PALM BEACH COUNTY
LIBRARY SYSTEM
3650 Summit Boulevard
West Palm Beach, FL 33406-4198

The Cambridge Guide to the Solar System

The Cambridge Guide to the Solar System provides a comprehensive and up-to-date description of the planets and their moons. Writing at an introductory level appropriate for high school and undergraduate students, Professor Lang leads the reader on a fascinating journey of exploration to the worlds beyond our home planet Earth. This is accomplished in a light and uniform style, including everyday metaphors and many spacecraft images.

A short introductory historical prologue traces the evolution of our understanding of planetary bodies, from Kepler and Galileo to modern telescopes and spacecraft. The major planets and their moons are then introduced by presenting common properties and similar processes such as impact craters, volcanoes, water, atmospheres and magnetic fields. This is then followed by chapters which focus on individual planets and other solar-system objects, including a comprehensive treatment of the various space missions: from the *Apollo, Clementine* and *Lunar Prospector* missions to the Moon, to the *Galileo* mission to Jupiter and the *Mars Pathfinder, Mars Global Surveyor* and 2001 *Mars Odyssey* missions. Subjects of topical human interest such as ozone depletion, global warming, the search for life on Mars, and asteroid collisions with Earth are also discussed. Each chapter begins with a set of bullet points and ends with a summary diagram describing the important information and discoveries presented in that chapter. Equations are kept to a minimum, and when employed are placed within set-aside *focus elements*. These *focus boxes* enhance and amplify the discussion with interesting details, fundamental physics and important related topics, but may be passed-over without interrupting the flow of the text.

Filled with vital facts and information, and lavishly illustrated in color throughout, this book will appeal to professionals as well as general readers with an interest in planetary science. The book is also supported by an Internet site for use by the instructor, student or casual reader. It contains all of the images in the book, together with their legends and some brief explanatory text. This site also includes similar material for the author's *The Cambridge Encyclopedia of the Sun*. The web site address is http://ase.tufts.edu/cosmos/

KENNETH R. LANG, Professor of Astronomy at Tufts University, has extensive experience in teaching about the solar system, at both the introductory and advanced level, and has served on recent NASA committees that plan its future activities. He is also a well-known writer, having published ten books translated into seven languages as well as more than 150 professional articles. Professor Lang's books include *Astrophysical Formulae*, published in a third enlarged edition in 1999, the prize-winning *Wanderers in Space* – Prix du livre de l'Astronomie in 1994, and *The Cambridge Encyclopedia of the Sun* recommended by the *Library Journal* as one of the best reference books published in 2001. Professor Lang has been a Visiting Senior Scientist in Solar Physics at NASA Headquarters and a Fulbright Scholar in Italy. He is a member of the International Astronomical Union, the Royal Astronomical Society, and the American Astronomical Society. In between traveling for work and pleasure, Professor Lang lives in Arlington, Massachusetts, with his wife, Marcella. They have three grown-up children: Julia, David and Marina.

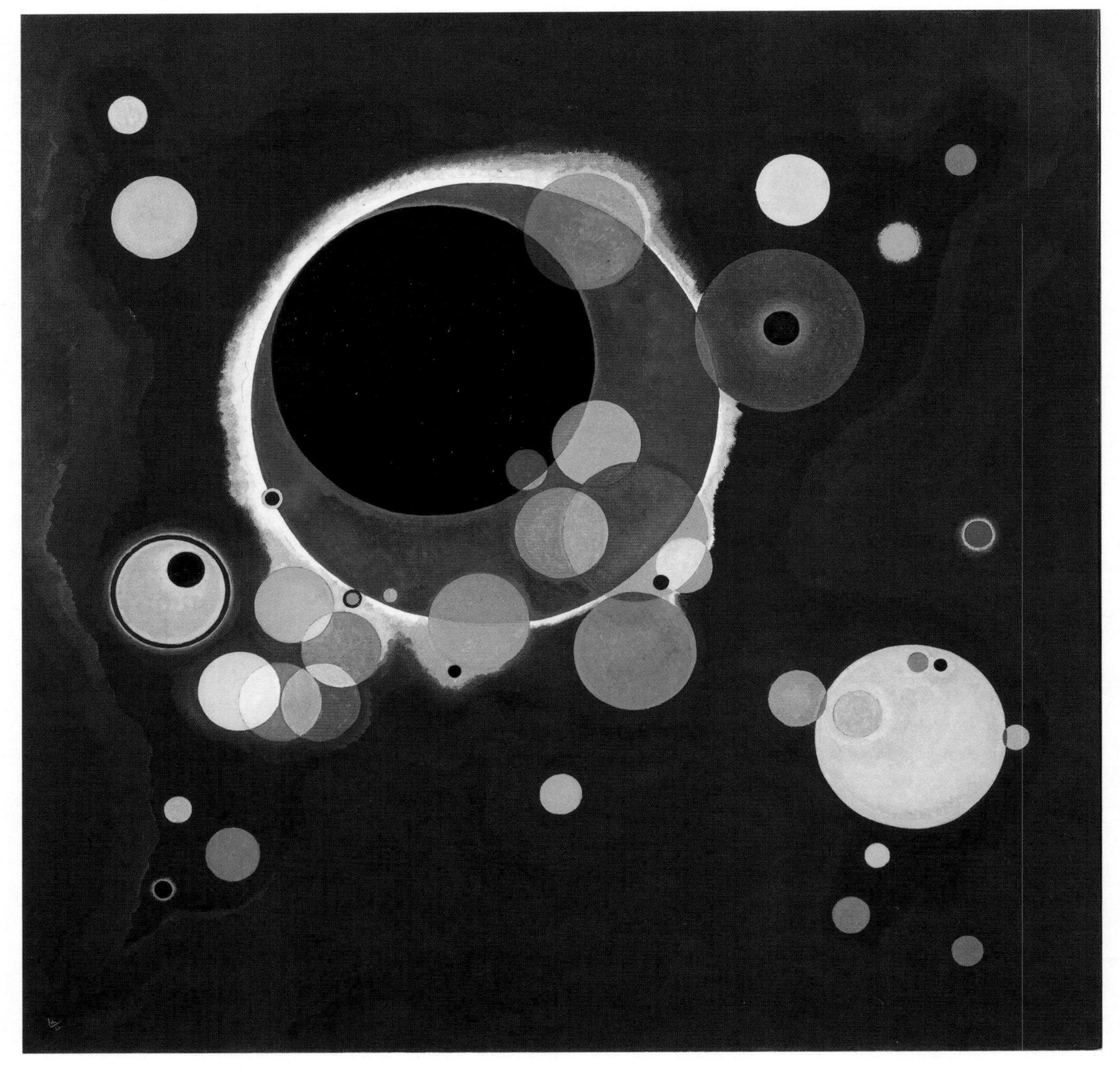

Several Circles. January–February 1926. The artist Vasily Kandinsky (1866–1944) seems to capture the essence of our space-age exploration of previously unseen worlds in this cosmic and harmonious painting. According to Kandinsky, "The circle is the synthesis of the greatest oppositions. It combines the concentric and the eccentric in a single form and in equilibrium. Of the three primary forms, it points most clearly to the fourth dimension." (Courtesy of the Solomon R. Guggenheim Museum, New York City, New York.)

The Cambridge Guide to the Solar System

KENNETH R. LANG
Tufts University, Medford, Massachusetts, USA

PUBLISHED BY THE PRESS SYNDICATE OF THE UNIVERSITY OF CAMBRIDGE
The Pitt Building, Trumpington Street, Cambridge, United Kingdom

CAMBRIDGE UNIVERSITY PRESS
The Edinburgh Building, Cambridge CB2 2RU, UK
40 West 20th Street, New York, NY 10011-4211, USA
477 Williamstown Road, Port Melbourne, VIC 3207, Australia
Ruiz de Alarcón 13, 28014 Madrid, Spain
Dock House, The Waterfront, Cape Town 8001, South Africa

http://www.cambridge.org

First published 2003

Printed in the United Kingdom at the University Press, Cambridge

Typefaces FF Scala 10/13 pt and Scala Sans *System* LaTeX 2ε [TB]

A catalogue record for this book is available from the British Library

Library of Congress Cataloguing in Publication data

Lang, Kenneth R.
The Cambridge guide to the solar system / Kenneth R. Lang.
p. cm.
Includes bibliographical references and index.
ISBN 0 521 81306 9 (hardback)
1. Solar system. I. Title.
QB501 .L24 2003
523.2–dc21 2002031562

ISBN 0 521 81306 9 hardback

Contents

Part 2 The inner system – rocky worlds

Preface

The planets have been the subject of careful observations and myth for millennia and the subject of telescopic studies for centuries. Our remote ancestors looked into the night sky, and wondered why the celestial wanderers or planets moved across the stellar background. They saw the planets as powerful gods, whose Greek and Roman names are still in use today. Then progressively larger telescopes enabled the detection of faint moons and remote planets that cannot be discerned with the unaided eye, and resolved fine details that otherwise remain blurred.

Only in the past half century have we been able to send spacecraft to the planets and their moons, changing many of them from moving points of light to fascinating real worlds that are stranger and more diverse than we could have imagined. Humans have visited the Moon, and robot spacecraft have landed on Venus and Mars. We have sent vehicles to the very edge of the planetary system, capturing previously unseen details of the remote giant planets, dropping a probe into Jupiter's stormy atmosphere, and perceiving the distant satellites as unique objects whose complex and richly disparate surfaces rival those of the planets. Probes have also been sent to peer into the icy heart of two comets, and robotic eyes have scrutinized the battered and broken asteroids.

The Cambridge Guide to the Solar System is a complete modern guide, updating and extending the prize-winning *Wanderers in Space* – Prix du livre de l'Astronomie in 1994. This book, written by the author and Charles A. Whitney, was completed before the *Clementine* and *Lunar Prospector* spacecraft were sent to the Moon, the *Magellan* orbiter penetrated the veil of clouds on Venus, the *Mars Pathfinder* landed on the red planet with its mobile roving *Sojourner*, the *Mars Global Surveyor* obtained high-resolution images of the surface of Mars, the *NEAR Shoemaker* spacecraft orbited the asteroid 433 Eros, the *Galileo* orbiter and probe visited Jupiter and its four large moons, *Deep Space 1* peered into the nucleus of Comet Borrelly, and Comet Shoemaker–Levy 9 collided with Jupiter. *The Guide* updates *Wanderers* to include the captivating results of all these missions, presenting more than a half century of extraordinary accomplishment.

The Cambridge Guide to the Solar System provides comprehensive accounts of the most recent discoveries, from basic material to detailed concepts. It is written in a concise, light and uniform style, without being unnecessarily weighed down with incomprehensible specialized materials or the variable writing of multiple authors. Metaphors, similes and analogies will be of immense help to the lay person and they add to the enjoyment of the material. Vignettes containing historical, literary and even artistic material make this book unusual and interesting, but at a modest level that enhances the scientific content of the book and does not interfere with it.

The book is at once an introductory text of stature and a thorough, serious and readable report for general readers, with much compact reference data. The language, style, ideas and profuse illustrations will attract the general reader as well as students and professionals. In addition, it is filled with vital facts and information for astronomers of all types and for anyone with a scientific interest in the planets and their satellites. The many full-color images, photographs, and line drawings help make this information highly accessible.

Each chapter begins with a set of pithy, one-sentence statements that describe the most important or interesting things that will be described in that chapter. A summary diagram, placed at the end of each chapter, captures the essence of our knowledge of the subject.

Set-aside *focus boxes* enhance and amplify the discussion with interesting details, fundamental physics and important related topics. They will be read by the especially curious person or serious student, but do not interfere with the general flow of the text and can be bypassed by the general educated reader who wants to follow the main ideas. Equations are kept to a minimum and, when employed, are almost always placed within the set-aside *focus elements*.

Numerous tables provide fundamental physical data for the planets and large moons. Many graphs and line drawings complement the text by summarizing what spacecraft have found. Guides to other resources are appended to the book as an annotated list of books for further reading, all published after 1990, and a list of relevant Internet addresses.

The Cambridge Guide to the Solar System has been organized into four main parts. The first introduces the planets and their moons, with a brief historical perspective followed by a discussion of their common properties. These unifying features include craters, volcanoes, water, atmospheres and magnetic fields. The second part discusses the rocky worlds found in the inner solar system – the Earth with its Moon, Mercury, Venus and Mars. The third part considers the giant planets, their satellites and their rings – worlds of liquid, ice and gas. The last part discusses the smaller worlds, the comets and asteroids, as well as collisions of these bodies with Jupiter, the Sun and Earth.

Chapter 1 traces our evolving understanding of the planets and their satellites made possible by the construction of ever-bigger telescopes. They resulted in the discovery of new planets and satellites, and resolved details on many of them. Here we include, in chronological order, the discoveries of Jupiter's moons, Saturn's rings, Uranus, Neptune, the asteroids, the icy satellites of the giant planets, tiny Pluto with its oversized moon, and the small icy objects in the Kuiper belt at the edge of the planetary realm. Other fundamental discoveries have been woven into the fabric of this chapter, including the realization that planets are whirling endlessly about the Sun, refinements of the scale and size of the planetary realm, and the spectroscopic discovery of the main ingredients of both the Sun and the atmospheres of the planets.

Chapter 2 begins with a description of how spacecraft have fundamentally altered our perception of the solar system, providing detailed close-up images of previously unseen landscapes and detecting incredible new worlds with sensors that see beyond the range of human vision. These new vistas have also resulted in a growing awareness of the similarities of the major planets and some moons. In the rest of Chapter 2 and in Chapter 3, they are therefore interpreted as a whole, rather than as isolated objects, by presenting comparative aspects of common properties and similar processes. This provides a foundation for subsequent examination of individual objects in greater detail.

Impact craters are found on just about every body in the solar system from the Moon and Mercury to the icy satellites of the distant planets, but in different amounts that depend on their surface ages and with varying properties. Ancient impacts on Venus have, for example, been erased by outpourings of lava, and the debris from subsequent impacts has been shaped by the planet's thick atmosphere. Numerous volcanoes have also been found throughout the solar system, including fiery outbursts on the Earth, towering volcanic mountains on Mars, numerous volcanoes that have resurfaced Venus, currently active volcanoes that have turned Jupiter's satellite Io inside out, and eruptions of ice on Neptune's largest moon, Triton.

Liquid water, which is an essential ingredient of life, covers seventy-one percent of the Earth's surface. Catastrophic floods and deep rivers once carved deep channels on Mars, and spring-like flows have been detected in relatively recent times. Water ice is ubiquitous in the outer solar system, including the clouds of Jupiter, the rings of Saturn, and the surfaces of most satellites. There is even evidence for subsurface seas beneath the water-ice crusts of Jupiter's satellites Europa, Ganymede and Callisto, and liquid water might also reside beneath the frozen surface of Saturn's satellite Enceladus.

Chapter 3 describes the atmospheres and magnetic fields that form an invisible buffer zone between planetary surfaces and surrounding space. Venus has an atmosphere that has run out of control, smothering this nearly Earth-sized world under a thick blanket of carbon dioxide. Its greenhouse effect has turned Venus into a torrid world that is hot enough to melt lead and vaporize oceans. Mars now has an exceedingly thin, dry and cold atmosphere of carbon dioxide. The red planet breathes about one-third of its atmosphere in and out as the southern polar cap grows and shrinks with the seasons. Jupiter's powerful winds and violent storms have remained unchanged for centuries, and Neptune has an unexpectedly stormy atmosphere. Saturn's largest moon, Titan, has a substantial Earth-like atmosphere, which is mainly composed of nitrogen and has a surface pressure comparable to that of the Earth's atmosphere. Temporary, rarefied and misty atmospheres cloak the Moon, Mercury, Pluto, Triton, and Jupiter's four largest moons.

Magnetic fields protect most of the planets from energetic charged particles flowing in the Sun's ceaseless winds, but some electrons and protons manage to penetrate this barrier. Jupiter's magnetism is the strongest and largest of all the planets, as befits the giant, while the magnetic fields of Uranus and Neptune are tilted. Guided by magnetic fields, energetic electrons move down into the polar atmospheres of Earth, Jupiter and Saturn, producing colorful auroras there.

Our description of individual planets begins in Chapter 4, with our home planet Earth. Earthquakes have been used to look inside our world, determining its internal structure and locating a spinning, crystalline globe of solid iron at its center. At the surface, continents slide over the globe, colliding and coalescing with each other like floating islands, as ocean floors well up from inside the Earth.

A thin membrane of air protects life on this restless world, and that air is being dangerously modified by life itself. Synthetic chemicals have been destroying the thin layer of ozone that protects human beings from dangerous solar ultraviolet radiation, and wastes from industry and automobiles are warming the globe to dangerous levels. The world has become hotter in the last decade than it has been for a thousand years, and at least some of this recent rise in temperature is due to greater emissions of greenhouse gases by human activity. The politicized debate over global warming is also described in Chapter 4, as are the probable future consequences if we don't do something about it soon.

This fourth chapter also discusses how the Sun affects our planet, where solar light and heat permit life to flourish. The amount of the Sun's radiation that reaches the Earth varies over the 11-year solar cycle of magnetic activity, warming and cooling the planet. Further back in time, during the past one million years, our climate has been changed by the recurrent ice ages, which are caused by variations in the amount and distribution of sunlight reaching the Earth.

An eternal solar gale now buffets our magnetic domain and sometimes penetrates it. Forceful mass ejections can create powerful magnetic storms on Earth, and damage or destroy Earth-orbiting satellites. Energetic charged particles, hurled out during solar explosions, endanger astronauts and can also wipe out the satellites that are so important to our technological society. Space-weather forecasters are now actively searching for methods to predict these threats from the Sun.

In Chapter 5 we continue on to the still, silent and lifeless Moon, a stepping stone to the planets. Most of the features that we now see on the Moon have been there for more than 3 billion years. Cosmic collisions have battered the lunar surface during the satellite's formative years, saturating much of its surface with impact craters, while lunar volcanism filled the largest basins to create the dark maria.

Twelve humans went to the Moon more than three decades ago, and brought back nearly half a ton of rocks. The rocks contain no water, have never been exposed to it, and show no signs of life. Yet, orbiting spacecraft have found evidence for water ice deposited by comets in permanently shaded regions at the lunar poles.

The fifth chapter also describes how the Moon generates tides in the Earth's oceans, and acts as a brake on the Earth's rotation, causing the length of day to steadily increase. The satellite also steadies our seasons by limiting the tilt of Earth's rotation axis. The story of the Moon's origin is given the latest and most plausible explanation: a glancing impact from a Mars-sized object knocked a ring of matter out of the young Earth; that ring soon condensed into our outsized, low-density Moon.

We discover in Chapter 6 that Mercury has an unchanging, cratered and cliff-torn surface like the Moon, but in a brighter glare from the nearby Sun. Although the planet looks like the Moon on the outside, it resembles the Earth on the inside. Relative to its size, Mercury has the biggest iron core of all terrestrial planets, and it also has a relatively strong magnetic field. Here we also mention tiny, unexplained motions of Mercury. As demonstrated by astronomers long ago, the planet does not appear precisely in its expected place. This discrepancy led Einstein to develop a new theory of gravity in which the Sun curves nearby space.

Chapter 7 discusses veiled Venus, the brightest planet in the sky. No human eye has ever gazed at its surface, which is forever hidden in a thick overcast of impenetrable clouds that are filled with droplets of concentrated sulfuric acid. Radar beams from the orbiting *Magellan* spacecraft have penetrated the clouds and mapped out the surface of Venus in unprecedented detail, revealing rugged highlands, smoothed-out plains, volcanoes and sparse, pristine impact caters. Rivers of outpouring lava have resurfaced the entire surface of Venus, perhaps about 750 million years ago, and tens of thousands of volcanoes are now found on its surface. Venus exhibits every type of volcanic edifice known on Earth, and some that have never been seen before. Some of them could now be active. Unlike Earth, there is no evidence for colliding continents on Venus, its surface moves mostly up and down, rather than sideways. Vertical motions associated with upwelling hot spots have buckled, crumpled, deformed, fractured and stretched the surface of Venus.

Our voyage of discovery continues in Chapter 8 to the red planet Mars, long thought to be a possible haven for life. Catastrophic flash floods and deep ancient rivers once carved channels on its surface, and liquid water might have lapped the shores of long-vanished lakes and seas. But its water is now frozen into the ground and ice caps, and it cannot now rain on Mars. Its thin, cold atmosphere lacks an ozone layer that might have protected the surface from lethal ultraviolet rays from the Sun, and if any liquid water were now released on the red planet's surface it would soon evaporate or freeze. Yet underground liquid water may have been seeping out of the walls of canyons and craters on Mars in recent times.

Three spacecraft have landed on the surface of Mars, failing to detect any unambiguous evidence for life. Corrosive chemicals have destroyed all organic molecules in the Martian ground, which means that the surface now contains no cells, living, dormant or dead. A meteorite from Mars, named ALH 84001, exhibits signs that bacteria-like micro-organisms could have existed on the red planet billions of years ago, but most scientists now think that there is nothing in the meteorite that conclusively indicates whether life once existed on Mars or exists there now. The future search for life on Mars may include evidence of microbes that can survive in hostile environments, perhaps energized from the planet's hot interior.

Chapter 9 presents giant Jupiter, which is almost a star and radiates its own heat. Jupiter radiates nearly twice as much energy as it receives from the Sun, probably as heat left over from when the giant planet formed. Everything we see on Jupiter is a cloud, formed in the frigid outer layers of its atmosphere. The clouds are swept into parallel bands by the planet's rapid rotation and counter-flowing winds, with whirling storms that can exceed the Earth in size. The fierce winds run deep and are driven mainly from within by the planet's internal heat. The biggest storms and wind-blown bands have persisted for centuries, though the smaller eddies are engulfed by the bigger ones, deriving energy from them. The little storms pull their energy from hotter, lower depths. Jupiter has a non-spherical shape with a perceptible bulge around its equatorial middle, and this helps us determine what is inside the planet. It is almost entirely a vast global sea of liquid hydrogen, compressed into a fluid metal at great depths. And above it all, Jupiter has a faint, insubstantial ring system that is made of dust kicked off small nearby moons by interplanetary meteorites.

Chapter 9 additionally provides up-to-date accounts of the four large moons of Jupiter, known as the Galilean satellites. The incredible complexity and rich diversity of their surfaces, which rival those of the terrestrial planets, are only visible by close-up scrutiny from spacecraft. Although the *Voyager 1* and *2* spacecraft sped by with just a quick glimpse of them, it was time enough for their cameras to discover new worlds as fascinating as the planets themselves, including active volcanoes on Io, smooth ice plains on Europa, grooved terrain on Ganymede, and the crater-pocked surface of Callisto. Then the *Galileo* spacecraft returned for a longer look, gathering further data on the satellites' surfaces and using gravity and magnetic measurements to infer their internal constitution. Changing tidal forces from nearby massive Jupiter squeeze Io's rocky interior in and out, making it molten inside and producing the most volcanically active body in the solar system. Jupiter's magnetic field sweeps past the moon, picking up a ton of sulfur and oxygen ions every second and directing them into a doughnut-shaped torus around the planet. A vast current of 5 million amperes flows between the satellite Io and the poles of Jupiter and back again, producing auroral lights on both bodies. There are no mountains or valleys on the bright, smooth, ice-covered surface of Europa. The upwelling of dirty liquid water or soft ice has apparently filled long, deep fractures in the crust. Large blocks of ice float like rafts across Europa's surface, lubricated by warm, slushy material. A subsurface ocean of liquid water may therefore lie just beneath Europa's icy crust, perhaps even harboring alien life that thrives in the dark warmth. Ganymede has an intrinsic magnetic field. As far as we know, it is the only satellite known that now generates its own magnetism. Callisto is one of the oldest, most heavily cratered surfaces in the solar system. Both Callisto and Europa have a borrowed magnetic field, apparently generated by electrical currents in a subsurface ocean as Jupiter's powerful field sweeps by.

Our voyage of discovery continues in Chapter 10 with Saturn, second only to Jupiter in size. Like Jupiter, the ringed planet radiates almost twice as much energy as it receives from the Sun, but Saturn is not massive enough to have substantial heat left over from its formation. Its excess heat is generated by helium raining down inside the planet. It is Saturn's fabled rings that set the planet apart from the other wanderers. The astonishing rings consist of billions of small, frozen particles of water ice, each in its own orbit around Saturn like a tiny moon. They have been arranged into rings within rings by the gravitational influences of small nearby satellites that generate waves, sweep out gaps and confine the particles in the rings. Saturn's rings are thought to be relatively young, less than 100 million years old. They may have originated when a former moon strayed too close to the planet and was torn apart by its tidal forces.

Saturn's largest satellite, Titan, has a substantial atmosphere composed mainly of nitrogen molecules, also the principal ingredient of Earth's air. Clouds of methane, raining ethane, and flammable seas of ethane, methane and propane could exist beneath the impenetrable haze. We should find out what lies beneath the smog when the *Cassini* spacecraft arrives at Saturn, in July 2004, and parachutes the *Huygens* probe through Titan's atmosphere four months later. Six medium-sized moons revolve around Saturn, each covered with water ice. They are scarred with ancient impact craters, and some of them show signs of ice volcanoes and internal heat. A number of small irregularly shaped moons of Saturn have remarkable orbits. The co-orbital moons move in almost identical orbits, the Lagrangian moons share their orbit with a larger satellite, and the shepherd moons confine the edges of rings.

Uranus and Neptune are treated together in Chapter 11, because of their similar size, mass and composition. Unlike all the other planets, Uranus is tipped on its side and rotates with its spin axis in its orbital plane and in the opposite direction to that of most of the other planets. No detectable heat is emitted from deep inside Uranus, while Neptune emits almost three times the amount of energy it receives from the Sun. This internal heat drives Neptune's active atmosphere, which has fierce winds and short-lived storms as big as the Earth. Both planets are vast global oceans, consisting mainly of melted ice with no metallic hydrogen inside. The magnetic fields of both Uranus and Neptune are tilted from their rotation axes, and are probably generated by currents in their watery interiors. The ring systems of both planets are largely empty space, containing dark narrow rings with wide gaps. One of Neptune's thin rings is unexpectedly lumpy, with material concentrated in clumps by a nearby moon. The rings we now see around these planets will eventually be ground into dust and vanish from sight, but they can easily be replaced by debris blasted off small moons already embedded in them. The amazingly varied landscape on Miranda, the innermost mid-sized satellite of Uranus, indicates that the satellite may have been shattered by a catastrophic collision and reassembled, or else it was frozen into an embryonic stage of development. Neptune's satellite Triton revolves about the planet in the opposite direction to its spin. The glazed satellite has a very tenuous, nitrogen-rich atmosphere, bright polar caps of nitrogen and methane ice, frozen lakes flooded by past volcanoes of ice, and towering geysers that may now be erupting on its surface. Triton may have formed elsewhere in the solar system and was captured into orbit around Neptune. Triton is headed for a future collision with Neptune as the result of tidal interaction with the planet.

Chapter 12 discusses the icy comets. They light up and become visible for just a few weeks or months when tossed near the Sun, whose heat vaporizes the comet's surface and it grows large enough to be seen. A million, million comets are hibernating in the deep freeze of outer space, and they have been out there ever since the formation of the solar system 4.6 billion years ago. We can detect some of them in the Kuiper belt reservoir at the edge of the planetary system, but billions of unseen comets reside in the remote Oort cloud nearly halfway to the nearest star. Two spacecraft have now passed close enough to image a comet nucleus, of Comet Halley and Comet Borrelly, showing that they are just gigantic, black chunks of water ice, other ices, dust and rock, about the size of New York City or Paris. When these comets come near the Sun, their icy nuclei release about a million tons of water and dust every day, from fissures in their dark crust. Some comets develop tails that flow away from the Sun, briefly attaining lengths as large as the distance between the Earth and the Sun, but other comets have no tail at all. Comets can have two kinds of tails: the long, straight, ion tails, that re-emit sunlight with a faint blue fluorescence, and a shorter, curved, dust tail that shines by reflecting yellow sunlight. They are blown away from the Sun by its winds and radiation, respectively. Meteor showers, commonly known as shooting stars, are produced when sand-sized or pebble-sized pieces of a comet burn up in the Earth's atmosphere, never reaching the ground. Any comet that has been seen will vanish from sight in less than a million years, either vaporizing into nothing or leaving a black, invisible rock behind. Some burned-out comets look like asteroids, and a few asteroids behave like comets, blurring the distinction between these two types of small solar-system bodies.

We continue in Chapter 13 with the rocky asteroids. There are billions of them in the main asteroid belt, located between the orbits of Mars and Jupiter, but they are so small and widely spaced that a spacecraft may safely travel though the belt. The combined mass of billions of asteroids is less than five percent of the Moon's mass. The Earth resides in a smaller swarm of asteroids, chaotically shuffled out of the main belt. Many of these near-Earth asteroids travel on orbits that intersect the Earth's orbit, with the possibility of an eventual devastating collision with our planet. The asteroids are the pulverized remnants of former, larger worlds that failed to coalesce into a single planet. The colors of sunlight reflected from asteroids indicate that they formed under different conditions prevailing at varying distance from the Sun. We could mine some of the nearby ones for minerals or water. An asteroid's gravity is too weak to hold on to an atmosphere or to pull most asteroids into a round shape. The close-up view obtained by passing spacecraft and radar images indicates that asteroids have been battered and broken apart during catastrophic collisions in years gone by. One spacecraft has circled the near-Earth asteroid 433 Eros for a year, examining its dusty, boulder-strewn landscape in great detail, obtaining an accurate mass for the asteroid, and showing that much of it is solid throughout. Other asteroids are rubble piles, the low-density, collected fragments of past collisions held together by gravity. Meteorites are rocks from space that survive their descent to the ground, and most of them are chips off asteroids. Organic matter found in meteorites predates the origin of life on Earth by a billion years; but the meteoritic hydrocarbons are not of biological origin.

The concluding Chapter 14 discusses colliding worlds, including pieces of a comet that hit Jupiter, comets that are on suicide missions to the Sun, and an asteroid that wiped out the dinosaurs when it hit the Earth 65 million years ago. The Earth is now immersed within a cosmic shooting gallery of potentially lethal, Earth-approaching asteroids and comets that could collide with our planet and end civilization as we know it. The lifetime risk that you will die as the result of an asteroid or comet striking the Earth is about the same as death from an airplane crash, but a lot more people will die with you during the cosmic impact. It could happen tomorrow or it might not occur for hundreds of thousands of years, but the risk is serious enough that astronomers are now taking a census of the threatening ones. With enough warning time, we could redirect its course.

The Cambridge Guide to the Solar System continues with an annotated list of books for further reading, all published after 1990, and a list of Internet addresses for the topics discussed.

The illustrator Sue Lee has combined artistic talent with a scientist's eye for detail in producing the fantastic line drawings and diagrams in this book. The text has been substantially improved by the careful attention of copy-editor Brian Watts.

This book was stimulated by the author's visit to the Jet Propulsion Laboratory, when the main results of the recent planetary missions were summarized by its director, Edward C. Stone, and the Project Scientists of many of them. Andrew P. Ingersoll, Torrence V. Johnson, Kenneth Nealson, R. Stephen Saunders, Donald K. Yeomans and Richard W. Zurek provided comprehensive scientific summaries matched only by the extraordinary accomplishments of the missions themselves. Planetary scientists with comprehensive knowledge have assured the accuracy, completeness and depth of individual chapters through critical review. I am grateful to my expert colleagues who have read portions of this book, and substantially improved it, either by thorough review or by expert commentary on some isolated sections. They include Reta Beebe, Doug Biesecker, Mark A. Bullock, Owen K. Gingerich, Torrence V. Johnson, Brian G. Marsden, Steven J. Ostro, Carl B. Pilcher, Roger A. Phillips, David Senske, Paul D. Spudis, David J. Stevenson and Donald K. Yeomans.

Kenneth R. Lang
Tufts University

Principal units

This book uses the International System of Units (Système International, SI) for most quantities, but with two exceptions. As is the custom with planetary scientists, we often use the kilometer unit of length and the bar unit of pressure. The kilometer now appears on most automobile speedometers. There are one thousand meters in a kilometer and a mile is equivalent to 1.6 kilometers. One bar corresponds to the surface pressure of the Earth's air at sea level. For conversion to the SI pressure unit of pascal, 1 bar $= 10^5$ pascal, or 1 pascal $= 10^{-5}$ bar.

Some other common units are the millibar (mbar), equivalent to 0.001 bar, the nanometer (nm) with 1 nm $= 10^{-9}$ meters, the micron or micrometer (μm) with 1 μm $= 10^{-6}$ m, the ångstrom unit of wavelength, where 1 ångstrom = 1 Å $= 10^{-10}$ meters, the nanotesla (nT) unit of magnetic flux density, where 1 nT $= 10^{-9}$ tesla $= 10^{-5}$ gauss, and the ton measurement of mass, where 1 ton $= 10^3$ kilograms $= 10^6$ grams.

The reader should also be warned that centimeter-gram-second (c.g.s.) units have been, and still are, widely employed in astronomy and astrophysics. The following table provides unit abbreviations and conversions between units.

Quantity	SI Units	Conversion to c.g.s. Units
Length	meter (m)	100 centimeters (cm)
Mass	kilogram (kg)	1000 grams (g)
Time	second (s)	
Temperature	kelvin (K)	
Velocity	meter per second (m s^{-1})	100 centimeters per second (cm s^{-1})
Energy	joule (J)	10 000 000 ergs
Power	watt (W) = joule per second (J s^{-1})	10 000 000 erg s^{-1} (= 10^7 erg s^{-1})
Magnetic Flux Density	tesla (T)	10 000 gauss (G) (= 10^4 G)
Force	newton (N) (= kg m s^{-2})	100 000 dynes (= 10^5 dyn)
Pressure	pascal (Pa) (= N m^{-2}) (= kg m^{-1} s^{-2})	10 dyn cm^{-2} (= 10^{-5} bar)

Part 1 Changing views and fundamental concepts

1 Evolving perspectives – a historical prologue

- Ancient Earth-centered models and modern Sun-centered ones have both been used to describe and predict the motions of the wandering planets, which can suddenly turn around, apparently moving in the opposite direction before continuing on their usual course.
- The stars seem to be revolving around the Earth each night, but the Earth is instead spinning beneath the stars. This rotation causes the Sun to move across the sky each day.
- Nicolaus Copernicus (1473–1543) argued in 1543 that the Earth is whirling endlessly about the Sun, completing one circuit each year.
- Almost four centuries ago, Johannes Kepler (1571–1630) used accurate observations, obtained by Tycho Brahe (1546–1601), to infer a precise mathematical relation between the mean orbital distance and period of each planet in its revolution about the Sun.
- Isaac Newton (1642–1727) introduced the laws of universal gravitation in his *Principia*, published in 1686, showing that the Sun's gravitational force holds the solar system together.
- In the early 17th century, Galileo Galilei (1564–1643) used his pioneering telescopic observations of the four large moons of Jupiter and the phases of Venus to support the Sun-centered model of the solar system.
- Two kinds of telescopes, the refractor and the reflector, enable astronomers to detect faint objects that cannot be seen with the unaided eye, and to resolve fine details on luminous planets that otherwise remain blurred.
- The known size of the solar system doubled when Uranus was discovered in 1781, and was enlarged by almost this amount once again with the discovery of Neptune in 1846.
- The asteroid belt between the orbits of Mars and Jupiter contains billions of asteroids, but it is largely empty space and has a total mass that is much less than that of the Moon.
- Estimates for the Earth–Sun distance were gradually refined over the centuries, setting the scale of the solar system and weighing the Sun from a distance.
- The nearest star is about 270 thousand times further away from the Earth than the Sun is.

- The composition of the Sun is encoded in the visible spectrum of sunlight, and some of the ingredients of planetary atmospheres can be inferred from the ways that they change that light.
- The lightest element, hydrogen, is the most abundant element in the Sun, and the next most abundant solar element, helium, was first discovered in the Sun.
- Saturn's rings are completely detached from the planet; they consist of innumerable tiny satellites each with an independent orbit about Saturn.
- Jupiter, Saturn and Uranus have a retinue of large satellites, Neptune has only one large moon, and Mars has two small ones.
- Pluto is a very small world with an oversized companion.
- A host of small, unseen icy objects can be found in the disk of the solar system, beyond the orbit of Neptune. This Kuiper belt is a reservoir for short-period comets.
- As many as a million million (10^{12}) tiny, invisible dirty balls of ice orbit the Sun in the Oort cloud, a spherical shell that lies at distances that are a quarter of the way to the nearest star. A small trickle of them are deflected, by passing stars or molecular clouds, into the heart of the solar system where they are seen as long-period comets.

1.1 Moving points of light

The ancient wanderers

Our remote ancestors looked up on any dark, moonless night, and saw thousands of stars. They identified the brightest by name and noticed patterns, now called constellations, amongst groups of them. The permanent stellar beacons are always there, firmly rooted in the dark night sky, and the constellations remain unchanged over the eons. Thousands of years ago, the component stars of the constellations, such as the Big Dipper, were in the same location with respect to each other as they are today.

As ancient astronomers watched the constellations sweep across the black dome of night, they focused attention on seven objects that did not move with the stars. Ranked in order of greatest apparent brightness, they are the Sun, Moon, Venus, Jupiter, Saturn, Mercury and Mars. Our ancestors called them *planetes*, the Greek word for "wanderers"; a designation we still use for all but the Sun and Moon. For much of recorded history, astronomy has been mainly occupied with describing and predicting the movements and trajectories of these ancient wanderers.

The Sun does not rise at precisely the same point on the horizon each day. Instead, the location of sunrise drifts back and forth along the horizon in an annual cycle that is tied to the seasons. Ancient astronomers used monuments to line up the limits of these excursions (Fig. 1.1). The length of the Sun's arc across the sky also changes with a yearly rhythm. The Sun rises highest in the sky every summer, with its longest trajectory and the most daylight hours (Fig. 1.2).

Like the Sun, the Moon rises and sets at different points along the horizon, and reaches varying heights in the sky. Since the full Moon always lies nearly opposite to the Sun, the winter full Moon rises much higher in the sky than the summer full Moon.

The Moon repeats its motion around the Earth on a monthly cycle, periodically changing its appearance (Fig. 1.3). Once each month, the Moon comes nearly in line with the Sun, vanishing into the bright daylight. On the next night the Moon has moved away from this position, and a thin lunar crescent is seen. The crescent thickens on successive nights, reaching the rotund magnificence of full Moon in two weeks. Then, in another two weeks, the Moon disappears into the glaring Sun, completing the cycle of the month and providing another natural measure of time.

Even the earliest sky-watchers must have noticed that the wanderers are confined to a narrow track around the sky, known as the *zodiac* from the Greek word for "animal". The Sun's annual path, called the *ecliptic*, runs along the middle of this celestial highway. The paths of the other

Fig. 1.1 Stonehenge Sunrise is framed by the ancient stone pillars of Stonehenge in southern England. The pillars stand about 4 meters above ground. This monument was used to find midsummer and midwinter four thousand years ago – before the invention of writing and the calendar. The Sun rises at different points on the horizon during the year, reaching its most northerly rising on Midsummer Day (summer solstice on June 21). After this, the rising point of the Sun moves south along the horizon until it reaches its most southerly rising on Midwinter Day (winter solstice on December 22). An observer located at the center of the main circle of stones at Stonehenge watched midsummer sunrise over a marker stone located outside the circle; midwinter sunrise and sunset were framed by other stones within the circle. (Courtesy of Owen Gingerich.)

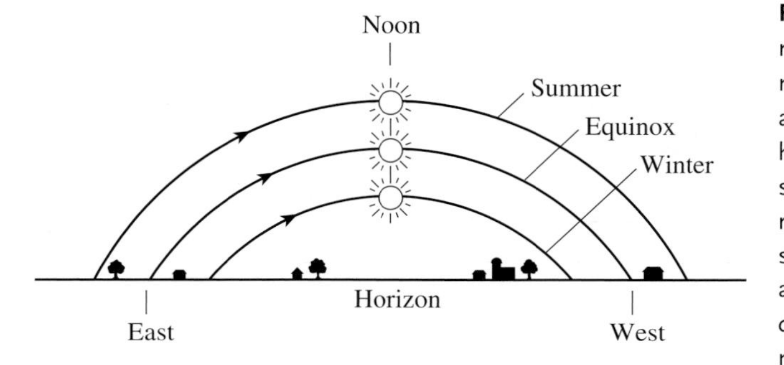

Fig. 1.2 The Sun's trajectory The Sun's motion across the sky looking south. The maximum height of the Sun in the sky, and its rising and setting points on the horizon, change with the seasons. In the summer, the Sun rises in the northeast, reaches its highest maximum height, and stays up longest. The Sun rises southeast and remains low in the winter when the days are shortest. The lengths of day and night are equal on the Vernal, or Spring, Equinox (March 20) and on the Autumnal Equinox (September 23) when the Sun rises exactly east and sets exactly west.

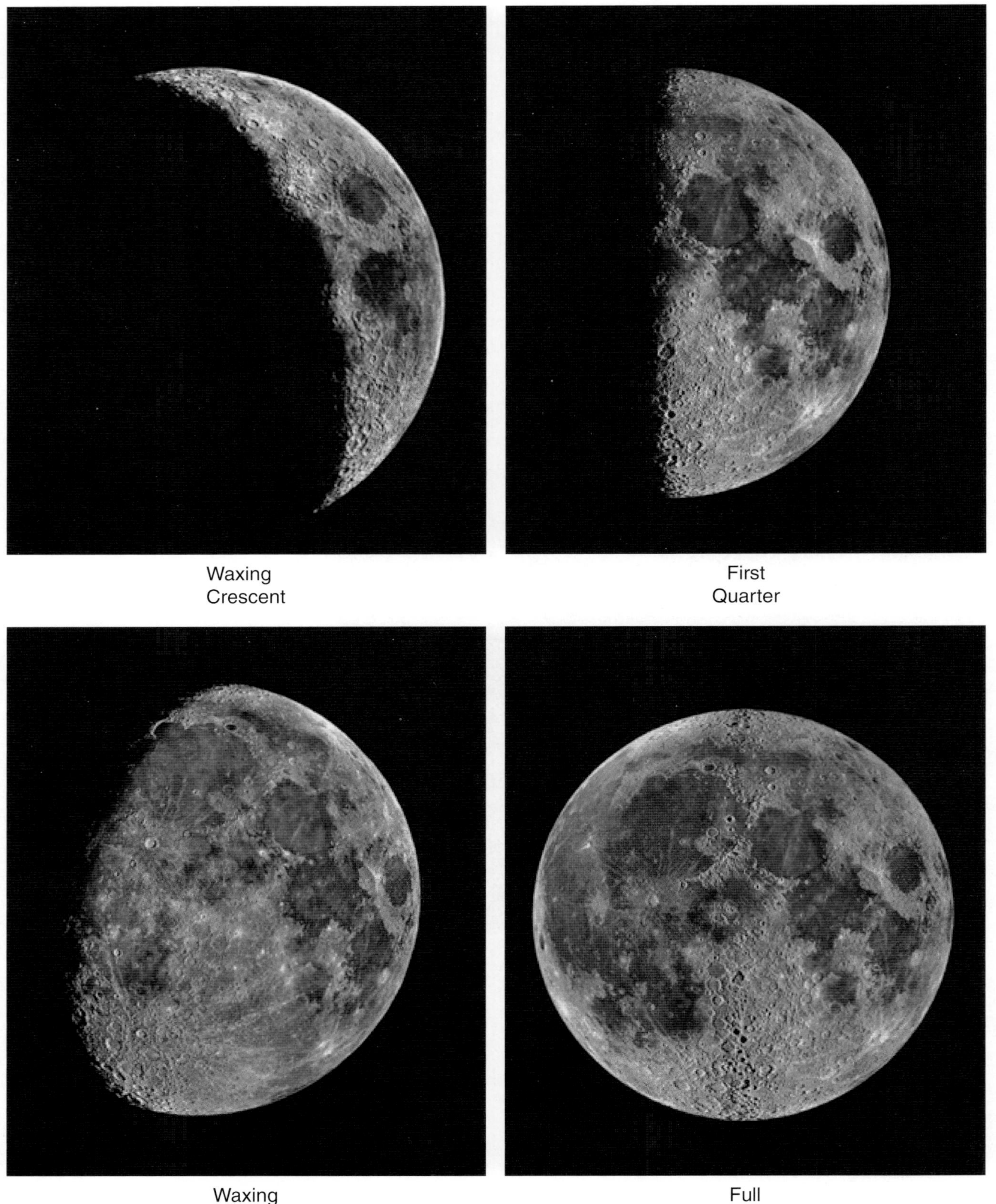
Waxing
Crescent
First
Quarter
Waxing
Gibbous
Full
Moon

Waning
Gibbous

Third
Quarter

Waning
Crescent

Fig. 1.3 The Moon's varying appearance During the monthly cycle, the Moon waxes (grows) from crescent to gibbous, and then after full Moon, it wanes (decreases) to a crescent again. The term crescent is applied to the Moon's shape when it appears less than half-lit; it is called gibbous when it is more than half-lit but not yet fully illuminated. The reason for the Moon's changing shape is described in Fig. 1.8 (Lick Observatory photographs).

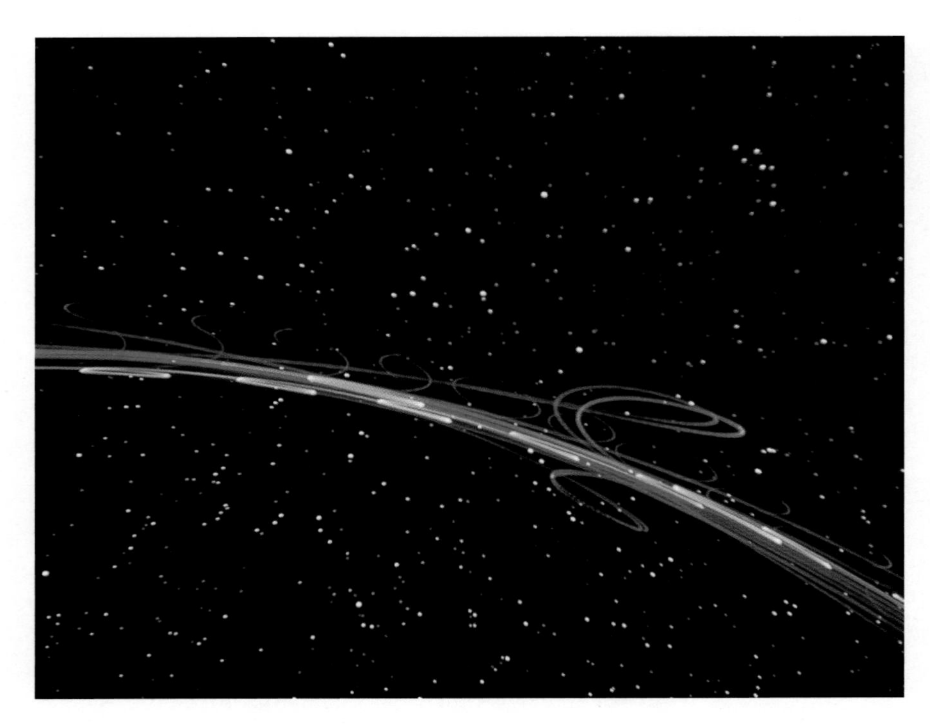

Fig. 1.4 Retrograde loops This photograph shows the apparent movements of the planets against the background stars. Mars, Jupiter and Saturn appear to stop in their orbits, then reverse direction before continuing on – a phenomenon called retrograde motion by modern astronomers. Ancient and modern explanations for this temporary backward motion are illustrated in Figs. 1.6 and 1.9, respectively. (Courtesy of Erich Lessing/Magnum).

wanderers also lie within the zodiac. Its narrowness is a sign that the planets move almost like marbles on a table because the planes of their orbits are closely aligned with each other.

It was obvious to astronomers from the earliest times that the wanderers do not move at uniform speeds or follow simple paths across the sky. When Mars, Jupiter and Saturn shine brightly in the midnight sky, each planet will gradually come to a stop in its eastward motion, move backward toward the west, and then turn around again and resume moving toward the east (Fig. 1.4). Although these planets travel eastward in the "prograde" direction most of the time, they sometimes appear to move in the westward "retrograde" direction before continuing on in their eastward course. Ancient and modern explanations of this looping, retrograde motion differ in their perspective on the locations and motions of the planets.

Circles and spheres

The ancient Greeks used geometrical models to visualize the cosmos, incorporating the symmetric forms of the circle and sphere. Their aim was to describe the regularities that underlay the planetary motions against the unchanging stellar background, and to thereby predict the locations of the planets at later times. They wanted to provide a reliable guide to the future, which is still the main point of science.

In arguments used by Pythagoras (572–479 BC), and subsequently recorded by Aristotle (384–322 BC), it was shown that the Earth is a sphere. During a lunar eclipse, when the Moon's motion carries it through the Earth's shadow, observers at different locations invariably saw a curved shadow on the Moon (Fig. 1.5). Only a spherical body can cast a round shadow in all orientations. The curved surface of the ocean was also inferred by watching a ship disappear over the horizon; first the hull and then the mast disappear from view.

In the Greek geocentric model, the central spherical Earth was supposed to be bounded by a much greater sphere, the imaginary celestial sphere of fixed stars. It wheeled around the central Earth once every day, with uniform circular motion and perfect regularity, night after night and year after year. Such a celestial sphere would also explain why people located at different places on Earth invariably saw just half of all the stellar heavens.

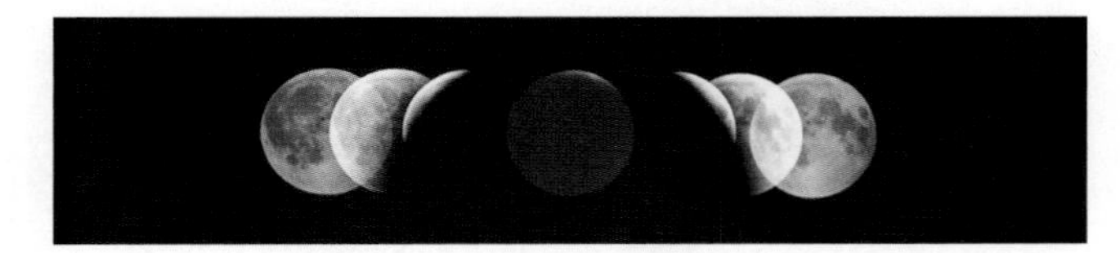

Fig. 1.5 Curved shadow of Earth This multiple-exposure photograph of a total lunar eclipse reveals the curved shape of the Earth's shadow, regarded by ancient Greek astronomers as evidence that the Earth is a sphere. Only a spherical body will cast the same circular shadow on the Moon when viewed from different locations on Earth or during different lunar eclipses. This photograph was taken by Akira Fujii during the lunar eclipse of 30 December 1982.

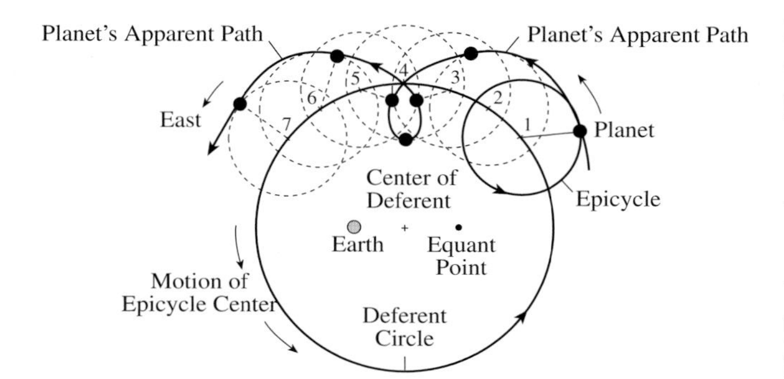

Fig. 1.6 Circles upon circles To explain the occasional retrograde loops in the apparent motions of Mars, Jupiter and Saturn, astronomers in ancient times imagined that each planet travels with uniform speed on a small circle, known as the epicycle. The epicycle's center moves uniformly on a larger circle, the deferent. A similar scheme was used by Ptolemy to explain the wayward motions of the planets in his *Almagest*. In the Ptolemaic system, the Earth was displaced from the center of the large circle, and each planet traveled with uniform motion with respect to another imaginary point, the equant, appearing to move with variable speed when viewed from the Earth.

The wayward planets were also supposed to move at constant speed on circular Earth-centered paths, but this contradicted observations. Since the planets move with changing speed across the sky, either the Earth had to be displaced from the center of motion or the planets were not moving at an unchanging rate. A planet in uniform circular motion about a center offset from the Earth would appear to a terrestrial observer to be moving with varying speed, faster when it is closest to Earth and slower when further away.

Combinations of uniform circular motion were additionally required to account for the looping, or retrograde, paths of the planets (Fig. 1.6). Each planet was supposed to move with constant speed on a small circle, or epicycle, while the center of the epicycle rotated uniformly on a large circle, or deferent. In this way Claudius Ptolemaeus (2nd century AD), known as Ptolemy, was able, in his *Mathematical Compilations* or *Almagest*, written about 145 AD, to predict the motions of every one of the seven wanderers, compounding them from circles upon circles.

In Ptolemy's model, the Earth was located to one side of the center of the deferent circle. An imaginary point, called the equant, was symmetrically positioned on the opposite side of the center (Fig. 1.6), and each planet was supposed to move uniformly with respect to the equant point. By selecting suitable radii and speeds of motion, Ptolemy could use this system of uniform motion around two circles to reproduce the apparent motions of the planets with remarkable accuracy. He succeeded so well that his model was still being used to predict the locations of the planets in the sky more than a thousand years after his death.

The Earth moves

The stars seem to be revolving about the Earth each night, but appearances can be deceiving. The Earth could instead be spinning beneath the stars. As the Earth rotates, the stars slide by and the celestial sphere just seems to be revolving once each day.

And the Sun might not be moving across the bright blue sky each day, for the Earth's rotation could produce this motion. Every point on the surface of a spinning Earth can be carried across the line of sight to an unmoving Sun, from sunrise to sunset, producing night and day (Fig. 1.7). Such a perspective involves a certain amount of detachment – the ability to separate yourself from the ground and use your mind's eye to look down on the spherical, rotating Earth, like a spinning ball suspended in space.

The Moon's nightly motion from horizon to horizon could also be neatly explained by the rotation of the Earth, and the Moon's monthly circuit against the background stars could be ascribed to its slower orbital motion around the Earth. This would also account for the Moon's varying appearance (Fig. 1.8). The Moon borrows its light from

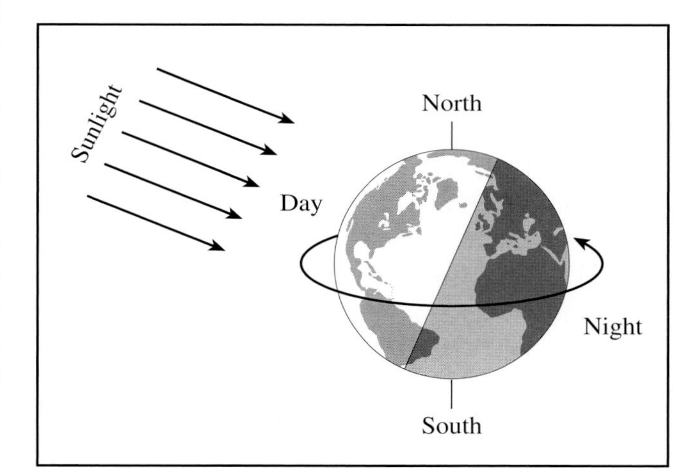

Fig. 1.7 Night and day The Earth rotates with respect to the Sun once every 24 hours, causing the sequence of night and day. Each point on the Earth's surface moves in a circular track parallel to the equator, and each track spends a different time in the Sun depending on the season. This drawing depicts summer in the northern hemisphere and winter in the southern hemisphere. Because the northern part of the Earth's rotational axis is tipped toward the Sun, circular tracks in the northern hemisphere spend a longer time in the Sun than southern ones.

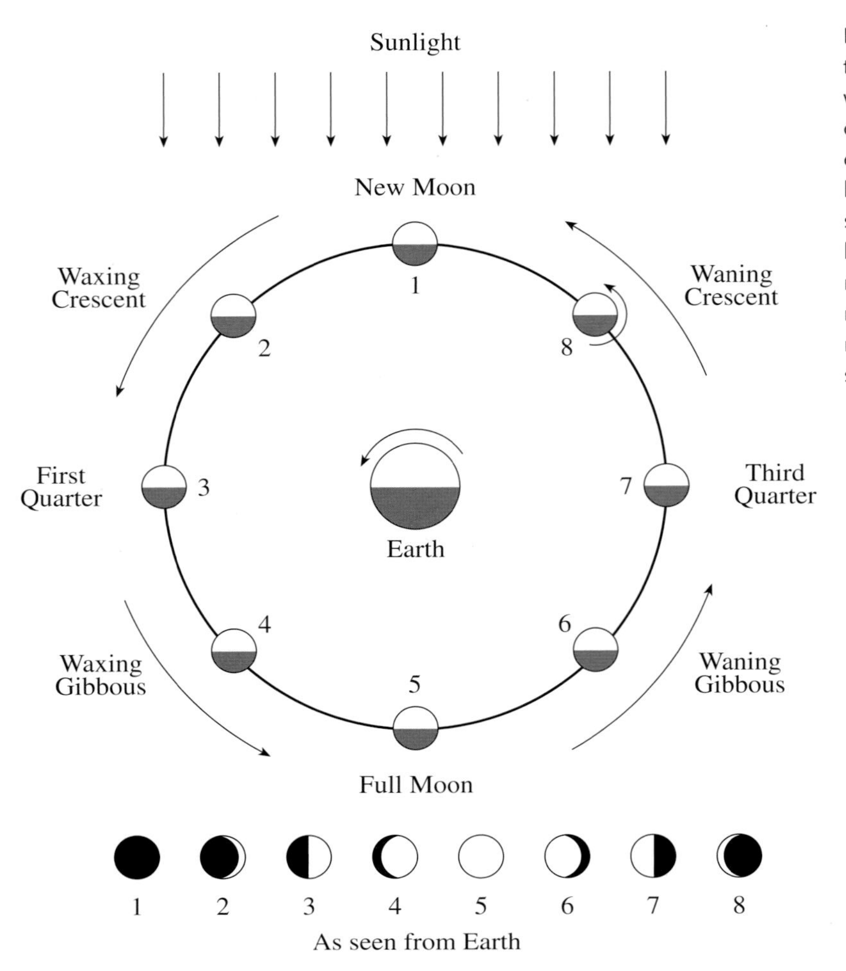

Fig. 1.8 Phases of the Moon Light from the Sun illuminates one half of the Moon, while the other half is dark. As the Moon orbits the Earth, we see varying amounts of its illuminated surface. The phases seen by an observer on Earth (*bottom*) correspond to the numbered points along the lunar orbit. The period from new Moon to new Moon is 29.53 days, the length of the month. As the Earth completes its daily rotation, all night-time observers see the same phase of the Moon.

the Sun, and the Sun illuminates first one part of the Moon's face and then another as the Moon orbits the Earth. On any given night, all observers on Earth will see the same phase of the Moon as our planet's rotation brings it into view.

The concept of a moving Earth nevertheless seems to violate common sense. The ground certainly seems to be at rest beneath our feet, providing the *terra firma* on which we carry out our daily lives. As Aristotle noticed, an arrow shot vertically upward falls to the ground where the archer stands, suggesting that the ground has not moved while the arrow was in flight.

Moreover, if the Earth is rotating, then it has to be spinning very rapidly. Using a radius of about 6.3 thousand kilometers, which is close to the value inferred long ago (by Eratosthenes, 276–194 BC), the Earth would have to be rotating at a velocity of about 460 meters per second to spin about its circumference once every 24 hours.

Yet, the globe on which we live might not only spin on its axis; it could also be whirling endlessly around the Sun, completing one circuit each year. This notion was presented by the Polish cleric and astronomer Mikolaj Kopernigk (1473–1543), better known as Nicolaus Copernicus, in his book *De Revolutionibus Orbium Coelestium*, or *Concerning the Revolutions of the Celestial Bodies*, published in 1543, the year of its author's death.

For Copernicus, the Sun was located at the heart of the planetary system, and the Earth was just one of several planets circling the Sun, in the same direction but at different distances and with various speeds. In order of increasing distance from the Sun, they are Mercury, Venus, Earth, Mars, Jupiter and Saturn. As Copernicus noticed, the further a planet is from the Sun, the longer it takes to complete a circuit.

There was no definite proof of this Sun-centered, heliocentric, hypothesis; it began as a bold leap in Copernicus' imaginative mind. But he could show that this new perspective provided natural explanations for observed phenomena. Venus and Mercury, for example, are never to be seen far from the Sun. They rise and set with the Sun, unlike Mars, Jupiter and Saturn, which can be seen at any time of night. Since the orbits of Venus and Mercury lie inside that of Earth and closer to the Sun, these planets are only seen around dawn or dusk. In contrast, the orbits of

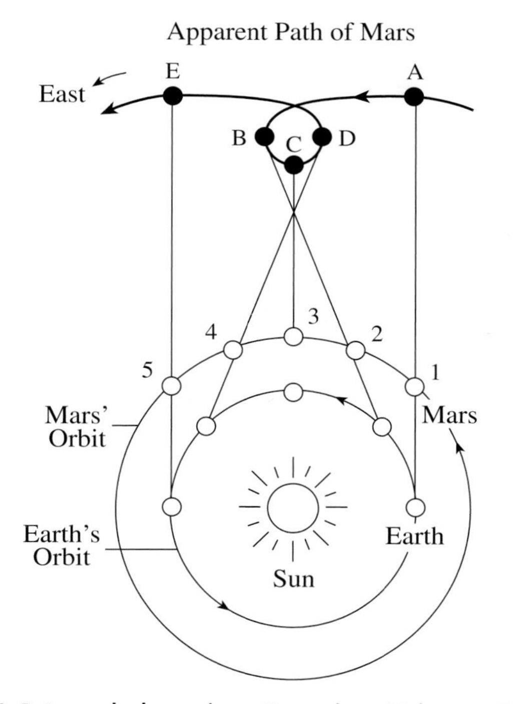

Fig. 1.9 Retrograde loops in a Copernican Universe A Sun-centered model of the solar system explains the looping path of Mars in terms of the relative speeds of the Earth and Mars. The Earth travels around the Sun more rapidly than Mars does. As Earth overtakes and passes the slower moving planet (*points 2 to 4*), Mars appears to move backward (*points B to D*) for a few months.

Mars, Jupiter and Saturn lie outside that of the Earth, so they are visible throughout the night.

The Sun-centered view also provides a simple explanation of the occasional backward, or retrograde, motions that were so hard to reproduce using an Earth-centered, or geocentric, model. Most of the time we see Mars, Jupiter and Saturn moving around the Sun in the same direction as the Earth, but during the relatively short time that the Earth overtakes one of these planets, that planet appears to be moving backward (Fig. 1.9). Moreover, one could confidently predict when a planet's apparent motion would come to a halt and turn around, and for how long it would seem to move backward.

We now realize that the tilt of the Earth's rotational axis and the annual orbit of the Earth can explain the seasons in the heliocentric model (Fig. 1.10). As the Earth orbits the Sun, its rotational axis points toward the same direction in the sky, at the star Polaris, but the northern and southern hemispheres are tilted toward or away from the Sun by up to 23.5 degrees. The greatest sunward tilt in a given hemisphere occurs in summer when the Sun is more nearly overhead and its rays strike the surface more directly. Winter occurs when that hemisphere is at its greatest tilt away from the Sun.

Copernicus' goal was to provide a geometric model that could replicate the planetary motions, but transforming their center to the Sun did not by itself improve the predictions. Proof of his Sun-centered model required improved observations and the introduction of non-circular motions. Yet, Copernicus' book did become a symbol for a new perspective of the heavens, a view that was ultimately to unite the Earth and planets in the domain of terrestrial physics. He opened the way to the study of not only how the celestial bodies move, but to an investigation of the forces that propel them and the underlying laws that govern their motion.

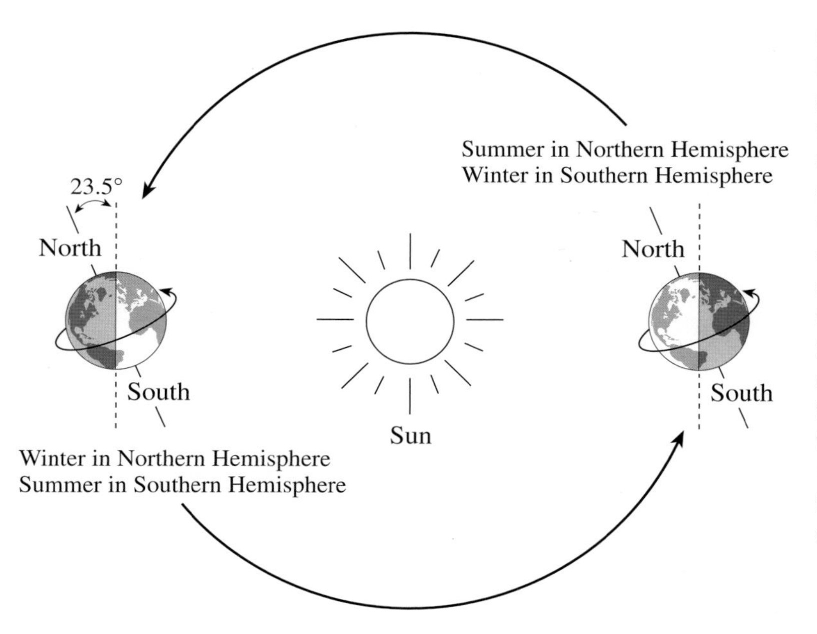

Fig. 1.10 The seasons As the Earth orbits the Sun, the Earth's rotational axis in a given hemisphere is tilted toward or away from the Sun. This variable tilt produces the seasons by changing the angle at which the Sun's rays strike different parts of the Earth's surface. The greatest sunward tilt occurs in the summer when the Sun's rays strike the surface most directly. In the winter, the relevant hemisphere is tilted away from the Sun and the Sun's rays obliquely strike the surface. When it is summer in the northern hemisphere, it is winter in the southern hemisphere and *vice versa*. (Notice that the radius of the Earth and Sun and the Earth's orbit are not drawn to scale.)

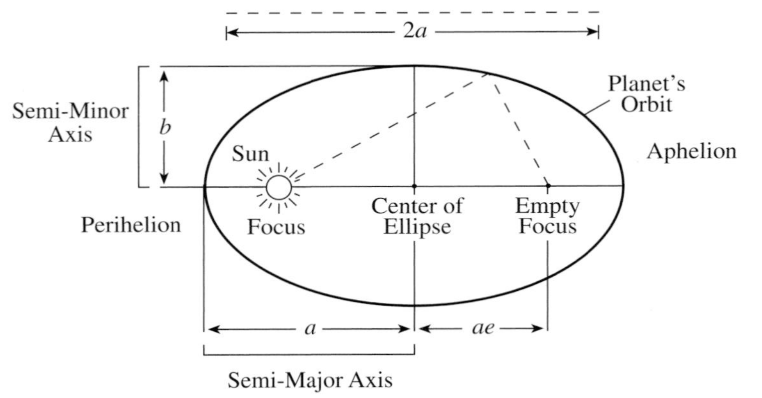

Fig. 1.11 Elliptical motion Each planet moves in an ellipse with the Sun at one focus. The length of a line drawn from the Sun, to a planet and then to the empty focus, denoted by the dashed line, is always $2a$, or twice the semi-major axis, a. The eccentricity, e, or elongation, of the planetary ellipse has been greatly exaggerated in this figure; planetary orbits are closer to being circular.

The harmony of the world

In the hope of developing a more precise description of planetary motions, the Danish astronomer Tycho Brahe (1546–1601) amassed a great number of observations that were more accurate and complete than any previous ones. This was before the days of telescopes, and he used ingenious measuring instruments that resembled large gunsights with graduated circles. These data were eventually interpreted by Tycho's assistant and successor, Johannes Kepler (1571–1630), who was able to determine precise mathematical laws from them.

Since circular motions could not describe Tycho's accurate observations of the planets, Kepler concluded that non-circular shapes were required (Focus 1.1). In 1605, after four years of computations, Kepler found that the observed planetary orbits could be described by ellipses with the Sun at one focus (Fig. 1.11). This ultimately became known as Kepler's first law of planetary motion.

So, a planet speeds up when it approaches the Sun, and slows down when it moves away from the Sun, and that accounts for the varying planetary speeds observed from Earth. Kepler was able to state the relationship in a precise mathematical form that can be explained with the help of Fig. 1.12. Imagine a line drawn from the Sun to a planet. As the planet swings about its elliptical path, the line (which will increase and decrease in length) sweeps out a surface at a constant rate. This is also known as the "law of equal areas". During the three equal time intervals shown in Fig. 1.12, the planet moves through different arcs because its orbital speed changes, but the areas swept out are equal.

Kepler labored another decade before publication of *Harmonice mundi* or *Harmony of the World*, in 1619, where he claimed to have listened to, and described mathematically, the music of the heavenly spheres. Kepler investigated arithmetic patterns between the periods and sizes of the planetary orbits, discovering the harmonic

Focus 1.1 Elliptical planetary orbits

According to Kepler's first law, the planets move in elliptical orbits with the Sun at one focus. The planet's closest point to the Sun, when the planet moves most rapidly, is called the perihelion; and its most distant point is the aphelion, where the planet moves most slowly. The distance between the perihelion and aphelion is the major axis of the orbital ellipse. Half that distance is called the semi-major axis, designated by the symbol a_P. The semi-major axis, a_E, of the Earth's elliptical orbit about the Sun is called the astronomical unit, abbreviated AU. It sets the scale of the solar system, and, when combined with the Earth's orbital period, permits the determination of the Sun's mass and the Earth's orbital velocity, but only after astronomers had found out how large an AU is (Section 1.3).

If a planet were viewed from the empty focus of its elliptical orbit, it would appear to move with almost uniform velocity, so it is analogous to the equant point in Ptolemy's description of planetary motion (Fig. 1.6). The close relationship of Ptolemy's equant point to the empty focus of an ellipse explains why his model was so successful.

The shape of an ellipse is determined by its eccentricity, e. If $e = 0$ its shape is a circle. The ellipse becomes more elongated and squashed as its eccentricity increases toward $e = 1.0$. The eccentricity of the planetary ellipse has been greatly exaggerated in Fig. 1.11, with an eccentricity of about $e = 0.5$. With the exception of Mercury, which is difficult to observe, all of the major planets have orbits that are nearly circular, with eccentricities of less than $e = 0.1$. This means that the Sun is very near the center of each orbital ellipse.

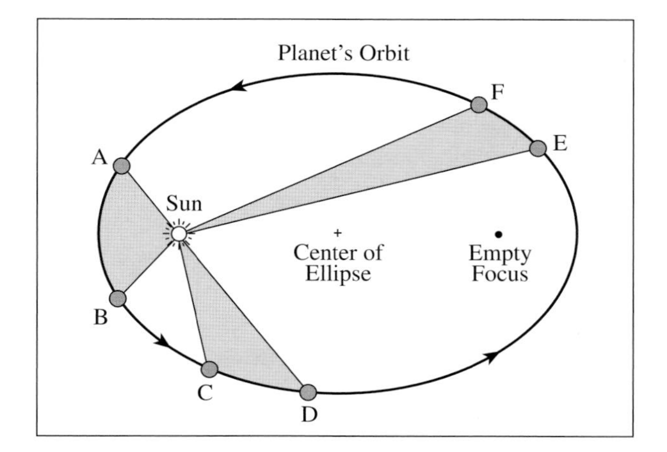

Fig. 1.12 Kepler's first and second laws Kepler's first law states that the orbit of a planet about the Sun is an ellipse with the Sun at one focus. The other focus of the ellipse is empty. According to Kepler's second law, the line joining a planet to the Sun sweeps out equal areas in equal times. This is also known as the law of equal areas. It is represented by the equality of the three shaded areas ABS, CDS and EFS. It takes as long to travel from A to B as from C to D and from E to F. A planet moves most rapidly when it is nearest the Sun (at *perihelion*); a planet's slowest motion occurs when it is farthest from the Sun (at *aphelion*).

relation that is now known as Kepler's third law. It states that the squares of the planetary periods are in proportion to the cubes of their average distances from the Sun.

These periods and distances are given in Table 1.1 with other mean orbital parameters of the planets. Here the periods are given in units of the Earth's orbital period of one Earth year, and the distances from the Sun are specified in units of the Earth's mean distance from the Sun, the astronomical unit which is designated AU. One Earth year is equivalent to 31.557 million (3.1557×10^7) seconds; the length of one AU, which is equal to 149.598 billion (1.49598×10^{11}) meters, took centuries to determine (Section 1.3).

If P_P denotes the orbital period of a planet measured in Earth years, and a_P describes its semi-major axis (Focus 1.1) measured in AU, then Kepler's third law states that $P_P^3 = a_P^2$, where the subscript "P" denotes the planet under consideration. This expression is illustrated in Fig. 1.13, for the major planets and for the brighter moons of Jupiter. The mean orbital velocity of each planet is proportional to the ratio a_P/P_P, so the velocity varies inversely with the square root of the distance or as $a_P^{-1/2}$.

In other words, the more distant planets have longer orbital periods and they move around the Sun with a slower speed. For example, Jupiter is 5.2 times as far away from the Sun as the Earth is, and it takes Jupiter 11.86 Earth years to travel once around the Sun. The Earth's mean orbital velocity is nearly 30 kilometers per second, while Jupiter's orbital velocity is about one-half that amount. Both planets are whirling around the Sun with awesome speed.

Kepler used precise observations to suggest that the Sun is at the hub of the solar system, with planets revolving around it in accordance with a precise mathematical relation between the mean orbital distance, period and velocity of each planet. He thereby began the transformation of astronomy from applied geometry to a branch of dynamical physics.

Table 1.1 Mean orbital parameters of the major planets[a]

Planet	Semi-Major Axis, a_P (AU)	Sidereal Orbital Period[b], P_P (years)	Eccentricity, e	Inclination to the Ecliptic, i (degrees)	Mean Orbital Velocity (km s^{-1})
Mercury	0.387 099	0.2409	0.2056	7.00	47.87
Venus	0.723 332	0.6152	0.0068	3.39	35.02
Earth	1.000 000	1.0000	0.0167	0.00	29.79
Mars	1.523 662	1.8809	0.0934	1.85	24.13
Jupiter	5.203 363	11.8626	0.0484	1.31	13.07
Saturn	9.537 070	29.4475	0.0541	2.49	9.67
Uranus	19.191 264	84.0169	0.0472	0.77	6.83
Neptune	30.068 964	164.7913	0.0086	1.77	5.48

[a] The dashed line divides the six planets known in Kepler's time from the two major, outer planets discovered later.
[b] The orbital elements are for the epoch 2000 January 1.5, and they were obtained from the web at http://ssd.jpl.nasa.gov/elem_planets.html and http://ssd.jpl.nasa.gov/phys_planets.html

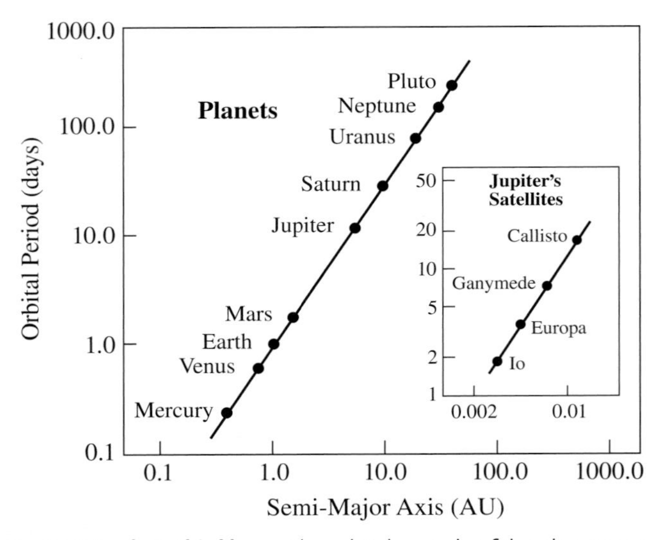

Fig. 1.13 Kepler's third law The orbital periods of the planets are plotted against their semi-major axes, using a logarithmic scale. The straight line that connects the points has a slope of 3/2, thereby verifying Kepler's third law that states that the square of the orbital periods increase with the cubes of the planetary distances. This type of relation applies to any set of bodies in elliptical orbits, including Jupiter's four largest satellites shown in the inset.

Newton and universal gravitation

Kepler told astronomers where the planets move, but not why. Although his laws could be used to accurately predict the motions of the planets, they did not explain why the planetary orbits are ellipses or what physical forces keep the planets whirling endlessly around the Sun. Both mysteries were solved after Isaac Newton (1642–1727) invented universal gravitation, with its unlimited range and capacity to act on all matter.

According to tradition, Newton was sitting under an apple tree when an apple fell next to him on the grass. This reminded him that the power of gravity, whose pull influences the motion of falling bodies, seems undiminished even at the top of the highest mountains. He therefore argued that the Earth's gravitational force extends to the Moon, and showed that this force can pull the Moon into an orbit. It was as if the Moon is perpetually falling toward the Earth while always keeping the same mean distance from it. The Sun's gravitational force similarly deflects the moving planets into their curved paths, so they forever revolve around the Sun.

Of all the known forces of nature, gravitation exerts its influence out to the greatest distance – except perhaps a hypothetical cosmological constant introduced by Albert Einstein (1879–1955) to keep a static Universe from collapsing. The enormous reach of gravity can be traced to two causes. In the first place, gravitational force decreases relatively slowly with distance, and this gives gravitation a much greater range than other natural forces, such as those that hold the nuclei of atoms together. In the second place, gravitation has no positive and negative charge, as electricity does, or opposite polarities as magnets do. This means that there is no gravitational repulsion between masses, and every atom in the Universe feels the gravitational attraction of every other atom. In contrast, the repulsive and attractive forces among like and unlike charges in an atom cancel each other, shielding the atom from the electrical forces of any other atom.

The basic principles of universal gravitation were described by Newton in his *Philosophiae Naturalis Principia Mathematica*, or *Mathematical Principles of Natural Philosophy*, commonly known as the *Principia*, published in 1686. He recognized that a force is required to alter the speed or direction of any body, and he used astronomical observations of orbital motions to provide constraints on the gravitational force imposed by one object on another. This force had to increase with the mass of an object and decrease with increasing distance from it. And from Kepler's third law, Newton inferred that the force of gravity must fall off as the inverse square of the distance.

Expressed mathematically, any mass, M_1, produces a gravitational force, F_{gravity}, on another mass, M_2, given by the expression:

$$\text{gravitational force} = F_{\text{gravity}} = \frac{G M_1 M_2}{D^2},$$

where the universal gravitational constant $G = 6.6726 \times 10^{-11}\ \text{m}^2\ \text{kg}^{-1}\ \text{s}^{-2}$, and D is the distance between the centers of the two masses. This is sometimes called the inverse-square law, since the force of gravity is inversely proportional to the square of the distance.

As a historical aside, Newton did not use his theory to show that the planets move in elliptical Keplerian orbits. He thought that no planet would continue to move in exactly the same ellipse, and that no calculation would exactly predict the locations of the planets. As it turned out, Einstein had to invent an extension of Newton's theory to describe gravity in the vicinity of a very massive object, including peculiarities in Mercury's orbital motion about the nearby Sun (Section 6.7). But for everyday effects on planet Earth and within most of the solar system, Newton's theory provides an exceptionally accurate description, and there is no noticeable difference between the two theories.

The concept of universal gravitation, and Newton's expression for the gravitational force, can be used to derive Kepler's third law in the form:

$$\text{Kepler's third law} = P_{\text{P}}^2 = \frac{4\pi}{G} \frac{a_{\text{P}}^3}{(m_{\text{P}} + M_\odot)}$$
$$= 5.9165 \times 10^{11} \frac{a_{\text{P}}^3}{M_\odot} \text{ seconds squared},$$

where a_P is the semi-major axis of the planet's orbital ellipse in meters, P_P is the orbital period in seconds, and m_P and $M_\odot$ respectively denote the mass of the planet and the mass of the Sun in kilograms.

Within the solar system, the dominant mass is that of the Sun, which far surpasses the mass of any other object there. That is why we call it a solar system, governed by the central Sun. The sum $(m_P + M_\odot)$ is therefore, to the first approximation, a constant equal to the Sun's mass, $M_\odot$, regardless of the planet under consideration.

Why doesn't the immense solar gravity pull the entire planetary system into the Sun? Motion opposes the force of gravity. Everything in the Universe moves, and there is nothing completely at rest. You might say that motion seems to define existence. When you stop moving it is all over.

The reason that the planets do not fall into the Sun is that each planet is also moving in a direction perpendicular to an imaginary line connecting it to the Sun, at exactly the speed required to overcome the Sun's gravitational pull. This orbital speed depends only on the Sun's mass and the planet's distance, but it is independent of the planet's mass.

Thus, gravity explains the moving points of light. It is the Sun's gravitational attraction that keeps the planets in their orbits, and it is the Sun's gravitational force that holds the solar system together. The Earth and other planets are strongly attracted to the Sun's huge mass, and it is their relentless motion that keeps the planets from falling into the Sun. And even before Newton's monumental work, Galileo had extended the human senses, using the telescope to discover previously invisible worlds and to provide evidence for a Sun-centered cosmos.

1.2 Discovery of new worlds

Galileo, the telescope, and the unseen cosmos

One of the most fascinating and lively books in astronomy, *Sidereus Nuncius* or *Starry Messenger*, was published in 1610. In it, the Italian astronomer and physicist Galileo Galilei (1564–1643) described how he turned the newly devised telescope toward the heavens, bringing the sky down to Earth and the Earth into the sky. In 1609 he found craters, rugged mountains and valleys on the Moon, perceiving another Earth-like world hanging unattached in space. In England, Thomas Harriot (1560–1621) made very poor telescopic maps of the Moon at about the same time; he greatly improved them only after he saw Galileo's results.

At least one cosmic object, the Moon, was no longer the polished, smooth and perfectly spherical body imagined by the ancients. Even the Sun was found to be spotty and impure under telescopic scrutiny by Galileo, Harriot, and other pioneering observers.

Galileo next used his rudimentary telescope to show that the Earth is not the only object with a satellite, our Moon, accompanying its motion through space. In 1610 he discovered four satellites that circle Jupiter. This meant that there was more than one center of motion in the Universe, and it contradicted Ptolemy's theory in which all astronomical objects move around the central Earth.

Galileo's observation that the innermost satellite of Jupiter revolves around Jupiter the fastest, and the outermost the slowest, meant that Jupiter is a miniature "solar system", mimicking the Copernican system in which the nearest planets to the Sun move around it with the fastest speeds.

Early telescopic observations by Galileo were also used to show that Venus goes through a complete sequence of Moon-like phases, from new to full, appearing at times as a thin crescent and thickening at other times into a round disk. This meant to Galileo that Venus had to circle the Sun. If nearby Venus orbited the Earth inside the Sun's orbit, then it could never appear completely illuminated, but Venus could appear in all its phases if it orbited the Sun (Fig. 1.14). Of course, Venus might orbit the Sun while the Earth remained at rest, so Galileo's persuasive evidence did not provide definite proof of the complete Copernican model.

Galileo's *Dialogo Massimi Sistemi Del Mondo, Tolemaico e Copernicano*, or *Dialogue on the Two Great World Systems, Ptolemaic and Copernican*, published in 1632, demonstrated the advantages of the Copernican cosmology, and provided telescopic evidence in its favor. His adoption of Copernicus' theory, in which the Earth moves around the Sun, was nevertheless opposed by theologians of the time, since a strict interpretation of the *Bible* (Psalm 104) indicated that "God fixed the Earth on its foundation, so it will never be moved".

After trial by Inquisition in 1633, the Roman Catholic Church forced Galileo to recant his support of the Copernican system as "abjured, cursed and detested". He was banished to confinement at his house in Arcetri, in the hills surrounding Firenze, where he spent his last years. Legend has it that as Galileo rose from kneeling before his inquisitors, he murmured, "eppur, si muove" – "even so, it does move", but he would hardly have been foolish enough to risk even greater punishment. Not until 1992, more than 350 years after his trial, did Pope John Paul II in effect apologize for the harshness of Galileo's sentence.

Galileo's use of the telescope to extend the human senses marked the beginning of a new age in

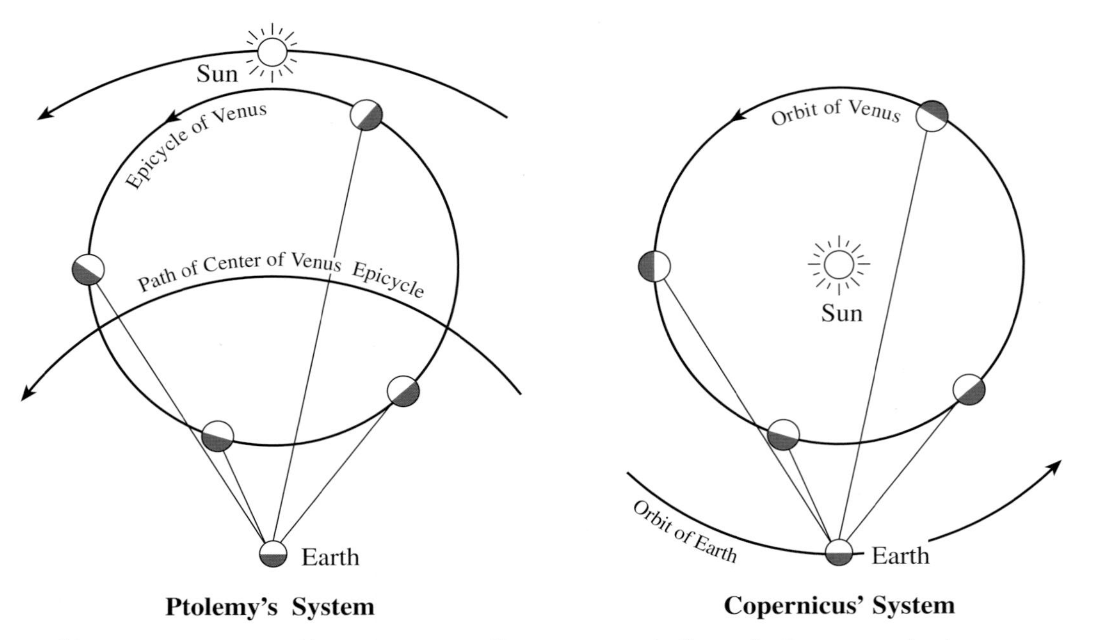

Fig. 1.14 Two world systems Among Galileo's most compelling arguments in favor of a Sun-centered solar system, previously advocated by Copernicus, was the fact that Venus displays phases like those of the Moon. In the long-held, Earth-centered Ptolemaic system (*left*) Venus always lies between the Earth and Sun, so it can never appear fully illuminated when viewed from the Earth. But in the Copernican system (*right*), Venus can show the whole range of illuminated phases.

astronomy – an age in which telescopes are used to view objects hitherto unseen and unknown, and to scrutinize known ones in greater detail. His observations led to the study of the planets as worlds like our own Earth, and not merely wanderers moving against the background stars. This era continues today, as we build new telescopes on the ground and send others into space, to discover new worlds and to investigate familiar ones in different ways.

Telescopes to extend our vision

Astronomical telescopes, which have been in use for about four hundred years, have enabled us to detect previously unknown objects, or to see known ones in greater detail, transforming our perception of the planets and their satellites. Telescopes extend our vision by collecting enough light to detect intrinsically faint sources or to resolve bright sources whose individual features are too near to each other to separate with the unaided eye. Until the Space Age (Section 2.1), the history of astronomical discovery was largely a matter of building bigger and better telescopes with increasingly sophisticated instrumentation.

There are two kinds of telescopes, the refractor, used by Galileo, and the reflector, initiated by Isaac Newton. As the names suggest, the refractor uses a lens to focus light, employing the principle of refraction, while the reflector uses a mirror to reflect and focus light (Fig. 1.15).

Modern refractors consist of a lens and a detector. The parallel light rays from a distant object are bent by refraction at the curved surface of a convex lens, known as the objective or the object glass. The objective lens brings the incoming light to a focus, where the light rays meet and an image is formed (Fig. 1.15). A detector placed at the focal plane, parallel to the objective lens, is used to record the image.

The distance from the objective lens to the focal plane, called the focal length, determines the overall size of an image. The greater the focal length, the bigger the image. The diameter of the objective lens is called the aperture, and the focal ratio of the lens is the focal length divided by the aperture.

The earliest refractors, such as the one Galileo used in 1609–1610, were slender tubes with a convex objective lens at one end and a second, smaller lens, termed the eyepiece, at the other end. The eyepiece was used to enlarge the image and to render the light rays parallel again so they could be focused by the eye to its retina. The magnification, or power, of such a telescope is equal to the ratio of the focal lengths of the objective and the eyepiece.

Galileo used a concave eyepiece placed in front of the focal plane, giving an erect image but a very narrow field of view; his telescopes had a magnification of about ×20. Kepler introduced a convex eyepiece behind the focal plane, which widened the viewing angle and inverted the image.

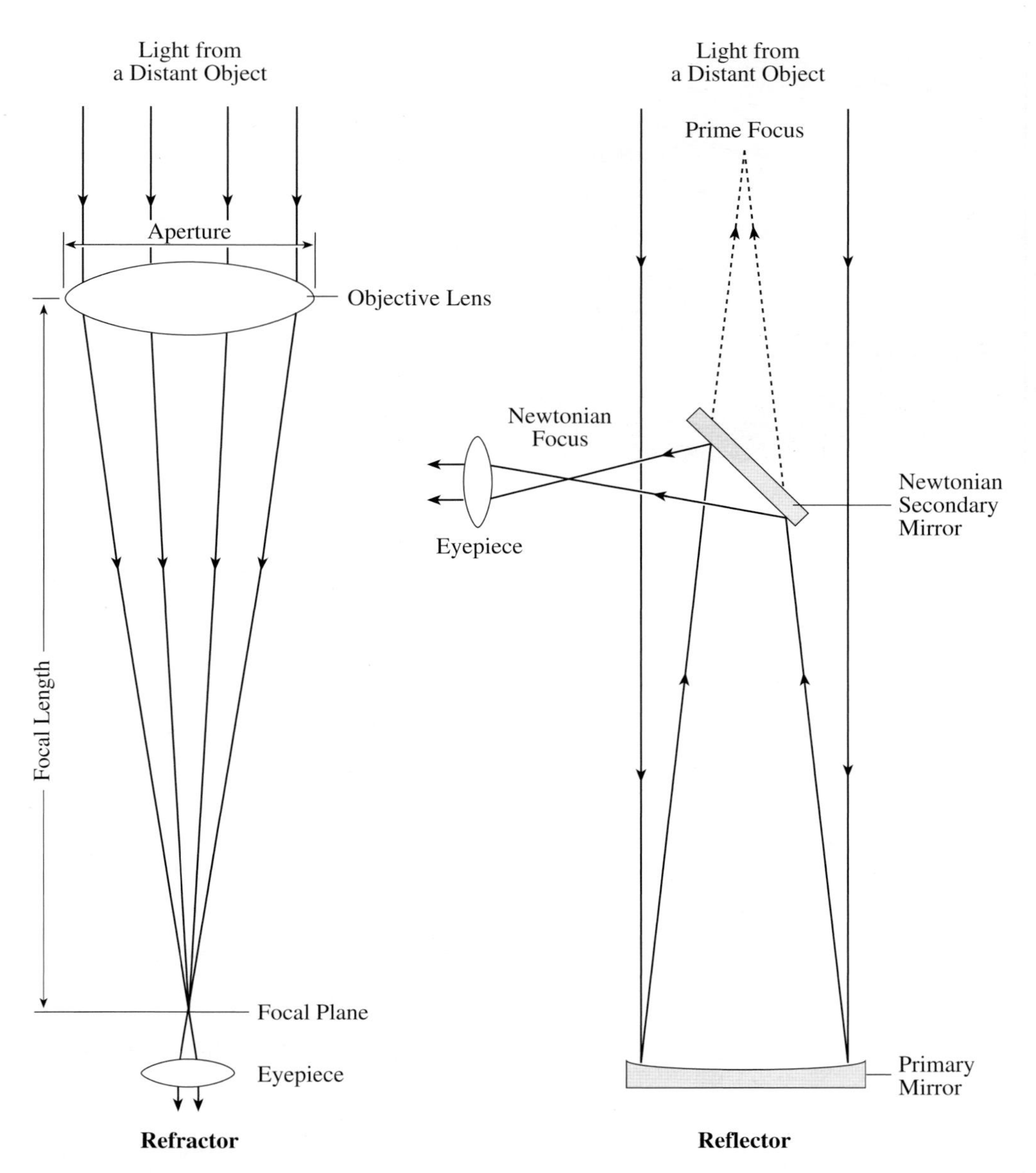

Fig. 1.15 Refractor and reflector Light waves that fall on the Earth from a distant object are parallel to one another, and are focused to a point by the lens or mirror of a telescope. The earliest telescopes were refractors (*left*). The curved surfaces of the convex objective lens bend the incoming parallel light rays by refraction, and bring them to a focus at the center of the focal plane, where the light rays meet and an image is created. A second, smaller lens, called the eyepiece, was used to magnify the image in the early refractors; later versions placed photographic or electronic detectors at the focal plane. The reflecting telescope (*right*) uses a large, concave, or parabolic, primary mirror to collect and focus light. A small, flat secondary mirror, inclined at an angle of 45 degrees to the telescope axis, reflects the light sideways, at a place now known as the Newtonian focus. Other light-deflecting mirror arrangements can be used to obtain any desired focal length, which varies with the curvature and position of small convex mirrors.

Many astronomers have now grown used to viewing this upside-down world.

Unfortunately, the objective lens in a refractor does not bring all parallel rays of light to a unique focus. As shown by Isaac Newton, sunlight is a mixture of all the colors seen in the prismatic display of a rainbow or in sunlight reflected by the crystals of new-fallen snow (Fig. 1.16). Each color has a definite wavelength, from the long red waves to the short violet ones. When sunlight passes through a glass lens, each wavelength or color is bent or refracted through a slightly different angle. This unequal refraction, known as chromatic aberration, produces a blurred image.

In 1668, Newton got around the problem by building an entirely different kind of telescope, the reflector, that uses a primary, concave mirror with a parabolic shape to gather the parallel light rays of a distant object and focus them to a point (Fig. 1.15). Light does not pass though

Fig. 1.16 Light painting This picture was made by using crystals to liberate the spectral colors in visible sunlight, refracting them directly onto a photographic plate, more than three hundred years after Isaac Newton used a prism, in 1672, to disperse sunlight into a spectrum of rainbow-like colors. The beautiful display shown in this picture was obtained in the rarefied atmosphere atop Hawaii's Mauna Kea volcano, where many of the world's best telescopes are located. (Courtesy of Eric J. Pittman, Victoria, British Columbia.)

a mirror, as it does through a lens, and the mirror concentrates light of all colors to the same focus, producing a sharp image. Newton placed a small, flat, secondary mirror just before the focal point of the primary mirror, to reflect the light to the side where an eyepiece was located. This Newtonian focus remains popular for many amateur telescopes, but professional astronomers use other light-deflecting mirror arrangements to obtain any desired focal length.

The critical parameter of a telescope is the diameter of the light-gathering lens or mirror. The larger the diameter, the more light is gathered, the brighter will be the image, and the fainter the objects that can be seen or recorded. The amount of light that can be gathered is proportional to the area of the lens or mirror, and consequently to the square of its diameter.

Although few of Galileo's many telescopes survive, he probably used one with an objective lens of about 0.05 meters (2 inches) in diameter to make his startling discoveries. The collecting area of such a lens is roughly fifty times that of the pupil of the unaided eye, which is about 0.007 meters across. This increase in light-collecting power allowed Galileo to see thousands of stars that had never been seen before, and to detect the four large moons of Jupiter. His primitive telescope could detect about half a million stars, over the entire heavens, compared with roughly two thousand visible with the unaided eye on a dark, moonless night.

Telescopes also provide angular resolution, which is the ability to detect the separation between things that are close together. Even Galileo's tiny telescope was able to resolve objects that remain blurred to the unaided eye, detecting previously unseen craters on the Moon and resolving the Milky Way into a myriad of stars.

This ability to discriminate fine details is called the resolving power of a telescope, and it depends on the diameter, D, of the light-gathering lens or mirror and the wavelength of observation (Focus 1.2). At a given wavelength, a bigger objective lens or primary mirror provides better angular resolution. The resolving power of a telescope operating at the wavelengths that we detect with our eye is about $0.13/D$ seconds of arc if D is in meters. Practically, however, the resolution at these wavelengths is limited by the Earth's atmosphere.

Focus 1.2 Angular resolution

The angular resolution, θ, of a telescope is determined by the diameter, D, of the objective lens or primary mirror, as well as the wavelength, λ, of observation. The mathematical expression is:

$$\text{angular resolution} = \theta = \frac{\lambda}{D} \text{ radians}$$
$$= 2.063 \times 10^5 \frac{\lambda}{D} \text{ seconds of arc,}$$

where one radian is equivalent to 206 265 ($2.062\,65 \times 10^5$) seconds of arc. This equation tells us that a bigger lens or mirror provides finer angular resolution at a given wavelength. A telescope with a mirror or lens that is 0.13 meters (5.1 inches) in diameter provides an angular resolution of about 1 second of arc at a red visual wavelength of 630 nanometers, or 6.30×10^{-7} meters.

Atmospheric effects limit the resolution of any telescope operating at visible wavelengths to about one second of arc, so you cannot improve the angular resolution by building a telescope bigger than about 0.13 meters in diameter. Nevertheless, a bigger telescope still gathers more light than a smaller one, permitting the detection of fainter sources. If a large telescope is placed in space, above our distorting atmosphere, greater angular resolution can also be achieved.

Our equation applies equally well at radio wavelengths where very big telescopes are required to achieve significant angular resolution. At a radio wavelength of 0.1 meters, an angular resolution of 1 second of arc requires a telescope with a diameter of 20 000 (2×10^4) meters. The advantage of radio signals is that the atmosphere does not distort them, or limit the angular resolution. We can observe cosmic radio sources on a cloudy day, just as your home radio works even when it rains or snows outside.

The resolution of ground-based telescopes operating at visible wavelengths is affected by turbulence in the Earth's atmosphere. It reduces the clarity of the image, limiting the angular resolution to about 1 second of arc. Similar variations cause the stars to twinkle at night. This atmospheric limitation to angular resolution is called *seeing*. The best seeing, of 0.2 seconds of arc in unusual conditions, is found only at a few sites in the world, and observatories are located at most of them. Better visible images with even finer detail can be obtained from the unique vantage point of outer space, using satellite-borne telescopes unencumbered by our atmosphere.

The serendipitous discovery of Uranus

The first planet to be discovered since the dawn of history was found accidentally, by a professional musician and self-taught amateur astronomer, William Herschel (1738–1822), using a home-made reflecting telescope in a systematic study of the stars from his home in Bath, England. While surveying the heavens on the night of 13 March 1781, Herschel came across an unusual object that was definitely not a star. It showed a disk, which no star can do, and it moved slowly from one night to another across the background of distant stars. This meant that it belonged to our solar system.

The object was soon lost in sunlight, but it was picked up a few months later. It was quickly recognized as a new planet, with a nearly circular orbit at twice the distance of Saturn, and Herschel became world famous – almost overnight. He was eventually appointed King's Astronomer with a pension, which permitted him to give up music as a career and devote himself full time to astronomy.

Herschel proposed that the new planet be named the "Georgian Planet" in honor of King George III, England's reigning monarch and a patron of the sciences. After some controversy, the new planet was instead named Uranus, the ruler of the heavens in Greek mythology. One consequence of this naming was that a newly discovered, heavy element was designated uranium in honor of the discovery of a new world.

Herschel succeeded where others had failed because of his skill as both an observer and a mirror maker. He was able to produce a 0.15-meter (6.2-inch) metal mirror of unsurpassed light-collecting area and resolution, making his Newtonian reflector, of 2.1-meter (84-inch) focal length, comparable to, or even better than, the refractor telescopes of his time.

The new planet Uranus orbits the Sun at 19.2 AU, or at 19.2 times the Earth's distance from the Sun. As predicted by Kepler's law, Uranus requires 84 Earth years to complete a round trip. Thus, Uranus has only completed two circuits around the Sun since it was discovered.

When he found Uranus, Herschel was apparently unaware of a numerical sequence that predicted its relative distance from the Sun. Known as the Titius–Bode law, after the last names of the first persons to state it, the sequence describes the regular spacing of the planets, suggesting that the next planet beyond Saturn would be located at 19.6 AU, or at about twice Saturn's distance (Focus 1.3). The so-called "law" also indicated a missing planet at 2.8 AU, in the gap between Mars and Jupiter, and suggested that another unknown planet would be located at 39 AU, or about twice the distance of Uranus. As it turned out, the asteroids were next discovered in the gap, and Neptune was eventually found close to the most distant location.

Focus 1.3 The Titius–Bode law

In the inner solar system, each planet's orbit is about 1.5 times the distance of its inward neighbor, and this ratio increases to roughly a factor of 2.0 in the outer solar system. This relative spacing of the planets is described by the Titius–Bode law, first noted in 1766 by Johann Daniel Titius (1729–1796), and brought to prominence by Johann Elert Bode (1747–1826) in his popular book on astronomy entitled *Anleitung zur Kenntnis des gestirnten Himmels*, or *Instruction for the Knowledge of the Starry Heavens*. It was Bode who subsequently named Herschel's new planet Uranus.

The law states that the relative distances of the planets from the Sun can be approximated by taking the sequence 0, 3, 6, 12, 24, . . . , adding 4, and dividing by 10. Mathematically, the semi-major axis, a_n, of the nth planet, in order of increasing distance from the Sun, is given by:

$$a_1 = 0.4 \quad \text{for } n = 1$$
$$a_n = 0.1[4 + 3 \times 2^{n-2}] \quad \text{for } n = 2, 3, \ldots, 9,$$

where a_n is the relative distance compared with that of the Earth and is given in AU.

A comparison of the semi-major axes, a_P, of the planets with this law is given in Table 1.2. The Titius–Bode law predates the discovery of Uranus at $n = 8$ by 15 years, the discovery of the first asteroid at $n = 5$ by 35 years, and the discovery of Neptune at $n = 9$ by 80 years. Although there is no well-accepted explanation for why this expression works so well, it probably has something to do with the dynamics, evolution or origin of the solar system.

Table 1.2 Comparison of measured planetary distances from the Sun with those predicted from the Titius–Bode law

Planet	n	Measured a_P (AU)	Predicted a_n (AU)
Mercury	1	0.387	$0.4 = (0 + 4)/10$
Venus	2	0.723	$0.7 = (3 + 4)/10$
Earth	3	1.000	$1.0 = (6 + 4)/10$
Mars	4	1.524	$1.6 = (12 + 4)/10$
Ceres (asteroid)	5	2.767	$2.8 = (24 + 4)/10$
Jupiter	6	5.203	$5.2 = (48 + 4)/10$
Saturn	7	9.539	$10.0 = (96 + 4)/10$
Uranus	8	19.19	$19.6 = (192 + 4)/10$
Neptune	9	30.06	$38.8 = (384 + 4)/10$

The ubiquitous asteroids

When Uranus was found to have an orbit near the place predicted by the Titius–Bode law, astronomers began a search for the missing planet that ought to be located between Mars and Jupiter, at a predicted distance of 2.8 AU from the Sun. The first object to be found in this location was nevertheless discovered quite unexpectedly by the Sicilian astronomer Giuseppe Piazzi (1749–1826) while he was preparing a catalog of accurate star positions.

On the opening night of the 19th century (1 January 1801), Piazzi found that the position of a supposed "star" had moved from its place noted when the stars in that region of the sky had been previously mapped. The object's motion was confirmed on the nights that followed, which meant that it was not a star and had to belong to the solar system. Piazzi gave it the name Ceres, honoring the goddess of Sicily.

Before other astronomers received word of Piazzi's discovery, the new object had moved too close to the Sun to be observed, and when it returned to dark skies, it could not be located. Hearing of the dilemma, a young mathematician, Karl Friedrich Gauss (1777–1855), devised a method for determining orbits from only three observations, and this led to its recovery about a year after it had been sighted.

By 1807 three more objects, named Pallas, Juno and Vesta, had been found, all located near 2.8 AU from the Sun. Yet, unlike the planets, none of them could be resolved into round disks with even the largest telescopes. They therefore became known as "asteroids" because they appear to be "star-like" points of light. Because they orbit the Sun like the much larger, major planets, asteroids are also known as minor planets.

The asteroids remained unresolved because they are very small and relatively nearby, rather than very large and distant like the stars. Even the largest asteroid, Ceres, has a radius of only 475 kilometers, which is less than one-third the radius of the Moon, less than one-tenth the radius of the Earth, and less than one-hundredth the radius of Jupiter.

No other asteroids were identified for 38 years, but the hunt for new ones became something of an astronomical sport in the last half of the 19th century. More than 300 asteroids had been discovered by 1891, and the pace of discovery subsequently increased by using long-exposure photographs of several hours to detect their motion against the stars.

Each asteroid is given a number corresponding to its chronological place in the discovery list, and a name that is usually provided by the discoverer. For instance, 433 Eros was the 433rd asteroid to be discovered. Nowadays several hundred asteroids are found annually, but they do not receive official numbers until their orbits are reliably known.

The list of asteroids with reliably known orbits reached the 2000 mark in 1977, and there were 100 000 known in the late 20th century. Astronomers estimate that there may be as many as half a million (500 000) faint asteroids smaller than one kilometer across, whose orbits have yet to be established.

Yet, despite their vast numbers, the combined mass of all the asteroids is estimated to be less then five percent that of the Moon, and nowhere near the mass of a single large planet.

Most of the asteroids with well-determined orbits lie in a great asteroid belt between the orbits of Mars and Jupiter (Fig. 1.17), at distances of 2.2 to 3.3 AU and with orbital periods of 3 to 6 Earth years. The asteroids are so little, and distributed across such a large range of distances, that the asteroid belt is largely empty space. This leaves plenty of room for spacecraft to pass though to the giant planets, undamaged by collision with any asteroid.

The asteroids are scattered around their orbits in a haphazard fashion, much like runners near the end of a long race on a small track. So, at first glance, the asteroids appear to fill the belt quite uniformly. But if the asteroids could be arranged along a line outward from the Sun – as though they had been placed on the starting line – a different pattern would emerge. Not all distances from the Sun are equally well represented. There are a few prominent gaps, and these are named the Kirkwood gaps after the American astronomer, Daniel Kirkwood (1814–1895), who discovered them in 1866 (Fig. 1.18).

According to Kepler's laws of planetary motion – which apply equally well to the asteroids as to the major planets – each orbital distance corresponds to a specific orbital period. This means that the arrangement of asteroids

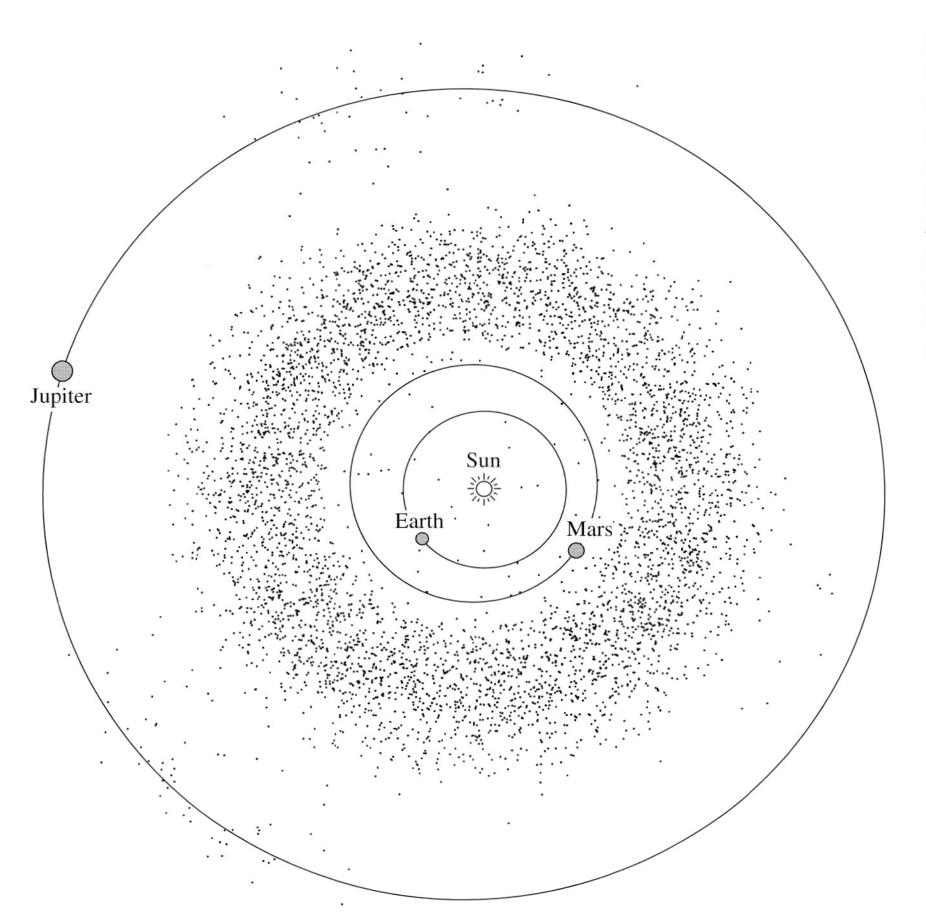

Fig. 1.17 Asteroid belt The locations of five thousand flying rocks, called asteroids or minor planets, whose orbits are accurately known. The vast majority of the asteroids orbit the Sun in the main belt located between the orbits of Mars and Jupiter. A few of them pass inside the orbit of Earth, while others move about 60 degrees ahead of and behind Jupiter in similar orbits. (Courtesy of Jeff Bytof, University of California at San Diego.)

according to distance can also be considered as an arrangement according to orbital period. As a consequence, the Kirkwood gaps imply that certain periods are missing.

A careful comparison with the period of Jupiter, 11.86 Earth years, shows that the missing periods are rational fractions of it. A rational fraction, such as 1/4, has integer numerator and dominator. The most prominent of the Kirkwood gaps are at periods given by 1/4, 1/3, 2/5, 3/7 and 1/2 times Jupiter's period (Fig. 1.18), suggesting to Kirkwood that Jupiter's gravity produces the gaps. Asteroids with these orbits always come nearest Jupiter in the same point of their orbit, so Jovian gravitational perturbations recur repeatedly at the same orbital position. The recurring gravitational jolts dislodge the asteroids from their orbits, but the details depend on modern concepts of chaos (Section 13.1).

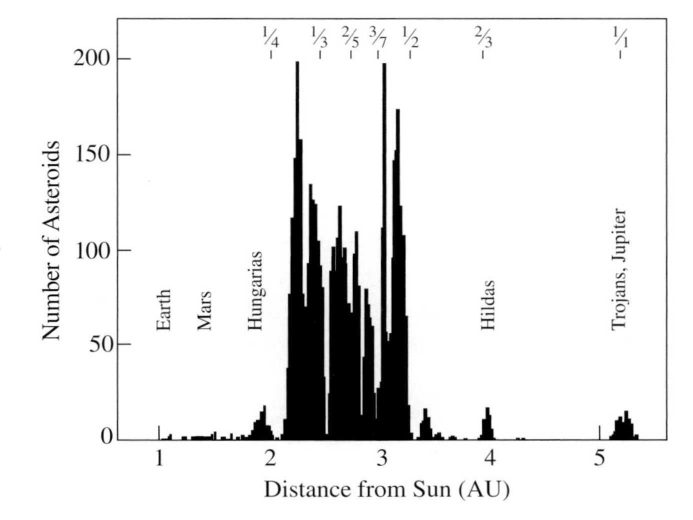

Fig. 1.18 Kirkwood gaps The number of asteroids at different distances from the Sun. Most of the asteroids are found in the asteroid belt that lies between 2.2 and 3.3 AU from the Sun. Repeated gravitational interactions with Jupiter seem to have tossed asteroids out of the Kirkwood gaps with orbital periods of 1/4, 2/7, 1/3, 2/5, 3/7 and 1/2 of Jupiter's orbital period. These fractions are placed above the relevant gap in the figure. In addition, there are several peaks corresponding to groups of asteroids with nearly the same orbital distance, such as the Trojan asteroids that have orbits that are identical in size to the orbit of Jupiter.

Neptune's discovery, triumph of Newtonian gravitational theory

Neptune's discovery was no accident, in contrast to those of Uranus and the first asteroid. It was a direct consequence of precise mathematical calculations of Uranus' motion.

Uranus had been detected by professional astronomers and mistaken for a star on no less than 22 occasions during the century that preceded the realization that it was a planet. These additional observations could be combined with the post-discovery ones to determine Uranus' trajectory and calculate its future position. Before long it was found that the planet was wandering from its predicted path.

A large, unknown world, located far beyond Uranus, was evidently producing a gravitational tug on Uranus, causing it to deviate from the expected location. Two astronomer–mathematicians, John Couch Adams (1819–1892) in England and Urbain Jean Joseph Leverrier (1811–1877) in France, independently located the planet by a mathematical analysis of the wanderings of Uranus.

Adams, a recent graduate from Cambridge University, finished his work first, deriving a precise position of the planet in mid-1845. He left a summary of his results with the then Astronomer Royal, George Biddell Airy (1801–1892), who did not feel compelled to look for the unknown world.

Leverrier finished his best calculations about a year later, and, unlike Adams, published his results. Both scientists had assumed that the undiscovered planet occupied the next place in the sequence of the Titius–Bode law, and they arrived at nearly identical locations for it.

When Leverrier's memoir reached Airy, he persuaded James Challis (1803–1882), Professor of Astronomy at Cambridge University, to make a search for the undiscovered planet. For a variety of reasons, Challis began the investigation slowly, and Leverrier had in the meantime sent his results to the Berlin Observatory where Johann Gottfried Galle (1812–1910) and his student Heinrich Louis d'Arrest (1822–1875) found the planet. They identified it on the first night of their search, on 23 September 1846, using a 0.23-meter (9-inch) refractor; it was located within a degree of both Adams' and Leverrier's predicted positions. Only later did Challis realize that he had previously recorded the planet twice when beginning his own search.

The discovery of the new planet, named Neptune after the Roman god of the sea, was acclaimed as the ultimate triumph of Newtonian science. It resulted from mathematical calculations, based on Newton's theories, of the effects of an unknown planet whose gravity was pulling Uranus from its predicted place. If proof were needed, this achievement certified the validity of gravitational theory.

Neptune is located at a mean distance of 30.06 AU, at 1.6 times the distance of Uranus and fairly close to the 38.8 AU predicted by the Titius–Bode law. Remote Neptune takes so long to travel around the Sun, about 165 Earth years, that it has not made a full orbit since it was discovered in 1846.

1.3 Physical properties of the Sun and planets

Distance to the Sun and other nearby stars

How far away is the Sun from the Earth, and how fast is the Earth moving through space? Kepler's model of planetary motion only provided a scale model for the relative distances of the planets from the Sun, and for a long time no one knew exactly how big the solar system was. Our planet's true distance from the Sun remained unknown for centuries. And since the distances were not reliably known, the velocity of the planet around the Sun could not be determined. The mean orbital speed is equal to the circumference of the orbit divided by one year, the time for the Earth to complete one trip around the Sun.

The crucial unit of distance for the planets is the Sun–Earth distance, known as the astronomical unit and designated AU for short. It can be determined by first estimating the distance between Earth and a nearby planet, and then inferring the Sun–Earth distance from geometry and Kepler's third law. The planetary distances are themselves determined by triangulation from different points on the Earth.

The triangulation method involves measurements of the angular difference in the apparent direction of a nearby cosmic object, as seen against the distant background stars, from two widely separated locations. This angular difference is known as the parallax.

The solar parallax, designated by the symbol $\pi_{\odot}$, is defined as half the angular displacement of the Sun as viewed from opposite sides of the Earth, or mathematically by $\sin \pi_{\odot} = R_{\mathrm{E}}/\mathrm{AU}$, where the equatorial radius of the Earth is $R_{\mathrm{E}} = 6.378\,140 \times 10^6$ meters. The ratio of R_{E} and the AU provides an angle in radian units, and one radian is equivalent to $2.062\,65 \times 10^5$ seconds of arc.

Giovanni Domenico (Jean Dominique) Cassini (1625–1712), the Italian-born French astronomer and first director of the Paris Observatory, obtained an early triangulation of Mars in 1672, combining his observations with those taken from Cayenne, French Guinea. The planet was then in opposition, at its closest approach to Earth. From the two sets of observations, made 7.2 thousand kilometers apart, it was possible to estimate the distance to Mars and to infer a solar parallax of 9.5 seconds of arc. This corresponds to a value of 139.2 billion (1.392×10^{11}) meters for the astronomical unit or about 7 percent less than the modern value of 149.6 billion meters.

Astronomers in the 18th century attempted to improve the measurement accuracy of the Sun's distance during the rare occasions when Venus crossed the face of the Sun, in 1761 and 1769. The method also involved comparison of observations from widely separated locations to determine the distance by triangulation. Difficulties in timing the beginning moment of transit led to uncertain and differing results, as did world-wide observations of the next Venus transits in 1874 and 1882. Values of the solar parallax between 8.6 and 8.9 seconds of arc were inferred.

The AU was established with increasing accuracy in the 19th and 20th centuries, by determining the distances of Mars and the nearby minor planet 433 Eros during their closest approaches to the Earth. The results (Fig. 1.19) converged toward a solar parallax of 8.80 seconds of arc.

The quest for accuracy in the mean distance of the Sun from the Earth culminated in the 1960s, when radar (radio detection and ranging) was used to accurately determine the distance to Venus (Fig. 1.20). The round-trip travel-time, T, for a radio pulse to travel from the Earth to Venus and back – about 276 seconds when Venus is closest to the Earth – was precisely measured. The distance to Venus was then obtained by multiplying half the round-trip time, $T/2$, by the speed of light, $c = 2.997\,924\,58 \times 10^8$ m s^{-1}. The radar measurements have determined the mean distance between the Sun and the Earth to an accuracy of about 1000 meters.

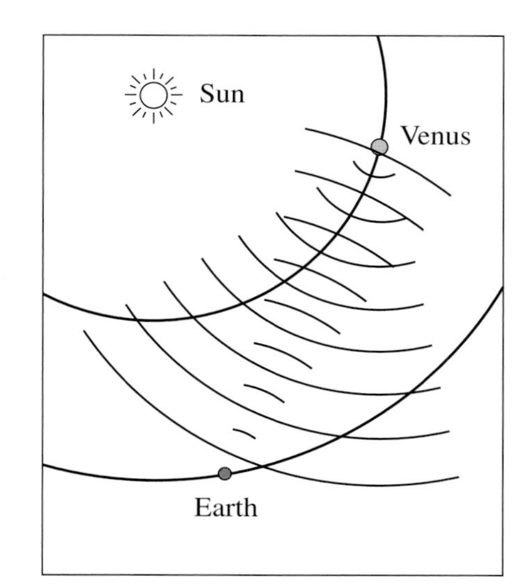

Fig. 1.20 Radar-ranging to Venus Accurate distances to the nearby planets have been determined by sending radio pulses from Earth to the planet, and timing their return several minutes later. The figure shows the emission of a pulse toward Venus; when it bounces from Venus the radiation spreads over the sky and we receive only a small fraction of the original signal, delayed by the round-trip travel-time. If T is the round-trip time and c is the speed of light, the total distance traveled is cT and the distance to Venus is $cT/2$. For Venus, the round-trip time is 4.6 minutes when the planet is nearest Earth and increases to 28.7 minutes when it is furthest away from us.

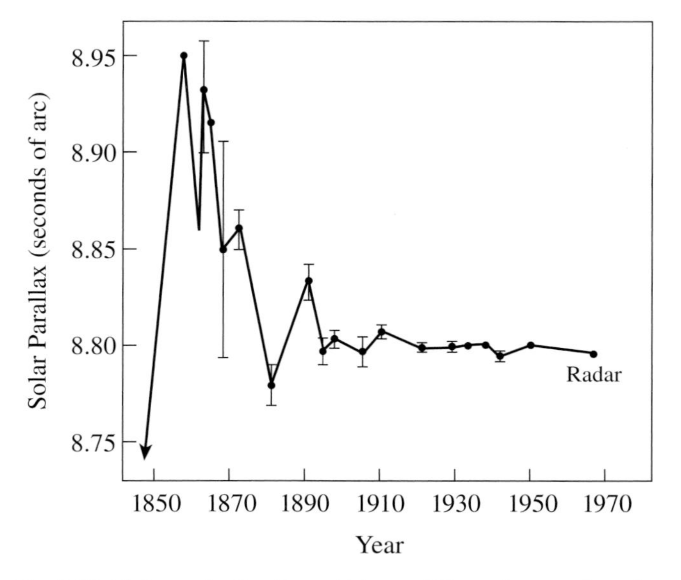

Fig. 1.19 Distance to the Sun Values of the solar parallax obtained from measurements of the parallaxes of Venus, Mars, and the asteroid Eros between 1850 and 1970. Here the error bars denote the probable errors in the determination, whereas the points for 1941, 1950 and 1965 all have errors smaller than the plotted points. In the 1960s, the newly developed radar (radio detection and ranging) technology enabled the determination of the Sun's distance with an accuracy of about 1000 meters. The radar value of the solar parallax is 8.794 15 seconds of arc.

Nowadays the accuracy of the mean Earth–Sun distance is fixed by the exact value for the speed of light. The time, τ_{AU}, for light to travel across one AU is given as a primary astronomical constant:

$$\text{Earth–Sun light travel-time} = \tau_{AU} = 499.004\,782 \text{ seconds},$$

with a derived value for the mean Earth–Sun distance of:

$$\begin{aligned}\text{astronomical unit} &= \text{AU} = c\tau_{AU} \\ &= 1.495\,978\,70 \times 10^{11} \text{ meters.}\end{aligned}$$

This value of the AU corresponds to a solar parallax of $\pi_{\odot} = 8.794\,148$ seconds of arc. Today the AU is technically defined as the unit of distance consistent with the units of time and mass of 1 day and the Sun's mass, and an application of Kepler's third law to the Earth's orbit. It is roughly the mean distance from the Earth to the Sun, accurate enough for most purposes.

Once you have an accurate value for the Sun's distance, the Earth's mean orbital velocity can be determined by dividing its orbital circumference by one year,

$$\begin{aligned}&\text{Earth's mean orbital velocity} \\ &= 2\pi \text{ AU}/P_E = 2.979 \times 10^4 \text{ meters per second}\end{aligned}$$

where $\pi = 3.141\,59$ and $P_E = 1$ year $= 3.1557 \times 10^7$ seconds.

And what about the distances to the other nearby stars? Even the closest stars, other than the Sun, are too far away for us to detect a shift in position from any two points on Earth. To triangulate their distances, astronomers needed a wider baseline, the Earth's annual orbit. Measurements separated by six months, from opposite sides of the Earth's orbit, can reveal a shift in the position of a nearby star with respect to the more distant ones. Half of this angular displacement, known as the annual parallax, π, is the ratio of the AU and the star's distance, D, or $\sin \pi = \mathrm{AU}/D$ radians.

If a nearby star lies in the plane of the Earth's orbit, or in the ecliptic, it will move backwards and forwards in this plane during the year by the amount of π. When the star's direction is perpendicular to this plane, it moves in a circle of angular radius π, and at other ecliptic latitudes the star appears to move in an ellipse with semi-major axis π.

The annual parallax of even the nearest star is so small that it could not be detected until telescopes were invented and refined. Until then, there was no direct measurement of the Earth's motion around the Sun. In the 16th century, the Copernicans argued that the failure to detect an annual parallax simply meant that the stars were too far away and that the angular shift in position was too small to detect. In other words, the Sun–Earth distance had to be much smaller than the distances of the stars.

James Bradley (1693–1762) placed the argument on a more quantitative basis, when telescopic astronomy was still in its infancy. In 1729 Bradley reported that his telescope failed to detect any annual parallax larger than one second of arc for the bright, and presumably nearby, stars. This meant that these stars had to be more than 206 thousand times more distant than the Sun. (At the same time, Bradley discovered a more subtle effect, called the aberration of light, in which the apparent position of a star is constantly altered by the ever-changing motion of an Earth-based observer.)

The stars are so far away, and the parallaxes so tiny, that the first convincing, reliable measurement of a star's annual parallax did not occur until 1838, when Friedrich Wilhelm Bessel (1784–1846) succeeded with the star 61 Cygni. Over the course of a year, the apparent position of this star shifted by a total angle of almost two-thirds of a second of arc, to give a parallax of half that amount. Bessel's parallax of 0.31 seconds of arc was close to the modern value of 0.292 seconds of arc, which corresponds to a distance of 706 thousand AU. Traveling at the speed of light, $c = 2.9979 \times 10^8$ m s^{-1}, it takes radiation 352 million seconds, or 11 years, to travel from 61 Cygni to the Earth.

Bessel's observation was a remarkable achievement, for astronomers finally knew the distance of at least one star, other than the Sun. It also provided the first definite proof that the Earth really is in annual motion about the Sun.

During the 20th century, telescopes of larger size and greater resolving power have been constructed, detecting the annual parallax of about 2000 nearby stars. The nearest one is Proxima Centauri with a distance of 135 million light seconds, or 4.29 light years. Since sunlight travels to the Earth in just 499 seconds, the nearest star is 271 thousand times further away from the Earth than the Sun is.

Distant stars with parallaxes smaller than 0.05 seconds of arc cannot be measured with Earth-based telescopes because of atmospheric distortion that limits their angular resolution. However, instruments aboard the *Hipparcos* satellite, which orbited the Earth above its atmosphere, obtained parallaxes for more than a hundred thousand stars in the 1990s with an unprecedented accuracy of 0.001 seconds of arc, leading to dramatic improvements in our estimates of the distances to the nearby stars.

Mass, radius and mass density of the Sun and major planets

Once an accurate value for the mean distance between the Earth and the Sun is known, one can use it with the orbital period of the Earth to infer the mass of the Sun from Newton's formulation of Kepler's third law. The precise distance to a planet can also be combined with the angular separation of one of its satellite to determine the orbital distance of that satellite from its planet, which can then be combined with the satellite's orbital period to establish the planet's mass using a similar mathematical expression.

When the mass of the Sun and planets are thus determined, we find that the Sun doesn't just lie at the heart of our solar system, it dominates it. Some 99.866 percent of all the matter between the Sun and halfway to the nearest star is contained in the Sun (Table 1.3). All of the objects that orbit the Sun – the planets and their satellites, the comets and the asteroids – add up to just 0.134 percent of the mass in our solar system. As far as the Sun is concerned, the planets are insignificant specks, left over from its formation and held captive by its massive gravity.

We can, for instance, weigh the Sun from a distance, determining its mass from the Earth–Sun distance, 1 AU $= 1.495\,978\,7 \times 10^{11}$ meters, and the Earth's orbital period, $P_E = 1$ year $= 3.1557 \times 10^{11}$ seconds using the expression (Section 1.1)

$$\text{Sun's mass} = M_\odot = 5.9165 \times 10^{11} \frac{(\mathrm{AU})^3}{P_E^2}$$
$$= 1.989 \times 10^{30} \text{ kilograms.}$$

Table 1.3 Distribution of mass in the solar system

Mass of the Sun	$M_\odot = 1.989 \times 10^{30}$ kilograms
Mass of Jupiter	$M_J = 1.899 \times 10^{27}$ kilograms
Mass of the Earth	$M_E = 5.974 \times 10^{24}$ kilograms
Total Mass of the Planets	2.668×10^{27} kilograms $= 446.6\,M_E$
Total Mass of the Satellites	6.2×10^{23} kilograms $= 0.104\,M_E$
Total Mass of the Kuiper-belt Objects	3.0×10^{23} kilograms $= 0.05\,M_E$
Total Mass of the Asteroids	1.8×10^{21} kilograms $= 0.0003\,M_E$
Total Mass of the Planetary System	2.669×10^{27} kilograms $= 446.7\,M_E = 0.00134\,M_\odot$

The linear radius of the Sun, $R_\odot$, can be determined from its angular diameter, θ, using

$$\text{Sun's angular diameter} = \theta = \frac{2R_\odot}{\text{AU}} = 0.0093 \text{ radians},$$

where a full circle subtends 2π radians and 360 degrees. Since there are 60 minutes of arc in a degree, we can express the angular diameter of the Sun in minutes of arc, $\theta = 60 \times 360 \times 0.0093/(2\pi) = 31.97$ minutes of arc, where $\pi = 3.14159$.

Precise solar diameter measurements during the 1980s resulted in a mean value for the solar near-equatorial radius, with an uncertainty of 2.6×10^4 meters:

$$\text{solar radius} = R_\odot = \frac{\theta \times \text{AU}}{2} = 6.95508 \times 10^8 \text{ meters}.$$

The mass density of the Sun, $\rho_\odot$, is obtained by dividing this mass by the Sun's volume, $4\pi R_\odot^3/3$, where the Sun's radius $R_\odot = 6.955 \times 10^8$ meters. It is $\rho_\odot = 1409$ kilograms per cubic meter, only about one-quarter of the mass density of the Earth, which is 5515 kilograms per cubic meter.

By the mid-19th century the eight major planets were known, and their distances established. The radius of each planet can be determined from its angular extent. For instance, Jupiter has an angular diameter of $\theta_J = 46.86$ seconds of arc when it is closest to Earth, or at a distance of $D_J = 4.2$ AU from us. That corresponds to a radius of $R_J = 7.14 \times 10^7$ meters. You can do the arithmetic yourself using $R = \theta\, D/2$ with 1 AU $= 1.496 \times 10^{11}$ meters and converting the angle to radians with 1 radian $= 2.06265 \times 10^5$ seconds of arc.

The radius of Jupiter is 11.2 times the Earth's radius, and its volume is more than one thousand times that of the Earth. It is therefore called a giant planet, as are Saturn, Uranus and Neptune.

The mass of a planet can be inferred from the orbital parameters of its satellites. The motion of a satellite is governed by the more massive planet, all in accordance with the inverse-square law of gravity. Newton used it centuries ago to infer the masses of the Earth, Jupiter and Saturn from their satellite motions, obtaining values comparable to modern ones.

The gravitational expression of Kepler's third law can be used to determine the planet's mass, M_P, using the orbital period, P_S, and distance, D_S, of the satellite. As an example, we can obtain Jupiter's mass from observations of any one of its four large satellites, discovered in 1610. The satellite Io's period is $P_S = 1.77$ days and Io's distance from Jupiter is $D_S = 4.22 \times 10^6$ meters, yielding a mass $M_J = 4\pi^2 D_S^3/(GP_S^2) = 1.9 \times 10^{27}$ kilograms, or about one-thousandth the mass of the Sun, where $\pi = 3.14159$, the Newtonian constant of gravitation $G = 6.672 \times 10^{-11}$ m^3 kg^{-1} s^{-2}, and there are 86 400 seconds in one day.

The masses of the other giant planets can be inferred from their satellites in the same way. The same technique can be used to obtain Earth's mass, from the motion of its Moon, and the mass of Mars from its two satellites.

Venus and Mercury have no satellites, but their mass can be obtained from detailed examination of the planetary orbits around the Sun. Since all objects exert gravitational forces on each other, any planet will be pulled ever so slightly out of its pure elliptical orbit by the other planets. The tiny competing tugs of a planet's neighbors therefore produce minute wiggles to a basic Sun-dominated motion, and observation of the small orbital meandering can be combined with gravitational theory to infer a planet's mass. Modern measurements of the exact, changing locations of the planets and of interplanetary spacecraft have detected these tiny perturbations, leading to precise determinations of the planetary masses.

The bulk properties of the planets are specified by their mass density – that is, the ratio of their mass to volume. For instance, the mean mass density of Jupiter is 1330 kilograms per cubic meter, abbreviated 1330 kg m^{-3}. That is comparable to the Sun's mean mass density of 1409 kg m^{-3}.

Contemporary values for the angular size at closest approach, radius, mass and mass density of the major planets are given in Table 1.4. They can be divided into two

Table 1.4 Angular diameter, radius, mass and mean mass density of the major planets[a]

Terrestrial Planets	Mercury	Venus	Earth	Mars
Angular Diameter (seconds of arc)[b]	10.9	61.0		17.88
Equatorial Radius, R_P (km)	2440	6052	6378	3396
Equatorial Radius, R_P ($R_E = 1.0$)	0.382	0.949	1.000	0.532
Mass, M_P (10^{23} kg)	3.302	48.685	59.736	6.4185
Mass, M_P ($M_E = 1.0$)	0.0553	0.8150	1.0000	0.1074
Mean Mass Density (kg m^{-3})	5427	5204	5515	3934
Giant Planets	**Jupiter**	**Saturn**	**Uranus**	**Neptune**
Angular Diameter (seconds of arc)[b]	46.86	19.52	3.60	2.12
Equatorial Radius, R_P (km)	71 492	60 268	25 559	24 766
Equatorial Radius, R_P ($R_E = 1.0$)	11.209	9.449	4.007	3.883
Mass, M_P (10^{23} kg)	18 986	5684.6	868.32	1024.3
Mass, M_P ($M_E = 1.0$)	317.710	95.162	14.535	17.141
Mean Mass Density (kg m^{-3})	1326	687	1318	1638

[a] The mass and mean mass density values are from http://ssd.pl.nasa.gov/phys_props_planets.html, the equatorial radii for the giant planets are those at the 1-bar pressure level in their atmospheres, and they are taken from *The New Solar System*, Fourth Edition (Cambridge University Press, New York 1999).

[b] The largest angular diameter seen from the Earth when the planet is at its closest approach to Earth.

groups: the four terrestrial planets, which resemble the Earth, and the four giant planets like Jupiter. The terrestrial planets, Mercury, Venus, Earth and Mars, are those closest to the Sun; they are relatively small, dense planets with solid surfaces. The remote giant worlds are all big and massive, but with low mass densities. Of these, Jupiter and Saturn consist largely of gas and liquid, while Uranus and Neptune are made up of a mixture of gas, liquid and melted ice. Many of the constituents of the Sun and giant planets can be determined by the technique of spectroscopy, which is discussed next.

1.4 Ingredients of the Sun and planetary atmospheres

The ingredients of the Sun and planetary atmospheres can be determined when the intensity of their radiation is shown as a function of its wavelength. Such a display is called a spectrum, and the study of spectra is known as spectroscopy.

Each chemical element or compound produces a unique set, or pattern, of spectral signatures at certain specific wavelengths and only at those wavelengths. They resemble a barcode or a fingerprint that can be used to identify the element or compound.

A hot, glowing body like the Sun emits radiation at all wavelengths, with a continuous spectrum. If this radiation passes through a cool, tenuous gas, such as the outer layers of a planetary atmosphere, part of the radiation is absorbed at discrete wavelengths. These spectral features are called absorption lines because they look like a line in the spectrum. If the gas is heated to incandescence, it will emit radiation at the same specific wavelengths, and the spectral features are called emission lines. The patterns of either the absorption or emission lines tell us the atoms or molecules that are present in the gas.

The technique of astronomical spectroscopy was first developed using the bright light of the Sun. When its spectrum is examined at high wavelength resolution, numerous fine, dark absorption lines are seen crossing the rainbow-like display (Fig. 1.21). The separate colors of sunlight are somewhat blurred together when coarser resolution is used, and the dark places are no longer found superimposed on its spectrum.

These dark gaps, or absorption lines, were first noticed in 1802 by the English chemist William Hyde Wollaston (1766–1828), and investigated in far greater detail by the Bavarian telescope maker Joseph von Fraunhofer (1787–1826). By 1815, von Fraunhofer had cataloged the wavelengths of hundreds of them, assigning Roman letters A, B, C, . . . to the darkest and most prominent of them, starting

Fig. 1.21 Visible solar spectrum A spectrograph has spread out the visible portion of the Sun's radiation into its spectral components, displaying radiation intensity as a function of wavelength. When we pass from short wavelengths to longer ones (*left to right* and *top to bottom*), the spectrum ranges from violet through blue, green, yellow, orange and red. Dark gaps in the spectrum, called Fraunhofer absorption lines, represent absorption by atoms or ions in the Sun. The wavelengths of these absorption lines can be used to identify the elements in the Sun, and the relative darkness of the lines helps establish the relative abundance of these elements. (Courtesy of the National Solar Observatory/Sacramento Peak, NOAO.)

from the long-wavelength, red end of the visible solar spectrum and progressing to its short-wavelength side.

The detailed explanation for the Sun's absorption lines was provided in the mid-19th century in a chemistry laboratory in Heidelberg, Germany. There the chemist Robert Wilhelm Bunsen (1811–1899), inventor of the Bunsen burner, and his physicist colleague Gustav Robert Kirchhoff (1824–1887) unlocked the chemical secrets of the Universe. When they vaporized an individual element in a flame, and heated it to incandescence, the hot vapor produced a distinctive pattern of bright emission lines whose unique wavelengths coincided with the wavelengths of some of the dark absorption lines in the Sun's spectrum, identifying that element as an ingredient of the solar gas. Solar lines designated by Fraunhofer by D, E and H and K were respectively ascribed to sodium, iron and calcium.

The lightest element, hydrogen, was also identified in the solar spectrum, in 1862 by the Swedish physicist Anders Jonas Ångström (1814–1874); it accounts for Fraunhofer's C and D lines. Subsequent investigations of the great strength of these lines indicated that hydrogen is the most abundant element in the visible solar gases.

Since the Sun was most likely chemically homogeneous, a high hydrogen abundance was implied for the entire star. This accounts for the Sun's low mass density. We now know that hydrogen accounts for 92.1 percent of the number of atoms in the Sun, and that hydrogen is the most abundant element in most stars, in interstellar space, and in the entire Universe.

Helium, the second most abundant element in the Sun, is so rare on Earth that it was first discovered in the Sun. A previously unknown emission line was first noticed during a solar eclipse on 18 August 1868, at a wavelength of 587.56 nanometers near the two yellow sodium lines. Since this feature had no known Earthly counterpart, a new chemical element had been discovered, which Norman Lockyer (1836–1920) named "helium" after the Greek Sun god *Helios*. Helium was not found on Earth until 1895, when William Ramsay (1852–1919) discovered it as a gaseous emission from a mineral called clevite. Helium accounts for 7.8 percent of the number of atoms in the Sun, and all of the heavier elements amount to only 0.1 percent.

Today, helium is used on Earth in a variety of ways, including the inflation of party balloons and in its liquid state to keep sensitive electronic equipment cold. Though plentiful in the Sun, helium is almost non-existent on the Earth. It is so terrestrially rare that we are in danger of running out of helium during this century.

The next most abundant elements in the Sun are carbon, oxygen and nitrogen, as well as the inert element, neon (Table 1.5). Like helium, they are not amongst the most prominent absorption lines noted by Fraunhofer. The large solar abundance of these elements was realized during spectroscopic studies in the early-20th century, when scientists understood the processes that produce spectral lines. The line intensities are related to the abundance and internal structure of the element, as well as the physical conditions of the gas in which the lines originate.

Table 1.5 The ten most abundant elements in the Sun

Atomic Number[a]	Element	Symbol	Number of Atoms (Silicon = 10.0)	Date of Discovery
1	Hydrogen	H	279 000	1766
2	Helium	He	27 200	1868[b]
6	Carbon	C	101	(ancient)
7	Nitrogen	N	31.3	1772
8	Oxygen	O	238	1774
10	Neon	Ne	34.4	1898
12	Magnesium	Mg	10.7	1755
14	Silicon	Si	10.0	1823
16	Sulfur	S	5.15	(ancient)
26	Iron	Fe	9.00	(ancient)

[a] The atomic number is equal to the number of protons in the nucleus of an atom.
[b] Helium was discovered on the Sun in 1868, but it was not found on Earth until 1895.

Spectroscopic observations of the planets are beset by two difficulties. One of these is the relatively cool temperature of the planetary atmospheres, and the other is the confusion caused by solar and terrestrial spectral features.

Temperature affects the physical state of an element or compound, and determines which spectral lines may be emitted by it. At the high temperatures in the solar atmosphere, at or above its visible disk, the gas consists of atoms or ions whose spectral signatures are located in visible sunlight or at shorter ultraviolet and X-ray wavelengths. In contrast, the outer planetary atmospheres have relatively low temperatures, in which there are no ions and atoms combine to form molecules. These molecules produce extremely complex, banded absorption features in the long-wavelength infrared part of the spectrum and are often unseen at visible wavelengths.

The sunlight that strikes a planet already contains the Sun's absorption lines, to which the planet's are added. Before the reflected light reaches a telescope on the ground, that light must pass through the Earth's atmosphere where it is further modified. Molecular oxygen and water vapor in our air produce strong absorption bands at red and infrared wavelengths. These terrestrial features hide, line by line, absorption of these same gases in the atmospheres of other planets. A spectrum gathered above the Earth's atmosphere would not have these confusing terrestrial lines.

Astronomers got around some of these difficulties by extending planetary spectra to the infrared wavelengths, and by comparing photographic spectra with those of the Moon, which has no substantial atmosphere and whose spectrum contains the spectral imprints of the Sun's outer atmosphere and the Earth's air. Although the spectra were crowded with solar and terrestrial absorption lines, the features of any substance that was very abundant in a planetary atmosphere, and of low abundance in our air, could be detected.

In 1932, Walter Sydney Adams (1876–1956) and Theodore Dunham, Jr. (1897–1984) detected the absorption lines of carbon dioxide, CO_2, in the spectrum of Venus. This was because there is a lot of carbon dioxide in the atmosphere of Venus and relatively little in the Earth's atmosphere; in fact, carbon dioxide accounts for 96 percent of the thick atmosphere of Venus (Section 3.1). Gerard Peter Kuiper (1905–1973) used ground-based spectroscopy in the 1950s to show that carbon dioxide is present in the Martian atmosphere, but the amounts were not known with certainty until the thin, cold atmosphere was investigated from space (Section 3.1).

Strong red and infrared absorption lines were also detected in early spectroscopy of the giant planets (Fig. 1.22). In the 1930s Rupert Wildt (1905–1976) showed that they could be explained by methane, CH_4, and ammonia, NH_3, noticing that methane must be enormously abundant on all four planets. Dunham verified his interpretation when he filled a long pipe with each gas and measured absorption lines at the same wavelengths as those found in the planetary spectra.

The presence of methane and ammonia would be expected if the planets formed together with the Sun. The massive giant planets would then approximate the solar composition, which would explain their low mean mass densities. The overwhelmingly abundant hydrogen, H, would combine with the abundant carbon, C, oxygen, O, and nitrogen, N, in the low-temperature environment far from the Sun, to form stable molecules of methane, CH_4, water vapor, H_2O, and ammonia, NH_3.

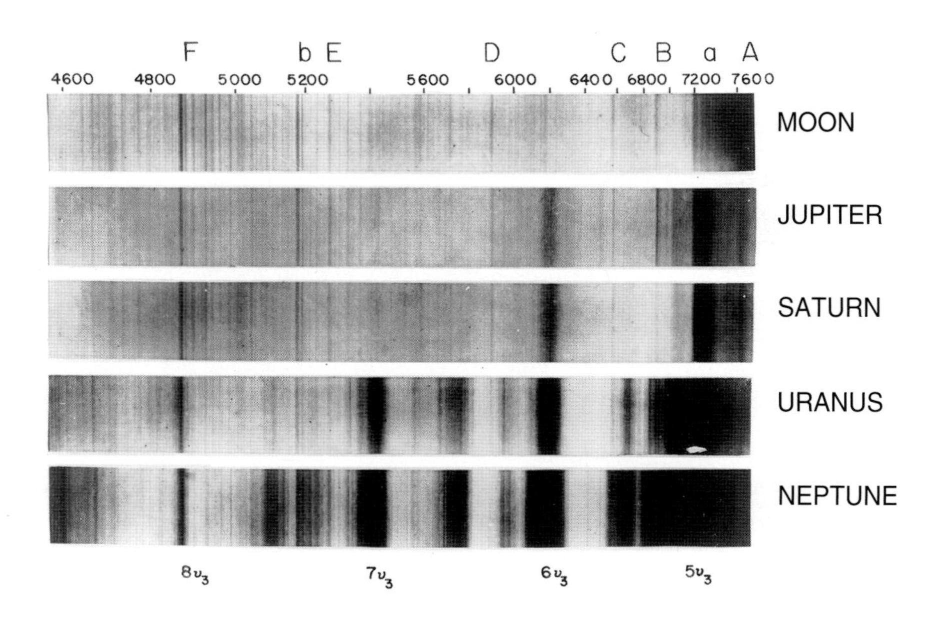

Fig. 1.22 Spectra of the giant planets The radiation spectra of the giant planets at visible wavelengths photographed by Vesto M. Slipher at the Lowell Observatory in 1907. Slipher's work was discussed by Rupert Wildt in 1931, who interpreted some of the bands of Jupiter as absorption by ammonia and methane, the natural gas we use for cooking and heating. Multiples of methane's vibration frequency, ν_3, are given along the bottom of this figure. The Moon's spectrum is also shown to illustrate the dark Fraunhofer lines found in reflected sunlight and those introduced by the terrestrial atmosphere. The wavelength scale at the top of the figure is in the ångstrom units that were in common use when Slipher made his observations; just divide these numbers by ten to get the wavelength in nanometers.

Nevertheless, the exact composition of the atmospheres of the giant planets had to await space-age infrared spectroscopy, which showed that Jupiter and Saturn are mainly composed of hydrogen (Section 3.2).

1.5 Satellites, rings and small, distant objects

Discovery of satellites and rings

For nearly half a century, the only satellites known in the solar system were the Earth's Moon and Jupiter's four largest satellites, discovered by Galileo in 1610 and now often called the "Galilean satellites" in his honor. They are named after four of the god Jupiter's lovers. Io is the innermost of the four Galilean satellites, succeeded by Europa, Ganymede, and Callisto, in order of distance from Jupiter.

In 1655 the Dutch astronomer–physicist, Christiaan Huygens (1629–1695), discovered Titan, the largest satellite of Saturn, named after Saturn's older brother. Within a few decades Giovanni Domenico (Jean Dominique) Cassini (1625–1712), at the Paris Observatory, had discovered four more moons circling Saturn; they are named Iapetus, Rhea, Tethys and Dione (Fig. 1.23). Like the Earth's Moon, Saturn's second-largest moon, Iapetus, always presents the same face to its planet. According to classical mythology, Gaia (Earth) gave birth to Uranus (Heaven), without the aid of any male, and coupled with him to conceive six male Titans, including Iapetus, and six female Titanesses, including Rhea, Tethys and Dione.

Both Cassini and Huygens used "aerial" refractors, with an objective lens placed on a tower or tall pole and a separate eyepiece near the ground. This novel arrangement permitted the use of an objective lens with slight curvature and long focal length to help correct for aberration and bring the image into sharp focus. Titan was, for example, discovered using a 0.05-meter (2-inch) objective lens with a focal length of 7 meters (276 inches), connected to an eyepiece by just a string for alignment.

Huygens turned his telescope toward Saturn itself, and explained its mysterious handle-like appendages. Galileo

Fig. 1.23 Satellites of Saturn Six of Saturn's large moons are shown in this one-minute exposure made with the United States Naval Observatory's 0.66-meter (26-inch) refractor. From left to right, the satellites are Titan, Dione, Enceladus, Tethys, Mimas and Rhea (on the other side). The faint image below the planet is that of a star. A partially transparent metallic film was used to weaken the light from Saturn and its rings. (Courtesy of Dan Pascu.)

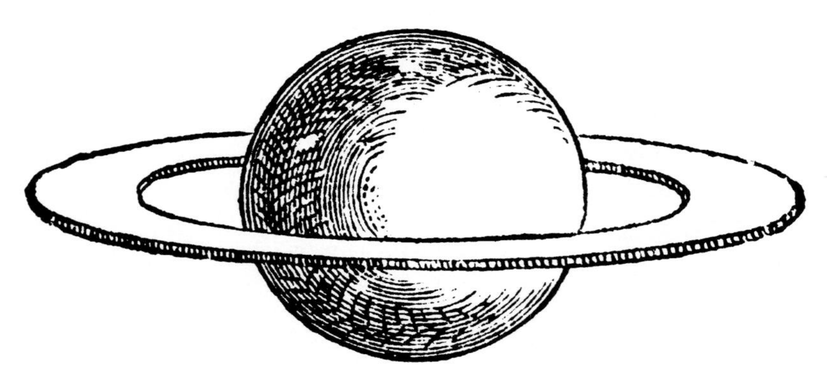

Fig. 1.24 Saturn's ring In 1659 Christiaan Huygens published this drawing of Saturn and its ring in his monograph *Systemia Saturnium*. Huygens recognized that a detached ring would explain the planet's ever-changing appearance, and announced his discovery in the form of an anagram, a succession of scrambled letters. The drawing shown here was accompanied by the deciphered anagram "Saturn is girdled by a thin flat ring, nowhere touching it, and inclined to the ecliptic".

had noticed that the planet was not round, but had blurry objects on each side. When these objects disappeared two years later, Galileo wondered if Saturn "had devoured his own children". In 1656 Huygens, then only 27 years old, realized that the planet was surrounded by "a thin flat ring, nowhere touching it, and inclined to the ecliptic" (Fig. 1.24). Because the ring is tipped with respect to the plane of the Earth's orbit around the Sun, it changes its shape when viewed from Earth, slowly opening up and then turning edge-on as Saturn makes her slow 29.5-year orbit around the Sun. When the ring is opened up, it resembles handle-like appendages, but when it is viewed edge-on the ring virtually disappears.

Cassini also observed Saturn's ring, suggesting that it is composed of swarms of satellites too small to be resolved individually, circling the planet with different velocities. In 1675 he discovered a dark separation in the ring, that is now known as the "Cassini Division".

In fact, there are three main rings visible from the Earth, the outer A, central B and inner C rings (Table 1.6). The A and B rings are separated by the Cassini division. The C ring is also known as the crepe ring since it is the most transparent of the three rings.

Then, in 1867, James Clerk Maxwell (1831–1879), only 26 years old, explained in detail just how Saturn's wide, thin rings could remain suspended in space, instead of being pulled into the planet by its gravity. His calculations indicated that the rings couldn't be solid. The side of the rings closer to the planet feels a stronger gravitational pull than the side further away, and the difference would rip a solid ring to pieces. Maxwell proposed that the rings are instead composed of a vast number of small particles, each pursuing its individual orbit in the plane of the planet's equator.

The innumerable particles that make up Saturn's rings act as tiny satellites that move in accordance with Kepler's third law, with the inner parts moving at a faster speed than the outer ones. James Edward Keeler (1857–1900) confirmed this observationally near the end of the century, in 1895 at the Allegheny Observatory in Pittsburgh. The inner rings of Saturn revolve at a faster speed than the planet rotates, suggesting that the ring particles are gradually falling into Saturn and must be continually replenished (Section 10.4).

In the meantime, more planetary satellites had been found, and for the next three centuries their discovery

Table 1.6 The main rings of Saturn

Name	Distance from Planet Center (Saturn radii[a])	Orbital Period[b] (hours)	Width (kilometers)
A ring	2.025 to 2.267	12.1 to 14.2	14 670
Cassini division	1.949 to 2.025	11.4 to 12.1	4585
B ring	1.525 to 1.949	7.9 to 11.4	25 580
C ring	1.235 to 1.525	5.8 to 7.9	17 490

[a] The equatorial radius of Saturn at the 1-bar pressure level is 60 268 000 meters or 6.0268×10^7 meters, nearly ten Earth radii.

[b] Saturn's rotation period is 10.6562 hours, so the inner B ring and all of the C ring move around the planet at a faster rate than the planet rotates.

progressed more or less in tandem with the development of increasingly powerful telescopes. William Herschel (1738–1822), the discoverer of Uranus, used his 0.15-meter (6.2-inch) reflector to identify four, two each of Uranus (Oberon and Titania in 1787) and Saturn (Mimas and Enceladus in 1789). It wasn't until 1851 that William Lassell (1799–1880) found two more Uranian satellites, Ariel and Umbriel, using a 0.61-meter (24-inch) reflector, and a fifth moon, Miranda, was found nearly a century later – in 1948 – by Gerard Peter Kuiper (1905–1973).

The five large satellites of Uranus are named for characters in literature. Oberon and Titania are the king and queen of the fairies in Shakespeare's *A Midsummer Night's Dream*. Inside their orbits is Umbriel, a "dusky, melancholy sprite" in Alexander Pope's *Rape of a Lock*. Close to the planet is Ariel, described by Shakespeare as "an airy spirit" in *The Tempest*. Closer yet is Miranda, named for Prospero's daughter in *The Tempest*.

Saturn's satellite Mimas has the name of one of the giants who fought against the gods in Greek mythology. The ringed planet's moon Enceladus is named for the Titan who was crushed in a battle between the Olympian gods and the Titans; earth piled on top of him became the island of Sicily. Two other satellites of Saturn, discovered in the 19th century, are named after Hyperion, a Titan, and Phoebe, a Titaness.

Lassell found Neptune's largest satellite in 1846, just a few weeks after the discovery of the planet. The satellite was named Triton – a god of the sea in Greek mythology – the son of Poseidon, the Greek equivalent of Neptune, the Roman god of the sea. Triton is the only large satellite in the solar system that travels in a retrograde orbit, opposite to the direction of the planet's rotation and the orbital direction of all planets and most satellites. A second, much smaller Neptunian satellite was not definitely known for more than a century; Kuiper located it on photographic plates in 1949. It was named Nereid for a sea nymph lured by Triton's conch-shell music in Greek mythology.

Neptune's satellites differ from those of Jupiter, Saturn and Uranus. Each of these planets has a group of large satellites that revolve in regularly-spaced, circular orbits in the same direction as the rotation of the planet and close to the planet's equatorial plane, presumably because they share the rotation of the material from which the planet and its satellites formed. In contrast, Neptune has just one large satellite, Triton, and it moves in the backward retrograde direction.

According to one hypothesis, Triton was born in its own independent orbit around the Sun, and was subsequently captured by Neptune into an eccentric, inclined retrograde orbit. While its orbit was evolving, Triton could have cannibalized satellites it collided with, thereby removing any other large satellites Neptune may have once had.

The large planetary satellites, with radii larger than 100 kilometers or 10^5 meters, were all discovered by the mid-20th century, and most of them were known by the end of the 19th century (Table 1.7). Altogether 21 of them are known – one for Earth, four for Jupiter, nine for Saturn, five for Uranus, and two for Neptune.

Careful telescopic scrutiny during the 19th and 20th centuries led to the discovery of relatively small satellites accompanying Mars and Jupiter. The two moons of Mars were, for example, located by Asaph Hall (1829–1907) in 1877, when the planet was closer to Earth than usual, using the 0.66-meter (26-inch) refractor of the United States Naval Observatory. The two Martian moons are named Phobos (fear) and Deimos (terror) after the attendants of the Greek god of war (Ares) in Homer's *Illiad*; Mars is the Roman god of war. These tiny satellites reflect so little sunlight that they must be comparable to a large asteroid in size, and they might have been captured from the nearby asteroid belt.

The inner Moon, Phobos, is a real maverick, moving around Mars at a faster speed than the planet rotates. Its orbit is steadily shrinking as the result of tidal interaction with Mars (Section 8.10), so Phobos is spiraling toward unavoidable destruction! It will either collide with the Martian surface or be torn apart by Mars' tidal forces to make a ring around Mars in about 100 million years.

Jupiter has eight small outer moons in eccentric tilted orbits that are so far from the planet that the Sun competes for their gravitational control. These eight outer moons fall into two widely separated groups (Fig. 1.25). The innermost four move in the same direction as the planet rotates, but in highly inclined orbits that are not in the plane of the planet's equator. The outermost four circle Jupiter in the backward, or retrograde, direction (Table 1.8). The Sun's gravitational perturbations would have dislodged the outermost satellites if they had direct orbits.

All eight of the outer moons of Jupiter are thought to be former asteroids that wandered near Jupiter and became captured into its family. The two groups of unusual moons may have formed in collisions, resulting in four fragments from each of two original parent bodies.

The list of small satellites more than doubled in the 1970s and 1980s when the *Voyager 1* and *2* spacecraft traveled close enough to the distant giant planets to detect numerous satellites smaller than 100 kilometers in radius. At about the same time, faint ring systems were also discovered girdling Jupiter, Uranus and Neptune (Sections 9.5, 11.4).

Table 1.7 Large planetary satellites[a]

Name	Mean Radius[b] (km)	Mass (10^{20} kg)	Mean Mass Density (kg m^{-3})	Distance from Planet Center (10^6 m)	Period of Revolution[c] (days)	Year of Discovery
EARTH						
Moon	1738	734.9	3344	384.4	27.3217	
JUPITER						
Io	1822	894.0	3528	422	1.77	1610
Europa	1561	480.0	3014	671	3.55	1610
Ganymede	2631	1482.3	1942	1070	7.16	1610
Callisto	2410	1076.6	1834	1883	16.7	1610
SATURN						
Mimas	199	0.38	1142	186	0.94	1789
Enceladus	249	0.8	1000	238	1.37	1789
Tethys	530	7.6	1006	295	1.89	1684
Dione	559	10.5	1498	377	2.74	1684
Rhea	764	24.9	1236	527	4.52	1672
Titan	2575	1345.7	1881	1222	15.9	1655
Hyperion	142		1250	1464	21.3	1848
Iapetus	718	18.8	1025	3561	79.3	1671
Phoebe	110		2300	12 944	550 R	1898
URANUS						
Miranda	236	0.71	1201	130	1.41	1948
Ariel	579	14.4	1665	191	2.52	1851
Umbriel	585	11.8	1400	266	4.14	1851
Titania	789	34.3	1715	436	8.71	1787
Oberon	761	28.7	1630	583	13.5	1787
NEPTUNE						
Triton	1353	214.2	2061	354	5.88 R	1846
Nereid	170		1500	5513	360	1949

[a] The physical parameters of the natural satellites are given on the web at http://ssd.jpl.nasa.gov/sat_props.html
[b] The radii are given in units of kilometers, abbreviated km, the mass is in kilograms, abbreviated kg, and the mass density in kilograms per cubic meter, abbreviated kg m^{-3}. By way of comparison, the equatorial radius of the planet Mercury is 2440 kilometers, so Ganymede and Titan are both bigger than Mercury.
[c] The letter R following the period denotes a satellite revolving about its planet in the retrograde direction, opposite to that of the planet's rotation and orbital motion about the Sun.

Pluto – a small world with an oversized companion

The discovery of Neptune in 1846 resulted from a mathematical study of the differences between the predicted and observed positions of Uranus, attributed to the gravitational pull of the then unknown planet. Astronomers hoped that similar irregularities in Neptune's motion would lead to the discovery of another remote planet; but because of Neptune's long 165-year orbit there were insufficient observations. Prediction of another unknown planet therefore had to be based upon perturbations in Uranus' motion, after corrections for the gravitational effects of Neptune.

Two astronomers used the corrected Uranus data to predict an undiscovered planet beyond Neptune. The first such prediction was made in 1909 when William Henry Pickering (1858–1938) argued that both Neptune and a remote Planet O were producing gravitational tugs on Uranus. Percival Lowell (1855–1916) made the next attempt in 1915; he called his unknown object Planet X.

Table 1.8 Small planetary satellites discovered before the space age

Name	Radius (km)	Distance from Planet Center (10^6 m)	Period of Revolution[a] (days)	Year of Discovery
MARS				
Phobos	9–13	9.4	0.32	1877
Deimos	5–8	23.5	1.26	1877
JUPITER				
Leda	8	11 094	239	1974
Himalia	90	11 480	251	1904
Lysithea	20	11 720	259	1938
Elara	40	11 737	260	1905
Ananke	15	21 200	631 R	1951
Carme	22	22 600	692 R	1938
Pasiphae	35	23 500	735 R	1908
Sinope	20	23 700	758 R	1914

[a] The letter R following the period denotes a satellite revolving about its planet in the retrograde direction, opposite to that of the planet's rotation and orbital motion about the Sun.

Lowell directed the most ambitious search for the trans-Neptunian planet, at his observatory in Flagstaff, Arizona, but no new planet was found at a variety of predicted locations between 1905 and 1919. The search from the Lowell Observatory continued a decade later using a new 0.33-meter (13-inch) photographic telescope. Once three photographs had been taken at intervals of several days, they were set in pairs in a blink microscope that would show the apparent motion of a planet, asteroid or comet against a background of nearly half a million stars on each photograph.

After months of painstaking work, Clyde William Tombaugh (1906–1997) discovered, on 18 February 1930, the sharp, faint, moving image of the elusive quarry (Fig. 1.26). The new object was named Pluto, for the Roman god of the underworld. It is a small frozen world at the outer fringe of the planetary system, with a highly elongated orbit that carries it between 29.7 and 49.3 AU from the Sun.

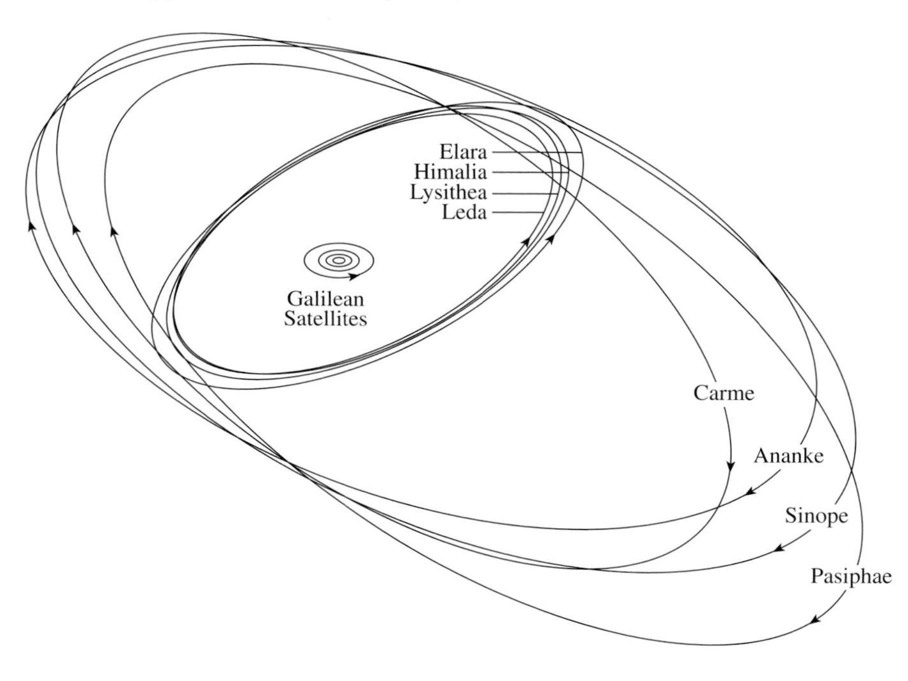

Fig. 1.25 Jupiter's outermost satellites Eight satellites of Jupiter have eccentric orbits that are inclined with respect to Jupiter's equatorial plane. The inner group of four moves around Jupiter in conventional direct orbits, in the same direction that the planet orbits the Sun, at distances between 11 and 12 billion (1.1 and 1.2 $\times 10^{10}$) meters. Their names all end in the letter "a" – Leda, Himalia, Lysithea, and Elara. The outer group of four has backward retrograde orbits that lie at distances between 20 and 24 billion (2.0 and 2.4 $\times 10^{10}$) meters from Jupiter. The outer group has names ending in "e" – Ananke, Carme, Pasiphae, and Sinope. In contrast, the four Galilean satellites, which are relatively near to the planet, move with direct orbits close to the planet's equatorial plane, as do four smaller satellites that are even closer to Jupiter.

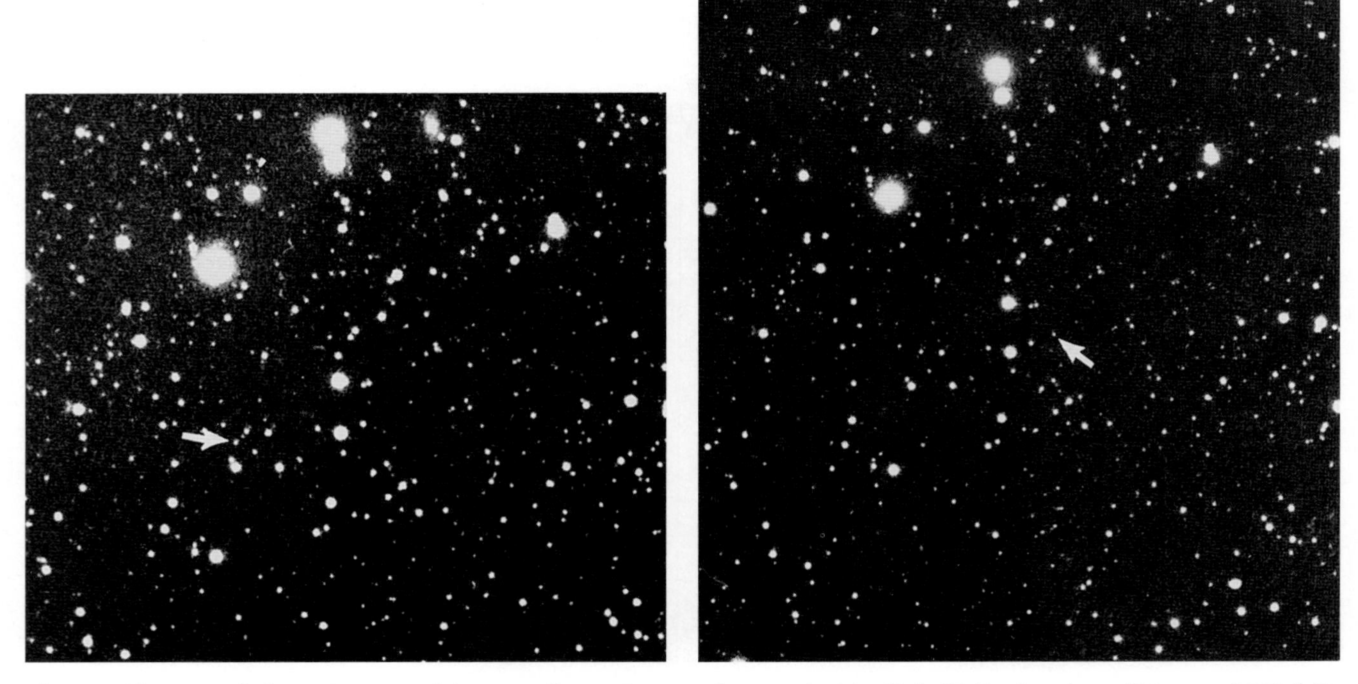

Fig. 1.26 Discovery of Pluto A region of the constellation Gemini, photographed by Clyde W. Tombaugh on 23 January 1930 (*left*), and the same region photographed six days later (*right*). When comparing the two plates on 18 February 1930 with a blink microscope, Tombaugh noticed an object (*arrows*) on the second plate that had changed its location with respect to the background stars since the first plate was taken. This was a previously unknown object that had to belong to the solar system. Because of its slow apparent motion across the sky, the planet images were separated by just 3.5 millimeters on the two photographs. (Courtesy of the Lowell Observatory.).

Pluto's orbit around the Sun is so far from circular that it crosses the orbit of Neptune. Pluto is, however, protected from possible collision with the large planet by a special orbital relationship, called a resonance. Simply put, for every three orbits of Neptune around the Sun, Pluto completes two.

Pluto is now known to be a double object, with a companion that is half as big as Pluto. This discovery was an accidental by-product of observations made for another purpose. In 1978, astronomers at the United States Naval Observatory were obtaining a series of photographs to improve the accuracy of Pluto's orbit, when several of the images appeared slightly distorted, from a round to oblong shape. The elongation seemed to disappear every few days, and careful examination showed that it is caused by another small world that orbits Pluto. The two objects are so close together and so far away that they remain blurred together when viewed with even the best telescopes on the ground, but they can be clearly resolved with the *Hubble Space Telescope* that orbits the Earth above its obscuring atmosphere (Fig. 1.27).

Pluto's companion is named Charon, after the boatman who ferried new arrivals across the river Styx at the entrance to Pluto's underworld, Hades. Penniless ghosts are said to have waited endlessly because Charon gave no free rides.

The announcement of this remarkable doubling was a happy surprise, for it permitted determining the mass of Pluto. Charon orbits Pluto at a distance of 19 640 kilometers, once every 6.387 Earth days. For comparison, a satellite that close to Earth would orbit in 7 hours. Charon's slow revolution about Pluto is a result of Pluto's small mass – only 0.2 percent (0.002) of the Earth's mass and only about one-sixth the mass of our Moon.

This means that Pluto was not found because it was correctly predicted. Its mass is far too small to have noticeably influenced the past motions of Uranus. In fact, the distant Earth exerts a larger gravitational influence on Uranus than Pluto does. The discovery of Pluto was the result of a meticulous and systematic search that was guided by an incorrect prediction, which merely happened to point in the general direction of Pluto.

When Pluto and Charon pirouette into an edge-on view from Earth, we see a series of mutual eclipses as the two objects alternately pass directly in front of each other. Timing the starts and ends of such occultations permits an accurate determination of their size. Pluto has a radius of just 1160 kilometers, or about one-fifth the size of Earth, whose equatorial radius is 6378 kilometers. Charon is nearly half the size of Pluto, with a radius of 635 kilometers.

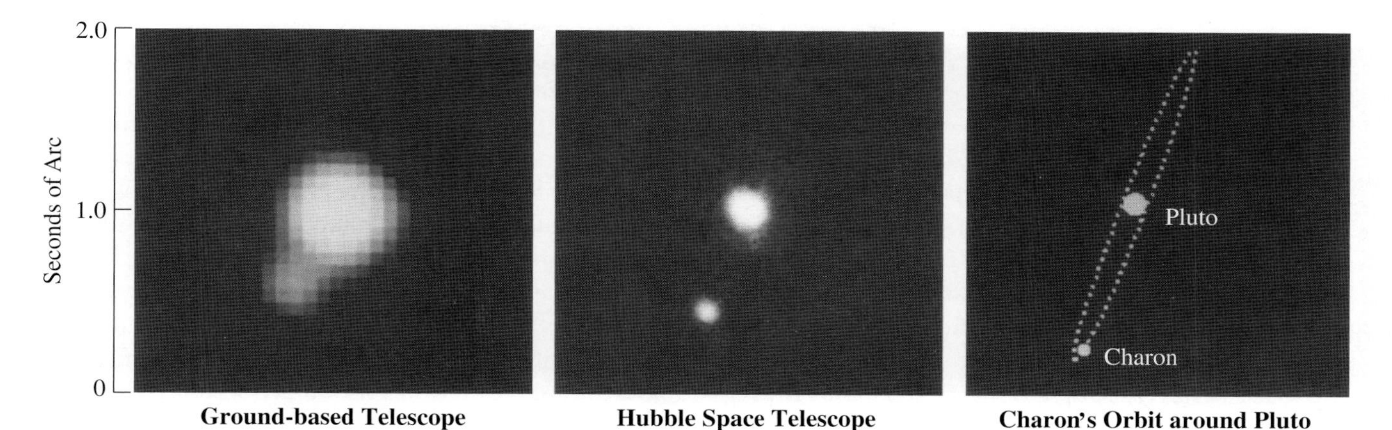

Fig. 1.27 A double object The *Hubble Space Telescope (HST)* distinguishes between Pluto, the bright object at the middle of the central image, and its companion Charon, the fainter object in the lower left of the central image. Observations from telescopes on Earth (*left*) were unable to clearly resolve the pair because of atmospheric distortions. At the time of the *HST* photograph, Charon's orbit around Pluto (*right*) was seen nearly edge on, and Charon was near its maximum angular separation from Pluto, a mere 0.9 seconds of arc. Pluto's diameter is 2320 kilometers, Charon is half that size, and the two objects are just 19 640 kilometers apart. The *HST*'s ability to distinguish Pluto's disk at its distance of 4.4 trillion (4.4×10^{12}) meters is equivalent to seeing a baseball at a distance of about 100 kilometers. (Courtesy of NASA.)

After Pluto's discovery, astronomers continued to speculate that some unknown, massive and remote planet was responsible for the apparent perturbations in the motion of Uranus. However, after accounting for the gravitational effects of Neptune, using a precise mass obtained when *Voyager 2* encountered the planet in 1989, the small unexplained differences between the predicted and observed locations of Uranus simply disappeared. This means that there is no massive trans-Neptunian planet, and that all of the sizeable planets have been discovered. Neptune is effectively the outermost major planet in the solar system, serving as a lonely distant sentinel to outer space.

Pluto is an anomaly. It is much smaller than the giant planets that occupy the outer parts of the planetary system, and is comparable in size to some of their satellites. Pluto is smaller than Saturn's satellite Titan and all four of Jupiter's largest moons. In many ways, Pluto is the twin of Neptune's largest satellite Triton (Table 1.9). They have almost the same size and mean mass density. Moreover, many other small worlds have now been discovered just beyond the orbit of Neptune. Pluto is more akin to this group of objects than to the families of terrestrial or giant planets.

Table 1.9 Triton and Pluto compared

	Triton	Pluto
Distance from Sun, at end of 20th century (AU)	30	29
Rotation Period (Earth days)	5.9	6.4
Rotation Direction	retrograde	retrograde
Radius (km)	1375	1160
Mean Mass Density (kg m^{-3})	2070	1840–2140
Temperature (K)	38	50

Small cold worlds in the outer precincts of the planetary system

As it turned out, Pluto shares a similar orbit and composition with numerous small balls of ice and rock, forming a distant, flat ring just outside of Neptune's orbit. It is now known as the Kuiper belt in recognition of Gerard Peter Kuiper's 1951 prediction of its existence. The name Edgeworth–Kuiper belt is also sometimes used to acknowledge Kenneth E. Edgeworth's (1880–1972) discussion of such a belt. Kuiper argued that the dark outer edge of the planetary realm is not empty, but is instead full of small, unseen bodies created from the leftover debris of the formation of the giant planets. The low-density material in these distant regions would have been spread out into such a large volume, and moving in such slow, ponderous orbits around the Sun, that it could not gather or coalesce into a body any larger than Pluto.

The first trans-Neptunian, Kuiper-belt object was discovered on 30 August 1992 by Jane X. Luu (1963–) and David C. Jewitt (1958–) using the University of Hawaii's 2.2-meter (87-inch) reflector on Mauna Kea. An electronic detector attached to a large telescope was required to detect the meager amount of sunlight reflected back from such a small object at such great distances. Within a decade of the first discovery, more than 340 objects were identified in the Kuiper belt. Every one of them is millions of times fainter than can be seen with the naked eye.

The largest known Kuiper-belt object has a radius of 650 kilometers, a little more than half the radius of Pluto, at 1160 kilometers, and comparable to the size of Pluto's companion, Charon, with a radius of 635 kilometers. The Kuiper-belt object, dubbed Quaoar after an American-Indian creator god, circles the Sun once every 288 Earth years, while Pluto takes 248 Earth years to complete the trip. The similarities between Quaoar and Pluto have reinforced the view that Pluto originated in the Kuiper belt, and was ejected into a Neptune-crossing orbit by some gravitational disturbance long ago. It is estimated that the Kuiper belt contains several objects about the size of Pluto, and many smaller ones, so Pluto should no longer be provided the status of a major planet.

Scientists estimate that the Kuiper belt may contain at least 35 000 objects with a radius of about 100 kilometers. Their combined mass could be as much as 3×10^{23} kilograms, or half of the total mass of all the planetary satellites in the solar system and roughly two hundred times larger than the total mass of all the asteroids in the asteroid belt between the orbits of Mars and Jupiter.

This speculation depends on the estimated size of the new-found objects and the extent of their belt. Since most of the newly discovered objects lie within 55 AU of the Sun, there could be an abrupt outer edge to the Kuiper belt. On the other hand, the more distant objects could have eluded discovery because they are too small and/or dark, reflecting too little light to be seen with even the most powerful telescopes and sensitive detectors. The important thing is that an entirely new class of small worlds has been discovered orbiting the Sun in the outer fringes of the planetary disk, making Pluto much less unique than had been thought previously.

Pluto, Charon and Triton could be the last survivors of a lost population of small icy worlds that once inhabited the early solar system. The oversized moon Charon may have joined Pluto during a past collision or very close encounter. This might also account for the substantial eccentricity and inclination of Pluto's orbit. All three objects may have been swept up by Neptune, which snared Triton in its backward orbit and locked Pluto into its resonant orbit. Other similar objects would have been thrown out of the solar system or ejected into the Kuiper belt by gravitational interactions with the newly formed giant planets. Pluto, its companion Charon, and Triton survive because they have found gravitational niches in the solar system where they remain in stable orbits.

Comet reservoirs

There are probably many more small objects than large ones in the Kuiper belt. Most of them have been hibernating in the deep freeze of outer space since the formation of the solar system. However, Neptune's gravity slowly erodes the inner edge of the belt, within about 40 AU from the Sun, and gradually pulls some of its members closer to the Sun. The Kuiper belt is therefore a likely reservoir for a certain class of comets, called the short-period comets, that become visible when they emerge from cold storage and move toward the Sun. When a Kuiper-belt object has been launched into the inner solar system, within a few AU of the Sun, the increased solar heat vaporizes the object's icy surface, forming an Earth-sized cloud of gas and dust that reflects enough sunlight to be seen as a short-period comet (Section 12.4).

The short-period comets have periods of less than 30 Earth years, typically 5 to 20 years, and conform somewhat to the pattern of planetary motion. They all move in the same prograde direction as the planets and their orbits are tilted only slightly from the orbital plane of the Earth, known as the ecliptic. This is consistent with an origin in the outer parts of the disk of the solar system, or in the Kuiper belt.

The other class of comets, those with long periods greater than 30 Earth years, come into the planetary realm at every possible angle – their orbits are inclined at all angles to the ecliptic. Roughly half of them orbit the Sun in the retrograde direction, opposite to the motion of the planets. These long-period comets approach the Sun on very elongated orbits, coming from enormous distances of 50 thousand AU or more.

The orientation and size of the orbits of long-period comets indicate an origin in a vast, remote spherical shell of icy objects. It surrounds the solar system and extends to interstellar distances of about 100 thousand AU, or about one-third the distance to the nearest star, Proxima Centauri at 271 thousand AU. This comet reservoir is named the Oort cloud, after Jan Hendrik Oort (1900–1992) who first postulated its existence in 1951.

No one has ever seen the Oort cloud, and no one ever will, but the trajectories of long-period comets indicate that they must have come from such a place. Even at these enormous distances, the Sun's gravity is usually powerful enough to hold the unseen comets in gigantic elliptical orbits. The small, invisible, dirty balls of ice circulate around the Sun at a slow, leisurely rate of about 40 meters per second, requiring millions of years to complete a loop. Just a few of them are dislodged from their distant home in the deep freeze of outer space.

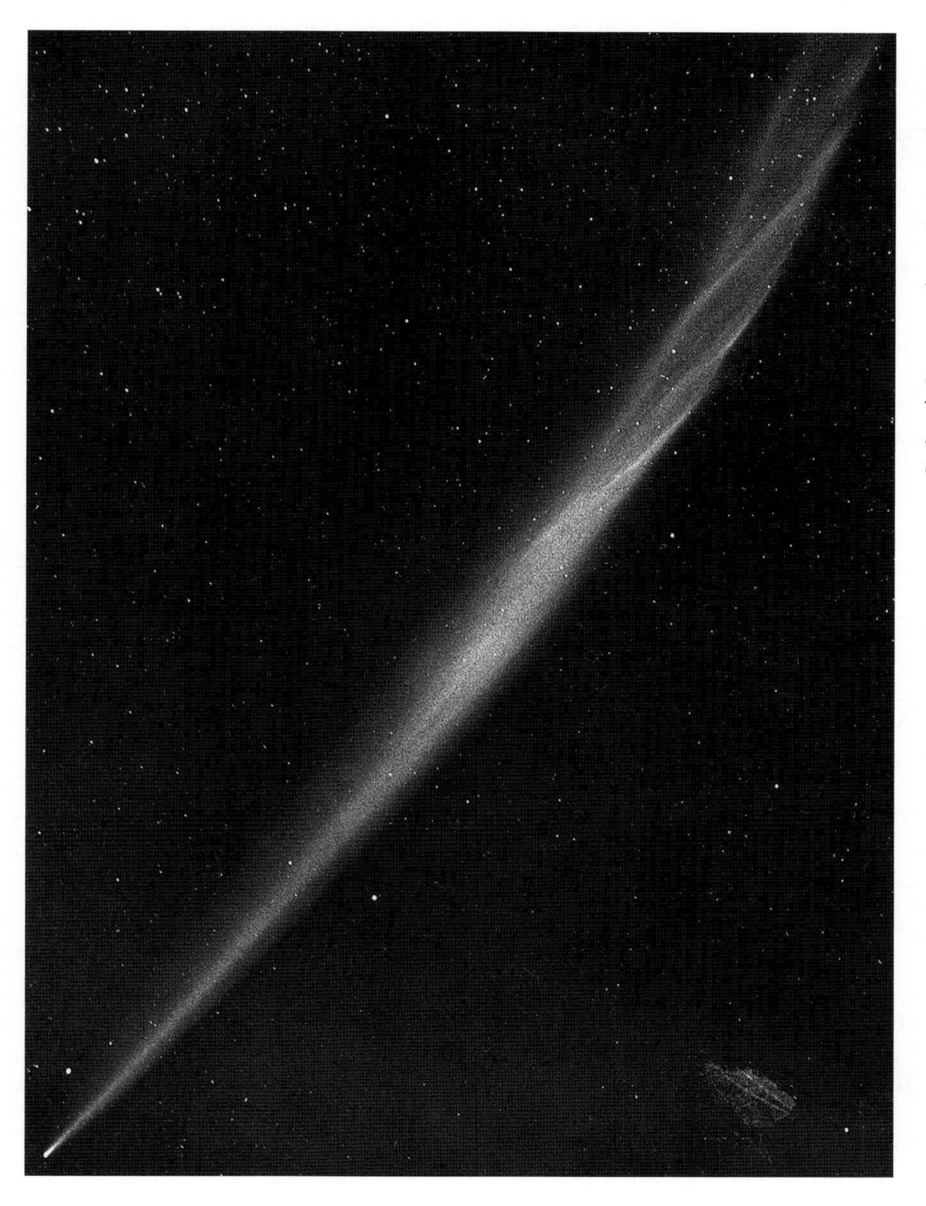

Fig. 1.28 A comet lights up When a comet travels close to the Sun, the solar heat vaporizes ice from the comet's surface, and solar forces bend the liberated material into comet tails that always point away from the Sun. The long tail of this comet stretches 120 billion (1.2×10^{11}) meters, or nearly the mean distance between the Earth and the Sun, at 1 AU = 1.496×10^{11} meters. It is named Comet Ikeya–Seki (1965 VIII) after the last names of its discoverers, Kaoru Ikeya (1943–) and Tsutomu Seki (1931–), and the year and order of its discovery. (Courtesy of the Lick Observatory.)

The random jostling of stars or giant molecular clouds passing near the Oort cloud can knock some of these icy objects from their stable orbits, sending them into the heart of the solar system where they can be seen as a long-period comet (Section 12.4). When tossed near the Sun, these dirty balls of ice become vaporized and light up with long tails and changing shapes that have inspired awe for centuries (Fig. 1.28).

This doesn't happen very often. In a million years, only about a dozen stars pass close enough to the Sun, within about 200 thousand AU, to stir the orbits of the objects in the Oort cloud and send a steady trickle of comets into the inner solar system on very elongated orbits around the Sun. The external perturbing action is so minute and uncommon that there ought to be as much as a million million (10^{12}) potential comets out there to account for the several hundreds of comets actually seen during all of recorded history.

The Oort cloud marks the outer boundary of the solar system, and completes our inventory of objects found within it, providing a background and foundation for an entirely new perspective obtained during the space-age exploration of the planets. Direct, close-up investigation by spacecraft in the second half of the 20th century has forever changed the moons and planets into fascinating, diverse worlds presented in the next chapter. Their distinctive characteristics and unique forms have been shaped and molded by processes that are also discussed in Chapter 2.

2 The new close-up view from space

- The first spacecraft flew past the planets and their satellites, providing an initial reconnaissance. This was followed by orbiting spacecraft that mapped out the global terrain, and returned for a close-up inspection.
- There have been six manned landings on the Moon, returning 382 kilograms of rocks for examination in the terrestrial laboratory.
- Some spacecraft have landed on the surface of Mars and Venus, and plunged into the atmosphere of Jupiter.
- Every solid planet or satellite contains impact craters, but in different amounts that depend on the ages of their surfaces.
- Impact craters on the Moon, Mercury and the icy satellites of the giant planets record an ancient, intense rain of meteorites about four billion years ago, and a continued cosmic bombardment since then.
- Numerous volcanoes are found on the surface of Earth and Venus; Mars has fewer volcanoes with some signs of relatively recent volcanic activity.
- The volcanoes on Jupiter's satellite Io have turned the satellite inside out; it is heated inside by the tidal flexing action of nearby massive Jupiter.
- Volcanoes of ice may have created some of the features now frozen into the bright smooth surface of Neptune's largest moon, Triton; dark geyser-like plumes have been observed in the process of eruption on the satellite.
- Seventy-one percent of the Earth's surface is covered with liquid water, and our bodies are largely composed of water.
- There could be water ice at the poles of Mercury and the Moon, within permanently shaded craters.
- Although Venus is now dried out, it may have once contained a small ocean.
- Catastrophic floods and deep rivers once carved channels on Mars, and an ancient ocean may have once covered the planet's northern lowlands.

- Layered deposits found on Mars might have been laid down long ago in numerous lakes and shallow seas when the red planet was able to sustain liquid water for long periods of time.
- Vast amounts of frozen water are now thought to exist in the polar caps of Mars and beneath the surface as permafrost.
- Substantial amounts of water ice have been detected just below the surface of the Martian northern and southern hemispheres at latitudes poleward of about 60 degrees.
- Spring-like flows have been detected coming out of the surface of Mars in relatively recent times.
- Jupiter's satellite Europa is covered with smooth, white water ice, which has cracked due to the contorting tidal effects of Jupiter's strong gravity. The warmth generated by tidal heating may have been sufficient to form an ocean of liquid water below Europa's icy covering.
- The release of water from just below the surface of Jupiter's satellite Ganymede is suggested by light regions separated by dark plates. Cracks and grooves in the bright, smooth ice on Saturn's satellite Enceladus similarly suggest the release of subsurface water.
- Magnetic measurements provide indirect evidence for oceans of salty, liquid water below the icy crusts of Jupiter's satellites Europa, Callisto and Ganymede.

2.1 Flybys, orbiters, landers and probes

We live at an incredible time, when all of the major planets, and most of their satellites, have been viewed close up with the inquisitive eyes of robotic spacecraft. They have perceived awesome, unanticipated features that are far beyond the range of human vision with even the best telescopes on the ground. No two of these fascinating new worlds are exactly the same (Fig. 2.1). Most of them have been investigated many times, with increasingly sophisticated instruments.

This captivating voyage of discovery began close to home, in the late-1950s, when the first artificial satellites were lofted into orbit around the Earth and the Russian *Luna 3* spacecraft swung once around the far side of the Moon, which had never been seen before. It is strongly deficient in the large, dark maria that characterize much of the Moon's near side facing Earth (Fig. 2.2).

The exploration of the Moon served as a stepping stone to the planets, and established a blueprint for subsequent planetary missions. For both our Moon and the planets, the initial reconnaissance was provided by spacecraft that flew by them, obtaining just a brief glimpse out of the corner of their eye. In the lunar case, three *Ranger* spacecraft were also sent crashing into the Moon's surface, transmitting high-definition pictures on the way down. This was followed by a more detailed exploration with orbiting spacecraft, such as *Lunar Orbiter 1, 2, 3, 4* and *5*, that can circle a moon or planet many times and map out its terrain. Like the explorers of new territories on Earth, the orbiters were sent to reconnoiter, to get the lay of the land, and to disclose possible dangers awaiting future visits. The next step involves landers, like *Luna 9* and *13* and *Lunar Surveyor 1, 3, 5, 6,* and *7*, that explore the surfaces, and probes that plunge into the atmospheres.

So far, humans have only visited the Moon. An estimated half-billion people watched the televised first visit, on 20 July 1969, when Neil Armstrong groped cautiously down the lander's ladder and stood firmly on the fine-grained lunar surface. An ancient dream had come true – man had set foot on another world.

In all, there have been six manned landings on the Moon, beginning with *Apollo 11* in July 1969 and ending with *Apollo 17* in December 1972. The actual landings were performed by the bug-like *Lunar Module* that separated from the main spacecraft, while in orbit around the Moon, and returned to it. At first the astronauts traveled on foot, staying near to the *Lunar Module*, but they subsequently moved to more remote locations in roving vehicles (Fig. 2.3). Altogether, 382 kilograms of rocks were brought back from the Moon for analysis in the terrestrial laboratory, determining the Moon's age, chemical composition, history and probable origin (Sections 5.7, 5.9).

Fig. 2.1 Space-age view of the planets A montage of planetary images taken with different spacecraft and presented, from top to bottom, in order of increasing distance from the Sun. They are Mercury (*top*) taken from *Mariner 10*, a Venus image from *Magellan*, the Earth seen by *Galileo*, with our Moon, a *Mars Global Surveyor* image of Mars, a *Cassini* image of Jupiter, a *Voyager 1* image of Saturn, and *Voyager 2* views of Uranus and Neptune (*bottom*). The inner planets, closest to the Sun (Mercury, Venus, Earth and Mars) are roughly to scale to each other, and the outer planets, furthest from the Sun (Jupiter, Saturn, Uranus and Neptune) are also shown roughly to scale to each other. (Courtesy of JPL and NASA.)

Fig. 2.2 The far side of the Moon A view of the Moon never seen before the Space Age, and not visible from the Earth. This picture of the back side of the Moon was recorded by cameras aboard *Apollo 16* in April 1972. The image is centered on the boundary between the lunar near side (*upper left*) and the Moon's hidden face (*lower right*). Three lunar maria are visible as dark patches on the near side. Clockwise from upper left: the Sea of Crises, the Border Sea and Smyth's sea. At lower right are the light-colored and heavily cratered highlands of the far side of the Moon. They contain almost none of the dark maria that characterize the side of the Moon that faces the Earth. (Courtesy of NASA.)

Fig. 2.3 Lunar rover The battery-powered lunar rovers, used in the last three *Apollo* missions, could carry two astronauts and all their equipment for thousands of meters across the lunar surface. The *Lunar Module* carried the astronauts back to their orbiting command module, while the rover remained on the Moon. Because there is no substantial atmosphere, water or weather on the Moon, both the rover and the footprints in the lunar soil may last for millions of years. By that time micrometeorites will have pitted the rover and erased the footprints. (Courtesy of NASA.)

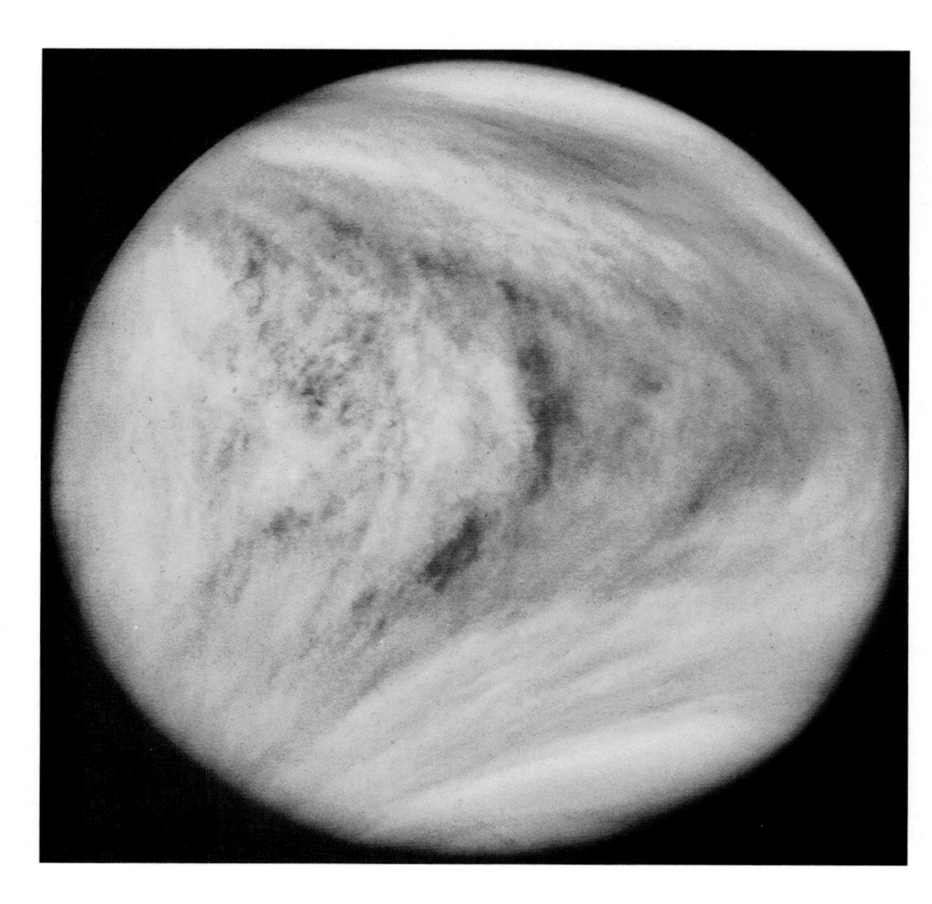

Fig. 2.4 Veiled Venus Bright opaque clouds of sulfuric acid wrap around Venus. The creamy yellow veil of clouds circulates once around the planet in only four Earth days, moving at speeds of up to 100 meters per second. Strong zonal winds combine with weaker poleward winds to carry the clouds in a slow spiral toward the poles. The thick cloudy atmosphere of Venus always blocks our view of the surface, and warms that surface to a torrid 735 kelvin. (Courtesy of NASA.)

Table 2.1 Some key events in the space-age exploration of the planets

Spacecraft	Launch Date	Encounter Date	Object	Discovery
Sputnik 1[a]	4 Oct. 1957	4 Oct. 1957	Earth	First artificial satellite.
Explorer 1	1 Feb. 1958	1 Feb. 1958	Earth	Radiation belts of charged particles.
Luna 3[a]	4 Oct. 1959	7 Oct. 1959	Moon	Photographed back side of Moon.
Mariner 2	26 Aug. 1962	14 Dec. 1962	Venus	First successful planetary flyby, measured solar wind.
Mariner 4	28 Nov. 1964	14 July 1965	Mars	Flyby, craters.
Luna 9[a]	31 Jan. 1966	3 Feb. 1966	Moon	Soft landing, Oceanus Procellarum.
Surveyor 1	30 May 1966	2 June 1966	Moon	Soft landing near Flamsteed.
Lunar Orbiter 1	10 Aug. 1966	14 Aug. 1966	Moon	Global photographs of lunar surface.
Apollo 11	16 July 1969	20 July 1969	Moon	First humans on Moon, Mare Tranquillitatis, sample return.
Venera 7[a]	17 Aug. 1970	15 Dec. 1970	Venus	Soft landing on surface.
Mariner 9	30 May 1971	13 Nov. 1971	Mars	Orbiter, global images.
Pioneer 10	3 Mar. 1972	3 Dec. 1973	Jupiter	Flyby, passage through asteroid belt.
Mariner 10	3 Nov. 1973	29 Mar. 1974	Mercury	Three flybys, heavily cratered surface.
Venera 9[a]	8 June 1975	22 Oct. 1975	Venus	Orbiter and lander, surface photograph.
Viking 1	20 Aug. 1975	29 June 1976	Mars	Orbiter and lander, surface photographs, life search.
Viking 2	9 Sept. 1975	7 Aug. 1976	Mars	Orbiter and lander, surface photographs, life search.
Pioneer Venus	20 May 1978	4 Dec. 1978	Venus	Orbiter and multi-probe, global radar images.
Voyager 1	5 Sept. 1977	5 Mar. 1979	Jupiter	Flyby, ring, volcanoes on satellite Io.
Voyager 1		12 Nov. 1980	Saturn	Flyby, Titan's dense atmosphere.
Voyager 2	20 Aug. 1977	9 July 1979	Jupiter	Flyby.
Voyager 2		25 Aug. 1981	Saturn	Flyby.
Voyager 2		24 Jan. 1986	Uranus	Flyby, rings, magnetic field.
Voyager 2		25 Aug. 1989	Neptune	Flyby, excess heat, winds, rings, magnetic field.
Giotto[b]	2 July 1985	13 Mar. 1986	Comet Halley	Flyby, close-up view of nucleus Comet 1P/Halley.
Magellan	4 May 1989	10 Aug. 1990	Venus	Orbiter, radar maps of surface, volcanic resurfacing.
Clementine	25 Jan. 1994	21 Feb. 1994	Moon	Mapped global surface composition and topography of Moon. Possible evidence for water ice at lunar poles.
Galileo	18 Oct. 1989	29 Oct. 1991	Asteroid	Flyby, image of 951 Gaspra.
Galileo		28 Aug. 1993	Asteroid	Flyby, image of 243 Ida and its moon Dactyl.
Galileo		7 Dec. 1995	Jupiter	Orbiter and probe, atmosphere and properties of four largest satellites.
NEAR Shoemaker[c]	17 Feb. 1996	27 June 1997	Asteroid	Flyby, image of 253 Mathilde.
NEAR Shoemaker[c]		14 Feb. 2000	Asteroid	First orbiter and in-depth study of an asteroid, 433 Eros, determining its composition, size, shape, and mass distribution.
Mars Pathfinder	4 Dec. 1996	4 July 1997	Mars	Lander, surface rover *Sojourner* examines dust, pebbles and rocks near mouth of outflow channel.

Table 2.1 (*cont.*)

Spacecraft	Launch Date	Encounter Date	Object	Discovery
Mars Global Surveyor	7 Nov. 1996	12 Sept. 1997	Mars	Orbiter, laser altimeter, magnetometer, high-resolution images, mapping began on 4 April 1999, recent water flow and volcanic activity, ancient magnetism.
Lunar Prospector	7 Jan. 1998	15 Jan. 1998	Moon	Mapped global elemental abundance, magnetic field and gravity, detected lunar core, evidence for substantial water ice at lunar poles.
Deep Space 1	24 Oct. 1998	22 Sept. 2001	Comet Borrelly	Tested advanced space technologies, image of the nucleus of Comet 19P/Borrelly.
2001 Mars Odyssey	7 Apr. 2001	24 Oct. 2001	Mars	Mapped the amount and distribution of chemical elements and minerals on the Martian surface, and provided evidence for substantial subsurface water ice.
Cassini[b]/*Huygens*	15 Oct. 1997	June 2004	Saturn	Orbiter and probe, atmosphere, rings, magnetic environment, Titan probe.

[a] Spacecraft launched by the former Soviet Union.

[b] *Giotto* was a mission of the European Space Agency, or ESA for short. All other spacecraft are missions of the United States National Aeronautics and Space Administration, abbreviated NASA, except the *Cassini* mission, which is a joint ESA and NASA venture.

[c] The acronym *NEAR* stands for *Near Earth Asteroid Rendezvous*.

In 1970, an unmanned entry probe was sent from the *Venera 7* spacecraft into the thick, carbon-dioxide atmosphere of Venus, measuring the temperature and pressure all the way down to the surface. These data showed that the surface is hot enough to melt lead and that the atmosphere is ninety times as heavy as our air (Sections 3.1, 7.3).

The first spacecraft to be launched on lengthy journeys beyond the Moon were flyby missions, the *Mariners, Pioneers* and *Voyagers,* that passed near the planets and their satellites to give us new vistas, unavailable from the ground, and making important discoveries in the process (Table 2.1). In 1962, instruments aboard the *Mariner 2* flight to Venus detected a perpetual flow of charged particles in interplanetary space, emanating from the Sun (Section 3.5). *Mariner 10,* launched in 1973, provided the first spacecraft photographs of Venus (Fig. 2.4), and traveled on to reveal the heavily cratered surface of Mercury (Fig. 2.5).

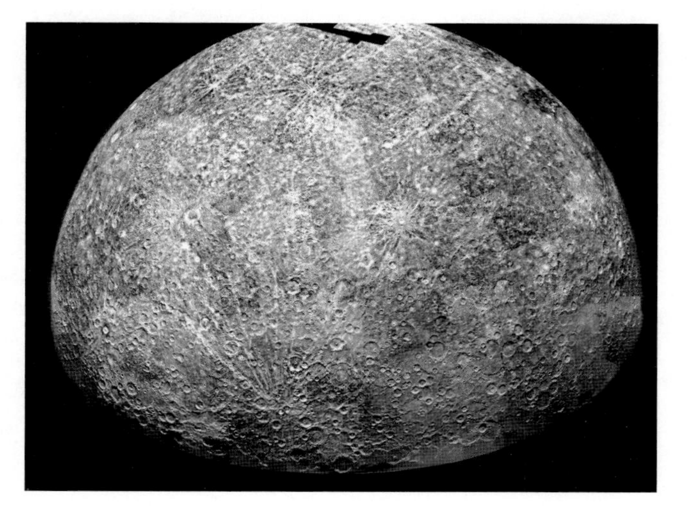

Fig. 2.5 Spaceshot of Mercury A photomosaic of Mercury's southern hemisphere produced from images acquired by *Mariner 10* during its first encounter with the planet in March 1974. Mercury has a heavily cratered surface that resembles the lunar highlands. Bright rayed craters are also present on Mercury, as they are on the Moon. (Courtesy of JPL and NASA.)

In 1972–74, the *Pioneer 10* and *11* missions to Jupiter showed that spacecraft could pass safely through the asteroid belt, blazing a trail for the extraordinarily successful *Voyager 1* and *2* flyby missions (Fig. 2.6). Their itinerary included Jupiter (1979, Fig. 2.7), Saturn

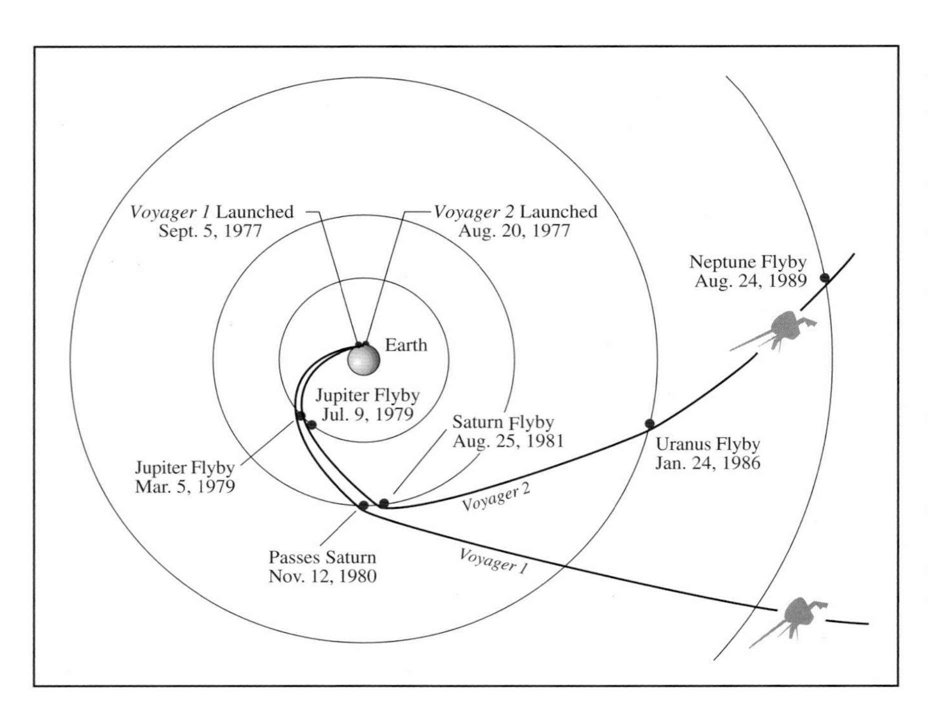

Fig. 2.6 Grand tour of *Voyager 1* and *2* The flights of the two *Voyager* spacecraft through the solar system. Both spacecraft were launched in 1977 and flew past Jupiter in 1979, transmitting remarkable details of the giant planet's weather and the surfaces of its four largest satellites. *Voyager 1* and *2* used the gravity of Jupiter to accelerate them on toward Saturn, providing close-up images of its rings and satellites in 1980–81. Saturn provided another gravity assist to propel *Voyager 2* on to Uranus, in 1986, and from there on to Neptune in 1989. *Voyager 1* was targeted differently at Saturn, sacrificing its grand tour for close views of the satellite Titan. Both spacecraft are now heading toward the edge of the solar system and will eventually leave it.

Fig. 2.7 Giant world Jupiter's clouded world with its alternating structure of light zones and dark belts. The two innermost Galilean satellites are also visible. Bright orange Io is seen just above the cloud tops, and icy-white Europa lies to the right. (Courtesy of JPL and NASA.)

(1980, 1981, Fig. 2.8), Uranus (1986) and Neptune (1989, Fig. 2.9).

Voyager 1 and *2* vastly improved our understanding of the atmospheres of the giant planets, and discovered unexpected rings, moons and magnetic fields. They also revealed the satellites of the giant planets to be unique and distinctive places with diverse surfaces and in some cases atmospheres or magnetic fields.

The initial explorations using flybys were followed by orbiters that greatly increased the time available for detailed study, often for years at a time. They revealed many features that previous flyby missions had missed, and forever changed our view of the planets and their satellites.

The extraordinary promise of planetary orbiters was first demonstrated in 1971–2 by the *Mariner 9* mission to Mars. The three previous, flyby missions, *Mariner 4, 6* and *7*, had discovered the ancient cratered terrain on Mars, but missed all of the younger geological features. The orbiting *Mariner 9* had sufficient time to completely explore the planet, revealing for the first time Mars' great volcanoes, the vast canyon system Valles Marineris, and evidence of ancient stream beds and water erosion (Sections 2.4, 8.7, 8.8).

The *Viking 1* and *2* orbiters amplified and enhanced our new perspective of Mars in the late-1970s (Fig. 2.10). Each *Viking* also had a 1-ton (1000-kilogram) lander that

Fig. 2.8 Saturn's realm The magnificent rings of Saturn encircle the planet, never touching its cloud tops. The prominent gap in the rings is named the Cassini Division. The yellow-brown atmosphere of Saturn, shown here in enhanced color, has a banded structure, but it lacks Jupiter's bright zones and belts. Three icy satellites (Tethys, Dione and Rhea) are visible as small white spots against the darkness of space (*left*), and another smaller satellite (Mimas) is visible against Saturn's cloud tops, just below the rings. Because of its rapid spin, Saturn has an oblong, egg-like shape, flattened at the poles and extended at the equator. (Courtesy of JPL and NASA.)

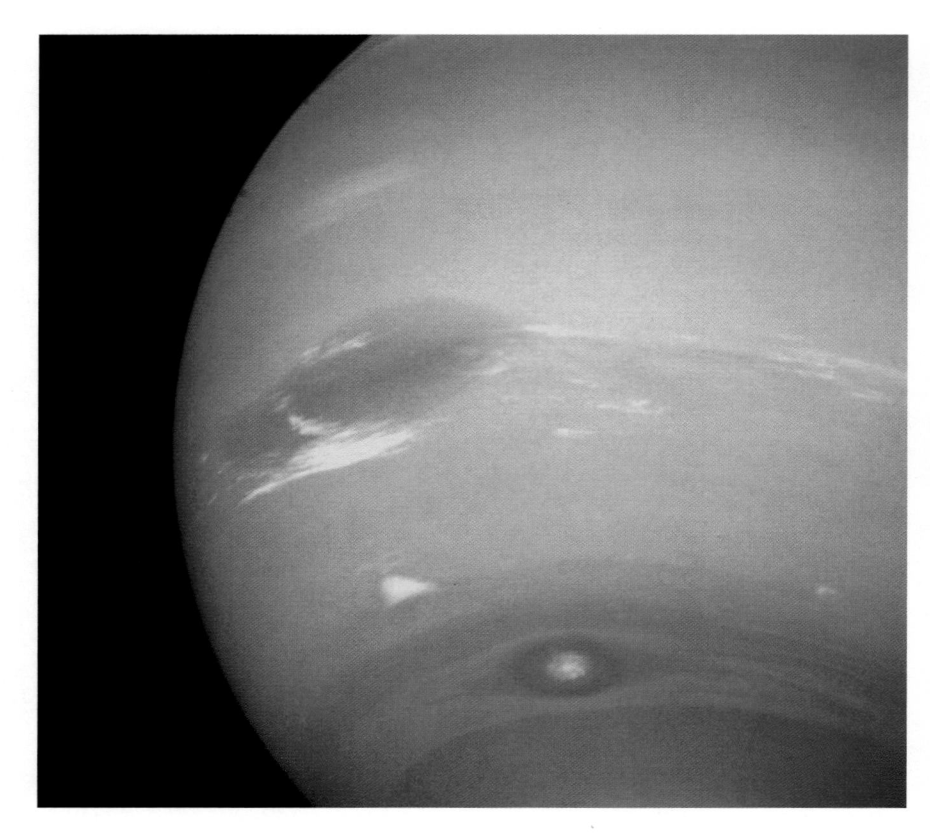

Fig. 2.9 Neptune's dynamic atmosphere After a journey of 12 years and 7 trillion (7×10^{12}) meters, the *Voyager 2* spacecraft captured this view of Neptune's Great Dark Spot (*left middle*), accompanied by bright, white clouds that undergo rapid changes in appearance. The spot is as large as the Earth and about one-third the diameter of the Great Red Spot of Jupiter. Another, smaller dark feature (*bottom center*) has a bright core. Internal heat drives strong winds on Neptune, moving at speeds of up to 450 meters per second. The *Voyager 2* images were sent by a small 20-watt transmitter. Traveling at the speed of light the radio signals took more than 4 hours to reach Earth. (Courtesy of JPL and NASA.)

Fig. 2.10 Mosaic of Mars This computer-generated mosaic of *Viking* orbiter images of Mars shows three volcanoes as dark spots to the west (*left*), while the bottom center of the scene shows the entire Valles Marineris canyon system, from Noctis Labyrinthus (*left*) to the chaotic terrain (*right*). Outflow channels are found in the north (*top*), and a variety of clouds and hazes are also visible, especially near the edge. (Courtesy of Alfred S. McEwen, U.S. Geological Survey.)

Fig. 2.11 Surface of Mars The Martian surface in western Chryse Planitia, as viewed from the *Viking 1* lander on 3 August 1976. Wind-blown dust clings against the eroded rocks, creates dust drifts and fills the sky. The gently rolling landscape resembles sand dunes on a rocky terestrial desert, but the dune drifts on Mars are composed of fine dusty material and the Martian dust drifts persist for much longer than sand dunes on Earth. These drifts were little changed during the six years they were observed from *Viking 1*. Mars is a cold and desolate world in which the silence is broken by the roar of winds, the hiss of dust, the rumble of mammoth landslides, and perhaps by outbursts of active volcanoes. (Courtesy of JPL and NASA.)

was sent safely to the planet's surface, obtaining beautiful panoramas of the Martian surface (Fig. 2.11) and measuring the properties of the thin, freezing atmosphere (Sections 3.1, 8.3). The *Viking* landers were also sent to search for extant life on Mars, but the results were inconclusive (Section 8.9).

On 4 July 1997 *Mars Pathfinder* landed near the mouth of a canyon system carved by massive floods of water long ago. The mobile roving *Sojourner* examined nearby rocks, dust and pebbles, which were all consistent with the downstream deposit of flowing water from this outflow channel (Section 8.9).

Close-up, high-resolution views of the surface of Mars were obtained from the *Mars Global Surveyor* at the end of the 20th century and the beginning of the 21st century. The images showed much finer detail than those obtained with the *Viking* orbiters, including layered deposits suggesting ancient lakes or shallow seas (Section 2.4), and evidence suggesting that water may have flowed on the surface in relatively recent times (Section 8.8).

Beginning in April 2001, the *2001 Mars Odyssey* mapped the amount and distribution of the chemical elements and minerals in the Martian surface, and obtained evidence for huge tracts of subsurface water ice.

In the 1990s, the *Magellan* orbiter used radar to penetrate the thick, cloudy atmosphere of Venus, mapping the entire planet with a clarity and resolution not available for much of Earth. Since Venus is perpetually shrouded in clouds, this was the only way to detect its surface. *Magellan's* radar images have revealed an unearthly world that was resurfaced long ago by rivers of outpouring lava, and disclosed numerous volcanoes that now pepper its surface (Fig. 2.12).

The *Galileo* orbiter–probe spacecraft, launched in October 1989, was so massive that no existing rocket had the power to launch it directly to Jupiter, its primary target. Instead, the spacecraft was placed on a looping trajectory that took it past Venus once and Earth twice (Fig. 2.13). The gravity of these planets was used to accelerate and propel the spacecraft in slingshot fashion toward its eventual rendezvous with the giant planet, somewhat like a pitcher winding up to throw a high-velocity strike. While the roundabout route took six years, in comparison to the direct, 21-month flights of *Pioneer 10* and *11*, it also took *Galileo* on close encounters with two asteroids along the way (Fig. 2.14).

Galileo carried an entry probe that penetrated Jupiter's kaleidoscopic clouds, obtaining the first direct, or *in situ*, sampling of a giant planet's atmosphere (Section 9.3). The main orbiting spacecraft looped around Jupiter for more than five years, until 2002, obtaining high-resolution images and analysis of the planet's stormy weather (Section 9.2) and its sparse ring system (Section 9.5). *Galileo* has provided new insights into the Galilean satellites (Section 9.4), including: volcanic activity on Io; beams of electrons that connect Io to Jupiter; compelling evidence for a global ocean beneath the ice crust of Europa; and the discovery of a magnetic field generated within Ganymede, the first such field to be found on a moon.

The *Cassini* spacecraft was launched on its seven-year journey to Saturn on 15 October 1997, with arrival expected in June 2004. It includes an orbiter, whose instruments will study the planet's atmosphere, rings, satellites, and magnetic environment. The spacecraft also carries the *Huygens Probe* that will be parachuted into the hazy, dense atmosphere of Saturn's intriguing moon Titan, determining the properties of its Earth-like atmosphere and its mysterious surface below.

Comets are so tiny and so far away that you cannot detect them until they come near the Sun, and their center is then buried within the brilliant glare of fluorescing gases and reflected sunlight. As a result, no one had ever seen the bare surface of a comet's nucleus until 1986,

Fig. 2.12 Venus unveiled A cloud-penetrating radar system on board the *Magellan* spacecraft has mapped the global landforms and features on Venus with a resolution of 120 meters, more completely than any planet including Earth. This hemisphere, centered at 180 degrees east longitude, shows the bright, planet-wide, equatorial highlands that contain towering volcanoes, long lava flows and deep faults and fractures. They run from lower left to upper right through Aphrodite Terra (*left of center*), a continent-sized highland, through the bright highland Atla Regio (*just right of center*) to Beta Regio (*far right and north*). Dark areas correspond to terrain that is smooth on the scale of the radar wavelength (0.13 meters); bright areas are rough. The orange tint, based on color images taken by the *Venera 13* and *14* landers, simulates the color of sunlight at ground level after being filtered through the planet's thick atmosphere and clouds. (Courtesy of JPL and NASA.)

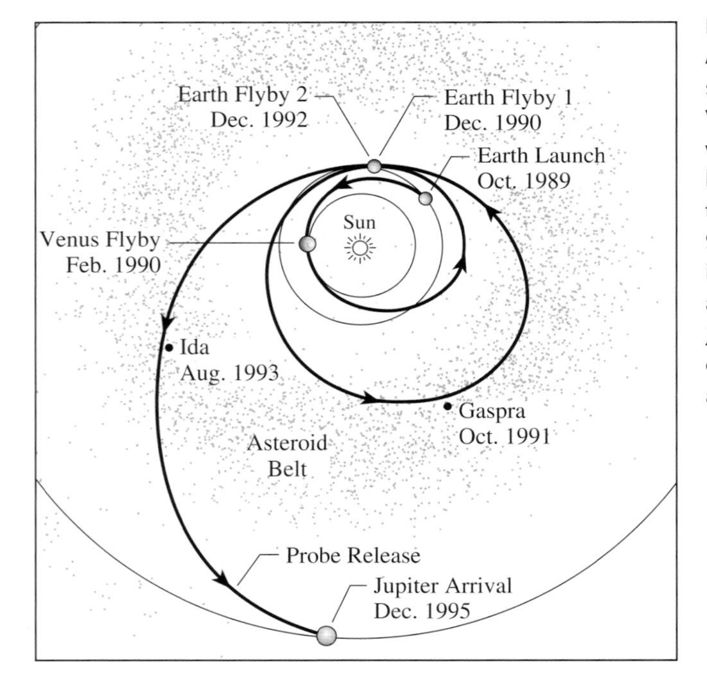

Fig. 2.13 *Galileo's* long flight to Jupiter After launch in October 1989, the *Galileo* spacecraft used the gravity of the Earth and Venus to accelerate it on to its encounter with Jupiter, six years after launch. In its long, indirect flight path, *Galileo* was able to fly past two asteroids at close range, 951 Gaspra in October 1991 and 243 Ida in August 1993 (see Fig. 2.14). It released an atmospheric probe just before arrival at Jupiter in December 1995; the main spacecraft continued to orbit the planet and examine its satellites for the next five years.

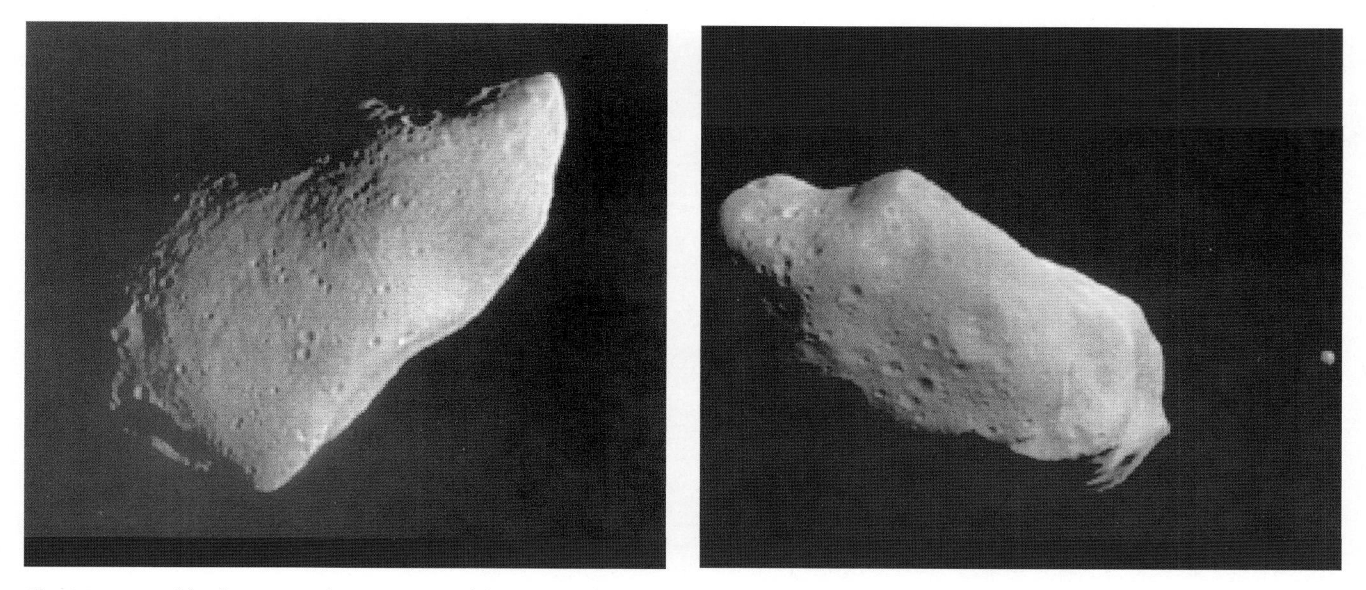

Fig. 2.14 Asteroids close up These images of the asteroids 951 Gaspra (*left*) and 243 Ida (*right*) were taken by the Jupiter-bound *Galileo* spacecraft on 29 October 1991 and 28 August 1993, respectively. Both objects have irregular, elongated shapes. The illuminated part of Gaspra is 18 kilometers long, from lower left to upper right, and it contains an abundance of small craters. Ida is about 52 kilometers in length, and is accompanied by a small moon (*right center*), named Dactyl. (Courtesy of JPL and NASA.)

Fig. 2.15 The black heart of Comet Halley The coal-black nucleus of Comet Halley is a dirty ball of ice, about the size of Paris or Manhattan. It is silhouetted against bright jets of water and dust that stream sunward (*right*) from at least three places that have been warmed by the Sun's radiation. In this projection, the nucleus measures 14.9 kilometers by 8.2 kilometers. This is a composite of images taken by the European Space Agency's (ESA's) *Giotto* spacecraft near its encounter with the nucleus of Comet Halley on 14 March 1986. (Courtesy of ESA.)

when the *Giotto* spacecraft peered into the core of Halley's Comet. It found the nucleus to be a black, oblong chunk of ice and dust, roughly the size of Paris or Manhattan (Fig. 2.15). At the moment of encounter, the comet was spewing out about 25 tons, or 25 thousand kilograms, of water every second, propelled into sunward jets by the vaporizing ice. So comets provide evidence that large quantities of water ice can be found in the outer solar system (Section 2.4). Some scientists think that former comets may have brought water to Earth.

At the turn of the 21st century, we can reflect in amazement at the incredible new worlds that have been discovered by the flybys, orbiters, landers and probes. Future spacecraft are now poised to continue the exploration in greater detail, focusing on issues such as the beginning of life on Earth, the search for life outside the Earth, and discovery of planetary systems around stars other than the Sun. Scientists will, for example, ultimately return samples of the surface of Mars for study in our Earth-bound laboratories back home, to examine them for fossil or recent evidence of life.

There is a growing awareness of the similarities of the major planets and some moons, despite the differences that make each of them unique. They all exhibit common properties and similar processes, such as impact craters, volcanoes, water and atmospheres, reminding us of the basic elements in ancient Greek philosophy – Earth, fire, water and air.

Fig. 2.16 Lunar crater Timocharis Astronauts on board the *Apollo 15* mission took this image of the medium-sized crater Timocharis, about 34 kilometers across, in August 1971. The deposits and ejecta have been thrown radially outward by the meteorite impact that created the primary crater with its circular rim. Smaller secondary craters are located beyond the radial ejecta (*lower left*). (Courtesy of NASA.)

2.2 Impact craters

The Moon and Mercury demonstrate the power of impact

The most distinctive features on the Moon are the circular craters that closely pepper its face. Comparatively recent ones still exhibit the details of the impact that created them (Fig. 2.16); older craters have been worn away by small particles that continuously bombard the Moon.

There are several lines of evidence that the lunar craters were formed by the explosive impact of interplanetary projectiles:

(i) The amount of material piled on a crater's raised rim is nearly equal to the material excavated from the interior, so if the rim was pushed back into the crater its depressed floor would rise to the level of the neighboring surface.
(ii) Nearly all of the Moon's craters are round. The explosive force of a large impacting object will produce round craters despite the fact that the projectiles that produced them must have arrived in a variety of directions – some nearly vertically, others at a glancing angle.
(iii) The rocks returned from the heavily cratered regions on the Moon consist of fragments of pre-existing rocks that have been welded together by the enormous pressures of impact.

So, the lunar craters must have been created by solid, rocky objects, named meteoroids, which came from interplanetary space and hit the Moon. When the meteoroids strike the surface of a planet or satellite they are called meteorites. Although the projectile vaporizes on impact, the explosion excavates material and hurls it outward, creating a raised rim, radial ejecta and secondary craters (Fig. 2.17). Meteorites of all sizes have hit the Moon, and its crust records the impact of more small meteorites than large ones.

Imagine the explosive impact of a colliding meteorite. When the Moon intercepts and stops the moving rock, its energy of motion is suddenly transformed to heat, creating intense pressure, high temperature, and two shock waves. The first shock engulfs the impacting object, vaporizes it, and compresses and melts rock at the immediate point of impact. The second shock wave travels out from the impact point, excavating the crater cavity and throwing up a rim of pulverized and melted rock around it. This material is carried out on ballistic trajectories in all directions from

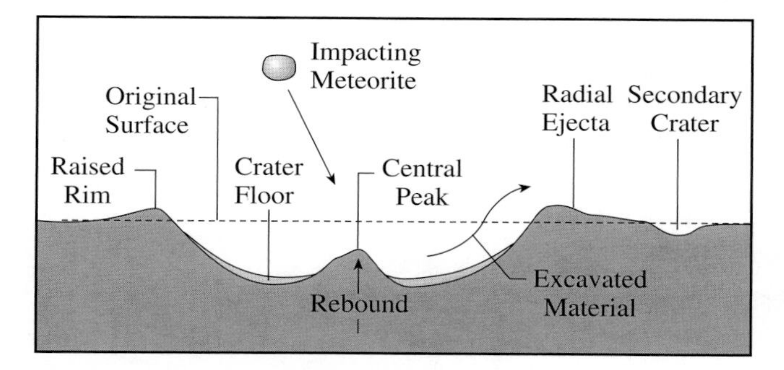

Fig. 2.17 Cross-sectional anatomy of a crater An impacting meteorite excavates a circular crater that is almost 40 times the diameter of the meteorite. The depth of the crater is roughly one-tenth its diameter, and the crater floor is depressed below the surrounding terrain. The explosion gouges out a circular hole, depositing material around its rim and ejecting debris radially outward in all directions. The surface rebounds from the impacting force of a large meteorite, creating a central peak in the floor of the biggest craters.

the point of impact; hence the circular craters and radial ejecta.

The rim is tossed out almost nonchalantly, something like flicking a particle off the end of a whip, but the excavated material is still many times as massive as the impacting projectile and about ten thousand times the volume. This is essentially due to the tremendous force of the impacting object, which can release energy equivalent to the explosion of tens of thousands of hydrogen bombs. Violent rebound of the crater floor, from the greater energy and shock of larger meteorites, gives rise to a central peak or peaks. In addition, many of the larger craters have terraced walls caused by rim material slumping in toward the crater center.

The largest lunar craters are the impact basins. A typical one is the Imbrium Basin, with a diameter of 1.5 thousand kilometers. Its outline can be seen with the unaided eye, forming an "eye socket" of the face of the "Man on the Moon". Its outer rim is defined by prominent mountain ranges, such as the Apennine mountains (Fig. 2.18). Such basins were created early in the Moon's history, and they were soon flooded and nearly filled with dark molten lava from the interior (Section 5.3).

Ejecta from the Imbrium Basin gouged out radial ridges and valleys that went a quarter of the way around the Moon, scattering a thick blanket of debris over most of the near side of the Moon. The energy of impact was so great that the floor of the crater rebounded, surging up and down and creating multiple rings as the lunar surface vibrated like the head of a drum.

Some of the rocks that have been returned from the Moon are the oldest rocks ever found. Their ages have been determined by the method of radioactive dating (Section 5.7), and they show that the Moon accumulated by the aggregation of rocky projectiles about 4.6 billion years ago (Section 5.9). When the Moon was very young, its outer layers were probably molten, but the crust cooled and became solid. The battered lunar surface that we see today remains a museum of impact scars that were created back then. It records an intense bombardment of leftover formation material that created the large impact basins and most of the lunar craters about 4.0 billion years ago.

Gradually the hail of impacting meteorites decreased, as most of the interplanetary meteoroids were swept up and pulled in, and the rate of cratering dropped rapidly during the subsequent billion years. A much lower, steady rate of crater production has persisted for the last 3 billion years, so relatively young craters are hard to find on the Moon. They are distinguished by white rays that splash across the lunar surface. The rays of older craters are darkened by eons of continued meteorite impact.

In their early years, all of the satellites and planets were the solar system's shooting gallery, subject to an intense bombardment of meteorites. We can see their scars on the Moon and on the planet Mercury, with its Moon-like surface that is similarly pockmarked with craters. Most of them were formed long ago and then eroded by subsequent meteorite impact. As on the Moon, there are small bowl-shaped craters on Mercury, smaller than 100 meters in diameter, and large impact craters up to a thousand kilometers (10^6 meters) across. Both worlds also contain a few, fresh young craters with bright rays as well as many older craters without rays, and both the Moon and Mercury have no significant atmosphere or any weather to erode their surface.

Mercury's surface also contains multi-ringed impact basins. The largest of these has been named *Caloris*, the Latin name for "heat", because it is located at a place on Mercury that faces the Sun when the planet is at the point in its orbit that is closest to the Sun. During the *Mariner 10* encounters with Mercury, half of the Caloris Basin was in shadow and the other half in sunlight. An irregular annulus of mountains cuts across the sunlit image, defining the edge of a huge excavation that is 1340 kilometers in diameter (Fig. 2.19).

The cataclysmic impact that created the Caloris Basin occurred an estimated 3.85 billion years ago when a meteorite roughly 150 kilometers across hit Mercury, like a cosmic bomb with an energy of a trillion 1-megaton hydrogen bombs. The violent explosion reverberated through the young planet, sending strong seismic waves along the surface and through the deep interior (Fig. 2.20). These waves converged to a focus on the side of Mercury opposite to

Fig. 2.18 The Moon's Apennine mountains The radial structure and steep inner slopes of these mountains (*lower right*) mark a section of the outer rim of the Imbrium Basin. The huge excavation was subsequently filled with lava to form the smooth Mare Imbrium and partially submerge the inner ring of mountains (*upper left*). The smaller circular craters include Timocharis (Fig. 2.16), about 34 kilometers in diameter (*middle left*) and the largest round structure Archimedes (*upper center*) with a diameter of 83 kilometers. (Photograph courtesy of UCO/Lick Observatory.)

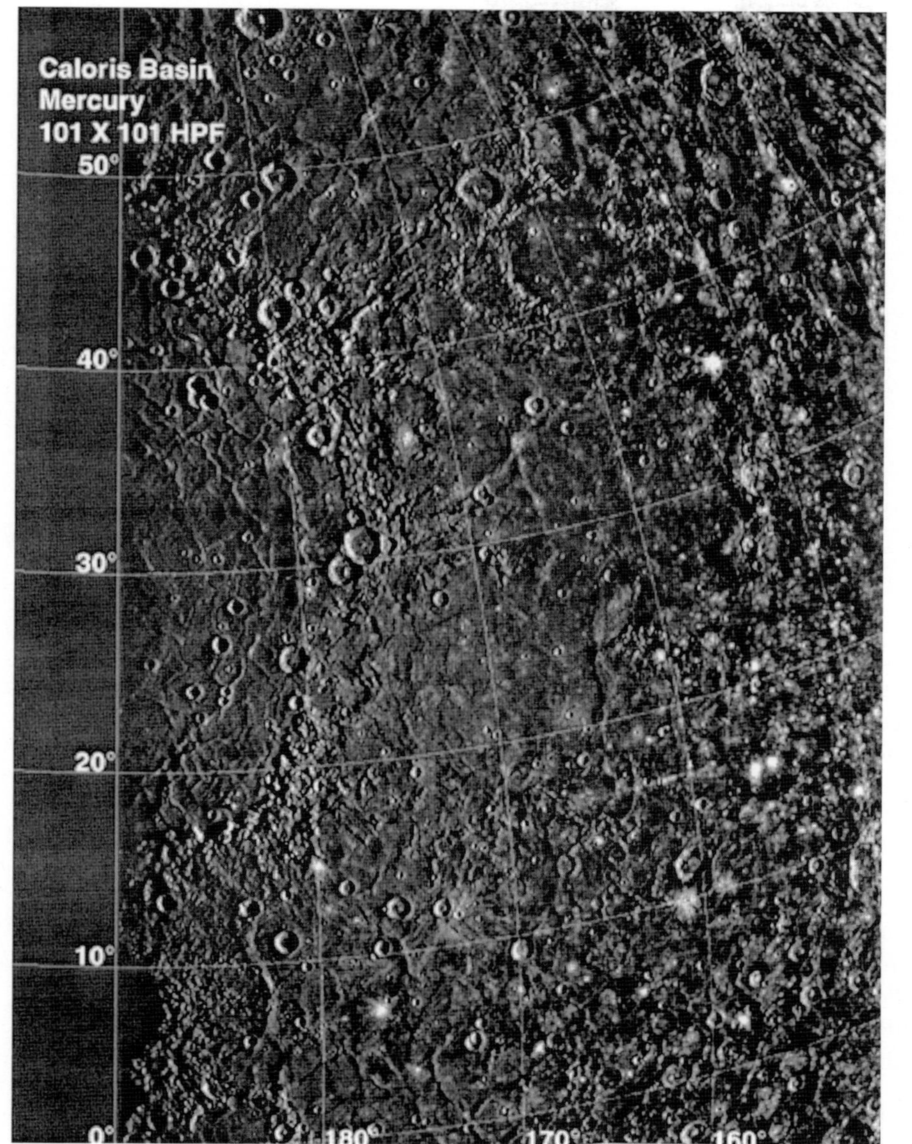

Fig. 2.19 Mercury's Caloris Basin A *Mariner 10* mosaic image of the sunlit portion of the Caloris Basin. This multi-ring feature spans 1.34 thousand kilometers, and its rim is marked by rough mountainous blocks that rise up to 1000 meters above the surrounding terrain. The collision resulted in a flat basin floor (*left*); it has been subsequently marked by smaller craters. (Courtesy of JPL and NASA.)

the Caloris Basin, producing a peculiar terrain of cracks, faults, hills and valleys.

The similarity of the surfaces of the Moon and Mercury, despite their differing masses and locations in the solar system, suggests that impacting objects were spread throughout the inner solar system during its early days. Mercury could have been bombarded at about the same time as the Moon, for scientists think that the entire solar system, with its Sun, planets and their satellites, formed 4.6 billion years ago.

The Moon itself probably formed from the remains of a collision. A Mars-sized object apparently struck the young Earth, melting the surface material at the point of impact and sending debris into orbit that eventually congealed to become the Moon (Section 5.9).

Ubiquitous impact craters – from Mars and Venus to Callisto and Miranda

When *Mariner 4* flew past Mars in 1965, snapping 22 close-up photographs, it revealed a wasteland riddled with the scars of an ancient rain of impacting meteorites (Fig. 2.21). Like the surfaces of the Moon and Mercury, the oldest Martian terrain probably bears the scars of the intense cosmic bombardment during the first 500 million years of the solar system, as well as the marks of a continual bombardment since then.

Yet, other parts of the surface of Mars have been extensively transformed by ancient, catastrophic floods and deep rivers (Sections 2.4, 8.8), and the pristine heavily cratered terrain also shows some of the wearing signs of passing

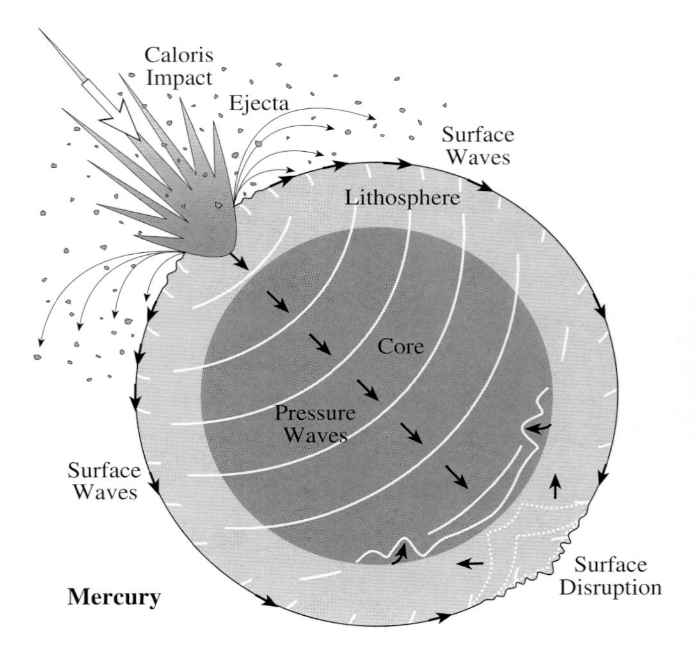

Fig. 2.20 Mercury's Caloris impact When an exceptionally large meteorite hit Mercury an estimated 3.85 billion years ago, it sent intense waves around the planet and through its core. They came to a focus on the opposite side of Mercury, disrupting the surface and producing hilly and lineated terrain there. The Caloris Basin was excavated at the impact site, and it now exhibits concentric waves that froze in place after the impact (Fig. 2.19).

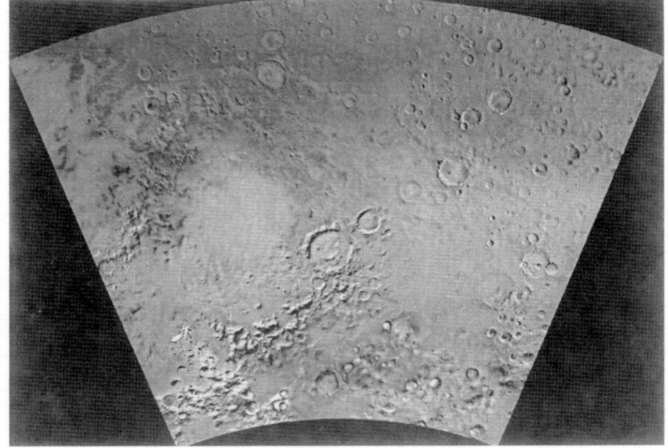

Fig. 2.21 Argyre region on Mars A mosaic of *Viking* images reveals the light-colored, circular floor of the Argyre impact basin (*middle left of center*), an 800-kilometer-wide hole blasted out of the surface of Mars by a giant meteorite impact. Mountains form part of the upraised rim structure of the impact crater, resembling the Apennine Mountains on the Moon (Fig. 2.18). In places, smaller, more recent, impact craters such as Galle (*lower left*) have obliterated the mountains. The smooth floor of the Argyre impact basin is apparently filled with lava, like the Imbrium Basin on the Moon (Fig. 2.18). The adjacent terrain shows the heavily cratered surface of Mars that is typical of the planet's southern hemisphere. Argyre is the name of the "silver" island at the mouth of the Ganges river on Earth. (Courtesy of JPL, NASA, and the U.S. Geological Survey.)

time. The largest impact craters on Mars are shallower than their lunar counterparts, with more subdued rims and flatter floors, and there are fewer smaller craters on Mars than there are on the Moon. These differences might be explained by enhanced erosion that modified the worn, old-looking craters and wiped out many of the existing small craters during the planet's early history, when the majority of craters were still forming. At that time the Martian atmosphere may have been more extensive than it is now, and Mars might have been warmer and wetter than at present (Section 8.8).

Large impact craters on Venus are relatively scarce when compared with the closely spaced, overlapping lunar craters. At one time Venus was probably as heavily cratered as the Moon, but the relatively small number and wide spacing of the craters now on Venus indicate that the surface we now see is much younger. When the Moon's cratering rate is scaled to Venus, the relative paucity of craters on its surface indicates an average age of about 750 million years, but the planet originated about 4.6 billion years ago. The relatively few craters we now see are due to meteoritic impact since the entire planet was resurfaced by rivers of outpouring lava about 750 million years ago (Sections 2.3, 7.5).

Following impact, large objects left craters on Venus that at first sight resemble those on the Moon, with central peaks, flat floors, and distinct circular rims. But the dense atmosphere on Venus affected both the incoming projectile and its ejected debris, creating features that are unlike any other craters in the solar system. The bright apron of debris that surrounds large craters on Venus often has a lobate, petal-like appearance with an unexpected asymmetry (Fig. 2.22). Material that was ejected from the crater became entrained in the hot, thick atmosphere, transforming it into a turbulent, fluid-like substance. The material flowed and spread out from the crater, creating patterns that resemble flowers or butterflies, rather than hurtling away from it to great distances in all directions.

Moreover, when the impact on Venus was oblique, the atmospheric wake of the incoming object prevented the ejecta from scattering back in the direction from which the meteorite came, so ejecta are missing in this region. Small incoming projectiles never made it to the ground, for they were burned up in the thick atmosphere. There are consequently no very small craters on Venus.

The surface of the Earth is sparsely cratered and therefore relatively young, usually less than 500 million years old although the planet is itself 4.6 billion years old. Erosion,

Fig. 2.22 Aurelia impact crater on Venus The unusual crater shapes on Venus are illustrated in this *Magellan* radar image of the Aurelia crater. Like the large impact craters on the Moon, it contains a circular rim, terraced walls and a central peak. But unlike lunar craters, lobate flows emanate from the ejecta, and a sector of the flow is missing, apparently due to an oblique impact from the upper-left. Interaction with the dense, thick atmosphere on Venus caused the ejected debris to act like a fluid, producing the lacy, rounded lobes of ejecta. Crater Aurelia, which is 32 kilometers in diameter, has been named in honor of the mother of Julius Caesar; apparently, Aurelia is also the name of Arnold Schwarzenegger's mother. (Courtesy of JPL and NASA.)

the deposition of sediments, the collisions of continents, and the internal churning of its rocks have erased any record of an ancient intense bombardment of the Earth. Impact collisions have nevertheless played an important role in shaping the Earth and affecting life on it. A meteorite of only 10 kilometers in diameter struck the Earth about 65 million years ago, wiping out the dinosaurs and many other species (Section 14.4).

Ancient impact scars are found on the rigid icy crusts of satellites in the cold outer parts of the planetary system. Jupiter's satellite Callisto, for example, has a rocky core surrounded by a deep layer of ice that is heavily scarred by impacts of meteorites (Fig. 2.23). The icy moons of Saturn, such as Mimas and Tethys, are also heavily cratered. Further out we find Miranda, the satellite of Uranus, with the most bizarre surface of all. It has regions of distinctly different terrain (Fig. 2.24). Some astronomers have argued that Miranda was once shattered into large fragments by a powerful collision, but that it managed to pull itself together again into a single body. Others reason that Miranda is a half-formed world, frozen when rock was still sinking into its interior and ice was rising to its surface (Section 11.5).

2.3 Volcanoes

Volcanoes on Earth and Venus

Volcanoes, another common aspect of the solar system, are driven by internal heat. For a large rocky planet, internal heat is continuously generated by the slow decay of radioactive material. Satellites can be heated by tidal interaction with their planet. Heat was also provided when the planets and satellites originated, as the result of high-speed collisions between smaller bodies.

Fig. 2.23 Jupiter's satellite Callisto This *Voyager 1* image shows the heavily cratered surface of Callisto. Its icy crust is as rigid as steel, and it therefore retains the scars of an ancient bombardment by impacting meteorites. An exceptionally large meteorite sent waves rippling across the surface, like a rock dropped into a pond. The extensive system of concentric rings, extending about 1.5 thousand kilometers from the impact site (*top center*), is named Valhalla after the home of the Norse gods. The impacting object apparently punctured the surface and disappeared. Today only the frozen, ghost-like ripples remain. Although there are very few large craters, the rest of Callisto is pockmarked with smaller impact craters that are flat for their size, and many of them have bright rims that resemble clean water ice splashed upon the dirtier surface ice. (Courtesy of JPL and NASA.)

Fig. 2.24 Uranus' moon Miranda The complex surface of Miranda, the innermost and smallest of the five major Uranian satellites, is seen at close range in this *Voyager 2* image. It contains a rugged, higher-elevation terrain (*right*) and a lower, complex terrain with ridges, grooves and jagged cliffs. The numerous craters on the rugged, higher terrain indicate that it is older than the lower terrain. The largest impact craters shown here are about 25 kilometers across. The other relatively young and complex terrain is characterized by sets of bright and dark bands such as the distinctive "chevron" feature (*top center*). (Courtesy of JPL, NASA, and the U.S. Geological Survey.)

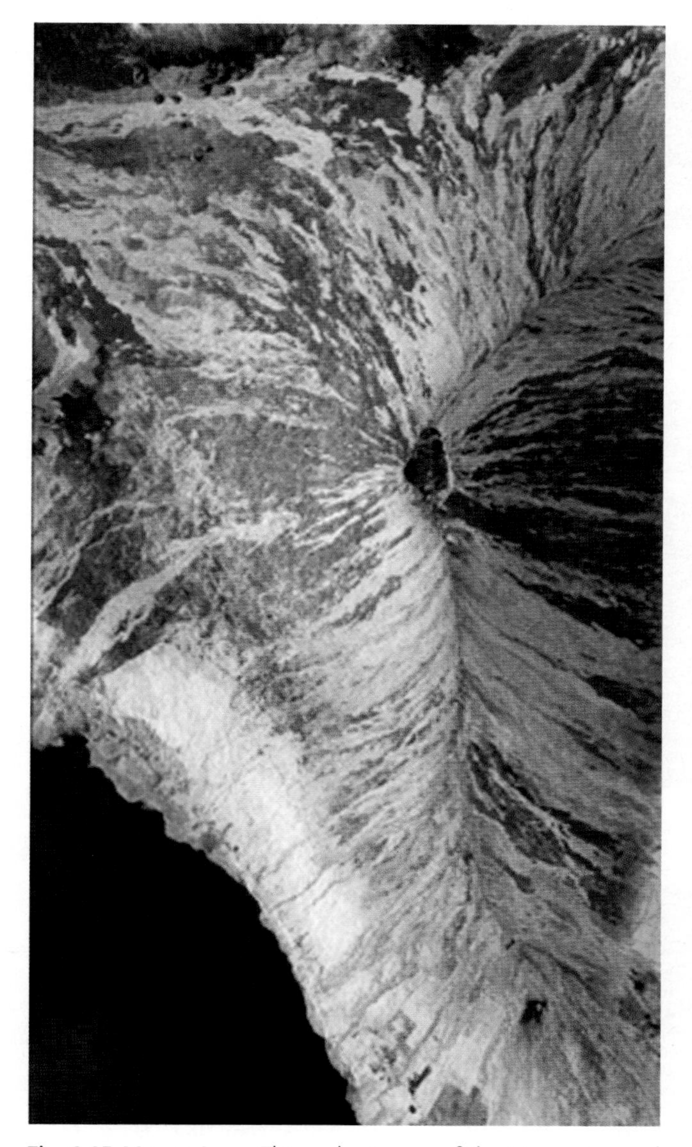

Fig. 2.25 Mauna Loa This radar image of the Mauna Loa volcano on the Big Island of Hawaii maps lava flows and other volcanic structures. Mauna Loa has erupted more than 35 times since the island was first visited by westerners in the early-1800s. The large summit crater, called Mokuaweoweo Caldera, is seen near the center of the image; this central depression was formed after lava flowed down the flanks of the volcano and magma was withdrawn inside it. If the height of the volcano is measured from its base on the ocean floor, Mauna Loa is the tallest mountain on Earth, rising almost 9 kilometers above the ocean floor. (Courtesy of JPL and NASA.)

Two planets now have a large number of volcanoes – the Earth and Venus. They have become hot enough inside to melt solid rock into liquid magma that is bottled up within their deep interior. The magma is swollen by heat, becoming lower in density, and rises through the cooler, higher-density material.

Eventually, the magma spreads out beneath the surface, pushes it up, and sometimes melts and punches holes in the crust, like a welder's torch. A volcano is formed and lava flows across the planet's surface. Molten rock trapped beneath a planet's crust is called magma; when molten rock issues from the crater of a volcano or a fissure in the crust it is called lava.

The Hawaiian Islands are giant volcanoes, formed when magma moved up from inside the Earth. Mauna Kea and Mauna Loa, on the big island of Hawaii, together form a mountain of lava that is much broader than it is tall; it is more than 120 kilometers across at its base and rises 9 kilometers above the ocean floor (Fig. 2.25). Such shield volcanoes have gentle slopes that have been built up from hundreds and even thousands of eruptions and individual flows of highly fluid lava. Mauna Loa is still erupting and growing, with repeated surges of lava that flow down its flanks.

Cone-shaped terrestrial volcanoes with steep slopes are formed when the lava is propelled out by hot gas. Examples are Vesuvius in Italy and Mount Fuji in Japan. The eruptions that formed these steep-walled mountains often expelled large clouds of volcanic ash.

The upwelling of pent-up heat and magma also forms rift valleys on Earth, with steep sides, sunken floors, and copious outpouring of lava. An example is the Great Rift Valley in Africa (Fig. 2.26), a long forking gash that crosses 4.5 thousand kilometers of the continent. It extends from Mozambique in the south to Ethiopia in the north, branching out through the Red Sea in one direction and diverging through the Gulf of Aden in another.

Tens of thousands of shield volcanoes have been identified on the face of Venus, by their round shapes and gentle slopes. They range in size from major, Hawaii-sized edifices that are hundreds of thousands of meters across (Fig. 2.27) to more numerous, smaller domes that pop up everywhere on the surface (Fig. 2.28). These shield volcanoes have been built up from runny lava that spreads out over great distances with the ease of spilt olive oil.

A smaller number of volcanic flows on Venus appear to be built from lava that is as stiff and thick as batter. In places, the sluggish lava has oozed onto the hot, flat surface of Venus, forming volcanic domes as round and flat as pancakes (Fig. 2.29). Each one has a dark feature almost precisely at the center, suggesting a vent from which the pasty lava flowed, like pancake batter on a hot griddle. Some of them even have little craters or pits on them that resemble bubbles that have burst in the batter. So, depending on the internal conditions when the magma formed in Venus, the resulting lava has the consistency and viscosity of either motor oil or toothpaste, and this helps determine the size and shape of the resulting volcanic formations.

Fig. 2.26 Nyiragongo The continent of Africa is being split apart by the pent-up pressure of hot, rising magma in numerous underlying hot spots along the Great Rift Valley. Volcanic outpourings like Nyiragongo fill the valley with lava as the rift slowly widens. (Courtesy of Bruce Coleman.)

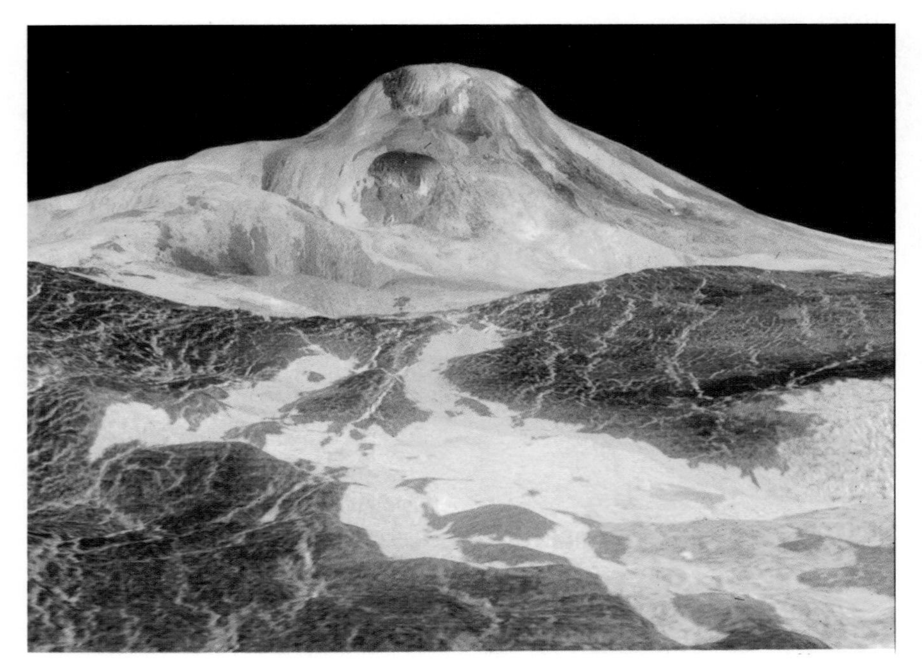

Fig. 2.27 Maat Mons This three-dimensional perspective of Maat Mons on Venus was obtained from radar data taken with *Magellan* in October 1991. It is 8 kilometers high, the second-highest peak on the planet. Fresh, dark lava extends for hundreds of kilometers in the foreground, perhaps flowing from a relatively recent eruption. Maat Mons is a giant shield volcano similar in size and shape to the big island of Hawaii (Fig. 2.25). *Maat* is the name of the ancient Egyptian goddess of truth and justice, and *Mons* is the Latin term for "mountain". The orange tint simulates the color of sunlight at ground level after filtering by the dense, thick atmosphere. (Courtesy of JPL and NASA.)

Unique, circular, domed features on Venus, named arachnids and coronae, are attributed to subsurface upwelling of magma that did not have quite enough force to completely break on through to the outside (Section 7.7).

Volcanoes on Mars

When *Mariner 9* neared Mars in 1971, the planet was engulfed in a dust storm. The eyes of the spacecraft – its

Fig. 2.28 Shield volcanoes on Venus Approximately 200 small volcanoes, ranging in diameter from 2 to 12 kilometers, can be identified in this *Magellan* radar image. They are shield-type volcanoes constructed mainly from eruptions of fluid lava flows similar to those that produce the Hawaiian Islands and sea-floor volcanoes on Earth. These small volcanoes are the most abundant geological features on the surface of Venus, believed to number in the hundreds of thousands and perhaps millions. (Courtesy of JPL and NASA.)

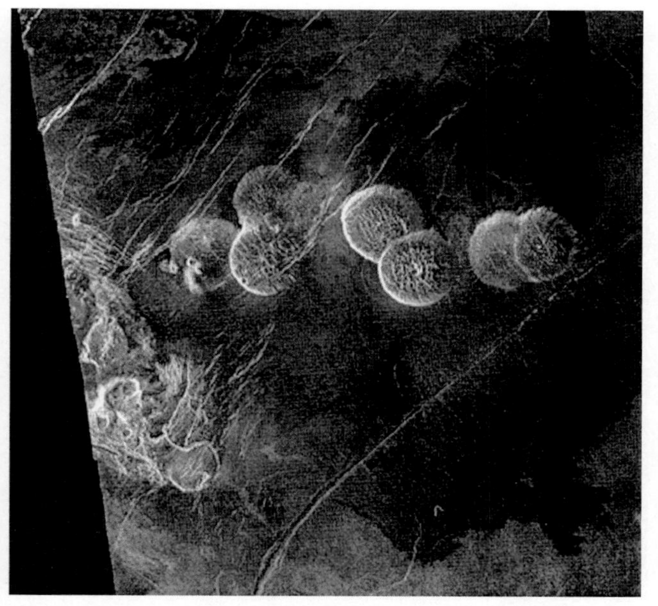

Fig. 2.29 Pancake domes on Venus These seven volcanic domes were discovered in *Magellan*'s radar images. They all have round shapes that are about 25 kilometers across, and steep sides that are less than 750 meters high. Their central vents may be lined up along a crack in the surface. These domes are interpreted as very thick, stiff and sluggish lava flows, rather than the fluid and runny type. Eruptions of the pasty, viscous lava, coming from a central vent on a relatively level surface, would form the circular, flattened shapes that resemble giant pancakes. Since there is little or no erosion by wind or water on Venus, newer pancakes look much the same as the ones on which they are superimposed. (Courtesy of JPL and NASA.)

cameras – could only peer at a disappointing, featureless ball, but as the dust storm began to settle four dark, round spots – the Tharsis volcanoes – poked out of the gloom. Even the thick blanket of dust could not cover these towering volcanic mountains. Thus, although Mars has just half the radius of the Earth, the red planet is still large enough to retain significant amounts of internal heat and to sustain long periods of volcanic activity.

The large volcanoes on Mars have the gentle slopes and rounded profiles of shield volcanoes on Earth, but the volcanoes on Mars stand higher. A striking example is Olympus Mons (Fig. 2.30) – *Mons* is a Latin term for "mountain". Olympus Mons is very much larger than any volcano on Earth. The major volcanic edifice is about 600 kilometers across at its base, rises more than 24 kilometers above its surroundings, and is rimmed by a cliff 6 kilometers high in some places. For comparison, the diameter of the base of Hawaii's Mauna Loa is just one-fifth that of Olympus Mons, and the height of the Hawaiian volcano is only a little over a third the height of the Martian one.

The impressive size of the volcanoes on Mars is attributed to the planet's thick outer shell, which remains fixed over the internal sources of magma. This gives the volcanoes on Mars a long time to grow, sometimes for billions of years. In contrast, the Earth's thinner crust is broken into pieces and moves over a source of magma, limiting the growth of individual terrestrial volcanoes and producing chains of smaller ones, such as the Hawaiian islands (Section 4.3).

Another type of large volcanic structure on Mars is the tholus, which is similar to the shield type of volcano but with somewhat steeper slopes, perhaps due to eruptions of more viscous lava or to a lower eruption rate. The Latin term *tholus* designates "a small domical mountain or hill". An example is Ceraunius Tholus (Fig. 2.31), with an estimated age of about 2.4 billion years.

A third kind of Martian volcano is the patera, with an exceptionally low summit and complex caldera. The name *patera* is Latin for "shallow dish or saucer". They

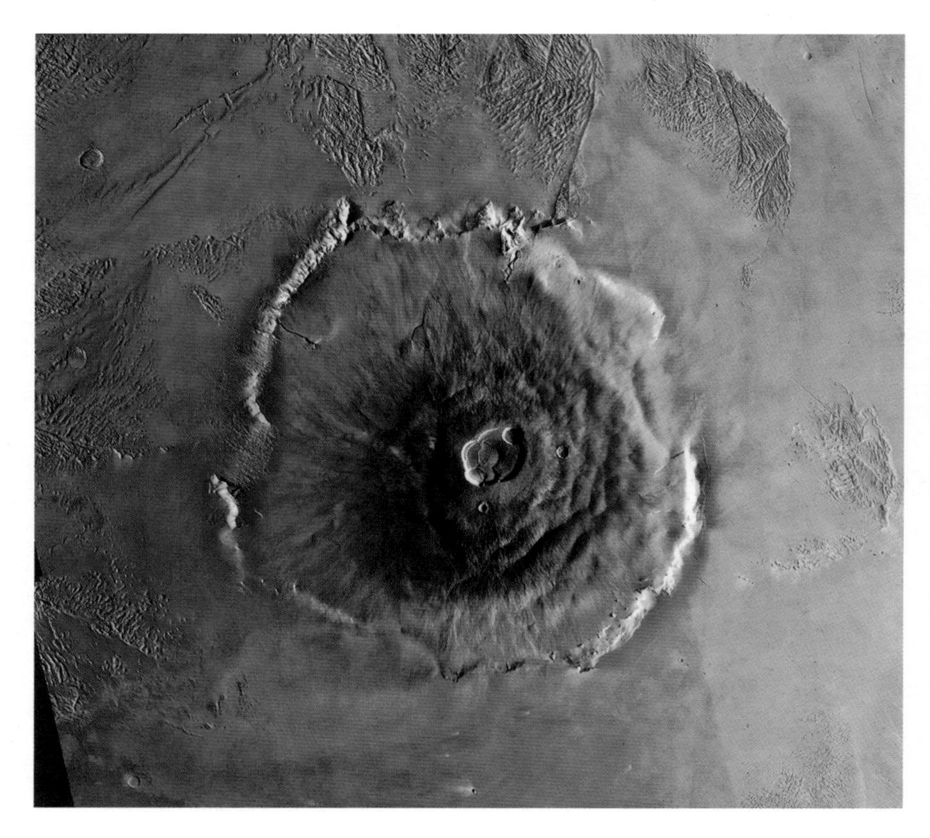

Fig. 2.30 Olympus Mons A mosaic of the towering Martian volcano Olympus Mons, using data obtained from the *Viking 1* orbiter in the late-1970s. It is the largest known volcano in the solar system, rising about 24 kilometers and spreading over 600 kilometers at its base. Counts of impact craters suggest that the lava flows on the gentle slopes of this volcano are relatively young, averaging only about 30 million years old. The summit caldera, or central depression, is a composite of as many as seven roughly circular depressions that formed by recurrent collapse when magma was withdrawn from within the volcano. The caldera is almost 3 kilometers deep and up to 70 kilometers across. The volcano is surrounded by a well-defined scarp, or cliff, that is up to 6 kilometers high. Many of the plains surrounding the volcano are covered by terrain containing ridges and grooves; it is called an *aureole*, the Latin term for "circle of light". *Mons* is the Latin term for "mountain". Mount Olympus, the highest mountain in Greece, is the home of the gods in Greek mythology. (Courtesy of JPL, NASA, and the U.S. Geological Survey.)

Fig. 2.31 Ceraunius Tholus The Tharsis region of Mars includes both volcanic and tectonic features. This *Viking 1* image portrays the shield volcano, Ceraunius Tholus (*middle right*), which is 115 kilometers in diameter. A 2-kilometer-wide channel extends from the summit caldera down the flanks of the volcano through a crater and into the adjacent plains. Smaller channels are just visible elsewhere on the flanks. The surrounding region (*left*) includes intensely fractured terrain. These structures are probably related to the uplift of the Tharsis region, causing fracturing and faulting. The term *tholus* means "small domical mountain or hill", and *ceraunius* means "thunderclap". Ceraunius Tholus is named for the Ceraunii mountains on the coast of Epirus, Greece. (Courtesy of JPL and NASA.)

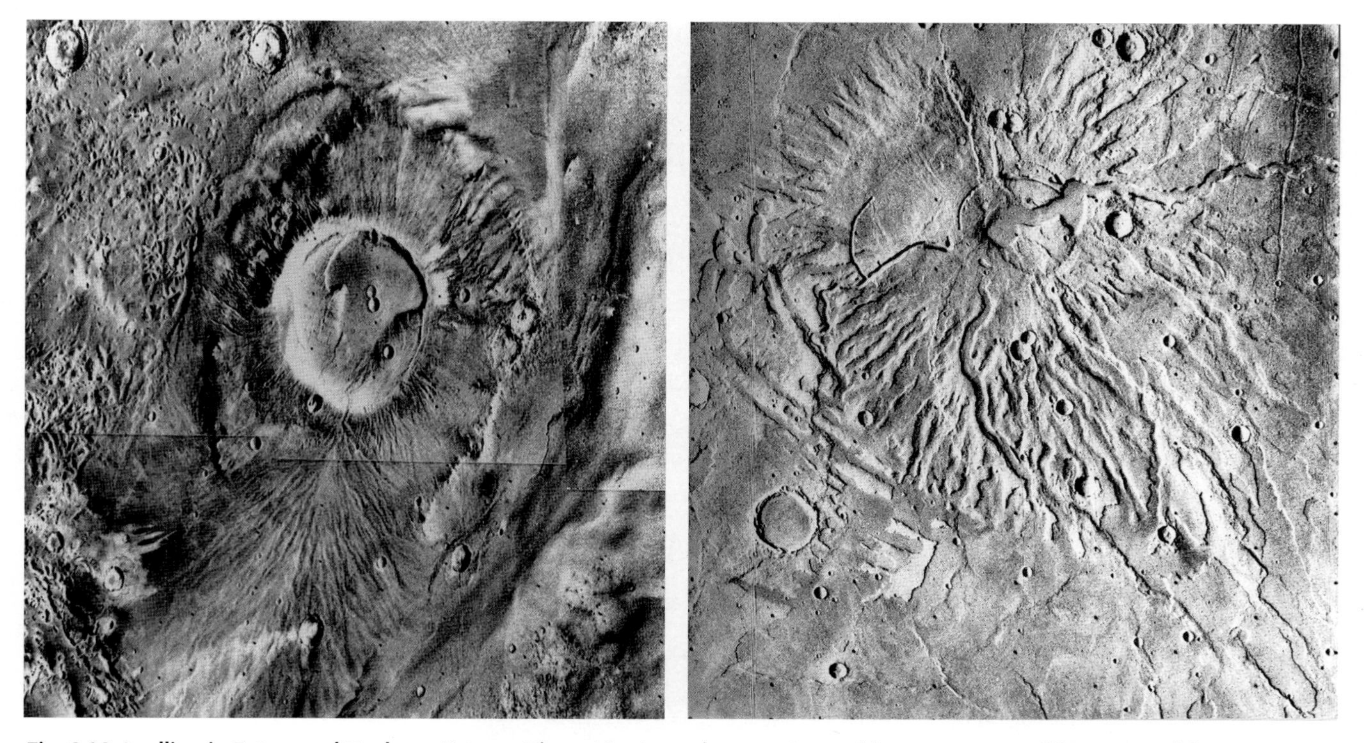

Fig. 2.32 Apollinaris Patera and Tyrrhena Patera These Martian volcanoes, imaged by cameras on *Viking 1*, are of the patera type, with low summits and broad flows. The summit caldera of Apollinaris Partera (*left*) is 80 kilometers wide, and the lava flows on its flanks were emplaced, on average, about a billion years ago. Tyrrhena Patera (*right*) is low-lying, fractured and highly eroded, with an estimated age of about 2 billion years. The caldera, about 12 kilometers across, is surrounded by a fracture ring about 45 kilometers in diameter. Several channels extend outward as much as 200 kilometers from the volcano center. The volcano could be comprised largely of ash flows, in contrast to the towering, younger, shield volcanoes which are mostly lava (Fig. 2.30). In Greek mythology, Apollo is the god of prophecy, sunlight, music and healing, and *patera* is an ancient Roman term meaning a "broad flat saucer or dish". The Tyrrhenian Sea is located between Italy and Sicily. (Courtesy of JPL and NASA.)

sometimes exhibit the worn-down appearance of old age (Fig. 2.32).

Images of the Martian surface suggest that volcanic activity might have persisted from the planet's youth into relatively recent times. Like the Moon and Mercury, the red planet bears the scars of a steady rain of meteorites. Relative ages of volcanoes and lava flows can be determined from the density of impact craters on them. While the most recent lava flows on Olympus Mons may be only a few million years old, and the average age is about 30 million years, lava could have been flowing out of this volcano for a long time before that.

Close-up images of other volcanic landforms, taken with cameras on *Mars Global Surveyor* in 1999–2002, indicate a distinct lack of craters and a fresh, young surface. The impact crater densities on some lava flows in the Martian plains, within the Elysium Planitia region, are up to a thousand times less than those on the lunar maria, suggesting ages of 100 million years or less. Volcanic cones that surround the Martian poles also appear to be geologically recent, with ages between 1 and 20 million years old. The presence of young lava flows and volcanoes implies that Mars may still be volcanically active today (Section 8.7).

Active volcanoes on Jupiter's satellite Io

There is one place in the solar system that is now more volcanically active than any other place; it is Jupiter's satellite Io. It is the hottest satellite in the solar system, so hot that you can see it melting before your eyes. Io is now spewing out 100 times more lava than all the volcanoes on the Earth. This is a totally unexpected discovery, made by the inquisitive camera eyes of *Voyager 1* in 1979 (Figs. 2.33, 2.34). Volcanoes are literally turning the satellite inside out, so parts of Io's surface are younger than your backyard. Because of the satellite's low gravity and lack of substantial atmosphere, the volcanic plumes spread out in graceful fountain-like trajectories, depositing circular rings of material up to 1.4 thousand kilometers in diameter.

What drives Io's continuous volcanism? The satellite is too small to now retain internal heat created during its

Fig. 2.33 Volcanic activity on Io *Voyager 1* captured this view of numerous volcanoes on Io on 13 July 1979. In contrast to Jupiter's satellite Callisto (Fig. 2.23), there are no impact craters on the surface of Io. Its volcanoes are continually resurfacing the satellite, keeping it young and erasing all signs of any impact craters. (Courtesy of JPL and NASA.)

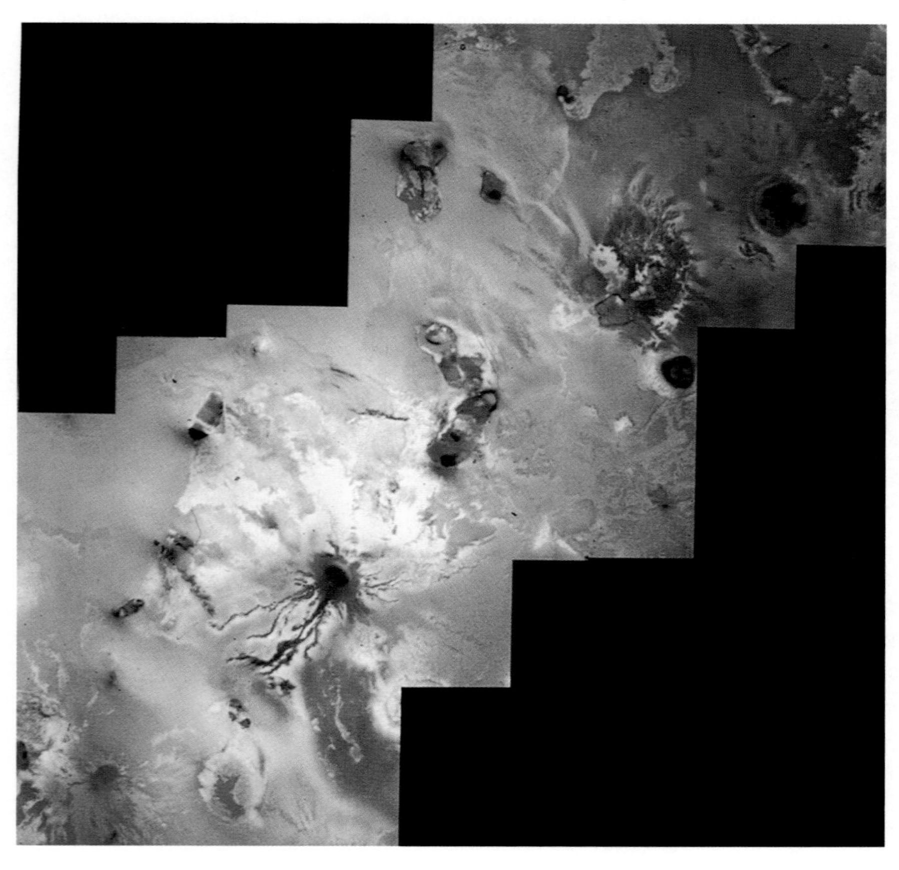

Fig. 2.34 Lava flows on Jupiter's satellite Io Numerous calderas and lava flows are located in this *Voyager 1* image mosaic of Io's volcanic plains. They are punctuated by dark snake-like tendrils of red, brown and black lava flows that are rich in sulfur. The bright whitish patches probably consist of freshly deposited sulfur-dioxide frost. This scene is about 2.1 thousand kilometers long, indicating that the slopes of active volcanoes stretch as far as 200 kilometers from the hot, black calderas into the cooler surrounding terrain. (Courtesy of JPL, NASA, and the U.S. Geological Survey.)

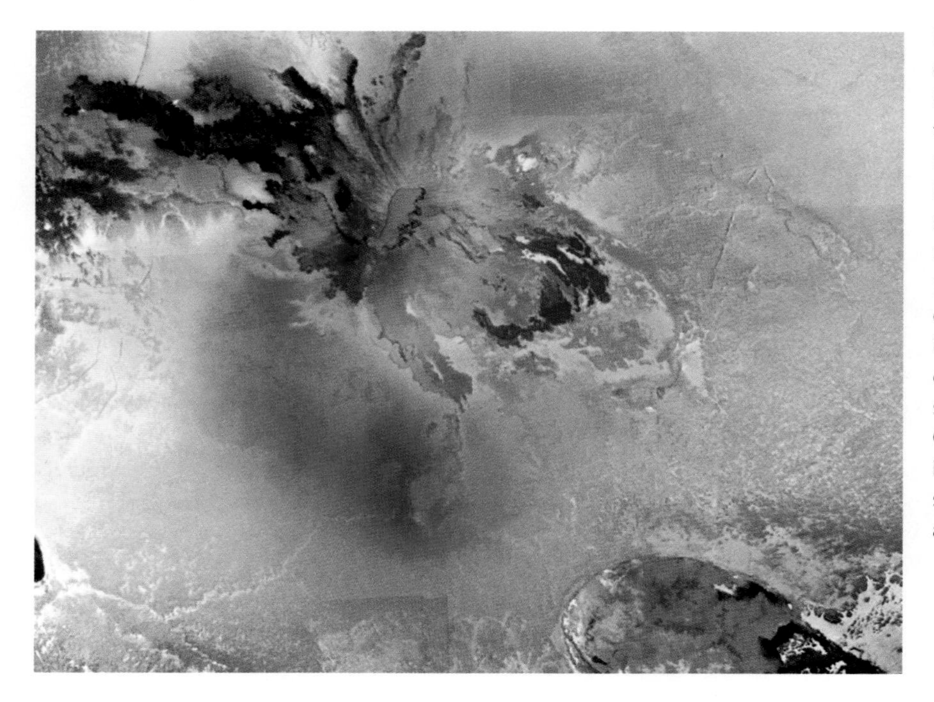

Fig. 2.35 Culann Patera One of the most colorful volcanic centers on Io is shown in this picture, constructed from images taken on 25 November 1999 through the red, green, and violet filters of a camera on board *Galileo*. Various colored lava flows spill out of the caldera on all sides. Unusual dark-red flows to the south east (*bottom right*) may be sulfur flows or silicate flows whose surfaces have been modified. The diffuse red material around the caldera is believed to be a compound of sulfur deposited from a plume of gas. Culann is the Celtic smith god, and *patera* is a Roman term meaning a "broad flat saucer or dish". (Courtesy of JPL, NASA, and LPL at the University of Arizona.)

formative years or to be significantly heated by radioactive rocks. The heat released during the satellite's formation and subsequent radioactive heating of its interior should have been lost to space long ago. After all, Io is about the same size as the Earth's Moon, which shows no signs of volcanism other than that associated with its earliest history about 4 billion years ago (Section 5.7). Io's internal heat is instead generated by tides that massive Jupiter raises in the solid body of the satellite.

The gravitational force of the nearby giant planet decreases with distance, so Jupiter pulls hardest on the side of Io facing it, and least on the opposite side; the center of Io is pulled with an intermediate force. These differences in the gravitational attraction of Jupiter on opposite sides of Io produce two tidal bulges in the solid body of the satellite – one facing Jupiter and one facing away. The giant planet thus effectively squeezes Io into the shape of an egg.

If Io remained in a circular orbit with the same face toward Jupiter, its tidal bulges would not change in height and no heat would be generated; but its orbit is not perfectly circular. Io's orbit has a forced eccentricity due to the combined gravitational interaction of Io with Jupiter and the other satellites of the giant planet. When the elongated orbit carries Io closest to Jupiter, the shape of Io is distorted more than when the satellite is further away. The resultant variation in the tides flex Io's surface, bending it in and out by as much as 100 meters during each orbit. Friction associated with this tidal flexing heats Io inside, melting its rocks and producing volcanoes at its surface.

Galileo returned for a close-up view of Io's volcanoes in 1999–2000 (Fig. 2.35), providing a better understanding of the sizzling world. Instruments on *Galileo* measured the temperatures of the volcanoes, showing that the lava is at 1700 to 2000 kelvin, up to twice the temperature of volcanoes on Earth. The high-temperature eruptions emit gaseous sulfur and sulfur dioxide; the bright surface flows are attributed to sulfur and the white surface deposits to sulfur dioxide. The very high temperatures apparently rule out liquid sulfur as a dominant volcanic fluid, and they have certainly driven off any water that might have been on Io. Other recent *Galileo* discoveries about Io are discussed in greater detail in Section 9.4.

Volcanism on Neptune's satellite Triton

Neptune's largest satellite, Triton, is the coldest moon ever recorded, with a temperature of just 38 kelvins, approaching absolute zero where all motion stops. It is so cold because Neptune is so far away from the Sun, therefore receiving little sunlight, and also because Triton reflects more of the incident sunlight than most satellites – only Enceladus and Europa are comparable. Yet, the frozen moon is a dynamic, alive world, molded by volcanic eruptions.

Triton's surface has a smooth, youthful appearance, with no large impact craters and few small ones. Global resurfacing by volcanoes of ice might have wiped out

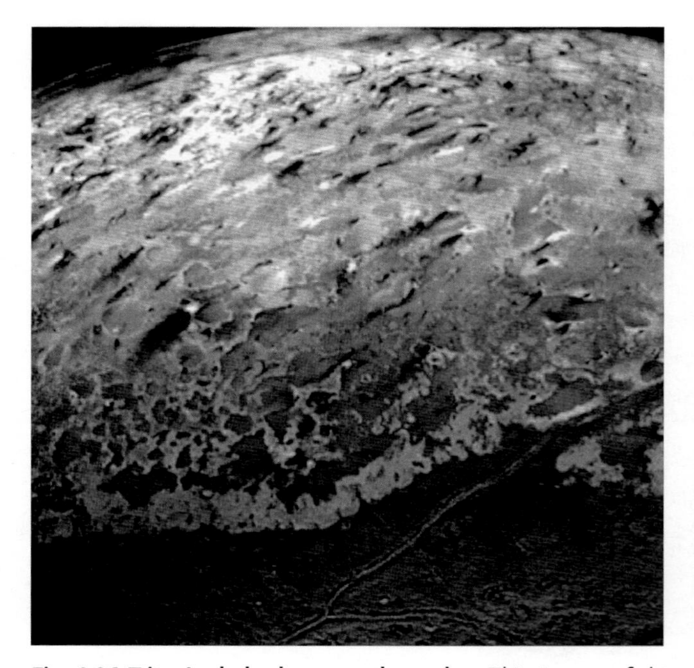

Fig. 2.36 Triton's dark plumes and streaks This image of the south polar terrain on Triton reveals about 50 elongated dark plumes, or "wind streaks" on the moon's highly reflective surface. The plumes originate at very dark spots, generally several kilometers across, probably marking vents where nitrogen gas was driven outward in geyser-like eruptions from beneath the surface. Winds in Triton's thin, nitrogen atmosphere may have carried the dark erupted material along, depositing it in the elongated streaks. This image was taken on 25 August 1989 from *Voyager 2*. (Courtesy of JPL and NASA.)

pre-existing craters on Triton, perhaps a few billion years ago when tidal flexing may have heated the satellite's insides. The deep underground heat may have turned the ice into liquid that rose to the surface, like a squeezed slush cone, filling the vast frozen basins that are now found there. These frozen lakes of ice look like inactive volcanic calderas; complete with smooth filled centers, successive terraced flows and vents.

Numerous dark plumes and streaks, found in the midst of the bright southern cap of Triton, suggest a different kind of volcanic activity, propelled by relatively recent eruptions of nitrogen gas (Fig. 2.36). Nitrogen boils at very low temperatures, at just 77 kelvin on Earth, and when it boils it expands, producing enormous pressures that can shoot gas and other material high into Triton's thin nitrogen atmosphere. Thus, geyser-like eruptions may have lofted the dark material outward from beneath the surface. The prevailing winds would then carry it across the satellite, depositing it on the ice as dark streaks.

Four active plumes were observed during the *Voyager 2* encounter with Triton. They rose in narrow straight columns to an altitude of 8 kilometers, where dark clouds of material were left suspended and carried downwind horizontally for over 100 kilometers, like smoke wafted away from the top of a chimney. Most of the dark streaks are probably remnants of such plumes.

Since the active plumes occur where the Sun is overhead, it is possible that sunlight produces the weak subterranean heat required to make the nitrogen boil and break through the overlying layer of ice. The sunlight would pass through the translucent ice and become absorbed by darker material encased beneath. The overlying nitrogen ice would trap the solar heat, for it is opaque to thermal infrared radiation, producing a solid-state greenhouse effect. Nitrogen gas, pressurized by the subsurface heat, might then explosively blast off the iced-over vents or lids, launching volcanic plumes of gaseous nitrogen and ice-entrained darker material, just as the water in an overheated car radiator is explosively released when the radiator cap is removed. The nitrogen geysers also resemble water-driven geysers on Earth, like Yellowstone's Old Faithful, but water boils at a much higher temperature of 373 kelvin. This brings us to water, another common property of many planets and satellites.

2.4 Water

Earth, the water planet

From space, our home planet Earth looks like a tiny, fragile oasis in space, a glistening blue and turquoise ball of water, flecked with delicate white clouds and capped with glaciers of ice (Fig. 2.37). Seventy-one percent of the Earth's surface is now covered with water. The oceans contain so much water that if the Earth were perfectly smooth the oceans would cover the entire globe to a depth of 2.8 kilometers. They contain about one billion trillion (10^{21}) kilograms of water.

Water is a marvelous substance. As a liquid, it will dissolve almost anything to some extent, and it can hold and release very large quantities of heat. When liquid water freezes, it expands and becomes less dense, in contrast to most substances. As a result, ice floats on the surface of lakes and oceans, so they freeze from the top down.

Water is crucial to life here on Earth. We, ourselves, are largely water. Just about anywhere there is water here on Earth, there is some sort of life – even thousands of meters inside the Earth. It follows that if liquid water was found on another planet, or on one of its satellites, that place might also be hospitable to life.

Fig. 2.37 Earth, the water planet Almost three-quarters of the Earth's surface is covered by water, as suggested by this view of the North Pacific Ocean. Earth is the only planet in the solar system where substantial amounts of water exist in all three possible forms – gas (water vapor), liquid and solid (ice). Here white clouds of water ice swirl near Alaska; the predominantly white ground area, consisting of snow and ice, is the Kamchatka Peninsula of Siberia. Japan appears near the horizon. From this orientation in space, we also see both the day and night sides of our home planet. (Courtesy of NASA.)

Water is made of the two most abundant, chemically reactive elements in the Universe, hydrogen and oxygen, and so ought to be very common. Yet, the Earth is the only place in the solar system where substantial quantities of water exist in all three possible states – as a gas (water vapor), liquid, and solid (ice). Ours is the only planet whose surface temperature matches the temperature of liquid water, between water's freezing and boiling temperatures of 273 to 373 kelvin, respectively.

When we look at our nearest neighbors, we see that Venus is too hot and Mars is too cold for significant amounts of liquid water to exist on their surfaces. Any water on Venus would now be in the form of steam, and water on Mars is now mainly locked beneath the surface in the form of ice and frost. The terrestrial planets were nevertheless colder or warmer in the distant past and they will have different surface temperatures in the future. During the last million years, the Earth's climate has, for example, been dominated by the recurrent ice ages, each lasting 100 thousand years, in which polar glaciers reached halfway to the equator, burying everything in their path under ice a kilometer thick. The ice ages are punctuated by brief intervals of unusual warmth, lasting about 10 thousand years. Human civilization developed during the most recent warm period, that began about 10 thousand years ago, and sooner or later we are headed for another ice age – unless global warming by human activity intervenes (Section 4.4).

It was too hot for liquid water to exist on any of the rocky inner planets in their very early history. Because of the energy released by the colliding rocks that merged to form these planets, they probably began as molten globes about 4.6 billion years ago. It was so hot that any surface water would have boiled off. Only when the initial bombardments slowed, and the glowing planets cooled, could water return to the surface. Oceans might then be liberated from their interiors by volcanoes or carried to the planets by icy comets arriving from the outer parts of the solar system.

Since the young, newly formed Earth would be very hot, water could not condense on its surface and it would start out dry. Our oceans were therefore probably supplied from outside the Earth at a later date, perhaps by small rock and ice bodies, similar to today's asteroids and comets, coming in from a cold outer reservoir. About one million comets would have to collide with the Earth in its youth to supply the amount of water now in its oceans, so there would have to be a lot more comets coming into the inner solar system in ancient times than there are now.

So, astronomers argue that the water now found in our oceans and lakes was supplied from outside the Earth by ancient impacts of comets and asteroids. Some geologists reason that the water was expelled from inside the Earth during intense bouts of volcanism. No one knows for sure whether external impact or internal volcanism resulted in most of our water. But since the Earth formed by the accumulation of colliding objects, even the water expelled by the volcanoes had to be originally supplied by cosmic impacts.

Possible water ice at the poles of Mercury and the Moon

As the planet nearest the Sun, Mercury bakes with midday temperatures on its sunlit side as high as 825 kelvin, but a strong tidal lock keeps the Sun directly over Mercury's equator at all times. This means that crater interiors near the poles are never exposed to direct sunlight. Calculations indicate that the permanently shadowed craters may have remained colder than 112 kelvin for eons, permitting substantial quantities of water ice to accumulate at the frigid crater floors near the poles.

When Mercury's poles tip toward the Earth, while never deviating from the north–south direction and remaining hidden from the Sun, astronomers have beamed radio signals at them and examined the echoes. The radar observations have revealed that the planet's north and south poles may contain substantial deposits of water ice, at least a couple of meters thick. Unusually strong radar echoes coming from the polar regions show prominent, radar-reflective caps that are plausibly attributed to water ice. These are return signals from pulses of radio radiation sent from the world's most powerful radio transmitter at the Arecibo Observatory in Puerto Rico.

The water in the dark and frigid places within polar craters could have been dumped on Mercury by icy comets that have been bombarding the planet since its formation until the present day. When the comet-loaded ice hits Mercury, the ice might vaporize and settle into the cold traps on the permanently shaded crater floors. A comet might also hit the polar regions directly, depositing water there. In either event, water ice will stay in the polar craters forever, never exposed to direct sunlight.

The rocks returned from the Moon are drier than a terrestrial desert. They resemble an Earth rock that has had all the water boiled out of it. For this reason, most scientists have assumed that there is no water on the Moon, and that there never was any.

Yet, if Mercury has reservoirs of frozen water at its poles, then the Moon might have some ice in craters at its polar regions that are also permanently in shadow, remaining eternally dark and cold. While in polar orbit around the Moon in 1994, the *Clementine* spacecraft bounced radar signals off the lunar surface and back to antennas on Earth, obtaining bright echoes that suggested ice might be present in a deep, cold, shadowed basin near the lunar south pole.

Then the *Lunar Prospector* spacecraft revived the prospects for lunar ice when its neutron spectrometer detected a signature of hydrogen atoms in both polar regions of the Moon. The instrument measures the neutrons dislodged from the Moon's crust by the impact of energetic particles called cosmic rays. The speed and energy of the neutrons depend on what they strike, and the observations suggested that they were bouncing off the nuclei of hydrogen atoms. The data indicate that there is hydrogen in the polar regions of the Moon at places that are almost perfectly correlated with dark areas, and that there are too many hydrogen nuclei to be solely implanted by the solar wind (Section 3.5). Scientists estimate that if the hydrogen atoms belong to water molecules, then 6 billion tons, or 6 trillion kilograms, of water ice is hidden in permanently shadowed regions near the Moon's north and south poles.

Astronomer's hoped that by crashing *Lunar Prospector* into a crater near the Moon's south pole, it would splash up a plume of water-bearing dust. More than a dozen large telescopes around the world were turned to watch the crash, on 31 July 1999, but they detected no plume of debris let alone the spectral signatures of water vapor. There are several explanations, including the possibility that the spacecraft missed its target.

The available evidence points toward the possibility of water ice at the lunar poles, but we won't know for certain until spacecraft land there. If such water reservoirs exist, they could help support human outposts on the Moon. Water could be purified to drink, or it could be chemically split into hydrogen, to burn as a rocket propellant, and oxygen to breathe. This would make it easier to establish a colony on the Moon, or to build a fueling station on it for interplanetary spacecraft, so the possibility of water ice at the Moon's poles is an important one.

A former ocean on Venus

Today the surface and atmosphere of Venus are exceptionally dry, which is what you would expect for a planet whose surface temperature is now a blistering 735 kelvin. The planet may nevertheless once have had vast quantities of liquid water. Models of planet formation predict, for example, that the Earth and Venus were once endowed with roughly equal amounts of water. When a runaway greenhouse effect heated Venus up, most of its water evaporated and was lost to space, while the Earth remained cool enough to keep its oceans, once they had originated. And if the very small quantities of water vapor now found in the atmosphere of Venus are a remnant of an ancient reservoir, then Venus has lost the equivalent of a very large lake or a small ocean.

Evidence that Venus once had an ocean is found in an excess of deuterium now in its atmosphere. Deuterium is an atom chemically identical to hydrogen but heavier and therefore more likely to be retained in the atmosphere. On Earth, it is found in heavy water, that comprises only about 0.016 percent (0.000 16) of the oceans. The natural explanation of the excess atmospheric deuterium on Venus is that the planet once had vast quantities of normal water, containing light hydrogen, and heavy water, containing deuterium. When these liquids were subsequently boiled away by the intense heat, the lighter hydrogen easily escaped from the planet, but some of the heavier deuterium remained behind as a residue. The amount of remaining deuterium suggests that Venus once had enough liquid water to uniformly cover the planet's surface with a global lake at least 4 meters deep, or just 0.12 percent of a full terrestrial ocean.

Water on Mars

Mars is a much drier, colder planet than the Earth. The *Viking 1* and *2* landers showed that the average surface temperature of Mars varies between 195 and 220 kelvin over a full Martian year of 687 Earth days, with daily variations between 10 and 60 kelvin. Over most of the surface of Mars, the temperature is below the freezing point of water at 273 kelvin, so nearly all of the water on the Martian surface is now frozen solid.

Liquid water cannot now exist for any length of time on the surface of Mars. It would immediately begin to boil, evaporate and freeze – all at the same time. Because of the low pressure of the thin Martian atmosphere, any liquid water would quickly vaporize. The surface pressure on Mars is at or below 1 percent (0.01) that of Earth's pressure at sea level (Sections 3.1, 8.3). And because it is now so cold on Mars, any liquid water or water vapor would soon freeze into ice.

Ninety-five percent of the Martian atmosphere is gaseous carbon dioxide (Sections 3.1, 8.3), and it contains only small quantities of water-ice crystals, in the form of clouds and haze. In addition, the red planet's polar caps contain large amounts of carbon-dioxide ice, known on Earth as dry ice. Since the freezing point of carbon dioxide is just 217 kelvin, the dry ice in a polar cap will sublimate, going directly from surface ice to atmospheric vapor, during the heat of the local Martian summer, and freeze back into the cap in the local winter.

Very little water vapor is present in the Martian atmosphere, making it drier than the driest of the Earth's deserts. If the water vapor were collected and condensed, it would amount to no more than a good-sized lake. Yet, despite the small amount of water vapor, the cold, thin atmosphere is close to saturation. It is about as wet as it can be. Consequently, the formation of clouds and fogs of water ice is a common feature of Martian weather. Such clouds have been observed along the flanks of volcanoes, above the polar caps, and in low-lying areas such as canyon floors.

Water is also frozen in the polar caps of Mars. When the summer heat evaporates the seasonal carbon-dioxide ice in the north, a massive underlying polar cap of water ice is detected (Section 8.5). The amount of water currently trapped in both Martian poles combined is about one-and-a-half times the amount of water frozen in the glaciers of Greenland. When melted, the water ice at the two poles of Mars could be enough to cover the planet with a small global ocean 20 to 30 meters thick.

Great quantities of water ice might also lie beneath the surface of Mars within its frozen soil, like the permafrost in Earth's arctic tundra. The layer of permafrost in the polar regions might be as much as a kilometer thick.

Spectroscopic instruments aboard the *2001 Mars Odyssey* spacecraft have found direct evidence of subsurface water ice buried in the Martian northern and southern hemispheres. The frozen reservoir is so pervasive at latitudes poleward of 60 degrees, an area much larger than Mar's permanent polar caps, that its total volume probably exceeds 10 thousand cubic kilometers. The concentration of ice in the upper meter of the ground is surprisingly high – one-fifth to one-third by weight and more than 50 percent water ice by volume. So if you heated one full bucket of this material it would be more than half a bucket of water. The amount detected so far is enough to fill Lake Michigan twice over, and there may be much more. Because *Mars Odyssey* can detect water to no more than about a meter under the surface, no one knows how far down the ice might extend.

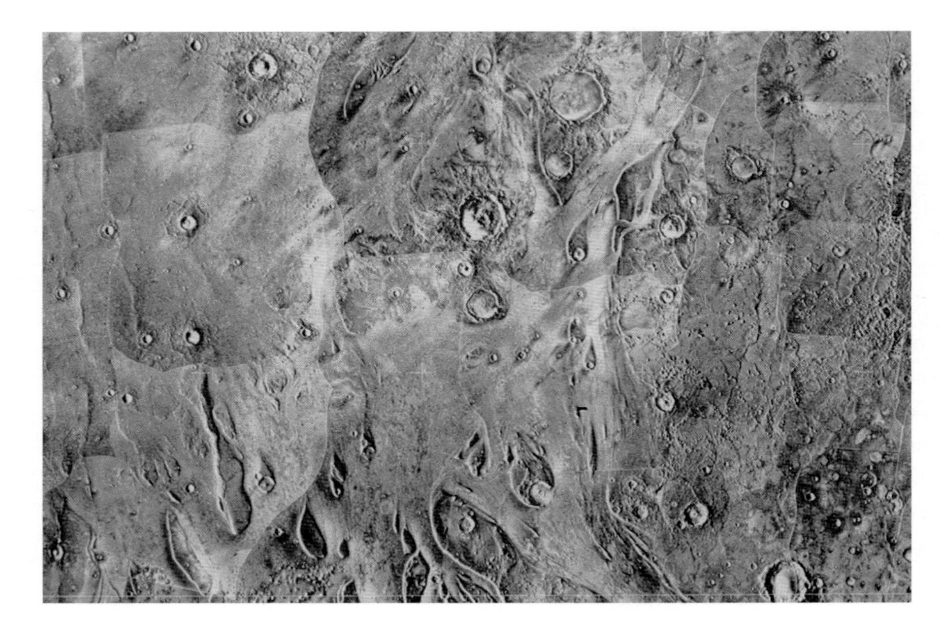

Fig. 2.38 Outflow channels A *Viking 1* image of the Ares Vallis region shows dry river beds produced by past flows of liquid water. Catastrophic floods must have carved out these outflow channels deep in the Martian surface, creating streamlined hills, scoured floors, and large teardrop-shaped islands where flowing water encountered an obstacle. Ares is the Greek god of war; the Roman equivalent is Mars. (Courtesy of Michael H. Carr, U.S. Geological Survey, and NASA.)

Vast quantities of liquid water flowed over the surface of Mars in the distant past. Huge, dry river beds and flood channels, imaged by *Mariner 9* and the *Viking 1* and *2* orbiters, provide unmistakable signs of former, water-charged torrents that cascaded across the Martian surface (Fig. 2.38). The flow channels that have been carved and etched into the surface of Mars are immense by terrestrial standards, as much as 100 kilometers wide and 2 thousand kilometers in length. The amount of water required to gouge out these river-like outflow channels is enormous, requiring catastrophic floods containing millions of tons, or billions of kilograms of liquid water (Section 8.8).

When and how all this water got to the surface is debated. The water may have come by outbursts from under the ground rather than by rain from above. On Earth, creeks flow into streams that converge into ever-larger rivers. There are no tributaries to the outflow channels on Mars. This suggests that the flows were caused not by runoff on the exposed surface, but rather by underground streams buried deep beneath the frozen ground. Ancient valley networks on Mars, however, form patterns similar to terrestrial watersheds.

High-resolution images from *Mars Global Surveyor* suggest that water has broken out of channel and canyon walls in relatively recent times, from underground where there was sufficient heat from volcanic activity to keep the water liquid. Stream-like seeps and flows have been detected coming out of the walls of canyons and craters, within the past few million years (Section 8.8). Many scientists suspect that liquid water is oozing out of a warm interior under the frozen surface of Mars. Others are more cautious, reasoning that fluid flows of mud or hot gas are responsible. Future investigations will determine what the correct explanation is, but in the meantime we can be reasonably certain that a deluge of water swept over much of the red planet a few billion years ago, creating massive floods, scouring the surface and carrying rocks and debris downstream.

As everyone knows, water flows downhill, so one clue to the water's fate is provided by the topography of Mars. The laser altimeter on board the *Mars Global Surveyor* made height measurements, with a precision as good as one meter, by bouncing laser beams off the planet's surface, determining land elevations by the time of the round trip. The dominant feature of the global topography is the six-kilometer difference in elevation between the low northern hemisphere and the high southern hemisphere. The northern plains have been resurfaced to a nearly billiard-ball smoothness, free of the immense canyons and valleys visible in the southern hemisphere (Fig. 2.39). This suggests that the floodwaters that cut the outflow channels drained northward and pooled in the vast northern lowlands at the ends of the channels.

Some scientists have argued that the northern lowlands were once the sites of an ancient ocean, covering up to one-third the surface area of the planet and up to 1.6 kilometers deep. Supporting evidence for this hypothesis includes hints of an ancient shoreline that extends around the southern boundary of the northern lowlands, like a bathtub ring. There are curved terraces parallel to the shoreline, attributed to either pounding waves or to receding water.

When the topographical maps are combined with measurements of the red planet's gravity, buried subsurface

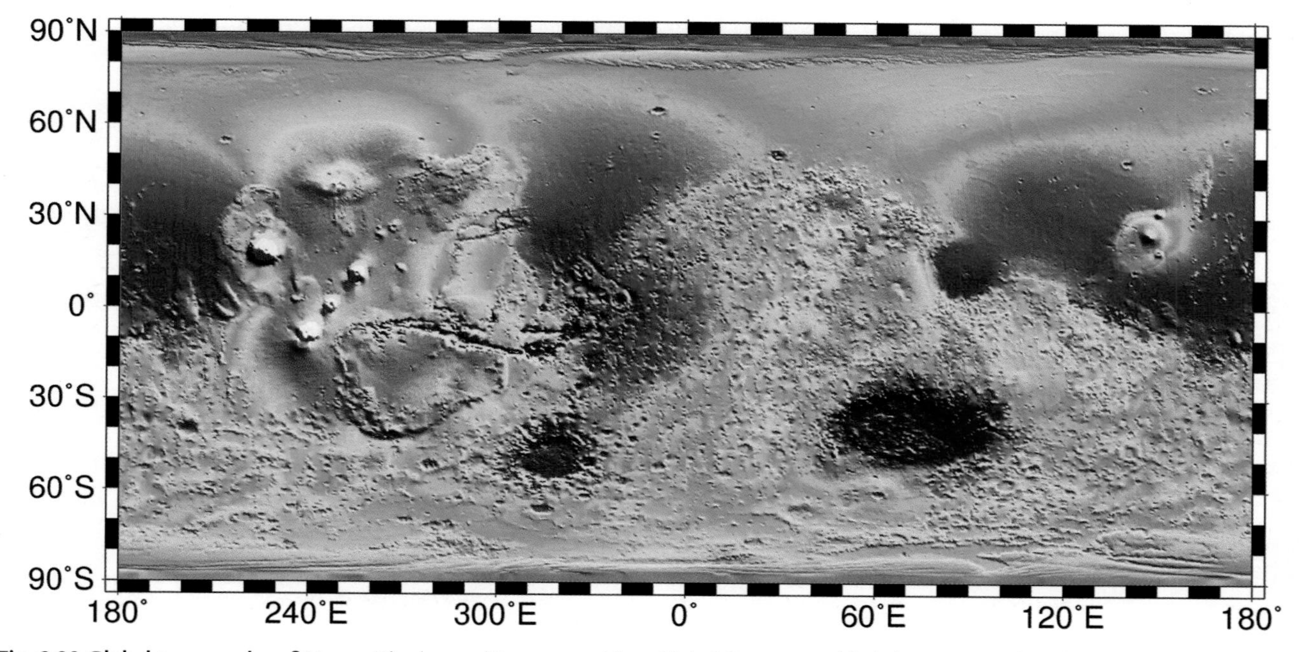

Fig. 2.39 Global topography of Mars The laser altimeter on *Mars Global Surveyor* yielded this precise global relief map of Mars, with the south pole (S) at the bottom and the north pole (N) at the top. By measuring the round-trip time of laser pulses bounced off the surface, the altitude was measured with a vertical accuracy of about one meter. The red areas in the southern hemisphere are high regions, about four kilometers above the average surface height, and the blue regions of the northern hemisphere are low places, about four kilometers below the average height. Dominant features also include the Hellas impact basin (45°S, 70°E) and the Tharsis region centered near the equator in the longitude range 220°E to 300°E. It contains the long, continent-sized, east–west Valles Marineris canyon system and several major shield volcanoes, including Olympus Mons (18°N, 225°E). Ancient floodwaters must have flowed downhill, from south to north (*top*), within the blue-colored outflow channels to the east (*right*) of Valles Marineris and to the west (*left*) of Olympus Mons, perhaps emptying into the northern lowlands and forming an ancient ocean there. (Courtesy of JPL and NASA.)

canyons are located beneath the ground, emanating from the visible outflow channels. This indicates that the transport of water continued far into some parts of the northern plains.

But if there was an ocean, where has all the water gone? Some of the water is now frozen into the north polar cap. The rest of the water, if it exists, may be frozen in the soil as permafrost.

The crisp, sharp images obtained with the cameras aboard *Mars Global Surveyor* have recorded deep and widespread layered deposits that might have been laid down long ago in numerous lakes and shallow seas when Mars was able to sustain liquid water for long periods. They are now seen just south of the equator, in exposed walls within ancient impact craters or basins and the chasms of the Valles Marineris (Fig. 2.40). These startling images remind us of the rock layers in the Grand Canyon. The sedimentary layers do not show evidence of any gullies, streams or channels that might have filled the layered regions with water, so the source of water would also likely be subterranean.

In one interpretation, the red planet was warmer and wetter in its early history, and was therefore much different from the cold, arid Mars we see today (Section 8.8, Focus 8.3). A thicker, warmer atmosphere would have permitted flowing water that filled low-lying areas with pools and lakes of liquid water, perhaps persisting long enough for layers of sediment to be compressed and cemented in place.

Another scenario for many of the Martian features attributed to water flow or deposit is strong winds that can mimic these effects. Raging winds, formed at a time when the Martian atmosphere was much denser than it is today, could have carved out flow-like features and lifted and deposited huge amounts of dust to form strata with no need for water at all.

However, most geologists agree that vast amounts of water flowed across the Martian landscape long ago, perhaps 3 to 4 billion years in the past. Mars may therefore have been hospitable to life a very long time ago. Life might even still be preserved below the surface if there is liquid water underground.

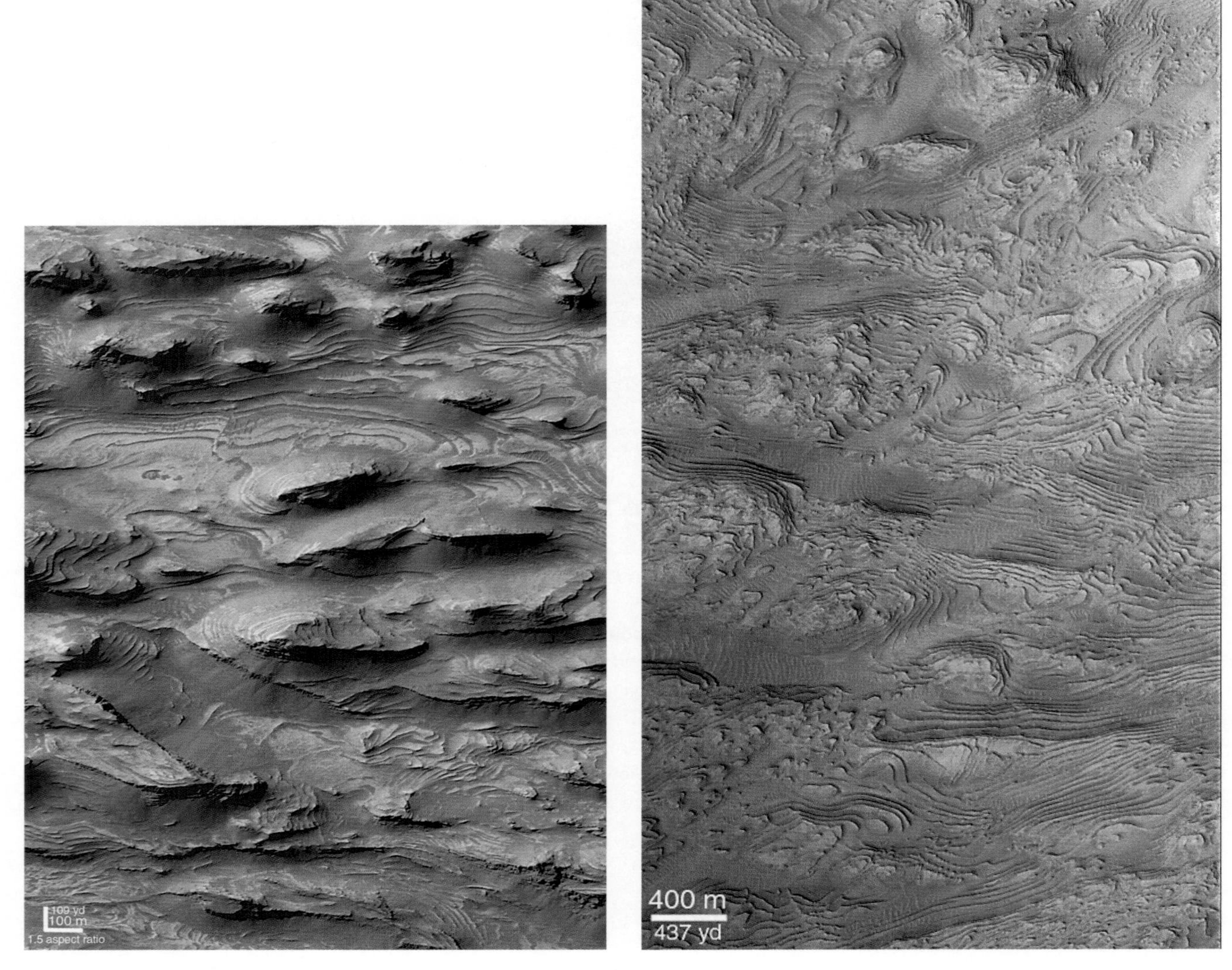

Fig. 2.40 Layered rocks on Mars These images from *Mars Global Surveyor* reveal hundreds of layers of similar thickness, texture and pattern that have been exposed in an impact crater in western Arabia Terra (*left*) and in a canyon located in southwestern Candor Chasma (*right*) of the Valles Marineris. The numerous, uniform deposits resemble regularly-layered, sedimentary rocks found on Earth. The Martian features could therefore be due to sediments that settled out of liquid water in ancient lakes or shallow seas a few billion years ago. The material might alternatively be due to deposits of airborne dust settling out of the atmosphere, that were later buried and compacted, or to layers of volcanic material. Arabia is a peninsula of southwestern Asia, bordering the Persian Gulf, the Arabian Sea, and the Red Sea; *terra* means "extensive land mass"; *candor* is Latin for "glossy whiteness or sincerity", and *chasma* is "a deep, elongated, steep-sided depression". (Courtesy of JPL, NASA, and Malin Space Science Systems.)

Water ice in the outer solar system

Dense rocky substances dominate the four terrestrial planets (Mercury, Venus, Earth and Mars) that are nearest the Sun, while the lighter gaseous and icy substances dominate the outer giant planets (Jupiter, Saturn, Uranus and Neptune). These compositional differences appear to result from the fact that the terrestrial planets formed close to the hot, bright, young Sun, and they suggest that water ice might be common in the colder, outer parts of the planetary system (Figs. 2.41 to 2.44).

The planets and their moons are thought to have formed in a swirling disk of material during the formation of the Sun about 4.6 billion years ago. This disk, called the solar nebula, probably had an elemental composition very similar to that of the Sun. The most abundant solar elements, after hydrogen, H, and helium, He, are oxygen, O, carbon, C and nitrogen, N (Section 1.4). Some of the oxygen could combine with less-abundant heavy elements, such as silicon and magnesium, to create rocky material throughout the solar nebula. When cooled to low temperatures in the outer parts of the solar nebula, the oxygen would

Fig. 2.41 Jupiter's satellite Europa Dark streaks mark Europa's smooth surface, forming a spidery, veined network in this *Voyager 2* image taken on 9 July 1979. In contrast to Jupiter's satellite Callisto (Fig. 2.23), Europa has very few impact craters; the absence of craters suggests that the ice crust is relatively young. Internal stresses apparently fractured its icy mantle, producing intersecting cracks that extend for millions of meters, but reach depths of less than 100 meters. The fractures may have been filled by liquid water gushing out from a global ocean in the satellite's interior, warmed by tidal heating. (Courtesy of NASA and JPL.)

also combine with hydrogen to make water, H_2O. Lesser amounts of methane, CH_4, and ammonia, NH_3, would also form in the frigid outer precincts of the planetary system.

In the inner regions of the solar nebula, the higher temperatures would vaporize water ice so it would not condense, leaving only high-density, rocky substances to coalesce and merge together to form the terrestrial planets. Further out, in the colder realm of the giant planets, large amounts of water ice could survive and not evaporate. The asteroid belt, that forms the great divide between the orbits of Mars and Jupiter, is thus known as the water line of the solar system; it is the first place that a 1-kilometer chunk of ice could form in the primeval solar disk and stay around. Both water ice and rock could condense from the solar nebula beyond the asteroid belt, and they are the most abundant ingredients of the satellites in the outer solar system. Because of their greater mass, the giant planets like Jupiter and Saturn were also able to retain abundant light gases such as hydrogen and helium that dominate their extensive atmospheres (Section 3.2).

Evidence for water ice in the outer solar system was first obtained by ground-based spectroscopy in the 1970s. The spectrum of sunlight reflected from Saturn's rings at infrared wavelengths between 2000 and 4000 nanometers was subtracted from that of the Moon, removing the features caused by absorption in the atmospheres of the Sun and the Earth. The resultant spectra indicated that the rings of Saturn are composed of pure water ice.

Recent ground-based infrared spectra of Saturn's rings indicate that their water ice is exceptionally pure. The spectra contain no detectable traces of rock, or silicate, dust, indicating that the rings contain less than one part silicate dust for somewhere between 1 million and 100 million parts ice.

Infrared spectroscopy also indicates the presence of water ice on the surfaces of Jupiter's large satellites,

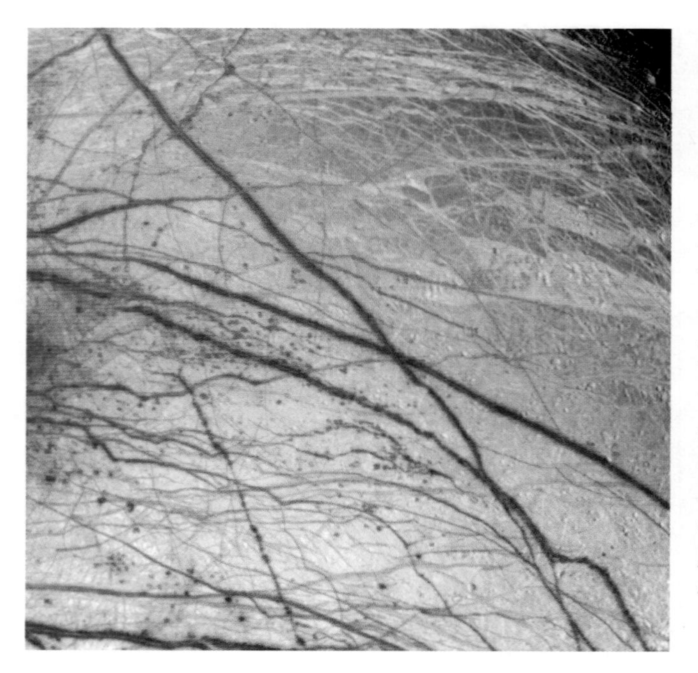

Fig. 2.42 Europa in color A composite, color-enhanced image of the Minos Linea region of Jupiter's moon Europa, taken on 28 June 1996 by imaging cameras on *Galileo*. The icy plains, shown here in bluish hues, reflect different amounts of light, probably as the result of differences in the sizes of the ice grains. The long red cracks in the ice could mark the sites of liquid water oozing out from the warm interior of Europa. The area covered in this image is about 1.26 kilometers across. In Greek mythology, Minos is the son of Zeus and the king of Crete, who kept a monster named Minotaur in a labyrinth. *Linea* is a "dark or bright elongate marking". (Courtesy of JPL, NASA, and PIRL at the University of Arizona.)

Fig. 2.43 Jupiter's satellite Ganymede Large dark blocks are frozen within the icy surface of Ganymede. They are believed to be part of the original crust of the satellite, resembling frozen-over continents floating on a background of translucent ice. The brilliant white material that surrounds some craters is probably clean water ice or bright snow that was splashed out from inside the satellite. The enhanced color of this *Galileo* image of Ganymede, taken on 29 March 1998, also reveals the two predominant terrain types on Ganymede, bright grooved terrain and older, dark, furrowed areas. The violet hues at the poles may be the result of small particles of frost. (Courtesy of JPL, NASA, and DLR – the German Aerospace Center.)

Ganymede and Callisto. They have low mass densities of 1940 and 1860 kilograms per cubic meter, respectively, indicating a composition of half rock and half ice; water ice and silicate rock have respective mass densities of 1000 and about 3000 kilograms per cubic meter. Both Saturn and Uranus have a retinue of icy moons with mass densities near that of water (Table 1.7), and the rings of Saturn could have formed from the break up a former satellite.

Jupiter's satellite Europa has a mass density of 2970 kilograms per cubic meter, so it could be mostly rock. Yet the surface of Europa is almost perfectly smooth and exceptionally bright, with no mountains and valleys in sight (Fig. 2.41). All of the light material in Europa must have once melted, floated to the top and froze into a crust of ice.

Very few impact craters are present on Europa's face, indicating that its smooth surface was formed relatively recently, geologically speaking. Some process must be keeping it young on time-scales of a few hundred million years or less. Liquid water or slush apparently oozes out within cracks in the ice, resurfacing the globe (Fig. 2.42). Some cracks in the icy moon could be as deep as the distance from Los Angeles to New York, and when you look down them you might see water rising.

Images from *Galileo* indicate that Europa almost certainly had a liquid ocean at one time, and they have considerably strengthened the evidence for an ocean of liquid water existing just beneath its icy surface at the present time (Section 9.4). The *Galileo* images show evidence for near-surface melting and movements of large blocks of ice, the first icebergs found outside planet Earth. In some cases, large ice rafts the size of cities have broken off and drifted apart, sliding away from each other, with edges that fit like the pieces of a jigsaw puzzle. Warm ice or even liquid water must lubricate the moving ice from below. After all, ice freezes from the top down, so you might expect Europa to be cold and desolate on the outside, with a warm ocean on the inside. It would be the first extraterrestrial ocean to be found and the first ocean to be discovered since Vasco Núñez de Balboa (1475–1519) discovered the Pacific Ocean in 1513.

What keeps Europa's subsurface ocean from freezing solid? The satellite may be kept warm inside by tidal heating from nearby massive Jupiter. A similar effect melts the inside of Io, producing its ubiquitous volcanoes (Sections 2.3, 9.4). The orbit of Europa is pulled slightly out of round by the gravitational action of Io, which is closer to Jupiter, and Ganymede, the next satellite out. Jupiter's varying gravitational pull as Europa moves along its eccentric orbit produces tides of different size, causing the satellite to stretch and distort, heating its interior and keeping the water liquid beneath its icy crust. These internal tides will also cause Europa's overlying ice shell to flex, producing cracks that open and close as Jupiter squeezes the moon in and out.

There is indirect magnetic evidence of a hidden ocean on Europa. Magnetic measurements from *Galileo* indicate that Europa interacts with Jupiter's magnetic field, even though the satellite does not have its own permanent magnetism. Internal electrical currents are generated as Europa sweeps through Jupiter's powerful magnetic field, and these electrical currents generate a temporary magnetic field that briefly alters Jupiter's field near the satellite.

For electrical currents to flow in Europa, some part of the satellite must conduct electricity. Ice is not a good conductor, but salty water is. A saltwater ocean about ten kilometers below the surface can produce the measured changes of Jupiter's magnetic field as it sweeps by in different orientations to the satellite.

To their surprise, scientists have also measured swings in the magnetic fields around Callisto and Ganymede, suggesting that they may also have salty water beneath their crust. Ganymede additionally has a molten core that generates its own permanent magnetic field (Section 9.4).

Ganymede is the largest moon in the solar system, with a radius that exceeds that of the planet Mercury, but the satellite's mass density is relatively low, at 1940 kilograms per cubic meter, so it probably contains substantial amounts of water. Its surface has large dark plates separated by lighter regions, and impact craters that are surrounded by bright material (Fig. 2.43). The dark regions are believed to be part of the original crust of Ganymede, which probably cracked and spread apart. The lighter

Fig. 2.44 Saturn's satellite Enceladus The bright, smooth surface of Enceladus, shown in this *Voyager 2* image obtained on 25 August 1981, reflects almost 100 percent of the incident sunlight, making it one of the most reflective objects in the solar system. When viewed up close, part of its surface is scarred with impact craters, and such impacts might have released liquid water from the satellite's interior. Other parts of the surface contain cracks and grooves, suggesting that internal stresses may have also discharged water that froze into smooth ice. (Courtesy of JPL and NASA.)

regions are most likely water ice that has moved in, replacing about half of the old, dark surface. The brilliant white material that surrounds some craters is probably clean water ice that splashed out from inside the satellite.

Saturn's moon Enceladus has a bright, smooth icy surface that contains cracks and grooves (Fig. 2.44), suggesting the release of water from below the surface within the last one or two billion years. This would be consistent with the satellite's low mass density of just 1240 kilograms per cubic meter, suggesting that it is just a big ball of water ice. Enceladus is caught in a gravitational tug-of-war between Saturn and the satellite, Dione, whose orbital period is about twice that of Enceladus. Dione's repeated gravitational tug produces Enceladus' eccentric orbit, and causes recurrent tidal flexing from Saturn that may warm the moon's interior.

Further out, Uranus and Neptune are largely composed of icy material that condensed out of the primeval disk at the time of their formation (Section 11.3). Comets in the distant Kuiper belt and the very remote Oort comet cloud are located beyond the planets where large amounts of ice are expected (Sections 1.5, 12.3). When one of these comets comes within the inner solar system, it expels about 25 tons, or 25 thousand kilograms, of water every second, propelled by water vapor generated when the Sun's heat melts the water ice (see Fig. 2.15, and Section 12.6). A typical comet contains one million billion (10^{15}) kilograms of water.

3 The invisible buffer zone with space – atmospheres, magnetospheres, and the solar wind

- A thin membrane of air protects, ventilates and incubates us.
- The surface of Venus now lies under a hot and heavy atmosphere of carbon dioxide; its greenhouse effect has raised the surface temperature on Venus to a torrid 735 degrees kelvin, hot enough to melt lead and zinc.
- High-velocity winds on Venus whip its highest clouds around the planet 60 times faster than the planet rotates.
- Mars now has an exceedingly thin, dry and cold atmosphere of carbon dioxide, with less than one-hundredth the surface pressure of our air, but global winds can stir up enough dust to cover the entire planet.
- The red planet breathes about one-third of its atmosphere in and out, as its southern polar cap grows and shrinks with the Martian seasons.
- The atmospheres of Earth, Venus and Mars may have originated after the planets formed, as the result of the collisions of comets and asteroids, and then evolved to their present states, primarily as the result of their varying distances from the Sun and the development of life on Earth.
- The Sun generated so little heat more than 2 billion years ago that the Earth's oceans should have been frozen solid; a thick carbon-dioxide atmosphere may have warmed the planet back then.
- Like the Sun, the most abundant element in the giant planets is the lightest element, hydrogen, and the next most abundant element is helium.
- Jupiter is all atmosphere, with no solid surface to rub against or continents to disturb the flow, so its Great Red Spot has existed for more than 300 years; the location and speed of powerful winds and violent storms on the giant planet have remained unchanged for a century.
- The composition of the atmosphere of a planet or satellite depends upon its temperature and mass. Hotter molecules tend to move faster, and are more likely to escape an object's gravitation. Smaller, less-massive objects have lower gravitational pull and are less likely to retain an atmosphere.

- Saturn's largest moon, Titan, has a substantial Earth-like atmosphere, which is mainly composed of nitrogen and has a surface pressure comparable to that of our atmosphere. An opaque smog-like haze hides the surface from view, but scientists speculate that ethane and propane may rain down to form flammable seas and lakes on Titan's surface.
- The *Cassini* spacecraft is expected to drop its *Huygen's Probe* into Titan's atmosphere and down to its surface in 2004.
- A thin film of gas cloaks the Moon, Mercury, Pluto, Triton, and Jupiter's largest moons, Io, Europa, Callisto and Ganymede, but these are temporary, rarefied and misty atmospheres that must be continuously resupplied. The few particles they have almost never hit each other.
- An energy-laden, electrically charged solar wind blows out from the Sun in all directions and never stops, carrying with it a magnetic field rooted in the star.
- Ancient magnetic rocks indicate that Earth's magnetic poles keep switching places, and that our planet's magnetic field may now be heading for a flip.
- The Sun's wind flows around the Earth's magnetic field. The magnetosphere of a planet is the volume of space from which the main thrust of the solar wind is excluded.
- Energetic electrons and protons have penetrated the Earth's magnetic defense. Some of them are confined within two doughnut-shaped radiation belts that encircle the Earth's equator but do not touch it.
- Jupiter's magnetosphere is the largest enduring structure in the solar system, more than ten times larger than the Sun. It was discovered by using Earth-based radio telescopes that unexpectedly detected the synchrotron radio emission of high-speed electrons trapped in the giant planet's immense magnetic field.
- The magnetic fields of Uranus and Neptune are tilted by enormous angles from their rotation axes.
- The Earth's aurora is a spectacular multi-colored light show that shines like a cosmic neon sign.
- When viewed from space, the aurora forms an oval centered on the magnetic poles of the Earth; similar aurora ovals have been detected in ultraviolet light at both the north and south magnetic poles of Jupiter and Saturn.

3.1 Atmospheres of the terrestrial planets

The breath of life

Our atmosphere forms an indispensable interface with nearby space, but it is often invisible. After all, you look right through the air in your room. Our atmosphere usually goes unseen on a warm, dry, windless day. Yet, the slow drift of floating clouds or the sight of birds and airplanes supported by their motion proves that there is something substantial surrounding us. We can sense the touch of the wind on a stormy day, and on cold days we feel the air against our skin.

We find a further clue in the rise of smoke above a candle or a group of hawks circling above a warm meadow. Hot air rises around the flame of the candle, and the flowing air replenishes the supply of oxygen required to keep the candle burning. The hawks are getting free rides in the rising currents of hot air above ground.

When astronauts look down at the Earth at sunrise or sunset, they detect the thin atmosphere that warms and protects us, and permits us to breathe (Fig. 3.1). It is only 10 kilometers from the ground to the top of

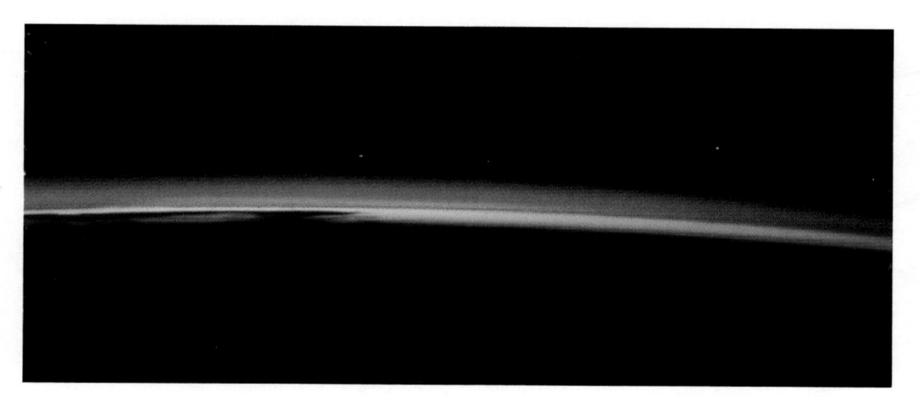

Fig. 3.1 A, thin colored line Brilliant red and blue mark the thin atmosphere that warms and protects us, as viewed from space at sunrise over the Pacific Ocean. Without this atmospheric membrane we could not breathe and water would freeze. (Courtesy of NASA.)

the sky, or no further than you might run in an hour. Everything beyond that thin layer of air is the black void of space. And everything below it is what it takes to sustain life.

If we were to weigh the surrounding air in a one-liter container we would find it tips the scales at slightly more than one gram. This is about one-thousandth the weight of the same amount of water. Determining its constituents is easy – just place the appropriate instrument in the air and see what is there. The major constituents of dry air on Earth are nitrogen molecules (77 percent), oxygen molecules (21 percent) that we breathe, and argon atoms (0.93 percent). Carbon dioxide is a miniscule 0.035 percent. There is no hydrogen in our air; most of the hydrogen on Earth is found in water. The water vapor in wet air is variable in amount, usually no more than 1 percent.

Almost all of the oxygen and carbon-dioxide molecules in our atmosphere are the breath of plants and animals, and they are continually being recycled in the photosynthesis and respiration processes. Animals breathe oxygen, and when they exhale they release carbon dioxide and water vapor. Conversely, green plants absorb carbon dioxide and water, use them in the photosynthesis of nourishment and then release oxygen into the atmosphere. This symbiotic relationship is one of the most remarkable features of life on Earth.

If plants did not continuously replenish the oxygen in our air, animals and humanity would exhaust the available supply in a mere 300 years. All the water on the Earth is split by photosynthesis and reconstituted by respiration every 2 million years or so. For millions of years, our ancestors breathed the same oxygen and drank the same water, binding them temporarily in their bodies and then releasing them again to the atmosphere.

The Earth is the only place in the solar system where we can stand naked and survive. The air brings oxygen to our lungs and refreshes our blood stream; sunlight provides just enough heat to prevent our fluids from freezing or boiling. There is a very different situation on the other terrestrial planets, where there are no plants to supply the oxygen and the temperatures are very different.

Earth's weather

Because the Earth is a sphere, there is an unequal distribution of the Sun's heat at the terrestrial surface, producing the climate differences that starkly distinguish one part of the globe from another. The equator, for example, receives more sunlight than the poles, so the equatorial regions, known as the tropics, are the hottest places in the world and the poles are the coldest. The extra warmth at tropical latitudes evaporates seawater and causes the air to expand and rise, carrying fresh-water moisture with it. As it rises, the tropical air cools and forms clouds of water ice. These clouds are moved over great distances by winds before condensing again to liquid water and falling to Earth as rain. When arriving on land, the water refreshes lakes and streams, and most of it eventually finds its way back to the sea.

The first to suggest a continuous, global circulation of the atmosphere was the English astronomer, Edmond Halley (1646–1742), best known today for the comet that bears his name (Section 12.2). Halley reasoned that the high-temperature air in equatorial regions would circulate toward the colder poles, and that the colder air from the north would move away from the poles to replace the warm tropical air. This movement of air in response to unequal temperatures is known as the wind, and the winds are blowing in an attempt to equalize global temperature differences. Since an increase in temperature produces higher pressure, the winds are also attempting to balance pressure differences, made unequal by different amounts of solar heating in various places.

Air currents tend to circulate from the tropics toward the poles and back, in the north–south direction, but they are deflected in the east–west direction by the Earth's rotation. Along the equator the near-surface air currents

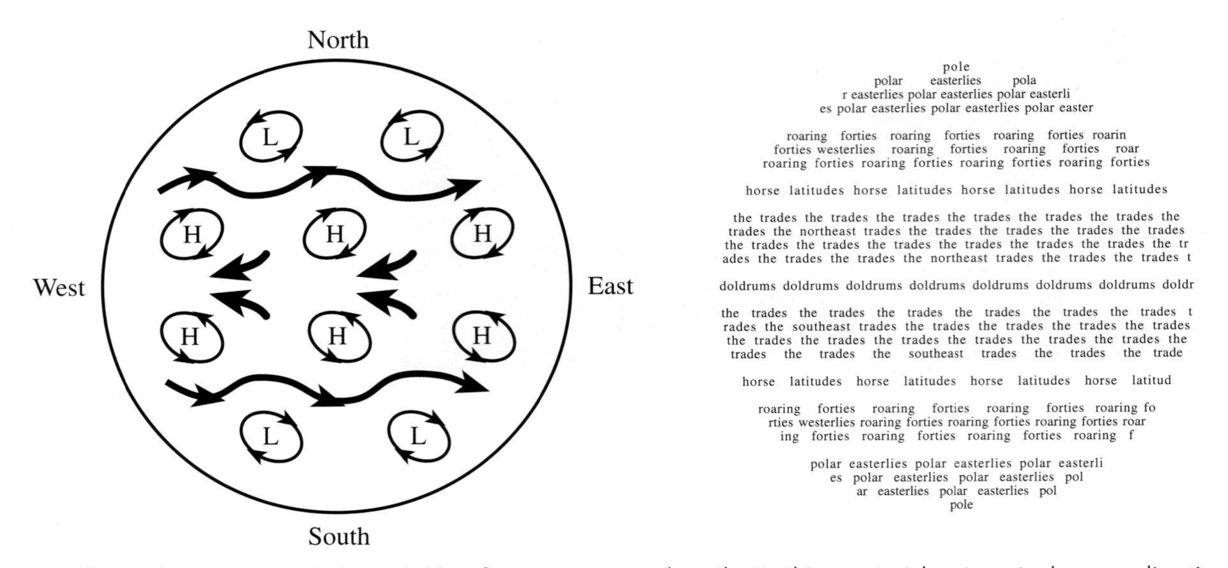

Fig. 3.2 Earth's weather patterns Trade winds blow from east to west along the Earth's equatorial regions, in the same direction that the planet rotates. At higher latitudes, there are high-altitude jet streams that move at speeds of up to 40 meters per second in the opposite direction to the trade winds. In the northern hemisphere there are high-pressure cyclones, denoted by H, and low-pressure anti-cyclones, L, rotating in the clockwise and counter-clockwise directions, respectively; in the southern hemisphere the cyclones and anti-cyclones rotate in the opposite direction. The prevailing winds are given in Annie Dillard's poem, "the windy planet" (*right*).

converge to form the trade winds, that blow mainly from east to west, almost every day of the year (Fig. 3.2). At mid-latitudes, the near-surface winds blow largely from the west and so are called westerlies. The high-altitude jet streams also blow eastward, at speeds of up to 40 meters per second, in a sinuous path that resembles the meandering of a river (Fig. 3.2).

The atmosphere's attempt to overcome the Sun's unequal heating of the Earth is never ending. The equator is always hotter than the poles, and the Earth is always spinning. The temperatures and pressures are never balanced, and the winds always blow.

Moreover, the winds do not blow in a straight line. The Earth's rotation creates vast swirling eddies in the air, just like those in a river or stream. One type of eddy is known as a *cyclone* from the Greek word for "wheel". It is a vast whirling mass of wind and precipitation.

A cyclone in the northern hemisphere is a low-pressure cell of air rotating counter-clockwise. An anti-cyclone is a high-pressure cell rotating clockwise in the northern hemisphere (Fig. 3.2). Between the trade winds and the westerlies, the pattern is primarily a series of high-pressure cyclones rotating clockwise in the north and counter-clockwise in the south. In the polar regions, the weather pattern is mainly a series of low-pressure cyclones rotating counter-clockwise in the north and clockwise in the south (Fig. 3.2).

As the cyclones pinwheel across the globe, they produce most of our stormy weather. Thunderstorms, blizzards, and tornadoes are examples. They tend to move from west to east across the North American continent in an almost steady parade. Cyclones also occasionally develop into hurricanes (Atlantic Ocean and Caribbean Sea) and typhoons (Pacific Ocean and China Sea).

Determining the detailed composition of planetary atmospheres

The ingredients of the atmospheres of the planets and satellites can be determined by spectroscopic observations at a distance, from the ground or space, and by space probes that are parachuted into the atmosphere. The planetary spectra contain absorption bands that fingerprint the molecules doing the absorption (Section 1.4). The probes obtain *in-situ* samples of the gas, measuring the constituents directly.

Ground-based spectroscopy has been successfully used to determine the abundance of the elements in the Sun, and some of the molecules in planetary atmospheres (Section 1.4). Carbon dioxide was identified as an ingredient of the atmospheres of Venus and Mars, and both ammonia and methane were found in the giant planets.

These early results were nevertheless incomplete and sometimes misleading. Since the planetary light has to pass through our atmosphere before reaching a telescope, the spectra contain strong, confusing absorption features caused by molecular oxygen and water vapor in the air. These molecules might therefore be present in the atmospheres of the other planets and almost impossible to detect. Moreover, some of the most abundant molecules in the atmospheres of the planets do not contain spectral signatures at visible wavelengths.

The main ingredients of most of the planetary atmospheres were therefore not definitely known until spectrometers aboard spacecraft were used to analyze their reflected sunlight for compositional information and to extend planetary spectra to invisible infrared and ultraviolet wavelengths.

Atmospheres of Venus and Mars

In many ways, Venus is Earth's twin sister, with almost the same weight and waistline. Her mass is 81 percent that of Earth, and her radius 95 percent, so the feel of gravity at the planet's surface is similar to that on Earth. The two planets also have nearly equivalent mean mass densities, of 5204 and 5515 kilograms per cubic meter respectively, indicating that their bulk composition must be nearly the same. Venus is also just a little closer to the Sun than the Earth, orbiting our star at a distance of 0.723 AU compared with the Earth's 1.000 AU. All of these similarities gave rise to the idea that the atmosphere of Venus might resemble the Earth's air, but with a more temperate climate.

Less than half a century ago, many astronomers believed that the surface of Venus was warm and wet, perhaps with steamy swamps and jungles and even living creatures. Then, in the late-1950s, it was earthbound radio astronomers that looked beneath the clouds, and discovered the extremely hot and inhospitable surface of Venus. Cloudy atmospheres are transparent to the long radio waves, so the radio radiation from Venus comes directly from its surface, and can be used to take its temperature. Beneath her gleaming clouds, the planet is an inferno with a temperature hot enough to melt lead or zinc.

Our knowledge of this torrid world was further enhanced when spacecraft directly measured the atmosphere. An entry probe parachuted from the Russian *Venera 7* spacecraft in 1970 transmitted measurements of the temperature and pressure all the way down to the bottom of the atmosphere, where the temperature reaches a sizzling 735 kelvin. Down there the thick, heavy atmosphere produces a pressure of 92 bars – that is, 92 times the air pressure at sea level on Earth. The oppressive heat and weight of the atmosphere of Venus were fully confirmed by atmosphere probes and landers released by seven *Venera* spacecraft from 1972 to 1983, as well as by the American multi-probe *Pioneer/Venus* mission in 1978. The thick, soupy atmosphere on Venus is uniformly hot and heavy, with neither geographic nor seasonal variations.

The principal constituent of the thick atmosphere on Venus is carbon dioxide. It was first identified in the upper atmosphere of Venus, by infrared absorption lines in the planet's spectrum obtained in 1932 (Section 1.4), and confirmed during the *Mariner 2* flyby in 1962. Three Russian space probes, *Venera 4, 5* and *6*, descended by parachute into the atmosphere in the 1960s, and obtained direct measurements of its principal constituents, showing that it consists of 96 percent carbon dioxide. The atmosphere of Venus contains about ten thousand times as much carbon dioxide as is present in our air.

A massive carbon-dioxide atmosphere is responsible for the high surface temperature of Venus through the greenhouse effect. This is the name given to the process by which an atmosphere traps heat near a planet's surface. The trapped heat warms the planet's surface to higher temperatures than would be normally achieved by direct sunlight in the absence of an atmosphere. On Earth, the surface temperature is raised by about 30 kelvin by this effect, resulting in the mild climate we enjoy today. But the greenhouse effect has raised the temperature of Venus' surface by hundreds of degrees.

Sunlight passes through the thick carbon-dioxide atmosphere to warm the planet's surface, but much of the surface heat is re-radiated in the form of long infrared waves that are absorbed by the atmospheric carbon dioxide. The atmosphere thus acts as a one-way filter, allowing the warmth of sunlight in, but preventing the escape of heat into the cold unfillable sink of space. So the heat is held close to the surface of Venus, elevating the temperature there.

Strong winds are blowing the highest clouds around Venus at speeds of up to 100 meters per second, racing around the planet's equator once every four Earth days (Fig. 3.3). Curiously enough, Venus' surface rotates in the same westward direction but with a much longer period of 243 Earth days. So the winds blow the entire outer atmosphere around the planet much more rapidly than the planet spins. Although terrestrial jet streams move at up to half the speed of the high-flying clouds on Venus, they are limited to narrow zones high in the Earth's atmosphere (see Fig. 3.2).

Near the surface, at the bottom of the massive atmosphere on Venus, the rapid winds have disappeared, and the atmosphere has become as sluggish and turgid as water at the bottom of Earth's oceans. Most of the atmosphere beneath the high-flying clouds probably rotates synchronously with the surface, just as most of the Earth's atmosphere does.

Fig. 3.3 Raging winds on Venus A high-velocity wind whips the upper layer of Venus' cloud deck around the planet's equator once every four Earth days, moving at speeds of up to 100 meters per second. This is illustrated by these photographs taken at ultraviolet wavelengths on consecutive days by the *Pioneer Venus* spacecraft in 1979. The Y-shaped clouds move towards the west (*left*). The winds of Venus are dominated by a zonal (east-to-west) circulation. The lower atmosphere and the planet's surface also rotate westward, but with the much slower period of 243 Earth days. (Courtesy of Larry Travis and NASA.)

Table 3.1 Percentage composition, surface pressure and surface temperature of the atmospheres of Venus, Mars and Earth

	Venus	Mars	Earth
Constituent			
Carbon Dioxide, CO_2	96	95	0.035
Nitrogen, N_2	3.5	2.7	77
Argon, Ar	0.007	1.6	0.93
Water Vapor, H_2O	0.010	0.03 (variable)	1 (variable)
Oxygen, O_2	0.003	0.13	21
Surface Pressure (bar)	92	0.007–0.010	1.0 (at sea level)
Surface Temperature (K)	735	140–300	288–293

The atmospheres of both Venus and Mars have the same ingredients as our air, but the proportions are different (Table 3.1).

The atmospheres of Earth, Venus and Mars each have one or more gases that can saturate. That is, the atmosphere fills up with as much of the vapor as it can hold, and then that substance condenses. In the Earth's atmosphere, water vapor condenses to form billowing white clouds of water ice. The same thing happens on Mars, where clouds of water ice are found near towering volcanoes. Carbon dioxide also condenses out of the thin, cold Martian atmosphere onto the surface. On Venus, our closest planetary neighbor, it is sulfuric acid that condenses to form the thick, unbroken layer of yellow clouds that always enshroud the planet (Section 7.2).

Mars, fourth planet from the Sun, was also long thought to resemble the Earth. The length of the day on Mars is only 37 minutes longer than our own, the rotational axes of both Earth and Mars are tilted by about the same amount, and the Martian year is 687 Earth days, or almost two Earth years. Both planets have four seasons – autumn, winter, spring and summer – although the Martian seasons are nearly twice as long. Mars has white polar caps that wax and wane with the seasons. Their alternate growth and recession mean that gases are being extracted from, and released into, an atmosphere. White clouds are also found on Mars, resembling those on Earth, and clouds are not possible without an atmosphere.

Thus, Mars has an atmosphere, clouds, polar caps and seasons, and these Earth-like qualities led astronomers to speculate that its atmosphere resembles Earth's. The red planet is nevertheless twice as far away from the Sun as the Earth, so it will be warmed less by solar radiation and ought to be colder – if there is no pronounced greenhouse effect on Mars.

The composition and extent of the Martian atmosphere wasn't understood until the Space Age. Early in the 20th century, astronomers incorrectly claimed to have detected the red spectral signatures of both water vapor and molecular oxygen on Mars. The same lines generated when the red planet's reflected sunlight passed though our moist, breathable air confused these astronomers. By mid-century Gerard P. Kuiper had identified the infrared lines of carbon

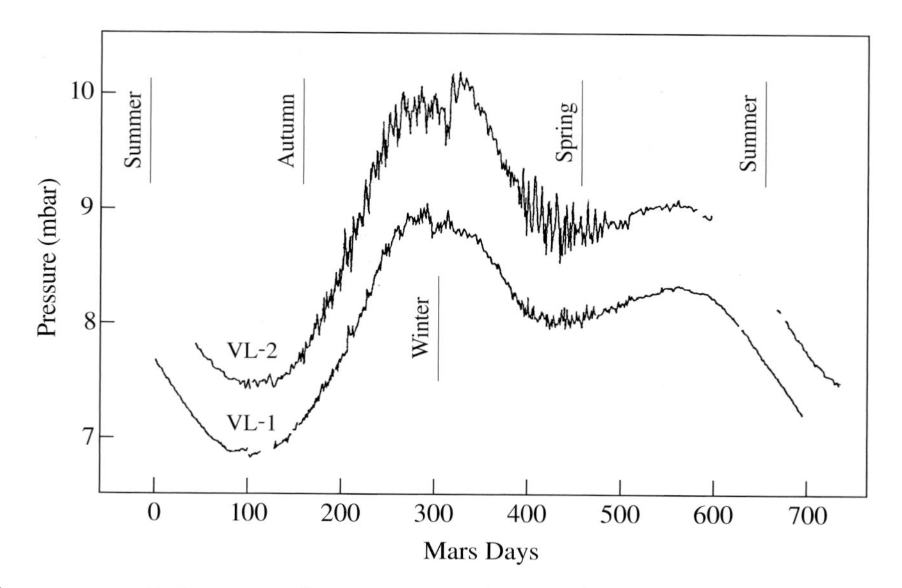

Fig. 3.4 Martian surface pressure Daily mean surface pressures at the two *Viking Lander* (VL) sites for one Martian year, showing that the red planet periodically removes and replaces carbon dioxide in its atmosphere. The seasons are for the northern hemisphere, and the pressure is given in millibars, abbreviated mbar, where 1 mbar = 0.01 bar and 1 bar is the sea-level pressure of the Earth's atmosphere. At the *Viking 1* site (bottom curve), the pressure ranged from 6.7 mbar during the northern summer to 8.89 mbar at the commencement of the northern winter. At the *Viking 2* site (*top curve*) the equivalent data were 7.4 mbar and 10 mbar. The higher values are probably due to the lower elevation; there was an approximate 1100-meter difference in the elevation of the two landing sites. Dust storms are thought to produce some of the smaller non-seasonal pressure variations. The seasonal pressure differences seem to be dominated by the southern polar cap, which is larger than the northern one. In the southern winter, and northern summer, carbon dioxide is condensed out of the atmosphere, enlarging the southern polar cap and reducing the total surface pressure of the planet. In southern summer, and northern winter, the carbon dioxide is released back into the atmosphere, with an increase in the total surface pressure.

dioxide on Mars. Yet, astronomers guessed that the atmosphere would be mainly nitrogen, like our own air, and imagined a surface pressure at least ten times greater than that eventually measured.

When the *Mariner 4* spacecraft passed behind Mars in 1964, its radio signal was sent through the Martian atmosphere, and a surface pressure of about 0.005 bar (5 mbar), or two-hundredths that of Earth, was inferred from the altered signal. So the atmosphere on Mars is exceedingly thin, with a surface pressure comparable to the pressure high in the Earth's rarefied stratosphere. In addition, by combining the spacecraft measurements of surface pressure on Mars with ground-based spectra of the planet, astronomers quickly concluded that carbon dioxide is the main ingredient of the Martian atmosphere.

The exact chemical composition of the atmosphere on Mars was determined by direct measurements in 1976 when the *Viking 1* and *Viking 2* landers arrived at the surface (see Table 3.1). Like Venus, the atmosphere on Mars is mostly carbon dioxide (95 percent) with just a whiff of nitrogen (2.7 percent). The landers confirmed that the local surface pressure is at or below 0.01 bar (10 mbar) (Fig. 3.4). They also showed that the surface temperature is almost always below the freezing point of water (Sections 2.4, 8.3).

The planet breathes its atmosphere in and out as its southern polar cap grows and shrinks, producing a seasonal change in atmospheric pressure by about 30 percent (Fig. 3.4). It is as if the planet was a giant lung that slowly breathes in and exhales the same gas, carbon dioxide. When the surface temperature drops during the southern winter, the atmospheric carbon dioxide condenses and freezes to enlarge the polar cap, resulting in a drop in atmospheric pressure. In the southern summer, the ice sublimates (goes directly from solid to vapor, without becoming liquid) back into the atmosphere, increasing the atmospheric pressure. The polar cap waxes and wanes due to this seasonal component of carbon-dioxide ice.

Powerful seasonal winds are driven by temperature differences between the northern and southern hemispheres. In each hemisphere, warm air rises in the summer and descends in the winter, when carbon dioxide is condensing to make the seasonal polar cap grow. The strong Martian winds also strip away light-colored dust in some areas and deposit it in others, accounting for the seasonal growth and decay of large dark areas seen from Earth (Section 8.4).

Martian winds stir up small, localized dust storms, and occasionally coalesce to produce a globe-encircling dust storm (Section 8.4). When substantial amounts of dust have been tossed aloft, sunlight is absorbed in the atmosphere rather than at the surface, and the storm can sustain itself by converting the Sun's energy into wind energy. The entire planet then becomes wrapped in an opaque yellow veil.

Evolution of the terrestrial planetary atmospheres

Nothing in the cosmos is fixed and unchanging, and nothing escapes the ravages of time. The atmospheres of the Earth, Venus and Mars are no exception, for they have been, and are being, slowly altered with the addition and removal of gases as time goes on. In fact, their atmospheres probably weren't even there when the planets formed. They had too little mass to attract and hold on to the abundant hydrogen gas that was around when they accumulated. The building blocks from which the terrestrial planets formed were primarily rocky and metallic objects, and these planets may have been initially too hot to retain substantial amounts of water vapor or carbon dioxide in their early atmospheres.

The present-day atmospheres of the terrestrial planets are thought to be secondary, having come from other sources, such as comets and asteroids. Because these objects formed further from the Sun's heat, they could retain water and carbon dioxide. Early in the history of the solar system, many more comets and asteroids were in orbits that intersected the orbits of Mars, Earth and Venus. The collisions would have released ices and gases, supplying the terrestrial planets with the volatile substances needed to form their early atmospheres and oceans. It is also possible that ancient volcanoes supplied some of the early gas and water from inside the young planets, but this internal material was originally supplied by the accumulation of impacting objects as the planet grew.

Initially these atmospheres were probably dominated by carbon dioxide, and oceans might have formed as the newborn planets cooled and the surface temperatures dropped below the boiling point of water. The atmospheres of Earth and Venus probably began with similar compositions 4 billion years ago, but their subsequent histories have been quite different. The Earth is now depleted of carbon dioxide, and has excessive oxygen, while Venus is depleted of water.

Being closest to the Sun, Venus was initially the warmest planet other than Mercury, which was too hot to retain any significant atmosphere. The atmosphere greenhouse effect raised the temperature of young Venus and boiled away any oceans that might have condensed. The increased water vapor blocked more heat, raising the surface temperature in a runaway greenhouse effect. This turned Venus into the torrid world we see today, with a surface temperature of 735 kelvin. Because Earth is a little further from the Sun than Venus, with a slightly lower initial temperature, the water on Earth remained liquid and it kept its oceans.

Life has been an important influence on the evolution of the Earth's atmosphere. Over the past 4 billion years, living things have produced a decrease in atmospheric carbon dioxide, while also providing an increase in atmospheric oxygen. The carbon dioxide was absorbed in ocean water where tiny marine creatures extracted it to manufacture carbonate shells. When these creatures died, their shells sank, producing carbonate sediments and rocks on the ocean floor. The result was a gradual depletion of carbon dioxide from the atmosphere.

Thus, the carbon dioxide on Earth probably came from the outside, and moved from the atmosphere to the oceans and into the rocks. We now know that there is as much carbon dioxide at the bottom of the ocean and in carbonate rocks as there is in the atmosphere of Venus, enough to exert a surface pressure of 70 bar if released into our air. During the same time, plants gradually supplied oxygen to the atmosphere, making it breathable by humans.

The balance is a delicate one. Because Earth is slightly further away from the Sun than Venus, the Earth evolved into a living world capable of sustaining a remarkable diversity of living things. If the Earth were placed in Venus' orbit, its atmosphere would get hotter and thus able to hold more water evaporating from the ocean. The additional water vapor would trap more heat from the Sun, raising the temperature further and evaporating more water. The runaway greenhouse would eventually raise the temperature to values as high as those on Venus today. All of Earth's water would be boiled away, and the Earth would turn into a dried-out, lifeless place like Venus.

Since Mars is further away from the Sun than any other terrestrial planet, it is warmed least by the Sun. The red planet has about half the size and a tenth of the mass of Earth, so its lower gravity would be less likely to hold onto a substantial atmosphere. Moreover, its smaller area would have intersected the orbits of fewer comets and asteroids. So we might expect a thin, dry and cold atmosphere similar to the one we see today on Mars. However, it also seem likely that at one time in the distant past, the

Martian atmosphere was warmer, denser and wetter than it is today, permitting torrents of water to flow across its surface (Sections 2.4, 8.8).

If the planet once had a thicker atmosphere, it could have slowly evaporated into space under the combined effects of energetic sunlight and the planet's weak gravitational field; ultraviolet light from the Sun breaks up the atmospheric molecules into lighter, more energetic atoms that can escape the relatively weak gravity. More recent climatic change on Mars, during the past few million years, has been induced by the changing tilt of its polar axis, causing atmospheric water and carbon dioxide to move into polar ice caps and back into the atmosphere again (Section 8.5).

The Sun is also changing as time goes on, growing slowly in luminous intensity with age, a steady inexorable brightening that is a consequence of the nuclear reactions that make it shine. As the Sun burns hydrogen into helium, in its energy-generating core, the increasing amounts of helium require a rise in temperature to sustain the nuclear burning, and hence an increase in the rate of the nuclear reactions and a slow brightening of the Sun. You couldn't detect the change over all of human history, but it has profound implications over cosmic periods of time.

Stellar evolution calculations indicate that when the Sun began to shine an estimated 4.5 billion years ago, it was 30 percent dimmer than it is today. Assuming an unchanging atmosphere, with the same composition and reflecting properties as today, the decreased solar luminosity would have caused the Earth's global surface temperature to drop below the freezing point of water at all times earlier than 2 billion years ago. The oceans would have been frozen solid, there would be no liquid water, and the entire planet would have been locked into a global ice age something like Mars seems to be in now.

However, sedimentary rocks, which must have been deposited in liquid water, date from 3.8 billion years ago. There is fossil evidence in those rocks for living things at about that time. Thus for billions of years the Earth's surface temperature was not very different from today, and conditions have remained hospitable for life on Earth throughout most of the planet's history.

The discrepancy between the Earth's warm climatic record and an initially dimmer Sun has come to be known as the faint-young-Sun paradox. It can be resolved if the Earth's primitive atmosphere contained about a thousand times more carbon dioxide than it does now. Greater amounts of carbon dioxide would enable the early atmosphere to trap greater amounts of heat near the Earth's surface, warming it by the greenhouse effect. That would keep the oceans from freezing.

Over time the Sun grew brighter and hotter. The Earth could only maintain a temperate climate by turning down its greenhouse effect as the Sun turned up the heat. Our planet's atmosphere, rocks, oceans, and life itself apparently combined to remove carbon dioxide from the atmosphere. Thus, the change over time in the solar energy striking Earth has been compensated by a steadily decreasing greenhouse effect, so that the surface temperature of the planet has remained almost constant for the past 4 billion years.

This convenient thermostat might soon be disrupted as human civilization dumps more and more carbon dioxide into the atmosphere by burning fossil fuels like coal and oil, raising the surface temperature during the past 100 years to higher values than in any century for the last thousand years (Section 4.4).

3.2 Atmospheres of the giant planets

The early photographic spectra indicated that the outer atmosphere of Jupiter is similar to that of Saturn, while the atmosphere of Uranus is more like that of Neptune. The observed features were identified in the 1930s as methane, the natural gas we often use for cooking and heating, and ammonia (Section 1.4), but they are minor constituents. The main ingredient of the giant planets is hydrogen, the most abundant element in the Sun and most stars; and like the Sun, the next most abundant element in the giant planets is helium.

The composition of the Sun (cooled, theoretically, to planetary temperatures) and the outer atmospheres of the giant planets are given in Table 3.2. Jupiter has very nearly the same composition as the Sun, made up mainly of the light gases hydrogen and helium. Saturn has about the same composition, with a bit more helium, but Uranus and Neptune are depleted of these two gases relative to the heavier hydrogen compounds like methane, ammonia and water.

If the massive giant planets were formed together with the Sun, as most scientists believe they were, then they ought to closely approximate the solar composition. This would help explain the low mass densities of the giant planets, between 710 and 1670 kilograms per cubic meter, which are comparable to that of the Sun at 1409 kilograms per cubic meter.

The overwhelmingly abundant hydrogen, H, would combine with the abundant carbon, C, oxygen, O, and nitrogen, N, in the low-temperature environment far from the Sun, to form stable molecules of methane, CH_4, water vapor, H_2O, and ammonia, NH_3.

Table 3.2 Percentage composition of the Sun and the outer atmospheres of the giant planets[a]

Constituent	Sun	Jupiter	Saturn	Uranus	Neptune
Hydrogen, H_2	84	86.4	97	83	79
Helium, He (atom)	16	13.6	3	15	18
Water, H_2O	0.15	(0.1)	–	–	–
Methane, CH_4	0.07	0.21	0.2	2	3
Ammonia, NH_3	0.02	0.07	0.03	–	–

[a] The percentage abundance by number of molecules for the Sun, cooled to planetary temperatures so that the elements combine to form the compounds listed, and for the outer atmospheres of the giant planets below the clouds. Dashes indicate unobserved compounds. (Courtesy of Andrew P. Ingersoll.)

But what about molecular hydrogen, H_2? The existence of large amounts of this substance in the giant planets was strongly suggested by the presence of methane and ammonia in their atmospheres, by their low mass densities, and by the exceptionally high solar abundance of hydrogen. However, because of its structure, molecular hydrogen does not produce absorption lines at visible wavelengths.

Definite spectroscopic proof that molecular hydrogen is the most abundant element in Jupiter and Saturn's upper atmospheres did not occur until the 1960s and late-1970s, when high-dispersion infrared spectroscopy, from both the ground and space, showed several weak absorption features due to molecular hydrogen (Fig. 3.5). Precise values for the ingredients of Jupiter's atmosphere were obtained in 1999 from the *Galileo* probe (Section 9.3).

Helium, the second most abundant element in the Sun, has no detectable spectral features to make its presence known in the cold outer atmospheres of the giant planets. Helium is chemically inert and does not combine with other atoms to make molecules. The presence and amounts of helium atoms have nevertheless been inferred from the hydrogen infrared spectral features.

Collisions between helium atoms and hydrogen molecules alter the latter's ability to absorb infrared light, an effect that the *Pioneer 10* and *11* and *Voyager 1* and *2* instruments could detect. When the infrared measurements were combined with measured changes in these spacecraft's radio signals, observed when they passed behind the planets, the helium abundance could be determined with an uncertainty of only a few percent. The *Galileo* probe obtained a more precise value of the helium abundance in Jupiter's atmosphere, accurate to better than one percent (Section 9.3).

The helium abundance for Uranus and Neptune is consistent with that expected from a solar composition, but helium has been significantly depleted from Saturn's upper atmosphere and somewhat reduced in Jupiter's (Table 3.2). Theoretical calculations indicate that helium rain has been falling toward the center of Saturn for the past two billion years, generating heat and producing lower amounts of helium in its outer atmosphere (Section 10.3). Helium rain must be similarly settling toward Jupiter's core, but in lesser amounts.

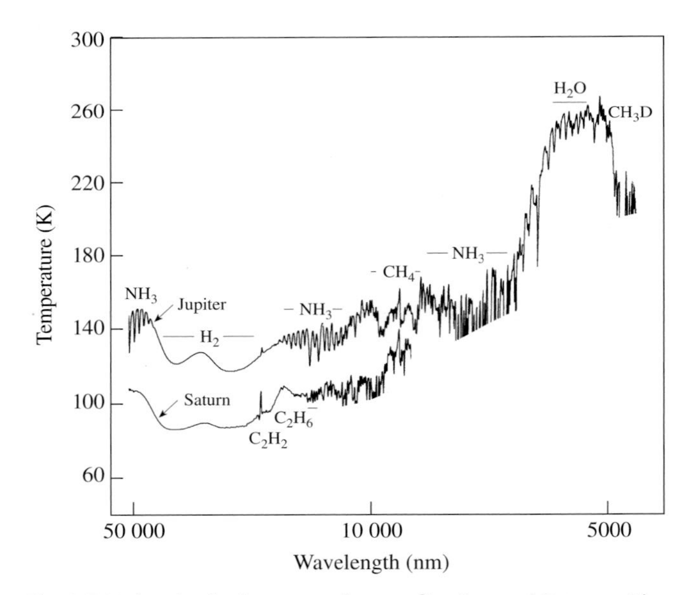

Fig. 3.5 Molecules in the atmospheres of Jupiter and Saturn The infrared radiation from the thin, cold, upper atmospheres of Jupiter and Saturn exhibit numerous features that have no counterpart in the spectrum of sunlight. Strong features are seen in Jupiter for molecular hydrogen, H_2, ammonia, NH_3, methane, CH_4, and water vapor, H_2O. Saturn's outer atmosphere is also abundant in hydrogen and methane, but the ammonia features are missing and those of acetylene, C_2H_2, and ethane, C_2H_6, are enhanced. These spectra were taken with instruments aboard *Voyager 1* and *2* during their Jupiter and Saturn flybys in 1979 and 1980, respectively. (Courtesy of Rudolf A. Hanel.)

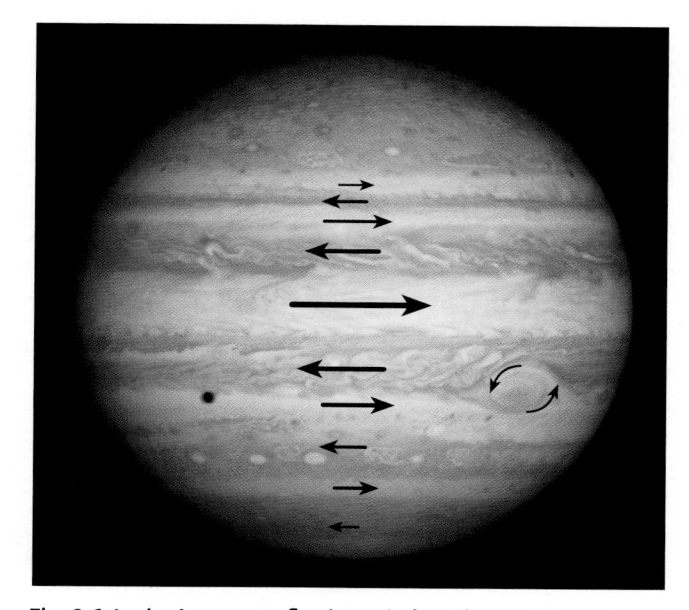

Fig. 3.6 Jupiter's counter-flowing winds The rapid rotation of Jupiter has pulled its winds into bands that flow east to west and west to east, shown in this image taken from the *Cassini* spacecraft on 7 December 2001. The windswept clouds therefore move in alternating light-colored, high-pressure zones and dark-colored, low-pressure belts. The arrows point in the direction of wind flow, and their length corresponds to the wind velocity, which can reach 180 meters per second in the equatorial regions (see Fig. 3.7). The Great Red Spot swirls in the counter-clockwise direction (*curved arrows*), like a high-pressure anticyclone in the Earth's southern hemisphere (Fig. 3.2), but it has lasted for more than 300 Earth years, much longer than terrestrial storms. Jupiter's moon Europa casts a shadow on the planet. (Courtesy of JPL and NASA.)

Raging winds on the giant planets

Astronomers have been using telescopes, both small and large, to scrutinize weather patterns on Jupiter for more than a century. They observe clouds of various ices, such as ammonia, that are formed in the cold outer atmospheres (Section 9.2). These clouds have been pulled into counter-flowing winds, moving in opposite eastward or westward directions and remaining confined to specific latitudes. These windswept clouds move in alternating light-colored bands, called zones, and dark ones, known as belts (Fig. 3.6). Since Jupiter's atmosphere has no solid surface to rub against or continents to disturb the flow, its winds are free to rage unabated in response to the planet's spin, with large-scale configurations that have remained unchanged for as long as they have been observed.

The weather patterns on Uranus and Neptune, where clouds of methane ice are observed, more nearly resemble those on Earth, which has low-latitude trade winds that blow westward and a meandering eastward current, the jet stream, in each hemisphere. Nevertheless, the Earth has the weakest winds in the solar system; its fastest jet streams move at speeds of about 40 meters per second. In contrast, Jupiter's winds move at constant speeds of up to 180 meters per second, and Uranus' fastest winds are just a little faster (Fig. 3.7). The high clouds on Venus and Mars also move at faster speeds than those on Earth, both with speeds of up to 100 meters per second. The winds on Neptune and Saturn move at speeds of up to 400 and

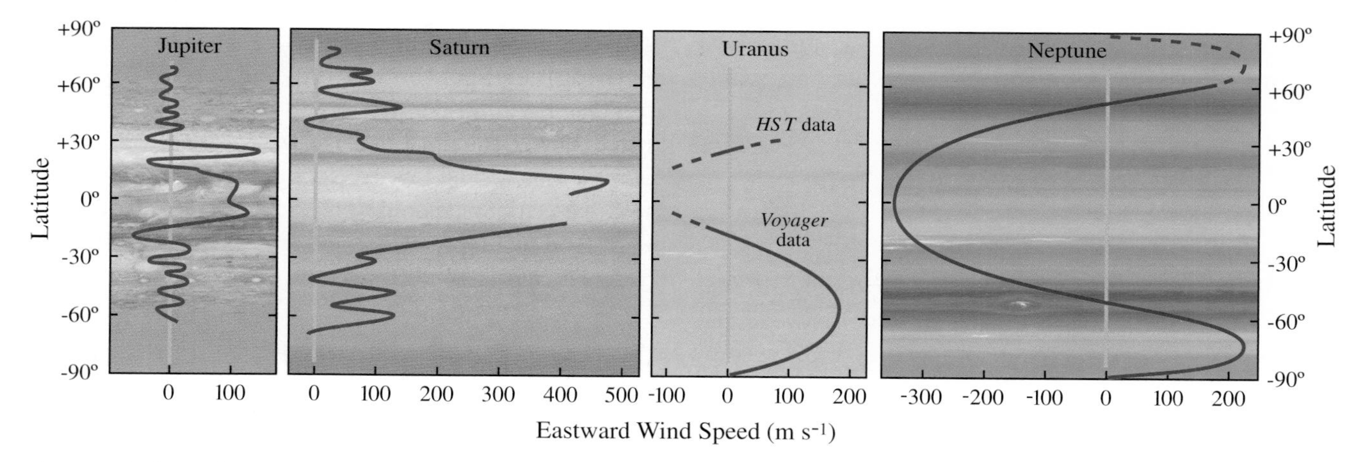

Fig. 3.7 Winds on the giant planets Variation of wind speed and direction as a function of latitude. Since a giant planet has no solid surface, the winds are measured relative to its internal rotation speed; the rotation period is determined from observations of the planet's periodic radio emission. Positive velocities correspond to winds blowing in the same direction but faster than the internal rotation; negative velocities are winds moving more slowly than the rotation. The winds are faster on Saturn than any other solar-system planet. (Courtesy of Andrew P. Ingersoll.)

450 meters per second, respectively, ten times the fastest winds on Earth. The speed of the Earth's winds are measured with respect to the rapidly rotating surface beneath them; for the giant planets that have no solid surface the internal rotation rate is inferred from the magnetic fields that are generated within their cores (Section 3.6).

Despite the slow motion of terrestrial winds, the energy available to drive them is greater than on any other solar-system planet. Because it is relatively near the Sun, the Earth intercepts more intense sunlight to power its winds than the more distant planets do. Although Venus is nearer the Sun than the Earth is, the bright clouds on Venus reflect most of the incident sunlight. The planet nearest to the Sun, Mercury, is too hot to retain a substantial atmosphere.

The circulation of the winds on the giant planets is powered by solar energy, as on Earth, plus internal energy left over from their formation. Even though Jupiter, Saturn and Neptune radiate about 1.67, 1.79 and 2.7 times more energy, respectively, than they receive from the Sun, the total power per unit area, from both sunlight and internal heat, is much less than that on Earth. The amount of power available to drive winds near the Earth's surface is 25 times that at the cloud tops of Jupiter and 400 times that in Neptune's atmosphere.

Small-scale turbulence in any planet's atmosphere dissipates the energy that is available to drive large-scale winds, and there is more dissipation for the planets that are nearer the Sun. Thus, both the global energy available to drive the winds and the amount of that energy that is dissipated decrease with increasing distance from the Sun, but at different ratios so there is more wind-producing power at the larger distances. The energy-dissipating turbulence is greatest at Earth, where the winds are weak, and least in Neptune's low-dissipation atmosphere, where the winds are stronger. Jupiter lies in between these two extremes.

Stormy weather on the giant planets

Giant anti-cyclones create continent-sized ovals on Jupiter that roll like ball bearings between the oppositely directed east–west winds. The large ovals revolve between the jet streams in the anti-cyclonic direction – clockwise in the northern hemisphere and counter-clockwise in the southern hemisphere. The smaller eddies are soon torn apart by the counter-flowing winds, lasting only about an Earth day or two, but the larger ones can persist for decades or centuries. Jupiter's Great Red Spot, about three times the diameter of Earth in size, has survived for more than 300 years.

How can the violent storms on Jupiter persist for such long times, whirling about for months, decades and even centuries? Some external source must be feeding energy into the spinning vortex, either from the sides or from below. Perhaps each whirling spot draws energy from the shearing motion of the clouds blowing in opposite directions on its sides. Both the biggest, long-lived storms and the powerful, banded winds on Jupiter may be energized by smaller eddies that merge into them. It is as if the larger features devour the smaller ones, consuming them to help maintain their flow and replenish their energy.

The little, short-lived storms may receive their energy from deep within the hot interior of the planet. In response to heating, gases in lower levels of the atmosphere will expand and thereby become less dense than the gas in the overlying layers. The heated material, due to its low density, rises, just as a hot-air balloon does, cooling at the cloud tops and sinking again. Such wheeling convective motions occur in a kettle of boiling water, and they produce towering thunderclouds here on the Earth. Lightning is also created in the thunderstorms on both the Earth and Jupiter (Section 9.2). The convective transport of energy can be described by complex mathematical equations that produce large-scale order out of a seemingly chaotic pattern of small-scale motions that never exactly repeat themselves.

Neptune also has a stormy atmosphere with raging winds, but the big storms are not as long lasting. The largest storm system on Neptune, observed when *Voyager 2* sped past the planet on 24 August 1989, is as broad as the Earth (see Fig. 2.9). It is called the Great Dark Spot because it resembles the Great Red Spot of Jupiter. Both of these giant-planet storms are found in the planetary tropics – at about one-quarter of the way from the equator to the pole, both rotate counter-clockwise, and both are about the same size relative to their planet. As *Voyager 2* watched, the Great Dark Spot contracted in and stretched out with a regular rhythm, like the mouth of a feeding fish, and drifted slowly equatorward. Yet, when the *Hubble Space Telescope* was turned toward Neptune in 1995, no trace of the Great Dark Spot could be found. Perhaps it simply ran out of food to supply its energy, or perhaps it moved into strong equatorial winds that could not support it.

Escape from an atmosphere

The large planets have atmospheres containing hydrogen and hydrogen compounds, the middle-sized planets have atmospheres containing oxygen compounds, and smaller objects, such as the Earth's Moon and the planet Mercury, have no atmosphere at all. George Johnstone Stoney (1826–1911) provided a simple explanation for these differences in 1898. It partly depends upon the mass and gravitational pull of the object, and it also depends on the temperature of the atmosphere.

Focus 3.1 Escape from the planets

Why don't terrestrial planets have hydrogen in their atmospheres, and why doesn't the Moon have any atmosphere to speak of? The ability of a planet or satellite to retain an atmosphere depends on both the temperature of that atmosphere and the gravitational pull of the planet or satellite. If the gas is hot, the molecules move about with a greater speed and are more likely to escape the gravitational pull of the planet. This is one of the reasons that Mercury, the closest planet to the Sun and therefore the hottest, has no atmosphere. The other reason is that Mercury has a relatively small size and mass, as far as planets go, and thus has a comparatively low gravitational pull. A planet with a larger mass is more likely to retain an atmosphere, which helps explain why massive Jupiter retains all the elements. Jupiter is also far away from the Sun's heat, so molecules in Jupiter's atmosphere move at a relatively slow speed.

An atom, ion or molecule moves about because it is hot. Its kinetic temperature, T, is defined in terms of the thermal velocity, V_{thermal}, given by equating the thermal energy to the kinetic energy of motion:

$$\text{thermal energy} = \frac{3}{2}kT = \frac{1}{2}mV_{\text{thermal}}^2 = \text{kinetic energy}.$$

Solving for the thermal velocity:

$$V_{\text{thermal}} = \left[\frac{3kT}{m}\right]^{1/2},$$

where Boltzmann's constant $k = 1.380\,66 \times 10^{-23}$ J K^{-1}, and the particle's mass is denoted by m. We see right away that at a given temperature, lighter particles move at faster speeds. Colder particles of a given mass travel at slower speed. Anything will cease to move when it reaches absolute zero on the kelvin scale of temperature. The temperatures at the surfaces or cloud tops of the planets are given in Table 3.3.

When the kinetic energy of motion of a particle of mass, m, moving at velocity, V, is just equal to the gravitational potential energy exerted on it by a larger mass, M, we have the relation:

$$\text{kinetic energy} = \frac{mV^2}{2} = \frac{GmM}{D} = \text{gravitational potential energy},$$

where the Newtonian gravitational potential is $G = 6.6726 \times 10^{-11}$ m^3 kg^{-1} s^{-2}, and D is the distance between the centers of the two masses. When we solve for the velocity, we obtain:

$$V_{\text{escape}} = \left[\frac{2GM}{D}\right]^{1/2},$$

where the subscript escape has been added to show that the small mass must be moving faster than V_{escape} to leave a larger mass, M. This expression is independent of the value of the smaller mass, m. The escape velocities at the surfaces or cloud tops of the planets are also given in Table 3.3.

A molecule will overcome the gravitational pull of a planet or satellite if the molecule's velocity exceeds the object's escape velocity, which increases with the object's mass (Focus 3.1). Small bodies with low mass, such as our Moon, have a very small escape velocity, and insufficient gravitational pull to retain any substantial atmosphere. Middle-sized planets, like the Earth, Venus and Mars, have moderate escape velocities and enough gravity to hold on to heavier, slower-moving molecules. Only the massive giant planets, like Jupiter and Saturn, have a high enough escape velocity that they can retain all molecules, including the lightest one, hydrogen.

Temperature also plays a role, for it helps determine if the molecules can move fast enough to escape an object's gravity. A planet will only retain molecules that are moving at velocities less than the planet's escape velocity, and a molecule's velocity increases with temperature (Focus 3.1). Hotter molecules dart about at faster speeds, and colder molecules move with slower speeds. Since the outer atmospheric temperature falls off with increasing distance from the Sun, molecules tend to have lower velocities out in the realm of the giant planets. This is the reason that some of

Table 3.3 Temperature and escape velocity of the planets

Planet	Temperature[a] (K)	Escape Velocity (km s^{-1})
Terrestrial planets		
Mercury	440	4.43
Venus	735	10.36
Earth	288–293	11.19
Mars	140–300	5.03
Giant planets		
Jupiter	165	59.54
Saturn	134	35.49
Uranus	76	21.29
Neptune	73	23.71

[a] Surface temperature for the terrestrial planets and atmospheric temperatures for the giant planets at the level where the atmospheric pressure is 1 bar.

the cold satellites of the remote giant planets are able to retain atmospheres in spite of their low mass and small escape velocity.

At a given temperature, a molecule's velocity increases with decreasing molecular mass (Focus 3.1), so lighter molecules move at faster speeds and are more like to escape a given planet or satellite than heavier ones. This explains why the Earth cannot retain light hydrogen atoms or molecules in its atmosphere, while the heavier oxygen and nitrogen molecules are held within our air.

3.3 Titan – a satellite with a substantial atmosphere

Saturn's largest moon, Titan, is the only satellite with a substantial atmosphere. Detailed investigations with instruments aboard *Voyager 1* in 1980 showed that the dominant gas surrounding Titan is molecular nitrogen, N_2, at 82 to 99 percent, similar to Earth (77 percent). In fact, the satellite is enveloped by about 10 times more nitrogen then we are, yielding a surface pressure 1.5 times greater than the sea-level pressure of Earth's atmosphere. The surface temperature on Titan is 94 degrees kelvin, as expected for a body so far from the Sun.

The spectrometers on *Voyager 1* showed that the next most abundant gas enveloping Titan is methane, CH_4, with an abundance between 1 and 6 percent. Gerard Kuiper had discovered signs of methane in Titan's spectrum as early as 1944.

Methane molecules rise up to high levels in Titan's atmosphere, where they are broken apart by ultraviolet sunlight and electrons coming from Saturn's magnetic environment. These molecular fragments recombine to form heavier hydrocarbon molecules such as ethane, C_2H_6, and familiar gases like acetylene, C_2H_2, propane, C_2H_8, and hydrogen cyanide, HCN – further details are given in Section 10.5.

It doesn't rain water on Titan, but it could rain fuel, in large drops that fall like snow. Given the known atmospheric composition and the temperatures, scientists speculate that thin clouds of methane-ice crystals may form in the lower atmosphere. Ethane and propane can rain all the way down to the surface, forming seas, lakes and ponds. The patchy reservoirs of liquid hydrocarbons could be driving the weather cycle on Titan, with towering clouds of methane and a rainy drizzle of ethane (Section 10.5).

We are not completely certain that there are any liquid seas on Titan, for we cannot see through its smog. The *Voyager 1* cameras showed that an opaque haze completely enshrouds the satellite (Fig. 3.8). The smog is unimaginably worse than a bad day in Los Angeles or Mexico City. Compared with any urban smog on Earth, there are

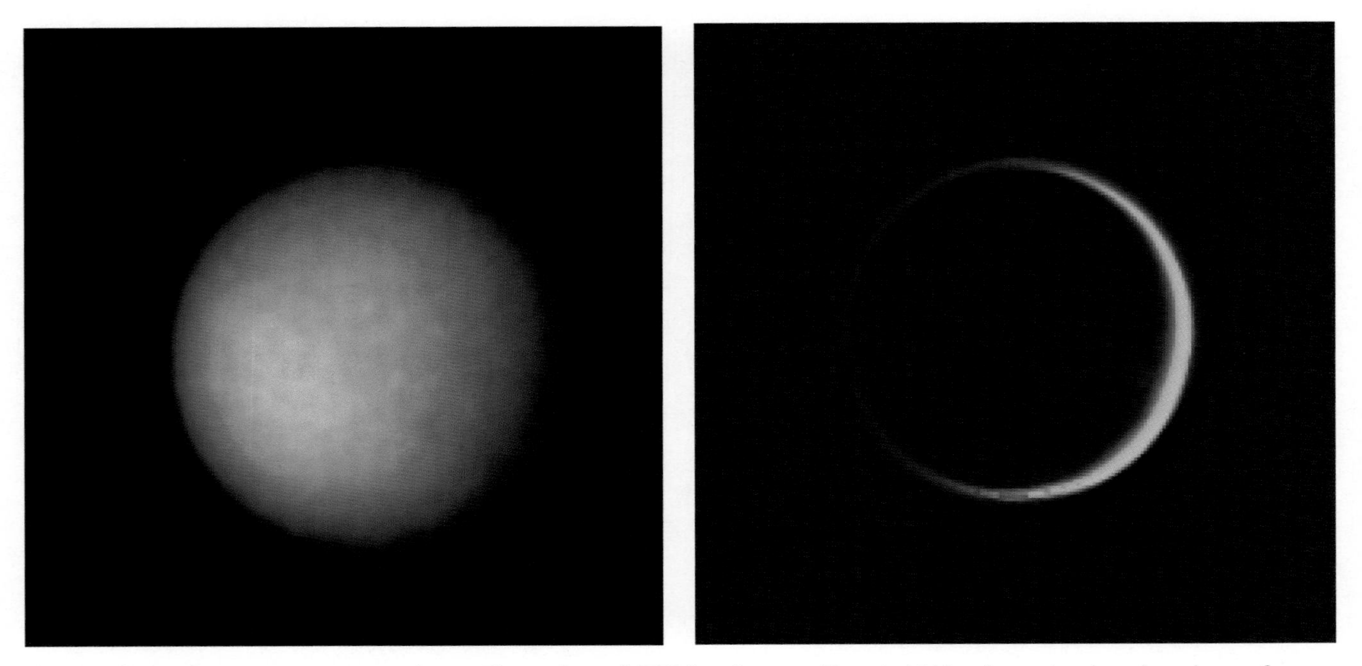

Fig. 3.8 Titan's dense, smoggy atmosphere The surface of 5150-km-diameter Titan is hidden from view by a hazy layer of smog, giving it a fuzzy, tennis-ball appearance in a *Voyager 1* image (*left*) taken on 4 November 1980. When illuminated from behind, the dense atmosphere forms a crescent several hundred kilometers above the satellite's surface (*right*) seen by *Voyager 2* on 25 August 1981. The extension of blue light around the moon's night side is due to scattering from smog particles in the sunlit portion. The *Cassini* spacecraft is expected to drop its *Huygens Probe* into Titan's atmosphere in 2004. (Courtesy of JPL and NASA.)

relatively few smog particles per unit volume of Titan's atmosphere, but the haze extends to an altitude of about 200 kilometers. This makes the smog thick enough to completely hide Titan's surface from view.

The orange smog must result from energetic sunlight that breaks simple molecules like methane apart and chemical reactions that create more complex substances from these fragments. The exact composition of this photochemical smog remains unknown, but its mere existence suggests that heavy compounds may be falling through the atmosphere and sinking into the hypothetical seas to form an organic sludge on their floors.

When the *Cassini/Huygens* spacecraft arrives at Saturn in mid-2004, it is expected to tell us what lies beneath Titan's obscuring veil of orange smog (Section 10.5). The *Cassini Orbiter* has a radar experiment to penetrate the smog and map the terrain, much as the *Magellan* radar looked through the perpetual clouds of Venus to image its surface (Sections 2.3, 7.4). Earth-based radar measurements, as well as infrared observations with the *Hubble Space Telescope*, already suggest a rough terrain with continent-sized features on Titan (Section 10.5). More detail will be obtained when the *Huygens Probe* descends slowly through the atmosphere, with cameras to image the surface and sensitive instruments that will measure the composition of the gases and particles around it.

As we shall next see, there are other satellites with sizes comparable to Titan that do not have a dense atmosphere, but are instead enveloped by very tenuous gas. Possible explanations for this difference are given in Focus 3.2.

Focus 3.2 Why does Titan have a dense atmosphere?

Why does Saturn's largest satellite, Titan, have such a substantial atmosphere when it is only slightly bigger than the planet Mercury and almost as big as Jupiter's largest satellite, Ganymede, which both have exceedingly tenuous atmospheres. The ability of a planet or satellite to retain an atmosphere is determined primarily by the body's mass and temperature (Focus 3.1), but because Titan, Ganymede and Mercury have nearly the same mass, differences in temperature should account for their atmospheric differences. Mercury is so hot that even the heaviest molecules move fast enough to escape the planet's gravity, while the temperature on Titan is so low that only the lighter molecules, like hydrogen, can escape.

But Ganymede is now sufficiently cold and massive to retain a Titan-like atmosphere. The difference between Titan and Ganymede is most likely a consequence of the temperature at the time of their "birth". Titan was born in the remote cooler regions of the solar system, and nearby Saturn never became as hot as Jupiter did during its birth. The low temperatures permitted ammonia, methane and water ice to form on Titan's surface when it was born, and these ices probably sublimated to form a primeval atmosphere of ammonia and methane, while water remained locked into its surface as ice. On Ganymede, water was probably the only ice that formed in the slightly warmer climate. If there was no ammonia or methane ice on Ganymede, and if the temperatures never became high enough for its water ice to sublimate, Ganymede would be left without any substantial atmosphere. The water ice is instead chipped away from outside, forming a tenuous oxygen atmosphere.

3.4 Tenuous atmospheres

By conventional standards, the Moon has no atmosphere. It has no clouds or weather. Its sky remains black in broad daylight, and there are no colorful sunsets on the Moon. In comparison, air molecules preferentially scatter blue sunlight down to us when the Sun is overhead, producing our bright blue sky, and the Sun is strongly reddened at sunrise or sunset because of scattering by dust and gas in our atmosphere. And since the Moon has no substantive atmosphere to transmit sound waves, there are no sounds to relieve the eternal silence on the Moon.

Yet, the Moon is surrounded by a tenuous mist of particles, a so-called atmosphere that is 100 trillion (10^{14}) times less dense than our air. The density of the Moon's tenuous atmosphere is so low, and its particles so far apart, that its atoms almost never hit and connect with each other. They just bounce around the surface.

Unlike the Moon, the planet Mercury is exceptionally hot on the side facing the nearby Sun. Because of the intense heat, any gas molecules on Mercury will move faster than the escape velocity of the planet, and move out into interplanetary space (see Focus 3.1). For this reason, it was also thought that Mercury does not have an atmosphere even though it is bigger and more massive than the Moon.

Small amounts of hydrogen and helium were nevertheless detected when *Mariner 10* flew past Mercury in 1973–74. Subsequent Earth-based observations found traces of sodium and potassium, which are also present in the Moon's tenuous atmosphere (Table 3.4).

The tenuous atmospheres on the Moon and Mercury could act as a source of ice in the polar regions that are never illuminated by sunlight. The atoms in the atmosphere just

Table 3.4 Number density of atoms in the atmospheres of the Moon and Mercury

Atom	Symbol	Moon (million atoms per m^3)	Mercury (million atoms per m^3)
Hydrogen	H	200	less than 17
Helium	He	6000	2000–4000
Oxygen	O	less than 40 000	less than 500
Sodium	Na	20 000	70
Potassium	K	500	16
Argon	Ar	less than 30 000 000	40 000

hop around the surface, like water on a hot skillet, until they reach the shadowed places. When they strike the cold surface there, the atoms will tend to stick and stay forever. Evidence for water ice at the poles of the Moon and Mercury was presented in Section 2.4.

The rarefied gas that envelops these objects is not permanent. Unlike the Earth's gaseous cloak, the atmospheres of the Moon and Mercury are constantly evaporating and being replenished. The gas could be supplied by comets that hit them, or created when micrometeorites or the solar wind strike the surface and release gases trapped in the rocks or soil. The solar wind is a perpetual flow of electrons and protons from the Sun (Section 3.5).

The gas atoms and the ions are picked up and removed by the Sun's wind. Lifetimes of atmospheric atoms on the Moon vary from about an hour for sodium up to a couple of hundred days for hydrogen.

The larger moons of the giant planets have sizes that are comparable to our Moon or Mercury, and these satellites are similarly cloaked in a thin film of gas (Table 3.5). These tenuous atmospheres are also temporary features, and must be continuously resupplied.

Io's active volcanoes eject sulfur dioxide, SO_2, gas into the vacuous space around the satellite, like exhaust from a rocket nozzle; the gas then flows down across the nearby ground and condenses to form white frost on the surface (Sections 2.3, 9.4). The resultant atmosphere has a maximum surface pressure one-millionth (10^{-6}) that of the Earth.

The volcanic gas on Io lingers near the active volcanic vents, forming dense localized patches of atmosphere, and the tenuous gas over one part of the surface apparently remains disconnected from any other part. Consequently, there are no global circulation patterns in Io's tenuous atmosphere.

A tenuous veil of oxygen molecules has been found around Europa, an ice-covered moon of Jupiter. The surface pressure of this atmosphere is barely a ten-millionth (10^{-7}) that of Earth. This extremely tenuous gas was

Table 3.5 Thin atmospheres[a]

Object	Mean Radius (km)	Atoms or Molecules	Symbol	Pressure[b] (bar)
Ganymede	2631	Molecular oxygen	O_2	10^{-7}
Mercury	2440	Helium, sodium, potassium	He, Na, K	–
Callisto	2410	Hydrogen, carbon dioxide	H_2, CO_2	–
Io	1822	Sulfur dioxide	SO_2	10^{-6}
Earth's Moon	1738	Hydrogen, helium, sodium, potassium	H, He, Na, K	–
Europa	1561	Molecular oxygen	O_2	10^{-7}
Triton	1353	Molecular nitrogen, methane	N_2, CH_4	1.5×10^{-5}
Pluto	1151	Molecular nitrogen, methane	N_2, CH_4	1.5×10^{-5}

[a] Listed in order of decreasing radius. Saturn's moon Titan has a radius of 2575 kilometers with a dense nitrogen atmosphere – see Section 3.3.
[b] The surface pressure is given in units of bars; the surface pressure of the Earth's atmosphere at sea level is 1.0 bar and the surface pressure of Titan's atmosphere is 1.5 bar. The number density of the tenuous atmospheres on the Moon and Mercury are given in Table 3.4.

discovered by ultraviolet spectroscopy from the *Hubble Space Telescope*. The same team of scientists subsequently found evidence for oxygen molecules enveloping Jupiter's moon Ganymede, but concentrated near the north and south poles of the satellite. The gas is likely to be as tenuous as that detected on Europa, with a comparable surface pressure.

Scientists believe that the atmospheric oxygen, O_2, comes from water ice, H_2O, on the surface of Europa and Ganymede. Charged particles bombard the water molecules, splitting off oxygen atoms, O, that combine to make oxygen molecules, O_2. Exposure to sunlight and meteor impacts could also create some of the gas. The relatively lightweight hydrogen, H, escapes into space, leaving the heavier oxygen molecules to accumulate to form an atmosphere.

There is plenty of ice on the surface of Triton, the largest satellite of Neptune. With a surface temperature of just 38 kelvin, Triton is the coldest body yet explored in the solar system. Nitrogen- and methane-ices are the dominant constituents of Triton's surface, feeding its tenuous atmosphere. Spectroscopic observations from the Earth and during the *Voyager 2* encounter with Neptune indicated that molecular nitrogen dominates the atmosphere of Triton, with a surface pressure of just 15 millionths (1.5×10^{-5}) of the surface pressure of the Earth's atmosphere at sea level. In spite of the low pressure, Triton's thin atmosphere can support clouds and haze that were detected by *Voyager 2*. Nitrogen ice sublimates, from ice to gas, in regions warmed by sunlight and then precipitates in the colder regions on the satellite. So Triton's nitrogen atmosphere comes and goes in response to seasonal temperature variations. The surface and atmosphere of Triton are discussed in greater detail in Section 11.5.

Pluto's atmosphere was discovered when astronomers watched a star move behind Pluto on 9 June 1988. If a star passes behind an airless planet, the star's light will be cut off abruptly and then quickly switch on when it reappears on the other side. Instead, the star dimmed gradually when it passed behind Pluto, indicating that greater amounts of sunlight were being absorbed as it passed through progressively deeper layers of the atmosphere. When the star emerged, it slowly brightened, so an atmosphere had to envelop Pluto.

Like Triton, the frozen surface of Pluto contains ices of nitrogen and methane, with smaller amounts of carbon-monoxide ice and water ice. Since nitrogen is the most volatile ice that has been identified, gaseous nitrogen molecules are probably the main constituents of Pluto's atmosphere; small amounts of methane and carbon monoxide are also likely atmospheric constituents as is hazy smog. The estimated surface pressure of Pluto's atmosphere is comparable to Triton's; variations in the surface temperature as Pluto moves along its eccentric orbit are likely to produce variations in the amount of gas and in the surface pressure.

3.5 The planets are inside the Sun's expanding atmosphere

The space just outside the Earth is not empty. It is filled with pieces of the Sun. Our star is expanding out in all directions, filling interplanetary space with electrically charged particles that are forever blowing from the Sun. This solar wind moves past the planets and engulfs them, carrying the Sun's rarefied atmosphere out to the space between the stars. So, we are actually living in the outer part of the Sun.

Unlike any wind on Earth, the solar wind is a tenuous mixture of charged particles and magnetic fields streaming radially outward in all directions from the Sun at supersonic speeds of hundreds of kilometers per second. It is mainly composed of electrons and protons, set free from the Sun's abundant hydrogen atoms, but it also contains lesser amounts of heavier ions. The seemingly eternal wind carries a magnetic field with it, with one end anchored in the Sun. This interplanetary magnetic field has a spiral shape due to the combined effects of the radial solar wind flow and the Sun's rotation.

The existence of the solar wind was suggested from observations of comet tails about half a century ago. When a comet is tossed into the inner solar system, the dirty ice on its surface is vaporized, forming two tails that always point away from the Sun (Fig. 3.9). One is a curved dust tail, pushed away from the Sun by the pressure of sunlight. The other is a straight ion tail that is affected by the solar wind.

The German astronomer, Ludwig Biermann (1907–1986) noticed, in the 1950s, that the ions in a comet's tail move with velocities many times higher than could be caused by the weak pressure of sunlight, and proposed that a perpetual flow of electrically charged particles pours out of the Sun at all times and in all directions, accelerating the ions to high speeds and pushing them radially away from the Sun in straight ion tails.

In 1958, Eugene N. Parker (1927–) of the University of Chicago showed how such a relentless flow might work, dubbing it the solar wind. It would naturally result from the expansion of the Sun's million-degree atmosphere, called the corona. He also demonstrated how a magnetic field would be pulled into interplanetary space from the rotating Sun, attaining a spiral shape.

Fig. 3.9 Comet tails Telescopic photograph of Comet Mrkos (1951 V) taken in August 1957, showing the straight, well-defined ion tail and the more diffuse, slightly curved dust tail. Both comet tails point away from the Sun. The electrically charged solar wind deflects the charged ions and accelerates them to high velocities, creating the relatively straight ion tail. The radiation pressure of sunlight suffices to blow away the un-ionized comet dust particles, forming a broad arc that can resemble a scimitar. (Courtesy of Lick Observatory.)

The existence of the Sun's ever-flowing wind has been fully confirmed by spacecraft that have measured its density, velocity and magnetic fields (Table 3.6). They have been making *in-situ* (Latin for "in original place", or literally "in the same place") measurements of the solar wind for decades.

The Soviet *Luna 2* measurements in 1959 indicated a solar-wind flux of 2 million million (2×10^{12}) ions (presumably protons) per square meter per second. This is in rough accord with all subsequent measurements. In 1962–63, the average solar-wind ion density was shown by the American *Mariner 2* to be 5 million (5×10^{6}) protons per cubic meter near the distance of the Earth from the Sun. Thus, the solar wind is exceedingly dilute, much nearer to a vacuum than that produced in any laboratory.

We now know that such a low density close to the Earth's orbit is a natural consequence of the wind's expansion into an ever-greater volume. Because of this expansion, the number density of particles in the solar wind diminishes as the inverse square of the distance from the Sun. Sporadic, gusty fluctuations, of a factor of ten in the number density of the solar wind, occur in response to varying solar activity.

Because the field lines are wrapped into a spiral form, the interplanetary magnetic field strength far from the Sun does not decrease as quickly as the number density of the solar wind. Instead, the field strength falls off as the inverse of the heliocentric distance.

The *Mariner 2* data also indicated that the solar wind has a slow and a fast component. The slow one moves at a speed of 300 to 400 kilometers per second; the fast one travels at twice that speed.

The radial, supersonic outflow creates a huge bubble of electrons, protons and magnetic fields, with the Sun at

Table 3.6 Mean values of solar-wind parameters at the Earth's orbit[a]

Parameter	Mean Value
Particle Density, N	10 million particles per cubic meter (5 million electrons and 5 million protons)
Velocity, V	400 km s^{-1} and 800 km s^{-1}
Flux, F	10^{12} to 10^{13} particles per square meter per second
Temperature, T	1.2×10^5 K (protons) to 1.4×10^5 K (electrons)
Particle Thermal Energy, kT	2×10^{-18} J $\approx$ 12 eV
Proton Kinetic Energy, $0.5\ m_p V^2$	10^{-16} J $\approx$ 1000 eV = 1 keV
Thermal Energy Density, NkT	10^{-11} J m^{-3}
Proton Energy Density, $0.25\ Nm_p\ V^2$	10^{-9} J m^{-3}
Magnetic Field Strength, B	6×10^{-9} T = 6 nT = 6×10^{-5} G

[a] These solar-wind parameters are at the mean distance of the Earth from the Sun, or at one astronomical unit, 1 AU, where 1 AU = 1.496×10^{11} m; the Sun's radius, $R_\odot$, is $R_\odot = 6.955 \times 10^8$ m. Boltzmann's constant $k = 1.380\,66 \times 10^{-23}$ J K^{-1} relates temperature and thermal energy. The proton mass $m_p = 1.672\,623 \times 10^{-27}$ kg.

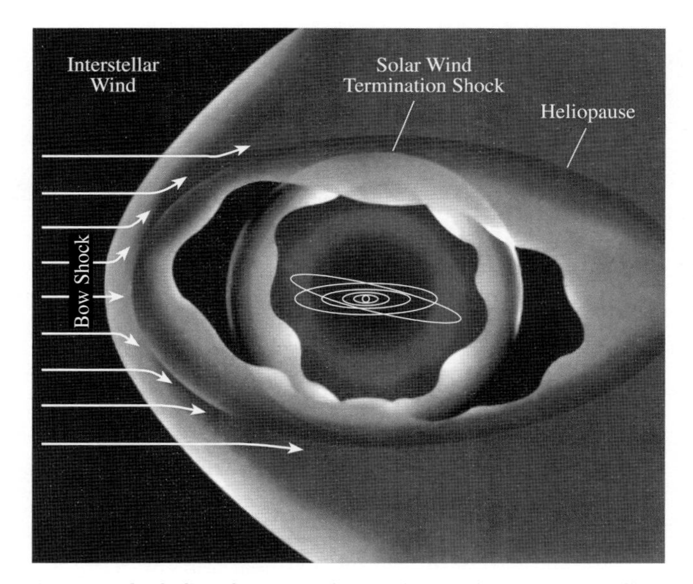

Fig. 3.10 The heliosphere With its solar wind going out in all directions, the Sun blows a huge bubble in space called the heliosphere. The heliopause is the name for the boundary between the heliosphere and the interstellar gas outside the solar system. Interstellar winds mold the heliosphere into a non-spherical shape, creating a bow shock where they first encounter it. The orbits of the planets are shown near the center of the drawing.

the center and the planets inside, called the heliosphere (Fig. 3.10), from *Helios* the Greek word for the "Sun". Within the heliosphere, conditions are regulated by the Sun. Its domain extends out to about 150 AU, or about 150 times the mean distance between the Earth and Sun, marking the outer boundary or edge of the solar system. Out there, the solar wind has become so weakened by expansion that it can no longer repel interstellar forces.

The perpetual solar gale carries the Sun's atmosphere out into interstellar space at the rate of almost a million tons (10^6 tons = 10^9 kilograms) every second. As the gas disperses, it must be replaced by material welling up from below to feed the wind. Exactly where this material comes from is an important subject of contemporary space research, but the main interest of planetary research is how the solar wind affects the planets.

3.6 Magnetospheres

Earth's magnetic dipole

In 1600, William Gilbert (1544–1603), physician to Queen Elizabeth I of England, authored a treatise in Latin, with the grand title *De magnete, magneticisque corporibus, et de magno magnete tellure* translated into English as *Concerning Magnetism, Magnetic Bodies, and the Great Magnet Earth.* In this work, which is still available in its English version, Gilbert showed that the Earth is itself a great magnet, which explains the orientation of compass needles. It is as if there was a colossal magnet at the center of the Earth.

At the equator, a compass needle points north and south, toward the poles. At each magnetic pole, it would stand upright, pointing into or out of the ground. And in between, at intermediate latitudes, the compass needles point north and south with a downward dip of one end, but not vertically as at a pole.

Since the geographic poles are located near the magnetic ones, a compass needle is aligned in the north–south direction. We usually put an arrow or point on the north end of the needle, and a terrestrial compass therefore points north. Since the Earth's rotation axis is inclined 11.7 degrees with respect to its magnetic axis, a compass needle does not point exactly toward the geographic North Pole, but within 11.7 degrees of it.

We can describe the Earth's magnetism by invisible magnetic field lines, which orient compass needles. These lines of magnetic force emerge out of the south magnetic pole, loop through nearby space and re-enter at the north magnetic pole (Fig. 3.11). The lines are close together near the magnetic poles where the magnetic force is strong, and spread out above Earth's equator where the magnetism is weaker than at the poles. You cannot see the invisible magnetic field lines, but compass needles point along them and other instruments can be used to measure their strength.

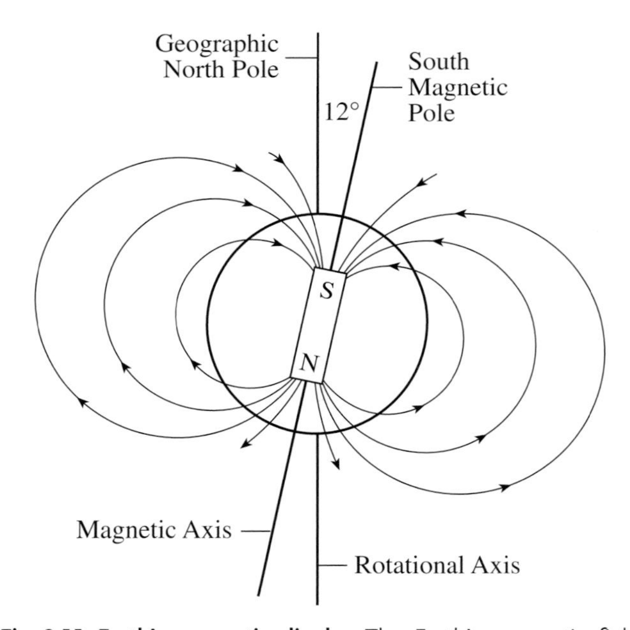

Fig. 3.11 Earth's magnetic dipole The Earth's magnetic field looks like that which would be produced by a bar magnet at the center of the Earth, with the north magnetic pole corresponding to the South (Geographic) Pole and *vice versa*. The Earth's magnetic dipole originates in swirling currents of molten iron deep in the Earth's core, and extends more than 10 Earth radii, or 63.7 thousand kilometers out into space on the side facing the Sun, and all the way to the Moon's orbit, at 384.4 thousand kilometers on the opposite side. Magnetic field lines loop out of the South (Geographic) Pole and into the North (Geographic) Pole. The lines are close together near the magnetic poles where the magnetic force is strong, and spread out where it is relatively weak. The magnetic axis is tilted at an angle of 11.7 degrees with respect to the Earth's rotational axis. This dipolar (two poles) configuration applies near the surface of the Earth, but further out the magnetic field is distorted by the solar wind (see Fig. 3.12).

The magnetic field strength at the Earth's magnetic equator is 0.000 030 5 tesla, or 0.305×10^{-4} T. Measurements of the surface magnetic fields of Earth show stronger fields near the poles where the magnetic field lines congregate, at roughly twice the strength of the field at the equator. The magnetic strength in both regions is several times weaker than a toy magnet, but the comparison is somewhat misleading since the Earth is a very big magnet.

Magnetism pervades the entire volume of the Earth. This global quality is expressed quantitatively by the magnetic dipole moment, equal to the product of the equatorial magnetic field strength and the cube of the planet's equatorial radius. For the Earth we have an equatorial radius of 6378 kilometers and a magnetic dipole moment of 7.91×10^{15} T m^3. In comparison, the magnetic dipole moment of a typical laboratory electromagnet is of the order of 10^{-5} T m^3 or more than 100 billion billion times weaker.

How is magnetism generated within the Earth's hot, molten interior? Heat cooks magnetism out of a permanent magnet, and liquefaction melts it away. But the compass needles are not ruled by a permanent magnet. Instead, the inside of the Earth is an electromagnet, generating magnetism by changing electric currents. Heat-driven circulation and the Earth's rotation combine to produce electrically conducting streams of molten iron, which generate the terrestrial magnetic fields by dynamo action.

Electric currents give rise to magnetic fields, and moving magnets generate electric currents. These two effects, in the churning outer core of the Earth, can amplify the small magnetic field that the planet captured from its surroundings when it formed. As opposing streams of molten iron, carrying tiny magnetic fields, sweep past one another, each induces currents in the other. This creates more magnetism, which induces more currents, and so on. In this way, the dynamo in the Earth's liquid outer core generates the magnetic fields detected at the terrestrial surface, taking energy from both the internal heat and the rotational energy of the planet. And because the currents deep down inside the Earth are always varying, the Earth's magnetism is a dynamic, changing thing.

Ancient magnetic rocks, found on the flanks of volcanoes, indicate that the Earth's magnetic field has not always been the same as it is today. When the molten volcanic lava flows to the surface and hardens into rock, its internal magnetism lines up with the Earth's magnetic field and freezes into position. These magnetic fossils record the direction and intensity of the terrestrial magnetic field when and where the lava solidified.

An inspection of magnetic fossils of differing ages from all parts of the world resulted in an amazing discovery. The great magnet of the Earth has flipped, or reversed its direction, many times in the past. During each flip, the north magnetic pole becomes the south one, and *vice versa*. The deep electric currents readjust, always remaining nearly aligned with the rotation axis, but with a swap in the magnetic poles.

An examination of volcanoes on land indicates that the Earth's magnetic field has reversed itself at least nine times over the past 3.6 million years. And the ordered succession of magnetized rocks on the spreading ocean floors records 100 and more full reversals of the direction of the magnetic poles during the past 200 million years. Tens of thousands of years separate some of the magnetic field reversals, while tens of million of years separate others.

So the terrestrial magnetic field is inevitably headed for a magnetic flip, but we don't know exactly when. The arrows on compass needles will then point south instead of north, reversing their direction. Animal species and satellites that depend on magnetic fields for guidance will lose their orientation, and will have to adapt to the changing field. The navigation systems of migrating birds and monarch butterflies, for example, depend in part on internal compasses that sense directions from the Earth's magnetic field. Honeybees, some wasps, some fish, sea turtles and even a species of mole rat take bearings magnetically.

The Earth's magnetic fields also influence the space near the Earth. They extend away from the Earth, decreasing in strength as the inverse cube of the distance. Yet, they remain strong enough to shield the Earth from the full force of the Sun's charged winds.

Earth's protective magnetosphere

Fortunately for life on Earth, the terrestrial magnetic field deflects the Sun's wind away from the Earth, like a rock in a stream or a windshield that deflects air around a car. It hollows out a protective cavity in the solar wind called the magnetosphere. The magnetosphere of the Earth, or any other planet, is that region surrounding the planet in which its magnetic field dominates the behavior of electrically charged particles such as electrons, protons and other ions. It diverts most of the solar wind around our planet at a distance far above the atmosphere, thereby protecting humans on the ground from possibly lethal solar particles.

The dipolar (two poles) magnetic configuration applies near the surface of the Earth, but further out the magnetic field is distorted by the Sun's perpetual wind. Although it is exceedingly tenuous, far less substantial than a terrestrial breeze or even a whisper, the solar wind is powerful enough to mold the outer edges of the Earth's magnetosphere into a changing asymmetric shape (Fig. 3.12), like a tear drop falling toward the Sun.

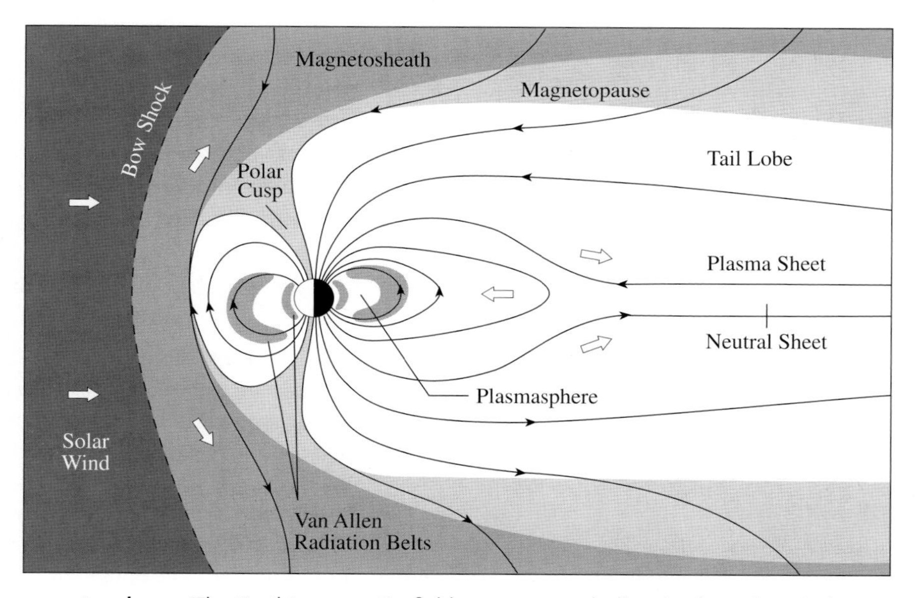

Fig. 3.12 The Earth's magnetosphere The Earth's magnetic field carves out a hollow in the solar wind, creating a protective cavity called the magnetosphere. The Earth, its auroras, atmosphere and ionosphere, and the two Van Allen radiation belts all lie within this magnetic cocoon. Similar magnetospheres are found around other magnetized planets. A bow shock forms at about ten Earth radii on the sunlit side of our planet. Its location is highly variable since it is pushed in and out by the gusty solar wind. The magnetopause marks the outer boundary of the magnetosphere, at the place where the solar wind takes control of the motions of charged particles. The solar wind is deflected around the Earth, pulling the terrestrial magnetic field into a long magnetotail on the night side. Electrons and protons in the solar wind are deflected at the bow shock (*left*), and flow along the magnetopause into the magnetic tail (*right*). Electrically charged particles can be injected back toward the Earth and Sun within the plasma sheet (*center*).

The hot, high-speed, magnetized solar wind confronts the Earth's magnetic field close to home, usually at a distance of about ten times the Earth's radius on the day side that faces the Sun. Here the solar wind pushes the Earth's magnetism in, compressing its outer magnetic boundary and forming a shock wave. It is called a bow shock because it is shaped like waves that pile up ahead of the bow of a moving ship. The bow shock results because the solar wind is supersonic, moving faster than sound and other waves that might propagate through the wind. The motion of the solar wind around the magnetosphere has therefore been compared to the flow of air around a supersonic aircraft.

After passing through the bow shock, the solar wind encounters and flows around the magnetopause, the boundary between the solar wind and the magnetosphere. The magnetic field carried in the solar wind merges with that of the planet, and stretches it out into a long magnetotail on the night side of Earth. The magnetic field points roughly toward the Earth in the northern half of the tail and away in the southern. The field strength drops to nearly zero at the center of the tail where the opposite magnetic orientations lie next to each other and currents can flow (Fig. 3.12).

Thus, the Earth's magnetosphere is not precisely spherical. It has a bow shock facing the Sun and a long magnetotail in the opposite direction. The term magnetosphere therefore does not refer to form or shape, but instead implies a sphere of influence.

Trapped particles

The Earth's protective magnetic cocoon is not perfect. Energetic charged particles flowing from the Sun can penetrate the magnetic defense and become trapped within the magnetosphere. This was realized in 1958 when James A. Van Allen (1914–) and his students used instruments aboard the *Explorer 1* and *3* satellites to unexpectedly discover a large flux of high-energy electrons and protons that girdle the Earth far above the atmosphere, moving within two belts that encircle the Earth's magnetic equator but do not touch it. They resemble a gigantic, invisible, torus-shaped doughnut. This was the first major discovery of the Space Age.

These regions are sometimes called the inner and outer Van Allen radiation belts. Van Allen used the term "radiation belt" because the charged particles were then known as corpuscular radiation; the nomenclature does not imply either electromagnetic radiation or radioactivity. The radiation belts lie within the inner magnetosphere at distances of 1.5 and 4.5 Earth radii from the center of the Earth, creating a veritable shooting gallery of high-speed electrons and protons in nearby space (see Fig. 3.12).

In 1907, about half a century before the discovery of the radiation belts, the Norwegian geophysicist Carl Størmer (1874–1950) showed how electrons and protons can be almost permanently confined and suspended in space by the Earth's dipolar magnetic field. An energetic charged particle moves around the magnetic fields in a spiral path that becomes more tightly coiled in the stronger magnetic fields close to a magnetic pole. The intense polar fields act like a magnetic mirror, turning the particle around so it moves back toward the other pole.

Thus, the electrons and protons bounce back and forth between the north and south magnetic poles (Fig. 3.13). It takes about one minute for an energetic electron to make one trip between the two polar mirror points. The spiraling electrons also drift eastward, completing one trip around the Earth in about half an hour. There is a similar drift for protons, but in the westward direction. The bouncing can continue indefinitely for particles trapped in the Earth's radiation belts, until the particles collide with

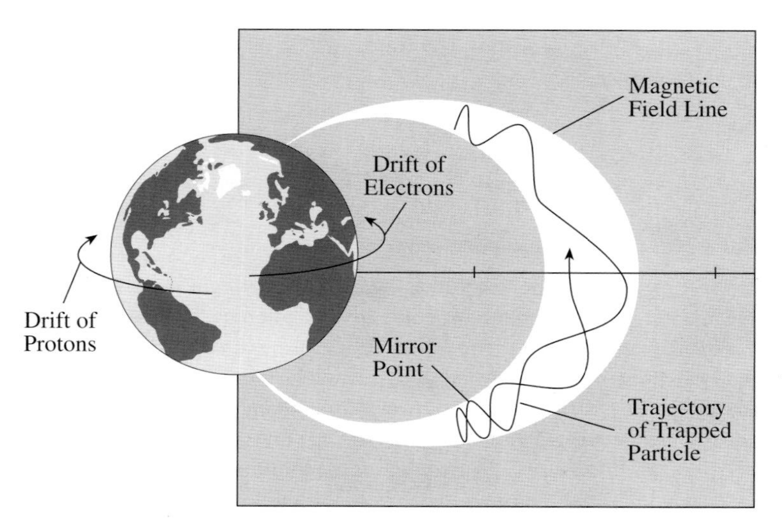

Fig. 3.13 Magnetic trap Charged particles can be trapped by Earth's magnetic field. They bounce back and forth between polar mirror points in either hemisphere at intervals of seconds to minutes, and they also drift around the planet on time-scales of hours. As shown by Carl Størmer in 1907, with the trajectories shown here, the motion is turned around by the stronger magnetic fields near the Earth's magnetic poles. Because of their positive and negative charge, the protons and electrons drift in opposite directions.

each other or some external force distorts the magnetic fields.

The problem at the time Størmer developed his theory was that there was no mechanism known to allow electrically charged particles into the dipolar magnetic field. After all, if electrons and protons cannot leave the magnetic cage, how could they get into it in the first place? Nevertheless, instruments aboard spacecraft have shown that energetic charged particles have entered the trap.

The high energy of the trapped particles may provide a clue to their origin. Protons within the radiation belts are up to ten thousand times more energetic than the usual energy of protons in the solar wind, so unusual solar activity could be responsible. Powerful explosions on the Sun, known as solar flares and coronal mass ejections, can accelerate charged particles to exceptionally high energies and produce violent gusts in the solar wind that can affect Earth (Section 4.5).

Once they arrive at the Earth's magnetosphere, electrons and protons from the Sun are deflected around the magnetopause, but it is not a perfect shield. The electrically charged particles can leak in across this boundary, or they can plunge down toward the Earth in the magnetically open polar cusps, like water poured into a funnel. However, the immense magnetic tail forms the bulk of the magnetosphere, and it provides the main location for breaking into the Earth's magnetic domain.

The magnetic fields of the solar wind and the Earth's magnetotail are sometimes pointing in opposite directions, and when this happens the two fields become linked, just as the opposite poles of two toy magnets stick together. When the solar and terrestrial magnetic fields touch each other, the magnetotail can be punctured, providing a back-door entry that funnels energy and particles into the magnetosphere (Fig. 3.14). Once inside the magnetic trap, the charged particles can be additionally accelerated to higher energies.

The magnetotail snaps like a rubber band that has been stretched too far. The snap catapults the outer part of the tail away from the Earth and propels the inner part back toward it. The solar wind is then plugged into the Earth's electrical socket, and our planet becomes wired to the Sun. Many of the electrons and protons in the Van Allen radiation belts, particularly the outer one, might originate in this way.

Particles within the inner magnetosphere close to the Earth can come from the upper terrestrial atmosphere, below the radiation belts. Solar ultraviolet and X-ray radiation create the ionosphere in the upper atmosphere, which can vary dramatically with solar activity (Section 4.4). Although most of the ionosphere is gravitationally bound to the Earth, some if its particles have sufficient energy to escape. The Earth's magnetic field traps oxygen ions, protons and electrons derived from the ionosphere, creating a plasmasphere that rotates with the Earth's magnetic field and extends outward to envelop the radiation belts. Further out, the magnetosphere is dominated by its interaction with the solar wind.

Energetic charged particles coming from interstellar space, known as cosmic rays, may also play a role in feeding the radiation belts, supplying the inner one with its high-energy protons. When cosmic rays bombard the Earth's atmosphere, which lies below the radiation belts, they collide with atoms in our air and eject neutrons from the atomic nuclei. These neutrons travel in all directions, unimpeded by magnetic fields since they have no electrical charge.

Once it is liberated from an atomic nucleus, a neutron cannot stand being left alone. A free neutron lasts only 10.25 minutes on average before it decays into an electron and proton. A small fraction of the neutrons produced in our atmosphere move out into the inner radiation belt before they disintegrate, producing electrons and protons in places they could not otherwise have reached. These electrically charged particles are immediately snared by the magnetic fields and remain stored within them, accumulating in substantial numbers over time. This mechanism might account for the energetic protons in the inner radiation belt.

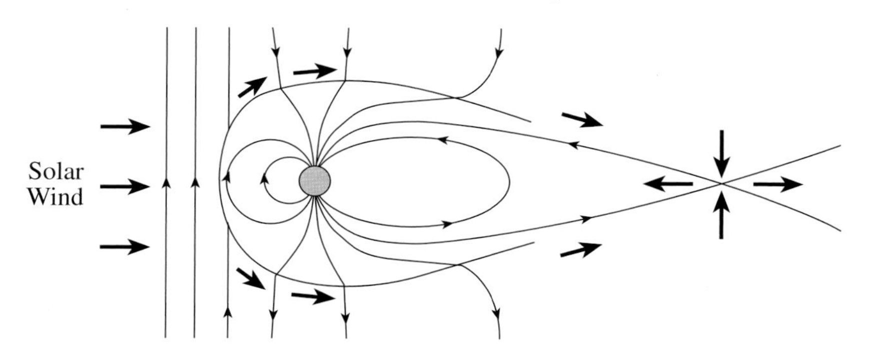

Fig. 3.14 Magnetic reconnection on the back-side The Sun's wind brings solar and terrestrial magnetic fields together on the night side of Earth's magnetosphere, in its magnetotail. Magnetic fields that point in opposite directions (*thin arrows*), or roughly toward and away from the Earth, are brought together and merge, reconnecting and pinching off the magnetotail. Electrically charged material is accelerated away from this disturbance (*thick arrows*), both away from the Earth and back toward it.

Table 3.7 Planetary magnetic fields[a]

Planet	Magnetic Dipole Moment[b] (Earth = 1)	Magnetic Field Strength at the Equator, B_0 (10^{-4} T)	Tilt[c] of Magnetic Axis (degrees)	Offset From Planet Center (R_P)	Bow-Shock Distance, R_{MP} (R_P)	Planet Equatorial Radius, R_P (10^6 m)
Mercury	0.0007	0.0033	+14°	0.05 R_M	1.5 R_M	$R_M = 2.439$
Earth	1	0.305	+11.7°	0.07 R_E	10 R_E	$R_E = 6.378$
Jupiter	20 000	4.28	−9.6°	0.14 R_J	42 R_J	$R_J = 71.492$
Saturn	600	0.22	< 1.0°	0.04 R_S	19 R_S	$R_S = 60.268$
Uranus	50	0.23	−58.6°	0.3 R_U	25 R_U	$R_U = 25.559$
Neptune	25	0.14	−46.8	0.55 R_N	24 R_N	$R_N = 24.766$

[a] The magnetic field strengths are given at the surface of Mercury and the Earth and at the cloud tops for the giant planets. Venus and Mars have no detected global, dipolar magnetic field, with respective upper limits of 2×10^{-9} and 10^{-8} tesla.
[b] The magnetic dipole moment, ($= B_0 R_P^3$) is given in units of the Earth's magnetic dipole moment of 7.91×10^{15} T m^3. Here we use the SI unit for magnetic field strength, the tesla, abbreviated T. The c.g.s. unit of magnetic field strength, the gauss or G for short, can be computed from one tesla = 1 T = 10^4 gauss = 10^4 G. The nanotesla or nT, is also used, with 1 nT = 10^{-9} T, and the nanotesla has historically also been called the gamma. A dipole moment of 1 T m^3 = 10^{10} G cm^3, where m and cm respectively denote meter and centimeter. The equivalent unit of 1 G cm^3 = 10^{-3} A m^2 is also used, where the current is in units of amperes, abbreviated A.
[c] The tilt is the angle between the magnetic axis and the rotation axis.

Planets with magnetospheres

Magnetic fields are ubiquitous in the solar system. Earth, Mercury and all the giant planets have strong magnetic fields generated within the planet, and Jupiter and Saturn have extensive magnetospheres. A magnetic field has been found on Mars, but the field's patchy nature suggests that it is not the result of an active internal dynamo. Instead, the magnetic field on Mars is probably a remnant of former times, frozen into expanses of solidifying lava (Section 8.6). Venus is the only major planet to have no detectable magnetic field. A magnetic field has even been found on at least one satellite, Jupiter's Ganymede (Section 9.4).

The magnetic fields surrounding Jupiter and Saturn dwarf Earth's magnetic environment. The strong magnetic fields on these giant planets produce large magnetospheres, and in addition the weak solar-wind pressures present in the outer solar system allow a given magnetic field to carve out a larger cavity. The magnetospheres of Jupiter and Saturn are dominated by planetary rotation, satellites are a major source of their charged particles, and internal forces shape those particles into an equatorial disk. Unlike Jupiter and Saturn, the magnetospheres of Uranus and Neptune are largely empty, and their magnetic fields are unexpectedly tilted.

As with the Earth, the magnetic fields near most of the other planets can be described by a magnetic dipole, and the best characterization of the planets intrinsic magnetism is the magnetic dipole moment. The dipole moment divided by the cube of the planet's radius yields the average strength of the magnetic field along the magnetic equator. Jupiter's magnetic moment is 20 000 times that of Earth. Saturn's magnetic moment is 600 times larger than the Earth's, but still 33 times weaker than Jupiter's. The magnetic dipole moments, equatorial magnetic field strengths and planetary radii are given in Table 3.7 for the six planets with known dipolar magnetic fields.

On the side facing the Sun, each of the planetary magnetic fields is compressed by the solar wind, forming a bow shock. It is located at the place where the pressure of the planet's magnetic field just equals the pressure of the solar wind (Focus 3.3). Such standoff distances for the six planets with detected magnetic fields are given in Table 3.7, but the varying solar-wind pressure can alter this distance by a factor of two.

The general form of the Jovian magnetosphere resembles that of the Earth (Fig. 3.15). but its dimensions are at least 1200 times greater. It is larger than the Sun in size, with a bow shock at more than 3 billion meters or at least 42 times the planet's radius. *Pioneer 10* and *Voyager 1* first encountered the Jovian bow shock at 95 and 86 Jupiter-radii, but the shock moved in and out due to the variable solar wind. Jupiter's largest satellites are all embedded within its magnetosphere and interact with it (Fig. 3.16).

Focus 3.3 Planetary magnetospheres

Six planets are known to have magnetospheres. The size of the magnetosphere, on the day side facing the Sun, is determined by the bow-shock distance, $R_{\rm MP}$, along the planet–Sun line at which the pressure of the planetary magnetic field balances the dynamic ram pressure of the solar wind. The magnetic pressure at the surface of the planet is given by $B_0^2/(2\mu_0)$, where $\mu_0 = 4\pi \times 10^{-7}$ is the permeability of free space. Since the dipole's magnetic field strength falls off as the cube of the distance from the planet, the magnetic pressure decreases as the sixth power of that distance. This means that the standoff point where the two pressures are equal occurs when:

$$\text{magnetic pressure} = \frac{R_{\rm P}^6 B_0^2}{2\mu_0 R_{\rm MP}^6} = m_{\rm p} N V^2 = \text{solar-wind pressure},$$

where the planet's radius is $R_{\rm P}$, the proton mass $m_{\rm p} = 1.67 \times 10^{-27}$ kilograms, N is the number density of the protons in the solar wind at the planet's distance from the Sun, and V is the solar-wind velocity at that distance. Solving for $R_{\rm MP}$ we have

$$R_{\rm MP} = \left(\frac{B_0^2}{2\mu_0 m_{\rm p} N V^2}\right)^{1/6} R_{\rm P}.$$

At the Earth's distance from the Sun, the number density of the solar wind is about $N = 5$ million protons per cubic meter and the solar-wind velocity is about $V = 400$ kilometers per second. The equatorial magnetic field strength of the Earth is $B_0 = 0.305 \times 10^{-4}$ tesla. With these numbers our equation gives $R_{\rm ME} = 10\,R_{\rm E}$, so the bow shock of the Earth is out at about ten times the Earth's radius. The bow-shock distance can be reduced to half this value when a powerful mass ejection from the Sun hits the Earth, producing extra ram pressure. Moreover, unusual drops in the solar-wind pressure have very occasionally inflated the leading edge of the magnetosphere five or six times further out in space, until it engulfed the Moon.

The values of $R_{\rm MP}$ for the other planets can be inferred by noting that the solar-wind number density, N, falls off with the inverse cube of the distance of the planet from the Sun, while the solar-wind velocity remains relatively constant. Values of $R_{\rm MP}$ are given in Table 3.7 for the six planets with detected dipolar magnetic fields.

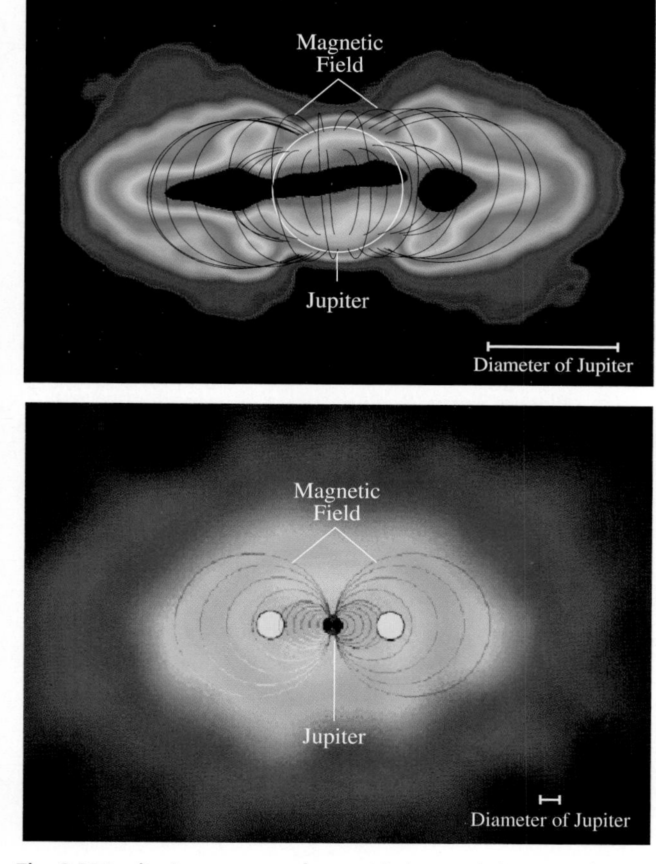

Fig. 3.15 Jupiter's magnetosphere High-speed electrons that are trapped in the Jovian magnetosphere emit steady radio radiation (*top*) by the synchrotron process; it is detected by ground-based radio telescopes such as the Very Large Array. An instrument aboard *Cassini* measures energetic atoms (*bottom*) created when fast-moving ions within the magnetosphere pick up electrons to become neutral atoms. The two open circles denote Io's orbital position on each side of the planet; Jupiter is denoted by the central black disk. (Courtesy of Imke de Pater, University of California at Berkeley (*top*) and NASA, JPL, and the Johns Hopkins University Applied Physics Laboratory (*bottom*).)

Jupiter's enormous magnetotail, driven outward by the solar wind, is almost a million million, or a trillion, meters long. It spans the distance between the orbits of Jupiter and Saturn, which is as great as the distance from the Sun to Jupiter itself. In contrast, the Earth's magnetotail barely flicks across our Moon's path, less than a half-billion meters from the Earth. Thus, Jupiter's magnetosphere is the largest enduring structure in the solar system, although it is occasionally exceeded in size by comet tails.

Jupiter's magnetic field was first recognized in 1954–5 by Earth-based observations of the planet's intense radio emission, and then directly measured by visiting spacecraft. High-speed electrons trapped within the planet's magnetic field generate the radio signals (Focus 3.4). The synchrotron radio emission is beamed in a direction nearly parallel to the magnetic equator (see Fig. 3.15), and it is therefore swept past an Earth-based observer as the planet

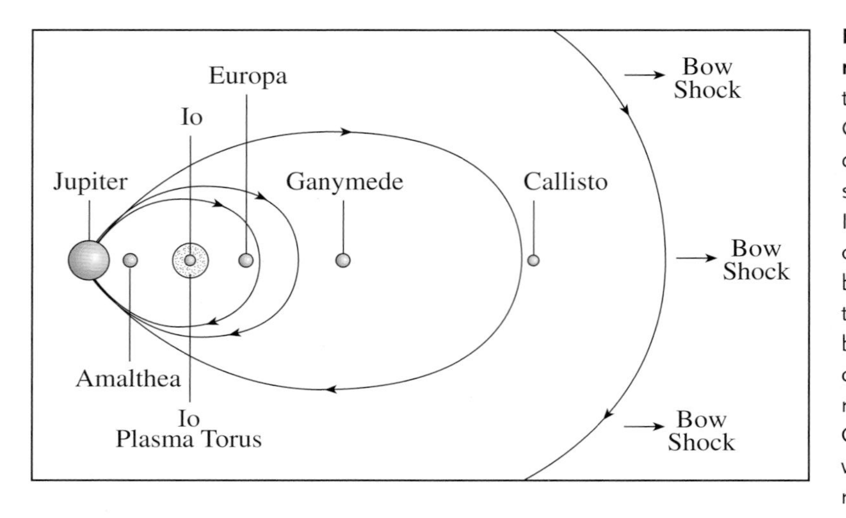

Fig. 3.16 Satellites within Jupiter's magnetic field This cross-section shows that the four Galilean satellites, Io, Europa, Ganymede and Callisto, are all embedded within Jupiter's magnetosphere. Small satellites orbit Jupiter within the orbit of Io, the innermost Galilean satellite. The outermost Galilean satellite, Callisto, orbits Jupiter near its bow shock. All of these satellites are being continuously bombarded with energetic charged particles that are trapped within Jupiter's magnetosphere. The distance from Jupiter to Callisto is 1.88 billion (1.88×10^9) meters, while the radius of the Sun is 0.7 billion meters, so Jupiter's bow shock is bigger than the Sun.

rotates and brings the magnetic equator in and out of alignment with the line of sight. Periodic variations in the strength of the radio emission indicate that the magnetic field rotates with a period of precisely 9 hours 55 minutes 29.7 seconds, or 9.9249 hours. Since the magnetic fields are generated deep within the planet, this is assumed to be Jupiter's rotation period; it differs from the rotation speed inferred from visible clouds that are blown in different directions and at various speeds by powerful winds (Section 3.2).

Focus 3.4 Radio broadcasts from Jupiter

The discovery of intense radio emission from Jupiter is one of the many examples of the accidental discovery of an unexpected phenomenon while looking for something else. In 1954–5 two radio astronomers, Bernard F. Burke and Kenneth L. Franklin, were using ground-based telescopes to observe the meter-wavelength radio emission of the Crab Nebula, a famous remnant of a stellar explosion in 1054 AD. They planned to study the changes in the Crab's radio signal, at a wavelength of 13.6 meters, as it passed behind the Sun, thereby determining properties of the outer solar atmosphere.

Their observations were hampered by radio bursts that resembled terrestrial interference, but the alert scientists noticed that they only appeared when the telescope was pointing in a certain direction in the sky. This meant that the radio bursts had an extraterrestrial origin. They were at first assumed to be coming from the Sun, an intense source of variable radio radiation, but the calculated position in the sky coincided with Jupiter.

Soon thereafter, a steady Jovian radio signal was found at shorter wavelengths of several centimeters. In 1959 George B. Field, and Frank Drake and Hein Hvatum, interpreted this emission as synchrotron radiation emitted by electrons trapped in the Jovian magnetic field and moving at relativistic speeds approaching the velocity of light. The electrons spiral about the magnetic field, emitting the radio radiation, named after the synchrotron particle accelerator on Earth where similar radiation was first observed at optical wavelengths detected by the human eye. This type of synchrotron radiation is not detected from any other planet.

The new theory for Jupiter's radio signals was confirmed by the observation of two characteristic signatures of synchrotron radiation. The radio broadcasts were weaker at shorter wavelengths, and stronger at the longer ones, unlike the thermal emission of a hot gas that is most intense at shorter wavelengths. The non-thermal radio radiation was also polarized, with a preferred orientation or direction, which ought to coincide with that of the magnetic fields. In addition, radio interferometer measurements indicated that the radio emission was much larger than the planet and roughly aligned with its equator (see Fig. 3.15).

As Jupiter rotates, it carries the magnetic fields and their trapped electrons with it. Since the radio signal of the electrons is beamed in a particular direction, it sweeps past the observer once every rotation. The varying radio noise has been used to define Jupiter's rotation period with a clock-like precision of 9 hours 55 minutes 29.7 seconds, or 9.9249 hours. This period is close to, but slightly different from, those of the Great Red Spot and other atmospheric features; the difference is used to measure the speed of Jupiter's powerful winds (Section 3.2).

Jupiter's strong, dipolar magnetic field and energetic trapped electrons were confirmed by *Pioneer 10* and *11*, which respectively passed within 2.8 and 1.7 Jupiter-radii from the planet's center in December 1973 and December 1974. *Voyager 1* and *2* subsequently approached within 5 and 10 Jupiter-radii, in March and July 1979 respectively, yielding information about the outer magnetosphere, as well as the magnetic interaction between Jupiter and Io (Section 9.4).

The *Pioneer* data showed that the magnetic field at the cloud tops is 4.28×10^{-4} tesla, or about 14 times stronger than the Earth's equatorial magnetic field strength. As is the case near the Earth's surface, the magnetic fields at Jupiter's cloud tops are strongest at the poles and weakest along the equator. The magnetic field can be described as a dipole with a magnetic axis that is titled at 9.6 degrees with respect to the rotation axis, similar to the Earth's tilt of 11.7 degrees; but the magnetic poles are reversed in comparison to those of the Earth, so a north-seeking terrestrial compass would point south in the vicinity of Jupiter. *Voyager* observations showed that the magnetic field of Jupiter extends to the orbit of Saturn and beyond.

Like its terrestrial counterpart, Jupiter's magnetosphere contains electrons and protons that are supplied from outside by the variable solar wind. Unlike the Earth, the giant planet's magnetosphere is also fed from within, by ions erupted from the active volcanoes on its innermost large satellite, Io. The dominant ions are sulfur and oxygen, both products of Io's unique volcanic activity (Sections 2.3, 9.4). Jupiter's belts of charged particles resemble the terrestrial Van Allen belts in shape, but the Jovian belts are up to a million times more densely filled with particles than those near the Earth are.

Ions and electrons within Jupiter's inner magnetosphere are accelerated by the spinning magnetic field of the planet, eventually reaching very high energies. The inner magnetosphere is a stiff, permanent structure that is tied to the planet and rotates with it. Once an electrically charged particle enters this region, the magnetic field picks the particle up and takes it for a long ride around the planet. As the powerful magnetic field spins, it extracts rotational energy from the planet, lashing and accelerating the charged particles to nearly the speed of light.

Thus, the power for populating and maintaining the magnetosphere of Jupiter comes principally from the rotational energy of the planet, as well as the tidal flexing of Io that results in its volcanoes (Sections 2.3, 9.4). In contrast, the power source of Earth's magnetosphere is principally the solar wind. The numerous high-energy particles in Jupiter's magnetosphere are capable of destroying sensitive electronic circuits in spacecraft that pass near the planet; they would also endanger humans should they ever venture close to it.

The high-speed charged particles exert an outward pressure on Jupiter's magnetic field, inflating it like an air-filled balloon. Because the field is weakest in its equatorial plane, the forces and pressures associated with the rapid rotation stretch the equatorial regions outwards in the form of a thin, elongated disk.

The varying solar-wind pressure buffets Jupiter's outer magnetosphere, so it expands and contracts. As varying solar activity pushed the magnetosphere in and out, approaching *Pioneer 10* and *11* and *Voyager 1* and *2* spacecraft crossed the bow shock several times. The changing shape and location of the magnetotail similarly caused the outward-bound spacecraft to cross it many times.

When gusts in the solar wind compress Jupiter's outer magnetosphere and push it back, some of its high-speed, electrically charged particles squirt out into interplanetary space with energies that exceed the typical energy of electrons and protons in the solar wind. The Jovian particles are continually replenished by acceleration within the planet's magnetosphere, spraying energetic electrons and protons throughout the solar system. Some of them reach the orbit of the Earth and even that of Mercury.

The discovery of Saturn's magnetosphere did not occur until September 1979 when *Pioneer 11* first crossed its bow shock, at 24 Saturn radii, closely followed by the *Voyager 1* and *2* encounters in November 1980 and August 1981, respectively. The magnetic field strength at the cloud tops is 70 percent of that at the Earth's equator, but spread over a much bigger volume. Saturn's dipolar magnetic field is almost precisely aligned with its poles of rotation. Like Jupiter, the magnetic poles of Saturn are reversed compared to those of the Earth. High-speed electrons in the spinning magnetic field give rise to periodic radio emission, and the inferred rotation period is 10 hours 39 minutes 22.3 seconds, or 10.6562 hours and just 44 minutes longer than Jupiter's rotation period.

Saturn's sizeable satellites and its rings control the population of trapped particles, absorbing energetic electrons and protons. Perhaps as a result, its magnetosphere does not contain a high density of high-energy electrons. The planet also does not have a volcanically active satellite to generate sulfur and oxygen ions. Its magnetic trap is instead permeated with low-energy ionized material, protons and oxygen ions, chipped or sputtered off of the water ice in the planet's rings and on its satellite surfaces. A vast, dense cloud of neutral, or un-ionized, hydroxyl, OH, molecules envelop the rings; it is also derived from the water, H_2O, ice.

The magnetic axes of Uranus and Neptune are tilted at a large angle with respect to their rotational

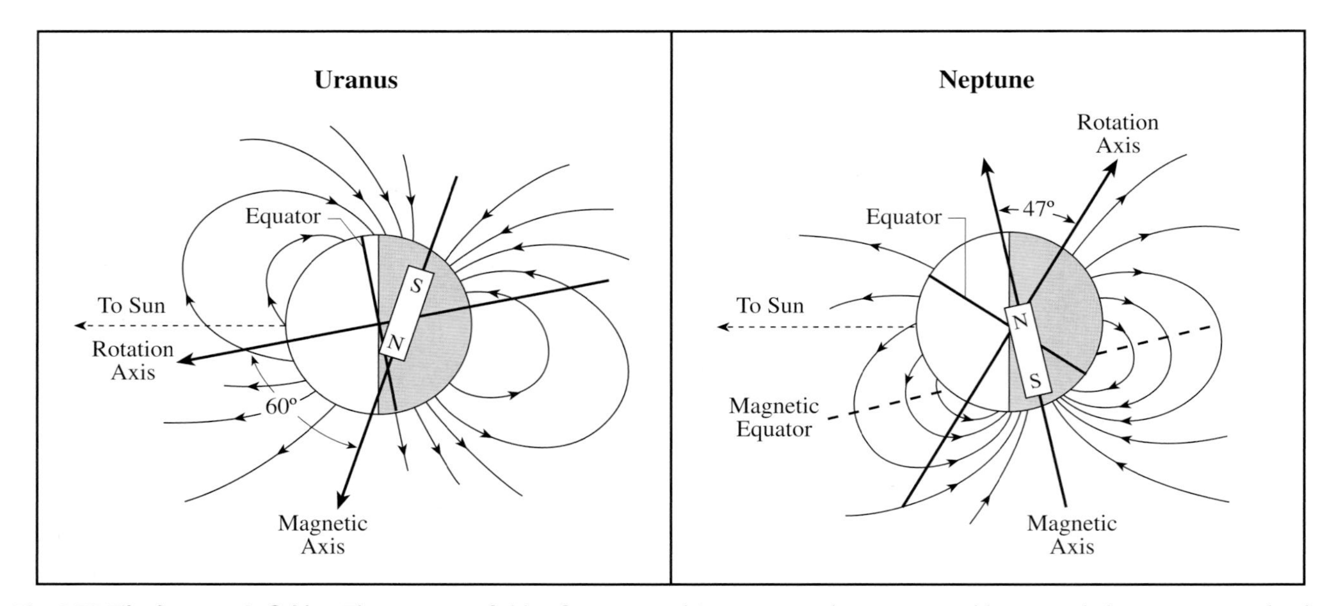

Fig. 3.17 Tilted magnetic fields The magnetic fields of Uranus and Neptune can be represented by a simple bar magnet, or dipole, embedded in the planet, but with a magnetic axis that is tilted with respect to the rotation axis. For Uranus the tilt is about 60 degrees; Neptune has a tilt of 47 degrees. In contrast, the magnetic axes of Jupiter, Saturn and Earth are much more nearly aligned with their rotation axes. The arrow of the rotation axis points from the geographic south towards geographic north, and the magnetic axis similarly points from magnetic south to magnetic north. On Uranus and Neptune a terrestrial compass would point toward the southern hemisphere of the planet, while on Earth it points toward the North (Geographic) Pole. In addition to the dipole part of their magnetic field, Uranus and Neptune have a large additional component known as the quadrupole part. A method of visualizing this is to imagine that the dipole has a magnetic center that is offset radially from the center of the planet. As shown here, the equivalent offset for Uranus is almost a third of the planet's radius, and there is a larger offset for Neptune of nearly half its radius. But such off-center dipoles are only useful as a picture of what the external field looks like and do not help in understanding how it is produced deep down in the planet.

axis, by 58.6 degrees for Uranus and 46.8 degrees for Neptune (Fig. 3.17), and these planets have fully developed magnetospheres. The rotation periods of the magnetic fields of Uranus and Neptune, inferred from their periodic radio emission, are 17.24 and 16.11 hours, respectively.

All planetary magnetic fields are generated by the dynamo action of moving electrically conducting material in their interior. Internal rotation and convection produce the motions, somewhat like a spinning and boiling pot of water. Vigorous internal convection is powered by the decay of radioactive elements in the Earth and Mercury; the giant planets have retained much of their primordial heat to drive the internal convection and power their dynamo. The combination of convection and rotation concentrates the magnetism, amplifying its strength and regenerating the magnetic fields.

Mercury and the Earth have cores of molten iron alloys. At the high pressures inside Jupiter and Saturn, their most abundant ingredient, hydrogen, behaves like a liquid metal (Sections 9.3, 10.3). Their strong magnetic fields are attributed to currents of electricity driven by the fast rotation of their liquid metallic interior. For Uranus and Neptune, water-rich material within their vast internal oceans most likely provides the electrically conducting medium.

3.7 Auroras

Curtains of green and red light dance and shimmer across the night sky in the Earth's polar regions, far above the highest clouds (Fig. 3.18). This light is called the aurora after the Roman goddess of the rosy-fingered dawn, a designation that has been traced back to Galileo Galilei (1564–1643). The auroras seen near the north and south poles have been given the Latin names *aurora borealis,* for northern lights, and *aurora australis,* for southern lights.

Most people never see the awesome lights, for auroras are normally confined to high latitudes in the north or south polar regions. But this does not mean that the auroras occur infrequently. Residents in far northern locations can see a faint aurora every clear and dark night.

Fig. 3.18 Aurora borealis Spectacular curtains of multi-colored light are found in these photographs of the fluorescent northern lights, or *aurora borealis*, taken by Forrest Baldwin in Alaska. (Courtesy of Kathi and Forrest Baldwin, Palmer, Alaska).

The northern lights have been documented for centuries. Ancient Vikings (500–1500 AD) thought they were the spirits of fallen warriors being carried to Valhalla, the home of the gods. The southern lights have never achieved a renown comparable to the northern lights, probably because the southern ones are not usually located over inhabited land and are instead seen from oceans that are infrequently traveled.

Rare, brilliant auroras can extend down to the Earth's equator, becoming visible as far south as Athens, Rome or Mexico City. They were noted by the ancient Greeks, Plutarch in 467 BC and by Aristotle in 349 BC, but auroras do not extend down to Greece very often, perhaps every 50 or 100 years.

Since auroras become more frequent as one travels north from tropical latitudes, it was thought that they would become brighter and occur most frequently at the highest northern latitudes. Arctic explorers were therefore surprised to see that the intensity and frequency of auroras did not increase all the way to the poles and instead peaked in an oval-shaped band that encircles the North Pole. This northern aurora oval has a radius of about 2.25 thousand kilometers and is centered on the Earth's north magnetic pole, with an inner and outer radius separated by about 500 kilometers.

Nowadays we can use spacecraft to view both the northern and southern lights from space (Figs. 3.19, 3.20). The *Space Shuttle* has even flown right through the northern lights. While inside the display, astronauts could close their eyes and see flashes of light caused by the charged aurora particles speeding through their eyeballs.

When viewed from above, the auroras form a luminous oval centered at each magnetic pole, resembling a fiery halo (Fig. 3.20). The aurora oval is constantly in motion, expanding toward the equator or contacting toward the pole, and constantly changing in brightness. Such ever-changing aurora ovals are created simultaneously in both hemispheres and can be viewed at the same time from the Moon.

Fig. 3.19 Aurora australis The eerie, beautiful glow of auroras can be detected from space, as shown in this image of the aurora australis, or southern lights, taken from the *Space Shuttle Discovery*. The colored emission of atomic oxygen extends upward to between 200 and 300 kilometers above the Earth's surface. (Courtesy of NASA.)

Visual auroras normally occur at 100 to 250 kilometers above the ground. This height is much smaller than either the average radius of the oval, at 2.25 thousand kilometers, or the radius of the Earth, 6.38 thousand kilometers. An observer on the ground therefore sees only a small, changing piece of the aurora oval, which can resemble a bright, thin, windblown curtain hanging vertically down from the Arctic sky.

The general correlation between the Sun's magnetic activity cycle and the northern lights has been known for more than two centuries. The number of sunspots, and related activity on the Sun, varies from a maximum to a minimum and back to a maximum in about 11 years. When the number of sunspots is large, bright auroras occur more frequently, and when there are few spots on the Sun the intense auroras occur less often. The increase in magnetic activity on the Sun, which causes the observed increase in sunspots, is somehow related to the greater number of exceptionally bright auroras.

The auroras are themselves caused by energetic electrons bombarding the upper atmosphere. The reason that auroras are usually located near the polar regions is that the Earth's magnetic fields guide the energetic electrons there. Most of these electrons come from the dark side of the Earth, in the magnetotail, and follow the magnetic fields back toward the Earth, up to the poles, and down into the oval. Electrical currents as great as a million amperes can be produced along the aurora oval, and the electric power generated during the discharge is truly awesome – about ten times the annual consumption of electricity in the United States.

When the electrons slam into the upper atmosphere, at speeds of about 50 kilometers per second, they collide with the oxygen and nitrogen atoms there and excite them to energy states unattainable in the denser air below. The pumped-up atoms quickly give up the energy they acquired from the electrons, emitting a burst of color in a process called fluorescence. It is something like electricity making the gas in a neon light shine or a fluorescent lamp glow. The process also resembles the beam of electrons that strikes the screen of your color television set, making it glow in different colors depending on the type of chemicals, or phosphors, that coat the screen.

The color of the aurora depends on which atoms or molecules are struck by the precipitating electrons, and the atmospheric height at which they are struck (Table 3.8). Low-altitude oxygen atoms produce green, a common aurora color. The high-altitude oxygen atoms account for rare, all-red auroras. Nitrogen molecules create low-altitude red light, below the oxygen's green, while nitrogen ions can produce violet-purple light at high altitudes. The green oxygen emission appears at about 100 kilometers and the red

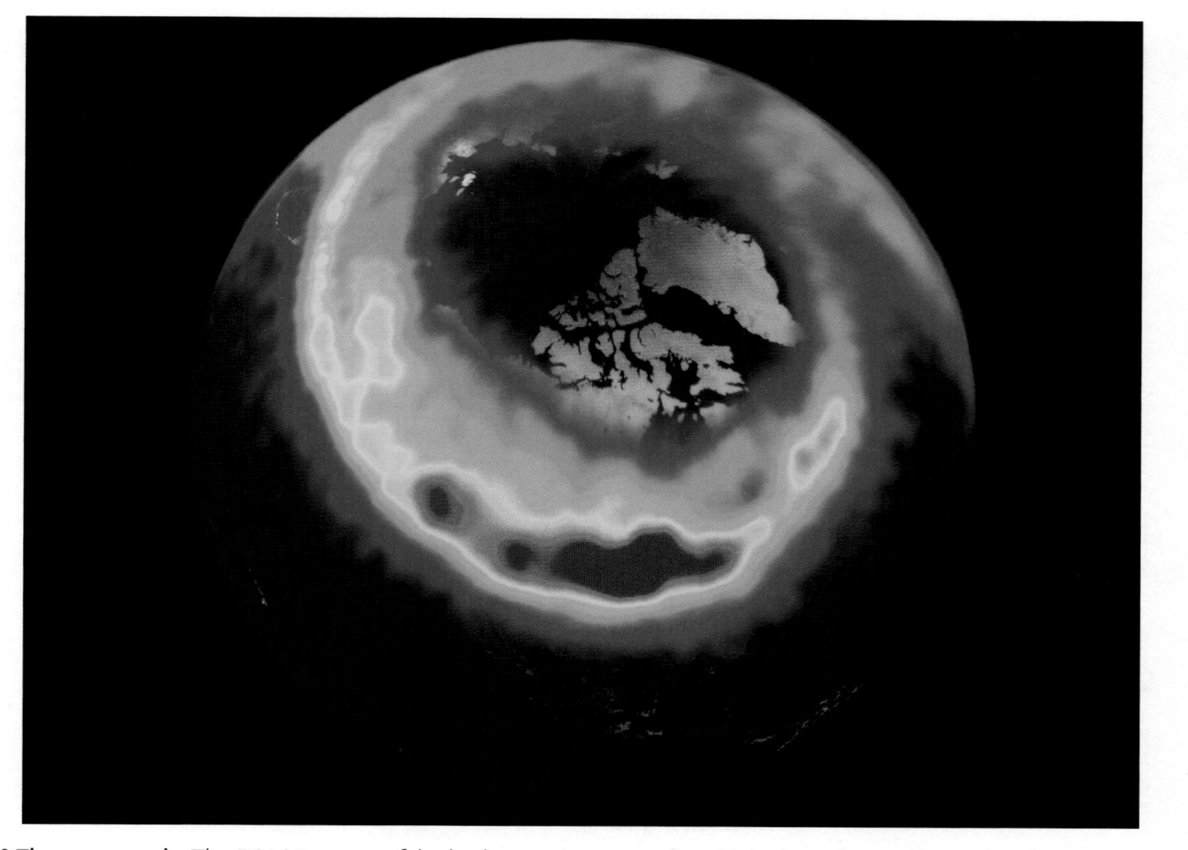

Fig. 3.20 The aurora oval The *POLAR* spacecraft looks down on an aurora from high above the Earth's north polar region in February 2000, showing the northern lights in their entirety. The glowing oval is 4.5 thousand kilometers across. The most intense aurora activity appears in bright red or yellow. The Earth's auroras are typically initiated by magnetic reconnection events in the Earth's magnetotail (Fig. 3.14), on the night side of the Earth. (Courtesy of the University of Iowa and NASA.)

oxygen light at 200 to 400 kilometers. At these heights, the auroras shine from the ionosphere, an electrically conducting layer in the Earth's upper atmosphere (Section 4.4).

So, energetic electrons colliding with oxygen and nitrogen atoms in the air cause the multi-colored aurora light show, but where do the electrons come from and how are they energized? Since the most intense auroras occur at times of maximum solar activity, it was once thought that the aurora electrons are energized during explosions on the Sun and hurled directly into the Earth's atmosphere from

Table 3.8 Frequent spectral features in the aurora emission

Wavelength (nm)	Emitting Atom, Ion or Molecule	Altitude (km)	Visual Color
391.4	N^+ (nitrogen ion)	1000	violet-purple
427.8	N^+ (nitrogen ion)	1000	violet-purple
557.7	O (oxygen atom)	90–150	green
630.0	O (oxygen atom)	>150	red
636.4	O (oxygen atom)	>150	red
661.1	N_2 (nitrogen molecule)	65–90	red
669.6	N_2 (nitrogen molecule)	65–90	red
676.8	N_2 (nitrogen molecule)	65–90	red
686.1	N_2 (nitrogen molecule)	65–90	red

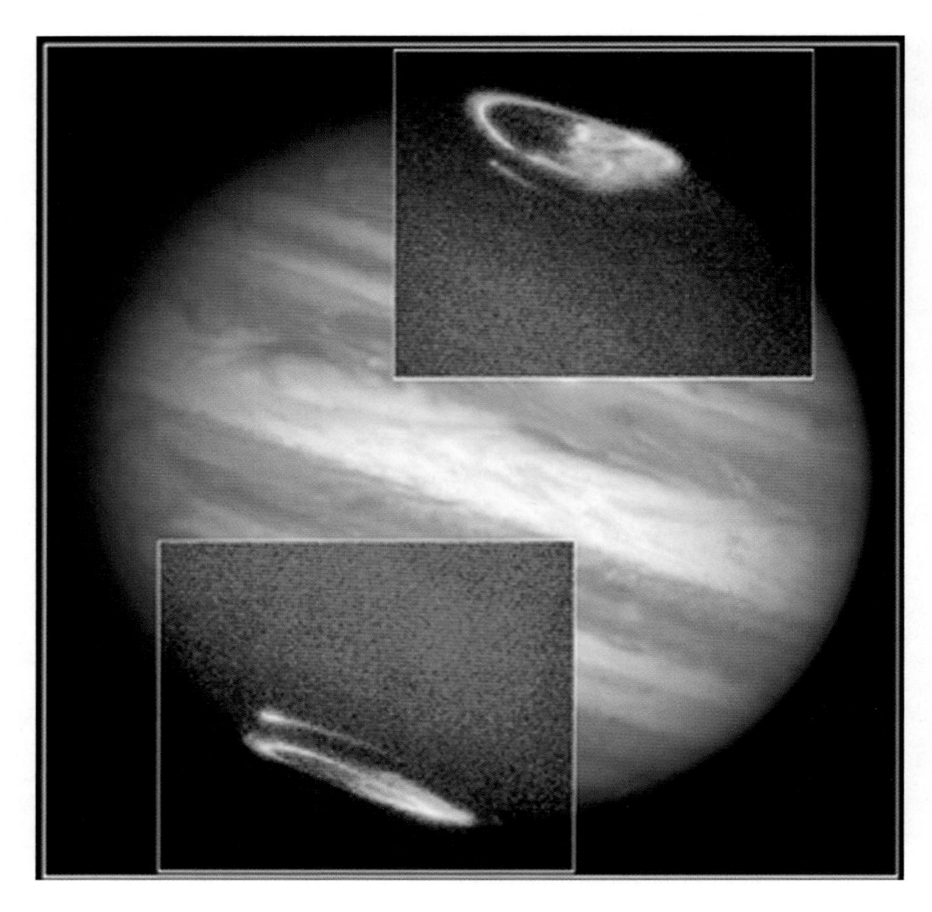

Fig. 3.21 Jupiter's auroras High-energy electrons and ions cascade into Jupiter's upper atmosphere and create bright auroras at ultraviolet wavelengths. This composite image, taken by the *Hubble Space Telescope* in September 1997, shows ovals in both the northern and southern regions. Elongated trails outside the ovals, starting from the far left below the ovals and moving to the right, are believed to mark the locations where powerful electrical currents from the volcanic moon Io enter the Jovian atmosphere. (Courtesy of John Clarke, University of Michigan, STSI, and NASA.)

the magnetic poles. After all, the intensity and frequency of the solar explosions also peak at the maximum of the solar magnetic activity cycle.

Even though changing conditions on the Sun may trigger the northern and southern lights, we now know that the electrons that cause the auroras arrive indirectly at the polar regions, from the Earth's magnetic tail, and that these electrons can be energized locally within the magnetosphere. Changing solar-wind conditions can temporarily pinch off the Earth's magnetotail, opening a valve that lets the solar-wind energy cross into the magnetosphere and additionally shoot energy stored in the magnetic tail back toward the aurora zones near the poles (see Fig. 3.14).

During this magnetic reconnection process, the magnetic fields heading in opposite direction – having opposite north and south polarities – break and reconnect at 140 to 160 thousand kilometers downwind of Earth on its night side. Electrons are pushed up and down the tail, and can be accelerated within the magnetosphere as they travel back toward the Earth and into its polar regions. The electrons that are thrown Earthward follow the path of magnetic field lines, which link the magnetotail to the polar regions and map into the aurora oval.

The rare, bright, auroras seen at low latitudes in more clement climates become visible during very intense geomagnetic storms that enlarge the magnetotail. The aurora ovals then intensify and spread down as far as the tropics in both hemispheres. Since these great magnetic storms are produced by solar explosions (Section 4.5), it is really the Sun that controls the intensity of the brightest, most extensive auroras, like the dimming switch of a cosmic light.

Brilliant curtains of aurora light are detected in Jupiter's upper atmosphere, just above the clouds. Like their terrestrial counterparts, the curtains of light on Jupiter are found in two oval-shaped regions circling the magnetic poles of the planet, just above the clouds (Fig. 3.21). The aurora glows are produced in these high-latitude regions because that is where the magnetic field directs electrically charged particles, electrons, protons and other ions. When these particles hit the planet's upper atmosphere, they collide with atoms and molecules there, leaving them in an excited state. As on Earth, the atoms and molecules release the extra energy in the form of light, and return to their normal state. But unlike Earth's colored light show, Jupiter's aurora ovals were first observed from space at ultraviolet wavelengths (Fig. 3.21); that is where most of the atoms and molecules radiate the most intense light.

Jupiter's aurora is the most powerful in the solar system. At about 10^{14} watts, it is typically one thousand times

Fig. 3.22 Auroras on Saturn High-energy electrons and ions are captured from the solar wind and funneled down into Saturn's upper atmosphere, creating aurora ovals at its northern (*upper left*) and southern (*lower right*) magnetic poles. This ultraviolet image was recorded by the *Hubble Space Telescope* in October 1997. The bright red aurora features are dominated by emission from atomic hydrogen, while the white regions within them are emitted by molecular hydrogen. (Courtesy of John T. Trauger, JPL, STSI, and NASA.)

more powerful than Earth's aurora. The Jovian lights are powered largely by energy extracted from planetary rotation, although there seems to be a contribution from the solar wind. Thus, internal processes seem to be the dominant source of power for Jupiter's aurora. This contrasts with Earth's aurora, which is mainly generated externally through the interaction of the solar wind and the terrestrial magnetosphere.

Jupiter's satellite Io affects the auroras on the planet. Electron and ions spewed out by volcanoes on Io are captured by the intense, rapidly rotating magnetic field and spiral inward at high energies toward the planet's polar regions. As the rotating magnetic field sweeps past Io, an invisible current of charged particles flows along Jupiter's magnetic field lines into the polar regions (Section 9.4), producing bright trails in the ultraviolet images (Fig. 3.21).

On Jupiter one can normally see a main ultraviolet oval, and in addition bright, swirling streaks are sometimes detected both within and outside the oval. They have been attributed to electric currents from three of the planet's large moons, Io, Ganymede and Europa. Very intense bursts of aurora activity have also been detected; they are apparently regulated by the variable solar wind, perhaps because it affects the size of Jupiter's magnetosphere.

Saturn's ultraviolet auroras (Fig. 3.22) are most likely caused when the gusty solar wind sweeps over the planet, perhaps like the Earth's aurora. But unlike the Earth, Saturn's aurora oval has only been seen from spacecraft in ultraviolet light, at least so far. It could not be detected from beneath the Earth's atmosphere which absorbs the ultraviolet.

In Chapter 4 we turn our attention to our home planet, Earth, third rock from the Sun.

Part 2 The inner system – rocky worlds

4 Third rock from the Sun – restless Earth

- Seismic waves generated by earthquakes have been used to look inside the Earth, determining its detailed internal structure.
- There is a crystalline globe of solid iron at the center of the Earth that spins faster than the rest of the planet. This inner solid core is suspended in a much larger, fluid outer core of molten iron, which is itself encased in a thick mantle of solid rock.
- The continents disperse and then reassemble, over and over again, roaming and wandering about the planet in an endless journey with no final destination.
- Sound waves and gravitational data have been used to effectively "empty" the oceans and see their floors, revealing an underwater range of active volcanoes that snakes its way along the middle of the ocean floor.
- The bottom of the oceans is continually renewed as new floor spills out of mid-ocean volcanoes and old floor is pushed back inside the Earth, but the water above the floors has remained for billions of years, shifting about the globe as oceans open and close.
- The outer part of the Earth is broken into a mosaic of large plates, like the cracked pieces of an egg shell; these plates move across the Earth at the rate of a few centimeters per year, or about as fast as your fingernails grow.
- Wheeling, churning motions deep inside the Earth's hot interior move continents sideways all across the planet; giant plumes of hot material can rise through the Earth's interior to lift entire continents up and down.
- The Earth's moving plates squeeze oceans out of existence, grind against each other to produce earthquakes, and dive into the Earth to produce volcanoes that make continents grow at their edges.
- Boston and Italy were once part of Africa, a glacier of ice once covered the Sahara Desert, and the Pacific Ocean once washed against the shores of Colorado.
- A colossal alp can erode away into a small, round knob of a hill in just a few hundred million years, while continents weld together to form new mountain ranges.
- The Earth's upper atmosphere is heated and ionized by the Sun's variable X-ray and extreme ultraviolet radiation.

- Ultraviolet radiation from the Sun creates the protective ozone layer in our atmosphere, and the amount of ozone it produces varies over the 11-year cycle of solar magnetic activity.
- Synthetic chemicals that have been destroying the ozone layer are now outlawed by international agreement.
- Invisible gases help to warm the Earth by trapping the Sun's heat and preventing some of it from being reflected back into space. This process is commonly known as the greenhouse effect.
- Warming of the Earth's surface and lower atmosphere by the greenhouse effect keeps the Earth from becoming a frozen ball of ice.
- Carbon dioxide and other heat-trapping gases, such as methane and nitrous oxide, have been increasing in the atmosphere for more than a century as wastes from industry and automobiles are added to the air.
- By burning coal and oil, humans have increased the amount of carbon dioxide in the Earth's atmosphere by 30 percent since the industrial revolution.
- Rising seas, retreating glaciers, melting ice caps, unseasonable warmth and cold, and unusually intense rains, snowstorms and floods are all recent signs of increased temperatures on the Earth.
- In the 1990s, our world became hotter than it has been for at least a thousand years. Most scientists attribute at least some of the recent rise in temperature to greater emissions of greenhouse gases by human activity.
- Supercomputer models forecast that the Earth will become noticeably hotter as the result of the unrestrained emission of heat-trapping gases by humans over the next 100 years. But uncertainties in the role of oceans, water vapor and clouds result in a wide range for the predicted temperature increase and possible climate change.
- The amount of the Sun's life-sustaining radiation that reaches the Earth, known as the solar constant, varies over the 11-year solar cycle of magnetic activity. Greater activity on the Sun makes our planet hotter, and lower solar activity results in a colder Earth, such as the Little Ice Age.
- Natural variations in the intensity of the Sun's radiation can explain many of the temperature fluctuations observed during past centuries, but global warming by heat-trapping gases, emitted by human activity, is required to explain the sharp rise in global temperatures during the 1990s.
- If current emissions of carbon dioxide and other greenhouse gases go unchecked over the next 100 years, global warming could produce agricultural disaster in the world's poorest countries, rising seas with coastal flooding throughout the world, and the spread of diseases carried by mosquitoes.
- The politicized debate over global warming is hampered by scientific uncertainties in predicting its future consequences, and by persons who selectively adopt the scientific forecasts that bolster their case.
- An international agreement to limit the human emission of heat-trapping gases, known as the Kyoto Protocol, is unlikely to be ratified by rich industrial nations because it is perceived as a threat to their economies and also because fast-developing poor countries are exempt from mandatory emission constraints.

- The ice ages, each lasting for about 100 thousand years, are caused by rhythmic changes in the Earth's distance from the Sun and in the wobble and tilt of its rotation axis, resulting in periodic variations in the amount and distribution of sunlight reaching the Earth.
- The Sun is slowly getting brighter as time goes on, and in 3 billion years it will boil the Earth's oceans away.
- High-speed protons and electrons are now being hurled out into space during explosions on the Sun, endangering space-walking astronauts and crippling some man-made satellites.
- Space-weather forecasters are now actively searching for methods to predict solar explosions, and one promising technique is to monitor the Sun for twisted magnetic fields.

4.1 Fundamentals

Table 4.1 Physical properties of the Earth

Mass	59.736×10^{23} kilograms
Mean Radius	6371 kilometers
Equatorial Radius	6378 kilometers
Mean Mass Density	5515 kilograms per cubic meter
Rotation Period	23 hours 56 minutes 04 seconds = 8.6164×10^{4} seconds
Orbital Period	1 year = 365.24 days = 3.1557×10^{7} seconds
Mean Distance from Sun	$1.495\,98 \times 10^{11}$ meters = 1.000 AU
Orbital Eccentricity	0.0167
Tilt of Rotational Axis, or Obliquity	23.27 degrees
Age	4.6×10^{9} years
Atmosphere	77 percent nitrogen, 21 percent oxygen
Surface Pressure	1.013 bar at sea level
Surface Temperature	288 to 293 kelvin
Magnetic Field Strength	0.305×10^{-4} tesla at the equator
Magnetic Dipole Moment	7.91×10^{15} tesla meters cubed

4.2 Journey to the center of the Earth

Looking inside the Earth's hidden interior

The internal structure of the Earth can be mapped with the help of earthquake waves. The Greek word for earthquake is *seismos*, meaning "to quake or tremor". Today, earthquake waves are often called seismic waves, and the study of earthquakes is known as seismology.

Earthquakes that originate in the planet's outer shell set the seismic waves in motion, and their velocities are determined by the density, temperature and chemical composition of the rocks they travel through. Waves become sluggish in hot, low-density rock, and they speed up in colder, denser regions. When moving between material of differing physical properties, the seismic waves change their speed and direction of movement sharply, enabling seismologists to determine boundaries between the Earth's internal layers.

The seismic investigations indicate that the Earth is layered inside like a peach. Its deeper layers are denser, and they are separated from one another in sharp transitions. There are three major parts: (1) the rocky crust, (2) a mantle of hot, plastic rock, and (3) the dense core (Fig. 4.1). They are the skin, pulp, and pit of the Earth, so to speak. The core has a liquid outer component and a solid inner one.

These different internal layers can be distinguished by the different chemical composition of their rocks. Most of

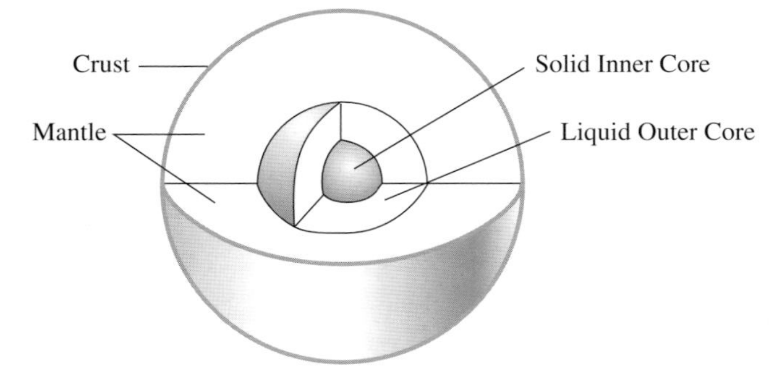

Fig. 4.1 Crust, mantle and core A relatively thin, rocky crust covers a thick silicate mantle. They overlie a liquid outer core, composed mainly of molten iron, and an inner core of solid iron. These nested layers have been inferred from seismic waves that travel through the Earth, changing velocity and direction at the layer boundaries (see Figs. 4.2, 4.3).

Table 4.2 The five most abundant elements in the Earth

Element	Symbol	Average Abundance (percent by mass)
Iron	Fe	34.6
Oxygen	O	29.5
Silicon	Si	15.2
Magnesium	Mg	12.7
Nickel	Ni	2.4

the rocks of the mantle consist of minerals in which silicon, denoted Si, and oxygen, abbreviated by O, are linked to other atoms. Such minerals are known as silicates. The core is composed mainly of iron, Fe, with some nickel, Ni. It is made up of an outer core of liquid, molten iron and an inner core of solid iron. The abundance of these ingredients in planet Earth is given in Table 4.2.

Most earthquakes occur just beneath the Earth's surface, when massive blocks of rock grind, lurch and slide against one another. The reverberations resemble ripples spreading out from a disturbance on the surface of a pond. These waves move in all directions and their arrivals at various places on the Earth can be detected by seismometers (Fig. 4.2). By comparing the arrival times at several seismic observatories, geologists can pinpoint the origin of the waves, the focus, and trace their motions through the Earth.

Rock layers of different density and stiffness will propagate the waves at different speeds, much the way that a tightened violin string will sound at a higher pitch than a looser one. As a result, the paths of seismic waves are bent and focused by their passage through the interior, as though they had passed through an immense lens. The lens of the human eye is a rough analogy; its concentric layers focus light onto the retina at the back of the eyeball in much the way that earthquake waves are focused by the interior of the Earth.

By careful mapping of the patterns of many earthquakes that travel to different depths, seismologists have peeled away the Earth's outer layers and looked at various levels within it (Fig. 4.3, Focus 4.1). It is similar to using an

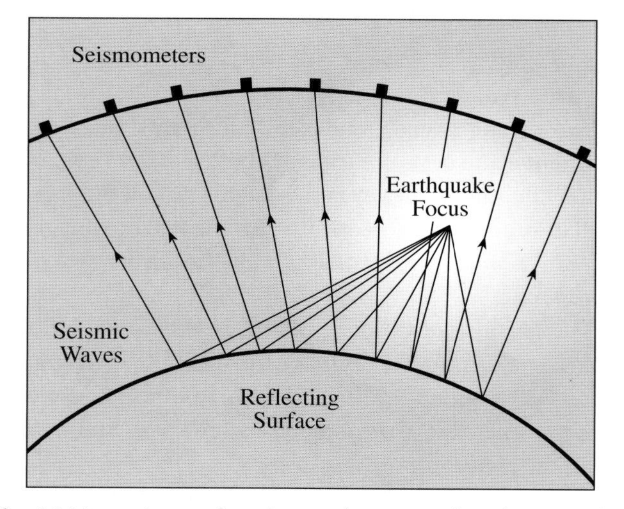

Fig. 4.2 Measuring earthquakes When an earthquake occurs beneath the surface of the Earth, it becomes the focus of seismic waves that travel through the Earth. Seismometers on the surface of the Earth record the arrival of the waves, and locate the position of the boundaries between internal layers of different composition, density, pressure and temperature (see Fig. 4.3). Seismic waves known as S, or shear and shake, waves cannot pass though a fluid, and are reflected by it. The reflected waves shown here mark the outer boundary of the Earth's liquid outer core.

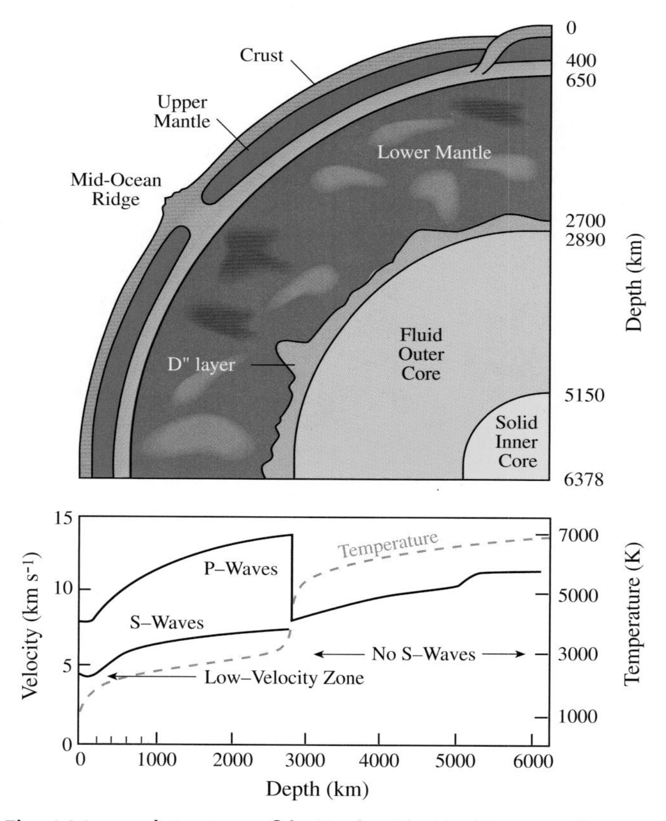

Fig. 4.3 Layered structure of the Earth The Earth's internal structure is determined by the varying velocity of earthquake waves. There are two types of waves that travel through the Earth. They are known as the compression P, or push and pull, waves and the shear S, or shake, waves. The P-waves move almost twice as fast as the S-waves, and the P-waves pass through the fluid outer core which the S-waves cannot do. The boundary between the mantle and core is marked by a precipitous drop in the velocity of the P-waves at a depth of about 2.9 thousand kilometers. The S-waves do not propagate beyond this boundary. The liquid outer core is separated from the solid inner core at a radius of 1.22 thousand kilometers where the P-waves increase in velocity.

Focus 4.1 Taking the pulse of the Earth

We cannot see the inside of the Earth and even our deepest mines are tiny dents in its surface. However, scientists have found a way to use earthquakes to illuminate the Earth's interior. Although the majority of earthquakes originate no deeper than 70 kilometers below the Earth's surface, they shake the planet to its very center, at a mean depth of 6371 kilometers, causing it to vibrate and ring like a bell. By combining the arrival times of different seismic waves that have traveled through the Earth's interior to various points on the surface, seismologists can determine the hidden interior structure of the Earth.

The earthquakes produce several types of seismic waves. There are the P- and S-waves that travel into the Earth, and the surface waves that propagate around it. The P-waves consist of compressional pulses through the Earth, expanding and compressing the rocks. They are analogous to sound waves in air, although the vibrations of the P seismic waves are much slower than audible sound. The S-waves are transverse deformations that set the Earth vibrating at right angles to the path of the waves, advancing like snakes The P stands for push and pull, while the S denotes shear or shake.

The P-waves can propagate through every part of the Earth, even its center. Some of them penetrate the deep interior and then re-emerge toward the surface on the other side. The S-waves do not travel in a fluid; they propagate only in resilient, solid substances that have elastic resistance to twisting.

Variations in temperature, pressure and composition within the Earth alter the seismic wave speed and cause the waves to bend or even to reflect. Reflections of the waves are particularly strong at the surface, the core–mantle boundary, and the top of the solid inner core. For instance, when the S-waves strike the liquid outer core of the Earth, they are reflected back toward the Earth's surface (Fig. 4.2). Studying the travel-times and velocities of waves taking different paths through the Earth therefore enables seismologists to deduce the layered structure of the deep Earth (Fig. 4.3).

In 1906, the British geologist Richard Dixon Oldham (1858–1936) found that at a certain depth the P-waves slowed sharply and the S-waves couldn't propagate (Fig. 4.3). These changes mark the bottom of the mantle and the top of the Earth's liquid outer core. The core–mantle boundary is sometimes called the Gutenberg discontinuity, after Beno Gutenberg (1889–1960), from the California Institute of Technology, who made the first accurate determination of its depth at an average of 2885 kilometers. In 1909, the Croatian geologist Andrija Mohorovicic (1857–1936) discovered that the velocity of seismic waves increases at about 35 kilometers below some continents (Fig. 4.3). This Mohorovicic discontinuity marks the place where the crust ends and the mantle begins. The inner core was discovered in 1936 by the Danish seismologist Inge Lehmann (1888–1993), and its boundary is located at a radius of 1225 kilometers.

ultrasonic scanner to map out the shape of an unborn infant in a mother's womb. Seismology is also somewhat like using Computed Axial Tomography (CAT) scans to derive clear views of the insides of living bodies from the numerous readings of X-rays that cross through the body from different directions.

The crust and mantle

The outer skin of the Earth, its crust, is a thin veneer of rocky material that covers the planet like the lumpy and split crust of an apple pie. At the base of the Earth's crust lies the Mohorovicic discontinuity, a tongue-twisting name shortened by most geologists to "Moho" – reminding us of the powerful "mojo" that can keep away evil spirits, and the jazz phrase "I've got my mojo working".

The Moho separates the dense mantle from the light crust. The boundary lies at a depth of about 5 kilometers from the ocean bottom and 35 kilometers below most places on continents; but as much as 60 kilometers below towering mountains. So, the crust is thinnest under the oceans and thickest under the continents. Because the continental and oceanic crusts are both less dense than the underlying material, they both tend to float on the mantle. High mountains have deep crustal roots that provide buoyancy and keep them afloat, much the way icebergs float on the ocean.

The Earth's buoyant crust is made up of two different materials. Oceanic crust is made up of basalt, while granite forms the basis of continental crust. Both of these low-density minerals are dominated by feldspars and quartz, but in different proportions.

The feldspars are the most abundant of all minerals in the crust, accounting for nearly half its volume. They are aluminum, Al, silicates with varying proportions of potassium, K, sodium, Na, and calcium, Ca. One main class is the orthoclase feldspars, with the formula $KAlSi_3O_8$, used extensively in the manufacture of porcelain and glass. The other main type of feldspar minerals is the plagioclase feldspar, ranging from pure sodium aluminosilicate to pure calcium aluminosilicate.

The oceanic crust is made of the black, shiny volcanic rock known as basalt. It covers more than half the Earth's surface. Basalt is a silicate mineral composed chiefly of feldspar of the pyroxene type, rich in calcium and aluminum. The basaltic magma originates at low depths in the mantle, and is less dense than the material in the upper mantle. This allows the basalt to rise up to the crust and erupt in volcanoes there. The ocean floor is covered with basalt, an outpouring of lava from volcanoes at the bottom of the sea, and volcanic islands like Hawaii and Iceland are largely composed of basalt.

The second most common mineral found in the crust is quartz, a hard, colorless crystalline mineral composed of silicon dioxide, or silica, with the formula SiO_2. The tough continental granites were formed in the fiery melts of magmatism, joining feldspars with at least 20 percent quartz.

The bulk of the Earth is in its mantle, the region that reaches down some 2890 kilometers, on average, from the thin crust to the core. The difference between the crust and the mantle is one of chemical composition. Material brought up by volcanic eruptions, as well as eroded mountains, indicate that the upper mantle is composed of dense minerals known as olivine $(Mg, Fe)_2SiO_4$ and pyroxene $(Mg, Fe)SiO_3$. These are silicates with a little magnesium, Mg, or iron, Fe, mixed in as minor constituents.

Lithosphere and asthenosphere

The outermost parts of the Earth, the crust and upper mantle, can be divided by their physical properties into the lithosphere and asthenosphere. The lithosphere is the solid region beneath our familiar oceans and mountains. It extends to depths of about 100 kilometers, which includes both the crust and upper mantle. The lithosphere, within the upper mantle, is crustal rock and mantle rock down to a zone in the mantle that is lubricious enough to move. Beneath the lithosphere lies the warm and plastic asthenosphere that reaches to a depth of about 300 kilometers. Its material is revealed by the slowness with which it propagates seismic waves.

The distinction between the lithosphere and asthenosphere is one of stiffness. The lithosphere takes the root of its name from the Greek *lithos*, for "stone". The lithosphere is the solid "plate" of the plate tectonic theory mentioned later in this section. The word asthenosphere comes from the Greek *asthenos*, meaning "without strength" or "devoid of force".

The radioactive elements responsible for the warmth of the asthenosphere are too weakly concentrated to melt the rock, but they cause it to soften and behave like putty. Rock in the asthenosphere flows slowly when strained for a long time, like applying slow pressure to an open tube of toothpaste, but the asthenosphere responds like a solid when it is struck by an earthquake.

Two cores

If you pick up a typical rock in the Earth's crust and determine its mass density, it will be roughly 3000 kilograms per cubic meter, or about three times that of water. By way of comparison, the mean mass density of the Earth is 5515 kilograms per cubic meter, which means that there must be high-density material located deep down inside the Earth. The fact that the Earth is not homogeneous, with its densest parts concentrated inside, has been known since the time of Isaac Newton (1642–1727), from the varying gravitational pull measured by pendulums located at different places on the Earth's surface. The material with the greatest density is concentrated in the planet's core, with a mass density of about twelve times that of water.

The Earth's core reaches about halfway to the surface, implying a volume that is one-eighth that of the entire Earth. If the mass density of the Earth were uniform, the core would have an equal share, one-eighth, of the mass of the Earth, but its actual mass is nearly three times greater. So the core's mass density is very high, and this points to iron as the most likely material. Iron has been identified as the main ingredient of the core because it is the most abundant heavy element in the Sun and in some meteorites. Laboratory measurements also show that the densities and seismic-wave velocities of the core are more closely matched by iron than any other element. The seismic evidence indicates that the core is less dense than pure iron would be at the pressures there. Although the core is mostly iron, it must consist of an alloy of iron that includes light elements, one of which may be hydrogen.

By weight, the Earth is mostly iron, but relatively little of the metal is found in the Earth's crust, which is principally made of lighter elements like silicon and oxygen. Billions of years ago, during the planet's very early history, the Earth must have been molten, permitting the iron to sink to the interior because of its enormous weight. Earth would have then cooled from the outside, forming solid rocks in the crust and mantle that consist of elements that are locked together and do not sink into the molten core.

Examination of earthquake waves has shown that there are two cores, an inner, crystalline solid core and an outer fluid one. The two cores are very different is size. The solid inner core has a radius of about 1225 kilometers, which is slightly smaller than the Moon whose mean radius is 1738 kilometers. The core-mantle boundary, which marks the outer edge of the outer fluid core is about 3485 kilometers in radius, or 55 percent of the Earth's mean radius of 6371 kilometers.

The seismic evidence indicates the presence of a rugged, interactive zone, known as the D″ layer, where the liquid-metal outer core meets the lowermost part of the rocky mantle. This turbulent, irregular boundary contains troughs and swells, deeper than the Grand Canyon and higher than Mount Everest, spreading across continent-sized areas. It may represent material that was once dissolved in the underlying fluid core, dense material that sank through the mantle but could sink no further down, or material formed as a result of chemical reactions between the core and mantle. Seismologists speculate that irregularities in the D″ layer at the core–mantle boundary may channel heat-flows to produce giant rising plumes of molten rock capable of penetrating the thick mantle and occasionally making their way to the Earth's surface.

The temperature of the deep core is difficult to determine, but we certainly know that the planet is hot inside. The central, inner core appears to be about 6900 kelvin, which is a bit hotter than the visible disk of the Sun, at 5780 kelvin. At first glance, this would seem to imply that the center of the Earth must be liquid, but this is contradicted by seismic evidence, which indicates a solid region in the deep interior. The clue to the apparent paradox is the high pressure at the center of the Earth, about 3.6 million times the pressure of the atmosphere at sea level. These pressures have been imitated in laboratory experiments, and they lead to a remarkable conclusion. At high pressures, iron can persist as a fairly rigid solid even at a temperature of thousands of degrees.

Most liquids will solidify if the pressures are high enough and the temperatures are relatively low. Probably the entire core was once molten, but the drop in temperature associated with a loss of heat permitted the inner portion to solidify under the high pressures. The pressures are low enough, and the temperatures still high enough, to sustain a liquid outer core. It also remains liquid because iron alloys melt at lower temperatures than most rocks.

As our planet grows older and colder, the solid inner core is growing continuously at the expense of the liquid outer core. The iron liquid at the base of the fluid core is freezing, solidifying and precipitating onto the surface of the solid core, making it grow at the rate of about 0.01 meters every 100 years. At the same time, the rocky mantle may be slowly dissolving into the liquid metal of the outer core.

The Earth's inner core is a solid lump of iron suspended at the center of the much larger, fluid outer core, something like a golf ball levitated in the middle of a fish bowl (Fig. 4.4). The outer core is itself curtained behind about

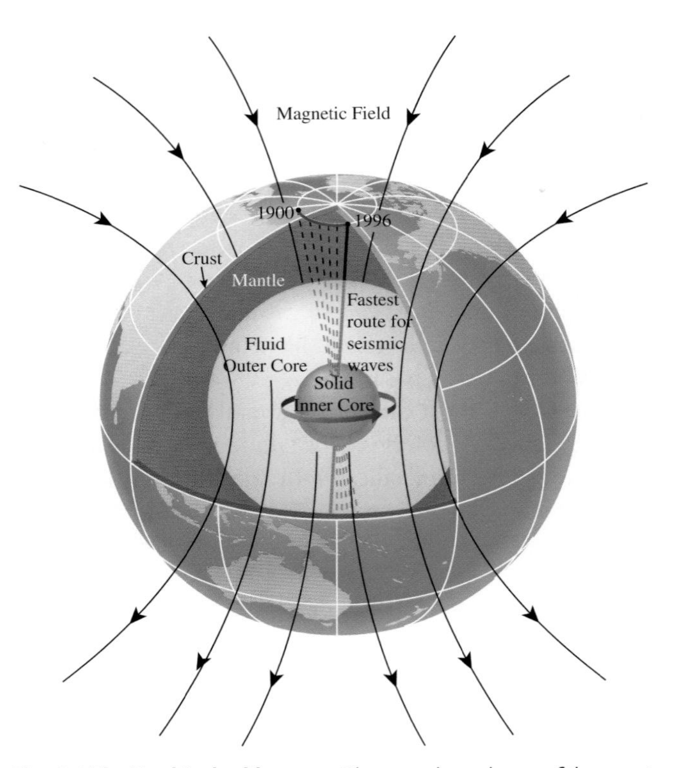

Fig. 4.4 The Earth's double core The mantle and part of the crust have been cut away here to show the relative sizes of the Earth's fluid and solid cores. The outer fluid core is about 55 percent of the radius of the Earth, and the inner solid core is slightly smaller than the Moon. The Earth's magnetic field is thought to be generated and sustained by moving currents in the planet's electrically conducting, fluid outer core, which is composed of molten iron. Geophysicists have discovered that the route of the polar (north–south) seismic waves through the Earth's interior is gradually shifting eastward because the inner core is rotating slightly faster than the rest of the planet. The fast rotation of the inner solid core may help explain how Earth's magnetic field reverses polarity (Courtesy of Paul Richards, Lamont-Doherty Earth Observatory.)

2890 kilometers of solid rock. So, the solid heart of iron is difficult to observe, but faint seismic vibrations in the ground have been used to sense it.

The seismic vibrations that pierce the inner core move through it at different speeds that depend on their direction, faster on polar north–south paths than equatorial east–west ones. This directional dependence of seismic-wave velocities is explained by the crystalline structure of the inner core. The crystals give the solid inner core a texture with a preferred orientation, like the grain in wood. By lining up along the Earth's spin axis, iron crystals make the inner core stiffer along this axis, thus letting sound waves travel faster in this direction. Some scientists have even speculated that the inner core may be just one, single, gigantic crystal of iron atoms rather than a mass of tiny crystals, like a huge diamond suitable for an interplanetary engagement. In either case, each crystal probably takes its direction from either the stress generated by Earth's rotation or from the terrestrial magnetic field.

Recordings of weak earthquake rumbles, which have traveled through the central core of the Earth, indicate that it spins faster than the outer Earth, but that they both rotate in the same direction. The fast lane for seismic waves is tipped slightly with respect to the Earth's north–south axis, and it moves around it (Fig. 4.4). This shift in orientation means that the crystalline globe at the center of the Earth is turning slowly within its solid rocky and liquid-metal enclosures. It is spinning with respect to the Earth's surface at between 0.2 and 0.3 degrees per year, completing one lap in between 1200 and 1800 years.

The Earth's magnetic field threads the solid inner core and may make it turn faster, much as a magnetic field turns the shaft of an electric motor. The magnetic coupling between the solid and liquid cores could also account for reversals in the Earth's magnetic polarity; the planet switches its north and south magnetic poles a few times every million years (Section 3.6). Currents in the electrically conducting, fluid outer core generate and maintain the magnetic fields, and turbulence in the fluid is always trying to toss the magnetism into a polarity reversal. The inner solid core exerts a stabilizing influence on this tendency, forcing the fields to stay in place, but the magnetic connection between the two cores is probably pulled apart as they spin with respect to each other. The coupling eventually gives way and the magnetism flips.

Differentiating the layered interior

The origin of the layered structure of the Earth's interior is still a geological mystery, but there are two extreme alternatives. According to one theory, the *hot* theory, the Earth accumulated rapidly (in 100 thousand to 10 million years), and the kinetic energy of the impacting material that coalesced to form the Earth kept the planet hot and molten as it formed. If the rocks were molten as the Earth grew, its constituents would separate, with the dense, heavy material sinking toward the interior, creating the dense core, and the light substances rising to the surface to form the low-density mantle and crust.

In an alternative scenario, the *cold* theory, the Earth gathered itself together relatively slowly, in 100 million to 1 billion years, and the planet started out cold, homogeneous and solid. The globe then became heated by emission from radioactive material that was uniformly distributed through the interior, and its temperature gradually rose to the melting point. When the planet melted, its gravitational field caused the heavy elements to sink toward the center, forming a dense core, while the lighter elements rose toward the surface, producing chemically distinct layers.

In both the hot and cold theories, the layered internal structure of the Earth results from a process known as differentiation, in which gravity separates elements in a molten state, pulling the heavier ones down. A similar thing takes place in a blast furnace or smelter. Slag-forming rock is loaded into the furnace, and molten metal is tapped periodically from the bottom. Thus, in both theories the Earth was once molten and after a process of gravitational separation or differentiation, the Earth began cooling from the outside. The solid crust formed and then the mantle, but the basic layered structures remained a feature of the Earth since its early history.

4.3 Remodeling the Earth's surface

Continents, oceans and ocean floors

There are two major types of terrain on Earth – the high, dry continents and the low, wet floor of the ocean (Fig. 4.5). Between them, and partially surrounding many continents, is a narrow strip of shallow ocean called the continental shelf. Today, the oceans cover 71 percent of the Earth's surface, and the world's continents amount only to scattered and isolated masses surrounded by water.

To those of us who are confined near the surface of the globe, the Earth seems rugged, with towering mountains rising several kilometers above the ocean, and deep trenches sinking as far beneath the ocean's surface. But a scale model of the Earth would have to be quite smooth. Those extremes reach only one-tenth of one percent, or 0.001, of the Earth's radius above and below the ocean surface. A basketball this smooth would have bumps no more than 0.0001, or 10^{-4}, meters high, roughly the size of the dot at the end of this sentence.

The smoothness of the Earth is due to the immense force of its gravity and the weight of its outer layers, which largely overcome the electrical force inside the solids making up the Earth and cause them to lie in concentric shells. In smaller bodies, such as asteroids less than a few hundred kilometers in diameter, the interior is strong enough to remain rigid and they retain their original irregular shapes and internal composition.

Moreover, ongoing erosion will wear down the world's highest mountains in just a few hundred million years, which is just a fraction of the Earth's age of 4.6 billion years. If the planet were a perfectly smooth sphere, the oceans would cover the entire globe to a depth of 2.8 kilometers. So, we can tell right away that high, dry

Fig. 4.5 Planet Earth from space As illustrated in this image of Africa, the Arabian peninsula, and the Indian Ocean, the Earth's surface consists of continents and oceans. The Antarctica ice cap is also shown, gleaming white at the bottom. Continents cover a little more than one-quarter of the Earth's surface, while ocean water covers almost three-quarters of the surface. This image was taken by *Apollo 17* astronauts in December 1972 as they left the Earth *en route* to the Moon. (Courtesy of NASA.)

land must be continuously recreated and pushed up out of the water.

Thus, the continents and oceans are not eternal, unchanging aspects of the Earth's surface. Their appearance of permanence is an illusion caused by the brevity of the human life-span. Just as an entire human lifetime is just a fleeting moment in the history of the Earth, today's map of the world is just a brief snapshot of its evolving, mobile, ever-changing surface. Over hundreds of millions of years, blocks of the Earth move about, producing drifting continents that completely alter our picture of the world.

Modern geologists have now pieced together the past, reconstructing the pieces of the Earth's moving jigsaw puzzle. They have shown that Boston and southern Florida are both former pieces of Africa, which were left behind when the Atlantic Ocean opened up. China used to be separated from Siberia by at least one ocean. Japan was once attached to Asia, and it may become part of Alaska in eight hundred million years.

Continental drift

The idea that continents have not always been fixed in their present positions was suggested more than three centuries ago, in 1587 by the Dutch mapmaker Abraham Ortelius (1527–1598) in his work *Thesaurus Geographicus*. However, the theory of moving continents was not developed into a thorough scientific hypothesis until the early 20th century, by the German meteorologist Alfred Lothar Wegener (1880–1930) in his influential and controversial book published in German in 1915 *Die Entstehung der Kontinente und Ozeane*, and first translated into English in 1924 as *The Origin of Continents and Oceans*. The English translation of the fourth edition, published in 1924, is still available from Dover Publishers in New York. Wegener noticed that the outlines of the continents themselves exhibit a number of remarkable symmetries. For example, the eastern edge of South America would fit snugly into the western edge of Africa, a remarkable fit noticed by Ortelius. In fact, much of the east and west shores of the Atlantic Ocean are as well matched as the opposite banks of a river (Fig. 4.6).

Wegener based his concept of continental drift not only on the similar shapes of the present continental edges, but also on the striking match of certain rocks and geological formations, fossil creatures, and ancient climates along the borders of continents on opposite sides of the Atlantic Ocean. He concluded that all of the continents were once

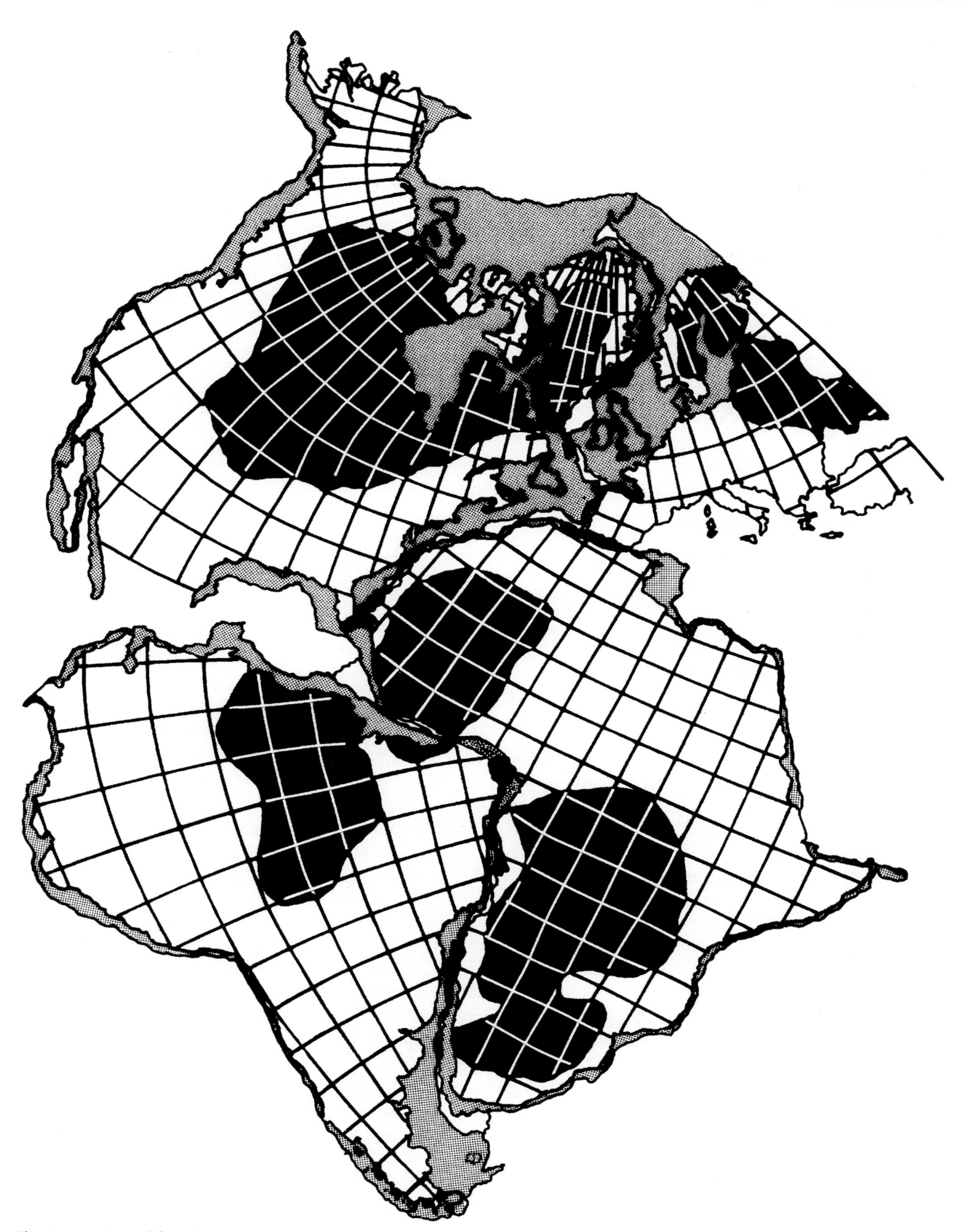

Fig. 4.6 Continental fit The continents fit together like the pieces of a puzzle. Here the fit has been made along the continental slope at a depth of 910 meters, or 500 fathoms (*gray areas*). Within the present continents are ancient terrains between 1.7 and 3.8 billion years old (*black areas*). The close fit of the shorelines of the continents suggests that they once formed a single landmass, known as Pangaea (see Fig. 4.7).

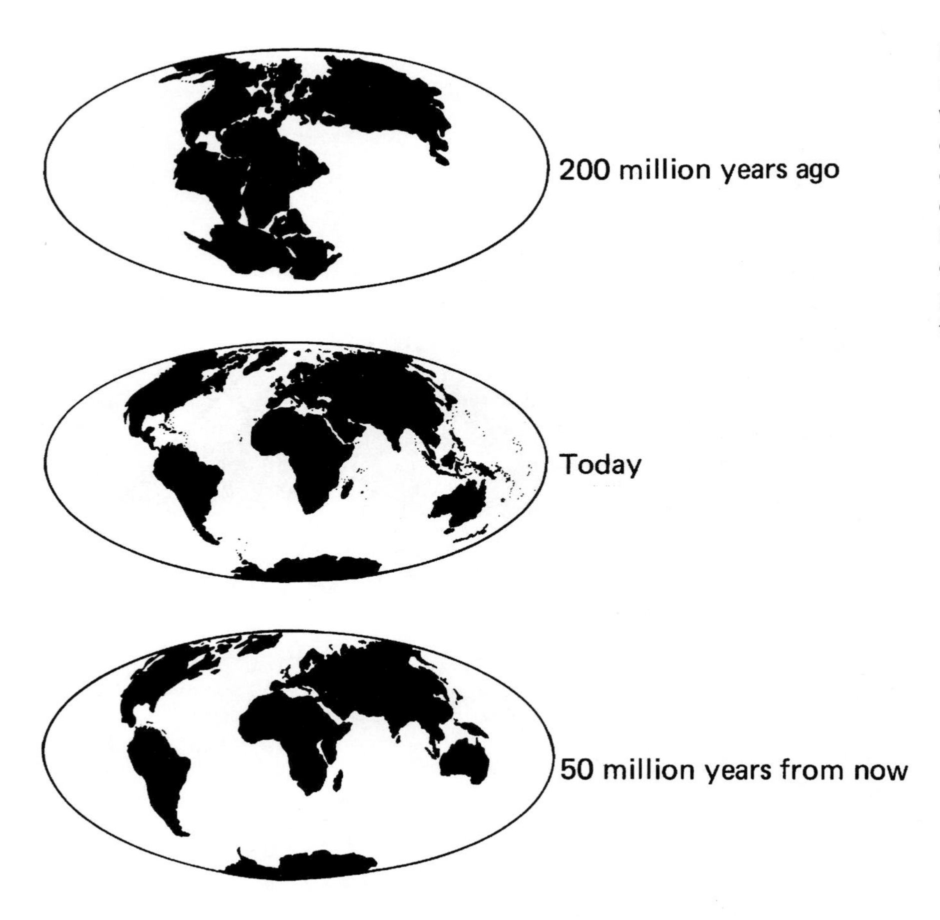

Fig. 4.7 Continental drift Two hundred million years ago all of the continents were grouped into a single supercontinent called Pangaea and the world contained only one ocean (*top*). The continents then drifted away from Pangaea, riding on the back of plates to the positions they now occupy (*middle*). The bottom diagram depicts the world geography 50 million years from now.

a part of a single landmass that fragmented and drifted apart (Fig. 4.7). If spacecraft had existed back then, their camera eyes would have seen one large continent and a single ocean surrounding it.

This hypothetical supercontinent is called *Pangaea*, a Greek word meaning "all Lands" and pronounced *pan-gee-ah*. After all, if today's continents spread apart from their obvious puzzle fit, they had to have been together in the first place. This would also account for the Earth's curiously asymmetric face, in which the ocean waters dominate the southern hemisphere while the continents dominate the northern hemisphere.

Pangaea broke into pieces about 200 million years ago when large amphibians and reptiles ruled the land. As the once-joined continents moved apart, the water rushed in to fill the gap caused by their separation. This led to the various smaller continents that we see today, but the arrangement is only temporary. The continents will continue to move apart, and since the globe is round, all lands will eventually converge again.

Thus, in about 200 million years from now a new Pangaea will be formed, and then, inevitably, another break-up will ensue as our restless planet continues to reform and reshape itself. In the process, the continents will continue on their endless journey, forever roaming and wandering about the planet with no final destination.

Wegener's theory of continental drift was disparaged, ridiculed and even scorned by most geologists for at least half a century. In retrospect, their objections are hard to understand. The discovery of glacial deposits in Africa and of fossils of tropical plants, in the form of coal deposits, in Antarctica certainly meant that these two continents had once been located at different parts of the globe with climates much different from their present ones.

The main difficulty was understanding how the continents could move across the Earth and plow their way through solid rock at the bottom of the ocean. A possible mechanism had been proposed by the English geologist Arthur Holmes (1890–1965), who noticed that both the Earth's surface and interior could be in motion. Internal heat could drive churning motions that might propel the continents from below. But these prescient ideas, developed in the 1930s and revitalized in Holmes' 1944 seminal textbook *Principles of Physical Geology* (Thomas Nelson and Sons, Ltd, London), were not widely accepted. Exploration of the ocean floor by sound waves provided the first evidence that Wegener and Holmes were on the right track after all.

Focus 4.2 Mapping the ocean floor from the top down

The ocean depths have recently been charted with great accuracy by measuring the height of the sea surface. Through the action of gravity on water, seabed mountains produce swells at the ocean surface, and canyons or valleys on the ocean floors produce dips in the surface waters. So the top of the ocean mimics the topography at its bottom.

A satellite is used to beam short pulses of micowaves, or short radio waves, down at the ocean, and to determine the time for the reflected pulse to bounce back. Since the microwave pulse travels at the speed of light, the distance between the satellite and the top of the sea is half the product of that speed and the round-trip travel-time. Because the separation between the satellite and the center of the Earth is known from the satellite's orbit, one can establish the distance between the Earth's center and the top of the ocean by subtraction.

The new technique was pioneered in 1978 with NASA's satellite *Seasat*. Although it worked for only three months, before unexpectedly failing, *Seasat* demonstrated the power of the technique (Fig. 4.8). An improved satellite mapper, named *Geosat*, was launched in 1985 by the United States Navy, to secretly gather similar information during the Cold War.

The topographical contours of the ocean top move up and down with features on the bottom, faithfully tracing out their highs and lows with an extraordinary precision of about one-tenth of a meter. This information helps submariners glide stealthily through the sea. Knowing the precise direction of gravity also improves the accuracy of guided missiles fired from the submarines.

With the Cold War ending in the mid-1990s, the Navy declassified the *Geosat* data, and it has been used with other satellite altimeter data, of the European Space Agency and NASA, to provide a full, detailed map of the sea floor.

Sea-floor spreading

The bottom of the ocean is not flat. It contains underwater mountains and valleys that are as grand as those on any continent. Although we cannot see these features in the inky darkness of the deep sea, we can use sound waves to reach down and touch them. Their distance is determined by recording the time it takes for electrically generated sound signals, called pings, to travel from a ship to the floor and back.

The German Navy used such echo-sounding measurements to reveal the rugged sea floor soon after World War I (1914–18). They showed that a chain of submarine mountains runs right across the middle of the floor of the Atlantic Ocean. Known as the Mid-Atlantic Ridge, it extends about 2 kilometers above the adjacent sea floor, which is at a depth of about 6 kilometers.

Nowadays, the United States Navy detects enemy submarines or ships with sonar, an acronym for sound navigation and ranging, transmitting a continuous rain of pulsed sound waves and using the same echo technique to measure distance. Many modern ships, including warships and some commercial fishing boats are also equipped with sonar to aid in navigation. Navy ships and research vessels equipped with sonar can now map a two-kilometer swath at the bottom of the ocean in a single ping of the sonar. Gravitational data, obtained from satellites that bounce radio beams off the sea surface (Focus 4.2), complement the sonar data, and they together result in highly detailed maps of the entire ocean floor (Fig. 4.8). They have shown that

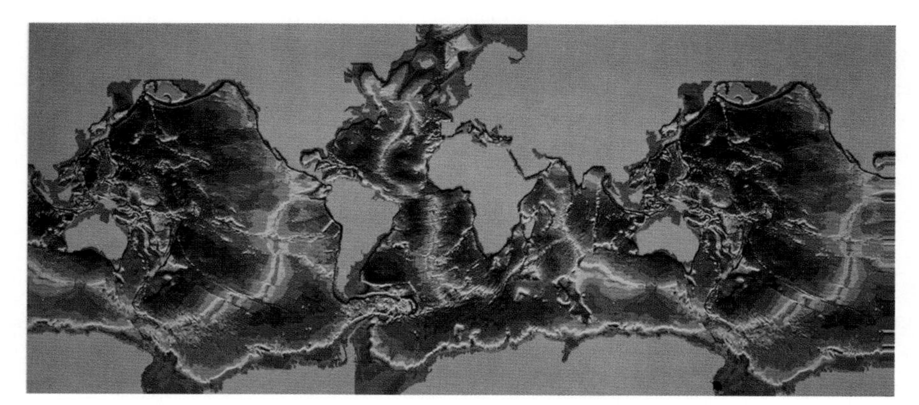

Fig. 4.8 Bottom of the ocean Map of the world's oceans floor as acquired by the *Seasat* satellite. The Mid-Atlantic Ridge runs down the middle of the ocean floor separating Africa from North and South America. As shown here, a succession of great ridges runs through all of the world's ocean floors, although not always in the middle. (Courtesy of William F. Haxby, Lamont-Doherty Geophysical Observatory, Columbia University.)

Fig. 4.9 Volcanic islands Lava erupting from the volcanic island Surtsey on 19 August 1966, almost three years after it rose out of the sea near the coast of Iceland. The volcanic island of Jolnir is in the background. It disappeared back into the sea about one month after this picture was taken, but Surtsey is still visited for research purposes today. All of these volcanic islands, including Iceland, mark points where a mid-ocean ridge has risen out of the ocean. (Courtesy of Hjalmar R. Bardarson, Reykjavik, from his book *Ice and Fire*.)

the Mid-Atlantic Ridge is just a part of a global mid-ocean ridge that snakes its way across the bottom of the world's oceans.

The global mid-ocean ridge is a gigantic network of underwater mountain ranges. The submerged mountains stand higher than the greatest peaks on land, and meander for more than 75 000 kilometers, creating the longest mountain chain on Earth. It is long enough to accommodate the total length of the Alps, Andes, Himalayas and Rockies. The mid-ocean ridge winds around the Earth, girdling the globe like the stitched seams of a baseball, not in simple lines but in offset segments. When the undersea mountains reach the surface they can form islands, like Iceland and its new neighboring island Surtsey, named after the Icelandic god of fire, Surtur (Fig. 4.9).

Even more remarkable are the deep canyons, collectively known as the Great Global Rift, that run along the mid-ocean ridge, splitting the ridge open as though it had been sliced with a giant's knife. Discovered in 1953, by the American scientists William Maurice Ewing (1906–1974) and Bruce C. Heezen (1924–1977), the rift marks a line where much of the Earth's internal heat is released. It is filled with hot molten rock, or magma, coming up from inside the planet.

Amazing creatures live down there in the eternal darkness at the bottom of the sea. Giant clams, tubeworms and crabs are warmed and fed by the superheated water. They thrive without light by digesting sulfur minerals emitted from the hot vents, substances that other animals would find poisonous.

The mid-ocean ridge is more accurately described as a tear in a sheet of paper rather than a cut, for it represents the place from which the ocean floor moves outward on both sides. It is as if the Earth was pulling itself apart and becoming unstitched.

Hot magma emerges from beneath the sea floor, and oozes into the canyons of the Great Global Rift, filling them with lava. As the lava cools in the ocean water, it expands and pushes the ocean crust away from the ridge. More lava then fills the widening crack, creating new sea floor that moves laterally away from the ridge on both sides, with bilateral symmetry.

If sea floor is continuously created in the middle of the ocean, where does it go? If all of the new material kept on piling up, the Earth would grow bigger as time goes on, and that is not observed. The size of the Earth has not changed significantly during the past 600 million years, and very likely not since shortly after its formation 4.6 billion years ago. The Earth's unchanging size implies that the ocean floor must be destroyed at about the same rate as it is being created. The floor disappears inside the Earth, where it is transformed by the heat and eventually recycles to rise again.

As it migrates away from the hot rift of its beginning, the new ocean floor grows colder and denser, subsiding to greater depths as it ages. After traveling across the Earth, in conveyer-belt fashion for many millions of years, the older heavier floor bends and descends back into the Earth, often at the edges of continents, creating a deep-ocean trench in the underlying rock. Such trenches are found all around the edges of the Pacific Ocean, and they can sink as far below sea level as the tallest mountains rise above it.

The over-all concept is known as sea-floor spreading, an idea introduced by the American geologist Harry Hammond Hess (1906–1969) in his 1962 paper entitled "History of Ocean Basins". In brief, new sea floor is formed at a rift in the mid-ocean ridge, turning cold and heavy as it spreads away from its source in two directions; the sea floor eventually sinks and disappears in a deep-ocean trench, where it is consumed. Its material is then recycled and born again as new floor emerges from the ridge.

The theory of sea-floor spreading accounts for, and unites, several unsolved mysteries of marine geology, including the very existence of the global mid-ocean ridge and deep-ocean trenches. Several lines of evidence now leave no doubt about the reality of sea-floor spreading.

First, it accounts for certain flat-topped submarine volcanoes, called guyots, under the Pacific Ocean; they look eroded away but now lie under 2 kilometers of water. It is now believed that guyots are once-active volcanoes, which rose above the water's surface before being eroded by the atmosphere down to sea level, and then carried beneath deep waters by the moving sea floor.

Second, the sediments on the floor of the ocean are relatively young. Research vessels have determined that the sediment layer is much thinner than it would be if it had accumulated over the lifetime of the ocean waters. The thickness of the sediments indicates that none of them have been accumulating for more than 200 million years. The oldest fossils found on the ocean floor are also no more than 200 million years old. From creation to extinction, from rift to trench, the ocean floor completely cleans house, erasing any record of its previous history in less than 200 million years.

In contrast, marine fossils in rock strata on land can be considerably older, including those found near the tops of the highest mountains, and the oldest known continental rocks have ages dating back to 3.96 billion years. Some tiny zircon crystals culled from rocks in Australia have even been dated at 4.4 billion years old, and they apparently required water for their formation. The sea most probably dates back to the formative stages of the young Earth, more than 4 billion years ago. Thus, young ocean floors have been replacing older ones, while the water above them has remained for billions of years.

Third, sea-floor spreading explains the ages of core samples recovered from the seabed during petroleum exploration. They show that the floor of the Atlantic Ocean is youngest in the middle, and grows progressively older with increasing distance from the mid-ocean ridge. Both the average age and the thickness of the sediment increase away from the ridge.

Perhaps the most decisive evidence for sea-floor spreading was the discovery of regular magnetic-field patterns in the ocean floor. Magnetic detectors towed behind ships and carried in aircraft could measure very small differences in the Earth's magnetic field from place to place, known as magnetic anomalies. Positive magnetic anomalies are places where the magnetic field is stronger than expected, and negative ones are weaker than anticipated.

One of the motivations of these studies was to detect perturbations in the magnetism caused by enemy submarines, but the results had far greater consequences. The pattern of magnetic anomalies was symmetrically placed, or mirrored, on each side of the mid-ocean ridge (Fig. 4.10). Frederick John Vine (1939–1988), then at Princeton University, and Drummond Hoyle Matthews (1931–1997), working at Cambridge University, compared the magnetic irregularities with the known history of magnetic-field reversals found on the flanks of volcanoes on land. These continental lavas show that every so often the Earth's magnetic field has changed its direction (Section 3.6), and the

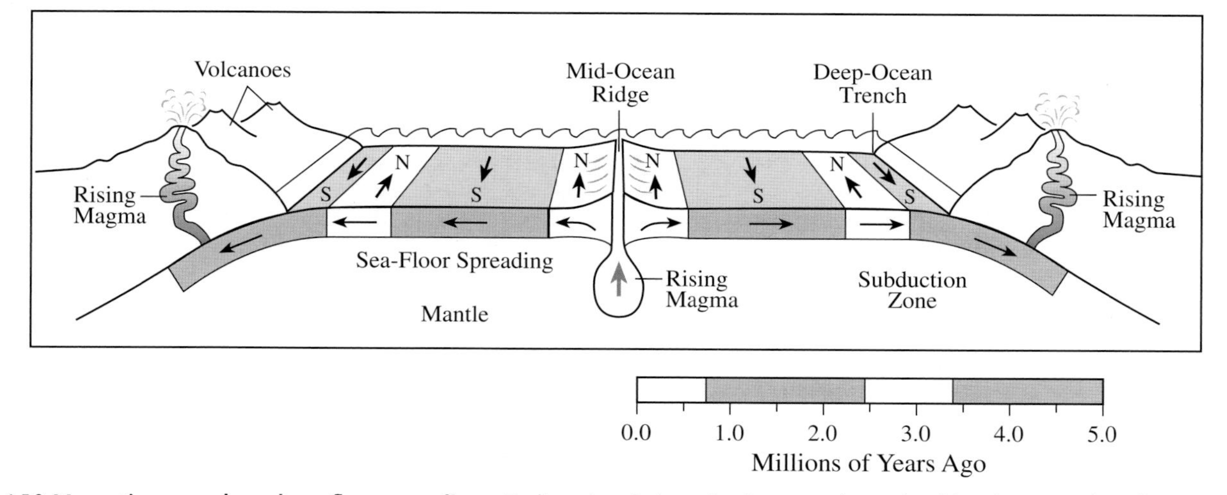

Fig. 4.10 Magnetic reversals and sea-floor spreading Radioactive dating of volcanic rocks on land has been used to determine the time-scale of magnetic reversals on the Earth (*bottom panel*). They indicate that the Earth's magnetic field has flipped, or changed direction, several times during the past five million years. The data describe normal epochs (*white*) when compasses would have pointed toward the geographic north, as they do now, and reversed epochs (*gray*) when compasses would have pointed south. The pattern of magnetic reversals on both sides of the volcanic mid-ocean ridge (*top panel*) is the same, indicating that sea-floor spreading has carried the solidified lava away from the central ridge. The sea floor is consumed at the other end, when it slides into a deep ocean trench at a subduction zone, shown here on a compressed time-scale. The sea floor can spread for 200 million years before diving into the Earth.

symmetric magnetic anomalies on the sea floor exactly match these polar reversals recorded on land (Fig. 4.10).

Each time the anomaly changes from positive to negative, the Earth's magnetic field turns upside down. Its magnetic poles flip, so the south magnetic pole switches from the North Geographic Pole to the South Geographic Pole or *vice versa*, and the north magnetic pole moves to the opposite geographic pole. Lava emerging at the present time would have a positive magnetic anomaly, with the Earth's south magnetic pole located at the North Geographic Pole.

The orientation of the Earth's dipolar magnetic field imprints itself on the volcanic rocks at the time they form, whether on land or under the sea. When fresh molten lava pours out of them, magnetic minerals within the cooling lava become aligned with the Earth's magnetic field, and this orientation or polarity remains as a fossil magnetic record, locked into the rock when it solidifies. Vine and Matthews proposed that the lava on both sides of the mid-ocean rift solidified and moved away, freezing-in the magnetic direction prevailing at the time, and when the Earth's poles flipped and reversed the lava flows preserved a set of parallel bands with opposite magnetic direction. Thus, the symmetric magnetic-anomaly stripes were recording the Earth's past magnetic field, and providing dramatic support for sea-floor spreading.

By radioactive dating of volcanic rocks on land, it is possible to tell when they solidified and to build up a chronology of the magnetic changes. This chronology can then put dates on the reversals found in the sea floor, and from the distances traveled it is possible to compute the rate of sea-floor spreading, assuming that the floor has moved at a constant rate. These rates have been independently calibrated astronomically by comparison of the seabed sediments with the orbital parameters that govern climate changes recorded in fossil organisms in the sediments (Section 4.4). They indicate that the sea floor has indeed been spreading at a constant rate for the past 5 million years.

The ocean floor moves away from the ridge at rates of 0.02 to 0.20 meters per year depending on the location, or just a little faster than your fingernails grow. When sustained for 200 million years, the spreading sea floor can push continents apart by between 4000 and 40 000 kilometers – entirely adequate to explain the widths of the great oceans. At the measured rate, it took just 150 million years for a slight fracture in an ancient former continent to widen into today's Atlantic Ocean.

Plate tectonics

The rind of the Earth, its outer shell known as the lithosphere, is subdivided into a mosaic of large plates, million of meters across (Fig. 4.11). They vaguely resemble the cracked pieces of an eggshell. The plates move horizontally atop a viscous layer of much hotter, softer, more malleable

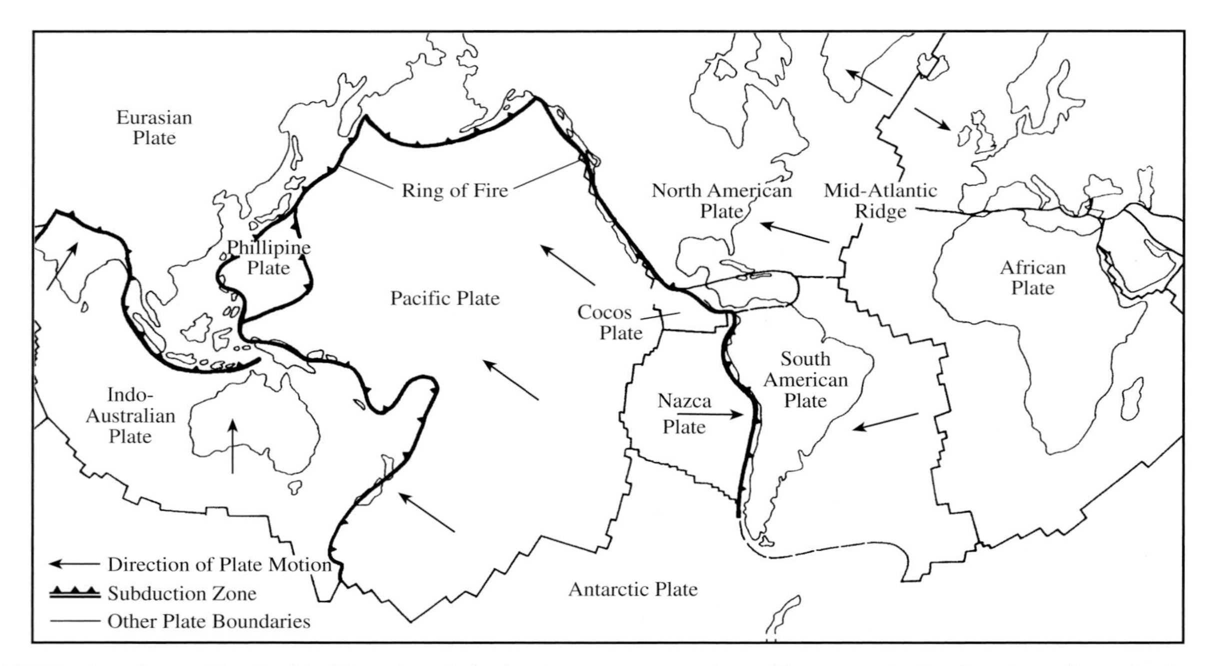

Fig. 4.11 Moving plates The Earth's lithosphere is broken into numerous plates. They move in the directions shown by the arrows at rates of a few-tenths of a meter per year. The lithosphere dives into the underlying asthenosphere at zones of subduction. They are denoted by the thick line with triangles, forming the famous Ring of Fire around the edge of the Pacific and Nazca Plates. Most of the Earth's earthquake and continental volcanic activity is concentrated along the subduction zones.

rock called the asthenosphere (Section 4.2). Because of the high temperatures and immense pressures found there, the uppermost part of the asthenosphere flows along at the base of the lithosphere.

Plates are composed of the Earth's crust and the rigid upper mantle just beneath it. Most plates contain both continental and oceanic crust, and they all include oceanic crust. Six of the nine major plates are named for continents embedded in them: the North American, South American, Eurasian, African, Indo-Australian, and Antarctic Plates. The other three are almost entirely oceanic: the Pacific, Nazca, and Cocos Plates. Accompanying them is a host of smaller plates.

Driven by heat from below, the plates move with respect to one another, accounting for most of our world's familiar surface features and phenomena, such as mountains, earthquakes and ocean basins. The continents are implanted within the moving plates, and continental drift is a consequence of the motion of plates carried along by the sea-floor spreading. So the moving plates carry the continents with them, on an endless journey with nowhere special to go.

The rigid plates are in continual, relentless movement, and they deform at their boundaries. Like drops of olive oil gliding across a warm frying pan, the continents sometimes collide and coalesce, sometimes slide and rub against each other, and at other times break up and scatter. The transformations produced by these interactive motions are known as plate tectonics, from the Greek word *tectonic* for "carpenter or building". They are forever reconstructing the face of the Earth.

A mid-ocean ridge is a crack in the sea floor that is filled in by magma from the mantle as two diverging plates separate. So, the ridge marks the boundary between two plates. As the plates move apart in opposite directions, a crack or rift opens up at the crest of the ridge, allowing more molten rock to move up and feed the spreading plates, like blood in an open wound that will not heal.

The spreading ocean floors are eventually pushed deep down into the planet's hot interior, at subduction zones where two converging plates meet. When a moving oceanic plate encounters a light continental plate, or a younger, lighter oceanic one, the heavier oceanic plate plunges steeply into the Earth along a subduction zone, like a down-going escalator, producing a deep-ocean trench. Because the continental rock has the lowest density, it remains on the top, while the ocean floor slides underneath (see Fig. 4.10). The buried material is consumed within the Earth, only to re-emerge, recycled and transformed at some other location.

Magma is generated at the sinking subduction zones where dense oceanic plates are pushed under the lighter continental ones, producing volcanoes that help us locate the edges of the plates. A dramatic example is the circular line of volcanoes that borders the Pacific Ocean. This active belt is known as the Ring of Fire because it is often the site of fiery volcanic eruptions (Fig. 4.11).

Scientists can track the plate motions using Very-Long-Baseline Interferometery, abbreviated VLBI, and by Satellite Laser Ranging, or SLR for short. In VLBI, radio receivers at widely separated telescopes record the strength and arrival times of cosmic radio signals from quasars, and the comparisons between the recorded data is used to determine the distance between the telescopes, with an accuracy of less than 0.01 meters. The SLR targets are satellites covered with tiny mirrors called corner cubes. Pulsed laser light, generated at stations on the ground, is bounced off the mirrors and the round-trip travel-time for the light to return is used to establish the distances to the satellites. When combined with precise orbital information for the satellites, these distances can be used to determine the changing separation of the ground stations and the rate of tectonic plate motion.

The VLBI and SLR measurements indicate that the plates move laterally, or horizontally, across the Earth at rates of 0.02 to 0.20 meters per year depending on the plate, which is consistent with sea-floor spreading. When scientists extrapolate the current plate motions into the past, like running a movie backwards, they find that all of the continents converge, joining together into a single supercontinent, Pangaea, about 200 million years ago (see Fig. 4.7). As suggested by the Canadian geophysicist John Tuzo Wilson (1908–1993) in the 1970s, the heat-driven plate motions cause the continents to disperse and then reassemble over and over again, as oceans periodically open and close.

The radio interferometer measurements indicate that the Pacific Plate is migrating with a northwestward velocity of 0.048 meters per year, carrying Los Angeles northward and producing earthquakes along the edge of the plate (Fig. 4.12). At this rate, Los Angeles will be a suburb of San Francisco in 10 million years. The interferometer observations also indicate that the Atlantic Ocean is widening by 0.017 meters per year, so it was 8.7 meters narrower when Columbus crossed it in 1492.

Earthquakes

An earthquake is a trembling or shaking of the ground caused by a sudden release of energy stored in the rocks below the Earth's surface. The devastating tremors and after-shocks can ravage large sections of the land, flattening entire cities, awakening dormant volcanoes and creating new ones, draining lakes and causing floods, avalanches

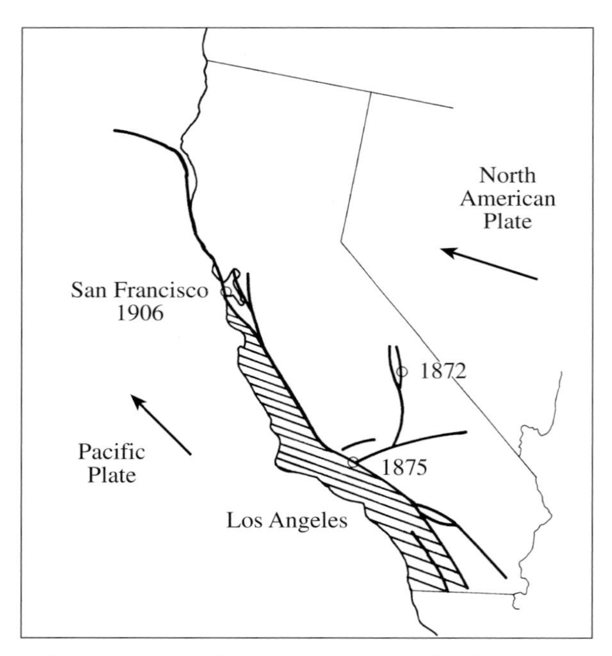

Fig. 4.12 Los Angeles is moving The Pacific Plate is moving northwestward at the rate of 0.048 meters per year, or about 5 meters every century, slowly carrying Los Angeles towards San Francisco. The North American and Pacific Plates strike and rub against each other like immense grindstones, producing earthquakes along their boundary known as the San Andreas fault. The dated circles denote places where very major earthquakes have occurred with magnitudes of 8 and over on the Richter scale.

and fires. Enormous tidal waves, called tsunami, can be generated by the quakes, racing across the ocean at high speed and wiping out everything on the shores they reach. Tens of thousands of people can be killed during a major earthquake, and hundreds of thousand more can be left homeless.

Like volcanoes, the world's earthquakes do not occur just anywhere, but usually along the edges of plates. They occur most often where the ocean floor is either being created or destroyed along a mid-ocean ridge and the deep-ocean trenches. Since an earthquake in the middle of the ocean floor is not likely to disrupt human life, we are naturally most interested in earthquakes that occur near the edges of continents where cities are located.

In addition to diverging plates, which are moving apart at a mid-ocean ridge, and convergent plates, which are heading toward a collision, there is a third type of plate boundary proposed in 1965 by John Tuzo Wilson. It is known as a transform fault, a place where plates move past one another, neither toward or away, and this is where earthquakes can occur. When the two plates meet along a transform fault, they "transform" their encounter into a slipping, sliding horizontal motion, and a sudden lurch in this motion can produce an earthquake.

The two plates on each side of a transform fault bump, crush, grind, rub and slide against each other, without creating or destroying crust, like two high-speed cars sideswiping each other, but in slow motion. A famous and visible example is the San Andreas Fault in California that marks the meeting of the Pacific Plate with the North American Plate (see Fig. 4.12). In 1906 a great earthquake devastated San Francisco, which is located at the edge of this fault.

The plates on each side of a transform fault build up stress along the line where they meet, and the stress is greatest where they are most tightly locked. As the friction and strain accumulate and rise over the years, a moment comes when the rock can't take it anymore, as when a festering problem of a family can surface into a screaming fight. In effect, the strain surpasses the strength of the rock. The stress is pushed to the limit, the two plates cannot slide further, and the accumulated energy is released as an earthquake. That part of the fault line then lurches back to its original equilibrium position, waiting for the next big one.

Since the time between major earthquakes in a given location can be a little longer than a human lifetime, many imagine that the danger is over, but the ground beneath their feet can remain unstable. Thus people living near the San Andreas Fault should no longer be concerned about *whether* another earthquake will occur, but *when* it will happen.

An instrument called a seismograph can measure the relative amount of energy released by an earthquake, its magnitude. The earthquake magnitude is given on a numerical scale, named after the American seismologist, Charles Francis Richter (1900–1985) who established it. Each increase of one unit on the Richter scale represents a 32-fold increase in the energy released by an earthquake. Major earthquakes usually measure between 6.0 and 9.1 (the highest ever recorded) on the Richter scale. The 1906 San Francisco earthquake would have measured 8.3 on the Richter scale, and the one that occurred there in 1989 measured 6.9. A magnitude 8 earthquake releases as much energy as detonating 6 million tons, or 6 billion kilograms, of trinitrotoluene, TNT.

The majority of earthquakes occur at depths of less than 70 kilometers. At greater depths, the rock turns from a solid into a flowing plastic due to the high temperatures and pressures, removing the frictional stress found in the solid-rock faults near the surface. Yet, almost 30 percent of the earthquakes occur at depths exceeding 70 kilometers and nearly 8 percent below 300 kilometers. These deep earthquakes occur when a tectonic plate plunges deeply into an ocean trench.

Something happens to the relatively cool, descending plate, triggering the earthquakes. It could be a transformation of the crystal structure of its minerals at high pressure, or it may be a result of old imperfections and zones of weakness in the descending slab. In either case, the likelihood of a powerful earthquake apparently depends on the angle of plate descent into a trench. When the angle is shallow, a lot of friction results and intense quakes occur; steep angles produce less rubbing and no great quakes.

The Earth's internal heat engine

What pushes the tectonic plates across the globe? Like humans, most of the driving forces that transform the Earth's face are hidden below its surface. Or, as the saying goes, it's what inside that counts. Heat, bottled up deep inside the Earth, produces internal currents that move the plates and propel the drifting continents.

The Earth's internal heat is left over from the time of the Earth's formation, within the liquid outer core, and augmented by the continued radioactive decay of elements such as uranium and thorium. As the internal heat tries to escape, it maintains a ceaseless, wheeling, churning motion, called convection, that turns and rolls over very slowly (Fig. 4.13). Convection occurs when molten rock becomes swollen by heat and rises through the cooler overlying material of lower pressure, like the currents in a pot of thick soup or oatmeal about to boil.

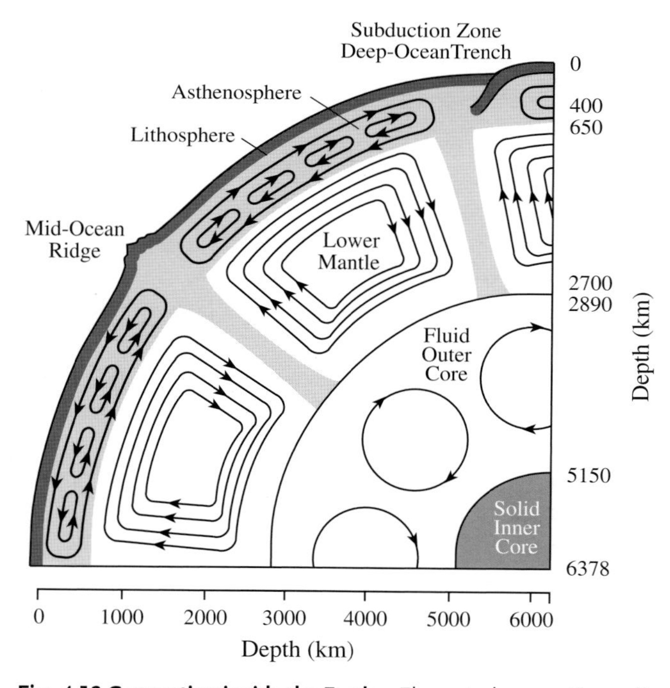

Fig. 4.13 Convection inside the Earth Elongated convection cells in the asthenosphere may be aligned in long cylinders that drive the overriding lithosphere plates along like a conveyor belt. A larger-scale circulation transports heat from the volcanic mid-ocean ridge to a deep-ocean subduction trench. Similar large-scale convection currents may lift and lower entire continents. Heat-driven convection in the fluid outer core probably generates and maintains the Earth's magnetic field.

The relatively low density of the hottest rock makes the material buoyant, so it ascends slowly; in contrast, the colder, denser rock sinks until heat escaping from the molten core warms it enough to make it rise again. Thus, the hot mantle rock flows in a circular pattern with hot rock rising in some places and cooler rock descending in others. The heated rock moves upward, spreads sideways, dragging portions of the lithosphere with it, and then cools and sinks, to be reheated and pushed upward again in an endless cycle. And the crust rides passively atop these giant convection cells, like dirt on a conveyor belt. Similar convective currents in the outer fluid core maintain the Earth's magnetic field. Thus, despite its ordered, layered structure, the inside of our planet is a dynamic, living mobile entity, with inner workings that can mix each layer up from the bottom to the top and back down again.

Powerful motions deep inside the planet not only push the plates horizontally, they also produce vertical changes at the surface. After all, the wheeling convection moves up and down, as well as sideways. Huge rising plumes of hot, buoyant, molten rock, originating and channeled at the rugged core–mantle boundary, can expand upward, piercing the mantle to lift and lower entire continents. The rising heat is now pushing South Africa up from below, and has been doing so for the past 100 million years.

The continents are poor conductors of heat, and therefore act like insulating blankets that try to block the heat's escape. The pent-up force of the trapped heat can even become powerful enough to split a continent apart. A gigantic plume of hot magma may even have played a major role in the break up of Pangaea.

Thus, the energy that drives the continents, spreads the sea floor, sets off earthquakes and ignites volcanoes is ultimately derived from the hot interior of the Earth. From the inner core all the way to the surface of the Earth, the dynamical activity of the Earth is driven by heat, and like any heat engine the Earth must be gradually running down. When the Earth's internal heat becomes totally depleted, it will become a geologically dead planet, and erosion will gradually flatten the mountains. In the meantime, continents continue to be renewed as the result of the Earth's internal heat engine.

The continents can grow in size by accumulating volcanic material when oceanic plates plunge under their edges, or by collisions of continental plates that can weld together and create some of the world's largest mountains. Underwater volcanoes can also rise to produce new islands.

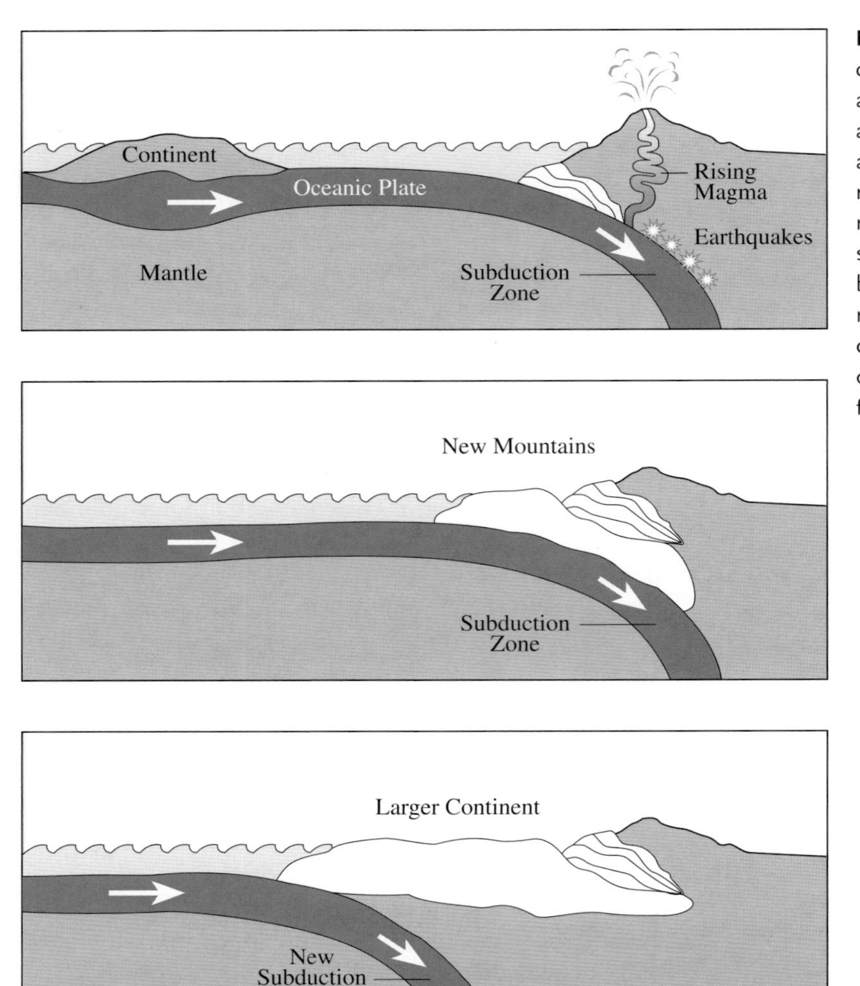

Fig. 4.14 Converging plates Magma, volcanoes and deep earthquakes are generated at a subduction zone (*top*) where a dense oceanic plate is pushed under a lighter continental plate. When continents on two moving plates meet head on, new mountains are generated (*middle*). In some situations, the advancing plate may become disrupted and the plate motion may stop. The two continents then become welded together forming a larger one, and a new subduction zone can be formed elsewhere (*bottom*).

An impermanent face

Moving plates provide the tools for sculpting the Earth's surface and altering its landscape. They have profoundly changed the way we view the world. Its entire surface is continuously shifting about and changing in shape and form. It is something like a rich old lady who keeps going in for face-lifts, in a futile attempt to resurrect her youth. Indeed, nothing on the Earth's face resembles itself as it was even several millions of years ago.

The transformations occur along the plate boundaries, not unlike the long, thin wrinkles in an aging face. High-standing belts of mountains and volcanoes are continuously being created when two plates converge along their borders, helping to hold the land above the sea.

As an ocean plate disappears into the trenches, great chains of towering volcanoes are created along the margins of continents (Fig. 4.14). The descending slab of lithosphere causes underground rock to melt, and the magma generated rises buoyantly to widen the continents at their edges. The Andes are still growing higher in this way, as the floor of the Pacific Ocean plunges beneath the west coast of South America.

The Pacific Ocean once reached to Colorado, and the western United States has been similarly grafted onto the continent. For most of the Earth's history, the land we call California did not exist. Where California has come to be there was only the deep blue sea reaching down to the spreading ocean floor. But the floor was moving into the Earth and being consumed under the ancient shoreline, creating volcanoes that rose to create new land.

On average, about two billion cubic meters of lava and ash are now being added to the continents by volcanoes each year.

The rising material also brings valuable metals and precious stones to the surface. All the gold in California originated in this way, as did the famous copper deposits in *Cyprus*, the Greek word for "copper", rising with volcanic magma and spewing out on the surface with the lava flows.

Diamonds were also forged in the crucible of the Earth's hot interior. The crushing pressures and blistering heat far within the mantle worked in unison to squeeze carbon into diamonds, and some of them have risen from the deep, entrained in explosive volcanic eruptions, even in the middle of continents.

Eventually, a moving continent reaches an open trench and jams it shut (Fig. 4.14), like trying to shove an eggplant down a garbage disposal. Continents are too light and thick to be subducted, and when they arrive at a trench it is closed up, like a sutured wound.

The violent collisions between continents have created the world's tallest mountains. When the continents meet, they buckle upward to form a range of mountains. Both land and oceanic sediment, built up over many millions of years, are tossed into the sky. The magnificent Himalayan range was formed this way (Fig. 4.15), when the Indo-Australian plate, with India firmly embedded, ran into the Eurasian plate, like a head-on collision of two cars. Slowly, the Himalayas shot up as India rammed into Asia, carrying the fossilized remains of ancient sea-creatures with them. Today the plate that carries India continues to slide beneath the Eurasian plate, widening the Indian Ocean and pushing the mighty Himalayas upward.

The European Alps have been fashioned in a similar fashion to the Himalayas, when the African Plate moved Italy up from Africa and collided with the Eurasian plate along Switzerland's former ocean shore. Today the African plate continues pressing Italy northward and raising the Alps.

Like adolescents, the relatively young mountains in the Alps and Himalayas just can't stop growing, but there is a limit to how high they can stand. Gravity opposes the upward forces, and erosion wears away mountain summits as they are being pushed up from below. As massive as it is, a range of mountains cannot resist eventual destruction by wind, water, and ice. Old mountain ranges, such as the Appalachians in the United States, once stood as high as today's Himalayas, but they have eroded into gentle undulations and rounded knobs.

Given the great age of the Earth, some 4.6 billion years, this erosion acting by itself would have worn away the continental mountains in a few hundred million years, and the globe would have long ago been covered by oceans. But the mountains are constantly being rebuilt while enlarging the continents. Many of the world's present-day continents have indeed been assembled as former continents welded together. Eurasia is still being put together as India and Australia are arriving from the south, and much of the eastern part of North American was once stitched together by similar collisions.

The edges of plates are not the only place that land rises above the sea. Some oceanic islands are located thousands of kilometers from the nearest plate boundaries. Chains or strings of such isolated islands are attributed to hot spots, an idea introduced by the prolific John Tuzo Wilson back in 1963. The hot spots are rising plumes of magma, liquid or molten rock anchored far beneath the ocean and deep within the mantle, even as far down as the core–mantle boundary. The relatively small, long-lasting and exceptionally hot regions provide a persistent source of magma capable of penetrating the mantle and piercing an overriding lithosphere plate, like a fixed blowtorch might melt holes in a steel plate moving by. As the plate glides slowly overhead at the rate of a few meters every century, it can leave a trail of islands that have risen out of the sea.

The Hawaiian islands were formed in such a way, as the Pacific Plate moved over a deep, stationary hot spot at the slow rate of 0.13 meters per year (Fig. 4.16). Stretching to the north and west of the big island of Hawaii, we find a string of smaller islands, including Oahu and Midway and submerged volcanoes, or seamounts, about 6 thousand kilometers long. Every one of these islands and seamounts was formed in the exact place where Hawaii now stands. The plume pushed the first Hawaiian island up above the ocean surface in this location about 70 million years ago.

Kilauea, the world's largest active volcano, is still rumbling because Hawaii has yet to move completely off the hot spot. At the same time, the underwater volcano, Loihi, is being formed as the Pacific Plate moves steadily on, continuing its relentless journey over the hot spot (Fig. 4.16). In about 50 thousand years Loihi should grow high enough to form the next Hawaiian Island.

But why aren't the Hawaiian Islands just one long extended island, eroded away on the oldest end and standing tallest at the youngest end? Although the lower parts of the hot-spot plumes are shaped like the thin stem of a wine glass, their tops flare out into mushroom-shaped reservoirs of molten rock that pools under the lithosphere and overriding crust, slowly melting them and gathering enough strength for penetration. It takes thousands of years to break on through to the other side, just as a welder's torch takes a while to burst through a steel plate. Sporadic bubbles rather than a continuous stream of hot rock burst through the ocean floor, forming a succession of oceanic islands.

Most of the hot spots lie under oceans and give birth to island chains, but some of them penetrate the mantle under the continents. Such a hot spot has created the hot springs, boiling mud and geysers of the Yellowstone National Park in Wyoming. As the North American Plate moved above this hot spot, it created a long line of volcanoes that are now extinct, with ages ranging from 0.6 to

Fig. 4.15 Mount Kailash This mountain, sacred to both Hindus and Buddhists, is called Kang Rinpoche (The Precious Snow Mountain) by the Tibetans. It was formed by the collision of two former continents, now welded together in a seam known as the Himalayan mountain range. Buddhists consider it to be the palace of the deity Chakasamvara (Wheel of Supreme Bliss) and Hindus consider it to be the dwelling place of Shiva. There are also many nearby caves where famous hermits like Milarepa meditated for years. Pilgrims have been visiting and circumambulating Mount Kailash, in western Tibet, for thousands of years. (Courtesy of Matthieu Ricard, Shechen Monastery, Katmandu, Nepal.)

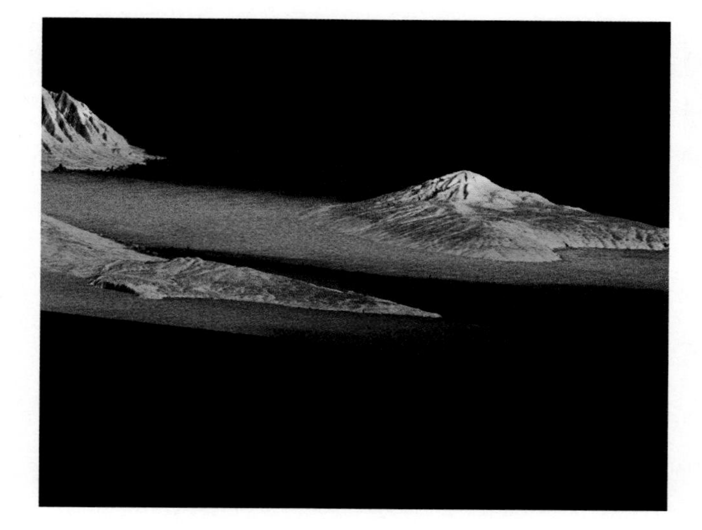

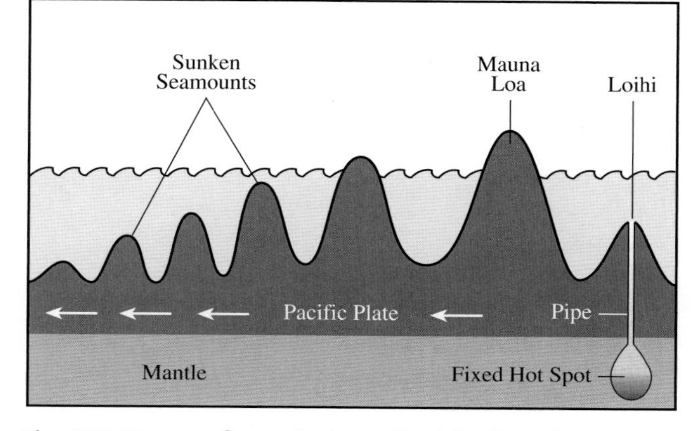

Fig. 4.16 Hot spot forms the Hawaiian islands A hot spot anchored deep within the Earth (*bottom*) has recently fed molten lava through a long pipe to Mauna Loa (Fig. 2.25) on the big island of Hawaii. The moving Pacific Plate has carried three other volcanic islands away from the hot spot; these extinct volcanoes are shown (*top*) in the accompanying radar image obtained from the *Space Shuttle* – Molokai (*left*), Lanai (*right*), and the northwest tip of Maui (*upper left*). As the plate moves on, wind and water erode the peaks, reducing the older ones to sunken islands known as seamounts. An underwater volcano, named Loihi, is now forming over the hot spot; it should rise above the ocean to become another Hawaiian island in about 50 thousand years.

6 million years. A semi-dormant volcano now rests under Yellowstone.

If a plate carrying a continent comes to rest over a hot spot, the heat and pressure from the upwelling magma will weaken and stretch the overlying continental crust. And when the crust is stretched beyond its limits, cracks or rifts will form in it. The magma rises and squeezes through the widening cracks, forming volcanoes. If the upwelling is short-lived, the result is merely a rift scar, such as the Rhine Valley. If it persists, the rift widens and a continent can literally be split in two. In time, the gap reaches a coastline, permitting sea water to flow in and a new ocean is created (Fig. 4.17).

Hot spots are now tearing Africa apart at its seams. A Great Rift Valley stretches from Ethiopia to Tanzania (see Fig. 2.26); as it widens the continent will break apart and the sea will eventually enter. The African and Arabian Plates have already pulled apart in another location, forming the Red Sea and the Gulf of Aden. They are developing into an ocean that may eventually rival the Atlantic Ocean in size (Fig. 4.18). At the same time, the Mediterranean Sea is narrowing as Africa moves toward Europe.

Thus a new dynamic picture of the Earth has emerged. Continents are growing at their edges; colliding continents are raising great mountain chains; earthquakes are bringing vast destruction as two plates grind together; chains of volcanic islands are rising from the ocean's depths; converging continents are squeezing oceans out of existence while new oceans open up where continents are splitting apart; and the ocean floor remains in eternal youth as new floor spills out of the mid-ocean ridge and old floor is pushed back into the Earth. As we shall next see, our atmosphere is also in a perpetual state of flux, forever changing in sometimes-dangerous ways.

4.4 The Earth's changing atmosphere

Our Sun-layered atmosphere

Our thin atmosphere is pulled close to the Earth by its gravity, and suspended above the ground by molecular motion. Its height is about 1 percent of the planet's diameter. Compared to the size of the Earth, the thickness of the air is something like the width of a window on a big building.

Because air molecules are mainly far apart, our atmosphere is mostly empty space, and it can always be squeezed into a smaller volume. The atmosphere near the ground is compacted to its greatest density and pressure by the weight of the overlying air. At greater heights there is less air pushing down from above, so the compression is less and the density and pressure of the air falls off into the near-vacuum of space.

Not only does the atmospheric pressure decrease as we go upward; the temperature of the air also changes. It decreases steadily with increasing height in the lowest region of our atmosphere, called the troposphere from the Greek *tropo* for "turning". The troposphere extends from the Earth's surface to about 12 kilometers above sea level (Fig. 4.19).

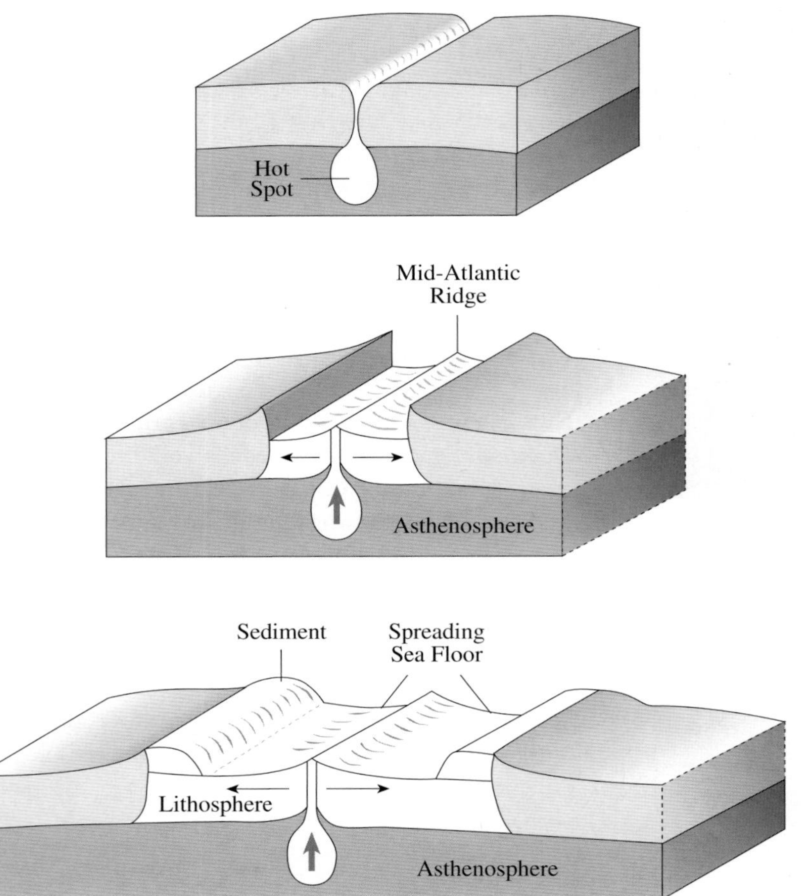

Fig. 4.17 The rifting of a continent A continental rift begins when molten lava rises up from deep in the Earth's interior and splits a continent open. As the fissure grows and widens, a future ocean floor spreads away from the ridge. Water should eventually flow into the cavity, making a new ocean.

All of the Earth's weather takes place in the troposphere, a thin planetary skin just one-thousandth the diameter of the Earth. Yet, about 80 percent of the atmosphere's total mass and almost all its water are concentrated in the troposphere.

The temperature falls at increasing heights in the troposphere because this layer of our atmosphere is heated from below by the greenhouse effect (described subsequently), and because the air expands in the lower pressure at higher altitudes and becomes cooler. The average air temperature drops below the freezing point of water (273 kelvin) about one kilometer above the Earth's surface, and bottoms out at roughly ten times this height.

But the temperature is not a simple fall-off with height. It falls and rises in two full cycles as we move off into space (Fig. 4.19). The temperature increases are produced by the Sun's invisible radiation.

Different types of radiation differ in their wavelength, though they propagate at the same constant speed – the velocity of light. They range from long radio waves, about a meter in length, to short X-rays whose wavelengths are roughly a billionth, or 10^{-9}, of a meter. The radiation at most of these wavelengths goes unseen by humans. Our eyes are only sensitive to a narrow range of visible colors, with wavelengths between 4 ten-millionths and 7 ten-millionths, or 4×10^{-7} and 7×10^{-7}, meters. Ultraviolet radiation is on the short-wavelength side of blue light, with wavelengths of about 2 ten-millionths, or 2×10^{-7}, meters. The wavelengths of X-rays are about a hundred times shorter than the ultraviolet rays.

The most intense radiation from the Sun is emitted at visible wavelengths, and our atmosphere permits it to reach the ground. This is the colored sunlight that our eyes respond to. The Sun emits lesser amounts of invisible, short-wavelength radiation, which is partially or totally absorbed in the atmosphere.

Even though the total amount of invisible solar radiation is substantially less than the visible emission, the individual short-wavelength rays are more energetic. That is why we get sunburns from the ultraviolet radiation that manages to get through the atmosphere, and need to be protected from the Sun when climbing at high altitudes where the air is thinner and more ultraviolet penetrates the atmosphere. The greater energy of radiation at

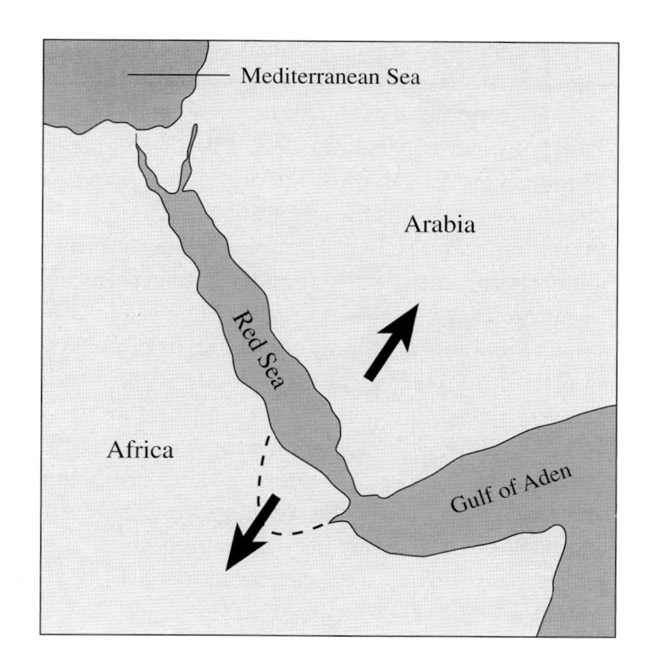

Fig. 4.18 An infant ocean An ocean is being born where the Arabian peninsula and the African continent are moving apart, a process that began about 20 million years ago. In a few hundred million years, the Red Sea could be as wide as the Atlantic Ocean is now. (Satellite photograph courtesy of NASA.)

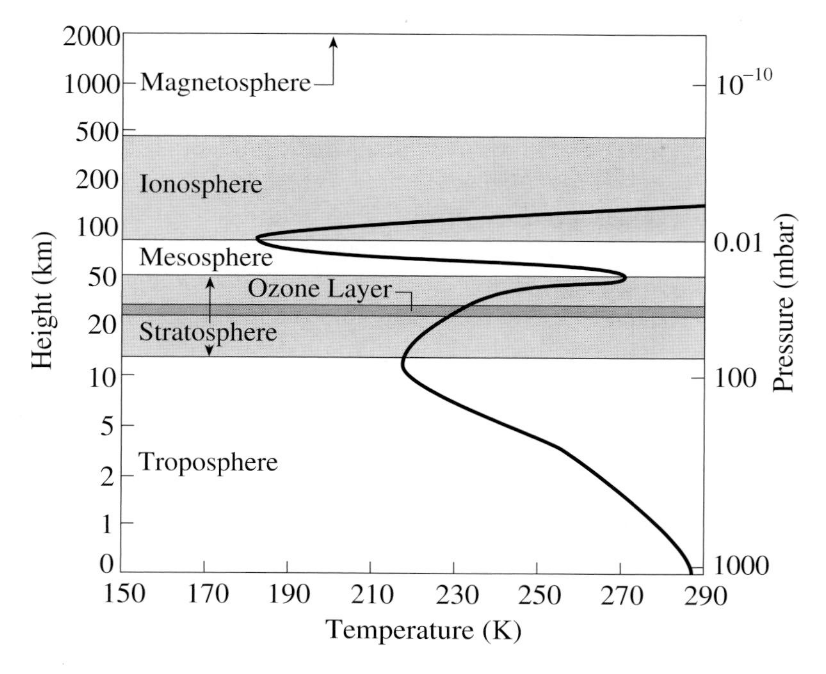

Fig. 4.19 Earth's layered atmosphere The pressure of our atmosphere (*right scale*) decreases with altitude (*left scale*). This is because fewer particles are able overcome the Earth's gravitational pull and reach higher altitudes. The temperature (*bottom scale*) also decreases steadily with height in the ground-hugging troposphere, but the temperature increases in two higher regions that are heated by the Sun. They are the stratosphere, with its critical ozone layer, and the ionosphere. The stratosphere is mainly heated by ultraviolet radiation from the Sun, and the ionosphere is created and modulated by the Sun's X-ray and extreme ultraviolet radiation.

shorter wavelengths also explains why X-rays, generated by machines, can see through your skin and muscles to detect your bones.

When absorbed in our air, the invisible short-wavelength radiation from the Sun transfers its energy to the atoms and molecules there, causing the temperature to rise. There is, for example, a gradual increase in temperature just above the troposphere, within the next atmospheric layer named the stratosphere (Fig. 4.19). This layer is located between 10 and 50 kilometers above the Earth's surface. Its name is coined from the words "stratum" and "sphere".

The mesosphere, from the Greek *meso* for "intermediate", lies just above the stratosphere. The temperature declines rapidly with increasing height in the mesosphere, reaching the lowest levels in the entire atmosphere. The main reason for the decreasing temperature is the falling ozone concentration and decreased absorption of solar ultraviolet radiation.

The temperature then begins to rise again with altitude in the ionosphere, a permanent, spherical shell of electrons and ions, reaching temperatures that are hotter than the ground. The ionosphere is created and heated by absorbing the extreme ultraviolet and X-ray portions of the Sun's energy. This radiation tears electrons off the atoms and molecules in the upper atmosphere, thereby creating ions and free electrons that are not attached to atoms.

The ionosphere was postulated in 1902 to explain Guglielmo Marconi's (1874–1937) transatlantic radio communications. Since radio waves travel in straight lines, and cannot pass through the solid Earth, they get around the planet's curvature by reflection from electrons in the ionosphere.

The Sun's invisible rays make ozone

Solar ultraviolet radiation is largely absorbed in the cold stratosphere, where it helps make ozone. When ultraviolet rays strike a molecule of the ordinary diatomic oxygen that we breathe, denoted by O_2, they split it into its two component oxygen atoms. Some of the freed oxygen atoms then bump into, and become attached to, an oxygen molecule, creating an ozone molecule, abbreviated O_3, that has three oxygen atoms instead of two. The Sun's ultraviolet rays thereby produce a globe-encircling layer of ozone in the stratosphere.

Although the ozone is present to the extent of only about 10 parts per million by volume, the ozone layer is critical to life below. It protects us by absorbing most of the Sun's ultraviolet emission and keeping its destructive rays from reaching the ground. If there were no ozone shield, plants, animals and humans could not even exist on land.

The amount of ozone in the stratosphere resembles the level of water in a leaky bucket. When water is poured into the bucket, it rises until the amount of water poured in each minute equals the amount leaking out. A steady state has then been reached. The amount of water in the bucket stops rising and it will stay at the same level as long as you keep pouring water in at the same rate. However, if you pour the water in at a different rate, or punch a few more holes in the bucket, the steady-state level of water in the bucket changes.

Solar ultraviolet radiation supplies ozone to the stratosphere from above, like pouring water into a bucket, at a rate that depends on the varying ultraviolet output of the Sun. We have recently been punching holes in the ozone layer from below, with chemicals used in our everyday lives.

Synthetic chemicals are destroying the ozone layer

Man-made chemicals, called chlorofluorocarbons, are consuming the protective ozone layer, eating holes in it and making it thinner. They are synthetic chemicals, entirely of human origin with no counterparts in nature. The name of the chemicals is a giveaway to their composition. Each molecule has been constructed in company laboratories by linking atoms of chlorine, fluorine and carbon.

The shorthand CFC notation abbreviates some of them. A number sometimes follows, providing a complex description of the number of atoms in each molecule, the most widely used being CFC-11 and CFC-12.

Beginning in 1930, the biggest producer of CFCs, the Du Pont Company, manufactured and marketed them under the name Freons. They have been widely used in refrigerators, plastic foams, spray-can propellants, automobile air-conditioning systems, and the cleaning of circuit boards used in televisions and computers.

These hardy chemicals don't interact chemically to form other substances. They are so inert and stable that once entering the atmosphere the CFC molecules can survive for more than a century, permitting them to drift and waft up into the ozone layer in the stratosphere. Although more than 20 million tons, or 20 billion kilograms, of CFCs have been released into the air, their combined concentration isn't very significant, only about one CFC molecule for every two billion molecules in the air. Yet, even these seemingly insignificant amounts can have enormous impact.

In 1974, Mario J. Molina (1943–) and F. Sherwood Rowland (1927–), two chemists who were then at the University of California at Irvine, showed that the chlorine in CFCs can destroy enormous amounts of ozone. Once arriving in the stratosphere, the Sun's ultraviolet rays will split chlorine atoms out of the CFCs, and the liberated chlorine sets off a self-sustaining chain reaction that destroys the ozone. A single chlorine atom will react with an ozone molecule, taking one oxygen atom to form chlorine monoxide; the ozone is thereby returned to a normal oxygen molecule and its ultraviolet-absorbing capability is largely removed. Moreover, when the chlorine monoxide encounters a free oxygen atom, the chlorine is set free to strike again. Each chlorine atom thus acts as a catalyst, destroying about ten thousand ozone molecules before it finally combines permanently with hydrogen.

Molina and Rowland were awarded the Nobel Prize in Chemistry in 1995 for their "contribution to our salvation from a global environmental problem that could have catastrophic consequences". They shared the prize with the German chemist, Paul J. Crutzen (1933–), who showed how the rate of ozone depletion could be accelerated by other chemical reactions in the atmosphere.

The ozone layer is itself invisible. But you can determine its ozone content by measuring the amount of solar ultraviolet radiation getting through the layer and reaching the ground. When there is more ozone, greater amounts of ultraviolet are absorbed in the stratosphere and less reaches the ground, and when the ozone layer is depleted, more of the Sun's ultraviolet rays strike the Earth's surface.

The British scientist G. M. B. Dobson (1889–1976) pioneered measurements of the air's ozone content about half a century ago. When his instrument was installed at Halley Bay, Antarctica, in 1957–8, Dobson found that the ozone abundance in polar spring (September–November) was noticeably less than that above other parts of the world.

Other British scientists continuously monitored the southern polar skies for 27 years; always detecting a springtime loss that became steadily larger as the years went on. By 1985 the ozone loss above Antarctica had nearly doubled when compared to the earlier measurements in the 1960s, and it extended all the way to the tip of South America, where another British monitoring station detected it. A continent-sized hole had opened up in the sky – the ozone hole (Fig. 4.20).

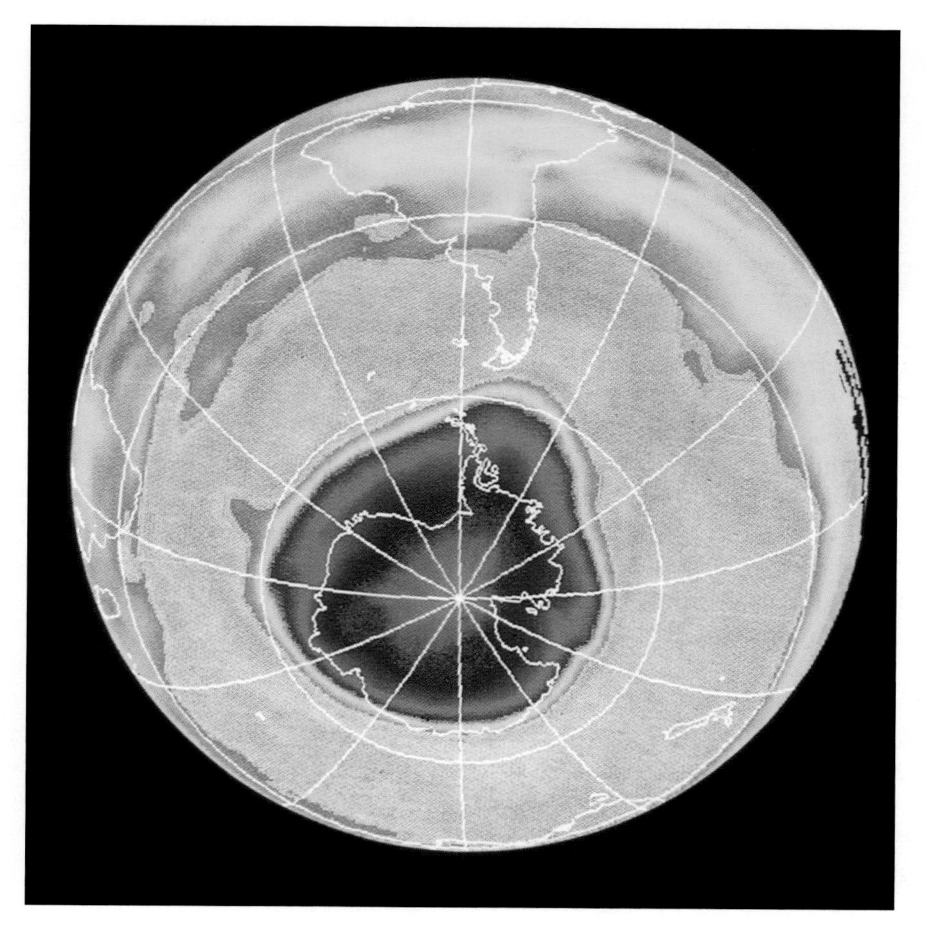

Fig. 4.20 Hole in the sky A satellite map showing an exceptionally low concentration of ozone, called the ozone hole, that forms above the South Pole in the local winter. In October 1990 it had an area larger than the Antarctic continent, shown in outline below the hole. Eventually spring warming breaks up the polar vortex and disperses the ozone-deficient air over the rest of the planet. (Courtesy of NASA.)

This unexpected discovery astounded space-age scientists who had not detected any ozone hole using satellites that had been monitoring the ozone layer from above. Their computers had been programmed to automatically reject large ozone depletions, apparently because their models did not predict such huge losses. So the now-famous ozone hole had been discarded as an anomaly, perhaps caused by an instrumental error. After re-analyzing the satellite data, the scientists confirmed the existence of an ozone hole in the local springtime above the South Pole.

We now know that whirling winds concentrate ozone-destroying chemicals, the CFCs, within a vast, towering vortex above Antarctica, resembling the eye of an immense hurricane. Each year the gaping hole opens up during Antarctic spring when the sunlight triggers ozone-destroying chemical reactions; the hole starts to close up in the early polar fall when the long sunless winter begins. Ozone-depleted air is dispersed globally, and the ozone is slowly restored, filling the hole until the cycle repeats in the following year.

Doing something about ozone depletion

The sudden and frightening discovery of an enormous ozone hole in 1985 sparked public awareness of the fragile ozone layer. The scientific community had been actively investigating Molina and Rowland's theory that synthetic chemicals, known as the CFCs, could be destroying the ozone layer. Although global models of the expected ozone depletion initially led to widely varying estimates of the potential threat, affecting the scientist's credibility and dampening public concern, a coordinated international investigation eventually led to a unified assessment of the problem.

A group of approximately 150 scientific experts reported in 1986 that atmospheric accumulations of CFC-11 and CFC-12 had nearly doubled from 1975 to 1985. The continued release of the synthetic chemicals at the 1980 rate could, they said, reduce the ozone layer by about 9 percent on a global average by the last half of the 21st century, with even greater seasonal and latitudinal differences. As a result, higher levels of dangerous ultraviolet radiation could reach heavily populated regions of the northern hemisphere.

Who cares if chemicals are punching a few holes in the sky and letting a little more sunlight reach the ground? The U.S. Environmental Protection Agency, or EPA, cared. In 1986 it published a report of the many serious consequences of ozone depletion. A thinner ozone layer lets more solar ultraviolet radiation through to the ground, where it can produce severe biological harm. The most energetic ultraviolet rays will reduce the effectiveness of the human immune system, increasing human vulnerability to infections and cancer.

The EPA estimated that there could be over 150 million new cases of skin cancer in the United States alone among people currently alive or born by the year 2075, resulting in over 3 million deaths. The dangerous ultraviolet radiation would also produce eye cataracts, distorting the vision of about 18 million people in the same population and blinding many of them. Added to this was the potential of widespread genetic damage to crops and forests, if nothing was done to stem the production of ozone-destroying chemicals.

Faced with the evidence of vanishing ozone, the global increase of atmospheric CFCs, and the prospect of widespread skin cancer and eye cataracts, international diplomats forged an accord in 1987 to limit and eventually ban the production of the substances that deplete the ozone layer (Focus 4.3). The treaty, known as *The Montreal Protocol*, has led to substantial reductions in ozone destroyers.

Although production of ozone-destroying substances has been substantially curtailed under international agreement, more than 20 million tons of them have already been dumped into the atmosphere, and this damage cannot be undone. Because of their long lifetime and slow diffusion into the stratosphere, the synthetic chemicals that are already in the air will keep on destroying the ozone layer for about a century.

The springtime hole that opens up in the ozone layer over Antarctica has, in fact, shown no signs of closing up, even though many of the chemicals producing them have long been banned. In the year 1990 the ozone hole was about as wide as the North American continent, covering approximately 14 million million (1.4×10^{13}) square meters. A decade later, in the year 2000, the ozone hole had grown to its largest size yet, measuring 28 million million (2.8×10^{13}) square meters despite the near abolishment of the CFCs.

The chemicals that were already in the air were apparently still wafting upward into the stratosphere. Hopefully their abundance in the ozone layer peaked at the end of the 20th century. After that, ozone loss should be stabilized and the trends reversed as the chemicals are gradually removed by slow washing-out processes.

The ozone layer will not regain full strength until well into the latter half of the 21st century when it should then recover to the natural levels that existed before the ozone hole was discovered. In the meantime, scientists will continue to monitor the ozone layer using a series of NASA satellites, while also keeping a close eye on the Sun's varying ultraviolet output which modulates ozone production by amounts comparable to human destruction of it.

Focus 4.3 *The Montreal Protocol*

Growing scientific, political and public awareness of ozone depletion eventually resulted in international negotiations to limit the manufacture and use of chemicals that destroy atmospheric ozone. A scientific consensus in the 1980s indicated that the destruction of the ozone layer by these synthetic chemicals had already begun, and that this would ultimately affect the health of large numbers of people if the production of the chemicals was not curtailed. The press and television also played a vital role in informing the public about the dangers, and the discovery of the ozone hole above Antarctica attracted added attention to the problem.

On 16 September 1987, representatives of 24 nations signed *The Montreal Protocol on Substances That Deplete the Ozone Layer*, including the United States which was the largest single producer and consumer of the suspect chemicals. The treaty agreed to a 50 percent reduction in important ozone destroyers, such as the chlorofluorocarbons designated as CFC-11 and CFC-12, below 1986 levels by mid-1998.

The ready acceptance of *The Montreal Protocol* was undoubtedly eased by the development of substitutes for CFCs in refrigerators, air conditioners, foaming, and cleaning solvents. In fact, the biggest producer, Du Pont, unilaterally stopped making the chemicals even before the Protocol required it.

The Montreal Protocol was strengthened in 1990, at a meeting in London, when it was agreed to a complete phase-out of CFC-11 and CFC-12 by the year 2000. The phase-out schedules of other ozone-depleting substances were accelerated at another meeting in Copenhagen in 1992, including the halons, which are another type of chlorofluorocarbon. At a meeting in Vienna in 1995 further controls were implemented. A total of 155 countries had ratified *The Montreal Protocol* by 1996, including the vast majority of both the producers and consumers of the dangerous substances.

The phase-out schedules differed for the rich, industrial countries and the poor, undeveloped ones. The industrial countries had to eliminate halon consumption as of 1 January 1994 and CFC consumption as of 1 January 1996. Developing countries were given a grace period, but had to complete their elimination by 1 January 2010. Several of these countries will complete their phase-out much before this date. A multilateral fund was also established to assist the developing countries in meeting their goals.

Further amendments were made at meetings in Montreal in 1997 and in Beijing in 1999, all aimed at reducing and eventually eliminating the emissions of all kinds of man-made ozone-depleting substances. For example, an amendment, that took effect on 1 January 2001, requires participating countries to stop the production of hydrochlorofluorocarbons (HCFCs) still used in refrigeration and cooling equipment.

The accord has accomplished far more than halting the production of dangerous synthetic chemicals. *The Montreal Protocol* was the first international agreement to protect the global environment. The treaty also marked the first time that the governments of the industrial nations agreed to help developing countries with environmentally safe substances and technology. It was further hoped that the precedent would pave the way for international agreement on global warming.

Heating by the greenhouse effect

Our planet's surface is now kept at a comfortable temperature because the atmosphere traps some of the radiant heat from the Sun and keeps it near the surface, warming the planet and sustaining living creatures. Jean Baptiste Joseph Fourier (1768–1830) first conceived the mechanism in the 1820s, while wondering how the Sun's heat could be retained to keep the Earth hot. The French mathematician had previously invented new mathematical tools to describe the transfer of heat. This branch of mathematics, which deals with waves, periodic motion and harmonic motion, is now known as Fourier analysis and it includes the Fourier series that is widely used by today's astronomers, engineers and physicists.

Fourier's idea, still accepted today, is that the atmosphere lets some of the Sun's radiation in, but it doesn't let all of the radiation back out. Visible sunlight passes through our transparent atmosphere to warm the Earth's land and oceans, and some of this heat is reradiated in infrared form. The longer infrared rays are less energetic than visible ones and do not slice through the atmosphere as easily as visible light.

So our atmosphere absorbs some of the infrared heat radiation, and some of the trapped heat is reradiated downward to warm the planet's surface and the air immediately above it. Fourier likened the thin atmospheric blanket to a huge glass bell jar, made out of clouds and gases, that holds the Earth's heat close to its surface.

The warming by heat-trapping gases in the air is now known as the "greenhouse effect", but this is a misnomer. The air inside a garden greenhouse is heated because it is enclosed, preventing the circulation of air currents that would carry away heat and cool the interior. Nevertheless, the term is now so common that we will also sometimes designate the heat-trapping gases as greenhouse gases, and let greenhouse effect designate the process by which an atmosphere traps heat near a planet's surface.

Right now, the warming influence is literally a matter of life and death. It keeps the average surface temperature of the planet at 288 kelvin (15 degrees Celsius or 59 degrees Fahrenheit). Without this greenhouse effect, the average surface temperature would be 255 kelvin (−18 degrees Celsius or 0 degrees Fahrenheit), a temperature so low that all water on Earth would freeze, the oceans would turn into ice and life, as we know it, would not exist.

The gases that absorb infrared heat radiation are minor ingredients of our atmosphere. In a series of careful laboratory experiments in the 1850s, the Irish physicist John Tyndall (1820–1893) showed that significant heat is absorbed by water vapor and carbon dioxide, with water vapor absorbing the most. Tyndall also explained why the sky is blue; although sunlight comes in all colors, air molecules preferentially scatter the blue light down to us.

Water vapor is by far the most powerful heat-trapping gas in our air. Sixty to seventy percent of the Earth's greenhouse warming is now caused by water vapor, and carbon dioxide provides just a few degrees.

The main constituents of the atmosphere, nitrogen (77 percent) and oxygen (21 percent) play no part in the greenhouse effect. The two atoms in these diatomic molecules are bound tightly together and are therefore incapable of absorbing significant infrared radiation. In contrast, water vapor and carbon-dioxide molecules consist of three atoms that are less constrained in their motion, so they absorb the heat radiation.

Why doesn't the atmosphere just keep heating up until it explodes? The greenhouse warming rises to a fixed temperature that balances the heat input from sunlight and the heat radiated into space. The level of water in a pond similarly remains much the same even though water is running in one end and out the other.

Of course, you can have too much of a good thing, like lying in the Sun all day in the summer or eating too much ice cream too fast. The greenhouse effect of a dense carbon-dioxide atmosphere on Venus has raised the planet's surface temperature to 735 kelvin, which is hot enough to melt lead (Section 3.1, 7.3). The high-temperature world has been boiled dry, like a kettle left on the stove too long. Mars has relatively little carbon dioxide or water vapor, and its surface is now frozen (Sections 3.1, 8.3). As Goldilocks might say, the surface temperature of Earth is just right, enough to keep most of our water in liquid form.

Humans are pumping increasing amounts of carbon dioxide into the air

For hundreds of years, humans have been filling the sky with carbon dioxide. The invisible waste gas is dumped into the air by burning fossil fuels – coal, oil and natural gas. When these materials are burned, their carbon atoms, denoted C, enter the air and combine with oxygen atoms, O, or oxygen molecules, O_2, to make carbon dioxide, abbreviated CO_2.

About a century ago, in the middle of the industrial revolution, it was mainly coal that fueled factory boilers and warmed city houses, releasing carbon into the atmosphere to make carbon dioxide. Since the gas is colorless and odorless, it could not be seen or smelled, but it was detected indirectly by the noxious fumes emitted as a byproduct of burning high-sulfur coal. Coal burning has blackened entire cities, such as London, described in 1854 by Charles Dickens (1812–1870) in *Hard Times.* His account is an eerie foreboding of the dark, polluted skies that are now found in Beijing and Mexico City, and the brown cloud overshadowing large areas of Asia.

In the 20th century, the perfection of the internal combustion engine and the mass production of automobiles made oil one of the most important sources of atmospheric carbon, and therefore of carbon dioxide. Around the globe, cars, sports utility vehicles and trucks have been releasing huge amounts of the potentially dangerous, heat-trapping gas into the air.

Every time we drive a car, use electricity from coal-, gas- or oil-fired power plants, or heat our homes with oil or natural gas, we release carbon into the lower atmosphere. The burning of forests has also contributed.

Just a few decades ago, no one knew if any of the carbon dioxide stayed in the atmosphere or if it was all being absorbed in the oceans. Then in 1958 Charles David Keeling (1928–) began measurements of its abundance in the clean air at the Mauna Loa Observatory in Hawaii. It is located at a remote high-altitude site in the midst of a barren lava field, far from cars and people that produce carbon dioxide and from nearby plants that might absorb it.

The sensitive measurements showed that the amount of carbon dioxide in the atmosphere above Hawaii increases and decreases in an annual cycle related to plant growth in the northern hemisphere. Every spring plants bloom, sucking CO_2 out of the air, and every fall CO_2 is released back into the air as plants either decay or lose their

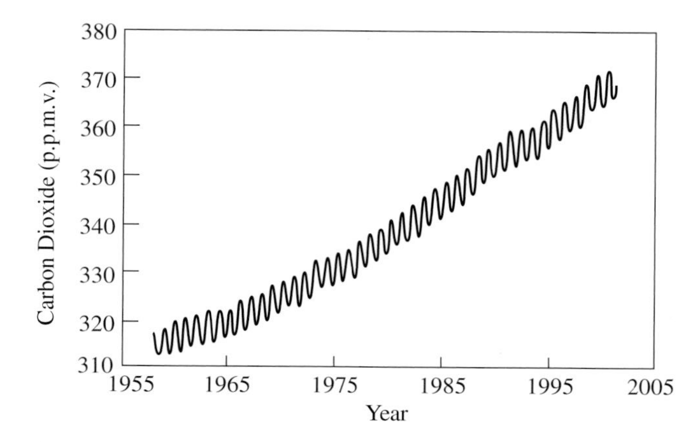

Fig. 4.21 Rise in atmospheric carbon dioxide The average monthly concentration of atmospheric carbon dioxide, or CO_2 for short, in parts per million by volume, abbreviated p.p.m.v., of dry air plotted against time in years observed continuously since 1958 at the Mauna Loa Observatory, Hawaii. It shows that atmospheric amounts of the principal waste gas of industrial societies, carbon dioxide, have risen steadily for more than forty years. The up and down fluctuations, that are superimposed on the systematic increase, reflect a seasonal rise and fall in the absorption of carbon dioxide by trees and other vegetation. Summertime lows are caused by the uptake of carbon dioxide by plants, and the winter highs occur when the plants' leaves fall and some of the gas is returned to the air. (Courtesy of Dave Keeling and Tim Whorf, Scripps Institution of Oceanography.)

leaves. The measurements had recorded the breathing of the plants all over the northern hemisphere.

But more importantly, Keeling's measurements showed that humans are also changing the composition of the atmosphere. Superimposed on the annual fluctuations, there was a systematic increase over the entire period of observation, continuing non-stop since 1958 (Fig. 4.21). Year by year the total measured concentration of carbon dioxide grew, as inexorably as the expansion of the world's population and human industry.

Since 1958, atmospheric concentrations of CO_2 have increased from 315 parts per million by volume, abbreviated 315 p.p.m.v., to 365 p.p.m.v. at the turn of the century.

Studies of ice deposits in Antarctica indicate that the amount of CO_2 has been increasing at an exponential rate ever since the beginning of the industrial revolution in the mid-18th century. Air bubbles that are trapped in the ice act like time capsules, conserving the atmosphere of the past. The air was sealed off in the bubbles when the ice was laid down, and extracted from cores drilled deep within the layered ice deposits. The ice cores show that the concentrations of the gas averaged 280 p.p.m.v. just before the industrial era. In the succeeding two-and-a-half centuries, a mere blink in the eye of cosmic time, the atmospheric concentration of carbon dioxide has increased 31 percent.

And that's not all. Air bubbles extracted from deeper layers of ice show that there is now more carbon dioxide in the atmosphere than there has been for 420 thousand years.

The atmosphere now contains almost 800 billion tons, or 800 trillion kilograms, of carbon dioxide. Humans continue to release about 7 billion tons of it each year. In other words, each person on Earth is, on average, dumping about a ton of carbon dioxide into the air every year, and there is no end in sight. (The world population in January 2002 was 6.202 billion, increasing at the rate of about 6 million people every month.)

Once added to the air, carbon dioxide spreads throughout the entire atmosphere. And it remains in the air for a long time, taking decades and even centuries to disappear. So future generations will have to contend with our present activities.

Since the oceans cover three-quarters of the Earth's surface, they can absorb great quantities of the carbon dioxide, eventually taking up about half of the amount that is released into the atmosphere by burning coal and oil. In the meantime, not all of the gas stays in the air. Some of it circulates through the atmosphere in one of nature's grand cycles, known as the carbon cycle.

During the spring and summer, trees and other vegetation take in carbon dioxide from the atmosphere, incorporating some of the carbon into the plant tissue and releasing oxygen into the air. Animals breathe the oxygen and return carbon dioxide to the atmosphere. Plants release some carbon when their leaves fall and decay in the autumn, and when the plants die they also sequester carbon in the soil.

Between 350 million and 65 million years ago, great quantities of carbon were stored deep underground, when plants and animals died and decayed and their remains were compressed into what eventually became coal, oil and natural gas. This carbon was removed from the carbon cycle, until humans started using fossil fuels and releasing it into the atmosphere again.

Every year, roughly half of the heat-trapping gas remains in the air, building up over the decades and centuries. But why should adding such small amounts of an invisible, non-toxic gas be a cause of concern? The total amounts are miniscule, but their consequences are significant. Even relatively small amounts of the gas can warm the Earth by the greenhouse effect, perhaps affecting the climate.

Roger Revelle (1909–1991) and Hans E. Suess (1909–1993) realized the threat decades ago. They argued that the oceans might not readily absorb all of the carbon dioxide being released into the air, and that the amount of atmospheric CO_2 would steadily increase as the fuel and power

Table 4.3 Greenhouse gases produced by human activity[a]

Greenhouse Gas	Pre-Industrial Concentration[b] (1860)	Recent Concentration[b] (2001)	GWP[c]	Lifetime (years)	Sources
Carbon Dioxide (abbreviated CO_2)	288 p.p.m.v.	369.4 p.p.m.v.	1	120	Burning coal, oil, and natural gas, deforestation, cement manufacture.
Methane (abbreviated CH_4)	848 p.p.b.v.	1782.5 p.p.b.v.	23	12	Livestock, rice growing, natural gas and oil production, coal mining.
Nitrous Oxide (abbreviated N_2O)	285 p.p.b.v.	314.5 p.p.b.v.	296	114	Nitrogen fertilizers, chemical manufacturing, waste treatment.
Chlorofluorocarbons					Refrigeration, aerosol cans,
CFC-11	zero	262.5 p.p.t.v.	3800	50	foam insulation,
CFC-12	zero	540.5 p.p.t.v.	8100	102	industrial solvents.
Sulfur Hexafluoride (abbreviated SF_6)	zero	4.0 p.p.t.v.	22 200	3200	Electrical equipment insulation, magnesium production, medical treatments.

[a] From http://cdiac.esd.ornl.gov/pns/current_ghg.html
Water vapor produces sixty to seventy percent of the Earth's global warming, but it is not included here because water vapor is not directly produced by human activity.
[b] Averages of the measured amounts, all by volume (v); p.p.m. = parts per million (10^6), p.p.b. = parts per billion (10^9), p.p.t. = parts per trillion (10^{12}).
[c] The GWP, or the Global Warming Potential, is used to contrast the radiative effects of different greenhouse gases relative to carbon dioxide. Values are the ratio of global warming from one unit mass of a gas to that of one unit mass of carbon dioxide over 100 years.

requirements of our world-wide civilization continued to rise. With prophetic insight, they wrote in 1957 that:

> Human beings are now carrying out a large-scale geophysical experiment of a kind that could not have happened in the past nor be reproduced in the future. Within only a few centuries we are returning to the atmosphere and oceans the concentrated organic carbon stored in sedimentary rocks over hundreds of millions of years. This experiment, if adequately documented, may yield a far-reaching insight into the processes determining weather and climate.

As subsequently documented by Charles Keeling, we are indeed altering the composition of our air in a unique global experiment.

Many scientists now think that humans are not only altering the composition of the atmosphere, but that the greenhouse effect of the increased amounts of CO_2 will alter the climate and weather. They also now realize that other greenhouse gases are being released into the atmosphere as the result of human activity, and that they additionally contribute to global warming and possible climate change.

New heat-trapping gases in our atmosphere

During the past few decades, several heat-trapping gases have been accumulating in the atmosphere as the result of human activity (Table 4.3). Although less common than carbon dioxide and water vapor, each molecule is far more powerful and potentially as significant for global warming. These other man-made greenhouse gases include methane, abbreviated CH_4, nitrous oxide, or N_2O for short, and the chlorofluorocarbons, known as the CFCs. Even though the total emissions of these molecules are quite small when compared with those of carbon dioxide, they are much more efficient at trapping infrared heat radiation. As a result, they can together contribute about as much global warming as carbon dioxide alone.

Methane is the same natural gas that we use at home for cooking and heating. Most of the atmospheric methane does not, however, come from gas wells. It is produced by agricultural activities such as growing rice and raising cattle. The gas is emitted by bacteria that thrive in oxygen-free places like rice paddies and the stomachs of cows.

Since pre-industrial times, the atmospheric concentration of methane has increased more than 110 percent. Over the same period atmospheric carbon dioxide has risen about 31 percent. Although methane is about 200 times less abundant than carbon dioxide, each incremental molecule of methane has about 23 times the heat-trapping power as each additional molecule of carbon dioxide.

When found in swamps, methane is known as marsh gas; it sometimes ignites spontaneously, producing flickering blue flares called will-o'-the-wisps. Some of it also escapes from coal mines, natural gas wells and leaky pipelines.

Nitrous oxide, or laughing gas, is also building up in the air, although not as rapidly as methane. The current rate of increase is about 0.2 percent a year, primarily as the result of nitrogen-based fertilizers but also from burning of fossil fuels in cars and power plants.

The chlorofluorocarbons, abbreviated CFCs, are very effective heat-trapping molecules. The addition of one CFC-12 molecule to the air can have the same greenhouse effect as the addition of 8100 molecules of carbon dioxide to the present atmosphere. Fortunately, the warming effect of these industrial chemicals may soon be leveling off since they have been banned on the basis of their ozone-destroying capability (Focus 4.3).

Contrary to popular misconception, however, ozone depletion and global warming are not the same thing. The CFC molecules that destroy ozone also trap heat, but the thinning of the ozone layer does not by itself make the Earth's surface hotter.

Signs of global warming

The term "global warming" has both a general and specific meaning. It is used in a general sense to indicate that the Earth is getting hotter. Global warming can also specifically imply that a warmer planet is the result of an increase of heat-trapping gases in the atmosphere, resulting from human activity. There is still scientific uncertainty in both the magnitude and timing of this type of global warming, but it is likely that some of the recent rise of the Earth's surface temperature is due to an increase in greenhouse gases emitted by human activity.

The possibility of global warming by humans was realized more than a hundred years ago, by the Swedish chemist Svante August Arrhenius (1859–1927). He proposed that an industrial increase of atmospheric carbon dioxide would gradually push the ground temperature upward. Arrhenius received the Nobel Prize in chemistry in 1903 for noticing that ions are formed in electrolytic solutions, conducting electricity and aiding chemical reactions, but this has nothing to do with his considerations of global warming.

Arrhenius proposed in 1896 that a doubling of atmospheric carbon dioxide beyond the amounts measured in the 1890s would boost the Earth's temperature by 5 to 6 degrees Celsius (9 to 11 degrees Fahrenheit), and that the temperature would drop by almost the same amount if the gas decreased by half. He thought that such temperature changes might account for the coming and going of the ice ages and the warm intervals between them. Although we now know that changes in the global distribution of sunlight bring on the ice ages, Arrhenius' estimate of the global warming by doubling the amount of carbon dioxide in our air is comparable to modern estimates.

Over the past century, the planet as a whole has warmed up in fits and starts, fluctuating between warm and cool periods. For example, the world heated up by about 0.5 degrees Celsius, or 0.9 degrees Fahrenheit, between 1920 and 1940; but the average temperatures dropped by about half that amount between 1940 and 1970, leading some experts to predict a coming ice age. Then the heat turned on again, rising by another 0.5 degrees Celsius in the last three decades of the 20th century. By the 1990s, the world had become hotter than any time in recorded history.

There are now many other signs that indicate a warming atmosphere. The oceans are rising; mountain glaciers are retreating; polar sea ice is melting; the very timing of the seasons is changing and erratic and severe weather is more common than just a few years ago. Taken alone, these events are no proof of global warming, but in combination they provide strong evidence for a warmer climate.

Rising seas provide additional evidence for a hotter world. As the water in the sea gets warmer, it will expand as most substances do when heated. The sea will then ascend to higher levels, in much the same way that heating the fluid in a thermometer causes the fluid to expand and rise up. This is because warm water or other fluids occupy a greater volume than cold ones. Measurements indicate that the global sea level increased somewhere between 10 and 25 centimeters (0.1 and 0.25 meters) during the 20th century. However, you could not have noticed the change, for the sea level was only rising between 1.0 and 2.5 millimeters (0.001 and 0.0025 meters) per year.

The melting of ice that now covers land, such as mountain glaciers, also contributes to the sea-level rise and most likely results from global warming. By the end of the 20th century, glaciers were retreating throughout the world, and those in Alaska were typically becoming about a meter thinner every 4 or 5 years.

As it melts and shrinks, the glacial ice releases water into streams and rivers that add to the sea. Such melt waters from mountain glaciers boosted the sea level between 2 and 5 centimeters (0.2 and 0.5 meters) in the 20th century.

Contrary to popular belief, the melting of floating icebergs will not raise the level of the surrounding sea. When ice cubes in your drink at home melt, they similarly do not cause any change in its level, for the melted ice produces the same volume of water that the ice cube displaced. However, because most of Antarctica's ice now covers land, it would add to the ocean's volume if it melted or broke off in blocks, in a process known as calving, so people are worried about the future meltdown of the Antarctica ice pack.

Long hot summers and warmer nights provide another mark of the warming trend. In the northern hemisphere, winter shortened by about a week on both ends during the last two decades of the 20th century. So spring arrives earlier than it used to, autumn lasts a bit longer, and the land stays greener. The longer, warmer summers are beginning to thaw the northern tundra, or permafrost, releasing methane that might further exasperate the build up of heat-trapping gases.

Something strange, it's claimed, is going on with the weather, providing other unwelcome signs that the heat is on, affecting our everyday lives. In fact, such events were predicted as an early symptom of a rise in the average global temperature, and now they are here. These warming signs include record rainfalls and severe snowstorms; extraordinarily destructive hurricanes and tornadoes; heat waves unique in weather annals; widespread droughts and devastating forest fires; unseasonable warmth and cold; and some of the worst floods in recorded weather history.

A hotter ocean surface produces such weather extremes by accelerating the water cycle. The cycle begins when some of the ocean evaporates, releasing warm fresh-water moisture into the air. This water vapor rises high into the colder atmosphere where clouds are formed. Winds drive the clouds for great distances over sea and land. Rain or snow from the clouds can then fall to land, refreshing streams, lakes or underground reservoirs. The water then runs down to the sea, where the cycle begins once more.

All the water in the oceans passes through this water cycle once in two million years. Yet, the ocean waters are at least 3.5 billion years old, so entire oceans have moved through the sky, completing more than a thousand such cycles.

John Updike (1932–) has captured this stately Sun-driven circulation in his 1984 poem *Ode to Evaporation:*

All around us, water is rising
  on invisible wings
to fall as dew, as rain, as sleet, as snow,
  while overhead the nested giant domes
of atmospheric layers roll
  and in their revolutions lift
  humidity north and south
from the equator toward the frigid, arid poles,
  where latitudes becomes mere circles
Molecular to global, the kinetic order rules
  unseen and omnipresent . . .

The rising temperatures heat the oceans more than they used to be heated, causing more water to evaporate. In addition, a warmed atmosphere holds more moisture than a cool one. As a result, there is more water in the air. When this extra water condenses, heavy rains, severe thunderstorms, blizzard-like snowstorms, and damaging floods become more frequent, intense and severe.

While the oceans are being heated, so is the land. It can become highly parched in dry areas, resulting in droughts that are more severe and persist longer than they used to. The rising temperatures also enlarge heat and pressure differences across the land, causing winds to develop and leading to more tornadoes and hurricanes.

Thus there are many indications that the Earth is getting hotter. Climate change has definitely arrived and the future is bound to be different from today. But it is still unclear how much of this warming and extreme weather is attributable to our accelerated release of greenhouse gases, and how much to natural causes. Many scientists think that both are involved and that human activity must be at least partly responsible. There is also scientific uncertainty about the future consequences of global warming.

How hot will it get in the future, and how fast?

There is almost no doubt that the heat is on, and that the climate is changing. The remaining scientific debates are over why the climate is changing, how fast it will change in the future, and the likely consequences of the change. Sophisticated computer models, known as general circulation models, try to answer these questions by analyzing the climate system and forecasting its future.

To assess the effects of global warming, one assumes that the dominant cause of climate change during the next century will be the release of greenhouse gases into the atmosphere by humans, and that the increasing rate of atmospheric build up of the waste gases will continue unabated. The supercomputer models then evaluate all the factors that push and pull the climate, setting the global thermostat.

The models tell of possible futures when atmospheric carbon-dioxide levels are about double the pre-industrial concentrations. Although we are almost one-third of the way toward reaching this state, it is going to take another 100 years to get there. And since the climate is very complex, the scientific pronouncements about the future climate vary widely. Some of the climate forecasts for the next century are scary and others are not. No one knows which one is true! And it's far enough in the future that none of the climate experts will be around to check the reliability of their forecasts.

A prestigious group of international scientists has been working on the problem for years. Known as the Intergovernmental Panel on Climate Change, or IPCC for short, it includes hundreds of climate experts. They evaluate the supercomputer model results and develop a coordinated assessment of future global warming under the auspices of the United Nations, including extensive reviews by both individual scientists and governments. In their report, entitled *Climate Change 2001: Impacts, Adaptation, and Vulnerability*, the group concluded that the burning of fossil fuels and other human activities are responsible for most of the rise in global temperatures during the last half of the 20th century.

If present trends in the emission of greenhouse gases continue for 100 years, the group concludes, then resultant human-induced global warming will raise the Earth's average surface temperature between 1.4 and 5.8 degrees Celsius (2.5 and 10.4 degrees Fahrenheit). The upper end of the IPCC forecast is essentially the same as Svante Arrhenius' 1896 computation of 5 to 6 degrees Celsius for a doubling of the carbon-dioxide content by industrial activity.

The worst-case scenario predicts a large temperature increase a century from now, comparable to the temperature rise since the last Ice Age, and there may also be dire consequences if the warming takes a middle course. Even a 2-degree-Celsius warming over the next century, near the bottom of the predicted range, will probably be the fastest warming in the history of civilization. As we shall see, some scientists predict apocalyptic consequences if it should occur, but others say these are exaggerated fears.

A vocal minority argues that no computer can adequately simulate the complexities of the real world. The computer models, they say, are flawed by massive uncertainties, making their predictions of future global warming about as reliable as a crystal ball or an ouija board. After all, the long-range forecasts of your local weather station are not all that reliable. Even with daily satellite images of weather patterns, we cannot reliably predict the local weather beyond a few days in advance. In Boston, for example, major snowstorms often fail to occur when predicted, and threatening hurricanes often veer off their expected course.

All of the models predict that the globe will warm as the result of the unrestrained emission of heat-trapping gases, but different temperatures are obtained under the same conditions and both modest and catastrophic climate changes are foreseen. The reason is that scientists don't understand the precise role of oceans, water vapor and clouds. They can all amplify the future global warming or cool it. So the experts carefully wrap their pronouncements in statistical properties, scientific uncertainties, and a range of possibilities. The most plausible outcome lies somewhere between the most extreme projections, between the "end of the world" scenarios and the "good for you" ones.

The mighty oceans are one of the biggest uncertainties. Current estimates suggest that their waters are now soaking up about half the carbon dioxide released into the atmosphere by human activity. Water absorbs less carbon dioxide when it is warmer, which is why you should always keep a carbonated beverage cold. So you might expect that the oceans will amplify the greenhouse warming, absorbing less of the waste gas when it gets warmer and leaving more of it in the air to warm the Earth. Yet, a precise knowledge of how carbon dioxide is buried deep within the oceans, and how the gas is released from them, is not available.

Scientists also do not fully comprehend how water behaves in a warming atmosphere and the amount by which it enhances the warming. Increased temperatures will evaporate more water from the oceans, and the additional moisture should increase the greenhouse warming. Water vapor, in fact, is the main heat-trapping gas in the lower atmosphere. Its greenhouse-amplifying effect is built into the supercomputer climate models, and their predicted average global temperature increase would be substantially less without it. Some critics argue that future warming may even dry out the upper atmosphere, tempering the warming effect of water vapor.

Our understanding of future global warming is additionally hampered by an uncertain knowledge of how clouds affect the Earth's temperature. At any given moment, clouds now cover at least half the area of the planet, and increased warming should produce even more clouds along with increasing water vapor. The high-flying clouds can cool the planet by reflecting more incident sunlight back into space. You may have noticed this cooling effect when a large cloud passed overhead. Clouds also produce a warming effect by absorbing some of the infrared heat radiation emitted by the ground. This heat is reradiated downward, keeping the planet warm. That accounts for warmer nights on a cloudy day. The net temperature effect of clouds depends on which effect dominates and how strong it is, but that is the difficulty, for no one seems to know for sure.

To add to the confusion, the existing supercomputer models offer only a blurred, myopic vision of the world, which isn't focused enough to resolve most clouds. Although they provide a plausible range of warming forecasts for a century from now, the calculations are reliable only on the broadest scales, such as average temperatures or seasonal changes across the entire world. No computer can possibly evaluate climatic changes everywhere on our planet and in the atmosphere over the next 100 years.

The finite capacity and speed of even the most advanced supercomputer limits its climate calculations to widely separated points, typically no better than 100 or 200 kilometers between adjacent points. This relatively crude resolution can blur distinctions between land and sea or mountains and plains. It also means that the models cannot zero in and pinpoint localized weather sources, such as clouds or even hurricanes. Thus, even the best global warming predictions represent a stripped-down version of the Earth's real climate, capable of approximating average conditions over entire countries a long time from now, but too crude to reliably forecast conditions within localized regions of the countries.

In addition, many of the predicted temperature changes from human-induced global warming pale in comparison to natural variations, from the annual seasons to the ponderous ice ages. The climate is naturally changing all the time, warming and cooling regardless of what people do. A prudent society should therefore examine if a hotter globe can be solely attributed to natural processes, as distinct from human ones, and determine how much of the recent rise in temperature is attributable to human activity and how much to natural causes.

Using past records to separate the warming effects of the Sun and humans

Changes in the amount or distribution of the sunlight illuminating the Earth can produce substantial variations in our climate, and we have proxy records of its variable output over past centuries. This data can be compared to global temperatures during the same period, to determine if the Sun is responsible for most of the warming and cooling of our Earth. Although natural variations in the solar output can explain most of the temperature variations over the past centuries, it appears that global warming by heat-trapping gases, emitted by human activity, is required to explain the sharp rise in global temperatures during the 1990s.

The Sun is, after all, the driving force for all climate and weather on Earth. Tropical regions of the Earth receive the greatest amount of heat because the Sun's vertical rays travel to the ground through the least amount of intervening air. Since higher latitudes are tilted toward or away from the Sun by 23.5 degrees, sunlight strikes them a glancing angle, penetrating a greater thickness of absorbing atmosphere and heating the ground less. The temperature difference creates winds that tend to carry water-filled clouds towards the North and South Geographic Poles, between the warmer tropics and the colder Poles, but they are also deflected in the east–west direction by the Earth's rotation (Section 3.1).

The annual seasons are also created by the tilt of the Earth's rotational axis. During the Earth's yearly orbit, a given hemisphere is tilted toward the Sun, producing summer, then away resulting in winter (Section 3.1, Fig. 1.10).

The word "climate" comes, in fact, from the Greek word *klima*, for "tilt". Nowadays climate denotes long-term changes, on time-scales of years, decades and centuries, while weather usually refers to short times of hours, days, weeks or months.

The annual change in solar radiation produces large seasonal temperature fluctuations in the northern hemisphere where most of the world's land is now found. The difference in the average surface temperature between northern winter and summer is an enormous 15 degrees Celsius (27 degrees Fahrenheit). In the tropics, the temperature changes little year round, since the amount of incoming solar radiation in those latitudes is least affected by the Earth's tilt.

The total amount of solar radiation striking the Earth can vary as the result of the Sun's changing brightness. The intensity of the Sun's radiation varies over an 11-year cycle of magnetic activity, first discovered as a cyclic variation in the total number and position of sunspots. The magnetized atmosphere in and around a sunspot, known as a solar active region, emits intense radiation at invisible ultraviolet and X-ray wavelengths, and the intensity of this radiation increases and decreases in step with the 11-year cycle. Visible sunlight also fluctuates in intensity with an 11-year period, but by lesser amounts.

The dark, Earth-sized sunspots come in pairs of opposite magnetic polarity, or direction, and they are connected by magnetic loops that rise above the visible solar disk. These loops of magnetism contain the hot, million-degree atmosphere of the Sun, which emits intense radiation at X-ray wavelengths. So, active regions shine brightly in X-rays, illuminating the thin magnetic loops that stitch the solar atmosphere together (Fig. 4.22). The number of active regions, with their bipolar spots and the magnetic loops that join them, varies in step with the 11-year sunspot cycle, peaking at the sunspot maximum. The sunspot cycle is therefore also known as the solar cycle of magnetic activity.

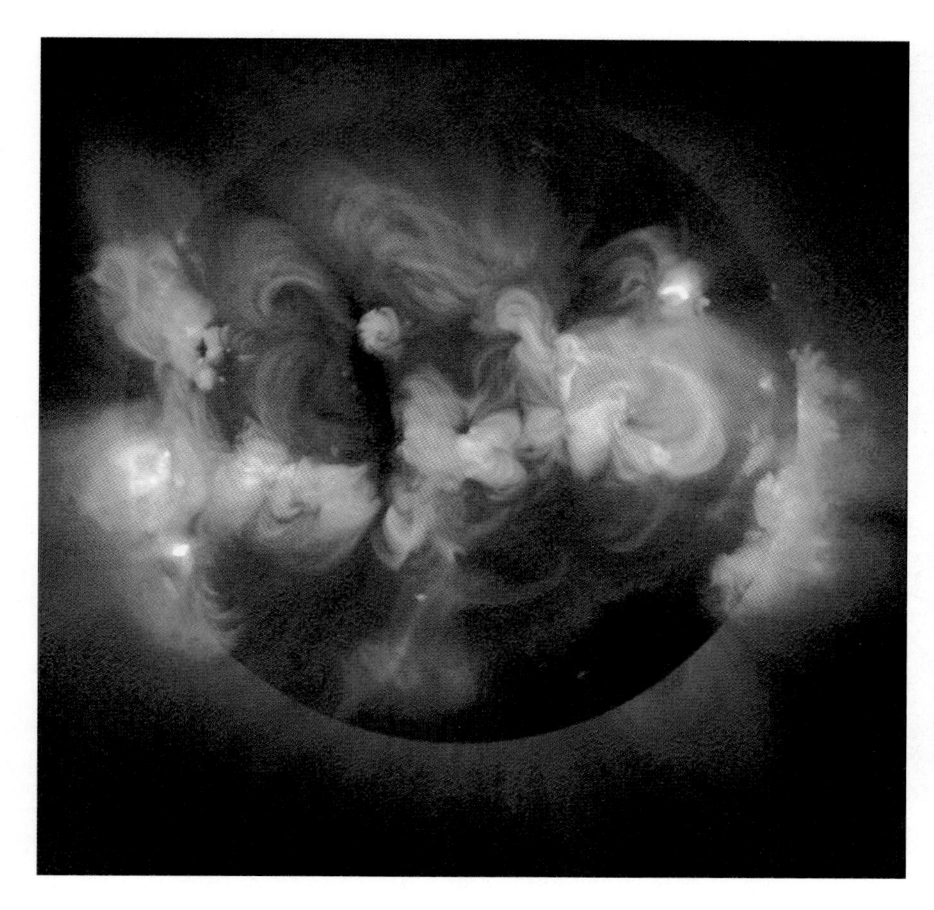

Fig. 4.22 The Sun in X-rays The bright glow seen in this X-ray image of the Sun is produced by ionized gases at a temperature of a few million kelvin. It shows magnetic coronal loops which thread the corona and hold the hot gases in place. The brightest features are called active regions and correspond to the sites of the most intense magnetic field strength. This image of the Sun's corona was recorded by the Soft X-ray Telescope (SXT) aboard the Japanese *Yohkoh* satellite on 1 February 1992, near a maximum in the 11-year cycle of solar magnetic activity. Subsequent SXT images, taken several years after activity maximum, show a remarkable dimming of the corona (see Fig. 4.23), when the active regions associated with sunspots have almost disappeared. (Courtesy of Gregory L. Slater, Gary A. Linford, and Lawrence Shing, NASA, ISAS, the Lockheed–Martin Solar and Astrophysics Laboratory, the National Astronomical Observatory of Japan, and the University of Tokyo.)

Because active regions emit intense ultraviolet and X-ray radiation, the Sun is brightest at these wavelengths during the peak of the magnetic activity cycle and dimmest at cycle minimum (Fig. 4.23). At the bottom of the 11-year cycle, the active regions are largely absent and the strength of the ultraviolet and X-ray emission of the Sun is greatly reduced. The ultraviolet emission doubles from activity minimum to activity maximum while the Sun's X-ray emission increases by a factor of 100.

At times of high solar activity, the Sun pumps out much more of the invisible rays, the air absorbs more of them, and our upper atmosphere heats up; when solar activity diminishes the high-altitude air absorbs less and cools down. At a given height in our upper atmosphere, the temperature, the density of free electrons, and the density of neutral, un-ionized atoms all rise and fall in synchronism with solar activity over its 11-year cycle. The Sun-induced changes in the content and structure of the ionosphere affects its ability to reflect radio waves, and the increased density of the upper atmosphere produces a frictional drag on artificial satellites, leading to their premature demise.

Although the sunlight that illuminates our days provides a seemingly reliable beacon, the Sun's visible luminosity also varies in tandem with the Sun's 11-year magnetic activity cycle and these changes could affect our climate. The colored sunlight passes right through to the ground, providing a direct warming or cooling of the lower atmosphere.

The total amount of the Sun's life-sustaining energy is called the "solar constant", perhaps because no variations could be detected in it for a very long time. Until the early-1980s, it was not known if the Sun's visible light was anything but rock-steady because no variations could be reliably detected from the ground. The required measurement precision could not be attained here on Earth because of the changing amount of sunlight absorbed and scattered by our atmosphere.

Stable detectors placed aboard satellites above the Earth's atmosphere have been precisely monitoring the Sun's total irradiance of the Earth since 1978, providing conclusive evidence for small variations in the solar constant (Fig. 4.24). It is almost always changing, in amounts of up to a few-tenths of a percent and on time-scales from 1 second to 20 years, and probably longer. This inconstant behavior can be traced to changing magnetic fields in the solar atmosphere.

Comparisons with other stars which resemble the Sun in mass and age indicate that the Sun could undergo more substantial variations in brightness than those observed so far by satellites.

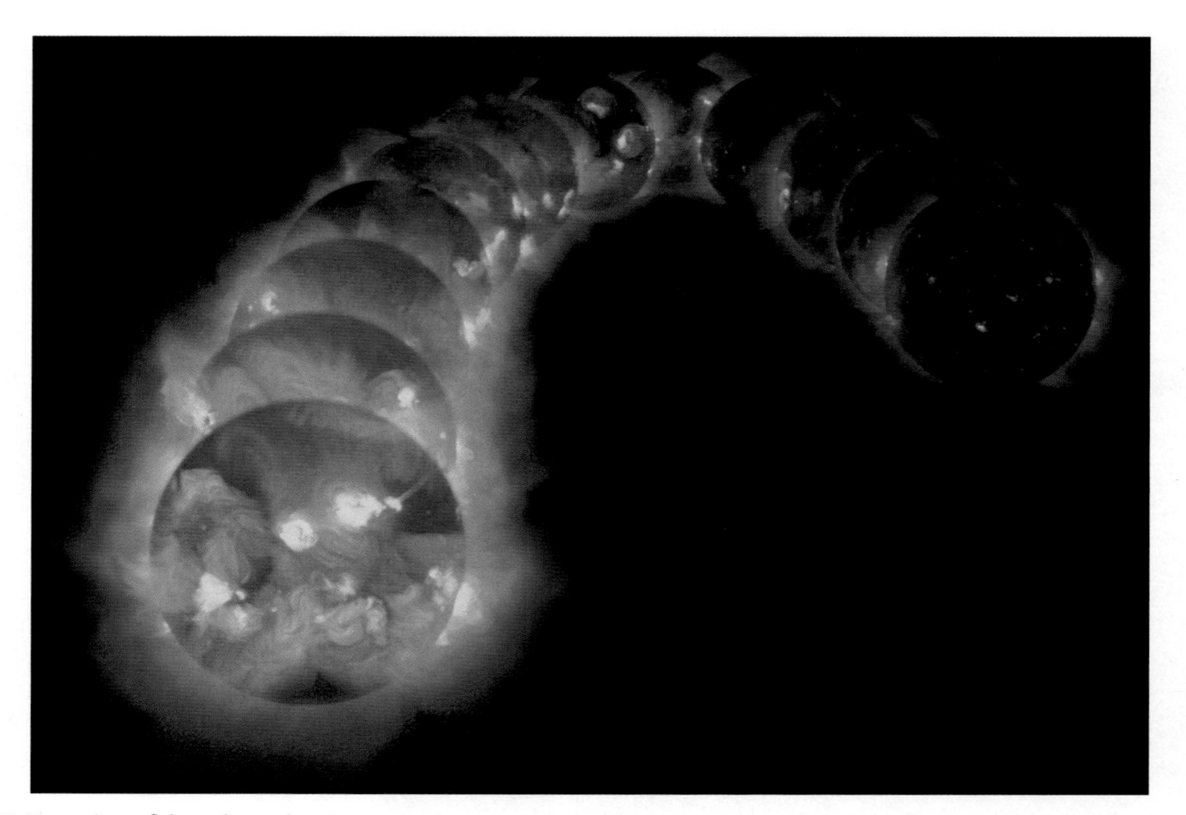

Fig. 4.23 X-ray view of the solar cycle Dramatic changes in the solar corona are revealed in this four-year montage of images from the Soft X-ray Telescope (SXT) aboard *Yokhoh*. The 12 images are spaced at 120-day intervals from the time of the satellite's launch in August 1991, at the maximum phase of the 11-year sunspot cycle (*left*), to late 1995 near the minimum phase (*right*). The bright glow of X-rays near activity maximum comes from very hot, million-degree coronal gases that are confined within powerful magnetic fields anchored in sunspots (Fig. 4.22). Near the cycle minimum, the active regions associated with sunspots have almost disappeared, and there is an overall decrease in X-ray brightness by 100 times. (Courtesy of Gregory L. Slater and Gary A. Linford, NASA, ISAS, the Lockheed–Martin Solar and Astrophysics Laboratory, the National Astronomical Observatory of Japan, and the University of Tokyo.)

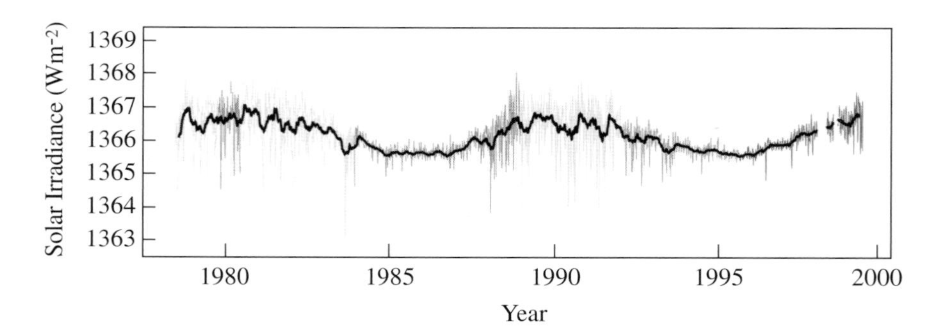

Fig. 4.24 Variations in the solar constant Observations with very stable and precise detectors on several Earth-orbiting satellites show that the Sun's total radiative input to the Earth, termed the solar constant or solar irradiance, is not a constant, but instead varies over time-scales of days and years. Measurements from five independent space-based radiometers have been combined to produce the composite solar irradiance over two decades since 1978. They show that the Sun's output fluctuates during each 11-year sunspot cycle, changing by about 0.1 percent between maximums (1980 and 1990) and minimums (1987 and 1997) in magnetic activity. Temporary dips of up to 0.3 percent and a few days' duration are due to the presence of large sunspots on the visible hemisphere. The larger number of sunspots near the peak in the 11-year cycle is accompanied by a rise in magnetic activity that creates an increase in luminous output that exceeds the cooling effects of sunspots. The solar constant, or irradiance, is given in units of watts per square meter, where one watt is equivalent to one joule per second. (Courtesy of Claus Fröhlich.)

In order to say that the climate is getting warmer, we cannot just extrapolate from recent measurements. The case for global warming must instead be based on century-long temperature records over a large part of the globe. We can then determine if the recent rise in temperature is a significant departure from long-term trends, and evaluate how much of the warming is attributable to human activity and how much to natural causes.

Scientists have therefore reconstructed variations in the climate of the past, comparing them to the Sun's changing output. Data extracted from tree rings and Antarctica ice cores indicate that solar activity has indeed fallen to unusually low levels at least three times during the past 1000 years, each drop corresponding to a long, cold spell of roughly a century in duration. As an example, sunspots virtually disappeared from the face of the Sun for the 70-year period between 1645 and 1715, when Europe experienced one of the coldest periods of the Little Ice Age. During that time, alpine glaciers expanded, the river Thames, England, and the canals of Venice, regularly froze over, and painters depicted unusually harsh winters in Europe (Fig. 4.25). The cold also sometimes brought widespread crop failures and famine to the continent, touching off waves of migration out of eastern Europe. The Sun was then about 0.25 percent dimmer, and the reduction in solar brightness produced an estimated drop of about 0.5 degrees Celsius (0.9 degrees Fahrenheit) in the global mean temperature.

It is very difficult to distinguish any human influence in the observed temperature record of past centuries. The natural ups and downs of the Sun's brightness or the cooling effect of volcanoes can explain almost all of it. Fine particles created during powerful volcanic eruptions, such as Mount Tambora in 1815, Mount Krakatau in 1883, and Mount Pinatubo in 1991, can spread out high above the ground, forming an invisible, umbrella-like shield that blocks some of the incoming solar radiation and causes temporary global cooling. Similar sulfate aerosols arise from fossil fuels burnt in power plants, factories and automobiles, partially offsetting the full warming effect of their carbon-dioxide emission.

The warning signal of human-induced global warming only rose above the confusing noise of the Sun and other natural effects in the 1990s, when the Earth became exceptionally hot. There is absolutely no evidence for such a decade in the historical temperature records going back 1000 years (Fig. 4.26). This sharp, unprecedented rise in the average global temperature during the last decade of the 20th century cannot be explained as a temporary swing produced by natural causes alone, and it is very likely that heat-trapping waste gases are at least partly responsible for it. The available evidence suggests that greenhouse gases emitted by industrial economies mainly caused this warming.

Humans have conquered the land, moved mountains, and redirected rivers. Airplanes have provided easy access to almost anywhere in the world, and communications satellites have connected us all in an electronic web. The globe has shrunk, and we might now be altering the entire atmosphere.

For most of history, we believed that climate and weather are governed by outside forces, beyond the influence of humans, but today we are no longer so sure. As the result of rapid, unprecedented population and industrial

Fig. 4.25 Hunters in the snow This painting by Pieter Bruegel the Elder (about 1525–1569) depicts a time when the average temperatures in Northern Europe were much colder than they are today. Severe cold occurred during the Maunder Minimum, from 1645 to 1715, when there was a conspicuous absence of sunspots and other signs of solar activity. This picture was painted in 1565, near the end of another dearth of sunspots, called the Spörer Minimum. (Courtesy of the Kunsthistorisches Museum, Vienna.)

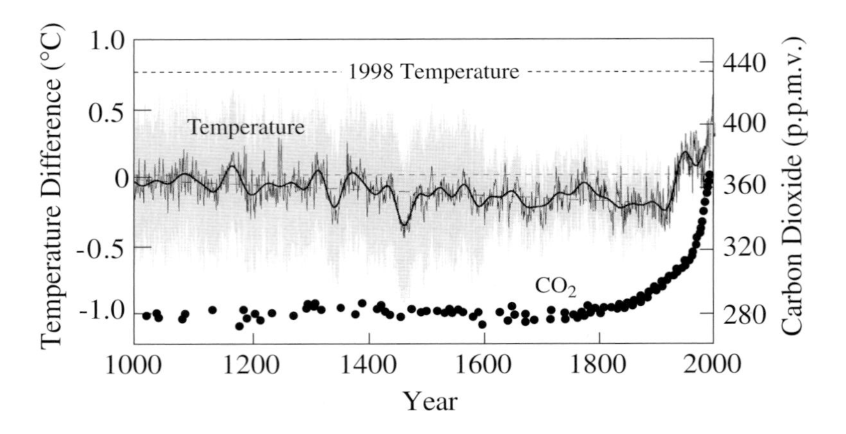

Fig. 4.26 Unusual heat The temperature in the northern hemisphere (*top curve, scale at left*) has been warmer in the latter part of the 20th century than in any other century of the last one thousand years, and the 1990s were the hottest decade during all that time. The sharp upward jump in temperature during the past 100 years was recorded by thermometers near the Earth's surface; earlier temperature fluctuations were inferred from tree rings, lake and ocean sediments, coral reefs, and ice cores. Temperature differences are from a reference zero level (*horizontal dashed line*) for the period 1902 to 1980. The heavy solid line is smoothed over a 40-year interval, and the light shading denotes the range of possible error in individual measurements. There has been an exponential rise in the amount of atmospheric carbon dioxide over the past two-and-a-half centuries (*bottom curve, scale at right in parts per million by volume, or p.p.m.v.*). (Courtesy of Michael E. Mann (*top curve*) and Charles D. Keeling (*bottom curve*).)

growth over the past century, humans can collectively alter the course of nature.

As the traditional Negro jubilee spiritual goes:

He's got the whole world in his hands,
The whole wide world in his hands, . . .

Here, the "He" is the God of creation – who handed over dominion of His Earth to us. The probable outcome of our present-day experiment with nature is a matter of controversial current debate, but we can make some educated guesses about it.

Likely consequences of global warming

If current emissions of carbon dioxide and other greenhouse gases go unchecked, temperatures should continue to rise and the climate will inevitably change. A significant warming ought to occur by the end of the 21st century, but the predicted consequences range from an ominous, catastrophic future to a mildly uncomfortable one. Two extreme arguments have therefore been advanced. One side, supported by environmentalists, warns of an imminent apocalypse and global catastrophe caused by our own tinkering with nature. In their perspective, belching smokestacks, gasoline-powered automobiles, power-generating stations and the voracious destruction of forests are turning up the heat on an overburdened environment, pushing it over the edge with raging storms, flooded cities and poisoned air. The alternate view, championed by some industrialists, is that doomsday predictions are exaggerated and hopelessly uncertain. And if the world does get a little warmer, they say, it won't be that bad. After all, most people like a warmer climate and many of us could use a little more heat in our lives.

There are both positive and negative aspects to even the worst-case forecasts of global warming. Not everyone and every place will be affected the same. Some regions will benefit, and others may suffer unbearable damage. Many aspects of a temperature increase would probably be welcome at high northern latitudes that are now too cold over much of the year. Regions that are near the equator are already hot, and a further increase in heat could be devastating. So there is bound to be a mixed verdict, and the outcome of the debate depends upon which kind of evidence you focus on.

Who should we believe? It is probably unwise to lapse into apocalyptic thinking, especially in view of the scientific uncertainty and long time-scales, but ostrich-like denial is also imprudent since a lot is at stake, from the wealth of nations to the future of the planet. The best we can do is to examine the full spectrum of potential, long-range consequences, from good to bad.

In the best-case projection, at the low end of the predictions, global warming will, by itself and independently of other influences, raise the average temperature of the world by 1.4 degrees Celsius (2.5 degrees Fahrenheit) over the next 100 years. If the warming takes this gradual, modest course, some parts of the world will benefit, and the normal resilience of society ought to accommodate the climatic change.

The modest increase in temperatures at mid-northern latitudes, where most people live, will be welcome. There will be longer summers, shorter winters and warmer nights. Residents of cities like Boston will suffer fewer colds, experience fewer heart attacks from shoveling snow, and spend less on heating, snowplowing and road salting. But summer air conditioning will cost more, the winter ski slopes may turn to slush, and the colorful fall foliage could disappear as trees move away from the heat to the north.

If global warming is at the upper end of the prediction, at 5.8 degrees Celsius (10.4 degrees Fahrenheit) in 100 years, many humans should also be able to adapt without much difficulty. After all, this rise in temperature is less than the average daily temperature difference between New York City and Atlanta, Georgia, or between Paris and Naples, and there is little evidence of greater risk to people who now live in the warmer southern climate. And those who live in the colder, northern regions are already used to a seasonal temperature increase between winter and summer that can be three times greater than the largest predicted heating over the next 100 years.

So, the good thing is that humans are adaptable. But the bad thing is also that humans are adaptable. As long as the climate changes occur slowly, we can adapt without realizing what is happening, but some very uncomfortable things can happen at the top part of the expected warming in 100 years.

There is an applicable proverb about frogs. If you put a frog in boiling water, it will jump out and save itself. But if you gradually increase the heat, with the frog in the water, it will die.

Very hot will be decidedly unwelcome in many places. Entire cities will be immobilized under the heat. The wealthy will move out of Palm Springs, and Las Vegas could become a ghost town. Residents in many other large cities should experience severe heat waves, making them feel like the world is melting down in a pool of sweat. As the climate becomes hotter and drier, drought will probably become more severe in areas prone to it, and supplies of fresh water will dwindle. More frequent bouts of extreme weather will also be expected, with widespread flooding and intense hurricanes.

On the one hand, agriculture in some regions will be better than other regions. The longer growing season and increasing carbon dioxide will foster plant growth, making much of the developed world greener. Agriculture will likely become more productive in Canada, northern Europe, Russia and the northern United States. As droughts turn some mid-American farms to dust, both agricultural production and population centers in the United States will shift north, and the same thing will probably happen in Europe.

The world's poorest countries are, on the other hand, highly vulnerable to agricultural disaster, for they are already located in arid and semi-arid regions. A further rise in temperature will almost certainly reduce crop yields in south Asia and sub-Sahara Africa, where expanding deserts will additionally claim more land.

As environmental conditions change over time, plants and animals will migrate, as they have throughout geological history – moving up and down in latitude as the globe warms and cools. The recent increase in temperatures has already caused some species to move north, and the accelerated heating could wipe out many of them in the future. Some plants and animals might not be able to move fast enough to keep pace with the rapid rate of temperature change, and climate-sensitive habitats could be destroyed altogether, hastening the extinction of some species.

No one can see much advantage in the rising seas, which are one of the most certain effects of the warming projected during the coming decades. There is just one indirect advantage – the meltdown of polar ice, that contributes to the sea rising, could result in permanent ice-free passage in the Arctic Ocean, providing a new shipping route between Europe and Asia.

The climate experts predict a rise in sea level of between 0.09 and 0.9 meters (3 inches to 3 feet) over the next 100 years if nothing is done to curtail the emission of greenhouse gases. The resultant flooding will seriously disrupt coastal areas where more than a quarter of the world's population now lives. In the worst-case increase, Venice and Alexandria will be inundated, as will many cities on the Atlantic and Gulf coasts of the United States, including Boston and New York City. Residents in South Florida will not have to worry about the sweltering heat; their homes will be flooded with seawater.

As with agriculture, the developing nations will get the short end of the stick. Thirty million people in Bangladesh could be displaced by a 0.9-meter increase in sea level, and the rising waters would most likely force the evacuation of 70 million Chinese. Salt water could move several kilometers inland at the mouths of rivers, invading coastal drinking-water systems. The Nile, Yangtse, Mekong and Mississippi deltas are all at risk. Island nations will suffer severe flooding or completely disappear under the rising waters; they include the Bahamas, many of the Caribbean islands, Cyprus and Malta in the Mediterranean, and several Archipelagos around the Pacific Ocean.

Even a modest rise in sea level will wipe many of the world's beaches out of existence. Flooding isn't the problem; it's the removal of sand by waves. Even a 0.3-meter (1-foot) rise in sea level creates wave action that erodes away up to 5 meters (16.4 feet) of some beaches. So, people who live by the seashore had better sell their homes, and you can forget winter vacations in parts of Florida and the Caribbean islands.

The last factor in our catalog of likely consequences of severe global warming is increased health risk – a topic close to the hearts of most people and nearly every politician. Many diseases might spread dramatically as the temperatures head upward, especially those borne by mosquitoes. As the world warms, mosquitoes will move north, into regions where the winter cold used to kill them, carrying malaria, dengue fever, yellow fever and encephalitis with them. Extremely hot weather may also directly terminate the lives of a lot of people, particularly the very old and the very young living in cities.

The list of possible consequences of global warming looks pretty grim, especially if the worst-case forecasts come true. Some of the threats are immediate and inevitable; others are remote and uncertain. Only time will tell for sure.

But that doesn't justify inaction. Even if world-wide emission of heat-trapping gases were capped at today's amounts, their concentrations in the atmosphere would continue to increase. And if we stopped burning fossil fuels altogether, the atmosphere would not immediately recover. The carbon dioxide already released in the air would stay there for at least a century, keeping the planet warm and the global temperatures high. So, it's time we acted to correct the problem – we may have waited too long already.

The heated debate about global warming

Whether we like it or not, global warming has become politicized, the subject of a contentious debate. It has entered the arena of world politics, a shadowy realm of diplomacy, economic interests, political alliances, and national security.

No ratified, international global-warming treaty exists, at least so far. Extremists on both sides of the issue have hired lobbyists to influence governments and mounted huge public relations campaigns to persuade the average person to accept their views. The conflicting information has confused the general public, often leading to an overall apathy. And since neither side in the debate will compromise, a consensus is impossible.

The climate scientists cannot resolve the contentious debate. When it comes to economic consequences, value judgments and policy considerations, they have no more authority than anyone else. No scientist should be permitted to influence public opinion outside his or her area of expertise. Some even say that the international climate panels understate caveats and focus on the harsher possible consequences of warming, perhaps to exaggerate their self importance and obtain more governmental funding.

When you strip away the rhetoric, the experts have little precise knowledge about the future severity of climate change and even less about the future physical impact in particular countries or regions. After decades of research, the model-builders cannot say precisely what will happen to the climate as the result of the atmospheric build up of heat-trapping gases. They just don't know enough about the atmosphere, clouds or the oceans to predict accurately the future global climate.

The uncertainty paralyzes discussion. Scientists have to generate a wide range of possible futures, some very threatening and others less so. Not all of these outcomes are likely to be true, and none is definitive, but people tend to latch onto those that fit their preconceptions. Especially the extremists, who selectively interpret the scientific forecasts to bolster their case – the liberals choose doom and gloom and the conservatives favor good times for all.

The environmental organizations and their allies insist that global warming is here, a harsh and inexorable reality, and that it is due to the rise in carbon dioxide caused by burning coal and oil. The Earth, they say, has entered a resulting widespread climatic disruption that is going to get a lot worse if quick actions are not taken to reverse the accumulation of heat-trapping gases in the atmosphere.

Advocates for this view employ the full range of potential environmental disasters, suggesting that they will almost certainly occur. Deserts will expand, drought will spread, water supplies will evaporate away, rising seas will flood coastal areas and cover island nations, widespread famine will ensue as farms dry up, and heat waves and tropical diseases will threaten our health. The future, some argue, could be apocalyptic as ice caps melt and the sea level rises. The frightened public often isn't aware that most of these catastrophes will not occur for 100 years, and then only if the upper end of the uncertain scientific predictions applies.

The industries considered most responsible for global warming are the most critical of its scientific validity. The coal and oil companies, as well as the oil-producing nations, argue that the computerized climate models are crude and approximate, incomplete, inconclusive, and so flawed that their predictions of future climate change cannot be used as a basis for taking action. The risks of global warming, they therefore argue, have been widely exaggerated, and there is no proof that anything catastrophic will happen.

Moreover, even if the global temperature is rising, they say, its cause is solely or mostly natural. The climate is always changing whether or not human beings have anything to do with it, and humans continue to be insignificant when compared to the natural forces that have determined the climate for millions of years.

The coal and oil companies insist that any future warming from carbon-dioxide emissions will be moderate and that rising levels of the benign gas will be a good thing. Far from being a pollutant, carbon dioxide is a powerful fertilizer that helps plants grow. If you reduced the amount of the gas in the air, the plants would be in real trouble; they could even disappear.

This side of the debate argues that any attempt to get rid of coal and oil will cause an economic disaster; it might even cripple the global economy. With sales exceeding two billion dollars a day and trillions of dollars a year, the oil industry is indeed a powerful economic force. A significant reduction in the use of coal and oil, some say, could eliminate millions of American jobs, reduce the United States' ability to compete globally, interfere with the free market, and endanger the lifestyle of every American.

The same arguments, they say, apply to most of the rich industrial countries, whose economies are dependent on the fossil-fuel industry. Any regulations to curtail emissions from burning coal and oil could slow economic growth in all of these countries.

With so much money at stake, it is perhaps not surprising that the oil and coal industries have spent millions of dollars to cast doubt on global warming. After all, they are just protecting their interests. Critics argue that they have also been suppressing the true implications of global warming, somewhat as the tobacco industry did about the dangers of cigarettes. Some lawyers have even threatened a class action suit to force a reduction in the emission of heat-trapping gases.

Not every industry supports the unrestrained use of fossil fuels. The insurance industry favors restraints, fearing that their profits will fall as extreme weather increases. Floods, hurricanes and other severe storms, attributed to global warming, have already caused annual insurance losses of billions of dollars, naturally passed on to the customers by higher rates, and have led to insurance policy exclusions for those living in storm-prone areas. Now entire countries are entering the fray, attempting international agreements with legally binding limits to the emissions of heat-trapping gases.

Doing something about global warming

Governments can blunt the feared global warming of the future by adopting energy policies that shift from coal and oil to natural gas, and eventually to energy sources that do not generate heat-trapping gases. Every time you turn a light on, the electricity most likely comes from burning coal. It still supplies 56 percent of the electricity in the United States. Yet, for the same energy production, coal burning releases more carbon into the air than burning oil, and natural gas releases even less. All that carbon combines with oxygen in the air to create carbon dioxide.

Some sources of electricity emit no carbon into the air and produce no heat-trapping gases. They include hydroelectric power, solar energy, and the power of the wind. In the United States, current subsidies and tax incentives for the development of oil, coal and natural gas amount to about 20 billion dollars a year. If these funds were shifted to the cleaner energy sources, it would make them more competitive. Of course, the largest carbon-free source of energy is nuclear power, which produces 20 percent of the United States requirement, and nuclear energy is subsidized by about 10 billion dollars a year.

Countries can avoid the clearing of their forests and plant a lot more trees. Each tree removes about a ton of carbon dioxide from the air, locking the carbon into its branches, trunk and leaves. Trees also tend to outlive humans and they prevent erosion. By protecting existing forests and planting new ones, countries could offset 10 to 20 percent of the expected carbon dioxide build up during the next century.

Mandates for limiting fossil-fuel emissions or protecting forests are nevertheless difficult to legislate, partly because the threat is uncertain. Policy makers like black-and-white issues, but future global warming effects are gray. There are pros and cons to a hotter world, winners and losers.

There is also a lack of immediacy. Very dramatic warming effects occur on a vast time-scale, over decades and centuries, so we are unlikely to witness them in our lifetimes. They certainly will not happen before the next re-election campaign of government leaders.

Global climate change is an issue that all countries have to deal with, both the rich industrial nations and the poor developing ones. But there are stark differences between the countries, blocking any substantive international agreement so far.

People in the poorer nations argue that the average person in the rich countries eats more food, consumes more energy and poisons the air more than they do. And the industrial countries became wealthy largely by burning the coal and oil that produced most of the heat-trapping carbon dioxide that is now in the air. They are thus responsible for most of whatever global warming is likely to bring. The rich nations are still responsible for most emissions today, about 80 percent of it. So, the poorer nations say, it is only right that the wealthy countries be the ones to cut back on their emissions.

The developing countries point out that they will endure greater damage from future climate disasters caused by the rich countries' emissions. Global warming, for example, is expected to increase crop yields in temperate northern regions, where the rich, industrial countries are located, while harming agriculture in the lower, warmer latitudes where most poor nations are.

Rich nations have the resources needed to adapt to climate changes; for some other countries this is not an option. The poorer countries are hard-pressed enough to assure the economic survival of their rapidly growing populations. As they produce more and more people and less and less food, these countries will not want to limit the economic growth required for survival.

Yet, the industrial countries insist that warming is a global concern, and that all countries must share in the solution. This is particularly so, they argue, since the developing countries' emissions are expected to surpass those of the rich nations in a few decades.

The United States has been cast as the wealthy villain, the most greedy, selfish and irresponsible of all. It is by far the biggest single producer of heat-trapping gases, both in total output and on a per capita basis, contributing 25 percent of the total with just 4 percent of the world's population. By way of comparison, the average European consumer, who also lives in an industrial country, consumes about half as much energy as the average American. So a little self-restraint and denial might be appropriate for the Americans, and it might help their strategic vulnerability.

Much of America's energy comes from oil-producing nations in the Middle East, and the United States spends at least 25 billion dollars each year in their military defense. Yet, most of them have deplorable human rights records and enormous gaps between rich and poor. These conditions have helped breed the terrorism that now threatens the United States.

And what about the poor developing nations? They are not about to adopt restraints that might slow their industrial growth just to keep the rich, industrialized nations a little cooler. Yet, if the developing countries take the same path as the wealthy ones, burning coal and oil to fuel their growth, then atmospheric carbon dioxide will soar.

If the poorer nations are forced to accelerate the burning of fossil fuels, to feed and house and employ their expanding populations, then their carbon-dioxide production will soon dwarf that of the rich industrialized countries. By 2015 the nations of Asia, led by China and India, will surpass even the unrestrained emissions of the richer nations.

So what's being done about the problem? In December 1997 representatives of the world's nations met in Kyoto, Japan, to establish, for the first time, specific legally binding targets and timetables for the emission of heat-trapping gases (Focus 4.4). The treaty, called *The Kyoto Protocol*, would require the United States and other industrial countries to reduce emissions by 2012 to an average of 5.2 percent below emission levels in 1990. The accord has been signed by more than 100 countries, but there is no way that it is going to cool the planet very soon. In order to take effect, *The Kyoto Protocol* has to be ratified by a substantial number of industrial nations, but none of them has ratified it, at least by 2001, and ratification is not expected.

The United States continues to find *The Kyoto Protocol* unacceptable because it unfairly requires only the industrialized countries to cut emission. The "one-sided" treaty is not likely to become acceptable until the fast-developing countries, like China and India, also face emission constraints. So the developing countries must eventually be persuaded to participate in any reduction of the emission of heat-trapping gases if the treaty is to survive.

Both the rich and poor nations want an agreement that will not cause serious harm to their national economy. Emission restraints will certainly be costly, most likely requiring the use of new forms of energy. Some wealthy countries fear it could cause a recession in their prosperous economies, while the poor ones are afraid that it will destroy the economic growth needed for the very survival of their people.

A global solution is needed, in which all nations participate in curtailing the emissions and agree on the best way to achieve it. Whatever the plan, its implementation should include a change in energy consumption habits to slow the inevitable global warming.

Individuals can reduce their consumption of the fossil fuels that electrify and heat their homes, offices and schools, power their vehicles, and fuel their factories. Ordinary people can use energy-efficient appliances and lighting, reduce their daily electricity use, drive their cars less, and insulate their homes and offices so they require less heat. Some of this energy conservation has already begun, but not enough is being done.

One-third of all greenhouse emissions come from automobiles. Burning a liter (0.264 U.S. gallons) of gasoline (petrol) in an average car produces 2.4 kilograms of heat-trapping gases, and over the course of a year that car dumps about 1000 kilograms of waste gas into the air – which is about equal to the weight of the car. A sports utility vehicle emits twice as much, and so does a pickup truck. So the world's population should buy fewer cars, at least those that are powered by gasoline, and they should especially avoid the gas-guzzling kind.

Focus 4.4 *The Kyoto Protocol*

In December 1997, at an International Climate Summit in the ancient Japanese capital of Kyoto, more than 100 nations agreed to reduce the emissions of heat-trapping gases that can warm the planet. Known as *The Kyoto Protocol to the United Nations Framework Convention on Climate Change*, or just *The Kyoto Protocol* for short, the agreement calls for reductions in the emissions of six greenhouse gases: carbon dioxide, methane, nitrous oxide, two fluorocarbons and sulfur hexafluoride.

The accord established different levels of reductions for individual countries. The United States would be committed to a reduction of 7 percent below 1990 levels, the European Union to 8 percent and Japan to 6 percent.

The developing countries are exempt from mandatory emission controls, since the vast majority of the emissions to date have not been caused by them. Even the fast-growing developing countries face no constraints, though their emissions are expected to surpass those of the industrial nations in two or three decades.

The protocol will take effect once it is ratified by at least 55 nations that collectively account for at least 55 percent of 1990 carbon-dioxide emissions. The terms become binding on an individual country only after its government ratifies the treaty. But no industrial country is expected to ratify the treaty.

The industrial countries cannot easily meet the limits of *The Kyoto Protocol* even if it is ratified. On the eve of the Kyoto negotiations in 1997, America's emissions were already up 10 percent from 1990 levels and they have risen about 1.2 percent a year since then. Other industrial nations, too, have recorded rising, not falling, emissions.

Powerful liberal groups have been calling for more ambitious limits to emissions, while more conservative interests argue against the drastic measures already required in *The Kyoto Protocol*. Both groups incorporate selected evidence about future global warming that agrees with their objectives, avoiding a balanced appraisal of the dangers.

The details of *The Kyoto Protocol* continue to be discussed in international bargaining sessions. It includes principles of international emissions trading which would allow a country to trade reductions if they fall below the country's limit. If an industrial nation were to exceed its emission quota, it could purchase the unused rights from lower-emitting countries, but the rulebook for these trades is still being worked out.

Under a joint implementation plan, a wealthy country could get credit towards its targets by investing in specific emissions-saving projects in developing countries, such as more efficient power plants. In another tradeoff, some countries in the former Soviet Union might be able to sell emission credits for reductions that occurred prior to the negotiations, primarily as the result of national economic problems.

Another provision under negotiation counts vast forests towards a country's emission-reduction credits, since trees absorb carbon dioxide from the air as they grow. Since forests can always be cut down, this is nevertheless just a temporary abatement, and not the same as a permanent prevention of carbon dioxide emitted by vehicles, power plants or factories.

As the diplomats and politicians continue their stately dance, *The Kyoto Protocol* will probably live or die on the basis of its economic repercussions. No nation will accept the treaty if it causes serious damage to the country's national economy, and economic incentives will be required to implement the planned reductions.

Every large automaker is now investing heavily in new engine technology to improve fuel efficiency. We already have hybrid cars that combine internal combustion with battery power, and we may eventually be able to purchase cars that use clean hydrogen gas as fuel.

And many of the fiercest corporate opponents to emission regulations are now voluntarily cutting their emissions of heat-trapping gases, perhaps to head off tougher regulations in the future. Oil giants like Shell and British Petroleum are taking steps to end the burning of natural gas at oil wells – some say to mollify environmentalists in their European markets where there is strong public interest in global warming. Large automakers in Europe, including Ford and General Motors, have reluctantly agreed to improve the fuel efficiency of their automobiles. Other corporate giants like Du Pont are lowering their output of certain chemicals, the CFCs, that contribute to global warming, an action that began years ago because some of the same chemicals damage the ozone layer.

So humans have modified the atmosphere, warming the globe, and we are starting to do something about it. But it's bound to be only a temporary fix. In the long run nature will take over the weather and climate once again. A hundred million years ago, when the dinosaurs roamed the Earth, there were no ice caps and tropical plants flourished near the South (Geographic) Pole. Deep cold nearly turned the Earth into a ball of ice about 10 thousand years ago, when the planet was in the depths of an ice age. In just a few million years from now, entire continents and oceans can be destroyed or created, changing the flow of air and ocean currents and altering global weather patterns. And even if it is pretty warm right now, the die is cast for the next glaciation and the ice will come again.

Ice and fire

During the past million years the Earth has undergone a series of warm and cold periods. During the cold periods, called ice ages, huge ice sheets build up on the continents and in the polar seas. The growing layer of continental ice flows toward the equator, scouring and covering large areas of land. Then the climate warms and the ice retreats.

Each glacial ice age lasts about 100 thousand years. There is a relatively short interval of unusual warmth between the ice ages that lasts 10 or 20 thousand years. During such an interglacial interval, the world's climate becomes more pleasant and serene. We now live in such a warm time, called the Holocene period, which has enabled human civilization to flourish.

The current Holocene interglacial, which has already lasted 11 thousand years, may not continue for more than a few thousand years, and we could then enter an ice age. The next time it happens, the advancing glaciers might bury Copenhagen, Detroit and Montreal under mountains of ice, and because of the drop in sea level people might then walk from England to France, from Siberia to Alaska, and from New Guinea to Australia.

The recurrent ice ages and warm intervals are caused by variations in the amount and distribution of sunlight reaching the Earth, but not by any intrinsic fluctuations in the amount of light radiated by the Sun itself. Three astronomical cycles combine to alter the angles and distance at which sunlight strikes the far northern latitudes of Earth, triggering the ice ages. This explanation was fully developed by Milutin Milankovitch (1879–1958) from 1920 to 1941, so the astronomical cycles are now sometimes called the Milankovitch cycles.

When there is less sunlight being received in far northern latitudes, the winter temperatures are milder there, but so too are summer temperatures. So, less polar ice melts in the summer, and over time the winter snows are compressed into ice to make the glaciers grow.

The varying gravitational forces produced by the other planets, whose distances from Earth change, produce a rhythmic stretching of the Earth's orbit. These planetary perturbations periodically change the shape of the Earth's orbit from circular to slightly elliptical and back again, over a period of 100 thousand years. As its path becomes more elongated, the Earth's distance from the Sun varies more during each year, intensifying the seasons in one hemisphere and moderating them in the other.

Shorter cycles are due to repetitive changes in the wobble and tilt of the Earth's rotational axis, which vary over 23 thousand and 41 thousand years respectively. The greater the tilt, the more intense the seasons in both hemispheres, with hotter summers and colder winters.

Successive layers of frozen atmosphere have been laid down in Greenland and Antarctica, providing a natural archive of the Earth's climate over the past 420 thousand years. Bubbles of air trapped in falling snowflakes and entombed in ice are deposited every year, building up on top of each other like layers of sediment. When extracted in deep ice cores, they reveal secrets about the ancient climate.

Such cores strongly support the idea that changes in the Earth's orbit and spin axis cause variations in the intensity and distribution of sunlight arriving at Earth, which in turn initiate natural climate changes and trigger the ebb and flow of glacial ice.

Air bubbles trapped in the ancient ice show that the Antarctica air temperatures and atmospheric heat-trapping gases rose and fell in tandem as the glaciers came and went and came again. The temperatures go up whenever the levels of carbon dioxide and methane do, and they decrease together as well (Fig. 4.27). Scientists cannot yet agree whether the increase in greenhouse gases preceded or followed the rising temperatures. And since the current level of greenhouse gases, recently deposited in our atmosphere by humans, far surpasses any natural fluctuation of these substances recorded during past ice ages (Fig. 4.27), we are not sure if the next ice age will dampen future global warming.

The increase in carbon dioxide in the distant past apparently resolves a difficulty in explaining the ice ages by the astronomical cycles. Although the largest climate variations occur every 100 thousand years, the corresponding variation in the intensity of incident solar radiation is far too small to directly create the observed temperature changes. The build up of greenhouse gas apparently amplifies effects triggered and timed by the rhythmic orbital change in the distance between the Sun and Earth.

Well-accepted models of stellar evolution indicate that the Sun began it life about 4.6 billion years ago shining with about 70 percent of the brightness it has today, and that it has been slowly increasing in brightness ever since. Assuming an unchanging atmosphere with roughly the same composition and reflecting properties as today, the faint young Sun could not have warmed the Earth above the freezing point of water until 2 billion years ago, and our oceans would have been locked in ice anytime before then. This contradicts geological evidence, in sedimentary rock layers, of liquid water at least 3.8 billion years ago, and fossil evidence of living things 3.5 billion years ago.

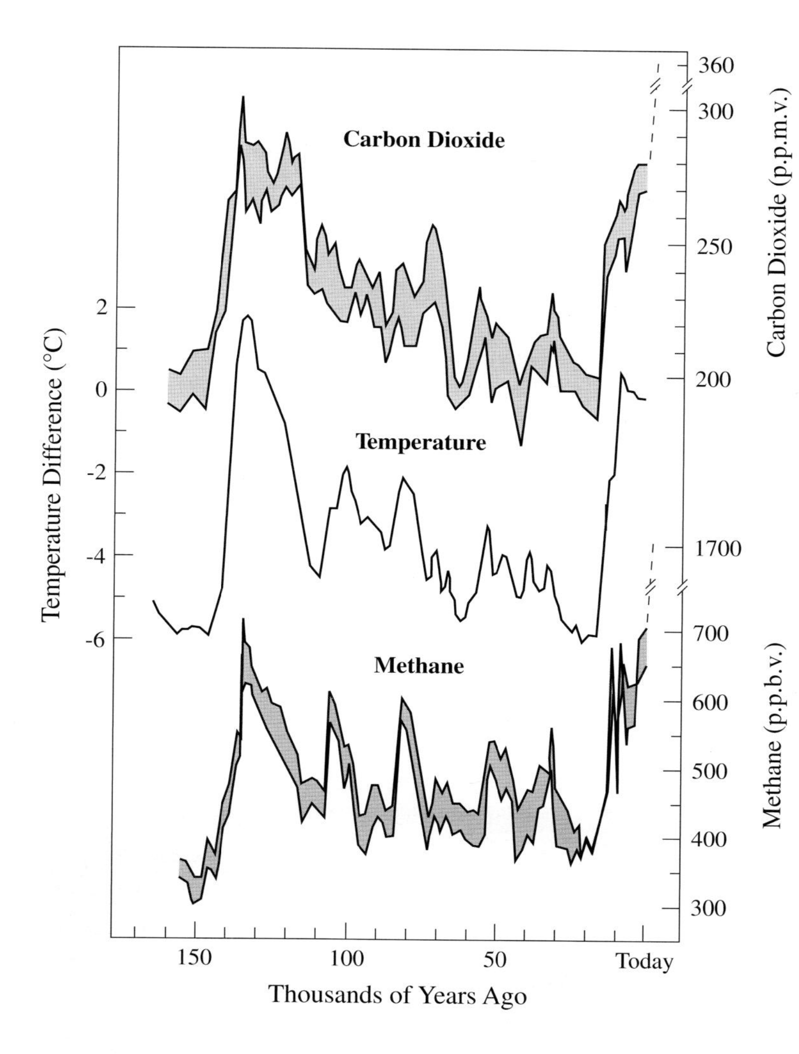

Fig. 4.27 Ice-age temperatures and greenhouse gases Ice-core data indicate that changes in the atmospheric temperature over Antarctica closely parallel variations in the atmospheric concentrations of two greenhouse gases, carbon dioxide and methane, for the past 160 000 years. When the temperature rises, so does the amount of these two greenhouse gases, and *vice versa*. This strong correlation has been extended by a deeper Vostok ice core, to 3623 meters in depth and the past 420 000 years. The carbon-dioxide and methane increases could amplify orbital forcing of climate change. The ice-core data does not include the past 200 years, shown as a dashed lines at the right. They show that the present-day levels of carbon dioxide and methane are unprecedented during the past four 100 000-year glacial–interglacial cycles.

The paradox can be resolved if the Earth's primitive atmosphere contained thousands of times more carbon dioxide than it does now. The greater heating of the enhanced greenhouse effect would have kept the oceans from freezing. The Earth could subsequently maintain a temperate climate only if its atmosphere, rocks and oceans combined to decrease the amount of carbon dioxide over time, turning down the planet's greenhouse effect as the Sun grew brighter.

Since the Sun keeps on getting hotter as time goes on, there is no escape from its eventual control over life on Earth. Calculations indicate that the Sun will become hot enough in 3 billion years to evaporate our oceans away, leaving the planet a burned-out cinder, a dead and sterile place. And if that doesn't do us in, the Sun will become hot enough to melt the Earth's surface in another 4 billion years (Fig. 4.28). Thus, our very remote descendants are destined to an end in fire, consumed by the Sun that once nurtured us. The only imaginable escape would then be interplanetary migration to distant moons or planets with a warm, pleasant climate.

But to get back to everyday reality, the Sun provides more immediate hazards, endangering humans and their spacecraft whenever they venture into space.

4.5 Space weather

There is danger blowing in the Sun's winds. Energetic charged particles and magnetic fields are being hurled into interplanetary space by explosions on the Sun, producing gusts and squalls in the solar wind that can wipe out unprotected astronauts and destroy man-made satellites. Down here on the ground, we are shielded from their direct onslaught by the Earth's atmosphere and magnetic fields, but out in deep space there is no place to hide.

Powerful explosions on the violent Sun come in two main varieties, known as solar flares and coronal mass ejections, or CMEs for short. Both kinds of solar activity are powered by the Sun's magnetic energy, and they both vary in step with the Sun's 11-year cycle of magnetic activity. Solar flares and CMEs are more frequent and tend

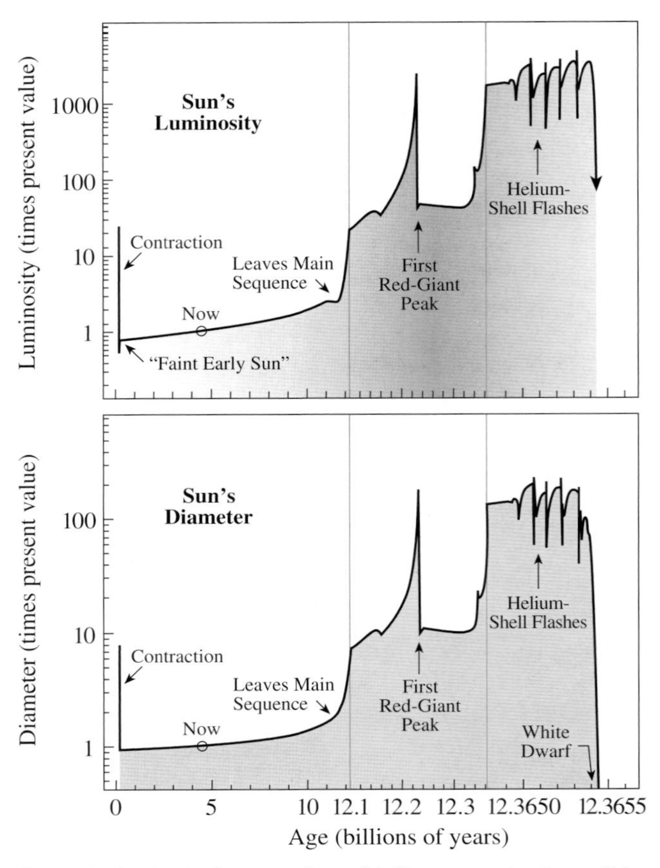

Fig. 4.28 The Sun's fate In about 8 billion years the Sun will become much brighter (*top*) and larger (*bottom*). Note the different time-scales, expanded near the end of the Sun's life to show relatively rapid changes. (Courtesy of I-Juliana Sackmann and Arnold I. Boothroyd.)

to be more powerful during the maximum in the activity cycle.

Solar flares are brief, catastrophic outbursts that flood the solar system with intense radiation and high-speed electrons and protons. In just a few minutes they can release an explosive energy of up to 10^{25} joule, equivalent to 20 million 100-megaton terrestrial nuclear bombs, raising the temperature of Earth-sized regions on the Sun to tens of millions of degrees. The other type of solar explosive activity, the CMEs, expand away from the Sun at speeds of hundreds of thousands of meters per second, becoming larger than the Sun and removing up to fifty billion tons, or 5×10^{13} kilograms, of the Sun's atmosphere (Fig. 4.29).

Energetic protons hurled out from intense solar flares are especially hazardous. They endanger any astronaut caught in space without adequate protection. The high-speed solar protons could even kill an unprotected astronaut that ventures into space. The electrically charged particles follow a narrow, curved path once they leave the Sun, guided by the spiral structure of the interplanetary magnetic field (Fig. 4.30). Solar astronomers therefore keep careful watch over the Sun during space missions, to warn of possible activity occurring at just the wrong place on the Sun, at one end of a spiral magnetic field line that connects the flaring region to wherever astronauts happen to be. They can then avoid making repairs to their space stations, and curtail any strolls on the Moon or Mars, instead moving inside storm shelters.

At any given phase of the solar cycle, intense solar flares are as much as 100 times more frequent than mass ejections, but the CMEs energize particles on a grand scale that covers large regions in interplanetary space. They move straight out of the Sun and flatten everything in their path, like a gigantic falling tree or a car out of control. Fast CMEs plow into the slower-moving solar wind and act like a piston that drives shock waves ahead of them, accelerating electrons and protons as they go, like ocean waves propelling surfers.

Humans in space also risk damaging exposure to galactic cosmic rays that originate outside the solar system.

Fig. 4.29 Coronal mass ejection A huge coronal mass ejection, abbreviated CME, is seen in this image. The white circle denotes the edge of the photosphere, so this mass ejection is about twice as large as the visible Sun. The black area corresponds to the occulting disk of the space-borne telescope, called a coronagraph. The disk blocks intense sunlight and permits the tenuous, million-degree corona to be seen. This coronagraph image was taken on 27 February 2000 with the Large Angle Spectrometric COronagraph (LASCO) on the *SOlar and Heliospheric Observatory* (*SOHO*). (Courtesy of the *SOHO* LASCO consortium. *SOHO* is a project of international cooperation between ESA and NASA.)

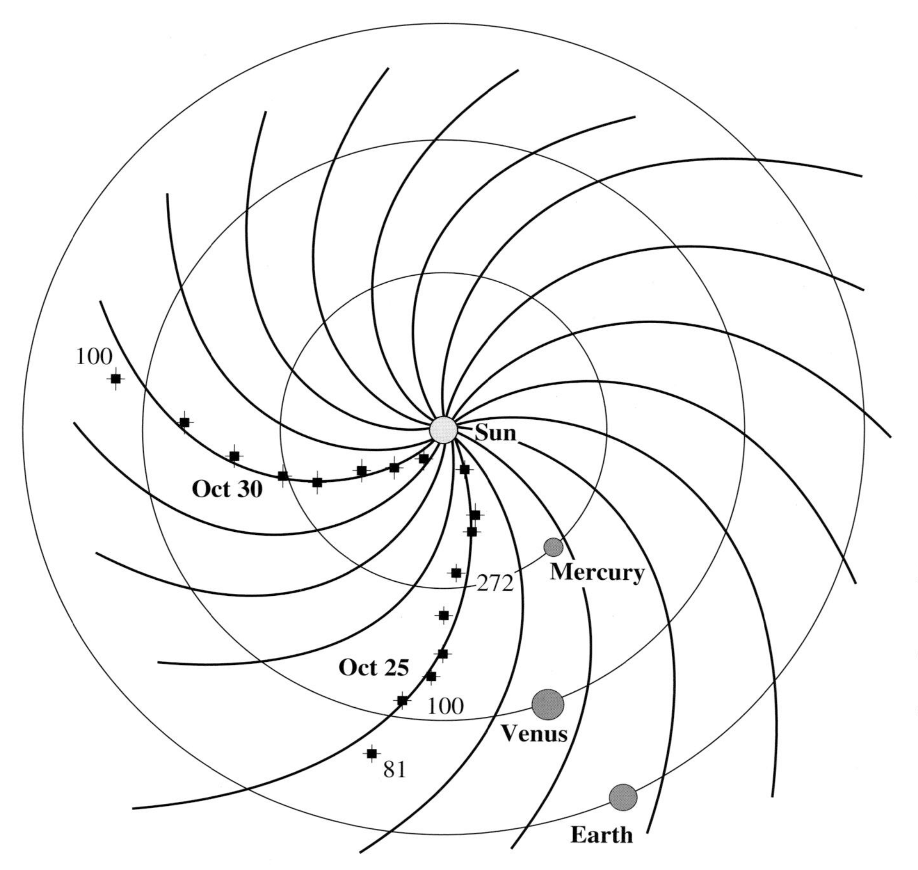

Fig. 4.30 Spiral path of interplanetary electrons The trajectory of flare electrons in interplanetary space as viewed from above the Sun's polar regions using the *Ulysses* spacecraft. As the high-speed electrons move out from the Sun, they excite radio radiation at successively lower frequencies; the numbers denote the observed frequency in kilohertz, or kHz. Since the flaring electrons are forced to follow the interplanetary magnetic field, they do not move in a straight line from the Sun to the Earth, but instead move along the spiral pattern of the interplanetary magnetic field, shown by the solid curved lines. The squares and crosses mark *Ulysses* radio measurements of type III radio bursts on 25 October 1994 and 30 October 1994. The approximate locations of the orbits of Mercury, Venus and the Earth are shown as circles. (Courtesy of Michael J. Reiner. *Ulysses* is a project of international collaboration between ESA and NASA.)

Although the flux of cosmic rays is much lower than Sun-driven particles, the cosmic-ray particles have much higher energy. At times of high solar activity, the frequency and intensity of solar explosions is greater, but the enhanced solar winds cut off the flow of cosmic rays into the solar system. The opposite conditions apply near the minimum in solar activity. So, there is some risk whenever someone takes a trip into space.

The dangers posed by cosmic rays are particularly daunting for a visit to Mars. With a projected launch date of 2020, NASA is planning to send astronauts on a three-year mission to the red planet – including six months in transit each way. As an example of the risks involved, cosmic-ray iron particles moving at nearly the speed of light can penetrate deeply into the body, even passing through your skull, ripping up the double helix of DNA molecules in the process. Long exposures to cosmic rays in space increase the risk of getting cancer, apparently to a forty-percent lifetime chance after a voyage to Mars and far above acceptable thresholds of government agencies.

Thus, there are some real potential hazards to travel in space, depending on the energy and flux of the particles out there, and the time and place of the voyage. Satellites near Earth can also be disabled. Storms in space can temporarily or permanently wipe out any one of the more than 1000 satellites now in operation, affecting the lives of millions of people.

Our networked society has become increasingly dependent on satellites that provide crucial information to corporations, governments and ordinary citizens. Geosynchronous satellites, that orbit the Earth at the same rate that the planet spins, stay above the same place on Earth to relay and beam down signals used for cellular phones, global positioning systems and internet commerce and data transmission. They can guide automobiles to their destinations, enable aviation and marine navigation, aid in search and rescue missions, and permit nearly instantaneous money exchange or investment choices. Other satellites move in lower orbits and whip around the planet, scanning air, land and sea for environmental change, weather forecasting and military reconnaissance.

High-speed, electrically charged particles, coming from solar flares or CMEs, scour exposed solar cells that are used to power spacecraft, and move right through the metallic skin of a spacecraft, producing erroneous commands. They have already destroyed at least one weather satellite and disabled several communications satellites.

The strong blast of X-rays from solar flares alters the Earth's ionosphere, disrupting radio contact with airplanes flying over oceans or remote countries. During moderately intense flares, radio communications can be silenced over the Earth's entire sunlit hemisphere.

The enhanced ultraviolet radiation and X-rays from solar flares also heats the atmosphere and causes it to expand, and similar or greater effects are caused by CMEs. The expansion of the terrestrial atmosphere brings higher gas densities to a given altitude, increasing the friction and drag exerted on a satellite and pulling it to a lower altitude. This consequence of rising solar activity has sent several satellites into a premature and fatal spiral toward the Earth, and future *Space Stations* will have to be periodically boosted in altitude to correct for the downward drag. Accurate monitoring of all orbiting objects depends on accurate knowledge of atmospheric change caused by the intense radiation of solar flares and CMEs.

When a CME slams into the Earth, the force of impact can push the bow shock down to half its usual distance of about 10 times the Earth's radius. Geostationary spacecraft, that stay over the same spot on Earth's equator, orbit our planet at about 6.6 Earth radii, moving around it once every 24 hours at the same rate that the planet spins. When the magnetosphere is compressed below their geosynchronous orbits, these satellites are exposed to the full brunt of the gusty solar wind and its charged, energized ingredients.

While altering the Earth's magnetic field, a colliding CME can produce strong electric currents in nearby space. If these currents connect to long-distance power lines on the ground, they can blow circuit breakers, overheat and melt the windings of transformers, and cause massive failures of electrical distribution systems. A CME can thereby plunge major urban centers, like New York City or Montreal, into complete darkness, causing social chaos and threatening safety. The threat is greatest in high-latitude regions where the currents are strongest, such as Canada, the northern United States and Scandinavia.

Thus, our technological society has become increasingly vulnerable to explosions on the Sun. They emit energetic particles, intense radiation, powerful magnetic fields and strong shocks that can have enormous practical implications when directed toward Earth. The solar emissions can disrupt navigation and communication systems, pose significant hazards to humans in space, destroy Earth-orbiting satellites, and create power surges that can black out entire cities. Recognizing our vulnerability, national centers and defense agencies continuously monitor the Sun from ground and space to forecast threatening activity. An example is the Space Environment Center, abbreviated SEC, of the United States National Oceanic and Atmospheric Administration. It collects and distributes space-weather data, using satellites and ground-based telescopes to monitor the Sun and interplanetary space.

With adequate warning, operators can power down sensitive electronics on navigation and positional satellites, putting them to sleep until the danger passes. Airplane pilots and cellular telephone customers can be warned of

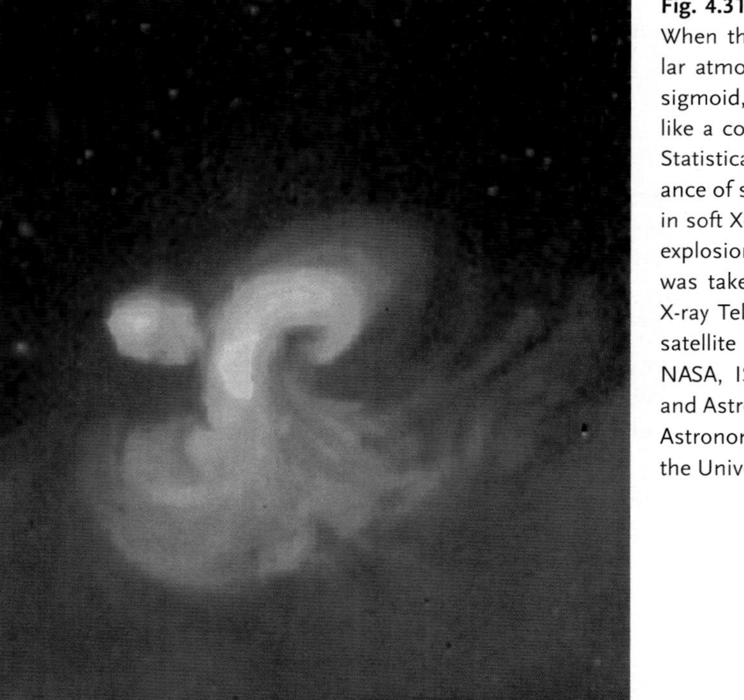

Fig. 4.31 The Sun getting ready to explode When the magnetic fields in the low solar atmosphere get twisted into an S, or sigmoid, shape, they become dangerous, like a coiled rattlesnake waiting to strike. Statistical studies indicate that the appearance of such a large S or inverted S shape in soft X-rays is likely to be followed by an explosion in just a few days. This image was taken on 8 June 1998 with the Soft X-ray Telescope (SXT) aboard the *Yohkoh* satellite (Courtesy of Richard C. Canfield, NASA, ISAS, the Lockheed–Martin Solar and Astrophysics Laboratory, the National Astronomical Observatory of Japan, and the University of Tokyo.)

potential communication failures. The launch of manned space flight missions can be postponed, and walks outside spacecraft or on the Moon or Mars might be delayed. Utility companies can reduce load in anticipation of induced currents on power lines, in that way trading a temporary "brown out" for a potentially disastrous "black out".

What everyone wants to know is how strong the storm is and when it is going to hit us. Like winter storms on Earth, some of the effects can be predicted days in advance. A CME arrives at the Earth one to four days after leaving the Sun, and solar astronomers can watch solar explosions happen. Solar flares are another matter. As soon as you can see a solar flare on the Sun, its radiation and fastest particles have already reached us, taking just 8 minutes to travel from the Sun to Earth. Dangerous, but less-energetic particles might take an hour to get here. The ultimate goal of space-weather forecasters is to predict when the Sun is about to unleash its pent-up energy, before a solar flare or CME occurs. One promising technique is to watch to see when the magnetism has become twisted into a stressed situation, for it may then be about to explode (Fig. 4.31).

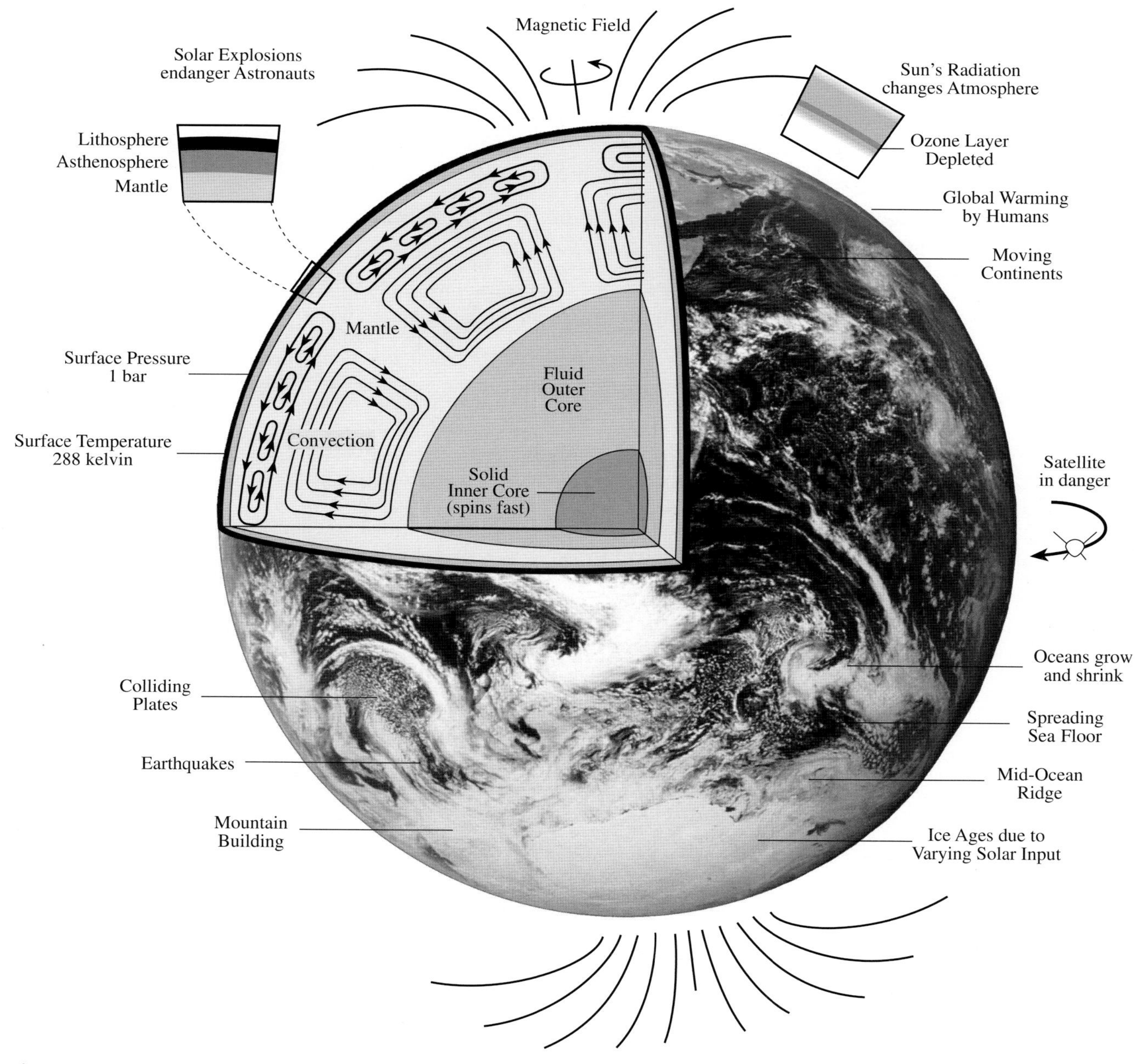

Fig. 4.32 Summary Diagram.

5 The Moon: stepping stone to the planets

- When the Moon moves into the Earth's shadow, the full Moon turns blood red; when the Earth travels into the Moon's shadow it can become dark during the day.
- The full Moon looks bigger near the horizon than directly overhead, but its changing size is an illusion.
- The Moon spins on its axis with the same period in which it revolves around the Earth, at 27.3 days, keeping its far side forever hidden to Earthbound observers.
- The near side of the Moon contains light, rugged cratered regions called highlands and dark smooth lava flows dubbed maria; the far side of the Moon is mostly highlands and has very few maria.
- For more than two centuries, lunar craters were attributed to volcanoes on the Moon, but they are now widely known to be due to the explosive impact of interplanetary projectiles.
- More than a quarter-century ago, twelve humans roamed the surface of the Moon and brought back nearly a half-ton of rocks.
- Because the Moon has almost no atmosphere, its sky remains pitch black in broad daylight and there is no sound or weather on the Moon.
- Two modest spacecraft, named *Clementine* and *Lunar Prospector*, have chalked up an impressive list of accomplishments, including evidence for water ice at the poles of the Moon.
- Humans might return to the Moon to create a unique astronomical observatory.
- Moonquakes, which are much weaker than earthquakes, indicate that the Moon has a small dense core, probably surrounded by a partially molten zone. The core has been confirmed by gravity measurements from the orbiting *Lunar Prospector* spacecraft, and the molten zone has been confirmed by laser-ranging measurements.
- There is no life on the Moon, and there apparently never was any.
- Rocks returned from the Moon contain no water and have never been exposed to it, but there is evidence for water ice in permanently shaded regions at the lunar poles.
- Earth rocks and Moon rocks are similar in their mix of light and heavy oxygen atoms, but the Moon rocks contain relatively little iron and few volatile elements common on Earth.

- Impact basins excavated by cosmic collision produce as much topographical relief on the Moon as there is on the Earth due to ongoing tectonic processes.
- Vast blocks of the lunar surface are magnetized, but they do not combine into an overall global dipole like the Earth's magnetism. Some of the ancient lunar magnetism has been concentrated on the side of the Moon opposite from large impact basins.
- Radioactive dating indicates that the Moon was assembled just 50 million years after the Earth and primitive meteorites were formed 4.6 billion years ago.
- During its early youth, between 4.4 and 4.6 billion years ago, a global sea of molten rock covered the Moon, but it is now covered by a layer of fine, powdery Moon dust.
- A heavy bombardment cratered the highlands between about 4.3 to 3.9 billion years ago, when the large impact basins were formed; lunar volcanism filled these basins to create the maria between 3.9 and 3.2 billion years ago.
- Most of the features we now see on the Moon have been there for more than 3 billion years.
- The Moon's gravity draws the Earth's oceans into the shape of an egg, causing two high tides as the planet's rotation carries the continents past the two tidal bulges each day.
- The Moon acts as a brake on the Earth's rotation, causing the length of the day to steadily increase and the Moon to move away from the Earth.
- The Moon provides a steadying influence to the Earth's seasonal climatic variation, anchoring and limiting the tilt of the planet's rotation axis.
- The Moon may have been created in a searing cataclysm during the ancient collision of a Mars-sized body with the growing Earth; the giant impact dislodged material that would become the Moon that we know.

5.1 Fundamentals

Table 5.1 Physical properties of the Moon[a]

Mass	7.349×10^{22} kilograms = 0.00123 M_E
Mean Radius	1737.5 kilometers = 0.2725 R_E
Mean Mass Density	3344 kilograms per cubic meter
Rotation Period	27.322 days = sidereal month
Orbital Period, the Sidereal Month	27.322 days = fixed star to fixed star
Synodic Month	29.53 days = new Moon to new Moon
Mean Distance from Earth	3.844×10^8 meters
Increase in Mean Distance	0.0382 ± 0.0007 meters per year
Mean Orbital Speed	1023 meters per second
Angular Radius at Mean Distance (Geocentric)	15 minutes 32.6 seconds of arc
Angular Radius at Mean Distance (Topocentric)	15 minutes 48.3 seconds of arc
Age	4.6×10^9 years

[a] Here M_E and R_E respectively denote the mass and radius of the Earth.

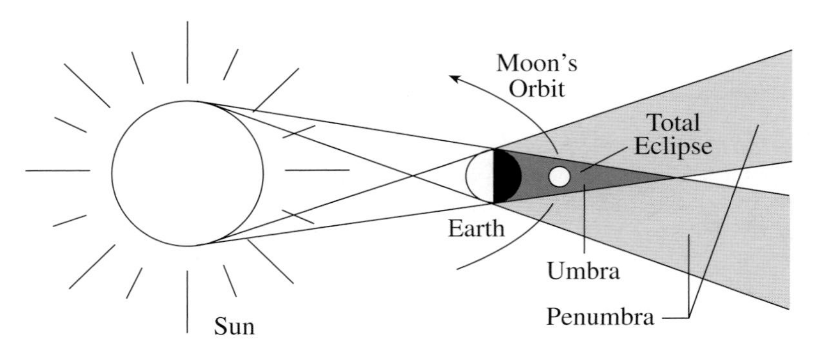

Fig. 5.1 Lunar eclipse During a lunar eclipse the initially full Moon passes through the Earth's shadow. A total lunar eclipse occurs when the entire Moon moves into the umbra. Because no portion of the Sun's visible disk can be seen from the umbra, it is the darkest part of the Earth's shadow. Only part of the Sun's disk is blocked out in the larger penumbra. A partial lunar eclipse occurs when the Moon's orbit takes it only partially through the umbra or only through the penumbra.

5.2 Eclipses

Once or twice in a typical year, the Moon's orbital motion carries it through the Earth's shadow. This is an eclipse of the Moon, and it can be seen from half of the Earth. There are two regions in the Earth's shadow at the time of a lunar eclipse: the umbral region where there is no direct sunlight, and the penumbral region where the Sun's light is partially shadowed (Fig. 5.1). The umbral shadow is darker, and it is in the shape of a narrow cone pointing away from the Earth.

The full Moon turns blood-red when in the umbral shadow of the Earth (Fig. 5.2). Ancient Hebrew writers often used this appearance as a metaphor to describe the end

Fig. 5.2 The blood-red Moon If the Earth had no atmosphere, the Moon would disappear in darkness during a total lunar eclipse. As shown here, the Moon actually becomes dark-red for an hour or so. This is because the Moon is illuminated by sunlight that is bent part way around the Earth and is reddened in passing through the Earth's atmosphere, just as the Sun is reddened at sunset. If the Earth is heavily clouded, the sunlight is obstructed and the Moon is particularly dark during a lunar eclipse. (Courtesy of Eric Mandon, Observatoire Populaire de Rouen.)

Fig. 5.3 Gossamer corona The Sun's corona as photographed during the total solar eclipse of 26 February 1998, observed from Oranjestad, Aruba. To extract this much coronal detail, several individual images, made with different exposure times, were combined and processed electronically in a computer. The resultant composite image shows the solar corona approximately as it appears to the human eye during totality. Note the fine rays and helmet streamers that extend far from the Sun and correspond to a wide range of brightness. (Courtesy of Fred Espenak.)

of the world. For instance, the prophet Joel declared that the Lord:

> ... will show wonders in the heavens and on the Earth, blood and fire and billows of smoke. The Sun will be turned to darkness, and the Moon to blood ...
>
> Joel 2:30, 31 – New International Version, Hodder and Stoughton

And then there are the lyrics to some of Bob Dylan's songs that include "the Moon rising like wildfire", and "when there was blood on the Moon".

A total eclipse of the Sun occurs when the Moon passes between the Earth and the Sun, and the Moon's shadow falls on the Earth. In an incredible cosmic coincidence, the Moon is just the right size and distance to blot out the visible solar disk when properly aligned and viewed from the Earth. In other words, the apparent angular diameter of the Moon and the visible solar disk are almost exactly the same, about 30 minutes of arc, so that under favorable circumstances the Moon's shadow can reach the Earth and cut off the light of the Sun.

The outer atmosphere of the Sun, known as its corona, becomes momentarily visible to the unaided eye when the Moon blocks the Sun's bright disk out and it becomes dark during the day. The corona is then seen at the limb, or apparent edge, of the Sun, against the blackened sky as a faint, shimmering halo of pearl-white light (Fig. 5.3). But be careful if you go watch an eclipse, for the light of the corona is still very hazardous to human eyes and should not be viewed directly. The million-degree corona can be seen all across the Sun's disk, and at any time, when viewing the Sun in X-rays with telescopes aboard satellites such as *Yohkoh* (see Fig. 4.22).

Since the Moon and the Earth move along different orbits whose planes are inclined to each other (Fig. 5.4), a total eclipse of the Sun does not happen very often. The Moon only passes between the Earth and the Sun about three times every decade on average. Even then, a total eclipse occurs along a relatively narrow region of the Earth's surface, where the tip of the Moon's shadow touches the Earth (Fig. 5.5). At other nearby places on the Earth, the Sun will be partially eclipsed, and at more remote locations you cannot see any eclipse of the Sun.

The Moon's orbital motion carries its shadow at about 3380 kilometers per hour with respect to the center of the Earth. But the Earth is rotating from west to east in the same direction as the eclipse shadow travels, at about 1670 kilometers per hour at the equator, slowing the

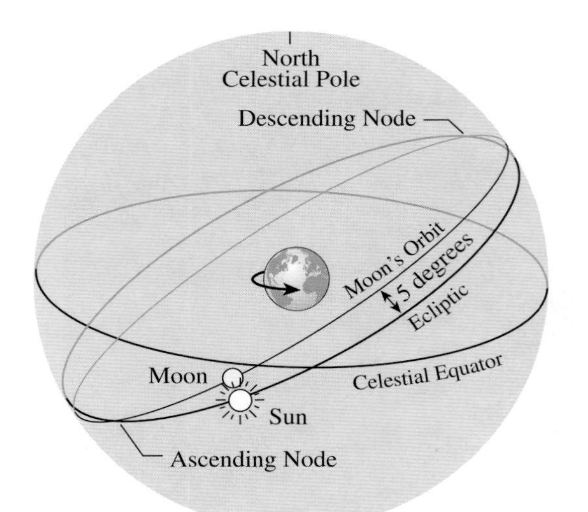

Fig. 5.4 Celestial paths of the Moon and Sun The Moon's orbit is tilted 5 degrees to the Sun's route across the sky, the ecliptic, allowing these paths to cross at two nodes. These are the only points at which eclipses can occur. During a lunar eclipse the Moon and Sun are located at opposing nodes, so that the Moon can move into the Earth's shadow cast by the Sun. A solar eclipse occurs when the Moon and Sun cross paths at the same node.

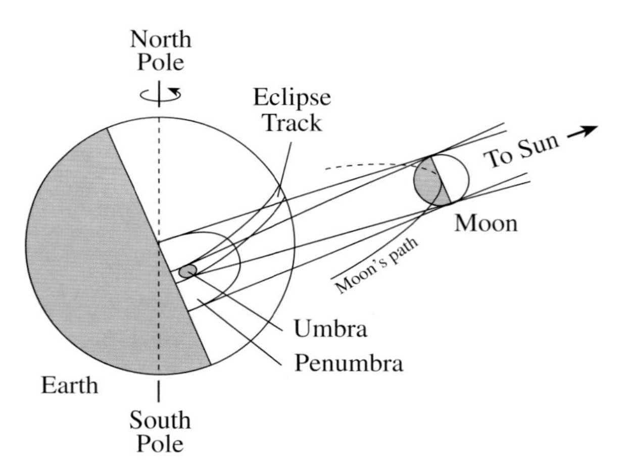

Fig. 5.5 Solar eclipse During a solar eclipse, the Moon casts its shadow upon the Earth. No portion of the Sun's photosphere can be seen from the umbral region of the Moon's shadow (*small gray spot*); but the Sun's light is only partially blocked in the penumbral region (*larger half-circle*). A total solar eclipse, observable only from the umbral region, traces a narrow path across the Earth's surface.

observed speed of the shadow by that amount. As a result, the longest total eclipse of the Sun observed at a fixed point on the ground lasts just under eight minutes.

If the Moon is at a distant part of its orbit at the time of solar eclipse, the Moon appears smaller than the Sun, and the tip of the Moon's shadow does not quite reach the Earth. The bright ring of the Sun's disk is then seen around the edge of the Moon. This is an annular eclipse, and it has none of the darkness and excitement of a total eclipse.

5.3 The Moon's face

When a full Moon rises or sets, it is a captivating sight. It looks huge, dwarfing everything in the foreground (Fig. 5.6). But appearances can be deceiving. The Moon is no bigger when it is close to the horizon than when it is high in the sky. Its changing size is an illusion caused by comparing the Moon to other objects when it is viewed along the ground.

This so-called Moon illusion arises from the way that the brain deals with apparent distance, not size. When people view the Moon near the horizon, there are large foreground objects, such as trees, buildings and hills, for comparison, so the Moon looks very far away and huge. When the Moon is overhead, alone in an otherwise empty sky, there are no other objects to gauge its distance; the Moon then appears to be closer and we think it is smaller than at the horizon (Fig. 5.7).

Artists often portray the Moon's face in all of its round fullness, and the Full Moon is the subject of all kinds of myths and superstitions (Focus 5.1).

Although it appears bright in contrast to the night sky, the Moon's face is as dark as the asphalt on highways and darker than most rocks on Earth. The lunar surface reflects just 10 percent of the sunlight that strikes it, which makes it one of the least shiny objects in the solar system.

Earthbound observers always see the same side of the Moon. We call this the near side of the Moon, in contrast to the far side, not visible from Earth. Our planet's greater gravitational pull on the near side of the Moon brakes the Moon's rotation and holds it in place like an invisible string, so the Moon's rotational period is precisely equal to its orbital period of 27.3 days. In other words, the Earth's gravity has synchronized the Moon's rotation with its orbital motion, so the Moon rotates on its axis once each orbit.

You can demonstrate synchronous rotation by holding a ball at arm's length and slowly turning around. As your body completes one rotation you always see the same side of the ball, but the ball has completed one rotation while revolving once about your body. But you can't watch someone else do this; you have to demonstrate it to yourself.

Although the same side of the Moon always faces the Earth, this doesn't mean that one side of the Moon is always dark. Like the Earth, the Moon gets its light from the Sun, and sunlight always illuminates one half of the Moon. As the Moon orbits the Earth we see varying amounts of its illuminated near side (Section 1.1, Figs. 1.3, 1.8). When we see a full Moon, the near side is in sunlight and the far side is dark. And when a new Moon is seen from Earth, the near side is dark and the far side is in full sunlight. Thus, although there is a "dark side of the Moon", it is not equivalent to the "far side".

Fig. 5.6 An enormous Moon In this awesome picture, a man and child seem enveloped by the huge Moon. Our brains trick us into thinking the Moon is much larger at the horizon than it is when viewed overhead - see Fig. 5.7. (Courtesy of der Foto-Treff.)

The Moon is the only planetary body that can be distinguished with the unaided eye as a globe, and even without a telescope you can tell that its surface is not uniform. Its face contains large, irregular features of light and dark material (Fig. 5.8), familiarly known as "The Man in the Moon".

It wasn't until Galileo Galilei (1564–1643) turned his primitive telescope to the Moon that it became clear that our satellite is rugged and mountainous like the Earth. When he looked closely at the division between light and shadow – day and night – on the globe of the Moon, Galileo discovered that the dividing line was ragged and

Focus 5.1 Full Moons

Ancient Greeks thought marriages consummated during a Full Moon would be prosperous and happy. In England, a distinction was made between lunacy and insanity; the former happened only during a Full Moon, while the latter was permanent. Yet, there is no scientific evidence that people become abnormally crazy at the time of Full Moon. Many Navajos believe that a woman is more likely to give birth during a Full Moon because of its pull on the amniotic fluid. It has even been rumored that a male child is more likely to be conceived when aided by the extra gravitational pull of a Full Moon. Of course, the pull of lunar gravity depends only on the Moon's mass and distance, and has no direct connection with the amount of sunlight that we see illuminating it.

A Full Moon is considered unlucky if it occurs on Sunday, the Sun's day, but lucky on Monday, whose name is derived from Moon Day. The phrase "once in a blue Moon" is based on the rare occurrence of a fourth Full Moon in a season, which normally has three. The last time a month elapsed without one Full Moon was in February 1866, an event that will not repeat itself for 2.5 million years.

At the time of Harvest Moon, the Full Moon rises at sunset, providing extra time for farmers to harvest their crops. According to folklore, the Harvest Moon also appears bigger, brighter and yellower than other Full Moons, but this is because it stays close to the horizon. All Full Moons look bigger near the ground than directly overhead, a visual effect known as the Moon illusion.

The yellow color of a rising Full Moon is due to scattering of light in the great thickness of air near the direction of the horizon; the haze and humidity of summer air can provide an orange color. It has even been suggested that the term honeymoon derives from the amber-colored Full Moons of June.

Fig. 5.7 Moon illusion We make judgements about size because of our perceptions of distance. The two black disks in this figure are the same size, but we see the bottom one as smaller because we think it is closer. The top disk seems larger because it appears to be farther away. The Moon on the horizon is similarly thought to be huge because comparisons with objects on the ground make us think it is far away – see Fig. 5.6. When people look straight up at the Moon, in an otherwise empty sky, they no longer have land clues to compute the Moon's distance and it is perceived as being closer and smaller.

that he could see high mountain peaks casting long pointed shadows. When sunlight strikes the lunar surface obliquely, every mountain, hill or valley is sharply delineated.

Since he had clear evidence that the Moon was not the perfectly smooth crystalline sphere that had been proclaimed in the writings of Aristotle (384–322 BC), Galileo could write in 1610 that:

> ... the surface of the Moon is not perfectly smooth, free from inequalities and exactly spherical, as a large body of philosophers considers with regard to the Moon and other bodies, but on the contrary, it is full of inequalities, uneven, full of hollows and protuberances. It is like the surface of the Earth itself, which is varied everywhere by lofty mountains and deep valleys.

But do not be deceived by these Earthly comparisons. The conspicuous mountain ranges on the Moon were thrown up about 4 billion years ago as rims of impact basins gouged out by immense cosmic collisions (Section 2.2), and the lunar mountains have nothing to do with the plate tectonics that created the much younger, terrestrial mountains (Section 4.3).

The Moon's rough terrain is mostly confined to the brighter regions that Galileo called *terrrae,* Latin for "lands"; they are now known as the highlands because they are higher than the dark regions (Fig. 5.9).

Galileo also discovered that the dark patches are smooth and level, resembling seas seen from a distance. He called them *maria,* the Latin word for "seas"; *mare,* pronounced "MAHreh", is the singular for "sea". However, we now know there is no water in the maria, and there apparently never has been. The dark maria cover about 17 percent of the lunar surface. When spacecraft were sent past the Moon to look at its averted face, they found that the far side contains very few maria (Section 2.1, Fig. 2.2).

Chemical examination of rock samples returned from the Moon has shown that the maria are ancient volcanic outflows composed of dark lava (Fig. 5.10). This material flowed out from inside the Moon to fill large impact basins that were formed at about the same time as the lunar highlands (Table 5.2). One of them, the Imbrium Basin that contains Mare Imbrium, now forms the "eyesocket" in the face of the "Man on the Moon"; it has a diameter of 1.5 thousand kilometers.

Craters form one of the most striking features of the Moon's landscape (Fig. 5.11). The word *crater* is derived from the Greek word for "cup or bowl", and it is a good

Table 5.2 Large impact basins and the lunar maria[a]

Maria (Latin)	Seas (English)	Basin Diameter (km)
Oceanus Procellarum	Ocean of Storms	3200
Mare Imbrium	Sea of Rains	1500
Mare Crisium	Sea of Crises	1060
Mare Orientale	Eastern Sea	930
Mare Serenitatis	Sea of Serenity	880
Mare Nectaris	Sea of Nectar	860
Mare Smythii	Smyth's Sea	840
Mare Humorum	Sea of Moisture	820
Mare Tranquillitatis	Sea of Tranquillity	775
Mare Nubium	Sea of Clouds	690
Mare Fecunditatis	Sea of Fertility	690

[a] Dating of rocks returned from the Moon indicate that the Imbrium, Serenitatis and Nectaris impacts occurred 3.85, 3.87 and 3.92 billion years ago.

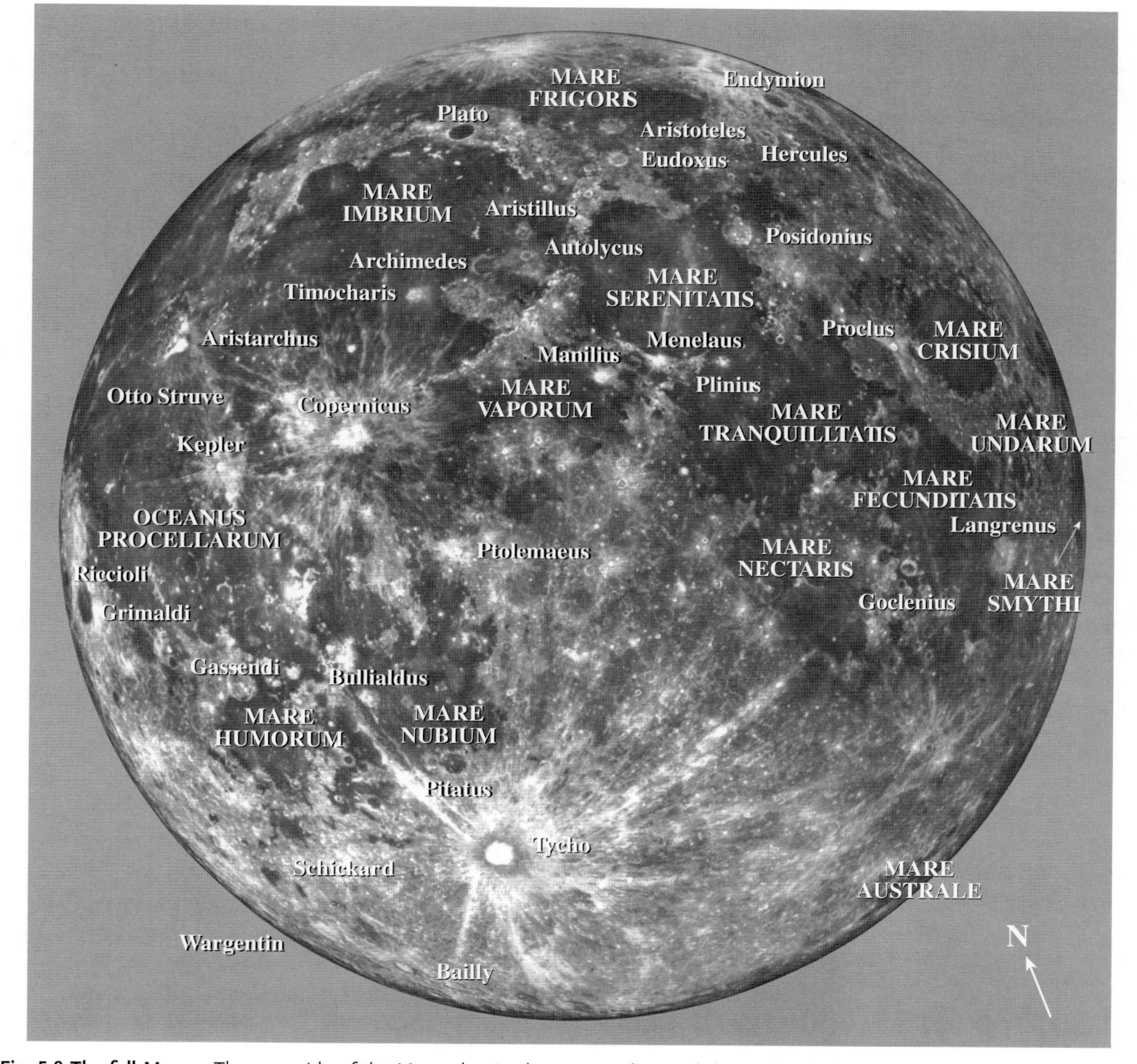

Fig. 5.8 The full Moon The near side of the Moon that is always turned toward the Earth. This Earth-based view of the full Moon enhances the contrast between the dark maria and the bright-rayed craters, such as Tycho (*near bottom center*), named after Tycho Brahe. The dark circular Mare Imbrium (Sea of Rains) is prominent in the northwest (*upper left*), immediately above the bright rays of craters Copernicus and Kepler (*middle left*). The dark circular Mare Serenitatis (Sea of Serenity) lies to the east (*right*) of Imbrium. (Photograph courtesy of UCO/Lick Observatory.)

description of the bowl-shaped depressions. They are just beyond the limit of visibility with the unaided eye, but a pair of binoculars will reveal a few of the larger ones. When seen through a telescope, the bright highlands are resolved into an enormous number of overlapping craters that have been visible to generations of observers.

At around the time of full Moon, a pair of binoculars will also show bright streaks that radiate from several craters like the spokes of a wheel. These are the lunar rays, the debris of crater formation. Some of the rays go more than one-quarter of the way around the Moon (Fig. 5.12).

The ubiquitous craters were thought to be of volcanic origin throughout the 19th century and well into the 20th century, but they are now widely known to be due to the explosive impact of interplanetary projectiles (Focus 5.2). Although the maria are filled with ancient volcanic outpourings of molten rock, they spread out

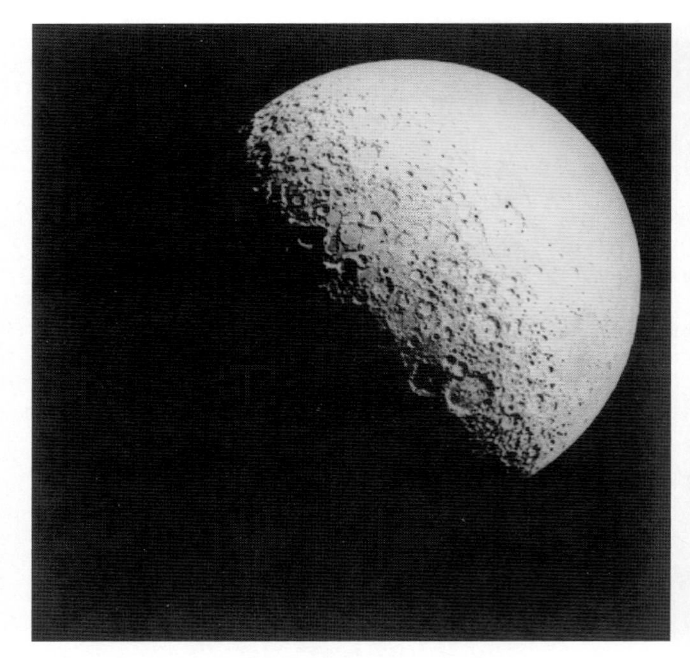

Fig. 5.9 Lunar highlands The heavily cratered lunar highlands are shown in this picture of the southern hemisphere of the Moon's near side. Humboldt crater is near the center of the image and Smyth's Sea is to the right. Impact craters of all sizes, including large impact basins measuring hundreds of kilometers across, were formed during an intense bombardment of the Moon about 3.9 billion years ago. This image was obtained during the *Apollo 15* mission in August 1971. (Courtesy of NASA.)

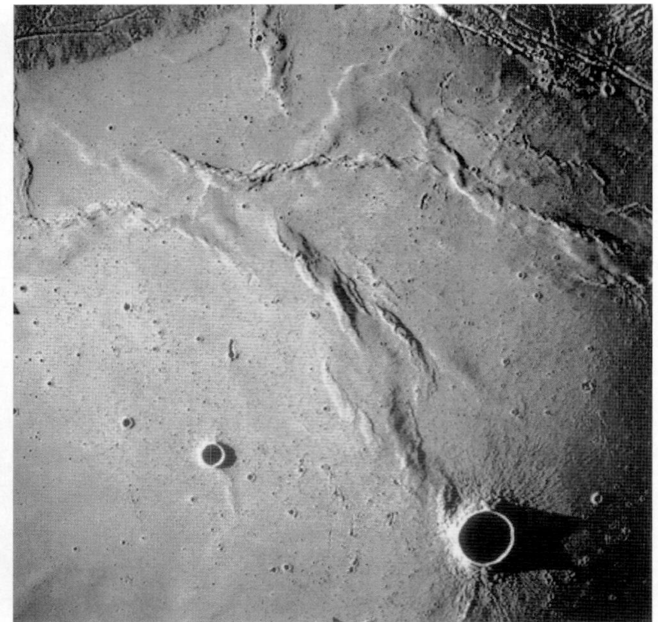

Fig. 5.10 Lava flows in a mare Lunar volcanism is seen frozen into place on the Sea of Serenity in this *Apollo 17* image taken in December 1972. The lunar maria contain relatively few craters when compared with the lunar highlands. The maria formed a secondary crust on the Moon, when lava filled the large impact basins over a period of several hundred million years ending around 3.2 billion years ago. The fluid spread rapidly, creating thin extensive sheets rather than piling up to form volcanoes. The largest crater shown here, named after the French mathematician Jean de Condorcet (1743–1794) is 74 kilometers in diameter. (Courtesy of NASA.)

rapidly and did not build up in one place. So there are no large volcanoes or calderas on the Moon.

Unlike the Earth, where erosion and tectonic processes tend to obscure the effects of impact, the surface of the Moon preserves a pristine record of an ancient bombardment extending back several billion years. Many of the planets and other satellites in the solar system bear the scars of a similar rain of debris, providing a common element in their history (Section 2.2).

5.4 Expeditions to the Moon

Race to the Moon

The Space Age began on 4 October 1957, when the Soviet Union launched the first artificial Earth satellite, *Prosteyshiy Sputnik*, the simplest satellite. Two years later the Soviets launched their *Luna 3* probe that was sent around the Moon, taking the first pictures of the normally invisible far side (Section 2.1, Fig. 2.2). And on 12 April 1961, cosmonaut Yuri A. Gagarin became the first human in space, orbiting the Earth in the *Vostok 1* capsule. Soviet officials used these accomplishments in a Cold War with the United States, citing them as evidence that communism is a superior form of social and economic organization.

The early Soviet capabilities in space underlined fears in the United States that a missile gap existed between it and its adversary, and seemed to verify the threat that the Soviet Union posed to world peace. After all, if you can send a spacecraft to the Moon, then you can surely launch a nuclear warhead on a guided missile to anywhere on the Earth.

Stimulated by the world-wide excitement generated by the first human flight in space, the visionary young President, John F. Kennedy, decided that the United States had to surpass the Soviet Union in some dramatic way in space. After expert advice, he concluded that there was a good chance of beating the Soviets to the first manned landing on the Moon. On 25 May 1961, just six weeks after the Gagarin flight, Kennedy delivered his now-famous address to a joint session of Congress, including the declaration: "I believe that this nation should commit itself to achieving the goal, before the decade is out, of landing a man on the Moon

Focus 5.2 Lunar craters – volcanoes or bombs?

Early interpretations of the lunar craters suggested that they were formed by volcanic activity. At the time, volcanic craters were the only Earth craters known, and no impact craters had been recognized on Earth. In addition, for more than two centuries reputable astronomers reported seeing smoke and even fire coming from volcanic eruptions on the Moon, especially near the craters Aristarchus and Alphonsus.

Gradually, the evidence began to favor the idea that the craters are formed by meteoritic impact. It was found that the floors of most lunar craters are slightly depressed below the surrounding level, in contrast to some of the Earth's craters that appear at the summits of towering volcanoes. In addition, large lunar craters contain flat floors and central peaks that are not often found in volcanic craters on the Earth. The central peaks of large craters on the Moon are created by the rebound of the underlying lunar surface following the collision of a big meteorite.

The round shape of nearly all the craters on the Moon can also be explained by the impact hypothesis. Because they strike at great speed, projectiles from space disintegrate in the explosive impact, producing a circular crater regardless of the direction at which the projectile struck. The lunar craters were also seen to resemble impact craters because the amount of material piled in the rims is nearly equal to the amount of material excavated from the interior.

Grove Karl Gilbert (1843–1918) was an early proponent of the impact theory for the origin of lunar craters, arguing that their size and shape closely resemble those formed by impacting objects. In his own words, written in 1892 when Gilbert was director of the U.S. Geological Survey,

> If a pebble be dropped into a pool of pasty mud, if a raindrop fall upon the slimy surface of a sea marsh when the tide is low, or if any projectile be made to strike any plastic body with suitable velocity, the scar produced by the impact has the form of a crater. This crater has a raised rim, suggestive of the wreath of the lunar craters. With proper adjustment of material, size of projectile, and velocity of impact, such a crater scar may be made to have a central hill.

Ralph B. Baldwin (1912–) provided additional evidence for the explosive origin of lunar craters in his influential book *The Face of the Moon* first published in 1949. He connected the relationship between the depth and diameter of craters on the Moon to the one describing the shell and bomb craters created during World War II, additionally noting that these man-made explosions have a circular form regardless of the angle of impact. Baldwin also argued that the dark, smooth maria occupy huge basins that were gouged out by rare, powerful impacts that punctured holes in the thin lunar crust, permitting lava to well out into them from the molten lunar interior.

But the arguments continued to rage. Former volcanic mountains might have sunk into the Moon's molten interior; the lunar craters could have been enveloped by lava lakes; and some astronomers continued to see glows, hazes and mists on the Moon that suggested volcanic eruptions. Most scientists had nevertheless dismissed the volcanic idea by the late-1960s, mainly because a large number of terrestrial impact craters had been identified.

The impact origin for lunar craters was confirmed when rocks were returned from the Moon. Samples from the highland craters and larger basins are conglomerates of preexisting rocks that have been welded together by impact. Dating of these rocks indicate that the battered highland crust is a museum of impact scars created during an ancient bombardment about 4 billion years ago (Section 5.7).

Although the lunar maria were filled during ancient episodes of volcanism, the Moon has apparently been volcanically inactive for 3 billion years. The supposed volcanic outbursts that were reported by several astronomers were evidently optical illusions.

and returning him safely to Earth". The president's call to action struck a responsive chord in the American public and galvanized their space program.

On 20 February 1962, John Glenn became the first American to orbit the Earth, and the race to the Moon was in full tilt. For several years the two superpowers traded accolades. The Russians sent the first woman – Valentina Tereshkova – into space, and they were the first to orbit three men in the same craft. On 18 March 1965 the Russian cosmonaut Aleksei A. Leonov was the first to walk in space, from the *Voskhod 2* capsule, closely followed by the American astronaut Edward H. White who took the first United States spacewalk on 3 June 1965 from the *Gemini 4* spacecraft.

During 1965 and 1966 the United States launched ten successful flights of the two-man *Gemini* spacecraft, including the first rendezvous of two spacecraft, *Gemini 6* and 7, and the nation was well prepared to embark on the *Apollo* program to land men on the Moon. It began with an ill-fated flight simulation on 27 January 1967, when faulty wiring ignited a flash electrical fire that asphyxiated and incinerated three astronauts on the ground. Yet, in just 22 months after this tragic setback, the manned *Apollo 8* spacecraft entered lunar orbit, all but ending the race to

Fig. 5.11 Craters Copernicus and Reinhold Bright ejecta radiate outward from the crater Copernicus near the lunar horizon. It is one of the youngest lunar craters on the near side of the Moon, with an estimated age of 900 million years and a diameter of 93 kilometers. The craters in the foreground are Reinhold A and B. (Courtesy of NASA.)

the Moon. It opened the way for the historic *Apollo 11* mission seven months later, when Neil Armstrong planted the American flag on lunar soil. The achievement was a spectacular triumph, the ultimate space first in the global geopolitical competition with the Soviet Union.

From 1960 to 1970 the Soviet Union sent several unmanned spacecraft to the Moon, including roving vehicles and sample returns to Earth, but they had lost the race to put a man on the Moon. Personal rivalries, shifting political alliances and bureaucratic inefficiencies had apparently bred failures and delays. After the triumphs of *Apollo 8* and *11*, the Russian lunar program faded into oblivion, and they turned their attention to long-duration missions in Earth-orbiting space stations.

The dissipation of the Soviet Union's lead in space tarnished the image of Soviet competence and diminished Soviet status in world affairs. In contrast, the *Apollo* program demonstrated the superiority of American purpose and technology in war or peace. Winning the race to the Moon and conquering the frontier of space taught us that nothing is impossible if we set our sights high enough; with resolve and willpower you can accomplish anything, especially in a democratic nation that stresses individual freedom.

The *Apollo* program

Before the United States accomplished manned landings, three types of robot spacecraft were sent to reconnoiter and answer two main questions for the proposed lunar landing. The first concerned the danger of encountering rocky terrain, where it would be impossible to land without capsizing. The second was the prediction, by some astronomers, that the lunar surface is covered by a thick layer of dust, perhaps as deep as a kilometer, that would make surface travel impossible. In fact, the astronauts might sink into the dust, suffocate and vanish into the Moon, like sinking into quicksand on Earth. The lunar surface was known to have been battered, churned and worn down by a hail of meteorites over the eons, creating loose debris of rocks, pebbles, grains, soil and dust.

To start resolving these uncertainties, three *Ranger* spacecraft crashed into the Moon, transmitting television pictures back to Earth as they rapidly approached the lunar surface. Watching these pictures was a dizzying experience, and the transmission of the final frames was interrupted by the crash itself. These were followed by five *Lunar Orbiters* that mapped most of the Moon's surface to locate potential landing sites, omitting only the polar regions. The final stage of preparation involved soft landings by the *Surveyors 1, 3, 5* and *7* that tested the detailed physical and chemical properties of the lunar surface and certified the safety of the initial *Apollo* landing sites. While the ground-control crews watched anxiously, the feet of the spidery three-legged *Surveyor* robots sank only a few centimeters into the lunar soil, showing that there was no thick dust layer and people could, indeed, walk on the Moon without sinking in over their heads.

The *Apollo* spacecraft was designed to carry three men into orbit around the Moon. A small, Spartan landing craft, the *Lunar Excursion Module* or *LEM* for short, would ferry two of the crewmen from lunar orbit to the Moon's surface and then back to the mother ship, while the third astronaut remained orbiting the Moon in the larger *Command Service Module*.

On 21 December 1968 three *Apollo 8* astronauts became the first humans to break free of the Earth's gravity. Although the crew would only orbit the Moon and not land on it, the unprecedented voyage provided the first sight of the Earth seen from afar – a radiant blue-and-white sphere rising beyond the battered face of the Moon in the dark void of space (Fig. 5.13). We then saw our home world in a new perspective, beautiful and vulnerable, a tiny, fragile oasis shimmering all alone in the vast, deep chill of outer space. The sheer isolation of the Earth became plain to every person on the planet. It stimulated a world-wide awareness of

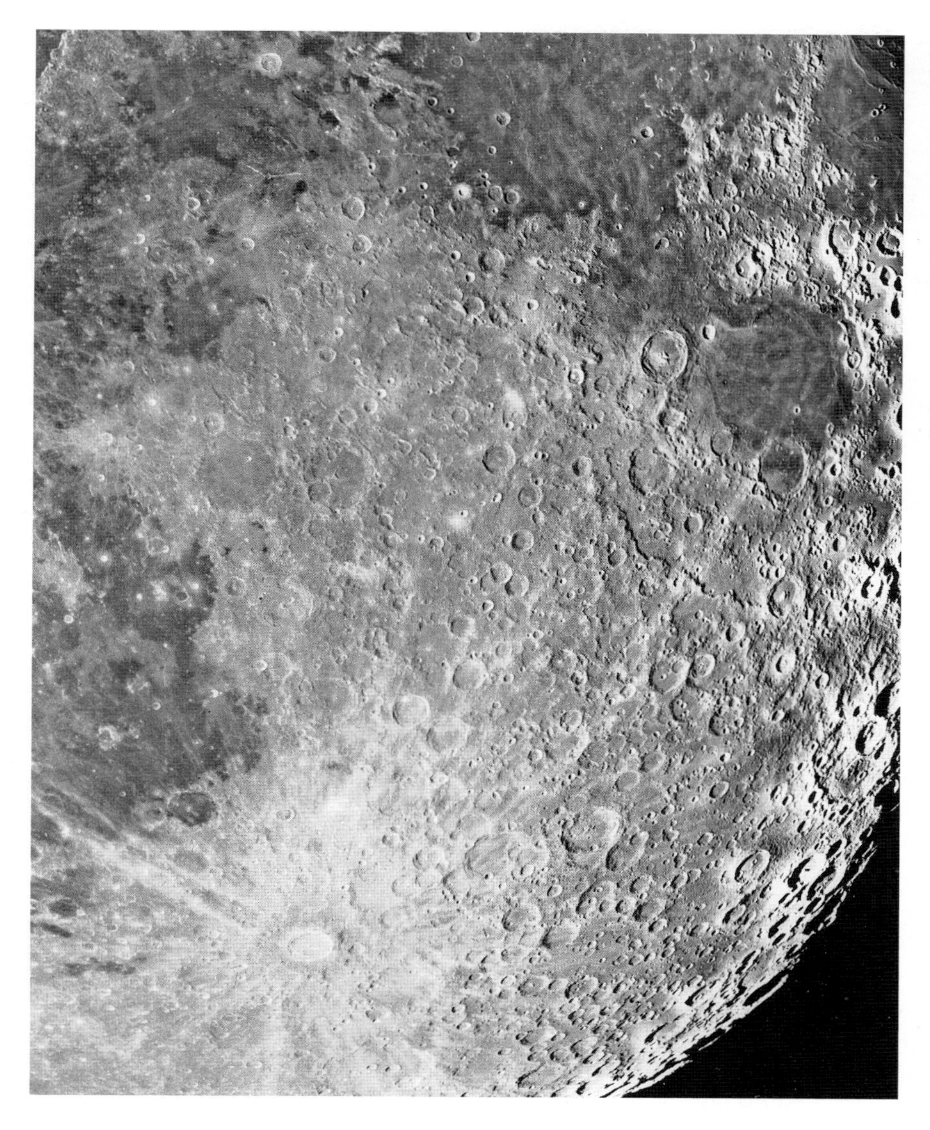

Fig. 5.12 Lunar rays White rays splash out across the Moon from crater Tycho (*bottom*). Tycho is a large, young crater with a diameter of 85 kilometers and an age of 107 million years. Only relatively recent craters retain their white rays, for those of older craters are darkened and worn away by continued meteorite impact. The dark, flat circular feature (*upper right*) is Mare Nectaris (Sea of Nectar). This clear image was produced using the unsharp masking technique that permits high contrast and fine resolution. (Anglo-Australian Telescope © 1976. Photograph prepared by David F. Malin.)

Fig. 5.13 Earthrise Expeditions to the Moon created a new image of the Earth as a blue and turquoise ball, light and round and shimmering like a bubble, flecked with delicate white clouds, suspended all alone in dark space. This image was taken from the *Apollo 8* spacecraft in 1968. (Courtesy of NASA.)

Table 5.3 *Apollo* missions to the Moon

Mission	Launch Date[a]	Landing Site	Accomplishments	Sample (kg)
Apollo 8	21 Dec. 1968	*Lunar Orbiter*	First humans to orbit Moon	
Apollo 10	18 May 1969	*Lunar Orbiter*	Test *Lunar Excursion Module*	
Apollo 11	16 July 1969	Mare Tranquillitatis	First human landing	21.6
Apollo 12	14 Nov. 1969	Oceanus Procellarum	First ALSEP[b]	34.3
Apollo 13	11 April 1970	Flyby	Landing aborted	–
Apollo 14	31 Jan. 1971	Fra Mauro, highland	First highland landing	42.6
Apollo 15	26 July 1971	Hadley–Apennine	First *Lunar Rover*	77.3
Apollo 16	16 April 1972	Descartes	Highland landing	95.7
Apollo 17	07 Dec. 1972	Taurus–Littrow	Last flight	100.5

[a] The spacecraft landed on the Moon four or five days after launch.
[b] ALSEP is an acronym for Apollo Lunar Surface Experiments Package.

our planet as a unique and vulnerable place, fostering the ecology movement and helping us to get a better feeling for the Earth's place in our lives and the Universe.

On 20 July 1969, the spindly-legged, *Lunar Module Eagle* carried two *Apollo 11* astronauts to the lunar surface. While an estimated half-billion people watched, Neil A. Armstrong took the controls to avoid a hazardous crater, and radioed the first words from another world "Houston, Tranquillity Base here. The *Eagle* has landed." It was enough to take your breath away.

With Buzz Aldrin at his heels, Armstrong groped cautiously down the ladder to the surface. He stood firmly on the fine-grained surface and an ancient dream had come true – man had set foot on another world and humans were no longer confined to their native world.

As Armstrong put it: "That's one small step for a man, one giant leap for mankind." Moments after his initial footstep, Aldrin gazed out at the Sea of Tranquillity and said simply "magnificent desolation". The next day, the Italian newspapers put it more succinctly: "Fantastico!"

Anyone who was alive then can still remember it; ask your parents or grandparents where they were when the first person landed on the Moon, and I am sure they can tell you. Even now, there is a sense of participation, a feeling that our lives were enriched and made memorable by the landing.

Twelve humans have walked on the lunar surface during the *Apollo* missions, to gather samples, take photographs and make other scientific measurements (Table 5.3). All the landing sites were on the near side and close to the lunar equator because these were the only places the astronauts could go safely (Fig. 5.14). Direct radio contact from Earth would be lost if they landed on the far side of the Moon. Sites near the equator were chosen to always be able to get astronauts back from the lunar surface quickly in case something bad happened down there. A landing near the edge or limb of the Moon, as viewed from Earth, was ruled out if the spacecraft was to return to Earth in daylight. Within these constraints, the landing sites were chosen to provide samples of a wide variety of terrain, from the smooth maria to the heavily cratered highlands.

Apollo 11 landed on the smooth plains of Mare Tranquillitatis, and *Apollo 12* settled down on a mare site near the edge of the vast Oceanus Procellarum. Rocks returned from these first missions confirmed the volcanic, basalt nature of the maria and established their antiquity, with ages greater than 3 billion years. The *Apollo 14* landing site was located in highland terrain near the Fra Mauro feature, an area thought to be covered with debris thrown out by the impact that formed the Imbrium basin. Dating of material obtained from this site indicated that the basin-forming impact occurred 3.85 billion years ago. *Apollo 15* was the first mission to employ a roving vehicle; it was sent to the Hadley–Apennine region containing both mare and highland units. The so-called Genesis rock found during this mission is a primitive chunk of highland material dating back some 4.5 billion years. *Apollo 16* landed on the Descartes highlands near the rim of the Nectaris basin, blasted out 3.92 billion years ago, and *Apollo 17* was sent to the Taurus–Littrow site, at the edge of the Serenitatis basin, excavated 3.87 billion years ago.

During its month-long cycle of day and night, the surface of the Moon experiences much greater temperature extremes than the Earth. At the near-equatorial *Apollo* sites, the maximum surface temperature is about 383 kelvin, higher than the boiling point of water, so the astronauts did

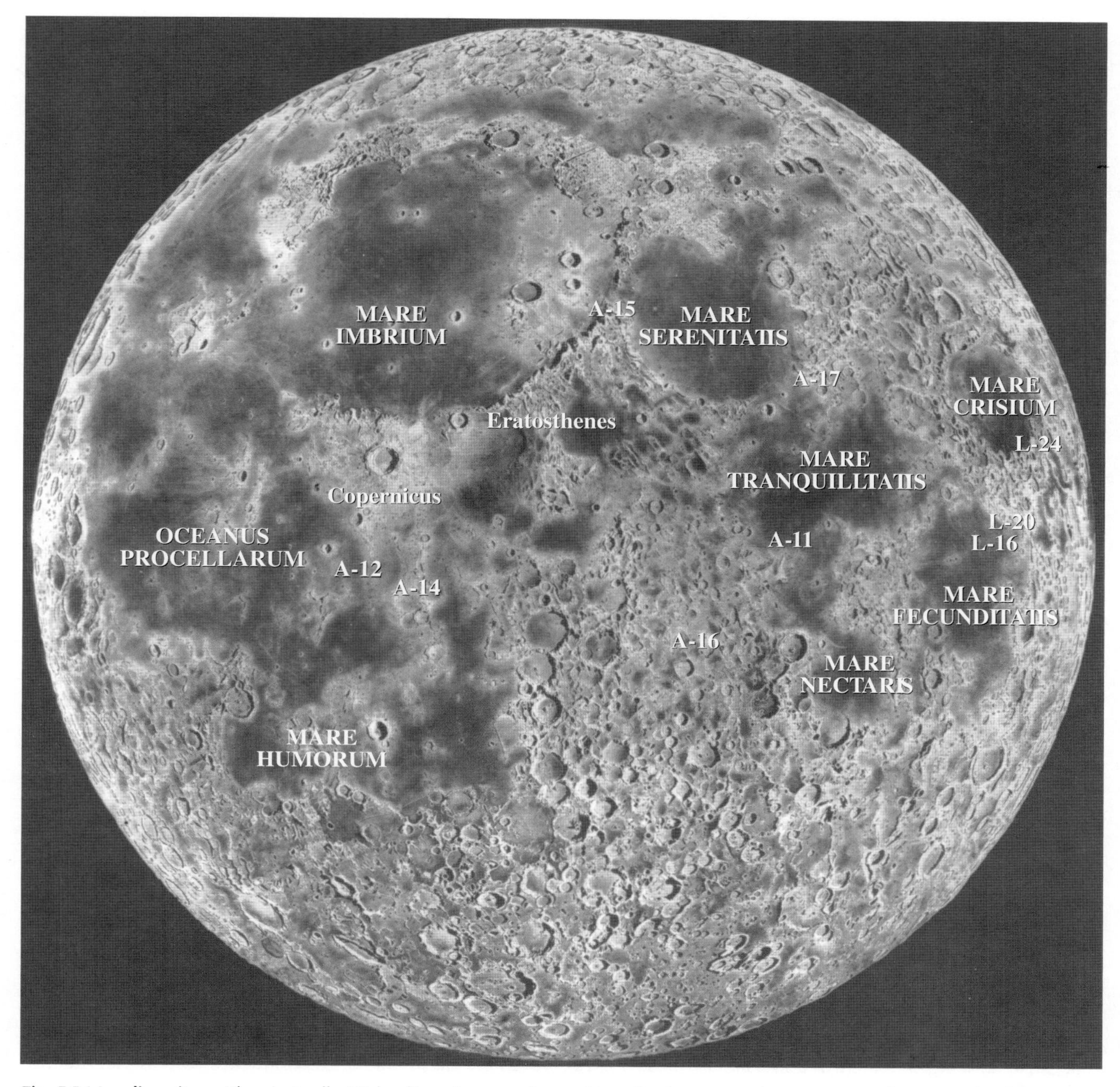

Fig. 5.14 Landing sites The six *Apollo* (A) landing sites were located in safe places near the equator on the near side of the Moon. Within this constraint, the sites were designed to obtain samples from a wide variety of terrain. *Apollo 11* and *12* respectively landed on Mare Tranquillitatis and Oceanus Procellarum. The spot chosen for *Apollo 14* was the Fra Mauro formation, which is covered with material ejected during the ancient impact that created the Imbrium Basin. By landing at a point just inside the Apennine Mountains, the *Apollo 15* astronauts could sample highlands, maria and the Hadley Rille. The *Apollo 16* mission sampled the highlands near crater Descartes, while *Apollo 17* landed near Mare Serenitatis. The location of the three Soviet *Luna* (L) unmanned, sample returns sites are also shown.

not want to land at high noon with the Sun overhead. They always landed in the lunar morning and stayed for just a few days, when the temperatures were similar to those on the Earth. This meant that the Sun was low, near the local horizon.

The astronauts' cameras recorded an eerie wasteland below a blackened sky (Focus 5.3), battered and scarred with craters of all sizes and covered with dust. It clung to the astronauts' clothing and equipment and showed the sharp outline of their footprints (Fig. 5.15); but there were

Focus 5.3 Black skies

The lunar astronauts stepped into a stark but beautiful world, where on a clear day you can see forever. With no clouds, dust, moisture or haze to obscure the view, distant details stood our clearly against the deep, black background. Because the Moon has no atmosphere to speak of, the sky was pitch black in broad daylight, there were no sounds to disturb the eternal stillness, and the Sun's true light could be seen, unfiltered by any air. By way of comparison, incident sunlight contains all the colors, but the Earth's air molecules scatter blue light more strongly than red light, making the overhead sky appear blue.

The twilight zone between the night's darkness and sunrise or sunset on Earth is caused by the Sun's rays being bent by the atmosphere. And when the Sun rises or sets, most of the blue light is scattered out before reaching us, so the light of the setting Sun is reddened and airborne dust helps this effect. In contrast, astronauts orbiting the Moon saw no twilight, and no colorful sunrise or sunset.

On the Moon there is no oxygen or water to weather the rocks or alter their chemistry. Erosion by micrometeorite bombardment of the Moon is very slow compared to erosion by weather and tectonic activity on Earth, so the footprints of *Apollo* astronauts will still be fresh and visible a million years from now. And since there is no air to breathe on the Moon, visiting astronauts were bundled in oxygen cocoons known as spacesuits.

How do we know that the Moon has no substantial atmosphere? The earliest evidence came from the abrupt vanishing of stars behind the edge of the Moon during lunar occultations. The word *occult* means "to hide". If the Moon had a thick atmosphere, the starlight would gradually dim during a lunar occultation, but this vanishing actually takes less than one second. It is in sharp contrast to the gradual fading of the Sun, stars, and planets when they set behind the horizon on Earth. The lack of a significant atmosphere also follows from the Moon's relatively small mass and gravity, together with its proximity to the warming Sun.

Instruments carried to the Moon by the *Apollo* astronauts identified the barest wisp of helium and argon atoms, and subsequent imaging observations from Earth revealed just a whisper of sodium and potassium atoms, emitting a detectable fluorescent glow when exposed to sunlight. But this is not a permanent atmosphere in the sense that ours is, for the lunar atmosphere is continuously being created, lost and replaced every few hours or weeks (Section 3.4).

The Moon's atmosphere is 100 trillion times more tenuous than the Earth's air, and so thin and rarefied that its constituent particles hardly ever hit each other. So, the Moon has no atmosphere in any practical sense.

no clouds of dust above the airless surface. Walking on the lunar surface was like walking on plowed soil or wet sand, and most of the finer dust had evidently been plowed down into the Moon by the churning of the meteorites.

Armstrong and Aldrin never strayed more than a hundred meters from their lander, like a timid child testing the water when entering a lake or sea for the first time. The astronauts of the next two missions (*Apollo 12* and *14*) had greater confidence and took longer Moonwalks (Fig. 5.16). During the last three missions (*Apollo 15, 16* and *17*) astronauts roamed as far as 7 kilometers from the landing site, visiting some of the most spectacular places on the Moon in a battery-powered car called the *Lunar Rover* (Figs. 2.3, 5.17). By the time of the *Apollo 17* mission, the lunar landings had become almost routine, so much so that Eugene A. Cernan muttered "let's get this mother out of here" as he blasted off the Moon in his *Lunar Module Challenger*.

Unlike the early landings on the smooth lunar maria, the last three *Apollo* flights visited mountainous areas: the Appenines, the Descartes highlands and the Taurus mountains. The tops of all the mountains were rounded off into gentle hills without sharp peaks or steep cliffs. Although it looked as if the Moon had been sandblasted smooth by eons of meteorite bombardment, the main reason for the gradual lunar slopes is that there is no water or ice erosion, as on Earth, to cut deep valleys and shape mountain crags, or tectonic activity to toss up crumpled mountain ranges.

The astronauts left behind the Apollo Lunar Surface Experiments Package abbreviated ALSEP. This nuclear-powered array of instruments included seismometers to monitor vibrations of moonquakes and meteorite impact (Section 5.5), magnetometers to measure possible magnetic fields (Section 5.6), and other instruments to analyze gases and charged particles streaming from the Sun to the Moon. The astronauts also brought lunar soil and rocks back home with them, altogether 382 kilograms and not an ounce of cheese (Fig. 5.18).

Mirrors were also left on the Moon, to reflect laser light sent from Earth. Every reflector contained 100 small mirrors, each in the shape of the three-sided corner of a box. These corner cubes reflect light directly back toward its point of origin. Observations of pulsed laser light, sent to the lunar mirrors and back, has permitted astronomers to measure the Moon's distance with an accuracy of two

Fig. 5.15 Boot prints on the Moon On 20 July 1969, Neil Armstrong became the first human to walk on the Moon. His boot print, shown here, reveals a thin layer of Moon dust, about 1 centimeter (0.01 meters) thick. Because there is no atmosphere, water or weather on the Moon, the footprint will probably remain for 1 or 2 million years. By that time, the constant rain of micrometeorites will have erased it. (Courtesy of NASA.)

centimeters, or to better than one part in 10 billion, showing that the Moon is moving away from the Earth (Section 5.8).

Sophisticated experiments were also performed from the *Lunar Command Module*, mapping the magnetic fields, chemical composition, surface radioactivity and terrain from a distance as the mother ship passed over the Moon.

Returning to the orbiting craft, the astronauts jettisoned the landing *Lunar Module* and headed for Earth, arriving home about three days later. Biologists felt there was a chance that the astronauts, or the returned rock samples, might infect the human race with some deadly lunar virus. The astronauts from the first three lunar landing missions, *Apollo 11, 12* and *14*, were therefore placed in quarantine for three weeks after their return. They remained in fine health and the crews of the last three missions, *Apollo 15, 16* and *17*, did not have to suffer quarantine.

The achievement of landing humans on the Moon was a spectacular American triumph in the Cold-War confrontation with the Soviet Union, in an incredible, warlike mobilization of scientists, engineers, and technology with an optimistic, can-do spirit. Imaginative thinkers viewed it as the stepping stone to permanent lunar bases, giant space stations and the colonization of Mars. But the dreams

Fig. 5.16 Moonwalk Charles Duke strolls across the lunar surface during the *Apollo 16* mission in April 1972. Small impacting particles have sandblasted the lunar surface, producing smoothed, undulating layers of fine dust and rounding the surfaces of lunar rocks. Larger meteorites have pounded and churned the surface, producing a layer of ground-up rocky debris. (Courtesy of NASA.)

Fig. 5.17 Moonride Eugene Cernan walks away from his lunar rover at the edge of the lunar highlands during the *Apollo 17* mission in December 1972. Sculptured, rolling hills are present in the background, and the South Massif is at the far right. The mountains of the Moon have smooth, rounded contours, primarily because there has been no water or ice erosion to sculpt them into steep peaks and valleys. The astronauts used roving vehicles like this one to travel across the Moon's rugged terrain, gathering rocks from a wide variety of locations. The rovers were left on the Moon. Free from wind, rain and rust, they will remain intact for millions of years; one might even imagine a returning astronaut using one that was discarded hundreds of years before. (Courtesy of NASA.)

Fig. 5.18 Moon rock Harrison Schmitt about to walk behind Split Rock during the *Apollo 17* mission in December 1972. Eugene Cernan had already scooped up samples from the debris on the front side of the boulder. The huge rock rolled down about a billion years ago, splitting into five pieces during the fall. The total length of the boulder, when reassembled is about 20 meters. (Courtesy of NASA.)

quickly dissipated and the sense of mission disappeared. The goal had been reached, the crisis was over, and the enemy had been conquered. The *Apollo* program lost its political appeal in the face of growing public indifference, its budget shriveled, and the program was cut short with the cancellation of the *Apollo 18, 19* and *20* lunar missions. Today NASA doesn't even have rockets big enough to send men to the Moon, and the *Apollo* program has not yet been followed by anything of comparable majesty.

Moreover, in the triumph of putting humans on the Moon we should not forget that space travel is a risky business. The first three *Apollo* astronauts died on the launch pad, and three others were almost lost when an oxygen tank exploded aboard *Apollo 13* on the way to the Moon. Sealed inside their spacecraft, the crew was in danger of dying by re-breathing their own carbon dioxide. They survived by converting the tiny *Lunar Excursion Module*, with its intact, breathable air and fuel, into a lifeboat, canceling the planned lunar landing and heading home. *Apollo 13* was very nearly a catastrophe. If the tank had exploded earlier, there would not have been enough electric power and water to go around the Moon and get home again. And if it had occurred later, when the astronauts were on their way down to land on the Moon, there would not have been enough fuel left in the *Lunar Excursion Module* to go home.

Return to the Moon

After the *Apollo* missions, no one had even a single glimpse of the Moon's far side for nearly two decades, and then it was obtained by the *Galileo* spacecraft on its way to explore Jupiter's realm. In order to reach the giant planet, *Galileo* pumped up its orbit and gained speed by swinging past the Earth, once on 8 December 1990, just 14 months after launch, and again on 8 December 1992, passing by the Moon in the process. It obtained images of the lunar limb and far side from vantage points not previously obtained. For instance, the Sun illuminated the western limb

Fig. 5.19 Rare side view of the Moon The *Galileo* spacecraft returned this image of the western limb and far side of the Moon on 8 December 1990, from a vantage point not possible from the Earth. It shows the bright far-side highland (*left*) and the near-side Oceanus Procellarum (*top right*). The dark spots near the center are Mare Orientale, about 900 kilometers across; it is barely visible from Earth at the western limb of the lunar near side. The huge South Pole – Aitken impact basin, a large circular depression about 2.6 thousand kilometers in diameter, is the large dark region at the lower left. (Courtesy of JPL and NASA.)

of the Moon during the 1990 *Galileo* flyby (Fig. 5.19), while sunlight brightened the opposite eastern limb during all of the *Apollo* missions.

Composites of *Galileo* images taken in three colors, violet, red and near infrared, have been used to depict compositional variations of the lunar surface (Fig. 5.20). They have been calibrated by *Apollo* sample returns that specify the chemistry at specific sites on the near side of the Moon. Some mare basalts are rich in titanium, while many others are relatively low in titanium but rich in iron and magnesium. The heavily cratered highlands are typically poor in titanium, iron and magnesium.

In early 1994, after a lapse of more than 20 years in lunar exploration, the United States Department of Defense placed a small spacecraft in orbit about the Moon. In sharp contrast to the eight-year, $25 billion *Apollo* program, the tiny, unmanned satellite required only two years and $75 million to build and launch. This is about one-tenth of the amount NASA spends on a single *Space Shuttle* flight, not including the cost of whatever is in the cargo bay.

Because one of the 1994 mission's byproducts was to prospect the surface mineral content of the Moon, the spacecraft was given the name *Clementine*, after the miner's darling daughter in the old Gold Rush ballad. However, this was not the main purpose of the spacecraft. It was built primarily as a military test of lightweight electronic imaging sensors that could detect the launch and track the flight of enemy ballistic missiles, possibly for use in a future star-wars missile shield.

Like its namesake, the spacecraft was "lost and gone forever" after orbiting the Moon for two months, but not without first chalking up an impressive list of accomplishments (Table 5.4). Unlike the *Apollo Command Modules*, that circled the Moon in low, near equatorial orbits, *Clementine* orbited across the lunar poles, permitting a global perspective as different regions rotated into view. The surface composition and topography of the entire satellite were mapped in unprecedented detail (Section 5.6), the South Pole–Aitken basin was completely mapped with high resolution for the first time (Fig. 5.21), and possible deposits of water-ice deposits were found in the cold permanently dark places within the poles (Sections 2.4, 5.6).

In the face of dwindling budgets and growing public interest in Earthbound problems, scientists and engineers found less-expensive ways of exploring space. NASA adopted a new "smaller, faster, cheaper" mode of operation, in which science is done by several cost-effective, high-risk spacecraft rather than a few major, expensive and

Table 5.4 The *Clementine* and *Lunar Prospector* missions to the Moon

Mission	Launch Date	Lunar Orbit	Accomplishments
Clementine	25 Jan. 1994	19 Feb. 1994 to 3 May 1994	Global surface composition, global topography, map of South Pole – Aitken basin, possible water ice at poles.
Lunar Prospector	6 Jan. 1998	15 Jan. 1998 to 31 July 1999	Global elemental abundance, global magnetic field maps, global gravity maps, detection of lunar core, water ice at poles.

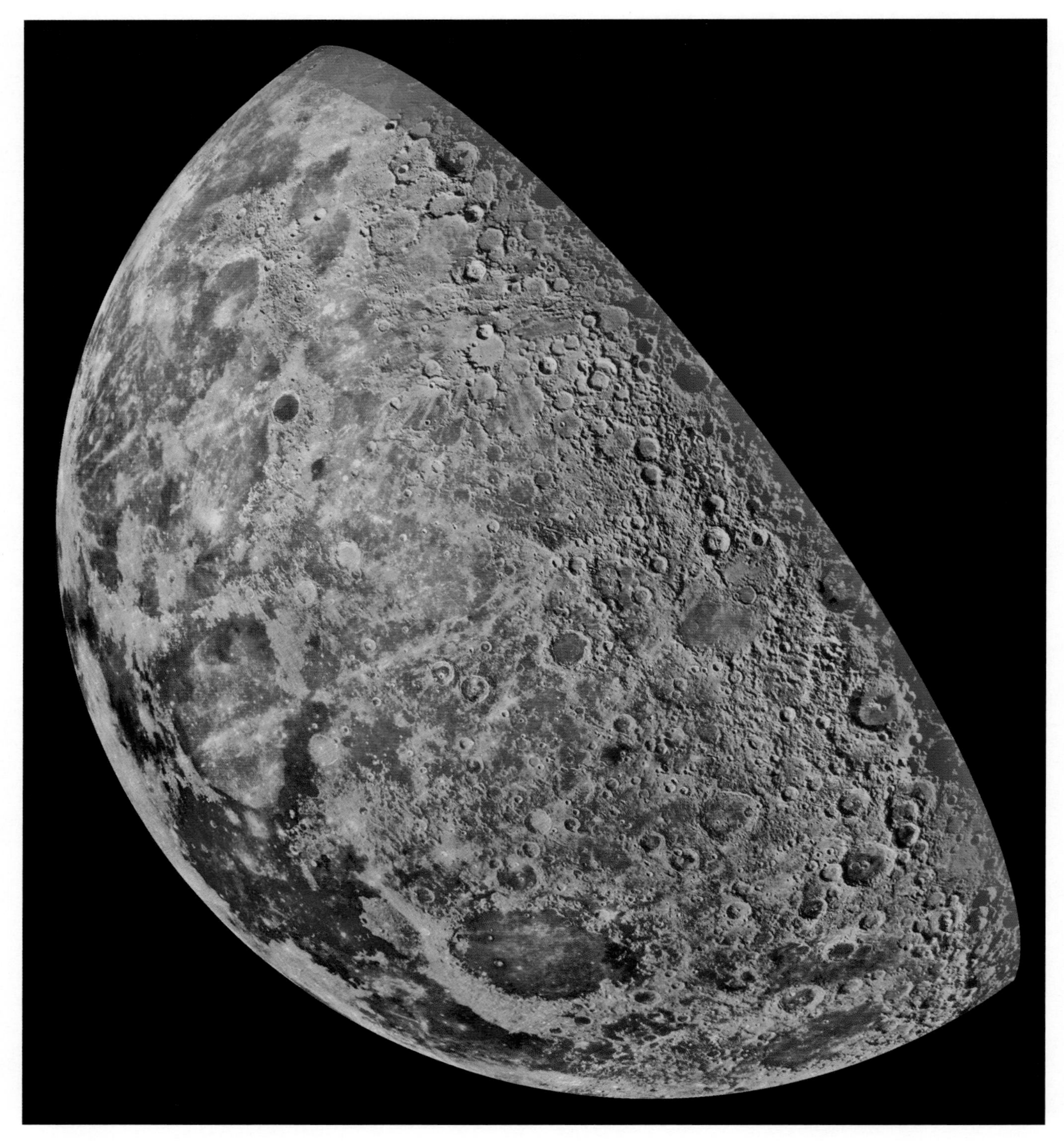

Fig. 5.20 Compositional variations This enhanced color mosaic shows volcanic flows with relatively high titanium content (*blue*), volcanic flows that are low in titanium but rich in iron and magnesium (*green, yellow and light orange*), and heavily cratered highlands that are typically poor in titanium, iron and magnesium (*pink and red*). In this view, taken by *Galileo* as it flew over the northern region on 7 December 1992, bright-pink highlands surround the lava-filled Crisium impact (*bottom*) and the dark-blue Mare Tranquillitatis (*left*) is richer in titanium than the green and orange maria above it. The youngest craters have prominent blue rays extending from them. (Courtesy of JPL and NASA.)

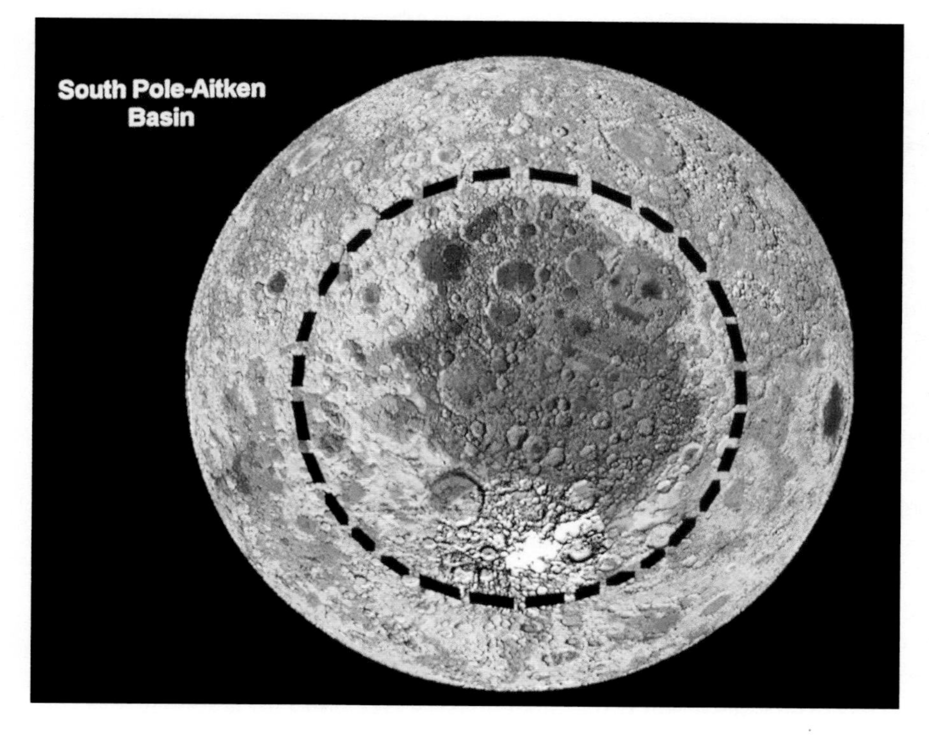

Fig. 5.21 Bottom of the Moon The laser altimeter on *Clementine* provided a detailed topographic map of the South Pole of the Moon for the first time, revealing the full extent of the South Pole–Aitken basin (*middle, center*). Colors indicate relative elevation, with orange being high and blue being low. The basin is about 2.6 thousand kilometers across and over 12 kilometers deep. It is the biggest, deepest impact feature known in the solar system, and dominates the relief of the far side of the Moon. Water ice might be preserved in the permanently shadowed regions at the Moon's south pole. (Courtesy of Paul D. Spudis, Lunar and Planetary Institute.)

low-risk ones. The *Lunar Prospector* spacecraft is an example. Launched on 6 January 1998, it was designed to obtain global data on elemental abundance, magnetic fields, and gravity fields, and it has also achieved an impressive array of accomplishments (Table 5.4), discussed in greater detail in Sections 5.5 and 5.6. A new space race, begun in the early-21st century, targets the Moon for future exploration (Focus 5.4).

Focus 5.4 Voyages to the Moon in the early-21st Century

The Moon will be bombarded in the early-21st century by orbiting, penetrating and crashing spacecraft from the United States, Japan and Europe.

The first commercial lunar spacecraft, *Trailblazer*, will be launched by the private U.S. company, TransOrbital, Inc., in 2003 from Baikonur Cosmodrome in Kazakhstan. *Trailblazer* carries video cameras to record, in unprecedented detail, separation from the launch vehicle, the Earth rising over the edge of the Moon, and the lunar surface. Its high-definition and stereoscopic imagery, with surface resolution of one meter, will be used in video games and movies, as well as in planning future landing sites. A crash landing of a hardened capsule will deliver personal items to the Moon, at the cost of about $2,500 a gram.

Japan plans to make the first soft-landing on the Moon in more than a quarter century, becoming the third nation in history to accomplish this feat. Their *Lunar A* spacecraft will land two penetrating probes that are equipped with seismometers to measure the Moon's internal structure, and devices to measure heat flow from the Moon. After the launch of *Lunar A* in 2003, Japan plans a launch in 2005 of the more ambitious orbiting lunar spacecraft dubbed *SELENE*, an acronym for *SELenological and ENgineering Explorer*. Its instruments include imaging cameras, a radar sounder, a laser altimeter, and spectrometers that will study the origin, evolution, and tectonics of the Moon.

The first European lunar spacecraft, ESA's *SMART-1*, an abbreviation for the first *Small Missions for Advanced Research in Technology*, begins its voyage to the Moon in 2003. The lunar orbiter will test new technologies, including novel propulsion schemes, as well as high-resolution imagery and spectrometry.

These spacecraft may pave the way for a renaissance in lunar exploration that could include mining the Moon, novel data-storage and high-vacuum experiments, and preparation for manned missions to Mars.

The new space race to the Moon may also be stimulated by China's apparent plans to put a man on the Moon by 2010, but the details of such a trip are shrouded in secrecy.

Astronomy from the Moon

A manned scientific station on the Moon would provide excellent opportunities for astronomy. With no significant atmosphere, every wavelength of radiation, from the longest radio waves to the shortest gamma rays, streams down without absorption to the lunar surface. Telescopes in artificial Earth satellites are now used to observe cosmic ultraviolet and X-ray emission that is partially or totally absorbed in the Earth's atmosphere, but such instruments are constrained by their weight, size, complexity and cost. Telescopes constructed on the rock-solid surface of the Moon would rest on a more stable platform than an orbiting satellite or space station, permitting high-resolution observations of the Universe at wavelengths that cannot be seen from the ground, including X-rays and gamma rays.

An astronomical observatory on the Moon would also avoid the problem of atmospheric turbulence that limits angular resolution at visual wavelengths to 0.2 seconds of arc at the best locations on Earth (Section 1.2, Focus 1.2). A lunar telescope as small as 1 meter in diameter would have a resolution of about 0.1 seconds of arc at the optical wavelengths detected with our eyes, while also avoiding interruption caused by cloudy weather and day time. Relatively small telescopes could be linked together electronically, using the techniques of interferometry to achieve angular resolutions that are thousands of times better than those currently available from either the space or the ground. The great increase in resolution might, for example, permit the detection of Earth-like planets around nearby stars; currently only much larger Jupiter-like planets can be observed from Earth.

The far side of the Moon would allow radio antennas to be shielded from terrestrial interference or noise. It could also be used to open a new window on the Universe at wavelengths longer than about 20 meters that are reflected by the Earth's ionosphere and do not reach the ground.

A manned lunar base could also serve as a support station for exploration of the rest of the solar system, including manned trips to Mars. Oxygen, hydrogen and metals might be extracted from the lunar soil to produce water, air, fuel and construction material. The Moon might even be mined for scarce resources on the Earth, such as helium-3, the gas that makes balloons float and might be used in fusion reactors in the future. The global reserves of helium-3 on the entire Earth amount to about 100 kilograms, but there is perhaps ten million times as much helium-3 on the Moon's surface, implanted there by the Sun's winds.

5.5 Inside the Moon

Moonquakes

As the Earth has earthquakes, so the Moon has moonquakes which were first detected by the sensitive seismometers placed by the *Apollo* astronauts at four widely spaced locations on the lunar surface – at the *Apollo 12, 14, 15* and *16* landing sites. Because winds, sea waves, and road traffic do not shake the Moon, the lunar seismometers can detect moonquakes that are much weaker than even the mildest earthquake. If you stood directly over the strongest moonquake you would not even feel your feet shake. They never exceed a magnitude of 2 on the Richter scale. Although earthquakes of this magnitude are recorded on Earth, they are not felt by humans and produce no damage to buildings.

The moonquakes are not only gentler than earthquakes, they also have distinctly different behavior. While tremors on the Earth start suddenly and persist for only a few minutes, the moonquake waves build up gradually and continue for more than an hour. Evidently the body of the Moon is an almost perfect medium for the propagation of seismic waves. Some of the moonquakes seem to emanate from the upper mantle, while others come from further down, more than halfway to the center of the Moon and much deeper than most earthquakes.

The lunar seismometers were able to record occasional impacts on the Moon – both from the meteorites and *Lunar Modules* that were sent crashing into the lunar surface near the end of some *Apollo* missions. Such records have permitted the construction of a model for the lunar interior, in much the way that geologists have modeled the internal structure of the Earth (Section 4.2).

The Moon is slightly asymmetrical in bulk form, with a thicker crust on the far side. Most of the volcanic maria occur on the near side where the crust is thinner (Fig. 5.22). On average, the lunar crust is about 70 kilometers thick. It is only a few tens of kilometers thick beneath the mare basins, but over 100 kilometers thick in the highlands and the far side of the Moon. As a result, the Moon's center of mass is offset from its geometric center by about 2 kilometers in the direction of the Earth.

A small lunar core

Scientists from the *Apollo* era were unable to agree whether the Moon has an iron-rich core; but they were certain that it had to be much smaller than the core of the Earth, which is 55 percent of the radius of the planet and 32 percent of

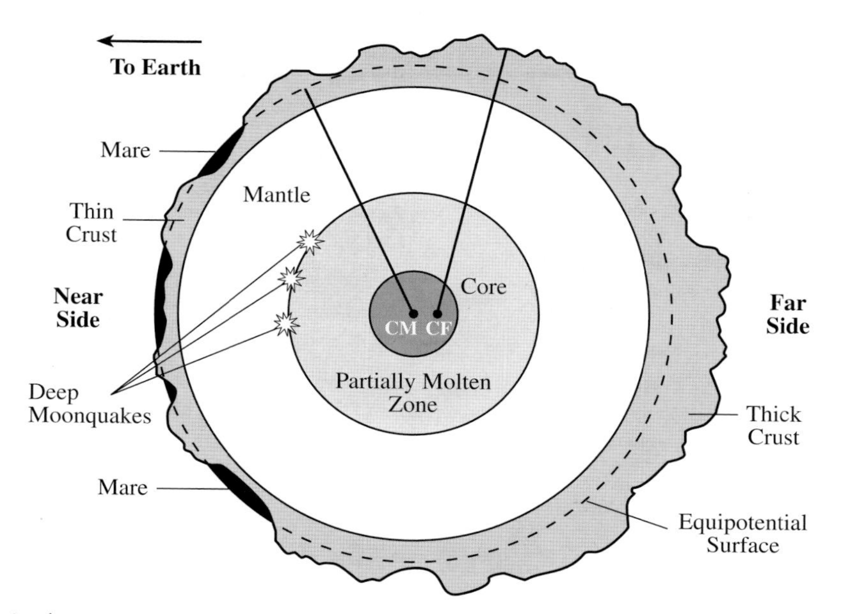

Fig. 5.22 Lunar interior A schematic cross-section of the Moon shows its internal structure. The lunar crust is thinner on the near side that faces the Earth, and thicker on the far side. Fractures in the thin crust have allowed magma to reach the surface on the near side, where the lava-filled maria are concentrated. The Moon has an iron-rich core with a radius of about 20 percent of the Moon's average radius of 1738 kilometers. A partially molten layer is believed to encircle the Moon's core, out to depths of about 1000 kilometers. The Moon's center of mass (CM) is offset by 2 kilometers from its center of figure, CF, so an equipotential surface, which experiences an equal gravitation force at all points, lies closer to the lunar surface on the hemisphere facing Earth. Therefore magmas originating at equipotential depths will have greater difficulty reaching the surface on the far side.

the planet's mass. But data from *Lunar Prospector* have now been used to gauge the size of the lunar core in two ways. They indicate that the Moon's core has a radius of about 350 kilometers, or 20 percent of the satellite's radius, and accounts for only about 2 percent of the body's mass.

Radio telescopes on Earth were used to measure small Doppler-effect changes in the *Lunar Prospector*'s radio signal as the spacecraft moved toward or away from the Earth, thereby identifying slight variations in the craft's velocity as it orbited the Moon. Since these velocity changes are caused by the varying gravitational pull of the Moon, they could be used to construct a full gravity map of the near and far sides of the Moon, from pole to pole. The resulting map revealed the distribution of mass within the Moon, and showed that it has a small, dense metallic core, about 350 kilometers in radius if mostly iron.

A second method studied the weak magnetic field induced within the Moon when it passes through the tail of the Earth's magnetosphere each month. This technique confirmed the presence of a lunar core of about the same size as that inferred from the gravity data.

The relatively small core of the Moon has profound implications for its origin. If the Moon and Earth coalesced independently, they should have more similar core sizes; instead the Moon seems to have coalesced from material blasted out of the young Earth (Section 5.9). The giant impact might have taken place when the Earth was still forming. In that case, most of the Earth's iron might have already sunk to its core, but there could have been enough iron-rich rock, expelled into space from the Earth and the impacting object, to build a lunar core.

Internal, partially molten zone

The *Apollo* seismic data suggested that the outer half of the Moon is cold and solid, but that it might be warm and partially molten in a lower zone. The moonquake waves lost energy if they went deeper than 1000 kilometers, about halfway to the center of the Moon at 1738 kilometers down. The *Apollo* scientists argued that the deep moonquakes might be generated at the boundary between the outer solid shell and the inner molten zone.

Evidence for a partially molten zone has been obtained from accurate measurements of the Moon's distance using laser ranging. The laser beams are sent to the Moon from telescopes on Earth and reflected from corner mirrors left on the Moon by the *Apollo* astronauts in 1969. Measurements of the round-trip travel-time of a pulse of laser light yields twice the distance between the Earth and the Moon, by multiplying the time by the velocity of light, with an accuracy of 0.02 meters. After more than 30 years of such determinations, scientists have concluded that the Moon's surface moves in and out by as much as a tenth of a meter, or 10 centimeters, every 27 days, in response to the shifting gravitational tugs of the Earth. This elastic yielding suggests that the interior is pliable, with a larger, partially molten layer surrounding the core.

Mascons

Precise radio tracking of *Apollo* spacecraft passing over the near side of the Moon showed that their orbits are gravitationally deflected toward the circular maria. The spacecraft acted as though the maria contained mass concentrations, abbreviated as mascons, which pulled at the spacecraft and changed their velocity when passing overhead. Virtually all the maria on the near side showed this unexpected feature, and the excess mass in each is about 10^{18} kilograms, or 1/100 000 the total mass of the Moon. Because radio tracking of the orbiting spacecraft was not possible when they passed to the far side, it was not known from the *Apollo* missions whether there are mascons on the far side of the Moon.

Both *Clementine* and *Lunar Prospector* have been used to detect mascons from gravity data for the entire Moon. *Lunar Prospector* created a more detailed version, discovering several mass concentrations beneath the floors of large impact basins, including at least four on the far side. The near-side basins have been filled with mare lava, but the far-side basins remain unfilled, so it isn't the lava that is providing the extra gravitational pull.

What are the mascons? The most likely explanation is that they represent an upward bulging of high-density mantle rocks that rose in the aftermath of basin-forming impacts. The impact that formed the largest basins has weakened the crust so much that the dense mantle has moved up beneath them, raising and fracturing the basin floors.

5.6 The lunar surface

Rocks from the *Apollo* missions

During the *Apollo* landings from 1969 to 1972 a dozen people roamed the Moon taking hundreds of rock samples, placing them in labeled bags, and returning 382 kilograms of Moon rocks to Earth in sealed containers. These specimens from another world have permitted scientists to decipher the composition of the lunar crust, and to reconstruct our satellite's history.

Since contact with the Earth's atmosphere would alter the composition of the lunar samples, they are kept in cabinets filled with a dry, oxygen-free atmosphere of nitrogen and are manipulated with long gloves sealed to the walls of the cabinets. When not under investigation, the rocks are kept in a massive steel-lined vault at the Lunar Receiving Laboratory of NASA's Johnson Space Center at Houston, Texas.

None of the rocks brought back from the Moon contain any moisture or hydrated minerals, and they show no signs of having been exposed to water. The oxygen that is on the Earth in rocks and water forms only rocks on the Moon. So, there is no water in the Moon rocks, and there hasn't been any for billions of years.

The Moon is lifeless, as we might expect from the lack of water. Extensive testing revealed no evidence for life, past or present, among the lunar samples. They contain no living organisms, fossils or native organic compounds. Thus, the Moon is a desolate place, devoid of life.

From its low mean mass density, we would expect the bulk of the Moon to be composed of silicates, or minerals in which atoms of silicon and oxygen are linked to other elements. Laboratory investigations of the lunar samples indicate that the lunar crust is indeed composed of such minerals, just as the Earth's crust is.

The surface of the Moon has been bombarded by meteorites for billions of years, breaking the lunar crust into rock fragments and fine-grained material known as the lunar regolith. It is the loose debris that has fallen back to the Moon after eons of meteoritic bombardments. The regolith is the Moon's version of soil, though it contains no organic material and there are no gardeners on the Moon.

The regolith covers the entire lunar surface to depths as great as 20 meters. It is thickest in the highland regions that have been exposed longest to meteoritic bombardment. The regolith in the maria is 2 to 8 meters deep.

The Moon rocks are roughly divisible into three types: anorthosites, basalts and breccias, and they all exhibit important differences from terrestrial rocks. The lunar samples are all much older than most rocks found on Earth, and they are composed of material that has been previously melted (anorthosites), erupted through magma–lava outflow (basalts), and crushed by meteoritic impacts (breccias).

The anorthosites are the oldest rocks ever found, dating back to 4.5 billion years. They are found in the light-colored lunar highlands, and contain a type of mineral known as plagioclase feldspar, commonly found on Earth, but with a difference. The lunar anorthosites were melted more than 4 billion years ago.

Basalts are dark lava that fills mare basins, forming a secondary crust. These thin volcanic veneers were created after heat from radioactive decay accumulated in the Moon, leading to the rise of magma and the eruption of basaltic lava about 3 billion years ago. The surface of Venus (Section 7.4) and the Earth's ocean floor (Section 4.3) are also secondary crusts formed in this way. But most of Venus was resurfaced about 750 million years ago, and the Earth's ocean floor is still being created.

The Moon contains none of the type of rocks that were generated deep inside the Earth during plate tectonic processes, such as the continental granites.

After the lunar rocks solidified, they were broken up, flung about and pulverized by meteoritic impacts. Energetic impacts, powerful enough to excavate meter-sized craters, have compacted and welded the regolith into aggregates called breccias. They are fossilized specimens that retain compositional information from the era in which they formed.

Scanning the surface

The *Apollo* rock and soil samples came from only six sites on the near side, chosen mainly to be safe and easy to get to. A global view of the Moon's surface composition therefore had to wait until the *Clementine* spacecraft surveyed the unexplored regions on both the near and far sides. Its ultraviolet, visible and infrared cameras took pictures at eleven different wavelengths used to identify different types of minerals. Since various rock-forming minerals reflect and absorb incident sunlight at different wavelengths, the *Clementine* data could be used to infer the chemical composition of most of the lunar surface, and rock samples could be used to calibrate the global data when it overlapped the *Apollo* landing sites.

The *Clementine* global data was used to map the abundance and distribution of iron on the Moon, showing that the dark, near-side maria consist of iron-rich lava, containing up to 14 percent iron by weight. In contrast, iron is practically absent in the near-side highland crust and across vast tracts of the far side, at about 3 percent iron by weight. These regions of very low iron content are dominated by aluminum-rich anorthosite (Fig. 5.23).

The highland crust on both the near and far sides is just what one would expect if the entire Moon was once covered in liquid rock at least several thousand kilometers deep. The heavy iron sank into this magma "ocean", while low-density feldspar (anorthosite) grains accumulated into floating "rockbergs" that coalesced and cooled to form the eventual constituents of the Moon's highlands. Some of the iron subsequently resurfaced when the Moon heated up inside and magma flowed up to the near-side maria.

Large meteoritic impacts dig holes into the lunar crust, exposing deeper material and revealing its composition. For instance, the floor of the huge South Pole–Aitken basin on the far side has an iron abundance of nearly 10 percent by weight.

Until *Clementine*, we also had no global map of the topography of the Moon. The laser altimeter on the spacecraft fired pulses of light at the Moon once every second and timed how long it took for the light beam to travel down to the lunar surface and back again. This enabled scientists to determine the distance to the surface, over and over again, with an accuracy of 50 meters. When these distances were combined with knowledge of the spacecraft orbit, maps of the elevation, or topography, of the entire lunar surface were obtained (Fig. 5.24).

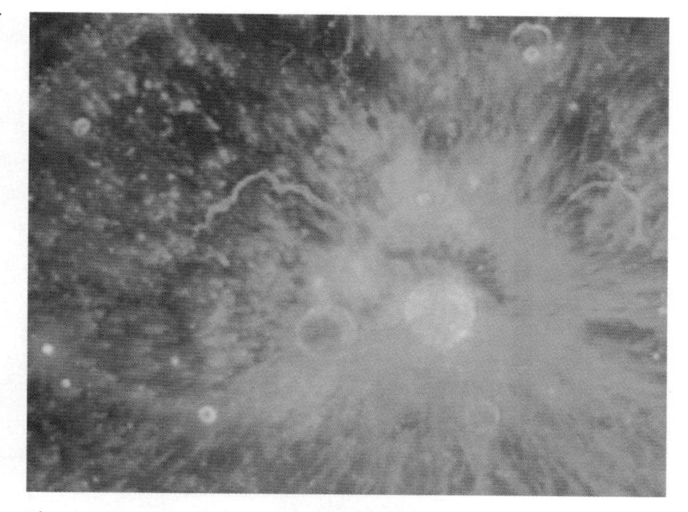

Fig. 5.23 Composition of Aristarchus plateau This colored *Clementine* mosaic illustrates the composition and mineralogy in the Aristarchus plateau, an elevated block of crust surrounded by the vast mare lava plains of Oceanus Procellarum. The fresh, excavated highland materials appear blue, mare lava flows are reddish purple, fresh basalt in the crater and rilles is yellow, and ash and volcanic glass is dark red. The two dark-blue spots in the center of Aristarchus are consistent with a composition of almost pure anorthosite, a primitive rock type that floated to the top of the ancient magma ocean. The sinuous rille is Schröter's Valley, or Vallis Schroteri, a large lava channel stretching for about 160 kilometers across the plateau. (Courtesy of JPL, NASA, and the U.S. Geological Survey.)

The new global maps showed an unexpected range of topography, over 16 kilometers and comparable to that seen in the geologically different Earth. The wide range of relief on the Moon is caused by the presence of huge impact basins. The huge South Pole–Aitken basin, which is over 12 kilometers deep and about 2.6 thousand kilometers across, dominates the far-side topography. The near side is relatively smooth, with typical relief of about 5 kilometers, primarily because its impact basins have been filled with mare basalt. As substantiated by *Clementine* gravity data, a thicker crust that helps block the outward flow of magma (Section 5.5) characterizes the far side.

Possible water ice at the lunar poles

Bright radar echoes, returned to *Clementine* from the south pole of the Moon, suggested that this region might contain

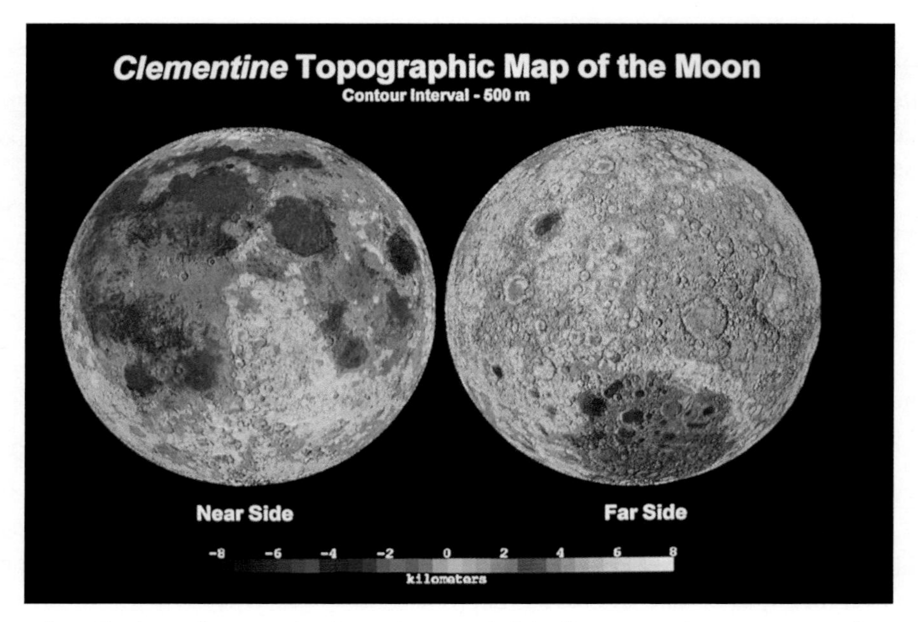

Fig. 5.24 Lunar topography The laser altimeter on *Clementine* provided the first comprehensive topographic map of the Moon. The near side (*left*) is relatively smooth and low (*blue* and *purple*), primarily because of the prominent impact basins, including Imbrium, Crisium and Nectaris, which are at least partly filled with mare basalt. In contrast, the far side (*right*) shows high relief (*red*) and extreme topographic variation comparable to that of the Earth. The Moon's wide range of relief is attributed to ancient impact basins that have been preserved for about 3.9 billion years, while the Earth's wide range stems from ongoing mountain building by colliding tectonic plates. The large circular feature on the southern far side (*right bottom*) is the South Pole–Aitken basin, 2.6 thousand kilometers in diameter and 12 kilometers deep. (Courtesy of Paul D. Spudis, Lunar and Planetary Institute.)

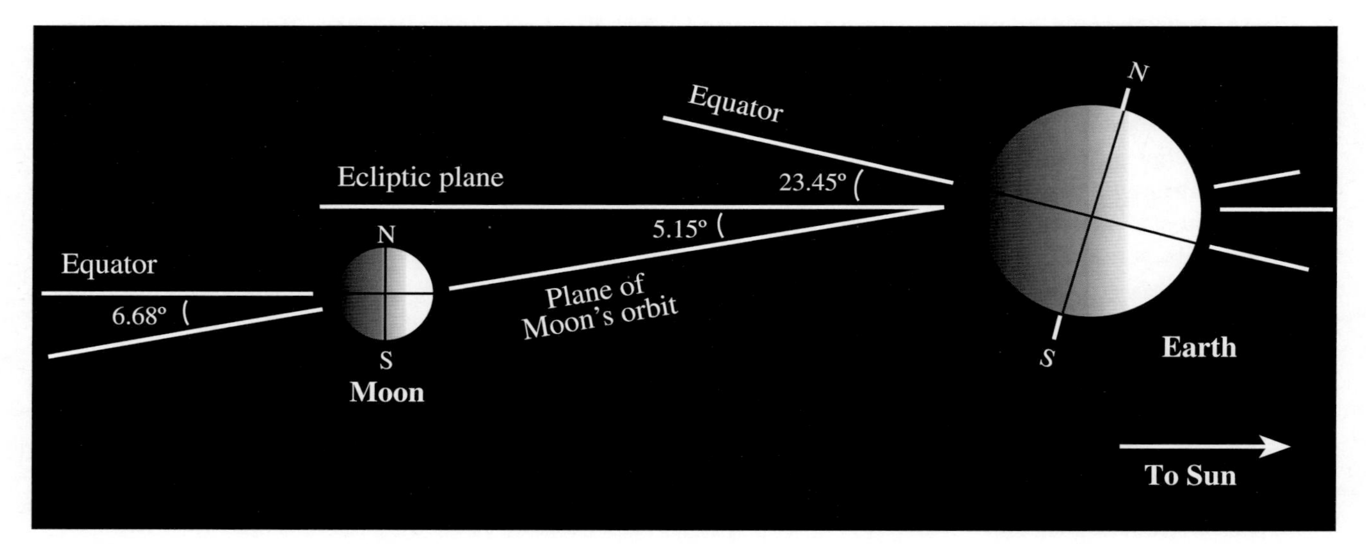

Fig. 5.25 Dark, cold lunar poles The near-vertical orientation of the Moon's north–south rotation axis to the ecliptic plane creates permanent night and a permanent deep freeze at the floor of craters located at the lunar poles. These regions might be reservoirs of water ice, delivered there by comets. The angle between the Earth's equator and the ecliptic, or the plane of the Earth's orbit around the Sun, is 23.5 degrees, and this tilt produces the seasons. The Moon provides a steadying influence for the Earth's tilt, keeping it from varying widely and producing dramatic climate variations. Also note that the plane of the lunar orbit falls neither in the Earth's orbital plane nor in the ecliptic.

radar-reflective water ice. *Lunar Prospector* then strengthened the possibility of water ice at the south pole, and also discovered what appears to be additional water ice near the north pole. During its passes over the poles, an instrument on *Lunar Prospector* detected substantial quantities of hydrogen, which mission scientists attributed to water ice found in permanently shaded areas near both poles. They estimate that there could be as much as 6 billion tons, or 6×10^{12} kilograms, of water ice located in the polar regions.

Since the Moon's rotation axis is oriented nearly perpendicular to the ecliptic plane (Fig. 5.25), the lunar poles are never tilted toward the Sun by more than a very small amount. This means that the bottoms of craters at the poles are in constant shadow and in a perpetual deep freeze of 50 to 70 kelvin. Any ice deposited in these frozen reservoirs would be preserved indefinitely in the eternal dark and cold.

But how could there be water ice on the Moon when the rocks returned from the Moon show no signs of ever being exposed to water? The rocks were taken from near-equatorial regions where intense sunlight would boil any liquid water or water ice away, and the Moon's fiery origin seems to have removed volatile elements during the satellite's formation (Section 5.9).

Water ice could have been delivered to the Moon by comets, which are essentially big balls of dirty ice (Sections 12.5, 12.6). Comet-borne water could have been deposited in the cold traps at the top and bottom of the Moon for more than 4 billion years, ever since the Moon's rocky crust formed, slowly accumulating in amount.

Experts assert that there must be water ice at the lunar poles, and that alternative explanations of the observations are unlikely. Other scientists disagree, so there is an ongoing controversy (Section 2.4). There may or may not be substantial amounts of water frozen into the lunar polar regions, and we may not know for certain until roving spacecraft visit them.

If there were a source of water on the Moon, it would make it more attractive for human outposts. Water could be purified to drink, or it could be chemically split into hydrogen, to burn as a rocket propellant, and oxygen to breathe. This would make it easier to establish a colony on the Moon, or to build a fueling station on it for interplanetary spacecraft.

The magnetized Moon

The Moon has no overall dipolar magnetic field, at least none that is strong enough to be detected. Its magnetic moment is at least ten million times weaker than the Earth's. Yet, some of the lunar rocks returned to Earth are magnetized. They have survived since the time that molten rocks covered the Moon and solidified 3 to 4 billion years ago, preserving fossilized remnants of ancient magnetic fields.

The *Apollo 15* and *16* missions each carried small subsatellites designed to measure the Moon's magnetism from lunar orbit. They found localized regions on the Moon with surface magnetic fields of between one-hundredth and one-thousandth that of Earth; the Earth's equatorial magnetic field strength is 0.3×10^{-4} tesla (Section 3.6). Large blocks of the lunar crust, as broad as 100 kilometers, are magnetized, but they do not combine into an overall global pattern.

Lunar Prospector measurements have shown that the largest concentrations of strong magnetic fields are on the lunar far side, located diametrically opposite to the Imbrium, Serenitatis, Crisium and Orientale impact basins on the near side. The basin rock is itself weakly magnetized, suggesting that the large basin-forming impacts demagnetized the crust at the impact side, while simultaneously magnetizing the crust on the opposite side of the Moon.

According to one explanation, the young Moon may have had its own global magnetic field generated early in its history when molten metal circulated in a small core. Large impacts, like the one creating the Imbrium basin, would create an ionized fireball racing around the Moon, piling the magnetic field up and concentrating it at the point opposite the impact. As the rocks cooled, the strongest, localized magnetic fields survived as fossils after the Moon lost its global magnetic field.

This is not the only explanation for lunar magnetism, for there are hundreds of smaller magnetized regions scattered over the entire Moon, in regions that are not located opposite to a large impact site. Perhaps they swept up, amplified and concentrated the magnetized winds that flow from the Sun. In an alternative explanation, the ancient magnetism may have been incorporated from surrounding material during the early stages of formation, or the Moon could have been magnetized by the Earth, at a time when the two bodies were closer together.

This is one of the remaining mysteries yet to be solved by the space-age exploration of the Moon.

5.7 The Moon's history

The age of the oldest rocks – Moon, Earth and meteorites

The time at which different features on the Moon originated can be determined from rocks returned from them. These relics have remained unaffected by the erosion that removed the primordial record from most Earth rocks. The ages of the lunar rocks can be determined by examining unstable radioactive elements and their stable decay products.

What matters are not the actual quantities of each element present, but the proportions – the ratio of stable elements, like lead, to unstable ones, like uranium and thorium. When this ratio is combined with the known rates of radioactive decay, the time since the rock solidified and "locked in" the radioactive atoms is found.

The method is known as radioactive dating, and it works this way. Certain types of nuclei, known as unstable parent isotopes, decay at a constant rate into stable lighter isotopes known as daughters. By measuring the amount of daughter material and knowing the rate of decay, the age of the rock can be estimated. The detailed mathematical

Focus 5.5 Radioactive dating

Radioactive elements can be used to clock the age of rocks on the Earth's surface, meteorites, and lunar rock samples. The number, N, of radioactive atoms in a rock changes with the time, t, since its solidification, according to the differential equation:

$$\frac{dN}{dt} = -\lambda N,$$

where λ is the decay rate. This equation integrates to give the number of radioactive atoms, N_t, at time t:

$$N_t = N_0 \exp(-\lambda t) = N_0 \exp\left(\frac{-0.693t}{\tau_{1/2}}\right),$$

where N_0 is the number of atoms at time $t = 0$, the time of solidification. The radioactive decay constant $\lambda = 0.693/\tau_{1/2} = \ln(2)/\tau_{1/2}$ and $\tau_{1/2}$ is the half-life of the radioactive species. Half-lives for the decay of radioactive isotopes are given in Table 5.5.

The number of radioactive atoms in the rock will be halved in a time equal to the half-life. Radioactive uranium, ^{238}U, decays, for example, into lead, ^{206}Pb, which is stable, with a half-life of about 4.47 billion years; so every 4.47 billion years the amount of uranium-238 in a rock will be halved.

We can apply the equations to ^{238}U, and express the abundance in terms of another kind of lead, ^{204}Pb, that is not a radioactive-decay product. If a terrestrial rock, lunar sample or a non-terrestrial meteorite became a closed system at time $t = 0$, then the present abundance of lead and uranium are related by the equation:

$$\left(\frac{^{206}\mathrm{Pb}}{^{204}\mathrm{Pb}}\right)_t = \left(\frac{^{238}\mathrm{U}}{^{204}\mathrm{Pb}}\right)_t [\exp(\lambda_{238}t) - 1] + \left(\frac{^{206}\mathrm{Pb}}{^{204}\mathrm{Pb}}\right)_0,$$

where the subscripts t and 0 denote the present and initial abundance, respectively.

If all of the rock samples have the same initial ^{206}Pb/^{204}Pb abundance, and if all of them have the same age, t, then a plot of $(^{206}\mathrm{Pb}/^{204}\mathrm{Pb})_t$ against $(^{238}\mathrm{U}/^{204}\mathrm{Pb})_t$ should lie in a straight line of slope $[\exp(\lambda_{238}t) - 1]$. Such a plot is called an isochron. If a system formed t years ago and initially contained no lead, then a curve of the ratios ^{207}Pb/^{206}Pb and ^{238}U/^{206}Pb also provides the age t.

These and similar methods have been used to show that the Earth, Moon and meteorites are 4.6 billion years old, with an uncertainty of no more than 0.1 billion years, where 1 billion years = 10^9 years = 1 Gyr (Also denoted as 1 Ga, where "a" is short for "annum", Latin for "year").

Table 5.5 Radioactive isotopes used for dating

Radioactive Parent [Name (Symbol) Mass No.]	Stable Daughter [Name (Symbol) Mass No.]	Half-Life (10^6 yr)
Rubidium (Rb) 187	Strontium (Sr) 87	48 800
Rhenium (Re) 187	Osmium (Os) 187	44 000
Lutetium (Lu) 176	Hafnium (Hf) 176	35 700
Thorium (Th) 232	Lead (Pb) 208	14 050
Uranium (U) 238	Lead (Pb) 206	4 470
Potassium (K) 40	Argon (Ar) 40	1 270
Uranium (U) 235	Lead (Pb) 207	704
Samarium (Sm) 146	Neodymium (Nd) 142	100
Plutonium (Pu) 244	Thorium (Th) 232	83
Iodine (I) 129	Xenon (Xe) 129	16
Palladium (Pd) 107	Silver (Ag) 107	6.5
Manganese (Mn) 53	Chromium (Cr) 53	3.7
Aluminum (Al) 26	Magnesium (Mg) 26	0.72

treatment is given in Focus 5.5. The method is something like determining how long a log has been burning by measuring the amount of ash and watching a while to determine how rapidly the ash is being produced.

The isotopes used for dating are given in Table 5.5 together with the half-life, the time for half of the radioactive substance to decay into a stable element. Many of the radioactive parent isotopes are still found on both the Moon and the Earth even though they have been decaying since the solar system originated about 4.6 billion years ago.

The daughter isotopes must be trapped in the rock and not escape or the estimated age will be too short. In fact, the daughters can escape quite easily when the rock is molten; only when it cools and solidifies do the daughters start to accumulate. For this reason, the ages determined for the rocks are really the times since the rock became solid. And if the rock is re-melted, say by the impact of a meteorite, its radioactive clock is reset, and the age will measure the time since the last solidification.

The radioactive dating method has been used to study rocks returned from the Moon. The oldest lunar samples, returned from the light, rugged highlands, indicate an age of nearly 4.6 billion years, and the lava flows that created the dark, near-side maria are dated at 3.9 to 3.2 billion years ago.

The oldest rocks on Earth are about 3.9 billion years old, but terrestrial crystals of zircon are as old as 4.4 billion years. Erosion by wind, water and geological processes has wiped out the oldest terrestrial rocks. The deep-ocean sediments, which are least affected by continuing geological activity on Earth, have ages of about 4.55 billion years.

Primitive meteorites, known as carbonaceous chondrites, have an age of 4.566 billion years, with an uncertainty of about 0.002 billion years. These meteorites are thought to date back to the earliest days of the solar system.

Rounding off the numbers and allowing for possible systematic errors, we can say that the Earth, Moon and primitive meteorites solidified at about the same time some 4.6 billion years ago, with an uncertainty of no more than 0.1 billion years. If the solar system originated as one entity, then this should also be the approximate age of the Sun and the rest of the solar system.

Formation of the highlands and maria

The retrieved lunar rocks have taken us back into time, to the formative stages of the Moon. They record events from the earliest history of the solar system that have been erased on Earth by water, wind and geological activity. Radioactive dating of Moon rocks and primitive meteorites indicates, for example, that the Moon was assembled a mere 50 million years after the solar system itself was born 4.6 billion years ago.

The cataclysmic bombardment associated with the final stages of the Moon's formation was so energetic that the globe was melted to depths of several hundred kilometers. In this global magma ocean, lighter mineral species floated to the top and formed the Moon's crust, and the denser material sank to the interior. As we have seen, the lunar highlands still contain the remains of these early low-density rocks, the anorthosites (Section 5.6).

The magma ocean gradually cooled and crystallized between 4.6 and 4.4 billion years ago, forming a thin, low-density lunar crust. Portions of this crust are today's highlands, on both the near and far sides of the Moon, rich in light elements such as aluminum and poor in heavy ones like iron.

When the radioactive dating method is applied to highland rocks retrieved from the Moon, there are two significant results. Highland rocks that crystallized from internally generated magmas, such as the anorthosite rocks, date from the earliest times, 4.6 to 4.1 billion years ago – a span of 500 million years after the origin of the Moon. In contrast, highland rocks that assembled from pre-existing rocks by impact, the breccias, all date from 3.9 to 3.8 billion years ago. The two results attest to prolonged igneous evolution on the early Moon, followed by a short, "cataclysmic" impact bombardment about 3.9 billion years ago. This hail of meteoritic debris cratered the lunar highlands and obliterated most of the direct evidence of the first half-billion years of lunar history.

The large impact basins were formed during the final stages of this heavy bombardment. The Nectaris basin was created 3.92 billion years ago, and the Imbrium impact took place an estimated 3.85 billion years ago. These impacts have been used to mark key events in the lunar timescale (Table 5.6). In pre-Nectarian time, the Moon's crust solidified. The major impact basins were gouged out during the heavy bombardment of Nectarian time. When the Imbrium basin was excavated, it marked the start of the Imbrian period of lunar volcanism that created the maria.

As the external cratering rate was declining rapidly (Fig. 5.26), internal processes set to work. The radioactive decay of long-lived unstable elements, such as uranium and thorium, produced heat that gradually warmed up the interior. There followed an era of volcanism, lasting for 700 million years, from 3.9 to 3.2 billion years ago. The outer zone of solid rock gradually cooled from the outside in, becoming thicker, and magma worked its way from deeper and deeper in the Moon. The magma flow may have stopped 3.2 billion years ago, since the youngest lunar lava samples are this old, but some mare basalts could be as young as 1 billion years old.

As molten basaltic rock welled up from the interior, it penetrated the thin crust beneath the great impact basins on near side of the Moon, flooding them with lava and producing the dark circular maria that can be seen today (Fig. 5.27). Successive lava flows set their marks in some maria, showing that they were not formed in a single quick pulse of volcanism, but by repeated outpourings that gradually filled the near-side basins.

The liquid lava moved quickly away from its vents, covering the sources rather than piling up, so no volcanic mountains were created. Flowing with about the consistency of motor oil, the lava spread for hundreds of

Table 5.6 Lunar time-scale

Period	Age[a] (10^9yr = 1 Gyr)	Characteristics
Pre-Nectarian	4.6 to 3.92	Moon accumulated 4.6 to 4.5 billion years ago in Earth orbit, newly formed Moon wrapped in molten rock, a magma ocean, until 4.4 billion years ago, solidification of lunar crust and formation of oldest impact basins.
Nectarian	3.92 to 3.85	Nectaris basin probably formed 3.9 billion years ago, Serenitatitis, Crisium and other impact basins are most likely this old, as are most rocks.
Imbrian	3.85 to 3.15	Period of lunar volcanism when most maria were formed, Imbrium basin excavated 3.85 billion years ago and the last big basin, Orientale, created 3.8 billion years ago.
Eratosthenian	3.15 to about 1.0	Formation of craters that are slightly degraded but without rays. Most of lunar surface remains unchanged, but some mare volcanism and large impacts.
Copernican	About 1.0 to present	Youngest craters formed, most of which have preserved rays. Crater Copernicus excavated 0.85 billion years ago and crater Tycho created just 0.10 billion years ago. Most of lunar surface remains unchanged.

[a] These ages are estimates based on inferences of the geological setting of the lunar samples.

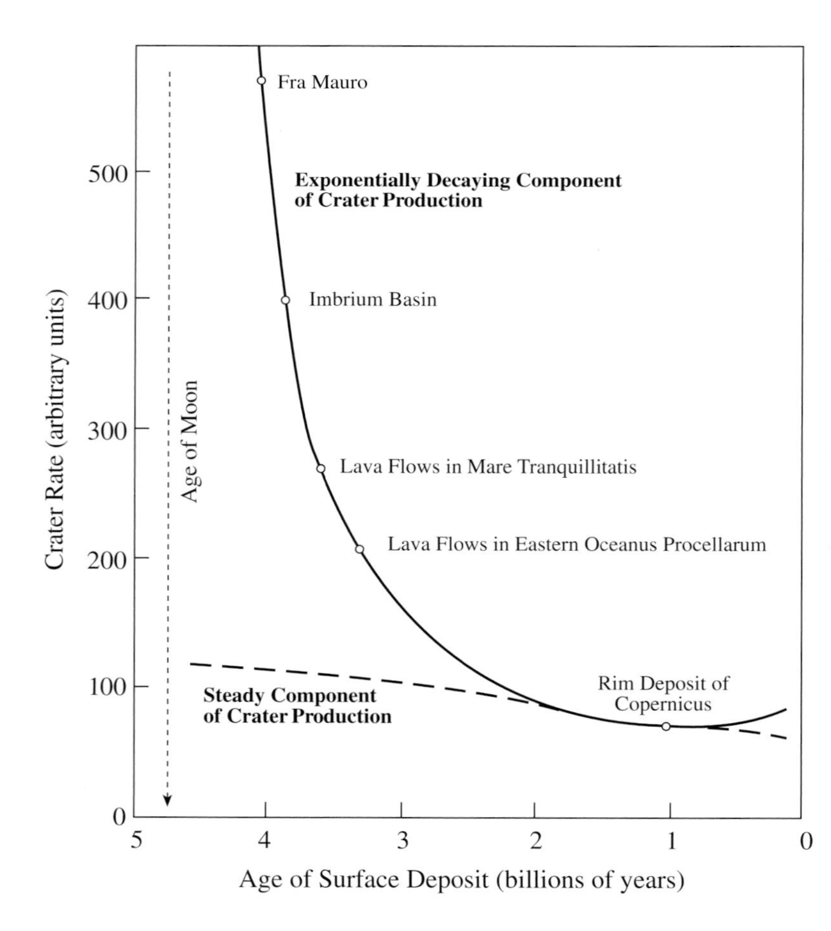

Fig. 5.26 Varying crater rate The rate of forming craters on the lunar surface is plotted against time. The circles denote the crater rate and rock ages at the various *Apollo* landing sites. The crater rate was very high during an intense bombardment that occurred 3.9 billion years ago. The rate dropped rapidly during the subsequent billion years, giving way to the lower steady rate of crater production that has persisted for the last 3 billion years. With such a curve, we can obtain approximate surface ages just by counting the number of craters in different parts of the Moon. Recent data have provided evidence of an increase in the cratering rate over the last 400 million years.

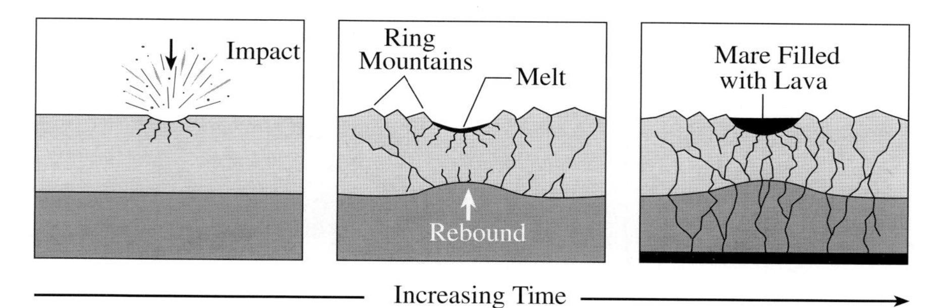

Fig. 5.27 Mare formation Disintegrating and vaporizing as it strikes, a meteorite blasts a huge impact basin out of the lunar surface (*left*), while the associated shock waves create fractures in the rock beneath the basin. The blast hurls up mountain ranges around the basin (*center*), and the underlying rock adjusts to the loss of mass above it by rebounding upward. The uplifted mantle causes additional fractures in the rock, while a pool of shock-melted rock solidifies in the basin. All the major impact basins on the Moon were created in this way between 4.3 and 3.9 billion years ago. Later, interior heat from radioactivity caused partial melting inside the Moon, and magma rose along the fractures, filling the basin with lava to form a dark mare (*right*). The lunar maria were filled by this volcanic outpouring between 3.9 and 3.1 billion years ago.

kilometers before hardening into a thin veneer, only a few hundred meters or less in thickness. And since the crust on the far side of the Moon was relatively thick to begin with, the molten rock had greater difficulty in penetrating it, explaining why there are so few maria on that side of the Moon.

The lava inundated all craters in its path, wiping the slate clean of previous impacts. preparing a fresh surface to record new impacts which, by this time, had greatly diminished in intensity. Thus the maria are relatively unscarred and most of their craters are small and relatively young.

Despite its violent beginnings, the Moon then settled down into a long quiet life. The face of the waterless, airless Moon has therefore remained largely unchanged for 3.2 billion years. By that time, the Moon had cooled so much that magma could no longer break through and erupt, and the pummeling by meteorite impacts continued at a much lower rate. Impacts occasionally blasted out craters like Copernicus, some 850 million years old, and Tycho, dated at 109 million years, but the ongoing rain of lesser impacts mainly churned up the lunar surface, covering it with rock fragments. This rubble forms a dusty covering to a magnificent, ancient museum of craters and basins formed billions of years ago.

5.8 Tides and the once and future Moon

The pattern of the tides

Walking along the ocean beach some morning, we might notice that the waves seem to be reaching further and further up the sand. The tide is flooding the beach. A few hours later, it hesitates and then begins to ebb, retreating onto the flats where the clams may often be found. The high tides occur simultaneously and symmetrically on opposite sides of the Earth; they return every 12 hours 25 minutes in each location, although not precisely to the same height. The time between consecutive high tides is slightly more than half a day because the Moon's revolution around Earth is in the same direction as the Earth's rotation on its axis, so Earth needs an extra 25 minutes of rotation to out-race the Moon and get into position.

The Moon creates two high tides because the gravitational force of the Moon draws the ocean out into an ellipsoid, or the shape of an egg. We can understand this by remembering that the gravitational force decreases with distance, so the Moon pulls hardest on the ocean facing it, and least on the opposite ocean; the Earth between is pulled with an intermediate force. As a result, the water directly beneath the Moon is pulled up away from the Earth's center, and the Earth's center is pulled away from the water on the opposite side, causing another high tide. Thus the differences of the gravitational attraction of the Moon on opposite sides of the Earth produce two tidal bulges – one facing the Moon and one facing away (Fig. 5.28).

As the Earth's rotation carries the continents past the tidal humps, we experience the rise and fall of water. In mid-ocean the tide is only 0.01 to 0.3 meters in height and usually goes unnoticed. But, when a shore blocks the tide, it often runs 2 or 3 meters high. Tides can also resonate in estuaries of the right shape, amplifying and building up the tides to 10 or 20 meters in height, as in the Bay of Fundy in Nova Scotia. You can create a similar effect by sloshing the water in your bathtub back and forth at just the right rate.

On a slowly rotating planet without continents, the tide would be highest along the line joining the centers of the

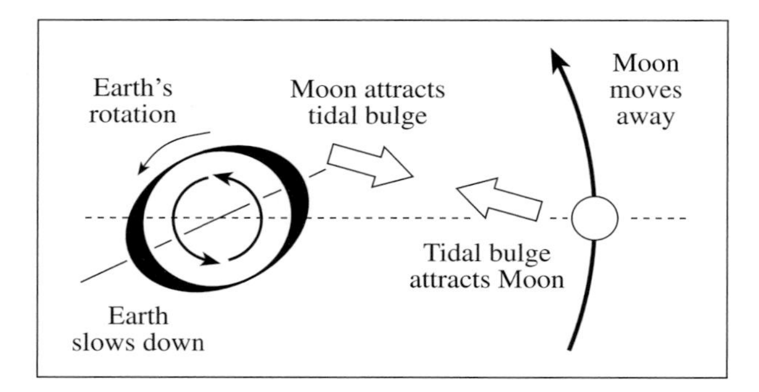

Fig. 5.28 Cause of Earth's tides The Moon's gravitational attraction causes two tidal bulges in the Earth's ocean water, one on the closest side to the Moon and one on the furthest side. The Earth's rotation twists the closest bulge ahead of the Earth–Moon line (*dashed line*), and this produces a lag in time between the time the Moon is directly overhead and the highest tide. The Moon pulls on the nearest tidal bulge, slowing the Earth's rotation. At the same time, the tidal bulge nearest the Moon produces a force that tends to pull the Moon ahead in its orbit, causing the Moon to spiral slowly outward.

Earth and Moon, that is, when the Moon is overhead. This is not the case for the Earth. The friction of the continents and the rapid rotation of the Earth carry the ocean's tidal bulge forward so it precedes the Earth–Moon line by about 3 degrees (Fig. 5.28). This means that in the open ocean the high tide actually occurs about 12 minutes after the Moon is overhead.

There are further delays for tides at the continental shores. When the flood tide moves in from the ocean, it may have to work its way among islands and peninsulas and along channels; this twisted path will delay the arrival by amounts that vary with location and with the time of the month. This time delay is called the "establishment" of the port, and the result is that high tide usually occurs an hour or two after the Moon is overhead, and occasionally more. Similar delays can be noted in tidal pools. They are lowest long after low tide in the ocean.

The Sun and the Moon both contribute to the formation of the tides, but the major portion of this rhythmic ebb and flood is driven by the Moon, whose contribution is 2.2 times that of the Sun. In the course of a month, the changing alignment of the Sun and Moon causes the tides produced by these two bodies to alternately reinforce and interfere, leading to the cycle of spring tides and neap tides. The spring tide occurs near new and full moon, when the Sun and Moon reinforce each other's tides, and the neap tide occurs near first and third quarter, when they interfere with each other (Fig. 5.29). The spring tides can be two or three times as high as the neap tides. The tides caused by the Sun vary by a small amount over the year as the Earth travels around its eccentric orbit and the Sun–Earth

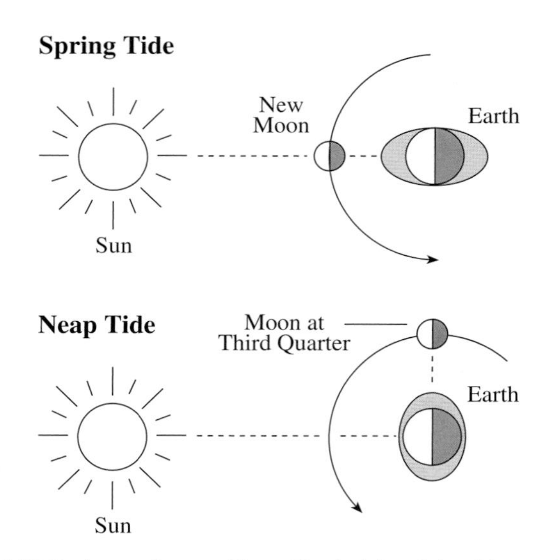

Fig. 5.29 Spring and neap tides The height of the tides and the phase of the Moon depend on the relative positions of the Earth, Moon and Sun. When the tide-raising forces of the Sun and Moon are in the same direction, they reinforce each other, making the highest high tides and the lowest low tides. These spring tides (*top*) occur at new or full Moon. The range of tides is least when the Moon is at first or third quarter, and the tide-raising forces of the Sun and Moon are at right angles to each other. The tidal forces are then in opposition, producing the lowest high tides and the highest low tides, or the neap tides (*bottom*). The heights of the tides have been greatly exaggerated in comparison to the size of the Earth.

distance changes, with the greatest solar tides when the Earth is nearest the Sun.

The days are getting longer

As the Earth rotates, the bulge raised on its surface by the Moon's gravity is always a little ahead of the Moon rather than directly under it (Fig. 5.28). The Moon pulls back on the bulge, and in the process it slows down the whole planet. In other words, our planet meets resistance in its daily rotation caused by the tidal interaction of the Moon with the Earth.

As the tides flood and ebb, they create eddies in the water, producing friction and dissipating energy at the expense of the Earth's rotation. The motion of the tides heats the ocean water ever so slightly and the Earth's rotation is slowed. The tides therefore act as brakes on the spinning Earth, slowing it by friction in much the way that the brakes of a car slow its wheels and become warm. The friction of tides dissipates energy at the rate of 5 billion horsepower (4×10^{12} watt). Tidal friction is slowing the rotation of the Earth, and the day is becoming longer at a rate of 2 milliseconds, or 0.002 seconds, per century. In other words, the days are getting longer at the rate of one second every 50 000 years.

This tidal effect on the Earth's rotation rate can help us understand the ancient astronomical records, and conversely those records also help us understand the effect of the tides. If the current, slow rotation rate is used to wind the Earth backward 2500 years to the time of an eclipse, the Earth would rotate about a quarter-turn too little, putting those predicted eclipse paths several thousand kilometers west of their actual locations. These paths would then conflict with the reported occurrences. As we shall see, the length of the month is also increasing; this reduces the discrepancy considerably, so the full story is a bit complicated.

Indirect historical measures of the Earth's rotation rate have been made by paleontologists through studies of fossil corals. The growth patterns of these corals consist of annual bands and fine daily ridges, produced by the effect of seasonal and daily changes of water temperature on the growth rate. The days were shorter in the past, but the length of the year was the same, so the number of days per year increases as we go back in time. Ancient corals confirm this, and they show a greater number of daily ridges per annual band than modern corals. Careful counting reveals that the day was only 22 hours long when we look back 400 million years. Studies of daily growth increments have been extended to fossilized algae called stromatolites, which indicate that the day may have been only 10 hours long 2 billion years ago.

Aside from such historical and paleontological determinations, this change of the Earth's rotation rate is imperceptible to humans, and it has not yet been measured directly. It is also mixed in with an erratic rate produced by the vagaries of the weather and the seasons, so all in all the Earth's rotation rate is no longer the best choice of clock. Astronomers now prefer to rely on atomic clocks due to their stability and continued accuracy.

Earth's tidal influence on the Moon

The Moon pulls the Earth's oceans, and the oceans pull back, in accord with Newton's third law that every action has an equal and opposite reaction. The net effect is to swing the Moon outward into a more distant orbit. This is because the tidal bulge on the side facing the Moon is displaced ahead of the Moon and this bulge pulls the Moon forward.

As the Earth slows down, the angular momentum it loses is transferred to the Moon, which speeds up in its orbit around us. It is not hard to see that this will swing the Moon away from the Earth if we look at the key equations (Focus 5.6). When we do the arithmetic, we find that the change of 0.002 seconds per century in the length of the

Focus 5.6 Conservation of angular momentum in the Earth–Moon system

One of the fundamental, unbreakable laws of physics is the law of conservation of angular momentum. It says that the product of mass, M, velocity, V, and radius, R, is unchanged in a closed system, that is, one not subject to an outside force. Thus:

$$\text{conservation of angular momentum} = M \times V \times R = \text{constant}.$$

For the Earth, the angular momentum is rotational, with $V = 2\pi R_E/P_E$, where the subscript E denotes Earth, and P_E is the Earth's rotation period of one day. So, we have:

$$\text{Earth's rotational angular momentum} = 2\pi M_E R_E^2/P_E.$$

Since the length of the Earth's day is increasing as time goes on, the Earth's rotational angular momentum is decreasing and the loss has to made up by an equivalent gain somewhere else in order to conserve angular momentum. This is done by an increase in the Moon's orbital angular momentum, which is given by:

$$\text{Moon's orbital angular momentum} = M_M \times V_M \times D_M,$$

where the subscript M denotes the Moon, M_M is the mass of the Moon, D_M is the Earth–Moon distance, and the orbital velocity of the Moon is given by the escape velocity of the Earth at the Moon's distance, or by

$$V_M = (2GM_E/D_M)^{1/2},$$

where G is the gravitational constant. Substituting this velocity expression into the angular momentum relation, we obtain:

$$\text{Moon's orbital angular momentum} = M_M \times (2GM_E D_M)^{1/2}.$$

Since the masses of the Moon and the Earth do not change to any significant extent, the Moon's distance has to increase to provide an increase in the angular momentum. The loss in the Earth's rotational angular momentum produced by a period increase of 0.002 seconds per century is equal to a gain in the Moon's orbital angular momentum that corresponds to a distance increase of 0.04 meters per year.

day implies an outward motion of the Moon amounting to about 0.04 meters per year. Small as it is, this value is just measurable with the laser reflectors planted on the Moon by the *Apollo* astronauts. The lunar laser-ranging data indicate that the Moon is moving away from the Earth at a rate of 0.0382 ± 0.0007 meters per year.

Will the Moon's outward motion carry it away from the Earth altogether? Probably not, because there is not enough energy in the Earth–Moon system for these bodies to overcome their binding energy and go their separate ways. Only the intrusion of a massive third body could achieve that, or some fantastic project to attach enormous rockets to the Moon and launch it into space.

What will ultimately happen is the following. The combination of the slowing Earth and the receding Moon means that the Earth's day will eventually catch up with the length of the month. When the day and the month are of equal length, the Moon-induced tides will cease moving; from then on the oceans will rise and fall much more gently, solely under the influence of the Sun. The Moon will hang motionless in the sky, and will be visible from only one hemisphere. At that stage the recession of the Moon will stop.

At that time, billions of years from now, the Sun's tidal action will take over, slowing the Earth's rotation even further, until the day becomes longer than the month. At this point, angular momentum will be drawn from the Moon, and it will begin approaching the Earth, heading on a course of self destruction until it is finally torn apart by the tidal action of the Earth. Perhaps it will form a ring around our planet. In any case, it will probably end its years where it apparently began – close to the Earth. By this time, however, the brighter Sun will have boiled the oceans away, and the Earth will have become a dry and barren place (Section 4.4).

Stabilizing the Earth

The orientation of the Earth's rotation axis causes the annual seasonal variations of our climate, and small variations in its orientation contribute to the advance and retreat of the ice ages. The obliquity of the Earth, the angle that its spin axis makes with the perpendicular to its orbital plane, is now a modest 23.5 degrees. This is sufficient to bring summer and winter as the northern or southern hemisphere is tilted toward or away from the Sun (Section 1.1, Fig. 1.10). Variation in the Earth's obliquity as small as $\pm$ 1.3 degrees, around a mean value of 23.3 degrees, may contribute to, or trigger, the ice ages (Section 4.4).

The climate forecast for a Moon-less Earth would be a lot bleaker. The gravitational pull of our large Moon acts as an anchor, limiting excursions in the Earth's rotation axis and keeping the climate relatively stable (Fig. 5.30). Without the Moon, the tilt of Earth's spin axis would vary chaotically between 0 and 85 degrees. Such large variations in the planet's obliquity would result in dramatic changes in climate. With an obliquity of 0 degrees, there would be no seasonal variation in the distribution of sunlight on Earth. At 85 degrees, the Earth's axis would be tipped completely over. The equatorial tropics could then be permanently in cold winter snows, and the poles would be alternately pointed almost directly at or away from the Sun over the course of a single year. Such wide climate changes might be hostile to many forms of life on Earth.

The nearby massive Sun holds the tilt of Mercury and Venus in place, but there is no help for Mars. Located far from the Sun and with two puny satellites, the obliquity of the red planet exhibits wild variations from 0 to 60 degrees on time-scales of about 5 million years, with profound changes in the Martian climate (Section 8.8).

5.9 Origin of the Moon

Constraints on models of the Moon's origin

There are several facts that must be explained by a successful account of how the Moon originated. Some of them have been known for more than a century, while others result from laboratory investigations of the rocks returned from the Moon.

Any origin theory must, for example, explain why the Earth has a relatively massive Moon when Mercury and Venus have no known moons, and Mars only has two miniscule ones that may be captured asteroids. In other words, our Moon is an unusual event in the formation process of the rocky terrestrial planets.

A satisfactory theory for the origin of the Moon must also explain the Moon's peculiar orbit, which lies neither in the plane of the Earth's orbit around the Sun, the ecliptic plane, nor in the Earth's equatorial plane. Our satellite revolves around the Earth inclined about 5 degrees to the ecliptic plane, which is itself tilted 23.5 degrees with respect to the Earth's equatorial plane (Section 5.2, Fig. 5.25).

Perhaps even more important is the mean mass density of the Moon, just 3344 kilograms per cubic meter, much lower than the Earth's mean mass density of 5515 in the same units. The Moon's overall mass density is much closer to the terrestrial mantle than that of the Earth as a whole, which includes its dense iron core.

The Moon is now moving away from the Earth at the rate of 0.038 meters per year, which means that the Moon was much closer to the Earth in the distant past. Assuming a constant rate of recession, the Moon was at least

Fig. 5.30 Steadying influence The Moon holds the Earth upright in space, stabilizing its orientation and keeping the planet from tilting over. Without the Moon's influence, chaotic forces could tip the Earth's rotation axis down so far that its poles are pointing at or away from the Sun, producing wild swings in the Earth's climate. This image of the Moon and Earth was taken from a distance of 6.2 million kilometers, by the *Galileo* spacecraft on 16 December 1992, soon after swinging around the Earth on its way to Jupiter. (Courtesy of JPL and NASA.)

171 thousand kilometers closer to the Earth 4.5 billion years ago; its current mean distance is 384 thousand kilometers and stronger tidal interaction probably propelled the Moon outward at a quicker rate in the past.

Comparison of the lunar samples to terrestrial rocks provides further constraints on the Moon's parentage (Table 5.7). The oldest rocks on the Moon solidified 4.5 billion years ago, which means that the Moon is about as old as the Earth. One important distinction comes from the similar quantities of oxygen isotopes, or light and heavy oxygen atoms, in Moon rocks and Earth rocks, indicating a close kinship and suggesting a common ancestry. By contrast, objects formed in other parts of the solar system, such as the asteroids or Mars, exhibit different oxygen isotope ratios. This indicates that the Earth and Moon formed in roughly the same part of the primeval solar nebula, unlike all of the meteorites and planetary samples found to date.

A second key constraint is in the compositional differences between the Earth and the Moon. The Moon rocks are bone dry and completely lack any water-bearing minerals. They are also missing other kinds of volatile elements, with low melting points, that could have been boiled out into space at high temperatures. Relative to Earth rocks, the Moon rocks are also highly depleted in siderophile, or

Table 5.7 Constraints on models for the origin of the Moon

Constraint	Implication
No massive moons on other rocky planets	Moon formation is an unusual process
Moon orbit tilted to Earth's equator	Origin process must explain peculiar orbit
Low mean mass density of Moon	Moon has no large iron core
Moon moving away from Earth	Moon once closer to Earth
Some Moon rocks 4.5 billion years old	Moon about as old as Earth
Oxygen isotope ratios	Earth and Moon formed near each other
Depletion of volatiles	Moon formed at high temperature
Enrichment of refractories	Condensation at high temperature
Depletion of metals	Removal of iron prior to formation

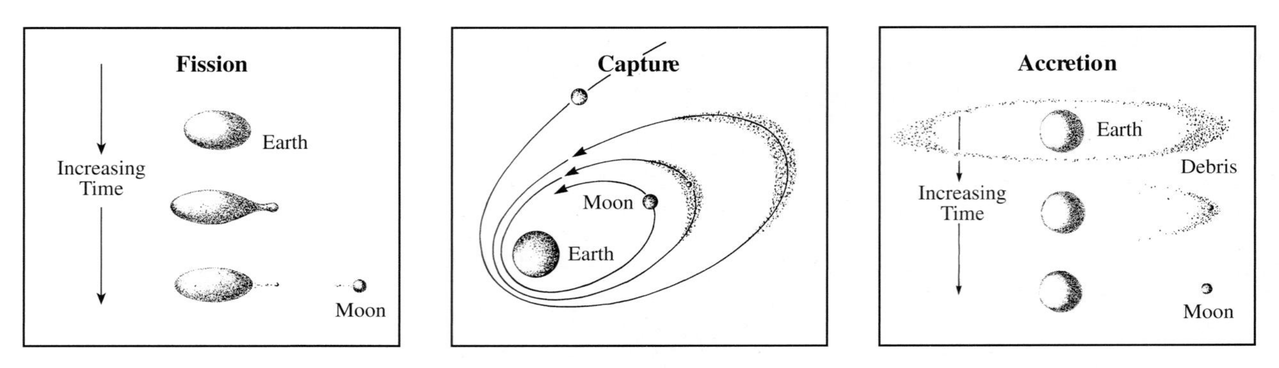

Fig. 5.31 Classical origin hypotheses According to the fission hypothesis (*left*), the rotational speed of the young Earth was great enough for its equatorial bulge to separate from the Earth and become the Moon. In the capture hypothesis (*center*), a vagabond Moon-sized object once passed close enough to be captured by the Earth's gravitational embrace. We have pictured disruptive capture, with subsequent accretion, but the Moon might have been captured intact. The accretion hypothesis (*right*) asserts that the Moon formed from a disk near the Earth.

"metal-loving" elements, such as cobalt or nickel, which tend to occur in rocks containing iron.

Yet, when compared to the Earth our satellite is enriched in non-volatile substances. Called refractories, these elements are the opposite of volatiles; they have high melting points and remain solid at high temperatures and require extraordinary heat to vaporize.

Early origin hypotheses

There are three classical hypotheses for the origin of the Moon that have been advocated for more than a century. They are the fission, capture and accretion models, nicknamed the daughter, pickup and sister theories (Fig. 5.31). But as Sherlock Holmes said in *The Adventure of Silver Blaze*, "I am afraid that whatever theory we state has very grave objections to it". So, we will briefly discuss the advantages and flaws of each of these hypotheses, and then move on to the more successful giant impact theory.

The fission hypothesis supposes that the Earth had no satellite in its earliest youth, but that it was once spinning so fast that a large fraction of its mass tore away to create the Moon. If this occurred after the Earth's iron had settled to the center, the Moon would naturally be depleted in metals and would have a low mean mass density characteristic of the outer layers of the Earth. Once the Moon had separated, tidal friction caused it to move slowly away toward its present orbit.

The fission hypothesis does not easily account for the compositional differences between the Moon and the Earth, such as the depletion of volatiles on the Moon. There are also two dynamical difficulties. First, the primordial Earth would have had to rotate exceptionally fast, once every 2.5 hours, if it were to throw off the material that became the Moon. Second, the Moon's orbital plane is tilted from the equatorial plane of the Earth; if the Moon spun off the planet's equator, the two planes ought to be coincident. There are ways around both of these problems, but the fission theory loses its attractive simplicity when it is doctored in these contrived ways.

If the Moon was not plucked out of the Earth, perhaps our satellite is a maverick that formed elsewhere, strayed too near the Earth, and was captured in orbit – either intact or as fragments torn apart by our planet's strong gravity. The principal advantage of this capture hypothesis is that it easily permits compositional differences between the Earth and Moon since they were formed in different locations within the solar system. The main obstacle is understanding how the capture could have taken place. A passing body would either collide with the Earth or receive a gravitational boost that would hurl it away from the Earth in slingshot fashion. In order to go into orbit about the Earth, the approaching Moon would have to slow down, and the chances of this happening are exceptionally low.

The third classical hypothesis suggests that the Moon and Earth formed concurrently from a cloud of gas and dust through a process not unlike the probable formation of the planets around the Sun. The raw materials for the Moon came from a disk of material in orbit around the Earth, and the planets originated in a similar accretion disk orbiting the Sun.

Such a model seems to apply nicely to giant gaseous planets, such as Jupiter, that have families of satellites resembling the solar system. But where does that leave the rocky terrestrial planets that have no known moons, Mercury and Venus? And what about Mars, with only two tiny satellites? If the process that formed our massive Moon is the natural way of things, we have difficulty understanding these other planets. And why should the

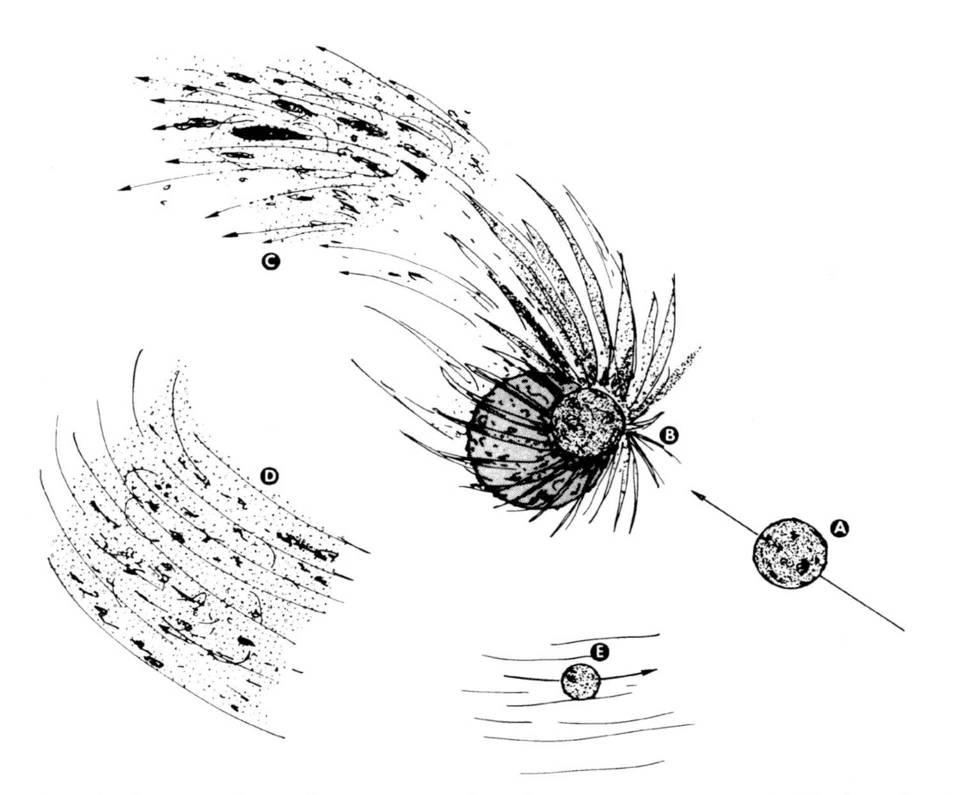

Fig. 5.32 Giant impact hypothesis According to the giant-impact hypothesis, a massive projectile (A), about the size of Mars, struck the young, still-forming Earth (B) in a catastrophic glancing blow nearly 4.6 billion years ago, resulting in a tremendous explosion and the jetting outward of both projectile and proto-Earth mass. Some fraction of this mass remained in Earth orbit (C), while the rest escaped Earth or impacted again on Earth's surface. A proto-Moon began to form from the orbiting material (D), accreting neighboring matter, and finally became the Moon (E). It may be mostly derived from the crust and mantle of the Earth and/or the impacting object, accounting for the Moon's relatively low mean mass density and lack of iron when compared with the Earth. The Moon accumulated so rapidly that the outer crust was molten, helping to account for the relative lack of water and other volatile elements. Then, as the crust cooled, the newborn Moon swept up the remaining objects nearby, blasting out impact basins and pock-marking the surface with numerous craters. (Courtesy of Alan P. Boss, Carnegie Institution of Washington.)

chemistry of the Earth and Moon be so different, and how were the volatile elements driven out of the Moon if it always orbited the Earth? Special assumptions can help extricate the accretion theory from its difficulties, but it loses its appeal when these special assumptions are introduced.

The giant impact hypothesis

Many astronomers now accept the idea of a violent and catastrophic lunar birth. According to this newer, giant impact hypothesis, a massive rogue projectile, perhaps 2.5 to 3.0 times the mass of Mars, sideswiped the Earth and dislodged the material that would become the Moon. This glancing, planet-shattering blow occurred almost 4.6 billion years ago, during a heavy bombardment that marked the last stages of the solar system's formative period. At this time, iron had already sunk to the core of the Earth, and a rocky crust was beginning to congeal around the partially molten planet.

The giant impact mechanism permits the Moon to form initially in the same part of the solar system as the Earth and to undergo a process that explains both the dearth of metals and volatile elements, as well as an enrichment of refractory elements, before solidification. The impact might have shattered the colliding object to smithereens and vaporized parts of the iron-poor upper layers of the Earth, blasting off a mix of terrestrial and impactor material into orbit where it coalesced to form the Moon that we know (Fig. 5.32). If the impacting object had a heavy iron core, it might have tumbled into the still-forming Earth, merging with the planet's core.

The glancing blow would have knocked the collision debris into the Moon's current tilted orbit, and if the satellite included material in the mantle of the impactor, this could also help explain the compositional differences between the Moon and Earth. The searing heat of such a collision would explain why the Moon holds no water and few volatile elements. All were boiled away. The new hypothesis therefore seems to explain the facts with a minimum of assumptions. As one astronomer stated, "it requires no magic, no special pleading, no extra twiddling and no *deus ex machina*".

Another reason that this model became at least imaginable was the discovery of extremely large impact basins on the Moon, such as the South Pole – Aitken basin on the far side. It was a short mental leap from very large impact basins to a planetary collision.

Thus, exploration of the Moon has resulted in the probable solution of the ancient mystery of the Moon's origin. It was most likely born in a fiery cataclysm out of the infant Earth, the result of an enormous, off-center collision during the early days of the solar system when such events were common.

The giant impact that gave rise to the Moon is a natural consequence of planet formation, making astronomers more aware of impact catastrophes in solar-system history. Similar collisions with larger or smaller projectiles could explain major planetary anomalies, such as the off-kilter, backward spin of Venus (Section 7.1), Uranus' bizarre, sideways orientation (Section 11.1), and even the demise of the dinosaurs that redirected the course of life on Earth 65 million years ago (Section 14.4).

The voyage to the Moon also opened a path to the rest of the solar system. Our satellite was the first port of call, the stepping stone in our captivating voyage to the planets, to which we now turn, beginning with Mercury whose composition probably resulted from a giant impact of its own.

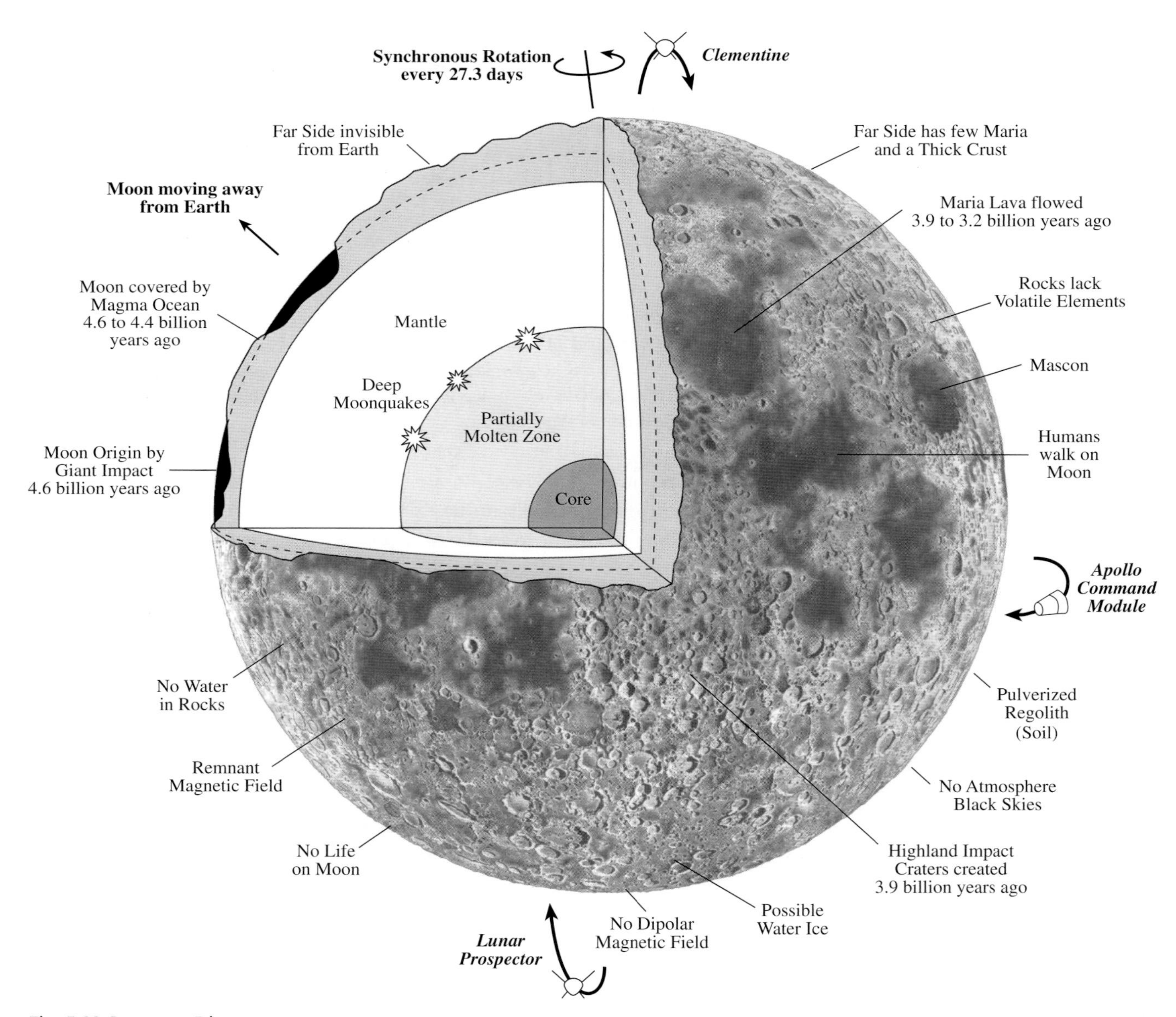

Fig. 5.33 Summary Diagram

6 Mercury: a dense battered world

- Because of its close proximity to the Sun, the innermost planet, Mercury, cannot be studied from Earth against the dark night sky; many astronomers and most people have never seen the elusive planet.
- During the daytime, Mercury's ground temperature reaches a blistering 740 kelvin; at night it plunges to a freezing 90 kelvin.
- Mercury has no atmosphere to speak of; just an exceedingly tenuous wisp of atoms that varies in location and time and is continually dislodged from the planet's surface.
- Because of spin–orbit coupling, produced by solar tides in the solid planet, there is a simple 2:3 ratio between Mercury's rotation period of 58.6 Earth days and its orbital year of 88 Earth days.
- The interval from sunrise to sunset at a given location on Mercury is 88 Earth days, and the night lasts 88 Earth days more, so the day on Mercury lasts 176 Earth days and twice Mercury's year.
- Mercury's rotation axis is aligned perpendicular to its orbital plane, so there are no seasons on the planet, and radar-bright spots suggest that water ice may reside in permanently shaded regions near its poles.
- Although Mercury is one of Earth's nearest neighbors, only one spacecraft, *Mariner 10,* has ventured near Mercury, viewing only about half of the planet's surface; the details of the other side of Mercury have never been seen.
- At first glance, Mercury's cratered surface resembles the Moon, but Mercury has unique surface features, including ancient intercrater plains, young smooth plains and no dark maria.
- Relative to its size, Mercury has the biggest iron core of all the terrestrial planets, and much larger than the core of the Moon.
- Mercury may have been blown apart by an ancient collision with a Moon-sized object.
- Mercury has a relatively strong magnetic field, the most powerful of all the terrestrial planets except Earth.
- Astronomers found long ago that Mercury did not appear in its expected place, leading Einstein to develop a new theory of gravity in which the Sun curves nearby space.

6.1 Fundamentals

Mercury is the smallest of the four rocky, terrestrial planets. It is the closest planet to the Sun, moving around it with the fastest speed and shortest year of any planet. Mercury spins with a slow rotation period of 58.6 Earth days, just two-thirds of its year. It has practically no atmosphere at all, but retains an unexpectedly strong magnetic field.

6.2 A tiny world in the glare of sunlight

A small, elusive planet

Mercury revolves closer to the Sun than any other known planet, with a mean distance from the Sun of just 0.387 AU. It therefore has the shortest year – 88 Earth days – and the highest orbital speed of any planet. Like a moth about a flame, Mercury races around the Sun at an average speed of 48 kilometers per second, pulled by the powerful solar gravitational field. Its rapid motion explains why Mercury is named after the wing-footed messenger of the gods in Roman mythology.

Table 6.1 Physical properties of Mercury[a]

Mass	3.302×10^{23} kilograms = $0.0553 M_E$
Mean Radius	2440 kilometers = $0.382 R_E$
Mean Mass Density	5427 kilograms per cubic meter
Rotation Period	58.646 Earth days
Orbital Period	87.969 Earth days
Mean Distance from Sun	5.79×10^{10} meters = 0.387 AU
Age	4.6×10^9 years
Atmosphere	Very tenuous (helium, sodium, potassium)
Surface Pressure	2×10^{-13} bar
Surface Temperature	90 to 740 kelvin
Magnetic Field Strength	0.0033×10^{-4} tesla at the equator = $0.01 B_E$
Magnetic Dipole Moment	5.54×10^{12} tesla meters cubed

[a] The symbols M_E, R_E and B_E respectively denote the mass, radius and magnetic field strength of the Earth.

As planets go, Mercury is a tiny world, with the smallest size of any terrestrial planet and slightly smaller than Jupiter's satellite Ganymede and Saturn's satellite Titan. Mercury's linear radius is easy to measure from its angular radius and distance. Its radius is 2440 kilometers or just 1.4 times the radius of our Moon. Mercury's mass has been a more elusive quantity to determine, because the planet has no satellites. The mass was first estimated from Mercury's gravitational influence on the orbital motion of Venus and passing comets; the estimate was improved by the planet's gravitational deflection of the space probe, *Mariner 10*. The planet's mass is 3.302×10^{23} kilograms, or just 0.055 times the mass of the Earth.

But Mercury is surprisingly massive for its size. Its volume is only slightly larger than the Moon's and yet it has four times the Moon's mass. This implies a mean mass density of 5427 kilograms per cubic meter, which is nearly as high as that of the Earth, 5515 in the same units and a little more than Venus at 5204.

Mercury's small apparent size and its proximity to the Sun make it difficult to see from Earth. The innermost planet never wanders more than 27.7 degrees in angular separation from the Sun (Fig. 6.1). This angle is less than that made by the hands of a watch at one o'clock. From Earth's perspective, Mercury's tight orbit never reaches into the dark night sky, and it can thus be observed only during the day.

The planet is visible to the unaided eye only in twilight when it is low in the sky and must be seen through a thick layer of air. Astronomers have therefore taken to observing Mercury near midday, when it is far from the horizon and can be seen through a relatively thin layer of air. At such times, Mercury can only be seen with telescopes.

But the terrestrial atmosphere limits the resolution of even the best Earth-based telescopes to features on Mercury that are just a few hundred kilometers across or wider – a resolution far worse than that for the Moon with the unaided eye. Moreover, space-borne telescopes with better resolution, such as the *Hubble Space Telescope*, cannot point at Mercury, because even stray light from the nearby Sun could damage their sensitive instruments.

Most people have never seen Mercury. Even Copernicus complained that it eluded him. So, it is not surprising that little was known about this enigmatic planet until it was explored by terrestrial radio detection and ranging, or radar for short, and by an unmanned spacecraft on an interplanetary flyby mission.

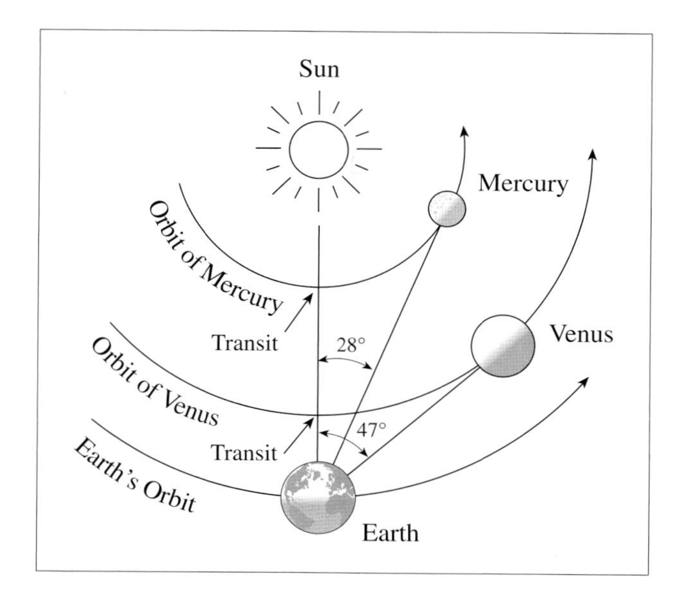

Fig. 6.1 Maximum elongation Unlike the planets which orbit the Sun beyond the Earth, Mercury and Venus can never be seen high in the sky in the dead of night. Because they are close to the Sun and inside Earth's orbit, these planets are always seen soon after sunset, or shortly before sunrise, and they show phases much as the Moon does. The elongation of Mercury and Venus, or their angular distances from the Sun as viewed from the Earth, never exceeds 28 degrees and 47 degrees, respectively.

Wild temperature swings in a world with no significant atmosphere

Solar radiation is ferocious on the surface of Mercury, the planet closest to the Sun. It is subject to the most intense sunlight, and experiences the greatest diurnal temperature variations, of any planet in the solar system. When Mercury is at the closest point in its orbit to the Sun, the noontime ground temperature on the side facing the Sun soars to 740 kelvin. This is hot enough to melt tin, lead, and even zinc. Because there is almost no atmosphere to hold in the heat, the ground temperature on Mercury plunges to 90 kelvin, or 183 degrees below zero, on the night side.

Like the Moon, the planet Mercury is, for all practical purposes, a world without any atmosphere. If it had any significant atmosphere, the intense solar heat on the side facing the Sun would have boiled it away into space. Sped up by the high temperatures, and held only loosely by Mercury's weak gravity, the atoms of any possible atmosphere would have little difficulty escaping into interplanetary space.

There is an exceedingly tenuous atmosphere on Mercury, consisting of hydrogen and helium atoms – discovered with instruments on the *Mariner 10* spacecraft – and sodium and potassium atoms – identified subsequently using Earth-based observations (Section 3.4). But it is constantly being evaporated away by the Sun's intense heat and replenished from below. The rarefied gas is so thinly distributed that its particles almost never touch each other, and they only hit the surface. The surface gas pressure is estimated at one-fifth of a million-millionth, or 2×10^{-13}, of the pressure at the Earth's surface. This thin atmosphere is a far better vacuum than can easily be produced in a laboratory on Earth.

6.3 Radar probes of Mercury

The halting spin of old age

Astronomers once supposed that solar tides in the body of Mercury would cause the planet to rotate on its axis once every 88 days, in step with its orbital period. Just as the Moon always presents the same face to the Earth, it was thought that one side of Mercury was always turned toward the Sun. To test this idea, the Italian astronomer Giovanni Schiaparelli (1835–1910) watched the surface markings seen though his 0.46-meter (18-inch) telescope, and concluded that the same side of Mercury did, indeed, always face the Sun. For three-quarters of a century, telescopic observers agreed with his conclusion. All of these astronomers were dead wrong!

In 1965, Mercury's true rotational period was determined with radio signals that rebounded from the planet. The world's largest radio telescope, located in Arecibo, Puerto Rico, was used to transmit megawatts of pulsed radio power at Mercury, and to receive the faint echo (Fig. 6.2). This technique is known as radio detection and ranging, abbreviated radar, and it is used to locate and guide airplanes near airports.

Each pulse was finely tuned, with a narrow range of wavelengths. Upon hitting the planet, its rotation detuned the pulse, slightly spreading the range of wavelengths (Fig. 6.3). One side of the globe was rotating away from the Earth, while the other side was rotating toward our planet. These motions produced slight changes in the wavelength of the echo, allowing the speed of the surface and the rotational period to be calculated, using the well-known expression for the Doppler effect (Focus 6.1).

The result came as an unexpected surprise. The rotation period was 58.6 days, or exactly two-thirds of the 88-day

Fig. 6.2 Arecibo observatory The world's largest radio telescope is nestled into the hills near Arecibo, Puerto Rico. Its metal reflecting surface has a spherical shape with a diameter of 305 meters. The reflected radio signals are focused to detectors suspended on the triangular structure hanging from the three towers. The facility can also transmit powerful radio pulses, sending them off the metal surface into space. Such pulses, sent and received from this giant telescope, first measured Mercury's rotation period in 1965. Until then it had been wrongly thought that Mercury kept one side permanently facing the Sun, with a rotation period equal to its orbital period.

period that had been accepted so long. Thus, with respect to the star background, Mercury spins on its axis three times during two full revolutions about the Sun. This relationship follows from $3 \times 58.6 = 2 \times 88$, and it is technically known as spin–orbit coupling.

But why does Mercury rotate with a period that is two-thirds of its orbital period? The answer is found in the Sun's varying tidal forces on the elongated planet as it moves along its eccentric orbit. If Mercury had a nearly round shape and a close to circular orbit, its rotation would have slowed to synchronism with its 88-day orbit. Like the Moon, it would have rotated with one face permanently toward the parent body. But the Sun tugs extra hard on the nearest, egg-shaped extension of the planet, and harder still when Mercury is closest to the Sun. This extra gravitational pull of the Sun gives an abrupt twist to Mercury's elongated body. These twists tend to speed up the rotation, and they may have knocked the planet into the shorter 58.6-day period.

Mercury was most likely slowed to this rotation rate within 500 million years of its formation, and has retained it ever since. If the planet tends to rotate a little too fast, the timing of the tidal twist is altered and the planet is slowed down; if it is rotating too slowly the twist speeds it up slightly, and the synchronism is re-established. Thus, the 2:3 resonance is likely to persist during the remaining history of Mercury.

The spin–orbit coupling has another curious effect that may have misled astronomers who reported the wrong rotation period. Because three times the true rotation period is equal to twice the orbital period, any surface markings on Mercury would have returned to the same side after two orbital revolutions (Fig. 6.4). Thus, astronomers could have been fooled, because looking at Mercury after two of its orbital periods they would see the same markings on the sunlit side and would find no disagreement with the 88-day period that they expected. Many of the conflicting observations were apparently ignored or missed. This is a striking example of curious observational circumstances and theoretical expectations that misled nearly everyone.

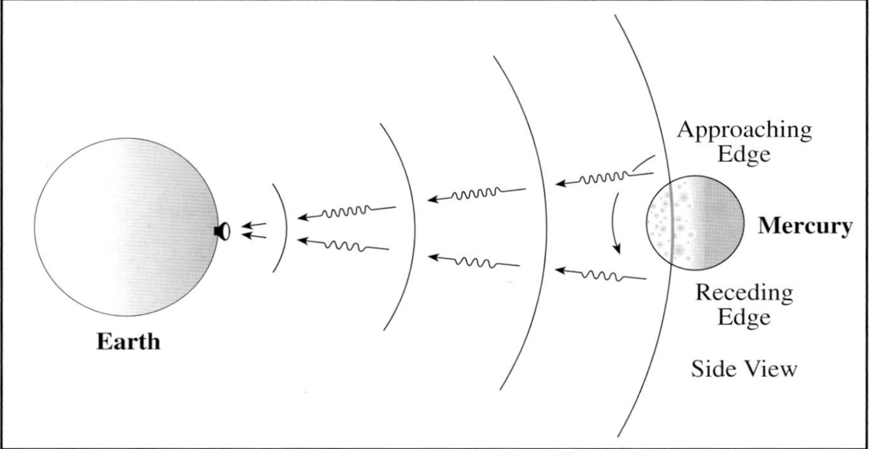

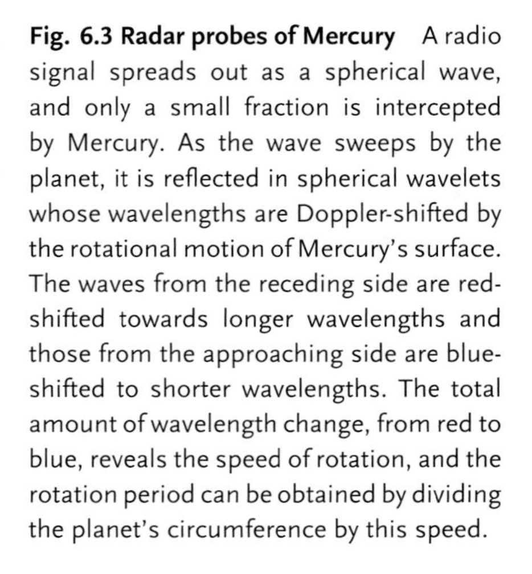

Fig. 6.3 Radar probes of Mercury A radio signal spreads out as a spherical wave, and only a small fraction is intercepted by Mercury. As the wave sweeps by the planet, it is reflected in spherical wavelets whose wavelengths are Doppler-shifted by the rotational motion of Mercury's surface. The waves from the receding side are red-shifted towards longer wavelengths and those from the approaching side are blue-shifted to shorter wavelengths. The total amount of wavelength change, from red to blue, reveals the speed of rotation, and the rotation period can be obtained by dividing the planet's circumference by this speed.

Focus 6.1 The Doppler effect

Just as a source of sound can vary in pitch or wavelength, depending on its motion, the wavelength of electromagnetic radiation shifts when the emitting source moves with respect to the observer. This Doppler shift is named after the Austrian scientist, mathematician and schoolteacher, Christian Doppler (1803–1853) who discovered it in 1842. If the motion is toward the observer, the shift is to shorter wavelengths, and when the motion is away the wavelength becomes longer. You notice the effect when listening to the changing pitch of a passing ambulance siren. The tone of the siren is higher while the ambulance approaches you and lower when it moves away from you.

If the radiation is emitted at a specific wavelength, λ_{emitted}, by a source at rest, the wavelength, $\lambda_{\text{observed}}$, observed from a moving source is given by the relation:

$$\text{redshift} = z = \frac{\lambda_{\text{observed}} - \lambda_{\text{emitted}}}{\lambda_{\text{emitted}}} = \frac{V_r}{c},$$

where V_r is the radial velocity of the source along the line of sight away from the observer, and the velocity of light $c = 2.9979 \times 10^8$ meters per second. The parameter z is called the redshift since the Doppler shift is toward the longer, redder wavelengths in the visible part of the electromagnetic spectrum. When the motion is toward the observer, V_r is negative and there is a blue shift to shorter, bluer wavelengths.

A rotating object will produce a blueshift on the side spinning toward an observer, and a redshift on that spinning away. Their combined effect will broaden a finely tuned radio pulse at wavelength λ by an amount $\Delta\lambda$ given by the expression

$$\frac{\Delta\lambda}{\lambda} = \frac{V_{\text{rot}}}{c}$$

where V_{rot} denotes the rotation velocity.

Long, hot afternoons on Mercury

The days are certainly long on Mercury, longer than the planet's year. At any given point on Mercury, the daylight interval between sunrise to sunset lasts 88 Earth days, and the night lasts 88 Earth days more. This means that at one location on the surface, successive sunrises occur every 176 Earth days (Fig. 6.4). So, an observer on Mercury would experience a full day that lasts two Mercury years.

In addition, when Mercury arrives closest to the Sun, and is moving at the fastest speed in its orbit, it moves quicker than the planet rotates. So, from the vantage of someone on the surface, the Sun would appear to stop in the sky and go backward until the planet's slow rotation catches up and makes the Sun go forward.

Possible water ice at the poles of Mercury

Despite the heat on the sunlit side of Mercury, radar astronomers have found evidence for ice at both the north and south poles. Both the intensity and orientation, or polarization, of the bright radar echoes suggests the presence of water ice. Their radar characteristics are similar to those seen on icy surfaces elsewhere in the solar system, such as the ice deposits on the Galilean satellites and the south polar ice cap on Mars.

The prospect of a planet as hot as Mercury having ice caps, or any water at all, seems preposterous. Yet, the ice could reside at the bottoms of polar craters. Mercury's rotation axis is very nearly perpendicular to the plane of its orbit, so the floors of deep craters near the poles are never exposed to sunlight. Any ice in these craters would remain in the cold, dark shade, evaporating very slowly. Perhaps the water was delivered to the crater floors by comets that slammed into them (Section 2.4).

Moreover, the radar-bright features have been matched with specific polar craters, located on *Mariner 10* images,

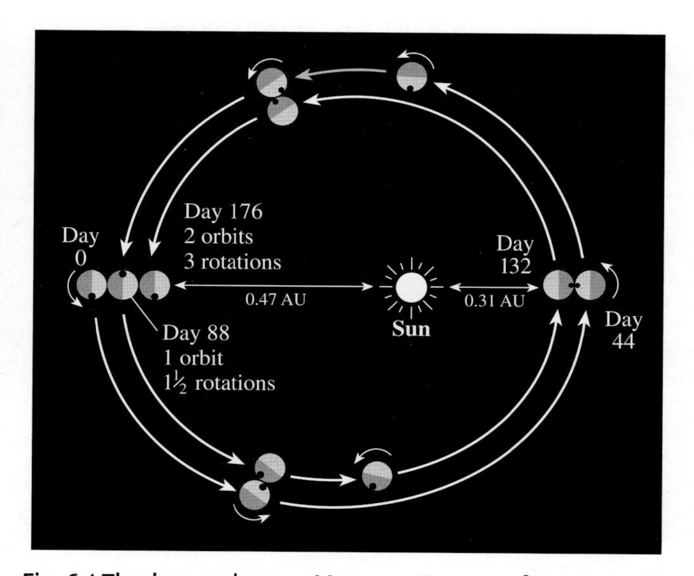

Fig. 6.4 The days are long on Mercury Because of its spin–orbit coupling, Mercury rotates once every 58.6 Earth days, and orbits the Sun in 88.0 Earth days. After two orbits the planet has rotated three times, but from Mercury's surface the Sun appears to have moved only once around the planet. So sunrise is repeated at a given point on the planet's surface (*black dot*) every two orbits, and a solar day on Mercury lasts two of the planet's years and 176 Earth days. The labels on this figure refer to Earth days.

Fig. 6.5 Mercury's south pole In this *Mariner 10* frame, south is down and the south pole is located on the right-hand edge of the large crater that has only its rim sticking up into the light. This crater is named Chao Meng-fu after the Chinese painter and calligrapher (1254–1322) and it is 167 kilometers in diameter. Radar-reflection data suggest that water ice might be present in the cold, dark, permanently shadowed floors of craters at the poles of Mercury. (Courtesy of JPL and NASA.)

and these craters should contain permanently shadowed interiors (Fig. 6.5). The shaded polar craters could, alternatively, contain other volatile substances, such as sulfur, which could produce strong radar echoes but have a higher melting point than water ice. So, we may not definitely know if there is water ice at the top and bottom of Mercury until inquisitive robot spacecraft land there and make the appropriate tests. In the meantime, we turn to the startling results of *Mariner 10.*

6.4 A modified Moon-like surface

Craters like the Moon

On 29 March 1974, the *Mariner 10* spacecraft penetrated the glare surrounding Mercury as it sped past the planet. The spacecraft had been flung toward Mercury during a close encounter with Venus seven weeks earlier. After passing Mercury's night side, *Mariner 10* went into orbit around the Sun, mimicking a tiny planet. As the spacecraft and Mercury looped around our star, they each returned to nearly the same place every six months, allowing repeated observations of the planet. Close-up photographs were taken during three such encounters, before the spacecraft's supply of maneuvering gas was depleted.

Altogether *Mariner 10* provided photographs of about half of Mercury's surface, and showed features thousands of times smaller than those seen from the best telescopes on Earth. These close-ups revealed a landscape that had never been seen before, and they gave us a glimpse of the planet's past. Still, the other face of Mercury remains unexplored, so a lot remains to be discovered when a spacecraft observes the unseen half of Mercury.

At first glance, Mercury closely resembles the Moon, for both worlds are small, heavily cratered, and without a significant atmosphere to cause erosion. Like the Moon, the planet Mercury has highlands that are pockmarked with impact craters (Figs. 6.6, 6.7), ranging in diameter from impact basins about a thousand kilometers across to craters only 100 meters in diameter, the limit

Fig. 6.6 Inbound mosaic of Mercury During its three encounters with Mercury in 1974–5, the *Mariner 10* spacecraft obtained thousands of images of the same sunlit hemisphere of the planet. Individual frames acquired by the approaching spacecraft have been joined together in this mosaic. From this perspective, Mercury looks similar to the Moon, with a heavily cratered surface, but close-up inspection of individual surface features reveals some fundamental differences. (Courtesy of JPL and NASA.)

Fig. 6.7 Outbound mosaic of Mercury After passing on the dark side of the planet, the *Mariner 10* spacecraft looked back at the sunlit hemisphere of Mercury and photographed these images that have been assembled into a mosaic. It shows large tracts of smooth plains, that may be due to extensive volcanism or the splashed-out debris from large impacts. Partially visible along the day–night terminator (*left*) is half of the Caloris basin (*above center toward the top*), a gigantic multi-ringed impact scar. (Courtesy of JPL and NASA.)

of *Mariner 10*'s photographic resolution. The ubiquitous craters on Mercury strongly resemble their lunar counterparts, indicating that they were formed by meteoritic impact (Section 2.2). As on the Moon, there are small bowl-shaped craters, larger craters with terraces and central peaks, relatively young craters with bright rays, and huge impact basins on Mercury.

There are also subtle differences caused by the planet's stronger gravity. Because the force of gravity on the surface of Mercury is about twice that on the Moon, material ejected from a crater on Mercury is thrown about half as far. Thus, secondary craters are closer to the primary crater rim on Mercury than they are on the Moon.

The impact craters on Mercury have been named after famous deceased artists, musicians, painters and authors (Focus 6.2). The bigger ones, in order of decreasing size, are named Beethoven, Dostoevskij, Tolstoj, Goethe, Shakespeare, Raphael and Homer; they range from 643 to 314 kilometers in diameter.

The multi-ringed Caloris basin is the biggest impact feature on Mercury. The huge excavation spans 1.34 thousand kilometers and occurred about 3.85 billion years ago. Only half of it was in sunlight during the *Mariner 10* encounter, and the focused effects of the explosion have been detected on the opposite side of the planet (Section 2.2). Since this feature is located at the hottest part of the planet, it has been named *Caloris*, Latin for "hot".

Intercrater highland plains

In many important respects, Mercury's resemblance to the Moon is superficial. The planet's most densely cratered surfaces are not as heavily cratered as the lunar highlands, and Mercury does not contain regions of overlapping large craters and basins. Also, unlike the Moon, the heavily cratered terrain on Mercury is interspersed with large regions of gently rolling, intercrater plains (Fig. 6.8).

Fig. 6.8 Heavily cratered highlands and intercrater plains The highlands of Mercury exhibit abundant craters, just as the lunar highlands do; but unlike the Moon, intercrater plains occur between large craters. This *Mariner 10* image of Mercury, which exhibits both highland craters and intercraters plains, is 400 kilometers across. Intercrater plains and heavily cratered terrain are typical of much of Mercury outside the area affected by the formation of the Caloris basin. (Courtesy of JPL and NASA.)

Focus 6.2 Naming features on a satellite or planet

Perhaps in an effort to bring the Universe a little closer to home, the places found on other worlds have been named after famous people or terrestrial surface features. The custom began in 1610, when Galileo Galilei (1564–1643) used the name *maria*, Latin for "seas", for the large dark regions on the Moon. Giovanni Battista Riccioli (1598–1671), an Italian theologian and astronomer, made a detailed telescopic study of the Moon in collaboration with Francesco Maria Grimaldi (1618–1663). The latter's excellent lunar map was inserted in Riccioli's *Almagestum novum* published in 1651, and the lunar nomenclature they adopted is still in use today. The lunar surface features were named after astronomers and philosophers, such as the crater Tycho for the Danish astronomer Tycho Brahe (1546–1601, Section 1.1). They also gave fanciful names to the maria, such as Mare Imbrium (Sea of Rains), Oceanus Procellarum (Sea of Storms) and Mare Tranquillitatis (Sea of Tranquillity), even though Riccioli personally believed there was no water on the Moon.

Lunar maps and catalogs continued to appear for nearly three centuries, often resulting in several different names for the same feature on the Moon. To deal with the confusion, the International Astronomical Union, abbreviated IAU, set up a standard lunar nomenclature in 1935, consolidating and unifying all previous names for 681 lunar craters, adopting the old custom of naming large craters for famous deceased scientists, scholars and artists and naming mountains for their terrestrial counterparts.

When spacecraft visited the Moon, a large number of new features were revealed, including those on the far side of the Moon that cannot be seen from Earth. In 1970, the relevant IAU group sanctioned hundreds of names for these new features, including the names of twelve American astronauts and Russian cosmonauts, half of whom were still living.

The IAU also standardized Martian nomenclature, incorporating many of the mythological names suggested by the Italian astronomer Giovanni V. Schiaparelli (1835–1910) for its light and dark markings. When spacecraft visited the red planet, an IAU task group was organized to review and sanction names for newly discovered features, and similar groups have subsequently been set up for all the planets and their satellites.

Anyone can suggest that an IAU task group consider a specific name, but investigators mapping the objects suggest many of them. All the suggested names are reviewed and upon approval eventually adopted by the full General Assembly of the IAU.

The naming has evolved into a complicated procedure, for there are now Latin descriptor terms that describe the feature types, such as chasma, labyrinth, montes, patera, planitia, rupes, terra and valles, and then names have to be suggested and adopted to go with each member of the type. The planitia, or low plains, on Mercury are, for instance, given the names for Mercury, either planet or god, in various languages, while the craters on Venus are all named after famous deceased women and its other features given the names of mythological goddesses – such as goddesses of fertility, home, love, sky, water and war (Section 7.4, Focus 7.1). A complete description of all the naming details can be found in the *Gazetteer of Planetary Nomenclature* of the IAU Working Group for Planetary System Nomenclature, available on the web at: http://planetarynames.wr.usgs.gov/

The lower number of craters on Mercury, as compared to the Moon, suggests that most of the planet was once covered by molten rock. This widespread volcanic resurfacing probably occurred between 4.2 to 4.0 billion years ago, obliterating the missing craters and lasting longer than the magma ocean did on the Moon – until 4.4 billion years ago. The planet's heavily cratered terrain was then excavated out of the cooled-and-solidified global lava at the tail end of the heavy bombardment recorded on the Moon (Table 6.2).

Smooth lowland plains

Widespread areas of Mercury are covered by relatively flat, sparsely cratered terrain called the smooth lowland plains (Fig. 6.9). They are younger than the intercrater highland plains, and they are about two kilometers lower. Unlike the dark maria on the Moon, the smooth lowland plains on Mercury are about the same brightness or color as the heavily cratered terrain and intercrater plains in the highlands of Mercury.

The smooth plains occur within and around the Caloris basin and on the floors of other basins, but a careful study of crater densities suggests that the smooth plains are younger than the Caloris impact. This age difference suggests that the lowland plains on Mercury are volcanic eruptions from the interior, rather than surface material melted and thrown out during basin-forming impacts. An investigation of the colors of sunlight reflected from the surface, and by implication the minerals it contains, support

Table 6.2 Mercury's early history

Period	Age[a] (10^9 yr = 1 Gyr)	Characteristics	Lunar Counterpart[b]
Pre-Tolstojian	4.2 to 4.0	Global volcanism and resurfacing, formation of intercrater highland plains, obliteration of oldest craters.	Pre-Nectarian
Tolstojian	4.0 to 3.9	Tolstoj impact basin and highland craters formed.	Nectarian
Calorian	3.85	Caloris basin impact, smooth lowland plains formed shortly thereafter.	Imbrian
Mansuarian	3.5 to 3.0	Period of diminished cratering.	Eratosthenian
Kuiperian	1.0	Young rayed craters such as Kuiper formed.	Copernican

[a] These ages are largely guesswork, for we have no samples of Mercury to calibrate its cratering time-scale.
[b] The Moon was covered by a magma ocean from 4.5 to 4.4 billion years ago, and most of its impact basins were filled with dark lava 3.9 to 3.2 billion years ago, forming the maria (Section 5.7, Table 5.6).

Fig. 6.9 Young smooth plains The smooth-plain terrain of Mercury is a possible example of a secondary crust of volcanic origin; but unlike the dark lunar maria the smooth terrain on Mercury is light in color. It was created after the heavy bombardment that excavated the older craters on the planet, but before the younger craters that are superposed on the smooth plains. The largest young crater (*top left, only one half showing*) is about 100 kilometers in diameter; such large craters have central peaks, flat floors, terraced walls, radial ejecta deposits and surrounding fields of secondary craters. The smooth plains shown here, in this *Mariner 10* image, have well-developed ridges. (Courtesy of JPL and NASA.)

the view that some of the smooth plains originated by volcanic outflow (Fig. 6.10).

Rupes – cliffs or scarps

The most remarkable geological features on Mercury are its winding cliffs or scarps, which are widely distributed over the planet. These unique global features have been named *rupes,* Latin for "rock or cliff", each preceded by the name of a ship of discovery or a scientific expedition. An example is Discovery Rupes, named after Captain James Cook's (1728–1779) ship on his third and last voyage to the Pacific from 1776 to 1780. The long cliff snakes its way across pre-existing craters and plains, attaining a length of 350 kilometers and rising to 4 kilometers, as high as the Pyrenees (Fig. 6.11).

The long cliffs on Mercury look like the wrinkled skin of a shriveled apple, and the shrinking of the planet as the interior cooled probably caused them. The total shrinkage implied by the cliff heights is about 0.1 percent (0.001) of the radius of the planet. Since the cliffs formed after most of the craters, but before some of the smaller craters found on them, Mercury probably shrank many hundreds of millions of years after the solidification of the crust, during the cooling of the planet's underlying mantle and partial solidification of its internal core. As the young planet solidified from the outside in, great blocks of its crust shifted up on one side and down on the other, thrusting the cliffs up into the sky.

There is no evidence of global shrinking on the Moon. This is probably because the Moon has a small metallic core, while Mercury has a large one. The entire globe probably shrank when at least some of its large iron core cooled and solidified.

The wrinkle ridges on the mare basalts do indicate local and regional shrinking on the Moon. Isolated cliffs, or scarps, in the lunar highlands also indicate some shrinking, but it is not likely to be global shrinkage as on Mercury.

All in all, Mercury seems to have undergone a series of events similar to the early history of the Moon, but on a slightly different time-scale (Table 6.2, and Section 5.7). Both objects, despite their different locations in the solar

Fig. 6.10 Surface minerals on Mercury This false-color composite was formed from recalibrated *Mariner 10* images at different wavelengths to highlight differences in surface minerals on Mercury. The color units in this and similar images are consistent with widespread plains formed by volcanism on the planet. The yellow splash at the lower right marks the location of a 60-km-wide rayed Kuiper crater believed to have excavated unusual material from below the surface about 1 billion years ago. Reddish areas contain fewer opaque minerals and may represent primitive crustal material; the dark and blue regions are consistent with enhanced titanium content. The composite also suggests that the planet has undergone differentiation with heavy minerals sinking inside and lighter ones rising to the surface. The Kuiper crater is named after the astronomer Gerard Kuiper (1905–1973), a *Mariner 10* team member. (Courtesy of JPL and NASA.)

system, appear to have begun life covered with a magma, subjected to an intense bombardment when the magma solidified, and followed by the rise of magma to the surface. But the volcanic vents on Mercury were probably then squeezed shut during a global compression of the crust while those on the Moon remained open.

6.5 An iron planet

Four of the terrestrial bodies – the Moon, Mars, Venus and Earth – exhibit a fairly linear relationship between size and mean mass density, in which bigger objects have greater density (Fig. 6.12). But Mercury does not conform to this relationship. Although it is less than half the size of the Earth and not much bigger than our Moon, the bulk mass density of Mercury, at 5427 kilograms per cubic meter, is typical of a far larger planet. The most natural explanation of Mercury's high mass density is that it contains an unusual amount of iron, which is cosmically the most abundant heavy element.

Most of this iron is probably concentrated in Mercury's interior, for no iron has been detected on its surface. Analysis of *Mariner 10* findings, as well as spectroscopic observations from Earth, has failed to detect any iron in Mercury's crustal rocks.

In order to account for the planet's large mass density, the apparent dearth of iron on the surface has to be balanced by a large iron core. Relative to its size, Mercury would have the largest metallic core of all the terrestrial planets (Fig. 6.12). The dense iron core takes up 75 percent of the planet's radius, so Mercury is mostly iron core surrounded by a relatively thin silicate mantle. In this respect, it is the opposite of the Moon that has a thick rocky mantle and a relatively small iron core (Section 5.5).

After its formation, Mercury probably remained molten long enough for the heavy elements to settle at its center, just as iron drops below slag in a smelter. The high weight of the iron atoms would have slowly carried them down into the interior, leaving the silicates to form a lighter rocky mantle floating on the surface.

As it descended, the iron released gravitational energy, much the way water gives up gravitational energy when it flows downward in a hydroelectric plant. The heat released this way was probably sufficient to melt a large portion of the planet. But Mercury would have then cooled quickly, and the crust would have formed as a single thick plate, eliminating the possibility of plate tectonics of the type found on Earth.

Why does Mercury have so much iron and so little rock? Some astronomers think the rock was blasted off long ago, when it collided with another massive object. Both Mercury and the object could have had thick rocky mantles that were completely vaporized in the collision, while their iron cores clumped together under their mutual gravitational pull. The vaporized rock particles might have spiraled into the massive Sun, while the iron planet remained in orbit. Such cataclysmic collisions are thought to have been common about 4.5 billion years ago, and one such giant impact with the nascent Earth created its Moon, with a lot of rock and little metal (Section 5.9).

A second possibility is that Mercury's close proximity to the Sun played a decisive role in its formation. Due to

Fig. 6.11 Discovery Rupes Mercury's surface is distinguished from the Moon and from the other terrestrial planets by having enormous cliffs, or scarps, that cut across its surface. The dark scarp cutting vertically across the center of this image, taken from *Mariner 10*, is about 350 kilometers long. The cliff transects two craters 35 and 55 kilometers in diameter, proving that it was created after these craters formed. These long cliffs, which are as much as 4 kilometers high, were probably created when the planet cooled and contracted. (Courtesy of JPL and NASA.)

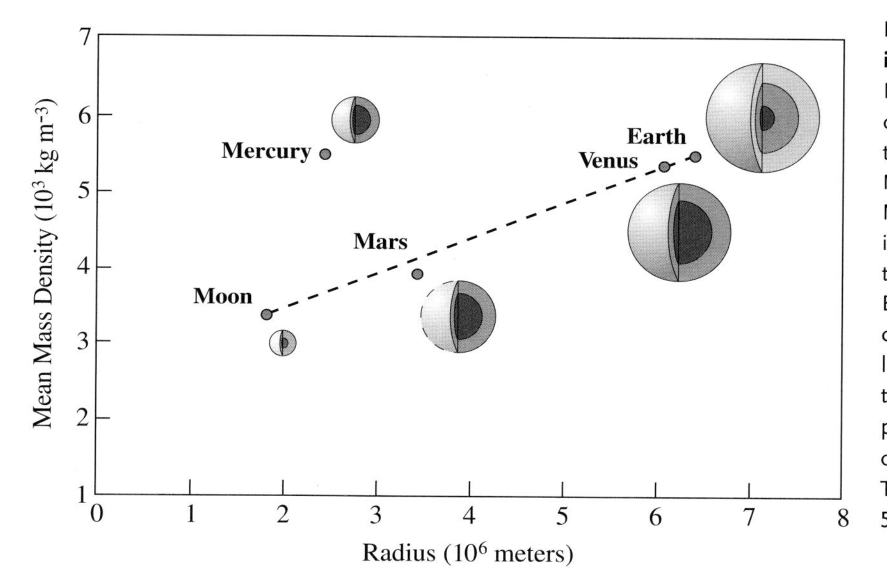

Fig. 6.12 Radius, mass density, and interior structure of terrestrial bodies Mercury is an anomaly in this comparison of the size and mean mass density of the Moon and the terrestrial planets. Mercury is just slightly bigger than the Moon and a little smaller than Mars, but its mean mass density is comparable to that of the larger terrestrial planets, Earth and Venus. As shown in the internal cross-sections, Mercury's unexpectedly large mass density is thought to be due to a large dense core extending up to 75 percent of its radius. The Moon's core occupies just 20 percent of its radius. The Earth's inner and outer core take up 55 percent of its radius.

the high temperatures near the young Sun, the heavy iron would have been more likely to accumulate than lightweight elements. The fall-off of temperature with increasing distance from the Sun does neatly explain the broader distinction between the dense terrestrial planets, formed in the realm of high temperatures near the Sun, and the distant giant planets that contain the more volatile, low-density elements that could condense at colder temperatures.

As a third possibility, the young Sun may have been so energetic that it vaporized Mercury's outer crust and mantle and drove them out of surrounding space. There is not enough evidence to decide between these three possibilities.

6.6 A mysterious magnetic field

Mariner 10 carried a sensitive device for detecting magnetic fields, a magnetometer. As the spacecraft moved toward Mercury, the magnetometer plotted the fluctuating magnetic field of the solar wind. However, when it reached the neighborhood of Mercury, it abruptly entered a new environment – a magnetic field that emanated from the planet. The increasing strength of this field as the spacecraft approached the planet indicated that the magnetic field at Mercury's surface would be 0.0033×10^{-4} tesla, or 0.01 times the Earth's surface magnetic field. Such a field was strong enough to carve out of the solar wind an elongated

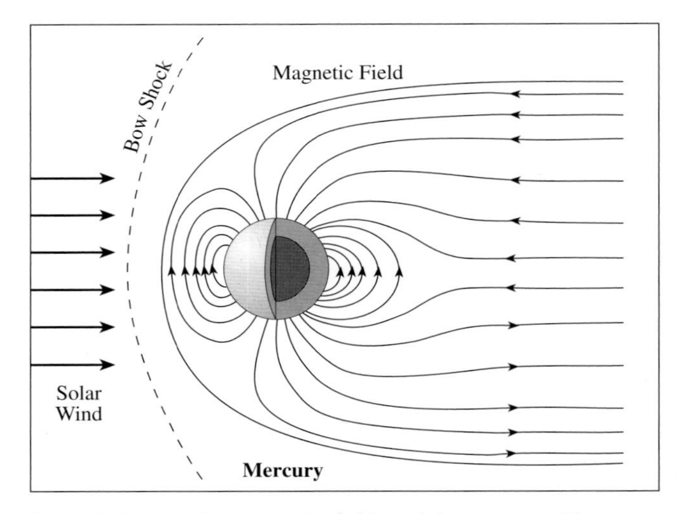

Fig. 6.13 Mercury's magnetic field and large core The magnetic field of Mercury is a miniature version of Earth's magnetic field, complete with bow shock, magnetosphere and magnetotail. Mercury's magnetic axis is closely aligned with its rotation axis, and its polarity is the same as the Earth's, with magnetic south corresponding to geographic north. This magnetic field is probably generated within the planet's large iron core, which takes up three-quarters of the planet's radius, but the exact mechanism for creating the field remains a mystery. The electrically charged solar wind compresses the magnetic field into a bow shock on the sunlit side and draws it out into a tail on the opposite side.

magnetic cavity with a tail, pointing away from the Sun (Fig. 6.13).

Near the planet, the dipole magnetic field has a shape similar to the field of a bar magnet, with the north pole along the rotational axis of Mercury. The planet's field and magnetic cavity, or magnetosphere, is a scaled-down, miniaturized version of the one surrounding Earth (Section 3.6), except that Mercury occupies a larger fraction of its magnetosphere.

Mercury's global magnetism is just strong enough to hold off the Sun's winds, and deflect them at about 1.5 Mercury radii from the planet center, on its sunlit side. But when the Sun is very active, its winds can compress the magnetism all the way down to the surface of Mercury.

In contrast, the remnant magnetic field on the Moon is not global; it occurs in localized concentrations of surface magnetism. Some of these magnetic bubbles are strong enough to deflect the solar wind, but in most places they are so weak that the solar wind impinges on the lunar surface and is absorbed there. The strongest remnant magnetism on the Moon could focus solar-wind gases around the margins of the magnetic bubbles, enabling future mining of solar gases there.

The magnetic fields of Mercury are too weak to maintain belts of trapped particles comparable to the Earth's Van Allen belts. Any charged particles initially trapped in Mercury's field quickly collide with the planetary surface, where they are absorbed.

The discovery of Mercury's magnetic field was completely unexpected and its source is a mystery. Rapid rotation of an electrically conductive, molten iron core, acting as a self-sustaining dynamo, was thought to be a prerequisite for generating a planet's magnetic field. But Mercury spins so slowly, taking 58.6 Earth days to rotate, that it was difficult to imagine how a dynamo could be sustained in its core. Moreover, most scientists thought that the planet's dense iron core should have cooled and solidified eons ago, so no molten metal should now be circulating down there.

The argument for a solid core goes something like this. Small objects like Mercury have a high proportion of surface area to volume; that proportion decreases with increasing size of the object. Since internal heat is generated throughout the volume of a satellite or planet, and radiated to space only from the surface, smaller bodies radiate their energy into space faster and cool more rapidly. Since the Earth and Mercury are both the same age, and the much larger Earth has a solid inner core, retaining only a liquid outer core, the interior of Mercury should have completely solidified over its 4.6-billion-year lifetime.

Nonetheless, Mercury does have a weak, global magnetic field, suggesting that at least some molten material might be circulating inside the planet. The best guess is that a thin, liquid outer core, enriched in sulfur to lower the melting point of iron, is a possibility. Other researchers have proposed that Mercury's magnetic field is a remnant of former times, when most of its core was liquid, and that this field is now frozen into the solid core. Yet another possibility is that the Sun's wind generated Mercury's magnetic field. No one knows for sure what the correct answer is, for we do not yet have enough information.

6.7 Einstein and the anomalous orbital motion of Mercury

For nearly two-and-a-half centuries, the dynamics of the solar system appeared to be explainable by Newton's law of gravitation, which was used to predict the paths of the planets with great precision. But Isaac Newton (1642–1727) himself expressed reservations, anticipating that his theory might not fully explain all the details of planetary motion. Thus, in 1684 he stated that:

> The planets neither move exactly in ellipses nor revolve twice in the same orbit. There are as many orbits of a planet as it has revolutions, as in the motion of the Moon, and the orbit

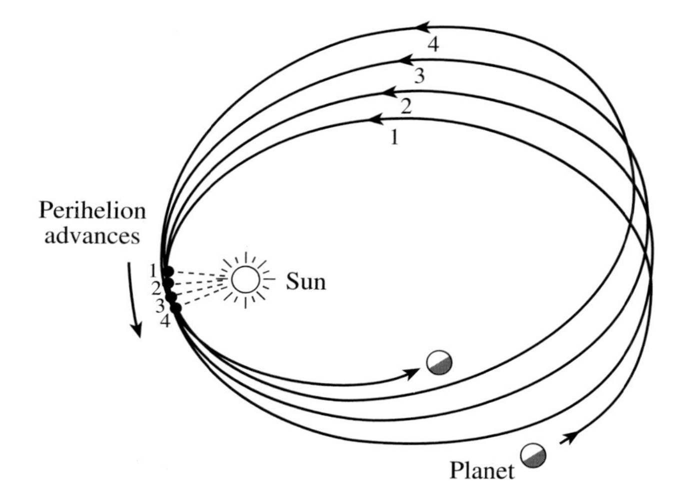

Fig. 6.14 Precession of Mercury's perihelion Instead of always tracing out the same ellipse, the orbit of Mercury pivots around the focus occupied by the Sun. The point of closest approach to the Sun, the perihelion, is slowly rotating ahead of the point predicted by Newton's theory of gravitation. This was at first explained by the gravitational tug of an unknown planet called Vulcan, but we now know that Vulcan does not exist. Mercury's anomalous motion was eventually explained by Einstein's new theory of gravity in which the Sun's curvature of space makes the planet move in a slowly revolving ellipse.

of any one planet depends on the combined motion of all the planets, not to mention the action of all these on each other.

By the end of the 19th century, laborious calculations had been used to describe the gravitational effects of all the planets on each other, but there was still something wrong with Mercury's motion. Newton's theory failed to provide the expected connection between old and new measures of the planet's position, and its trajectory could not be precisely specified.

Instead of returning to its starting point to form a closed ellipse in one orbital period, Mercury moves slightly ahead in a winding path that can be described as a rotating ellipse. As a result, the point of Mercury's closest approach to the Sun, the perihelion, advances by a small amount, 43 seconds of arc per century, beyond that which can be accounted for by planetary perturbations using Newton's law (Fig. 6.14). This unexpected effect is known as the anomalous precession of Mercury's perihelion.

The unexplained twist in Mercury's motion was discovered more than one-and-a-half centuries ago, as the result of watching the planet pass in front of the solar disk every decade or so. This resulted in accurate determinations of the planet's position relative to the Sun. An analysis of these transits led the French mathematician, Urbain Jean Joseph Leverrier (1812–1877), to report the anomalous precession to the French Academy of Sciences in 1849. A decade later,

Table 6.3 Important differences between Newton's and Einstein's theories of gravitation

Newton's Theory	Einstein's Theory
Mass produces a force called gravity	Mass distorts spacetime causing gravity
Gravity acts instantly at all distances	Gravitational effects propagate at the speed of light

he tried to resolve the problem using the same strategy that led to his prediction, in 1846, of Neptune from unexplained motions of Uranus (Section 1.2).

Leverrier attributed the anomalous motion to the gravitational pull of an unknown planet orbiting the Sun inside Mercury's orbit and moving ahead of it. The hypothetical planet was named Vulcan, and extensive searches were conducted for it. No such planet was ever reliably detected. The Newtonian theory of gravity was therefore unable to provide any physical source for the anomalous precession of Mercury, and hence it appeared not to make physical sense within the context of that theory. The cause remained a mystery until 1915 when Albert Einstein (1879–1955) explained it using a new theory of gravity in a paper entitled "Explanation of the Perihelion Motion of Mercury by Means of the General Theory of Relativity".

According to Einstein's theory, known as the General Theory of Relativity, space is distorted and curved in the neighborhood of matter, and the distortion is the cause of gravity. In the absence of matter, space is not distorted and is described by the geometry developed by the ancient mathematician, Euclid (around 300 BC). In the presence of matter, space becomes curved and it must be described by non-Euclidean geometry (Fig. 6.15).

The result is a gravitational effect that departs slightly from Newton's expression near a very massive object, and the planetary orbits are not exactly elliptical. This produces an advance of the perihelion, and the amount predicted by Einstein for Mercury was 43 seconds of arc per century – exactly the observed amount. Because the amount of space curvature produced by the Sun falls off with increasing distance, the perihelion advances for the other planets are much smaller than Mercury's.

Some of the differences between Newton's and Einstein's theories of gravitation are given in Table 6.3. But Einstein's theory is really an extension of Newton's theory and this extension is applicable only in the vicinity of very massive objects like the Sun. For everyday effects on planet Earth, there is no noticeable difference between the two theories.

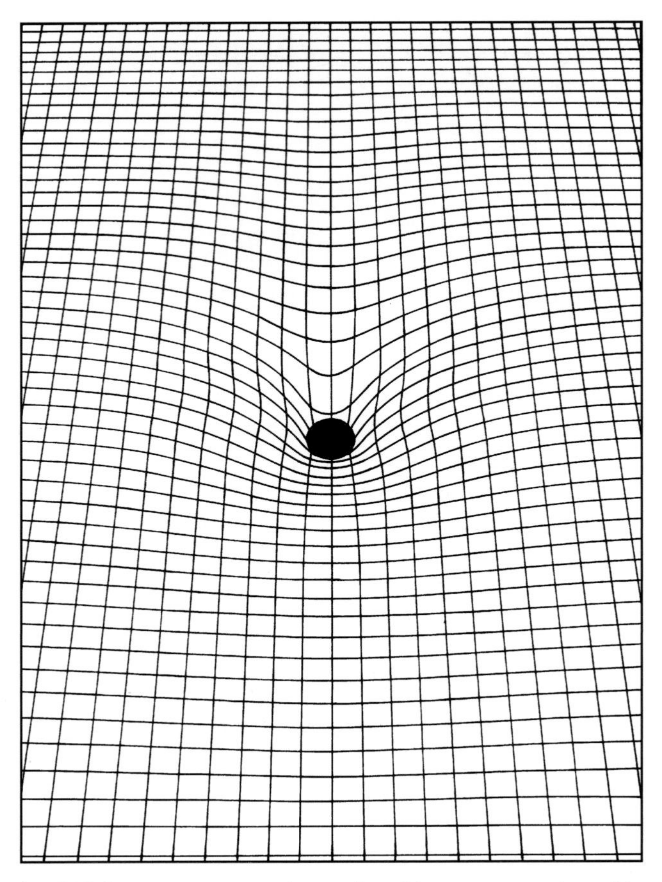

Fig. 6.15 Space curvature A massive object creates a curved indentation upon the flat Euclidean space that describes a world which is without matter. Notice that the amount of space curvature is greatest in the regions near the object, while further away the effect is lessened.

The accord between Einstein's calculations and the observed motion of Mercury depends on the assumption that the Sun is a nearly perfect sphere. If the interior of the Sun is rotating very fast, it will push the equator out further than the poles, so its shape ought to be somewhat oblate rather than perfectly spherical. The gravitational influence of the outward bulge will provide an added twist to Mercury's orbital motion, shifting its orbit around the Sun by an additional amount and lessening the agreement with Einstein's theory of gravity. Fortunately, scientists have used sound waves, detected as pulsations of the Sun's visible disk, to see inside the Sun and show that the internal rotation rate is not fast enough to produce a substantial asymmetry in the Sun's shape. So, we may safely conclude that measurements of Mercury's orbit confirm the predictions of the General Theory of Relativity.

Unlike some of today's theoretical physicists, Einstein realized that his new theory needed to be verified by definitive predictions of other consequences. He noticed that the path of light passing near a massive object will be bent by the curvature of space, and predicted that the apparent positions of stars would therefore be displaced when passing near the Sun. When Einstein's predicted value, of 1.75 seconds of arc for a Sun-grazing light ray, was confirmed by observations of stellar positions during a solar eclipse on 29 May 1919, it brought him international recognition.

The new theory was tested with greater precision by measuring the extra time-delay of radar pulses to Venus when they travel along the curved path near the Sun. The measurement requires an extremely precise clock, for the delay caused by the Sun's curvature of nearby space amounts to only two ten-millionths of the total round-trip travel-time of 1000 seconds.

Einstein's general theory has now been confirmed with a precision of up to one part in a thousand by a variety of observations, so it is widely accepted as a brilliant contribution to our understanding of nature – begun by his attempts to account for Mercury's unexplained motion.

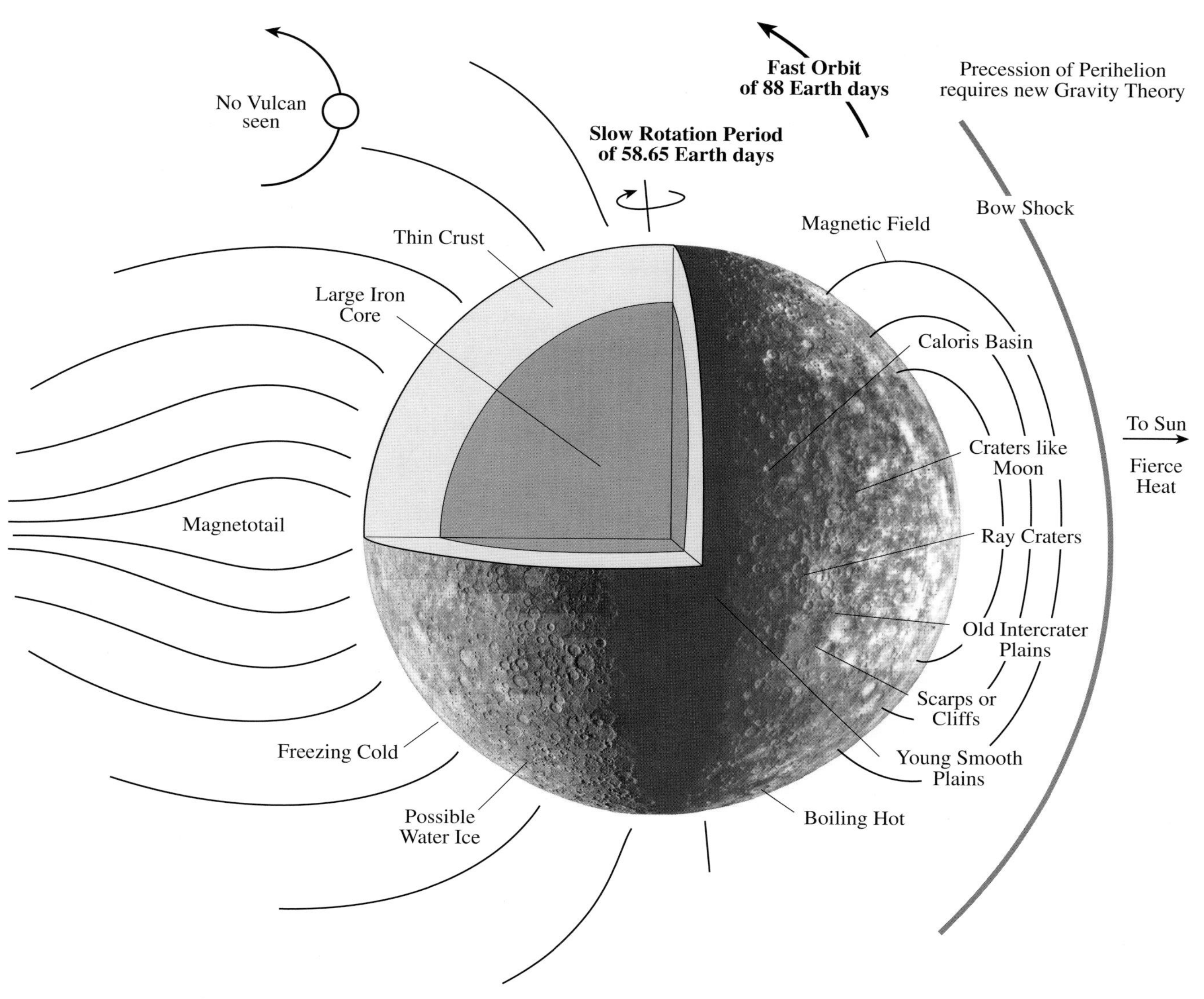

Fig. 6.16 Summary Diagram

7 Venus: the veiled planet

- Venus is the brightest planet in the sky. It orbits the Sun inside Earth's orbit, appearing in the evening or morning hours and never in the middle of the night.
- No human eye has ever gazed on the surface of Venus, which is forever hidden by a thick overcast of impenetrable clouds.
- Venus has a day longer than her year. The planet rotates once every 243 Earth days, in the opposite, retrograde direction from other planets except Uranus, and it takes 224.7 Earth days for Venus to orbit once about the Sun.
- In size, density and composition, Venus is almost identical to the Earth, but space probes and radar have penetrated its clouds to reveal an unearthly surface without a trace of liquid water or life.
- *Venera* spacecraft have dived through the clouds of Venus, surviving long enough to measure the properties of its torrid surface and even photographing it.
- The deadly efficient greenhouse effect of a thick, carbon-dioxide atmosphere has scorched Venus' surface, raising its temperature to 735 kelvin.
- The surface of Venus lies under a crushing atmosphere whose surface pressure is 92 times that on Earth.
- The circulation of the dense atmosphere on Venus evens out temperature variations on the ground, and seasons are absent as well.
- The pale-yellow clouds of Venus are composed of concentrated sulfuric-acid droplets.
- It takes only 4 Earth days for the high-flying clouds to move once about Venus, blown by fierce, rapid winds, but the slow winds near the surface rotate with the planet, once every 243 Earth days.
- There is no detectable magnetic field on Venus, but its dense atmosphere deflects the solar wind.
- The radar instrument aboard the *Magellan* spacecraft spent more than four years mapping out the surface of Venus in unprecedented detail, revealing rugged highlands, smoothed-out plains, volcanoes, and sparse, pristine impact craters.

- Venus is a smorgasbord of volcanism. About 85 percent of its surface is covered by smooth, low-lying plains of lava, and much of the remaining 15 percent is high standing with towering volcanoes.
- Meandering rivers of molten rock have gouged long, narrow channels through the plains of Venus.
- The entire surface of Venus was probably resurfaced by rivers of outpouring lava about 750 million years ago, wiping out all previous craters and about 90 percent of the planet's history; volcanic activity has continued at a reduced level up to the present.
- Tens of thousands of volcanoes now pepper the surface of Venus; some of the towering volcanoes could now be active.
- High volcanic rises on Venus are kept up by active motions below.
- Vertical motions associated with upwelling hot spots have buckled, crumpled, deformed, fractured and stretched the surface of Venus.
- Venus exhibits every type of volcanic edifice known on Earth, and some, called arachnids and coronae, that have never been seen before.
- The surface of Venus moves mostly up and down, rather than sideways.
- Liquid water is non-existent on Venus, and the lack of water could be why Venus does not have moving plates similar to those found on Earth.

7.1 Fundamentals

Venus is the planet most like the Earth in size and mass. Its radius of 6051.8 kilometers is just 5 percent less than the Earth's radius, and its mass is 19 percent less than the Earth's mass. Like the Earth, the planet Venus is a dense, rocky world, one of the four terrestrial planets. Venus spins in the backward direction, and so slowly that its day is longer than its year. The planet's surface lies under a hot and heavy atmosphere, with a high temperature and pressure, but there is no detectable magnetic field on the planet.

Table 7.1 Physical properties of Venus[a]

Mass	48.685×10^{23} kilograms $= 0.815\,M_E$
Mean Radius	6051.84 kilometers $= 0.949\,R_E$
Mean Mass Density	5204 kilograms per cubic meter
Rotation Period	243.025 Earth days, retrograde
Orbital Period	224.7 Earth days
Mean Distance from Sun	$1.081\,57 \times 10^{11}$ meters $= 0.723$ AU
Age	4.6×10^{9} years
Atmosphere	96 percent carbon dioxide, 3.5 percent nitrogen
Surface Pressure	92 bar
Surface Temperature	735 kelvin
Magnetic Field Strength	less than 3×10^{-9} tesla or $10^{-5}\,B_E$

[a] The symbols M_E, R_E, and B_E denote respectively the mass, radius and magnetic field strength of the Earth.

7.2 The veiled planet

Bright torch of heaven

Venus is the most brilliant of the planets; it is the brightest object in the night sky, after the Moon. The stunning beauty of Venus must have been known since the dawn of human history. Our name Friday is, for example, derived from the Anglo-Saxon "Frigadaeg", combining Friga, or Venus, and daeg, or day.

A female association has been common since the beginning of civilization. As Euripides (484–407 BC) put it: "Venus, the eternal sway, all race of men obey". In another example, the Chinese named the planet T'ai-pe – "the Beautiful White One".

The name Venus is that of the ancient Roman goddess of love and beauty; the Greek equivalent was Aphrodite. The Greek's worshipped Aphrodite on the island of Cythera, and therefore the adjective "Cytherean" has often been applied to the planet.

The oldest recorded observations of Venus are those of the Babylonians, who called the planet Ishtar, "the bright torch of heaven – the embodiment of all things womanly, the Mother of the Gods, and the goddess who evoked the power of dawn."

There are other non-female names for the planet. The Judeo–Christian Devil is also known as Lucifer, which was originally a Latin name for Venus as a morning star. For the Mayan civilization, which flourished between 300 and 900 AD, Venus was the Sun's brother, the male God named Kukulkan, who preceded the Sun in rising from the underworld of night. The Mayan astronomer–priests could accurately predict Venus' appearance for over a hundred years, but they also got a bit carried away and made human sacrifices to the planet.

The view from Earth

Venus is visible at the edges of night, lingering near either dawn or dusk. It hangs low and bright in the morning or evening sky, sometimes near the crescent Moon. Because Venus' greatest angular distance from the Sun, known as its maximum elongation, is 47 degrees, it appears as the "evening star" just after sunset or as the "morning star" just before sunrise, but never as both an evening and morning star on the same day.

The brightest planet is never seen in the middle of the night or at midday. In the dark black of midnight, we are on the opposite side of the Earth from the Sun and Venus. At noon the planet is hidden in the full, bright glare of the Sun.

Venus is the second planet from the Sun, the world next door and the planet nearest to us in space. Every 19 months the planet swings to within one hundred times the distance of the Moon.

Venus moves around the Sun once every 224.7 Earth days, like a runner on the inside track, at a mean distance of 0.723 AU. Since the Earth orbits the Sun at 1.00 AU in the same direction as Venus and with a slightly slower rate, it takes about 19 months for Venus to catch up with us. That is, about every 19 months Venus passes between the Sun and us.

During each 19-month circuit, Venus is visible from the Earth for approximately 260 days as an evening star on one side of the Sun, and for about 260 days as a morning star on the other side of the Sun. Between its evening and morning appearances, Venus disappears from view; it then passes between us and the Sun or moves behind our star.

The approximate 260-day length of a Venus appearance in the morning or evening coincides closely with the average length of a human pregnancy. So the cosmos certainly seems to be in tune with female cycles. Most of us are familiar with a woman's monthly lunar cycle, and there is another Sun-related one that is not so well known. The Sun pulsates, moving in and out, every 5 minutes. This is the average length of a woman's contractions during childbirth; at least it was when my kids were born.

When viewed through a telescope, Venus brightens and fades, and also changes in apparent size, during its dance around the Sun (Fig. 7.1). As noticed by Galileo Galilei (1564–1643) in 1610, the planet exhibits a complete sequence of Moon-like phases. Its apparent illumination goes from a full round disk to a narrow crescent and back to rotundity again every 19 months. This was one of the earliest indications that the planets move around the Sun rather than the Earth (Section 1.2). Venus also appears to grow when it approaches us in its orbit and shrinks as it recedes. When Venus is furthest from the Earth, on the opposite side of the Sun, it is fully illuminated and smallest, with a disk that is only 10 seconds of arc across. As the planet comes closer to Earth, it looks partly illuminated and larger, subtending about 64 seconds of arc.

Venus' cloud-laden atmosphere reflects nearly 80 percent of incident sunlight back into space, making it the brightest of planetary worlds. Yet, the clouds that create this luminous beacon also perpetually hide the planet's surface from view. No features can be seen beneath the unbroken layer of clouds by the human eye.

Spectroscopic observations have been more rewarding than casual visual inspection, showing in the 1930s that the planet's upper atmosphere is mainly composed of carbon

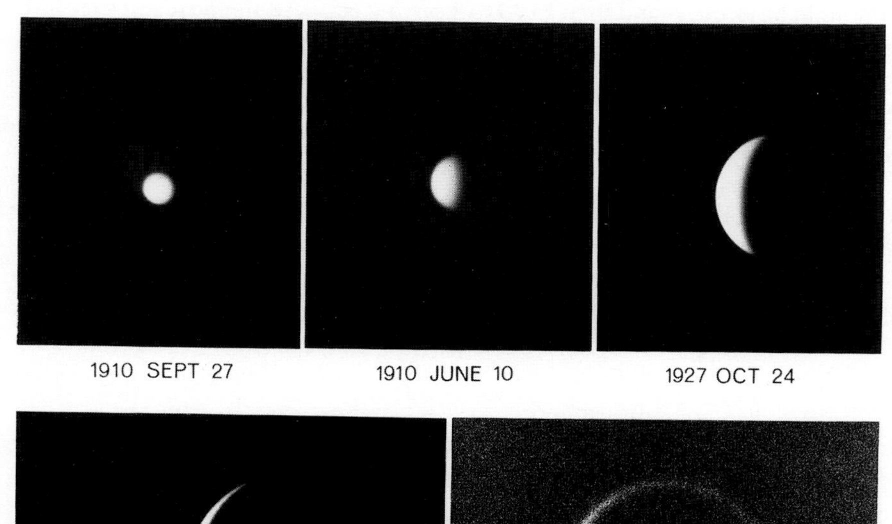

Fig. 7.1 The phases of Venus When fully illuminated, Venus looks small and far away; its apparent size is about seven times larger when the crescents are narrowest and Venus is nearest the Earth. After observing similar phases with his small telescope in 1610, Galileo wrote "Cynthiae figuras aemulatur mater amorum", or "The mother of love (Venus) emulates the figure of Cynthia (Moon)". The phases and variation in the apparent size of Venus provided important early evidence that the planets revolve around the Sun rather than the Earth. (Lowell Observatory photographs.)

dioxide, abbreviated CO_2. We now know that its thick, massive atmosphere is 96 percent CO_2, and that it contains about three hundred thousand times as much CO_2 as is present in our air.

Venus rotates backward, at an exceptionally slow rate

Although no human eye has ever seen the surface of Venus, radio waves can penetrate its obscuring veil of clouds and touch the landscape hidden beneath. By bouncing pulses of radio radiation off the surface, radar astronomers discovered in 1967 that Venus spins in the backward direction, opposite to that of its orbital motion. That is, unlike the other terrestrial planets, Venus does not rotate in the direction in which it orbits the Sun.

The radar observations also showed that Venus spins with a period longer than any other planet, at 243.025 Earth days. This rotation period is even longer than the planet's 224.7 Earth-day period of revolution around the Sun, so the day on Venus is longer than its year. The method of determining this slow rotation is the same Doppler technique used to discover Mercury's slow 58.6-Earth-day rotation period (Section 6.3), and both planet's have probably been slowed down by the Sun.

Tides raised by the Sun in the planet's thick atmosphere may explain why Venus turns very slowly and in the wrong direction. The Sun's gravitational force produces two tidal bulges; the rotating planet drags these bulges along with it, causing them to twist out of alignment with the Sun. As a result, the Sun's gravitational attraction tends to oppose the rotation, slowing the planet down. Friction between the atmospheric tides and the planet's surface also helped to gradually slow Venus' early rotation to the point where it stopped and reversed. Tidal friction in the tides that the Moon produces in the Earth's oceans similarly slows down our planet's rotation (Section 5.8).

7.3 Penetrating the clouds

A hot and heavy atmosphere

In many ways, Venus is the Earth's twin sister, with almost the same weight and waistline. So the feel of gravity on the planet's surface is similar to that on Earth. And because Venus is just a little closer to the Sun than Earth, the climate beneath Venus' clouds was thought to be warm and temperate.

Since no one could see the surface under the clouds, scientists were free to speculate about what might be found there, and fascinating creatures were imagined to flourish in the warm, wet environment. Throughout the 19th and most of the 20th century, our sister planet was pictured with a verdant, swampy surface, perhaps with oceans of water. It might even be teeming with life.

But space-age scientists have drastically altered this romantic vision. Their Earthbound radio telescopes and space probes have penetrated the clouds and uncovered the elusive planet's hidden secrets. They revealed a truly hellish and sterile surface, without a trace of flowing water or a sign of life. Beneath her pure, gleaming clouds, the planet of love is a torrid inferno!

The first hint of the inhospitable environment came in 1958 from ground-based radio astronomers, who measured unexpectedly large amounts of microwave radiation emitted by Venus. If the radiation was coming from the planet's surface, it had a temperature of hundreds of kelvin, hotter than a microwave oven and hot enough to boil away any oceans on Venus.

This explanation was not universally accepted. Some scientists thought the microwave radiation might be coming from high in the atmosphere, where lightning could generate the emission. But then the United States launched the first interplanetary spacecraft, *Mariner 2*, which flew by Venus in 1962. Instruments aboard *Mariner 2* pinpointed the source of the microwave radiation, showing that it comes from the scorching surface of Venus rather than from its atmosphere. As one scientist put it, Venus is no place to raise the kids.

Diving into the inferno

Of course, the only way to be certain about what lies beneath the clouds was to send a space probe through them. After some initial failures, the former Soviet Union mastered the technique. In 1967 their *Venera 4* spacecraft was the first to enter the atmosphere of another planet, make measurements there, and radio home the results. It confirmed that the atmosphere of Venus is about 96 percent carbon dioxide, and recorded increasing temperatures and pressures on its way down, at least until the spacecraft was either burnt up or crushed to pieces.

The day after *Venera 4* made its historic entry into the atmosphere of Venus, *Mariner 5* flew by the planet. Measurements of the way the atmosphere changed *Mariner 5*'s radio signal, when it passed behind the planet and reappeared on the other side, provided a clear profile of the temperature and pressure and confirmed their high values near the ground.

The next two probes, *Venera 5* and *6*, penetrated further, and then in 1970 *Venera 7* survived the heat and pressure of descent to become the first spacecraft to land on the surface of another planet. *Venera 8* repeated this feat in 1972.

Venera 7 and *8* measured the atmosphere's temperature and pressure all the way down to the bottom, showing that the surface temperature is a sizzling 735 kelvin. That is about as hot as a self-cleaning oven, hot enough to boil the ground dry, and to incinerate any humans that might visit the planet. The thick atmosphere traps the Sun's heat, raising the ground temperature by the greenhouse effect to about three times what it would be without an atmosphere.

The massive atmosphere imposes a pressure that is 92 bar, or 92 times the air pressure at sea level on Earth. It would crush you out of existence. The surface pressure is comparable to that experienced by a submarine 500 fathoms, or 1000 meters, below the surface of our terrestrial oceans. So, it's a marvel that *Venera 7* and *8* could withstand the pressure and send back information.

In 1975, *Venera 9* and *10* obtained photographs of the surface, and the Soviets subsequently parachuted seven more entry probes down there, determining among other things the composition of the rocks. Two more landers, *Venera 11* and *12*, descended to the surface in December 1978.

Two American spacecraft, together known as the *Pioneer Venus* mission, arrived at Venus in the same month as *Venera 12*. The *Pioneer Venus Orbiter*, also known as *Pioneer 12*, circled the planet and sent back useful data for 14 years. By sending down radio signals and measuring their echoes, its radar instrument made the first topographic map of the surface, revealing an exceptionally smooth world with just a few high places. Other instruments aboard the *Orbiter* scrutinized the atmosphere, clouds and ionosphere, the electrically charged layer between the atmosphere and outer space.

The *Pioneer Venus Multiprobe*, also called *Pioneer 13*, carried four craft, one large probe and three small ones, that plunged into the atmosphere at both high and low latitudes and on both the daylight and night-time sides, providing a comprehensive picture of the atmospheric structure.

In the late-1960s, the Soviet Union began to hurl spacecraft toward Venus at 19-month intervals, every time the planet in its orbit moved nearest to the Earth (Table 7.2). A spacecraft launched near that time requires the least energy and fuel to reach Venus, taking about four months. The Soviets sent them in steadily increasing numbers for two decades; the American launches were fewer, but they included more technologically sophisticated instruments.

Table 7.2 Some important American (*Mariner*, *Pioneer* and *Magellan*) and Soviet (*Venera* and *Vega*) missions to Venus

Mission	Arrival Date[a]	Accomplishments
Mariner 2	4 Dec. 1962	Flyby of Venus, first successful planetary flight, confirmed intense microwaves from surface of Venus, measured solar wind in interplanetary space.
Venera 4	18 Oct. 1967	First entry probe of another planet's atmosphere, confirmed that carbon dioxide is its main ingredient.
Mariner 5	19 Oct. 1967	Flyby, confirmed high temperature and pressure at surface of Venus.
Venera 5 and *6*	16, 17 May 1969	Penetrated further than *Venera 4*.
Venera 7	15 Dec. 1970	First probe to land on the surface of another planet, measured surface temperature, pressure and radioactive elements.
Venera 8	22 July 1972	Landed on surface.
Venera 9 and *10*	23, 26 Oct. 1975	First photographs of surface.
Pioneer Venus	4 Dec. 1978	First global radar maps of topography, first map of gravity field, five atmospheric probes.
Venera 11 and *12*	21, 25 Dec. 1978	Landers.
Venera 13 and *14*	1, 5 March 1982	Chemical analysis of rocks.
Venera 15 and *16*	10, 14 Oct. 1983	Orbiters, radar images of surface.
Vega 1 and *2*[b]	11, 15 June 1985	Landers, balloon atmosphere probes.
Magellan	10 Aug. 1990	Radar-imaging orbiter, global high-resolution maps of features.

[a] The *Venera* spacecraft were sent to Venus when the planet was near the Earth, taking about four months to get there.
[b] The main *Vega 1* and *2* spacecraft continued on to Comet Halley after dropping probes at Venus.

Clouds of concentrated sulfuric acid

The billowing white clouds on Earth are composed of crystals of water ice, formed when water vapor rises into the cold atmosphere. But because of the high surface temperature on Venus, there can be no liquid water on its surface and its atmosphere is extremely dry compared to the Earth. It possesses only a hundred-thousandth, or 10^{-5}, as much water as the Earth has in its oceans. If all of Venus' water could somehow be condensed onto the surface, it would make a global puddle only a couple of centimeters deep. So there are no water clouds on Venus, and no water rain.

What accounts for the unbroken layer of pale-yellow clouds that covers Venus? A detailed study of the sunlight reflected from the uppermost clouds indicates that the reflecting cloud particles have a spherical shape, implying that the particles are liquid droplets rather than ice crystals. Water and other plausible liquids were ruled out because they have the wrong reflecting and refracting properties.

Baffled astronomers found the answer in the 1970s. A combination of spectroscopy and polarimetry, or how the cloud droplets polarize light, showed that the clouds of Venus are composed of concentrated sulfuric acid! That is the same sulfuric acid that is commonly used in car batteries and contributes to the eye-stinging quality of smog in some industrial cities, especially near smelters.

But where does the acid come from? Instruments aboard *Pioneer Venus* showed that it is derived from gaseous sulfur dioxide and water vapor in the atmosphere. Chemical reactions involving sulfur dioxide, abbreviated SO_2, and water vapor, or H_2O, form the sulfuric acid, denoted H_2SO_4. A similar series of chemical reactions forms terrestrial smog.

When the concentrated sulfuric-acid droplets fall into the warmer atmosphere on Venus, they evaporate, and the gas rises again to the cloud layers. Thus, the acid rain on Venus never reaches the surface, and it is not removed from the atmosphere.

The *Pioneer Venus* investigations also showed that related substances contribute to the greenhouse effect on Venus. Although carbon dioxide is the most significant greenhouse gas on the planet, its action is enhanced by the presence of sulfur dioxide, water vapor, carbon monoxide and cloud particles. This mixture prevents most of the heat radiation from escaping into space, yielding the torrid surface temperatures.

Venus is certainly too hot for liquid water to exist on its surface, but large amounts of water vapor are bound up with the sulfur dioxide in its clouds. Volcanoes probably supply these gases and maintain the clouds. Solar ultraviolet radiation will decompose any water vapor or sulfur-dioxide molecules that rise to high levels in the atmosphere. The hydrogen is continuously lost into space, after water is torn apart. Sulfur and oxygen are too heavy to escape and they react with other atmospheric constituents. The thick clouds that are observed today could only be present if volcanoes supplied their constituents within the past 30 million years. Such volcanoes probably stay active for tens of millions of years, so they may still be active today.

On Earth, water rain efficiently removes sulfur dioxide and other sulfur gases from the atmosphere, so they are only present in very small amounts. Any sulfur gases that are discharged into the air by factories or volcanoes dissolve in our white clouds of water ice, to form droplets of sulfuric acid that are quickly washed out by water rain, sparing us the corrosive acid clouds on Venus. But the sulfuric acid that does rain down to the Earth's surface can damage forests and lakes, so it is still of concern.

Fig. 7.2 Surface rocks on Venus On 5 March 1982 the *Venera 14* lander touched down on Venus at 13 degrees south latitude and 310 degrees east longitude, where it survived for just one hour before succumbing to the planet's heat. That was long enough to radio back these photographs of the surface of Venus, which include part of the lander. The thin, plate-like slabs of rock, that are 0.5 to 1.0 meters across, could be due to molten lava that cooled and cracked. The composition and texture of these rocks is similar to terrestrial basaltic lava. (Courtesy of Iosif Shklovskii.)

Glimpse of a volcanic surface

Venera 9 and *10* touched down in October 1975, surviving long enough to send back one picture each, the first from the surface of any other planet. Six years later, *Venera 13* and *14* transmitted more images, and analyzed the rocks. Since the sulfuric-acid clouds reflect most of the incident sunlight, and absorb almost all of the rest of it, Soviet astronomers once thought that there might not be any light at the surface. They therefore equipped their earlier spacecraft with floodlights to illuminate the scene when they reached the ground, but it turned out that there was enough natural light to take the historic pictures.

The surface of Venus is always bathed in a diffuse light under a heavy overcast. So the landers found that daylight on Venus resembles a dark, smoggy day in Los Angeles or Mexico City. The pictures are sometimes colored orange, for this is the main color to make it through the thick clouds.

Some scientists thought that the high temperatures and pressures would melt, flatten and chemically weather the surface into a featureless plain. However, the surface photographs showed fresh-appearing rock without eroded edges (Fig. 7.2).

This suggested that active volcanic processes might be replacing the old eroded surface with fresh, young material. Moreover, the slab-like appearance of many of the rocks might be attributed to flowing lava. As the molten lava spread across the surface and cooled, it would produce the thin, fractured layers of rock that we see in the surface photographs.

Analysis of the surface rocks showed that they have a composition and mass density that resemble basalt, the type of dark, fine-grained lava that lines the Earth's ocean floors and fills the Moon's maria basins. The basalt rocks were identified at several landing sites on Venus, suggesting that much of the planet is encrusted by lava, covering its original surface. The basaltic crust was most likely extracted from a differentiated interior, when molten rock rose up through the crust, forming volcanoes that once poured lava over much of the planet's surface.

Circulation of the atmosphere

At ground level, the heavy atmosphere is sluggish and turgid, hardly moving at all. The speeds measured from the *Venera* landers and the *Pioneer Venus* probes are between 0.3 and 1.0 meters per second, or about the walking speed of a tired old man. A fast wind would have a tremendous force in the dense atmosphere, disrupting the surface. Because of the slow winds, the bottom of the atmosphere moves along with the surface, completing one circuit every 243 Earth days.

The wind speed increases with altitude, rising to about 100 meters per second in the clouds at about 60 kilometers in height (Fig. 7.3). The high-flying clouds race around the planet once every four Earth days, from east to west in the same backward direction that the planet rotates (Section 3.1). So, the top of the atmosphere is blown around Venus more than 50 times faster than the planet rotates; such a rapid motion is sometimes called super-rotation. These high-speed zonal (east–west) winds are partly driven by the rotation of the solid planet beneath them, but the exact mechanism for maintaining the flow is not well understood.

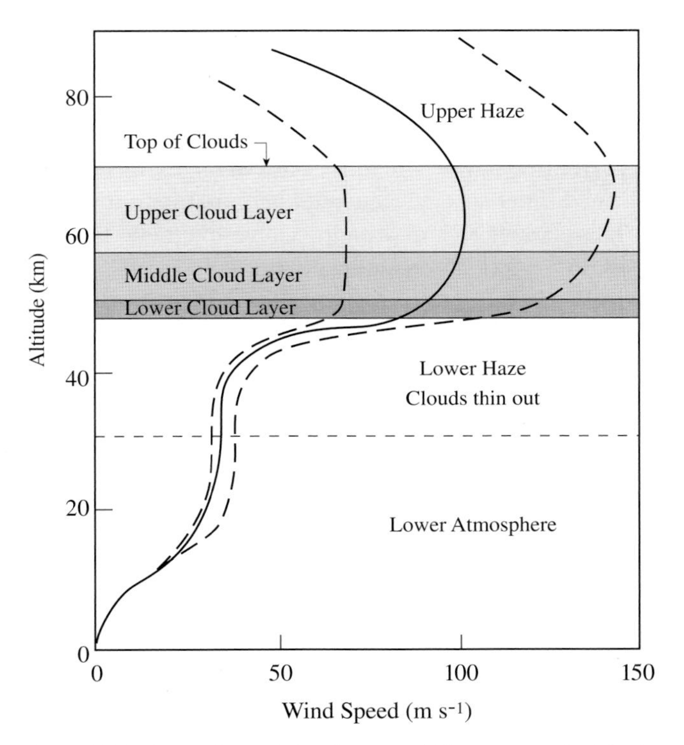

Fig. 7.3 Wind speeds and cloud layers of Venus At all altitudes in the thick atmosphere of Venus, the dominant winds blow westward with a speed that increases with height. From a gentle breeze near the ground, the wind speed increases to 100 meters per second at great heights. Space probes that penetrated the planet's clouds and balloons that floated in its atmosphere have been able to detect three distinct layers in the opaque sulfurous clouds. The top layer contains small droplets of sulfuric acid; the middle layer contains larger but fewer particles. The bottom layer is the densest and contains the largest particles; it is comparable to bad city smog in visibility. Beneath the lowest layer, the atmosphere is hot enough to vaporize all particles, so it is relatively clear down there.

The atmosphere and winds have transformed the impact craters on Venus, which are unlike those seen on any other world. The dense atmosphere affects the impact debris, changing it into fluid-like flows (Section 2.2), and the material ejected during impact is moved by the winds. Some fresh craters are surrounded by radar-bright haloes, streamlined hoods and tail-like wind streaks that act like wind vanes, pointing downwind at the time of impact (Fig. 7.4). The wind streaks indicate that the winds just above the surface were blowing toward the equator from the northern and southern hemisphere.

The atmosphere redistributes heat from one part of Venus to another, thereby moderating temperature differences. Most of the sunlight falling on Venus is either reflected by the clouds or absorbed in them. And because the Sun's rays fall directly on the equator and obliquely at the poles, the equatorial clouds are initially warmer than the polar ones. But this temperature difference generates winds that transfer heat in a single large Hadley cell (Fig. 7.5), named after George Hadley (1685–1768) who first proposed such a circulation for the Earth's atmosphere.

As on the Earth, the warm air rises at the equator to the cloud tops, where winds propel it toward both poles. After warming the poles, the circulating atmosphere sinks and flows back toward the equator at lower levels near the base of the clouds, completing the cell. The stronger zonal (east to west) circulation on Venus combines with this weaker Hadley (north–south) circulation, giving rise to a wind vortex that carries the clouds in a slow spiral toward the poles.

The dense lower atmosphere transports heat so efficiently from one part of the globe to another that the ground temperature varies by no more than a few degrees between the equator and poles or from day to night, so there is no place to escape the heat.

The changing distribution of sunlight on the Earth causes stormy weather and produces the changing winds that propel clouds across the globe. In contrast, Venus has a steady atmospheric circulation pattern almost devoid of weather. Since the planet's orbit is nearly a perfect circle and its rotation axis is hardly tipped at all, there are no noticeable seasons on Venus. Still, controversial evidence for localized thunderstorms was provided when some spacecraft detected radio signals that might be associated with lightning.

Missing magnetism

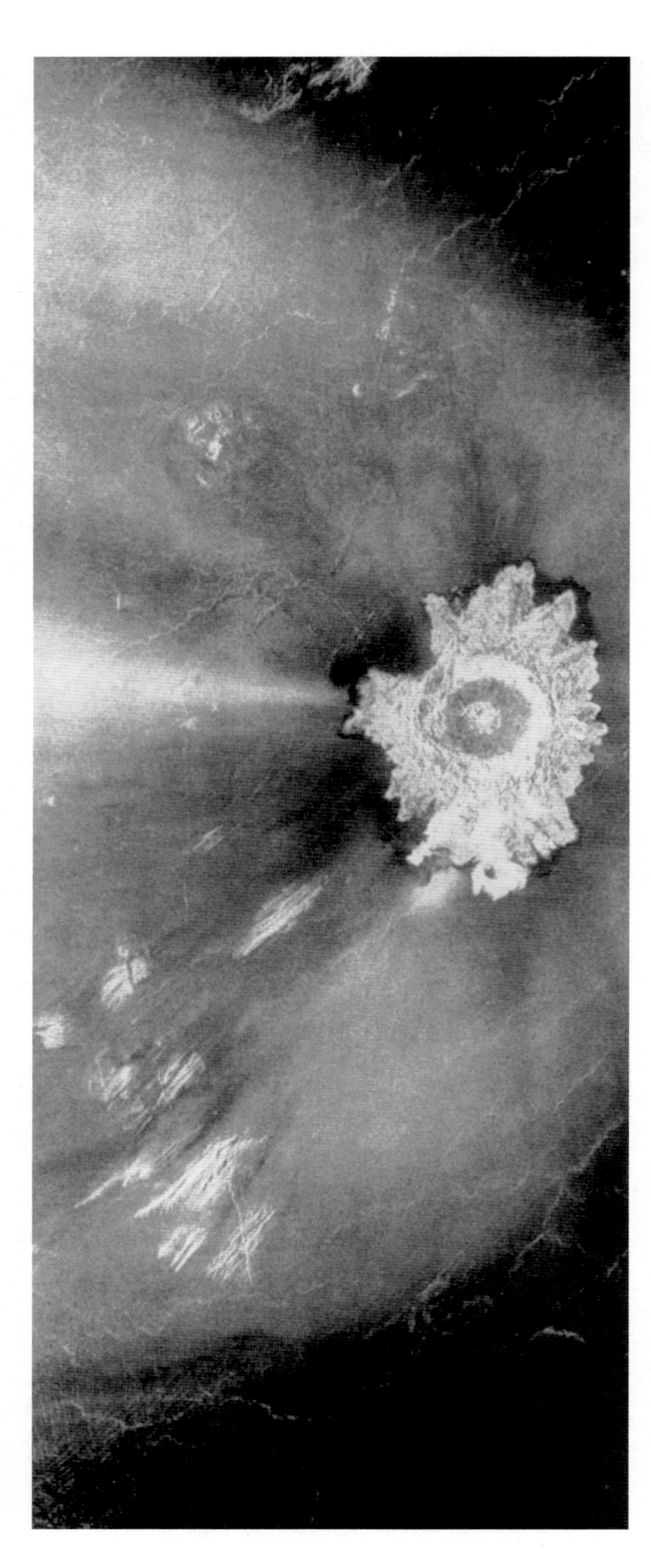

Unlike the Earth and most other planets, Venus has no significant global magnetic field, for reasons not fully understood. Measurements from *Pioneer Venus Orbiter* showed that if Venus has any magnetic field its strength is less than one-hundred-thousandth, or 10^{-5}, that of Earth, or less than 3×10^{-9} tesla. Lacking a magnetic field, Venus has no belts of trapped particles such as the Van Allen belts near Earth.

The weakness of this magnetic field is rather surprising because Venus and the Earth have a similar size and mass, and they might be expected to have similar interiors. The Earth has a molten outer core and a solid inner one, and by analogy Venus ought to possess them. However, Venus does not show the magnetic field that would be produced by currents within a molten outer core.

Contrary to popular belief, the slow rotation of Venus is not responsible for the lack of a significant magnetic field, but something in its core is out of whack. Some investigators argue that the core is now completely solid, while others suggest that core solidification has not yet commenced. In either case, Venus would lack both a molten outer core for currents to flow in and the heat given off by the freezing of a solid inner core which would help maintain the flow.

The planet's lack of an appreciable magnetic field exposes the upper atmosphere to the continuous hail of charged particles from the Sun. The intrinsic magnetic fields of the Earth and other planets fend off and deflect this electrically charged solar wind (Section 3.6). Nonetheless, the solar wind is prevented from reaching the surface by Venus' dense atmosphere and ionosphere, which create an obstacle scarcely bigger than the planet itself.

Energetic ultraviolet sunlight ionizes some of the atoms and molecules in the outer atmosphere above the clouds, forming an electrically charged layer similar to the Earth's ionosphere, and this layer helps shield the ground from the solar wind. The ions provide conduction paths for electrical currents that produce forces that counter the wind. As a result, the solar wind slows down and is deflected around the

Fig. 7.4 Winds blow crater ejecta A radar image of Adivar Crater exhibits a bright, jet-like streak (*middle left*) that extends over the surrounding plains. This unusual feature, seen only on Venus, is believed to result from the interaction of ejected debris and high-speed winds in the upper atmosphere, blowing from the east (*right*). The crater is 30 kilometers in diameter, and is named for the Turkish educator and author Halide Adivar (1883–1964); it is located at 9 degrees north latitude and 76 degrees east longitude, just north of the western Aphrodite highland. (Courtesy of JPL and NASA.)

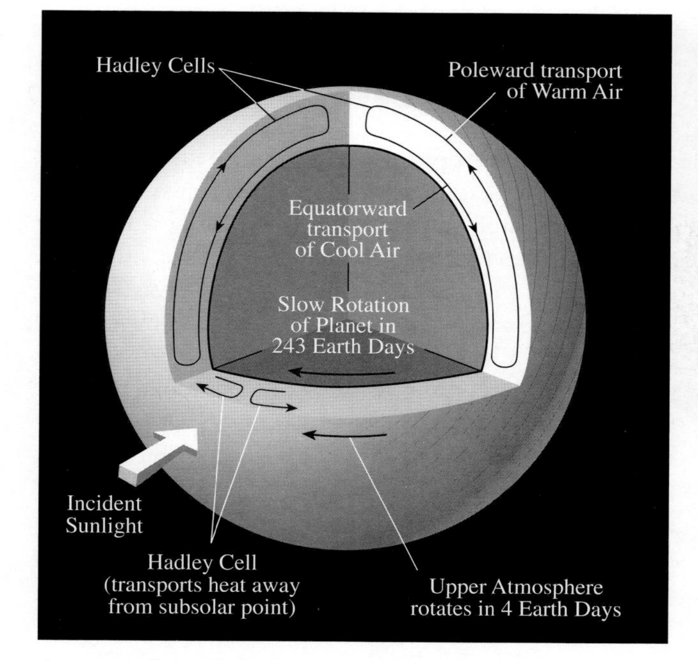

Fig. 7.5 Hadley cells in the atmosphere of Venus Incident solar energy drives a Hadley-cell circulation of the atmosphere, which keeps heat from building up at any one location. Air rises in warm regions near the equator on the sunlit side of Venus, where the planet is heated most by the Sun. Some of the warm air flows towards cooler zones near the poles, and sinks and returns to warm equatorial regions again. Strong winds also blow around the planet in the direction that Venus rotates, from east to west.

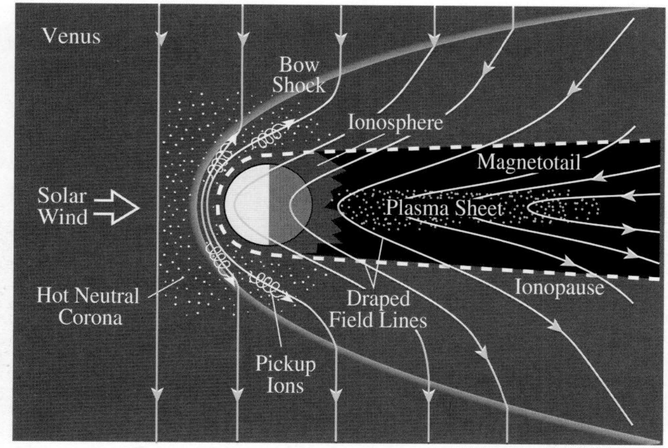

Fig. 7.6 Interaction with the solar wind Even though Venus has no appreciable magnetic field, the solar wind is prevented from reaching the surface by Venus' dense atmosphere and by electrical currents induced in its conducting ionosphere. The planet has a well-developed bow shock, but it does not have belts of trapped particles. Both the ionosphere and the extended corona of hot gas, derived from the upper atmosphere, help to divert the solar wind.

planet in a bow shock, and the interplanetary magnetic field is draped back to form a magnetotail (Fig. 7.6). The *Pioneer Venus Orbiter* found that the solar wind's interaction with Venus changes on times-scales of hours to years, depending on the vagaries of the wind, with a bow shock that expands and contracts in step with the 11-year cycle of solar magnetic activity.

7.4 Unveiling Venus with radar

Magellan and its predecessors

No human eye has ever gazed on the surface of Venus; it can only be sensed by radio transmissions. Radar, an acronym for radio detection and ranging, uses its own source of radio radiation, and does not need sunlight to probe the planet, gathering data day or night. Only radar is capable of piercing the thick clouds of sulfuric acid that blanket Venus.

The planet's topography is inferred from the length of time it takes for a radar pulse to reach a particular part of the surface and return an echo. The surface roughness is determined from the strength of the echo. Rough surfaces and slopes tilting toward the radar reflect more energy and return a stronger echo, thus appearing bright in a radar image. Surfaces that are smooth or tilt away from the incoming radio signal send less energy back, and appear dark in a radar image, somewhat like a wet road seen in the headlights of a car at night.

The surface features on Venus have been probed with radar systems of increasingly greater resolution over the decades. Blurred images were first obtained from powerful, ground-based radio telescopes, such as the one in Arecibo, Puerto Rico. The *Pioneer Venus Orbiter* then zoomed in to take a closer look in the 1970s, acquiring a global map of Venus at a coarse resolution of about 100 kilometers. This was followed by the twin spacecraft *Venera 15* and *16* whose radar instruments mapped much of the planet's northern hemisphere in the mid-1980s, resolving features as small as 1 kilometer across.

The surface of Venus was next scanned by a radar system aboard *Magellan*, for more than four years in the early-1990s. It mapped details as small as 120 meters across, producing the most complete global view available for any planet, including Earth. The spacecraft is named after the Portuguese explorer Ferdinand Magellan (1480–1521) whose expedition first circumnavigated Earth.

Magellan's radar images revealed a rich and varied landscape with stunning and unprecedented clarity, describing a surface whose nature and history turned out to be quite different from those of the Earth. The surface of Venus is covered by massive, global outpourings of lava, punctuated by unique volcanic constructs never seen before, scarred by sparse impact craters surrounded by beautiful outflows,

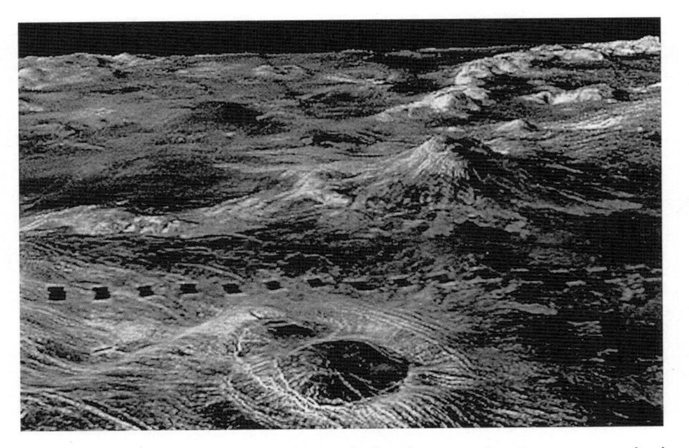

Fig. 7.7 Surface terrain The side-looking radar images and altimetry data from *Magellan* have been merged to obtain this three-dimensional view of the terrain on Venus. It includes radar-dark, lowland plains and radar-bright highlands. A hot upwelling of magma is believed to have formed the 200-kilometer-wide Nagavonyi Corona in the foreground; just behind the corona is a shield volcano that is 2 kilometers high. Nagavonyi is a Ganda (Uganda) crop goddess; the distant highlands include Phoebe Regio, a region named for the Greek Titaness. (Courtesy of JPL and NASA.)

and fractured, stretched, crumpled and split open by upwelling magma (Fig. 7.7). Even hardened professional astronomers were inspired with a sense of wonder at these discoveries.

Perhaps the most stirring aspect of the *Magellan* images is the fresh, pristine nature of the features they reveal. Although no one expected to see significant evidence of erosion and weathering on the dry planet, observers were struck by the detailed sharpness of its craters, fractured plains, volcanoes and crumpled landmasses. Most of these surface features have been preserved for hundreds of million of years. Everything is totally exposed and largely preserved – at least between periods of intense volcanic activity or internal upheaval.

As Aladdin said from his flying carpet, it's:

A whole new world.
A dazzling place, I never knew.
But when I am way up here,
It's crystal clear

Magellan had one instrument, a Synthetic Aperture Radar, abbreviated SAR, capable of imaging the surface, mapping its topography as an altimeter, and measuring electromagnetic radiation from the surface. Most typical radar systems send out one pulsed radio signal at a time, and process each echo by itself before sending out another pulse. In contrast, *Magellan* sent out several thousand radar pulses each second, and its SAR used fast computers to accumulate multiple echoes received from many locations simultaneously. The combined data simulated a large antenna and provided the superb resolution and fine detail.

The SAR used pulses of microwave radiation, finely tuned to a wavelength of 0.126 meters, to penetrate the clouds and illuminate a long, narrow strip down and off to the side of the spacecraft. The pulsed signals bounced off the planet's surface, and both the return wavelengths and the round-trip travel-times to receive the echoes were recorded and relayed back to Earth, resulting in an image of a swath of the terrain. As the spacecraft looped around the poles of Venus, the slowly rotating planet turned beneath it, exposing the entire globe to scrutiny during each 243-day rotation. Every second, the computers took in 36 million bits of data and relayed them back to Earth. Over four years, they obtained more than a million billion bits of information, more than had been obtained from all previous lunar and planetary spacecraft combined.

To complement the wide-angle, side-looking antenna, there was a second, narrow-beam antenna that looked straight down and used a pulsed signal and its echo to measure the topography. Elevations were measured with an accuracy of 30 meters with a surface resolution of about 5 kilometers.

A smoothed-out world

Radar data from the *Pioneer Venus Orbiter* and *Magellan* showed that Venus is an extraordinarily smooth world, largely at one low level and quite different from the Earth (Fig. 7.8). About 85 percent of the surface lies within one kilometer of the average planetary radius, 6051.84 kilometers. A coating of lava has smoothed these vast low-lying plains. Without its water, the topography of the Earth occurs at two distinct elevations, which correspond to the continents and ocean floors.

The lowest point of Venus is about 6048.0 kilometers from its center and the highest point at 6062.57 kilometers. Thus the variation of topography on Venus is almost 14.6 kilometers. The surface temperature and pressure vary with height, between 653 and 766 kelvin and between 45 and 119 bar at the highest and lowest elevations.

Although most of Venus' terrain consists of smooth, low-lying, volcanic plains, about 15 percent of the planet's surface consists of highlands that tower above the plains, rising an average 4 to 5 kilometers above the mean planetary radius. There are two large-scale elevated regions that punctuate the smoothed-out surface; they are Aphrodite Terra in the equatorial region and Ishtar Terra in the far north (Fig. 7.9).

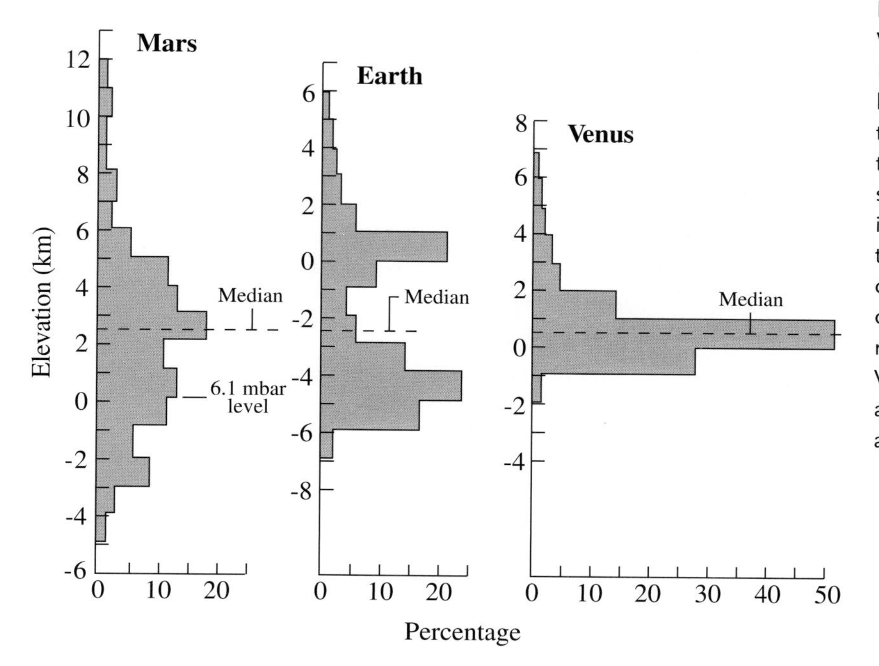

Fig. 7.8 Surface elevations for Earth, Venus and Mars Most places on Earth (*center*) stand near one of two prevailing levels, the high-standing continents or the low-lying ocean floors. In contrast, the surface of Venus (*right*) is unusually smooth and flat. A small percentage of its terrain consists of elevated highlands that are comparable in height to many continents on Earth. The surface features on Mars (*left*) are spread over a broader range of elevation than most of those on Venus; but the Martian surface elevations are not double-peaked as those on Earth are.

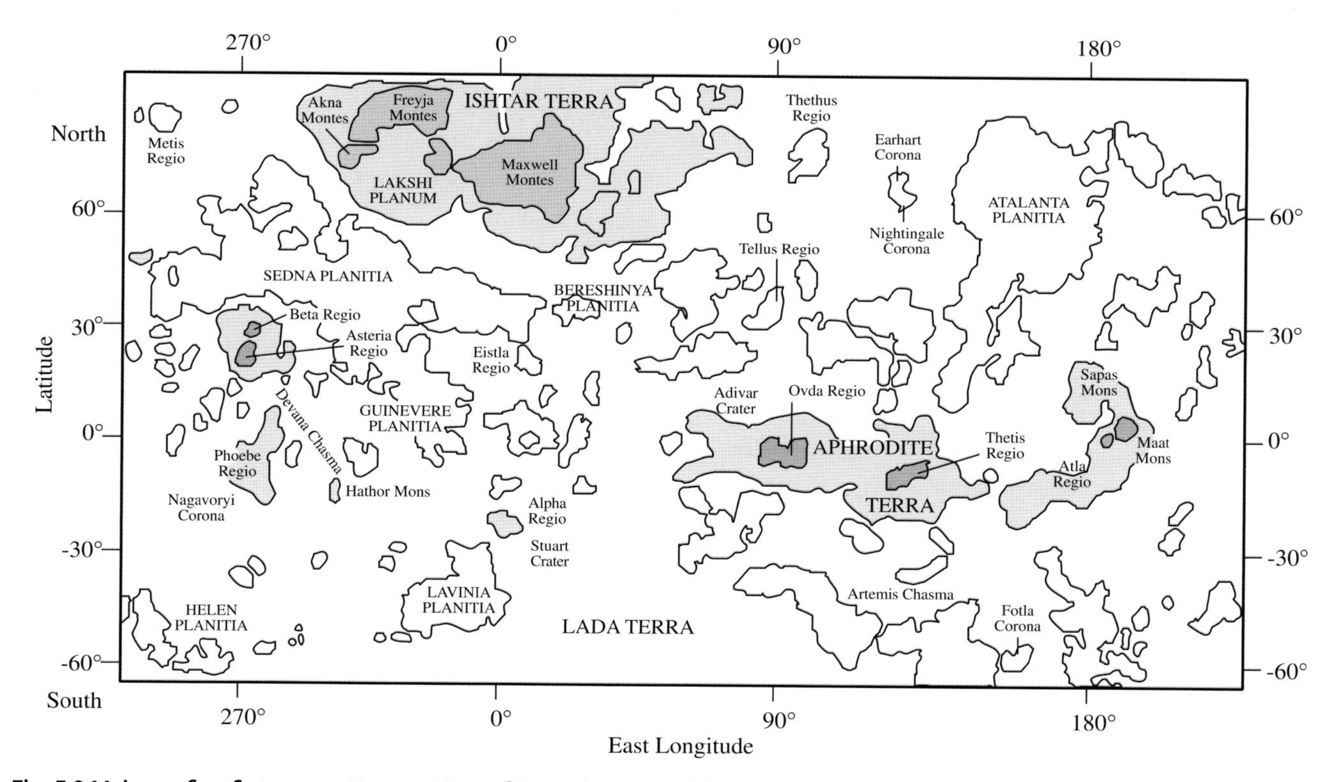

Fig. 7.9 Major surface features on Venus Most of Venus lies at roughly the same radius, in the lowland, volcanic plains (*white areas*). The highland massifs (*darker areas*) include Aphrodite Terra, an elongated feature extending along the equator, centered at about 105 degrees east longitude. Many of the elevated regions near the equator mark the sites of volcanism, such as Maat Mons (Fig. 2.27) and Sapas Mons (see Figs. 7.14, 7.15). The other main elevated region is Ishtar Terra in the north, centered at about 0 degrees longitude (*top*). Ishtar Terra is roughly the size of Australia; the elevated plateau in its western part, Lakshmi Planum, is bounded on three sides by mountains (see Fig. 7.16), including Maxwell Montes. This mountain's 11-kilometer height above the average radius exceeds by two kilometers the height of Mount Everest above sea level.

Table 7.3 Common features on the surface of Venus and the category of women used to identify them

Descriptor Term (singular, plural)	Feature Type	Category of Identifying Female
Chasma, Chasmate	Canyon, steep-walled trough	Goddesses of hunt or the Moon
Corona, Coronae	Oval-shaped feature	Fertility and Earth goddesses
Crater, Craters	Circular depression	Famous women
Dorsum, Dorsa	Ridge	Goddesses of the sky
Linea, Lineae	Elongate marking	Goddesses of war
Mons, Montes	Mountain	Miscellaneous goddesses[a]
Patera, Paterae	Irregular crater, shallow	Famous women
Planitia, Planitiae	Low plain, level surface	Mythological heroines
Planum, Plana	Plateau or high plain	Goddesses of prosperity
Regio, Regiones	Large area, broad region	Giantesses and Titanesses[b]
Rupes, Rupes	Cliffs or scarps	Goddesses of hearth and home
Terra, Terrae	Highland, extensive landmass	Goddesses of love
Tessera, Tessarae	Tile-like, polygonal terrain	Goddesses of fate or fortune

[a] The one exception is Maxwell named after James Clerk Maxwell.
[b] The two exceptions are Alpha Regio and Beta Regio.

Aphrodite Terra is over 10 000 kilometers long and covers a quarter of the planet's circumference at the equator. It contains tall volcanoes, long lava flows and deep faults and fractures. Western Aphrodite is built from the massifs Ovda and Thetis Regiones; the eastern part of Aphrodite is occupied by Atla Regio. Ishtar Terra is 5600 kilometers across. It consists of an elevated plateau encircled by narrow mountain belts (Fig. 7.9).

Each feature on Venus is described (Table 7.3) by two names – a woman's name that identifies it and a feature designation (Focus 7.1). As examples, Aphrodite and Ishtar are respectively the Greek and Babylonian goddess of love, and a terra is an extensive landmass. The name of a mythological goddess or a famous mortal woman identifies every feature on Venus. The only exceptions on Venus are Maxwell Montes, named in honor of the British physicist James Clerk Maxwell, whose 19th-century theories of electromagnetism made radar possible today, and Alpha Regio and Beta Regio, using the first two letters of the Greek alphabet. They were first seen in Arecibo radar images in the 1960s, before the female naming convention.

7.5 Volcanic plains

Planet-wide covering of lava

Venus has been extraordinarily volcanic. Extensive lava flows emerge from cracks in the crust and towering volcanoes, and rivers of lava snake their way through solid rock. Outpouring lava has covered much of the surface of the planet, and tens of thousands of small volcanoes are now found all across its face. Venus exhibits every type of volcanic edifice known on Earth, and some that have never been seen before.

The spreading lava has flooded and filled the low-lying regions of Venus, creating extensive smooth plains that cover about 85 percent of Venus' surface. The volcanic nature of these lowland plains, each designated by the term planitia, is demonstrated in the *Magellan* images. You can practically see the molten rock spreading like heavy cream across these plains, often running for hundreds of thousands of meters down gentle slopes. The magma has risen from within canyons as the crust pulled apart, cooling and solidifying into lava flows that look like frozen river currents (Fig. 7.10).

In other places, the molten material has burned paths in the pre-existing lava deposits, following a narrow, sinuous smoothly curving course. They can meander for millions of meters across the planet's surface (Fig. 7.11). Many end in outflows that look like river deltas. These river-like channels were formed not by water, but by lava that was hot enough to carve through solid rock, remaining hot and liquid over distances that are longer than the Nile, the longest river on Earth. The high surface temperature on Venus probably kept the lava liquid, and prevented the cooling flow front from damming up the molten rock behind it.

Focus 7.1 Naming features on Venus

There are two names that describe every feature on Venus. They are a particular name that identifies it, plus a descriptive name that says what it looks like. As indicated in Table 7.3, each class of feature has a specific type of associated identifying female, either real or mythological.

The identifying names can be suggested by anyone, but eventually they have to be approved by the International Astronomical Union, or IAU for short. The lists of approved names for the planets and their satellites are fascinating, and you can review them in the IAU's *Gazetteer of Planetary Nomenclature* located on the web at: *http://planetarynames.wr.usgs.gov/*

Craters larger than 20 kilometers in diameter are identified by the last name of a famous woman, who has to be dead at least three years at the time of naming. Some of my favorites include the singer Billie Holiday, the anthropologist Margaret Mead, the writer Flannery O'Conner, the ballet dancer Anna Pavlova, the singer Edith Piaf, and the abolitionist Harriet Tubman. Smaller craters are given common female first names, as are hurricanes on Earth.

The names for the major topographic features on Venus and surface features shown in other parts of this chapter are:

Advidar Crater	Halide, Turkish educator, author (1883–1964)
Akna Montes	Yucatan goddess of birth
Alpha Regio	First letter in Greek alphabet
Aphrodite Terra	Greek goddess of love
Artemis Chasma	Greek goddess of hunt and Moon
Asteria Regio	Greek Titaness
Atla Regio	Norse giantess, mother of Heimdall
Atlanta Planitia	Greek huntress associated with golden apples
Bereghiny Planitia	Slavic water spirit
Beta Regio	Second letter in Greek alphabet
Danu Montes	Celtic mother of God
Devana Chasma	Czech goddess of hunt
Eistla Regio	Norse giantess
Fotla Corona	Celtic fertility goddess
Freyja Montes	Norse mother of Odin, Nordic goddess of fertility and love, who weeps tears of gold as she seeks her lost husband
Guinevere Planitia	British queen of king Arthur
Helen Planitia	Greek, "the face that launched 1000 ships"
Ishtar Terra	Babylonian goddess of love
Lada Terra	Slavic goddess of love
Lakshmi Planum	Indian goddess of prosperity, consort of Lord Krishna
Lavinia Planitia	Roman, wife of Aeneas
Maat Mons	Ancient Egyptian goddess of truth and justice
Maxwell Montes	James C. (1831–1879), British physicist famous for his theory of electromagnetism
Metis Regio	Greek Titaness
Nagavonyi Corona	Ganda (Uganda) crop goddess
Ovda Regio	A violent, ill-tempered spirit who wandered the Finnish forest naked, looking for trespassers to tickle to death
Phoebe Regio	Greek Titaness
Sapas Mons	Phoenician goddess
Sedna Planitia	Eskimo, her fingers became seals and whales
Sif Mons	Scandinavian grain goddess whose long golden hair is the autumn grass, and lover of Thor, the god of thunder
Stuart Crater	Mary, Queen of Scots (1542–1587)
Tellus Regio	Greek Titaness, Roman goddess of the Earth
Tethus Regio	Greek Titaness
Thetis Regio	Greek Titaness

The meanings for the surface features shown in other figures of this chapter are also given in the figure captions.

A relatively young surface

Venus, like all planets, has been subjected to a continual rain of meteoritic bombardment over the eons. The plains of Venus are uniformly peppered with impact craters (Fig. 7.12), the scars of this bombardment, though nowhere near as liberally as on the surfaces of the Moon, Mars, and Mercury. On Venus the craters of a given size are far fewer in number and more widely spaced than on the Moon. At one time Venus was probably as heavily pock-marked with large craters as the Moon's ancient surface is, but the scarcity of the craters now on Venus indicates that the surface we now see is much younger than the lunar surface.

We can estimate when the lava flowed by counting the number of craters of a given size on the plains and

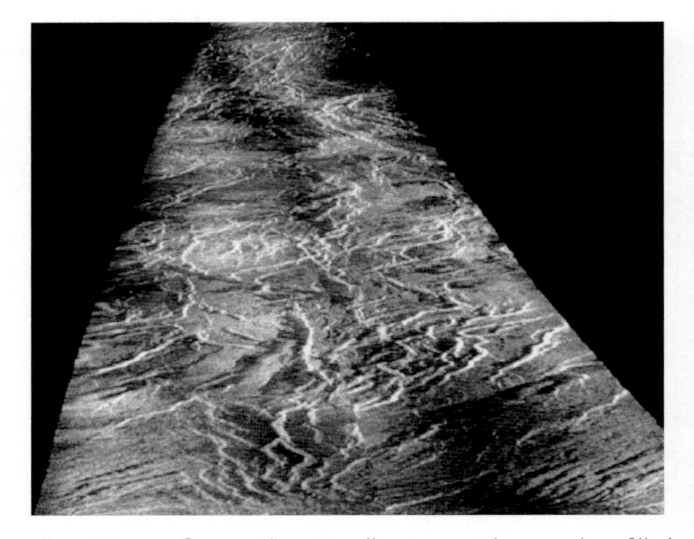

Fig. 7.10 Lava-flows This *Magellan* image shows a lava-filled canyon produced when Venus' crust was pulled apart and magma rose within the gap. The lowland plains have many similar canyon systems, typically about 10 kilometers wide and up to 1000 kilometers long, apparently containing solidified lava-flow. This region, located near the equator between Asteria Regio and Phoebe Regio, is about 10 kilometers wide and stretches 600 kilometers to the north (*top*). (Courtesy of JPL and NASA.)

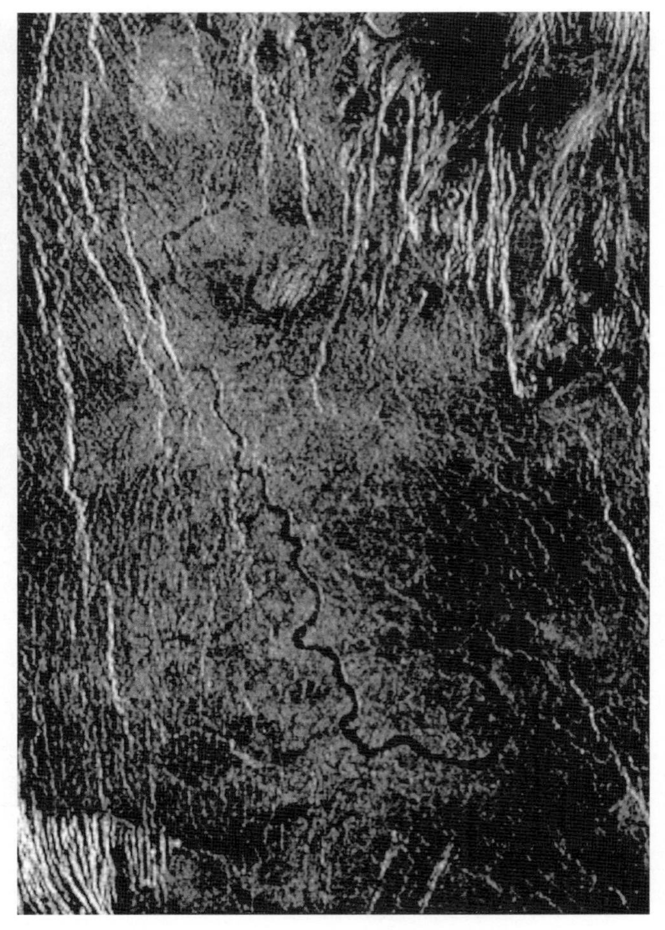

Fig. 7.11 Sinuous channel This segment of a sinuous lava channel in Lavinia Planitia is approximately 200 kilometers long and just 2 kilometers wide. Such channel-like features are common in the lowland plains of Venus. They meander across the plains, like terrestrial rivers, but apparently formed by fluid lava that melts its way across the surface. The very flat surface of Venus may explain why the channels remain within narrow boundaries without forming any lava lakes or tributaries. (Courtesy of JPL and NASA.)

comparing it to the number on the Moon – ignoring the smallest ones that are not found on Venus because the incoming projectiles burned up in the thick atmosphere. The Moon's cratering record tells us the number of impact craters that should appear in a given time.

Magellan has logged about 1000 impact craters, which when compared to the lunar record indicates an average surface age of about 750 million years. At that time molten rock emerged from inside the planet, spreading across the surface, eradicating all previous craters, and wiping out all traces of the first 90 percent of Venus' history. That is a relatively young age, only about 10 percent of the age of the solar system and of the planet Venus itself. No matter how you look at it, the surface of Venus is practically new born compared with the Moon's surface which still bears the scars of a heavy bombardment about 3.9 billion years ago.

Theories for the volcanic makeover

Everyone agrees that the smooth plains covering most of Venus came from volcanic floods, emanating from the planet's interior, but the experts disagree over when and how it occurred. According to one "global catastrophe hypothesis", planet-wide volcanism wiped the face of Venus clean about 750 million years ago, resurfacing the entire globe and drowning any existing craters in a flood of lava.

There are two equally likely catastrophe interpretations that cannot be distinguished. One is a single resurfacing at about 750 million years ago. The second is that there was continuous resurfacing of the planet over most of its earlier history, and that this resurfacing slowed down sharply at about 750 million years ago. In either interpretation, the planet switched over to a low rate of localized volcanism about 750 million years ago, but it has not disappeared.

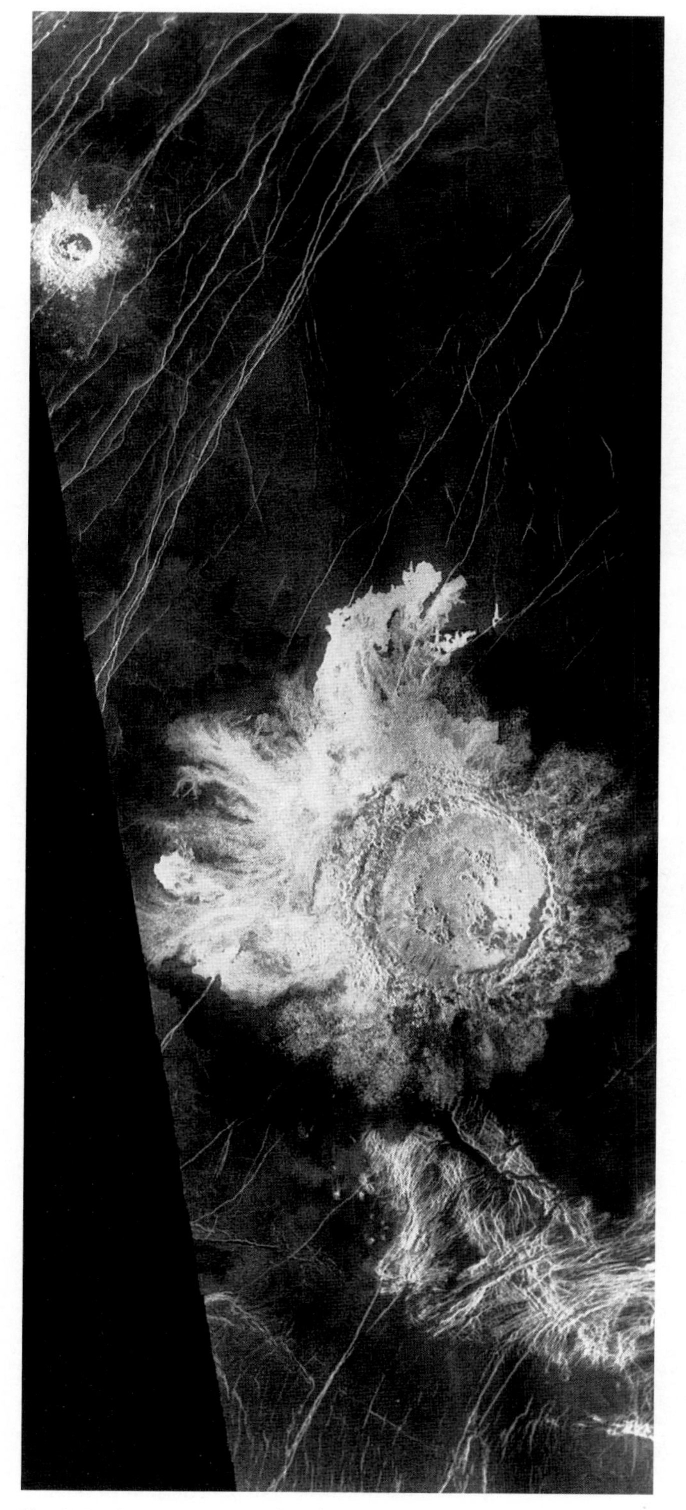

Fig. 7.12 Stuart Crater This beautiful crater exhibits asymmetric radar-bright ejecta attributed to an oblique impact and interaction with the dense, thick atmosphere (Section 2.2). Its rim, which is 67 kilometers in diameter, encircles a bright floor that may have unique physical properties. This impact crater's name honors Mary Stuart, Queen of Scots (1542–1587). (Courtesy of NASA.)

Most of the craters on Venus have lava on their floors, and some of them show exterior volcanism affecting their ejecta or breaking their rims, so there has been at least a modest level of ongoing volcanism.

A different model that has observational geological support is a series of discrete volcanic plain-forming events extending over a considerable period of geological time, with the volume of lava declining from one event to the next. In this view, volcanism is an ongoing process, occurring at different places and times, further in the past and in the future from 750 million years ago. In fact, volcanic uplifting and rifting suggest that volcanism on Venus may extend up to the present.

The intense volcanism that repaved Venus may have transformed its climate and roasted its atmosphere. When massive volcanoes erupted about 750 million years ago, they should have ejected a lot of greenhouse gases into the atmosphere in a relatively short time. Calculations indicate that the volcanic release of carbon dioxide, water vapor and sulfur dioxide might have raised the temperatures well in excess of the "present inferno". Still, this volcanism is not the cause of the present intense greenhouse environment on Venus, which probably existed long before the volcanic resurfacing.

7.6 Highland massifs

Towering volcanoes

Fifteen percent of the Venus surface comprises highlands, where the largest volcanoes are found, concentrated on or near the equatorial highlands (Fig. 7.13). Large-scale plumes of rising magma have probably pushed up this globe-encircling, elevated region from below. When the molten rock pierced the surface, lava flowed out to form towering volcanoes that are perched atop the raised highlands. Some of the volcanoes are found in Beta Regio on the western side of the equatorial highlands. Several rise out of Atla Regio in the eastern end of Aphrodite Terra, including Maat Mons and Sapas Mons (Fig. 7.14).

One of the highest volcanoes, Maat Mons, rises 9 kilometers above the surface, and spreads 200 kilometers across it (see Fig. 2.27). Sapas Mons is shorter and broader (Fig. 7.15). Both peaks are known as shield volcanoes, because they have the shape of a shield or an inverted plate. Similar giant shield volcanoes are found in the Hawaiian Islands, each with a broad base and gentle slopes.

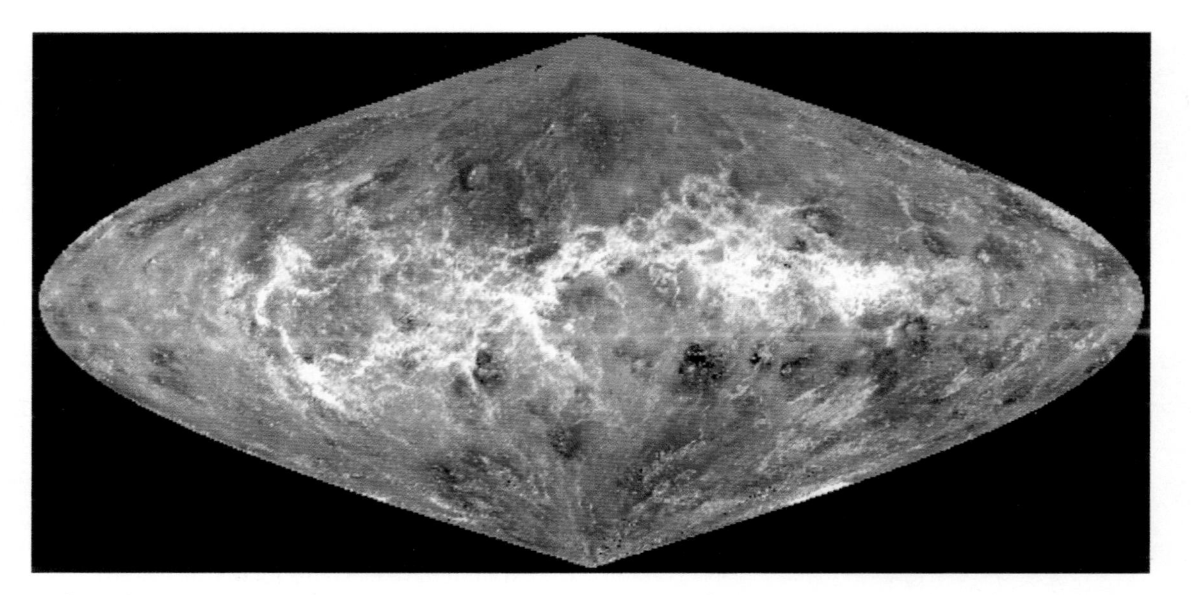

Fig. 7.13 Surface of Venus A radar system on board the *Magellan* spacecraft has mapped the surface of Venus to a resolution of 120 meters, producing this global view of the planet's terrain. A vast equatorial system of highlands and ridges runs from the continent-like feature Aphrodite Terra (*left of center*) through the bright highland Atla Regio (*just right of center*) to Beta Regio (*far right and north*). This image is centered at 180 degrees longitude, and drawn using a sinusoidal projection that does not distort the area at different latitudes. A more traditional projection of the *Magellan* global data was shown in Fig. 2.12. Dark areas correspond to terrain that is smooth at the scale of the radar wavelength of 0.126 meters; bright areas are rough. (Courtesy of Mark A. Bullock, JPL, and NASA.)

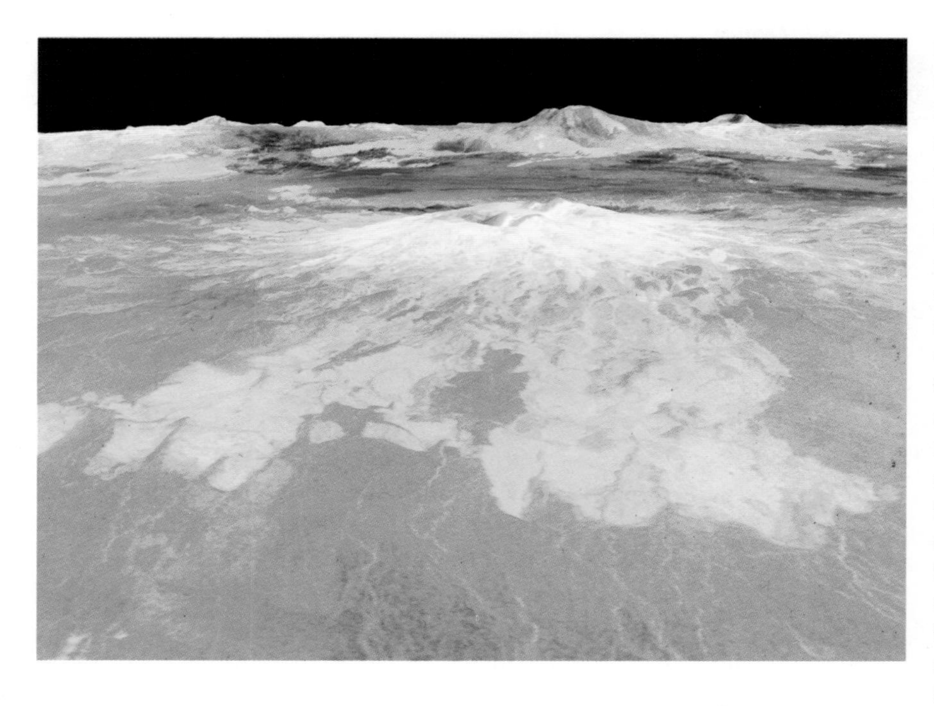

Fig. 7.14 Volcanoes on Venus Radar-reflectivity and topographic altitudes from *Magellan* have been combined to construct this three-dimensional perspective view of the surface of Venus. Lava-flows extend for hundreds of thousands of meters in the foreground at the base of the volcano Sapas Mons (*center*). Another volcano, Maat Mons, appears on the horizon; a close-up view of Maat Mons was shown in Fig. 2.27. The volcanic origin of these peaks is suggested by the fiery orange and yellow colors, but there can be no fires on Venus without oxygen. The artificial tints are based on color images taken by the *Venera 13* and *14* landers, simulating the color of sunlight that filters through the thick atmosphere. Sapas is named after a Phoenician goddess, Maat is the name of an ancient Egyptian goddess of truth and justice, and *Mons* is the Latin term for "mountain". (Courtesy of JPL and NASA.)

Scientists suspect that some volcanoes on Venus may still be active. Because sulfur dioxide and water vapor are continuously lost, the thick clouds of sulfuric acid must be maintained by ongoing volcanic release of these gases (Section 7.3). Variations in the observed amounts of atmospheric sulfur dioxide have been attributed to volcanoes that have been belching forth the gas. In addition, some of the volcanic peaks in the highlands reflect radar pulses with unexpected intensity, perhaps related to recent volcanic activity.

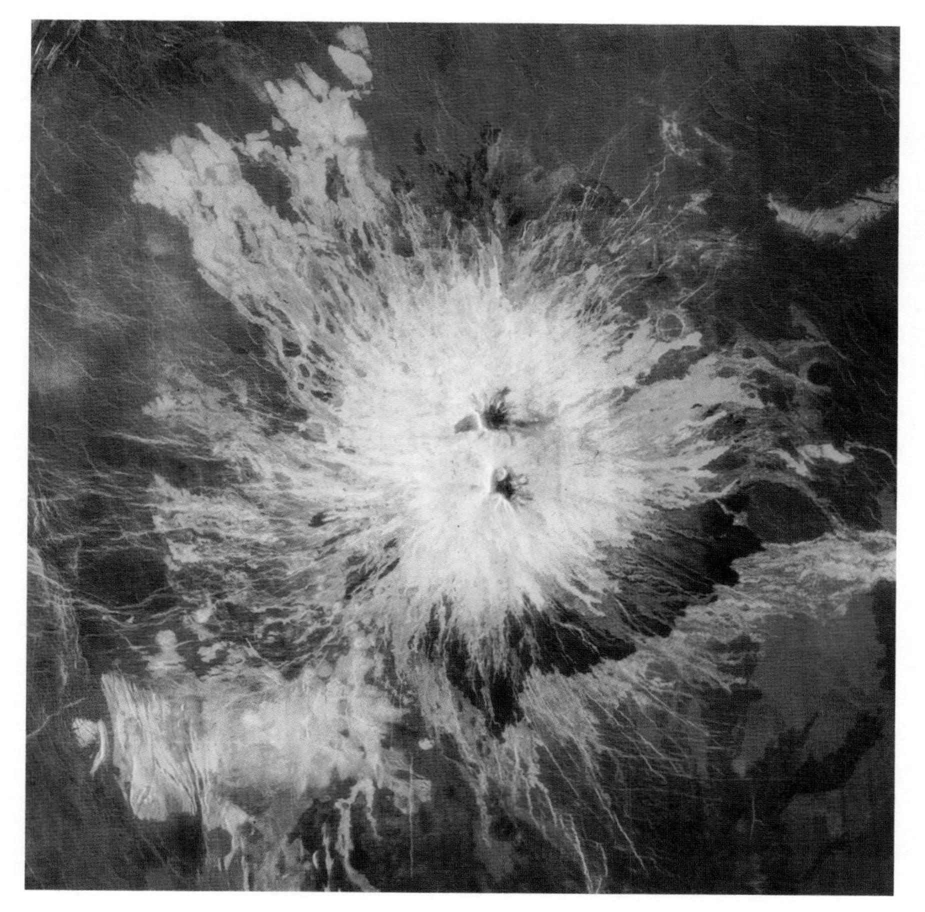

Fig. 7.15 Sapas Mons Located in an equatorial highland called Atla Regio, the volcano Sapas Mons is about 400 kilometers in diameter and only 1.5 kilometers high; it is named after a Phoenician goddess. The sides of the volcano are composed of numerous overlapping lava-flows from flank eruptions similar to terrestrial volcanoes such as the Hawaiian shield volcanoes. The summit contains two smooth, radar-dark mesas, as well as groups of pits, some as large as one kilometer across. They may have formed when underground chambers of magma were drained, causing the surface to collapse above them. (Courtesy of NASA.)

Mountain ranges

The region of Venus that most closely resembles terrestrial mountains is Ishtar Terra, located at far northern latitudes. It consists of an elevated plateau, Lakshmi Planum, which is bounded on three sides by mountain belts – the Danu, Akna and Freya Montes (Fig. 7.16). Lakshmi just drops off into the surrounding plains on the fourth side, forming an immense cliff. The belts of mountains, with their banded ridges and narrow valleys, resemble mountain ranges on Earth, and the loftiest peak, Maxwell Montes, rises to Himalayan altitudes, standing over 11 kilometers above the surrounding terrain.

The raised plateau and surrounding mountains closely resemble the Tibetan Plateau and the Himalayan Mountains on Earth, which were produced by the collision of India with Asia. However, since Venus has no colliding continents, the intensely deformed mountains must have been pushed up into the sky by a different process. One possible explanation suggests that the material beneath the mountains was cooler than surrounding areas and sunk. The crust would then be pulled together, bunching up into mountains and compressing the surface features. Or perhaps that part of the crust was compressed and folded after the surrounding plains were formed, like a carpet pushed against a wall. In either case, Ishtar Terra is a unique type of raised feature on Venus, and different processes hold up other highland regions.

Gravity's highs and lows

How can the high places on Venus stay up when its rock temperatures are halfway to their melting point? Something has to be holding the mountains up. So, spacecraft measured local variations in gravity and used them to look underneath the highlands and see what is there.

The movements of the orbiting spacecraft were tracked by recording small changes in the wavelength of its radio signal, and the corresponding small changes in the orbital speed, inferred from the Doppler effect, were used to specify local variations in the gravitation produced by the underlying mass and density.

It turned out that some of the highest places on Venus exert the strongest gravity, a correlation first noticed by tracking *Pioneer Venus Orbiter* and confirmed by sending

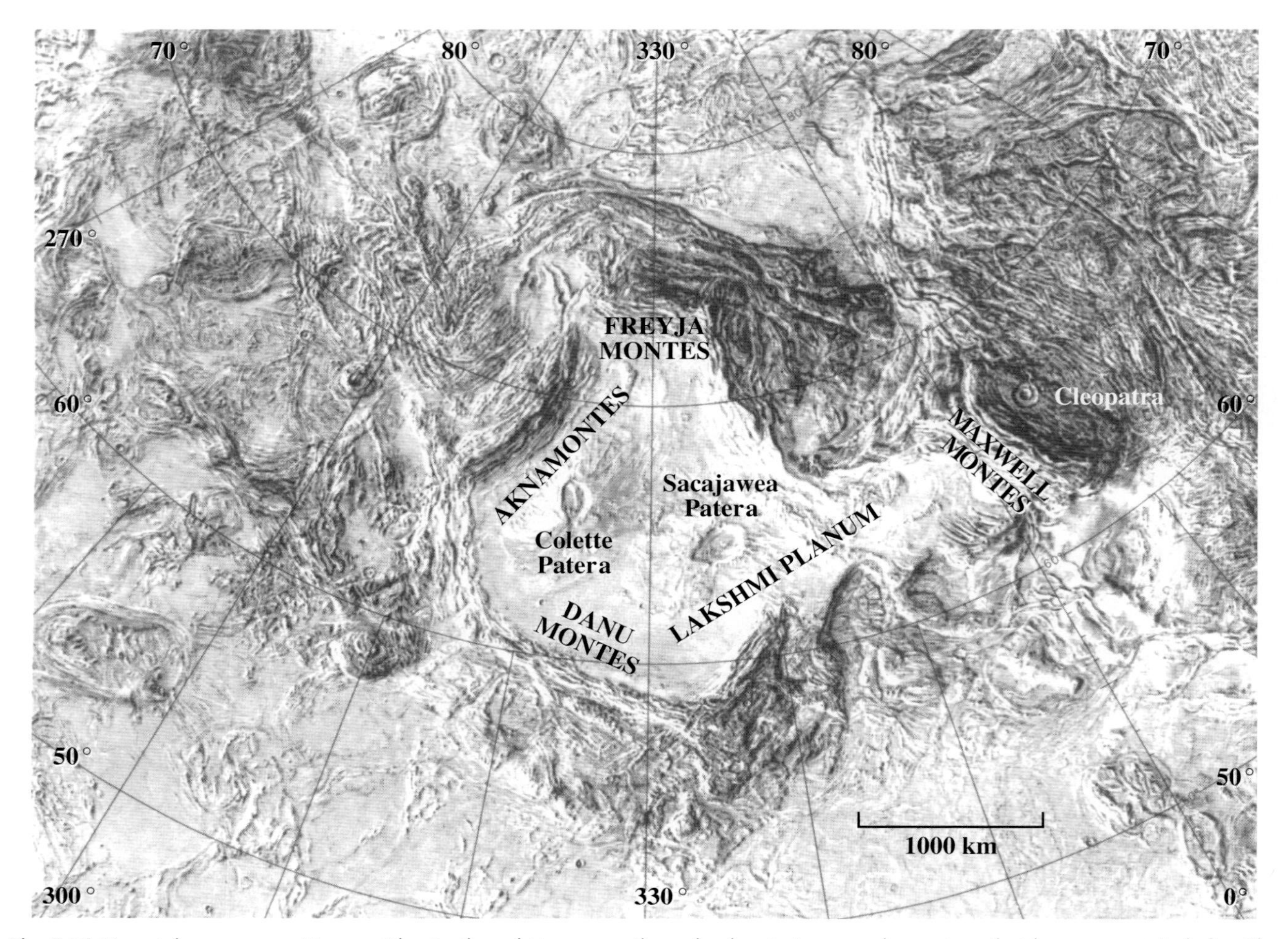

Fig. 7.16 Mountain ranges on Venus The Earth and Venus are the only planets in our solar system that have mountain belts. The highland massif of Ishtar Terra in the northern hemisphere of Venus includes a huge plateau, named Lakshmi Planum. It is about the size of Africa, rises 3.5 kilometers above the surrounding terrain, and is bordered by the Akna, Danu, Freyja and Maxwell mountains, each about 1000 kilometers in extent. Maxwell Montes stands 11 kilometers above the mean radius. Akna and Freyja are the respective names of the Yucatan goddess of birth and the Norse mother of Odin. Danu is the Celtic mother of God. Maxwell is named after the British physicist James Clerk Maxwell, while Cleopatra is the Egyptian queen who had affairs with Julius Caesar and Mark Anthony. The Blackfoot Indian woman, Sacajawea, guided the Lewis and Clark expedition to the Pacific Northwest, and Claudine Collette was a French novelist. (Adapted from a U.S. Geological Survey map using Arecibo, *Pioneer Venus* and *Venera 15* and *16* radar data.)

Magellan into a low-altitude orbit. These volcanic rises, such as Atla Regio and Beta Regio, are probably held up by active motions below, fed by rising plumes of sluggish, upwardly buoyant flow. The hot, low-density material wells up from deep within the planet, with glacial slowness in spite of its heat, eventually spreading out beneath the surface, and pushing the volcanic rises up. As the plumes near the surface, the lower pressures induce partial melting of the plume that sometimes punches holes in the crust, providing a conduit that permits lava to flow out from volcanoes and fissures in the planet.

The volcanic rises on Venus are thought to be very similar in origin to "hot spots" on Earth, such as the Hawaiian Islands. Other highland regions on Venus, such as the crustal plateaus Ovda Regio and Thetis Regio, are more akin to continental mountains on the Earth. They have low-density roots that extend down under the high-standing regions and balance the mass excess. The combination is in "isostatic" equilibrium, like an iceberg floating on the ocean.

7.7 Tectonics on Venus

Trapped heat deforms the surface

All planets have heat to get rid of. Gravity compressed and heated the planet's interiors when they formed, and colliding bodies also brought heat to them in their early stages. Then the decay of radioactive elements added more internal heat. It is still being generated within Venus by radioactive decay, producing molten rock inside the planet.

Size is the main factor in determining how much heat remains inside a planet or satellite, for larger bodies lose heat more slowly and remain internally active for longer times. Thus, the relatively small Moon has not been volcanically active for 2 or 3 billion years, even though lava flowed into its impact basins to form the lunar maria before then. The larger, rocky planet Venus must have had vast churning reservoirs of hot material beneath its crust for a much longer period, and probably still does.

The heat trapped inside a planet wants to get out, so it rises to crack and deform the planet's surface or spills out in volcanoes. In technical terms, the molten rock becomes swollen by heat and lower in density, rising through the cooler, overlying high-density material and carrying heat upward, like the convective bubbles in a pot of boiling water. The upward-flowing material wants to crack open or puncture a hole in the overlying material, breaking on through to the other side to release all that heat. The crustal deformations caused by the pent-up heat, and the internal changes affecting them, are known as *tectonics*, the Greek word for "carpenter or building".

The hot rising material has buckled, fractured and stretched the crust on Venus, like a crumpled piece of paper or a face seamed and thickened by age (Fig. 7.17). It has split the crust open and spread it apart, forming rift valleys with steep sides and sunken floors. Some of them are found in Alta region of Venus alongside its volcanoes (Figs. 7.18, 7.19, and 7.20). The linear rift zones in the equatorial highlands can extend for thousands of kilometers, but are cracked apart by just a few kilometers. In contrast, rifts that split open the Earth's continents can, because of moving plates, widen up to make way for its biggest oceans (Section 4.3).

When a bubble of hot material rises to just below the surface, it presses against the crust, causing the ground to bulge and crack. Circular and radial fractures are created around the edges of the rising dome, forming a network of radar-bright features reminiscent of a spider's web (Fig. 7.21). Some of them have therefore been nicknamed *arachnids,* from the Greek and modern Latin words for "spider". The term *coronae,* the Latin word for "crown" is used for the larger, elevated, circular structures that are also pushed up from below by rising molten rock trying to get out. Both arachnids and coronae are unique to Venus and have not been found on any other planet.

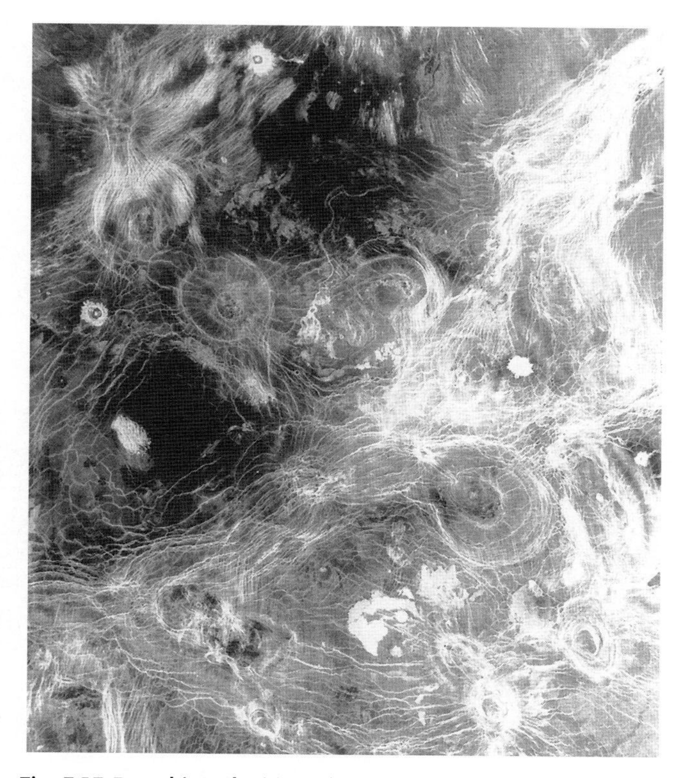

Fig. 7.17 Bereghiny Planitia This high-resolution *Magellan* mosaic shows an area in the low-lying plain Bereghinya Planitia, named for the Slavic water spirit. The image is 1.84 thousand kilometers wide, and centered at 45 degrees north latitude and 11 degrees east longitude. Its fractured surface is attributed to plumes of magma that rise from inside Venus, pushing against the planet's crust. The most prominent features are the circular and oval structures that are sometimes called arachnids for the web-like appearance of their fractures (see Fig. 7.21). Also visible are lava-flows, impact craters, and volcanic domes. (Courtesy of NASA.)

Coronae have concentric ridges and fractures that are hundreds of kilometers across, and large volcanic outpourings have occurred within them (Fig. 7.22). When enough lava spills out into a corona, the upwelling subsides and it is no longer supported from below. The bulge will deflate and buckle the surrounding terrain, producing an annulus of ridges and troughs that often surrounds coronae, like the moat around a castle. Or else, the magma cools and retreats as it ages and the molten rock drains back down the vent from whence it came. Then the dome will collapse like a giant fallen soufflé, creating ring-like fractures and a crumpled, cracked surface.

The increasing pressure of the upwelling magma can stretch the planet's skin until it bursts, like the broken cheese bubbles in a pizza or a split in an overcooked hotdog. Small volcanic domes, known as pancakes, are sometimes formed when pasty, sluggish lava breaks through and flows

Fig. 7.18 Atla Regio Fractures, seen as bright, thin lines, criss-cross the volcanic deposits in part of the Atla region of Venus. The fractures are not buried by the lava, indicating that the convulsive tectonic activity post-dates most of the volcanic activity. This *Magellan* radar image is approximately 350 kilometers across, and centered at 9 degrees south latitude and 199 degrees east longitude. Several circular volcanoes, surrounded by radar-bright lobes, are also present. This region is named Atla after a Norse giantess, mother of Heimdall. (Courtesy of JPL and NASA.)

Fig. 7.19 Alpha Regio This *Magellan* image mosaic reveals the radar-bright, highly fractured southeastern portion of Alpha Regio (*left*), forming an intricate pattern of ridges and valleys. The complexly deformed, high-standing terrain is called a tesserae. The image, which is approximately 600 kilometers across, also shows dark, smooth volcanic planes that border the eastern edge of Alpha Regio. Within these dark plains are eight rounded, pancake domes up to 35 kilometers across and up to 1 kilometer high (*right of center* and *bottom right*). Alpha is the first letter of the Greek alphabet. (Courtesy of NASA.)

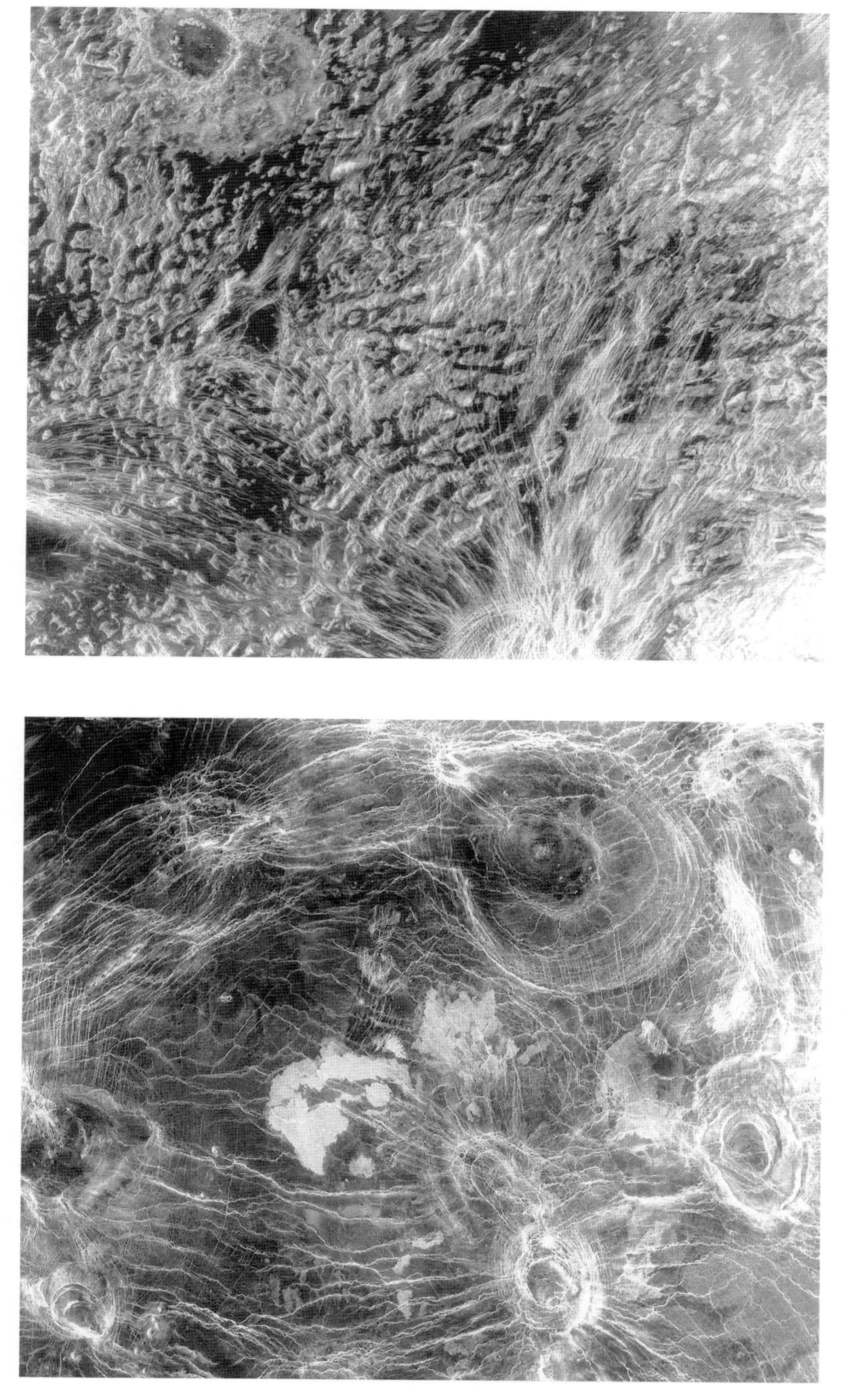

Fig. 7.20 Ovda Regio The surface of Venus is laced with fractures, ranging from elaborate networks of fine cracks to giant canyons millions of meters long. Episodes of fracturing and folding have created the high-standing domes and ridges shown in this highland area known as Ovda Regio; it forms the western part of Aphrodite Terra. The low-lying valleys between these ridges were then flooded by lava, creating this surrealistic radar image of bright, closely packed islands in a dark sea. An impact crater 60 kilometers across (*top left*) is superimposed on the older flood lava. An extensive fracture system extends outward like spokes from the circular feature at bottom right; these fractures also cut across older volcanic features. This *Magellan* radar image is 600 kilometers wide. The region is named Ovda after a Titaness having supernatural powers. (Courtesy of NASA.)

Fig. 7.21 Arachnids An enlarged view of a *Magellan* mosaic of Bereghinya Planita (Fig. 7.9), showing circular structures of up to 230 kilometers in diameter. Their central domes are surrounded by concentric and radial fractures. Such features have been informally dubbed arachnids for their spider-like appearance. They are similar in form, but usually smaller than, the circular volcanic structures known as coronae (Fig. 7.13). (Courtesy of NASA.)

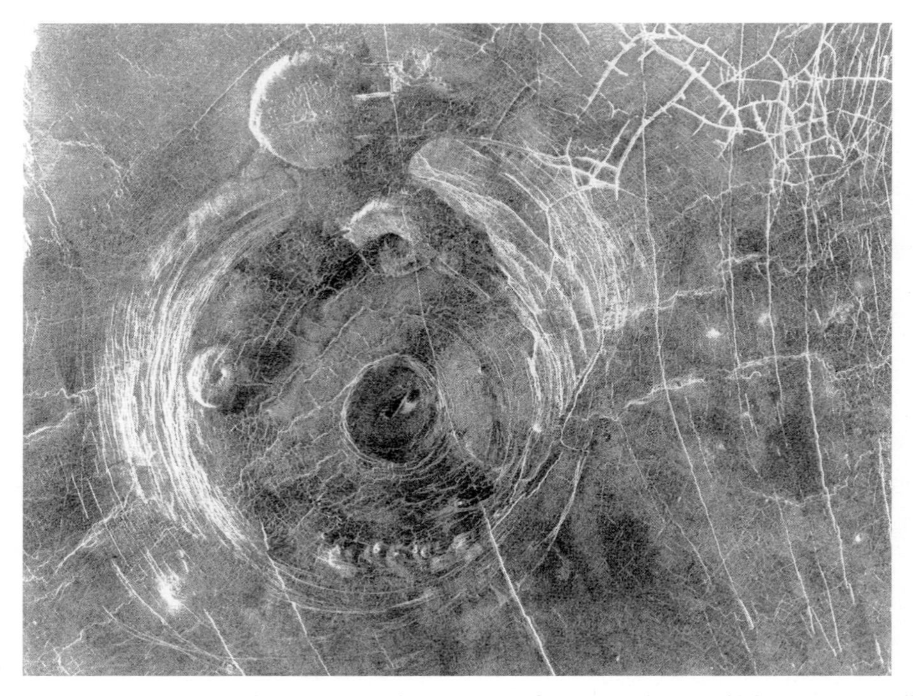

Fig. 7.22 Fotla Corona Large circular and oval structures with diameters of 0.2 to 1.0 thousand kilometers are called coronae. The one shown at the center of this *Magellan* image is known as Fotla Corona, about 200 kilometers across and named after the Celtic fertility goddess. It is located in a vast plain to the south of Aphrodite Terra, and is centered at 59 degrees south latitude and 164 degrees east longitude. Molten rock rising from the interior of the planet most likely explains the corona's circular shape, raised topography, complex fractures, and associated volcanism. Just north (*top*) of this corona is a flat-topped pancake dome, about 35 kilometers in diameter, thought to have formed by the eruption of sluggish, pasty lava. Another pancake dome is located inside the western (*left*) part of the corona. There is also a smooth, flat region in the center of the corona, probably a relatively young lava flow. Complex fracture patterns like the one in the northeast (*top-right*) of the image are often observed in association with coronae. (Courtesy of JPL and NASA.)

along the surface like toothpaste (Fig. 7.22). In other places the crust breaks and spreads open and lava flows into the gap like olive oil.

Just about everywhere that *Magellan* looked, it found intersecting ridges, cracks and grooves, readily visible in the absence of overlying soil or vegetation and the absence of erosion. The intricate, tortured networks are found over the entire globe, ubiquitous throughout the volcanic plains and within the highlands. Internal forces associated with the pent-up heat have been pushing and pulling the crust, producing these patterns. They are known as *tesserae*, the Latin word for "tiles".

As the surface moves up in some locations and down in others, the associated stresses pull the surface apart or push it together. Over time, these stresses can be created in different directions, producing a regularly spaced, gridded pattern of fractured terrain that is only found on Venus at this scale (Fig. 7.23). Some of the cracked patterns of the tesserae have regular six-sided shapes that can be attributed to global heating and cooling of the surface. Repeated episodes of surface deformation in some highlands have additionally created a chaotic network of ridges, troughs and depressions with linear and curved structures (Fig. 7.24).

Fig. 7.23 Tesserae Because there is little soil and no vegetation or erosion to confuse us, we can see tectonic patterns on Venus much more easily than on our own planet. This region, covering an area of 37 kilometers by 80 kilometers long, is located at 30 degrees north latitude and 333 degrees east longitude, on the low rise separating Sedna Planitia and Guinevere Planitia. It shows a criss-crossed pattern of radar-bright lines, which appear to be faults or fractures. The orthogonal system of ridges and grooves is formed in elevated terrain (by 1 or 2 kilometers), and spaced at regular intervals of 1 to 20 kilometers. Known as *tesserae*, from the Latin word for "tiles", the features suggest repeated episodes of intense surface fracturing that may be unrelated to volcanic activity. This type of terrain is not seen on any other planet. It covers about 8 percent of Venus's surface. Sedna is an Eskimo goddess whose fingers became seals and whales, and Guinevere is the legendary Queen of the British King Arthur. (Courtesy of NASA.)

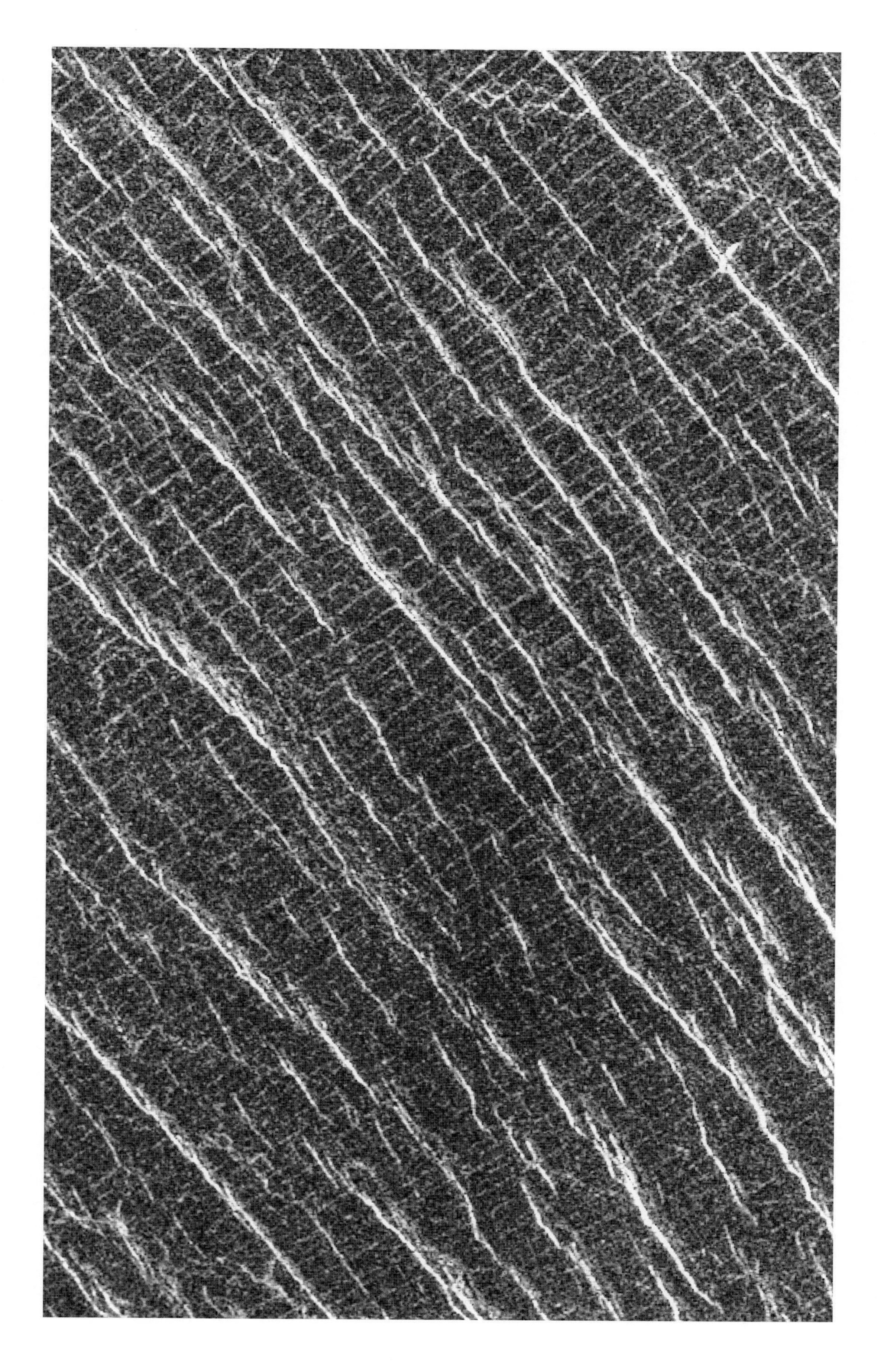

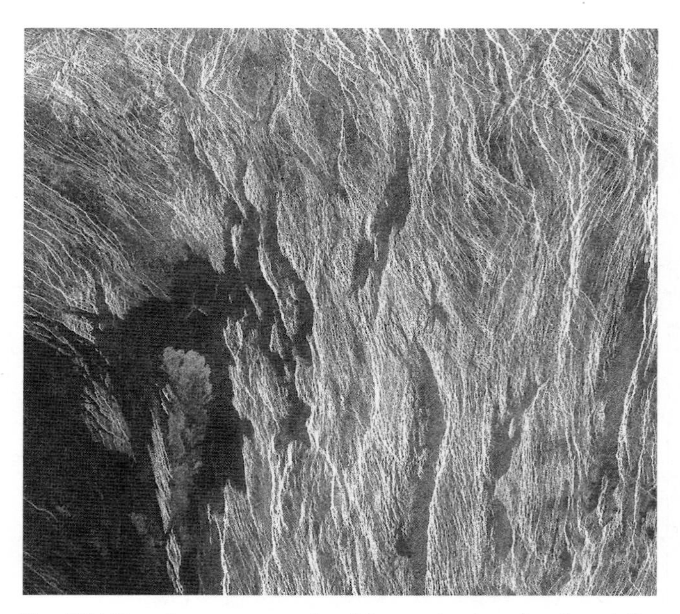

Fig. 7.24 Chaotic tessera terrain Many episodes of surface deformation apparently created this complicated version of the more ordered tessera shown in Fig. 7.23. The complex pattern of ridges and valleys was probably formed by several episodes of faulting, folding, shearing, compression and extension. The chaotic tessera terrain shown here has been embayed by smooth volcanic plains to the west (*left*); the *Magellan* image shown here is located at about 25 degrees south latitude and 357 degrees east longitude within Alpha Regio. The largest ridges and troughs are about 10 kilometers wide and less than 70 kilometers long. Alpha is the first letter of the Greek alphabet. (Courtesy of NASA.)

Up, down and sideways

On Earth, the surface manifestation of trapped heat is the sideways motion of large plates, each millions of meters across, which move laterally across its surface. They separate at a mid-ocean ridge, descend into deep-ocean trenches where long arcs of volcanoes are made, collide with each other to form the great mountain ranges, and grind horizontally together to set off earthquakes (Section 4.3). The plates are rigid, so any movement on one side also happens at the other; in between, the plates move sideways carrying the continents with them.

The Earth's crust is also recycled laterally. New ocean floor rises out of the interior at a mid-ocean ridge, moves horizontally for great distances across the planet, and is destroyed by diving into the deep-ocean trenches.

Because Venus and Earth are about the same size, and composed of the same rocky material, the rates of internal heat generation and the energy available to drive internal motions should also be similar. Thus, there ought to be enough heat trapped inside Venus to push plates around its surface.

Yet, there is no convincing evidence for a system of plates that slide horizontally about the surface of Venus, as they do on Earth. On Venus there are no features comparable to the Earth's extensive mid-ocean ridge or to its

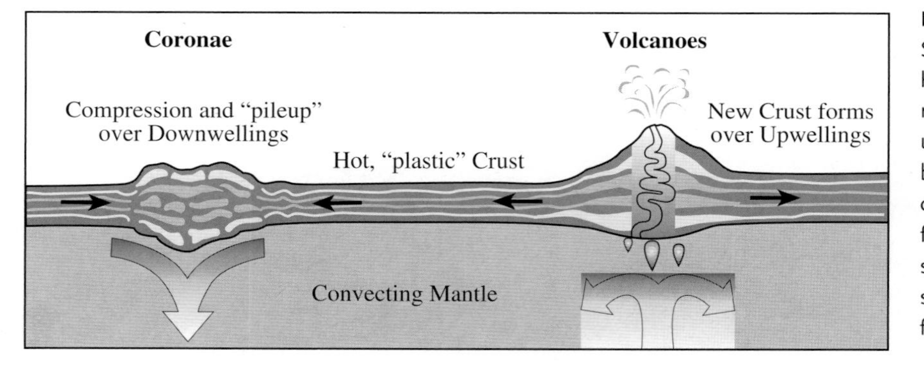

Fig. 7.25 Vertical motions on Venus Some of the surface features on Venus have been formed by vertical, up and down motions driven by hot material welling up from the planet's interior. When rising bubbles of hot rock press against the crust, they can form domed, cracked features known as coronae. Larger plumes support volcanic rises on Venus, while sinking regions can lead to mountain formation.

Fig. 7.26 Impact craters, terrain type, and terrain ages on Venus Impact craters (*top diagram*) are randomly scattered all over Venus. Most are pristine (*white dots*). Those modified by lava (*red dots*) or by faults (*triangles*) are concentrated in places such as Aphrodite Terra. Areas with a low density of craters (*blue background*) are often located in highlands. Higher crater densities (*yellow background*) are usually found in the lowland plains. The terrain type (*middle diagram*) is predominately volcanic plain (*blue*). Within the plains are deformed areas such as tessarae (*pink*) and rift zones (*white*), as well as volcanic features such as coronae (*peach*), lava floods (*red*) and volcanoes of various sizes (*orange*). Volcanoes are not concentrated in chains as they are on Earth, indicating that plate tectonics do not operate. Terrain age data (*bottom diagram*) indicate that volcanoes and coronae tend to clump along equatorial rift zones, which are younger (*blue*) than the rest of the Venusian surface (*green*). Tesserae, ridges and plains are older (*yellow*). In general, however, the surface lacks the extreme variation in age that is found on Earth and Mars. (Courtesy of Mary Beth Price and NASA.)

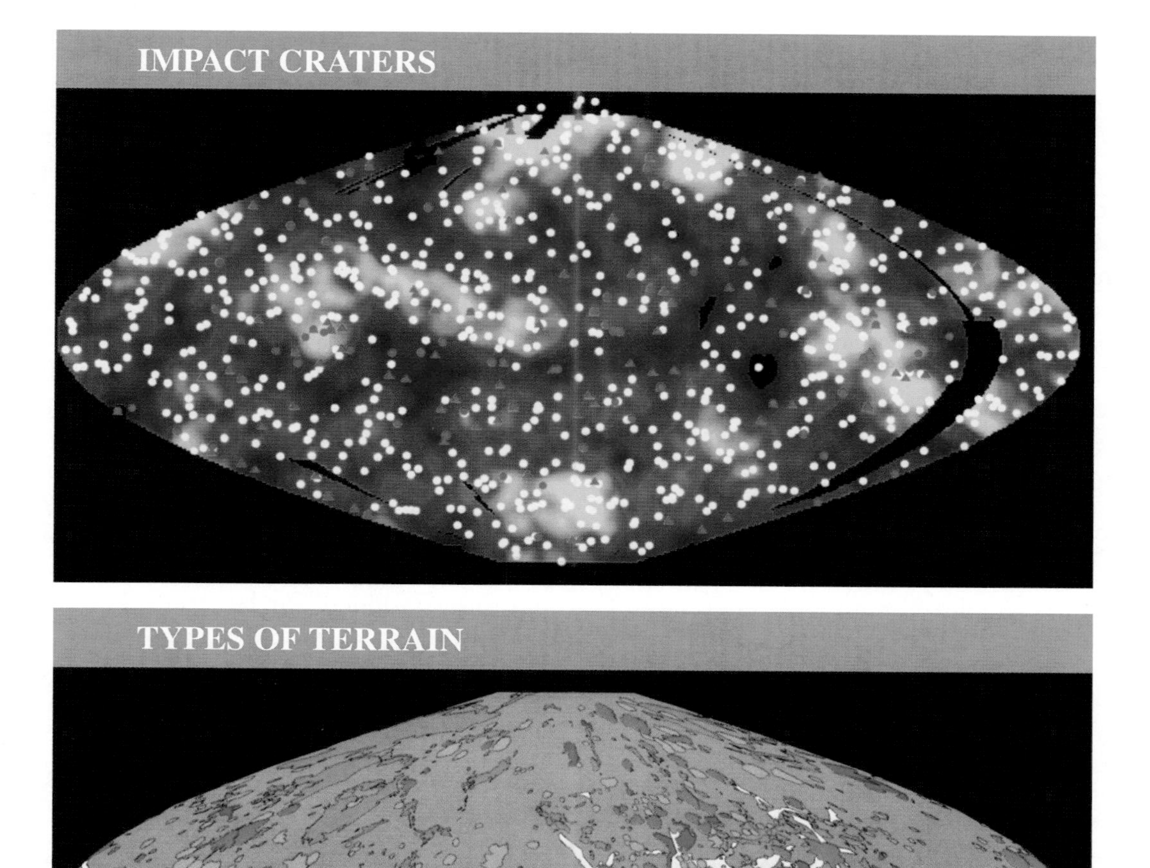
IMPACT CRATERS
TYPES OF TERRAIN

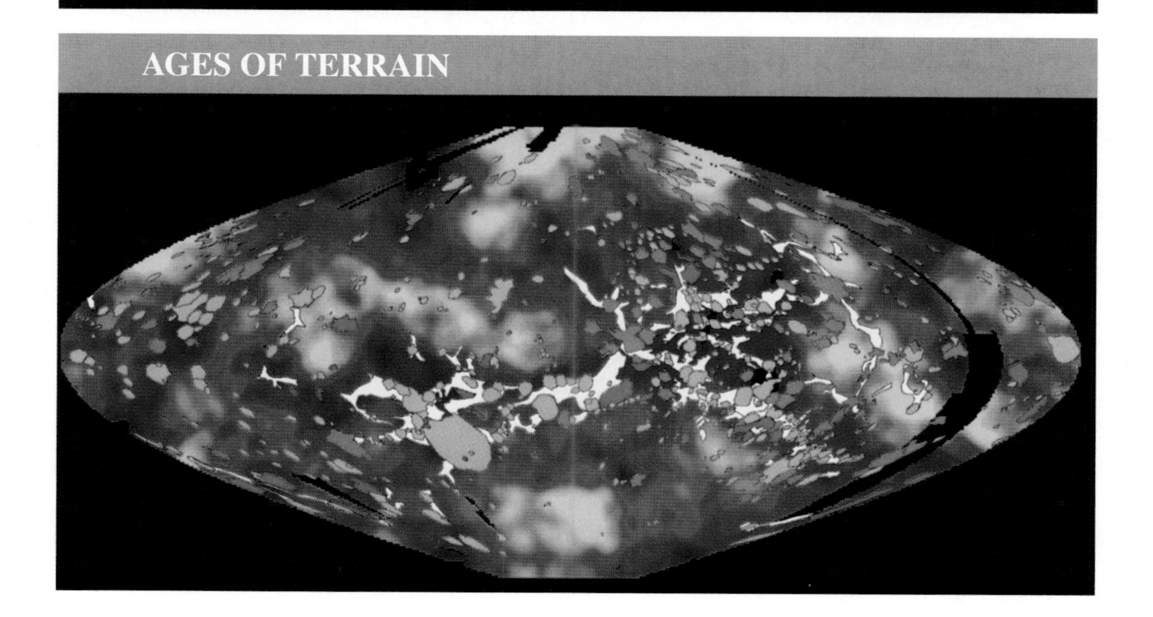
AGES OF TERRAIN

deep-ocean trenches. Thus, it is unlikely that the crust of Venus is recycled in the same way as the Earth's is.

Perhaps the reason Venus has no sliding plates is because there is no water to lubricate the kinds of fault motions necessary for plate tectonics. There can be no liquid water on the torrid surface, and the planet may also be exceptionally dry inside. The outer shell of Venus is probably seized up tight, like a car engine without oil. But this may not have been true in the past when greater amounts of internal heat might have overcome the lack of lubricating water.

Whatever the exact explanation, the outer solid shell of Venus seems to consist of one thick, rigid plate, not many shifting plates as on Earth. Such a thick, strong and unbroken lithosphere would also help support the high plateaus on Venus over long time-scales, and it might also explain why most of the surface is at one low level.

On Venus, the dominant movement is often vertical, or up and down (Fig. 7.25). Upwelling material pushes against the ground, creating arachnids, coronae and tesserae, and punctures the surface to form volcanoes. Volcanic rises are held up by the hot rising material, and long mountain ranges may have been built during sinking, downward compression. Moreover, vast regions of the planet consist of flat, lowland plains with no substantial motion, either vertically or horizontally.

Although astronomers know virtually nothing about the first 4 billion years on Venus, they have been able to piece together a sequence of events during the last 750 million years (Fig. 7.26). At the beginning of the record, they see isolated surface deformation giving rise to a locally fractured crust. Widespread lava flooding created the flat lowland plains soon after this episode of tesserae formation. After this brief but intense period of global volcanic floods, the style and rate of volcanism changed. Localized volcanoes grew on top of the vast plains and coronae were formed within them, but primarily in the equatorial regions where extensive rifts are also found.

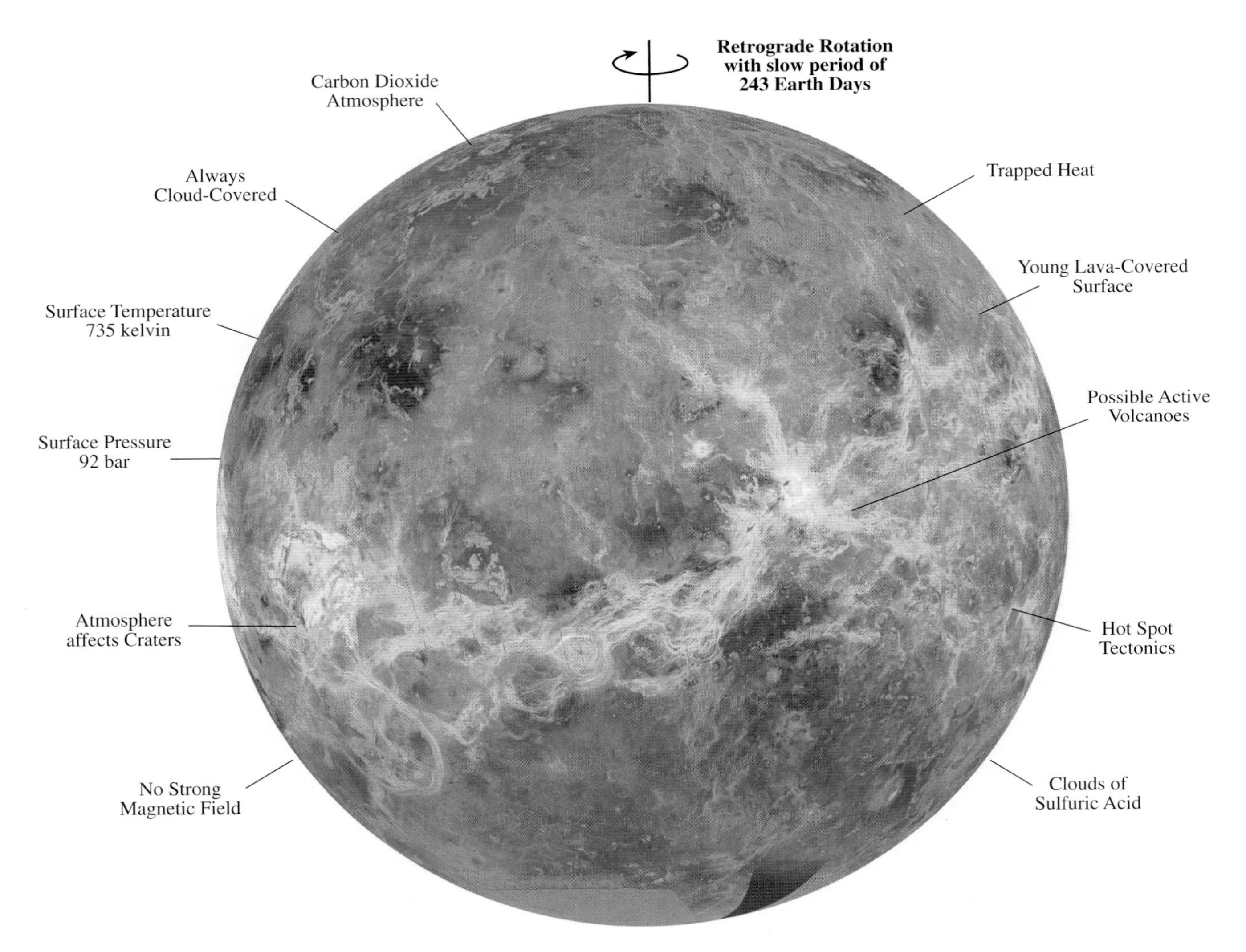

Fig. 7.27 Summary Diagram

8 Mars: the red planet

- Mars now has a thin, cold and dry atmosphere that is composed almost entirely of carbon dioxide.
- The low surface temperature and pressure of the Martian atmosphere are comparable to those in the Earth's stratosphere.
- The Martian atmosphere contains virtually no oxygen, so it has no ozone layer. The planet's surface is therefore exposed to the full intensity of the Sun's ultraviolet rays.
- It cannot now rain on Mars, and liquid water is now unstable on its surface. If any liquid water was now released on the Martian surface, it would survive for just a brief time before freezing or evaporating.
- Powerful and pervasive winds blow across Mars, sweeping up vast dunes of sand and fine-grained dust, creating tornado-like dust devils, and occasionally producing global dust storms that hide the entire planet from view.
- Polar caps of frozen carbon dioxide, or dry ice, wax and wane with the seasons at both the north and south poles of Mars. Because winter in the southern hemisphere is colder and longer than winter in the northern hemisphere, the southern winter cap is the largest.
- The northern residual polar cap, which remains in the summer heat, is composed of water ice, but the residual cap in the south consists of frozen carbon dioxide, or dry ice. If melted, the water in the northern cap would amount to much less than a planetary ocean.
- Mars is divided into two strikingly different hemispheres; in the south there are the older, elevated, heavily cratered highlands that resemble the lunar highlands. In the north there are the younger, lower-lying, smoother volcanic plains.
- The oldest terrain on Mars exhibits bands of magnetized material with alternating senses of polarization, most likely originating long ago when the red planet might have had a global magnetic field.

- Mars does not currently have a global magnetic field to deflect lethal energetic cosmic and solar particles, and it does not have a thick atmosphere with an ozone layer to absorb solar ultraviolet radiation. As a result, the surface of Mars is sterile and life cannot survive there unless protected in some other way.
- Volcanoes have persisted on Mars for a substantial part of its 4.6-billion-year history, and the planet may still be volcanically active.
- The dry tracks of past flowing water are etched into the surface of Mars, marking the site of ancient rivers and floods that occurred billions of years ago.
- Water might have lapped the shores of long-vanished lakes and seas on Mars.
- Huge amounts of water once flowed on the Martian surface, but exactly where all that water came from and what happened to it are still uncertain.
- Though cold and dry today, Mars might have been wetter and warmer long ago, with a thick, dense atmosphere.
- The Martian climate may go through huge swings triggered by periodic variations in its orbit and spin axis.
- Liquid water may have been seeping out of the walls of canyons and craters on Mars in recent times, creating small gullies and depositing the debris in fan-like deltas.
- Three spacecraft have successfully landed on the reddish-brown surface of Mars. The *Viking 1* and *2* landers failed to detect any unambiguous evidence for life on Mars. The *Mars Pathfinder* lander and its *Sojourner Rover* found that their landing site has been untouched by water since it flowed across the region more than two billion years ago.
- There are no detectable organic molecules in the Martian surface examined by the *Viking 1* and *2* landers, which means that this part of the surface now contains no cells, living, dormant or dead. The highly oxidized material has also rusted the Martian surface red.
- Cosmic impacts with Mars are capable of ejecting surface rocks into space, and some of them eventually arrive at the Earth. One such meteorite from Mars, named ALH 84001, exhibits several signs that bacteria-like micro-organisms could have existed on the red planet billions of years ago. The evidence includes carbonates, organic molecules, mineral features, and structures interpreted to be fossils. Nevertheless, most scientists now think that there is nothing in the meteorite that conclusively indicates whether life once existed on Mars or exists there now.
- If life did once exist on the surface of Mars, it might have survived when things got tough, not on, but beneath the surface, within rocks or deep underground in the wet and more temperate part of the planet's interior. In support of this conjecture, we know that terrestrial microbes live in complete darkness within rocks, inside the Earth's crust or at the bottom of its oceans, energized by the Earth's internal heat.
- The future search for life on Mars may include fossil or extant evidence of tough and tenacious, one-celled microbes that can survive in hostile environments, perhaps energized from the planet's hot interior by chemical processes that may involve inorganic compounds.
- Mars has two small moons, named Phobos and Deimos. Phobos is heading towards eventual collision with Mars.

8.1 Fundamentals

Table 8.1 Physical properties of Mars[a]

Mass	6.4185×10^{23} kilograms $= 0.1074 M_E$
Mean Radius	3389.92 kilometers $= 0.532 R_E$
Mean Mass Density	3933.5 kilograms per cubic meter
Surface Area	1.441×10^{14} square meters $= 0.2825 A_E$
Rotation Period or Length of Sidereal Day	24 hours 37 minutes 22.663 seconds = $8.864\,27 \times 10^4$ seconds
Orbital Period	686.98 Earth days = 668.60 Mars solar days
Mean Distance from Sun	2.2794×10^{11} meters = 1.52366 AU
Orbital Eccentricity	0.0934
Tilt of Rotational Axis, the Obliquity	25.19 degrees
Distance from Earth	5.6×10^{10} meters to 3.99×10^{11} meters
Angular Diameter at Closest Approach	14 to 24 seconds of arc
Age	4.6×10^9 years
Atmosphere	95.32 percent carbon dioxide, 2.7 percent nitrogen, 1.6 percent argon
Average Global Surface Pressure	0.0056 bar = 560 pascal
Surface Pressure at *Viking 1* and *2* sites	0.0067 to 0.0088 bar and 0.0074 to 0.010 bar
Surface Temperature Range	140 to 300 kelvin
Average Surface Temperature	210 kelvin
Magnetic Field Strength (Remnant)	$\pm 1.5 \times 10^{-6}$ tesla $= \pm 0.05 B_E$
Magnetic Dipole Moment	less than 10^{-4} that of Earth

[a] Adapted from H. H. Kieffer, B. M. Jakosky, C. W. Snyder and M. S. Matthews (eds.): *Mars*, University of Arizona Press, Tucson, Arizona 1992. The symbols M_E, R_E, A_E, and B_E respectively denote the mass, radius, surface area, and magnetic field strength of the Earth.

8.2 Telescopic observations, early speculations and missions to Mars

An Earth-like planet

Mars, fourth planet from the Sun, ranks third in brightness as seen from Earth – after Venus and Jupiter. But Mars is brighter than most stars, and it has intrigued humans since prehistoric times because of its reddish color and slow looping movement across the starry background (Section 1.1). The ancients associated the blood-red color with warfare and the Greeks and Romans named the planet after their gods of war – Ares and Mars, respectively.

Mars spins on its axis with a rate and tilt that are almost identical to the Earth's. The day on Mars is only 37 minutes longer than our own, and the polar axis on Mars is tilted at 25 degrees, about the same as Earth with its 23-degree tilt. Both planets therefore have four seasons – autumn, winter, spring and summer – although the Martian seasons last about twice as long since the Martian year is nearly two Earth years.

When it is summer in Mars' southern hemisphere it is winter in the north and *vice versa*, just like our home planet. But unlike Earth, which travels in a nearly circular orbit, Mars moves in a more elliptical orbit, which means that its distance from the Sun, and the amount of solar heating of its surface, varies noticeably. During the southern summer, Mars is 20 percent closer to the Sun than during the northern summer and the extra sunlight makes the southern summer hotter than the northern one.

Mars is a relatively small planet, about half the size and one-tenth the mass of Earth (Fig. 8.1). The total surface area of Mars is about equal to the land area of Earth, owing to the extensive oceans that occupy most of the terrestrial surface. Still, Mars is substantially larger than the Moon or Mercury, so we might expect it to be intermediate in many properties between Earth and the Moon.

Fig. 8.1 Earth and Mars This composite image demonstrates the relative size, similarities and main difference of Earth (*left*) and Mars (*right*). Mars is about half the size of the Earth. Both planets exhibit clouds and polar caps. Bluish-white clouds of water ice hang above volcanoes on Mars (*center right*) and large dark areas extend across its red surface (*top right*). The residual north polar cap on Mars (*top*) is made of water ice, and is circled by dark dunes of sand and dust. The Earth also has clouds and polar caps composed of water ice. However, about 75 percent of the Earth is covered with oceans, while liquid water cannot now exist for long times on the surface of Mars. The Earth image was taken from the *Galileo* spacecraft in December 1990, and the Mars image was acquired by the *Mars Global Surveyor* in April 1999. (Courtesy of NASA and JPL (*left* and *right*) and Malin Space Science Systems (*right*).)

It is indeed the Earth-like appearance of Mars in a telescope that has intrigued humanity during the past few centuries. Its orbit is closer to the Earth than any other planet except Venus, enabling us to discern polar caps that change in size with the Martian seasons and dark markings that seasonally distort its ruddy face (Figs. 8.2, 8.3). A white polar cap grows in the local winter at each pole and recedes with the coming of local spring. Large grayish-green regions flourish in the summer and become dormant in winter, as many plants do on Earth. Their seasonal growth on Mars has been called the "wave of darkening" since a dark band moves from a polar cap toward the equator as the cap shrinks.

White clouds repeatedly form at certain locations on Mars, and clouds are not possible without an atmosphere. Both the Earth and Mars have relatively clear atmospheres, heated seasonally by varying sunlight. Thus, Mars is in many ways the planet most closely resembling the Earth. Both planets have an atmosphere, clouds, polar caps and seasons.

Detailed telescopic observations of Mars are difficult. One reason is that the red planet is out of view for prolonged periods when it is on the other side of the Sun from the Earth. The other reason is that turbulence in the terrestrial atmosphere limits the angular resolution of even the most powerful telescope on Earth to about one second of arc (Section 1.2, Focus 1.2). As a result, the finest ground-based telescopes provide only a blurred vision of Mars.

Even when Mars is closest to the Earth, the planet subtends an angle of roughly 20 seconds of arc, and the best telescope can only separate or resolve details no smaller than about one-twentieth of the Martian diameter or about 300 kilometers across. Thus, most hypothetical Martian lakes, mountains, valleys or craters cannot be seen from Earth. To see more, one must use a spacecraft to approach Mars more closely.

Space vehicles have the shortest distance to travel when the Earth moves between Mars and the Sun. This alignment occurs every 780 days, or 26 months, and is known as an opposition, since Mars is then opposite the Sun in our sky (Table 8.2).

Early speculations about intelligent life on Mars

Over the past century, our fascination with Mars has been stimulated largely by the prospect that life may exist there, either in the past or the present. Large, seasonally varying, dark regions seemed to suggest life, since water melting from the polar caps might cause hypothetical vegetation to grow and progress from the poles to the equator.

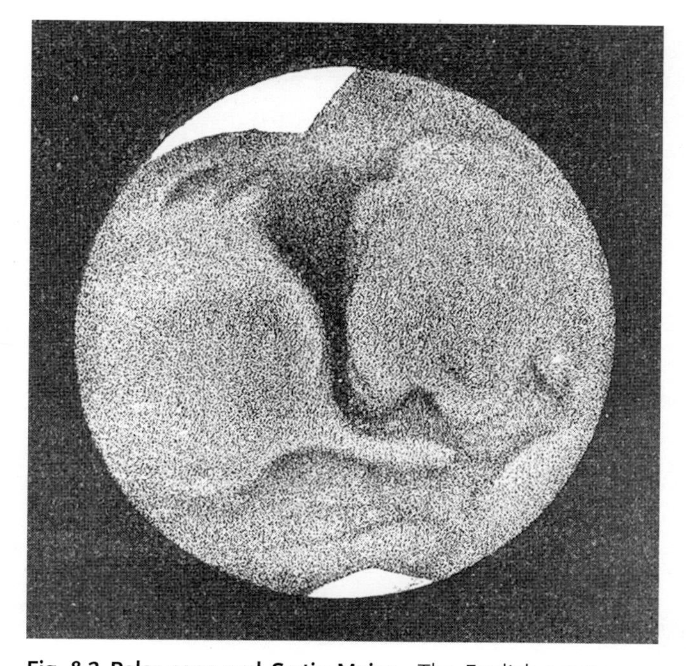

Fig. 8.2 Polar caps and Syrtis Major The English amateur astronomer Warren De La Rue (1815–1889) made this drawing of Mars on 20 April 1856, using a 0.33-meter (13-inch) reflector telescope. It shows bright polar caps and a dark, triangular feature now known as Syrtis Major Planitia (Gulf of Sirte Plains). The Dutch astronomer Christiaan Huygens (1629–1695) first sketched this feature in 1659. From his observations of Syrtis Major, Huygens concluded that the rotation period of Mars is about 24 hours. This drawing is reproduced from Camille Flammarion's 1892 book *La Planete Mars et ses Conditions d'Habitabilité*, Gauthier-Villars et Fils, Paris.

In the early-20th century, it was even widely believed that advanced civilizations had developed on Mars. These speculations resulted from ground-based telescopic observations apparently showing oases and canals stretching across the dusty plains of Mars, but often glimpsed at the limit of telescopic visibility.

During a particularly favorable opposition in 1877, when Mars was even closer to the Earth than during most other oppositions, the Italian astronomer Giovanni Virginio Schiaparelli (1835–1910), director of the Milan Observatory, reported that a maze of dark, narrow straight lines traverses the planet's surface (Fig. 8.4). He called them *canali*, the Italian word for "channels" or "canals", assuming that they were natural features. Schiaparelli mapped them and gave the broadest ones the names of large terrestrial rivers, such as the Ganges and Indus.

The French astronomer Camille Flammarion (1842–1925) subsequently wrote in 1892 that the channels resemble man-made canals, redistributing scarce water across a dying Martian world. Flammarion was also convinced that the Martian inhabitants might be more advanced than terrestrial humans. These ideas were eloquently described in his book *La Planéte Mars et ses Conditions d'Habitabilité*, published by Gauthier-Villars et Fils, Paris in two volumes in 1892 and 1909, including the passage:

> Anyone with an open mind would not reject the possibility that the canals are the result of intelligent action by the inhabitants of Mars. The present physical conditions of that globe do not rule out the possibility that it is inhabited by a human species whose intelligence and means of action could be much higher than ours is. It would be anti-scientific to deny the possibility that they could have changed the course of primitive rivers, and gradually redistributed the increasingly scarce water by a system of canals.

At about the same time, a wealthy Bostonian, named Percival Lowell (1855–1916), convinced much of the

Fig. 8.3 *Hubble Space Telescope* views Mars This perspective of Mars was obtained from the *Hubble Space Telescope* (*HST*) on 10 March 1997 – the last day of spring in the Martian northern hemisphere. The red planet was near its closest approach to Earth and a single picture element of the *HST* spanned 22 kilometers on the Martian surface. The image shows bright and dark markings observed by astronomers for more than a century. The large dark feature seen just below the center of the disk is Syrtis Major Planitia, first seen telescopically by Christiaan Huygens in the 17th century. To the south of Syrtis Major is the large circular impact basin Hellas (*center bottom*) filled with surface frost and shrouded in bright clouds of water ice. The seasonal north polar cap (*center top*) is rapidly sublimating, or evaporating from solid dry ice to carbon-dioxide gas, revealing the smaller residual water ice cap with its collar of dark sand dunes. (Courtesy of David Crisp, NASA, JPL, and STSI.)

Table 8.2 Oppositions of Mars 2001 to 2035[a]

Opposition Date	Right Ascension (hours minutes)		Declination (degrees minutes)		Diameter (seconds of arc)	Distance[b] (10^{10} m)
2001 June 13	17	28	−26	30	20.5	6.82
2003 August 28	22	38	−15	48	25.1	5.58
2005 November 7	02	51	+15	53	19.8	7.03
2007 December 28	06	12	+26	46	15.5	8.97
2010 January 29	08	54	+22	09	14.0	9.93
2012 March 3	11	52	+10	17	14.0	10.08
2014 April 8	13	14	−05	08	15.1	9.29
2016 May 22	15	58	−21	39	18.4	7.61
2018 July 27	20	33	−25	30	24.1	5.77
2020 October 13	01	22	+05	26	22.3	6.27
2022 December 8	04	59	+25	00	16.9	8.23
2025 January 16	07	56	+25	07	14.4	9.62
2027 February 19	10	18	+15	23	13.8	10.14
2029 March 25	12	23	+01	04	14.4	9.71
2031 May 4	14	46	−15	29	16.9	8.36
2033 June 27	18	30	−27	50	22.0	6.39
2035 September 15	23	43	−08	01	24.5	5.71

[a] An opposition occurs when the Earth moves between Mars and the Sun, and the two planets are closest. Adapted from William Sheehan: *The Planet Mars*, University of Arizona Press, Tucson Arizona 1996.
[b] The distance between the Earth and Mars at opposition in units of ten billion (10^{10}) meters. Because the Martian orbit is more elliptical than Earth's, the distance between the two planets at different oppositions varies as much as 50 billion meters.

American public that intelligent life existed on Mars. Rich enough to do as he pleased, Lowell built an observatory in the clear air of Flagstaff, Arizona with the specific intention of observing and explaining the Martian canals. When Lowell turned his telescope toward Mars in 1894, he found what he expected to see – a vast network of canals bordered by vegetation.

Lowell enthusiastically advocated the possibility of intelligent life on Mars in his books entitled *Mars*, published in 1895 by Houghton Mifflin, Boston, and *Mars and its Canals*, published in 1906 by Macmillan, New York. He argued that the Martian channels mark a vast, planet-wide irrigation network, constructed by intelligent beings to transport water away from the melting polar caps to parched equatorial deserts.

Most astronomers, however, could not see the canals, which had been glimpsed at the limit of telescopic detection, concluding that they were some sort of optical illusion if they existed at all. And no one ever succeeded in photographing the canals using telescopes on Earth.

Still, the idea of an intelligent species on Mars, perhaps facing extinction, was so powerful that belief in canals on Mars persisted until spacecraft took close-up photographs of the planet. They showed that there are no canals on Mars. As it turned out, the "canals" have no objective reality, beyond the tendency of the human mind to seek order in chaos.

The canals are an illusion created when the eye arranges minute, disconnected details into lines. Moreover, the wave of darkening is not a sign of the seasonal revival of life, but is instead caused by seasonal winds shifting bright dust from one region to another (Section 8.4). Thus, the basis for early speculations about intelligent life on Mars was illusory.

The space-age odyssey to Mars

Due to distortion caused by the Earth's atmosphere, the details of Mars remained hidden from view until spacecraft

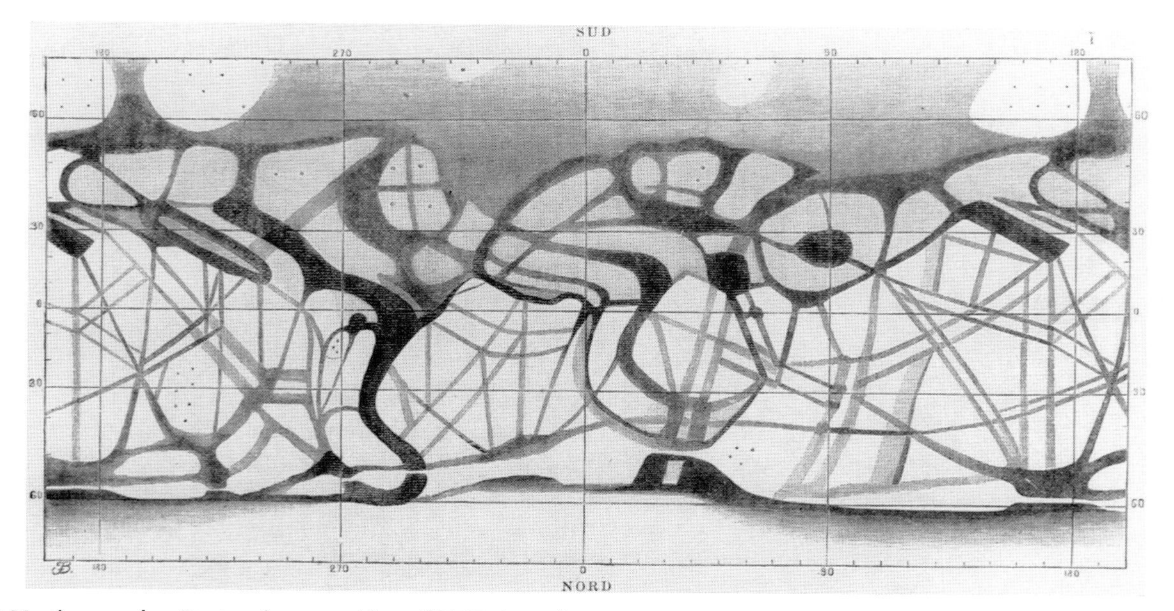

Fig. 8.4 Martian canals During the opposition of 1877, the Italian astronomer Giovanni Schiaparelli (1835–1910) mapped features he thought he saw on Mars, including a vast network of long, thin, straight lines criss-crossing the planet's surface. Some of the canals have apparently doubled, or divided in two, in this Mercator projection drawn by Schiaparelli during the opposition of 1881 using an 8.6-inch (22-centimeter) refractor. At this time, the apparent diameter of Mars was only 16 seconds of arc. Schiaparelli named these features canali, and they were likened to man-made water canals by subsequent observers, including Camille Flammarion (1842–1925) and Percival Lowell (1855–1916). Nevertheless, most astronomers failed to see the canals, and spacecraft have not detected them.

flew past it, and were then sent to orbit the red planet and land on its surface (Table 8.3). The search for life on Mars has been one of the main driving forces behind many of these missions.

8.3 The Martian atmosphere

A carbon-dioxide atmosphere

Astronomers have long known that Mars has an atmosphere. It is required for the formation and support of the clouds that have been observed telescopically since the 19th century. The seasonal waxing and waning of the polar caps also suggests the presence of an atmosphere on Mars. Gases are released into the Martian atmosphere when a polar cap warms up during the local summer and the cap becomes smaller; gases are extracted from the atmosphere during the winter growth of the cap.

During the first half of the 20th century, astronomers guessed that the Martian atmosphere was mostly nitrogen, like that of the Earth. Carbon dioxide had been spectroscopically identified in the planet's atmosphere by mid-century, but it was thought to be a minor ingredient. Then experiments aboard *Mariner 4* showed in 1965 that carbon dioxide is the primary gas, and nitrogen is a minor constituent.

The *Viking 1* and *2* landers made detailed measurements of the composition of the Martian atmosphere (Table 8.4). Carbon dioxide is indeed the principal constituent, amounting to 95.32 percent of the atmosphere at ground level, followed by nitrogen (2.7 percent) and argon (1.6 percent). When compared with Earth, there is much more argon in the Martian air.

Small amounts of oxygen, ozone and water vapor

Oxygen molecules are present in the Martian atmosphere, but in a miniscule amount of just 0.13 percent. In contrast, the Earth's atmosphere is filled with breathable oxygen, amounting to 21 percent of our air. Detection of significant amounts of oxygen would have argued for the possibility of plant life on Mars, since oxygen is unstable in a planetary atmosphere and plants are needed to continuously supply it. The small amount of free oxygen that is now present on Mars is the byproduct of the destruction of carbon dioxide by energetic sunlight. This process also results in the production of exceedingly small amounts of ozone.

Table 8.3 Important missions to Mars[a]

Mission	Launch Date	Encounter Date	Discovery and/or Accomplishments
Mariner 4	28 Nov. 1964	14 July 1965	Flyby, cratered southern hemisphere, cold, thin, carbon-dioxide atmosphere.
Mariner 6	24 Feb. 1969	1 July 1969	Flyby, confirmed *Mariner 4* findings.
Mariner 7	27 Mar. 1969	5 Aug. 1969	Flyby, confirmed *Mariner 4* findings.
Mariner 9	30 May 1971	13 Nov. 1971	Orbiter, volcanoes, canyons, outflow channels.
Viking 1	20 Aug. 1975	19 June 1976	Orbiter and lander, surface photographs, water ice in north polar cap, meteorology, life search.
Viking 2	9 Sept. 1975	7 Aug. 1976	Same as *Viking 1*.
Phobos 2	12 July 1988	Jan. 1989	Orbiter and lander, composition of surface of Syrtis Major.
Mars Pathfinder	4 Dec. 1996	4 July 1997	Lander, surface rover.
Mars Global Surveyor	7 Nov. 1996	12 Sept. 1997	Orbiter, laser altimeter, magnetometer, high-resolution images, recent water flow and volcanic activity, ancient magnetism.
2001 Mars Odyssey	7 Apr. 2001	23 Oct. 2001	Discovery of substantial amounts of subsurface water ice in northern and southern hemispheres at latitudes poleward of about 60 degrees.

[a] Between 1971 and 1974 the Russians sent four *Mars* spacecraft into orbit around the red planet, with *landers* that either crashed, missed the planet or failed within seconds of touchdown.

Since there is so little ozone in the Martian atmosphere, it has no ozone layer, and the planet's surface is exposed to the full intensity of the Sun's ultraviolet radiation. The lethal rays would destroy any exposed micro-organisms, so there might not be any live ones left on Mars. By way of comparison, the Earth has a thick ozone layer high in its atmosphere, which absorbs most of the dangerous ultraviolet sunlight and keeps it from reaching the ground (Section 4.4).

Any human visitor to Mars would have to wear a spacesuit to provide protection from the Sun's harmful radiation, as well as to supply oxygen to breathe. An unprotected person on the surface of Mars would receive a radiation dose as high as one hundred times our daily exposure on Earth. It would be comparable to being exposed to the fallout of a nuclear bomb.

There is now very little water vapor in the Martian atmosphere, about 0.03 percent near the surface. And that is

Table 8.4 Composition of the atmosphere at the surface of Mars[a]

Species	Abundance	Species	Abundance
Carbon Dioxide (CO_2)	95.32 percent	Water Vapor (H_2O)	0.03 percent[b]
Nitrogen (N_2)	2.7 percent	Neon (Ne)	2.5 p.p.m.v.
Argon (Ar)	1.6 percent	Krypton (Kr)	0.3 p.p.m.v.
Oxygen (O_2)	0.13 percent	Xeon (Xe)	0.08 p.p.m.v.
Carbon Monoxide (CO)	0.07 percent	Ozone	(0.04 to 0.2)[b] p.p.m.v.

[a] Composition by volume in percent or in parts per million, denoted p.p.m.v. Because carbon dioxide varies seasonally due to condensation at the polar caps, all percentage abundances will vary seasonally. Adapted from H. H. Kieffer, B. M. Jakosky, C. W. Snyder, and M. S. Matthews (eds.): *Mars*, University of Arizona Press, Tucson Airzona 1992.

[b] The abundance of water vapor and ozone vary with season and location. The annual global average of water vapor is 0.016 percent by volume.

Fig. 8.5 Frost on Mars Atmospheric water vapor freezes onto the surface of Mars, producing a thin coating of water ice on rocks and soil. These white patches of frost were photographed from the *Viking 2* lander at its Utopia Planitia landing site on 18 May 1979; the frost remained on the surface for about 100 days. (Courtesy of NASA and JPL.)

about as much of the vapor that the atmosphere can hold. It is practically saturated with water vapor. When the temperature drops, water can condense and freeze out of the saturated air, forming low-lying mists or ground fogs in canyons and frosts on the surface (Fig. 8.5).

White clouds commonly form in the vicinity of towering peaks on Mars, which attract clouds just like mountains of our own planet. One recurrent white spot was visible from Earth more than a century ago, acquiring the name Nix Olympica, or Snows of Olympus. Now we know that the white spots are not snow, but clouds of water-ice crystals that freeze out of the high, cold air. And Nix Olympica turned out to be the tallest volcano in the solar system, now called Olympus Mons.

The concentrations of Martian water vapor vary with location and season, but they are always low. If all the water vapor above a given place on Mars could rain down to the surface as a liquid, the average depth would be 0.000 015 (1.5×10^{-5}) meters. By comparison, the Earth's atmosphere normally contains ten thousand times as much water vapor, capable of raining onto the ground with depths of several centimeters. Moreover, the pressures and temperatures on Mars are such that it cannot now rain on Mars.

The thin, cold Martian air

Since Mars has only a tenth of the Earth's mass, the red planet has less gravitational pull. It might thus be expected to retain a less substantial atmosphere. And because Mars is 50 percent further from the Sun than the Earth is, the red planet receives about half as much sunlight and ought to be colder. The ground-level pressure and temperature on Mars are, in fact, lower than those outside the highest-flying jet airplane on Earth. The atmosphere near the surface of Mars is almost as thin as our best laboratory vacuums, and the temperatures are usually below the freezing point of water.

The surface pressure on Mars was first accurately determined when *Mariner 4* passed behind the planet, and its radio signal penetrated the Martian atmosphere in order to reach Earth. From the manner in which the signal was altered, a surface pressure of about 0.005 bar was determined, compared to 1.000 bar at sea level on Earth. The measurement was repeated when the spacecraft emerged from behind Mars, also indicating that the atmosphere on Mars is 200 times thinner than on Earth. The Martian air is so tenuous that it would not be breathable, even if it were made of pure oxygen – which it is not.

The Martian atmosphere is not substantial enough to significantly trap heat emitted from the planet's surface, which means that Mars is a frozen world. Warmed only by direct sunlight, without any pronounced greenhouse effect, the surface temperature on Mars averages 210 kelvin, well below the freezing point of water at 273 kelvin. Only the midday summer Sun near the equator is warm enough to thaw water ice during much of the year.

Under present conditions on Mars, liquid water is unstable and cannot stay on the surface of Mars. Because the temperature and pressure are so low, water on Mars is now stable only as ice or vapor. Over most of the surface, the temperature is usually below the freezing point of water, and when it warms above freezing the water turns almost directly into vapor. If liquid water was released onto the surface from the warmer interior, that water would survive for just a brief time before freezing into ice or evaporating explosively, turning into water vapor.

8.4 The winds of Mars

Why the winds blow

Even with its thin atmosphere, Mars experiences substantial winds driven by temperature differences in the atmosphere. As on Earth, the Martian atmosphere is warmed by sunlight during the day and cools at night, but due to the absence of a thick moderating atmosphere or oceans, these daily temperature variations are extreme on Mars (Focus 8.1). Since the Sun also warms the summer hemisphere more than the winter one, Mars has global temperature differences that depend on the season, just as the Earth does.

The Martian atmosphere responds to both the daily and seasonal temperature differences by generating winds that blow from hot to cold regions, transporting heat and trying to equalize the temperatures. And since a rise in temperature is equivalent to an increase in pressure, it is high-pressure air that rushes toward low pressure. This rush is the wind. The sharper the temperature or pressure difference, the stronger the wind.

The influence of the winds on Mars is pervasive. They create a restless world of constant change, stirring up dense, billowing clouds of dust and scouring the surface, creating time-varying light and dark patterns, and carrying dust from one place to another. These are called aeolian effects, after Aeolus, the Greek god of the wind.

Focus 8.1 Weather report from Mars

When *Viking 1* and *2* landed on Mars, they each deployed instruments that provided daily weather reports for years. Winds were usually light and variable, blowing with speeds of several meters per second along the ground, but they occasionally increased to ten times that speed in gusts. Although the temperatures rarely climbed above freezing, the day to night temperature excursions were large. The *Mars Pathfinder* lander took ground temperature readings at a more rapid pace, showing that the temperatures can go up or down by 20 kelvin in just seconds.

The two *Viking* landers found that the surface pressure on Mars can change by 30 percent, while always remaining less than one-hundredth of that on Earth. The frozen world becomes so cold during the southern winter that almost a third of its atmosphere freezes, dropping out of the sky and enlarging the southern polar cap. When the temperature rises in southern summer, the gas is released back into the atmosphere and the surface pressure rises again. This enormous seasonal change in the mass of the atmosphere has no counterpart on Earth, where pressure changes of a couple percent are cause for concern, often indicating a major storm.

Windblown dust and sand

Since Mars now lacks liquid water to wash its surface clean, the red planet is covered with dust and sand. Dust particles are typically about one-millionth of a meter (10^{-6} m = 1 μm) across, comparable in size to cigarette smoke and roughly one-hundredth the width of a human hair. Sand particles are about one-thousandth of a meter (10^{-3} m = 1 mm) across, a thousand times bigger than the dust.

The dust and sand behave differently in the winds of Mars. The smaller and lighter dust particles get picked up and carried by the wind. The bigger, heavier sand particles bounce across the surface when strong gusts of wind come along. The same thing happens at a windy beach on Earth, where the bouncing sand grains sting your face or legs. But because the Martian air is so thin, the wind that moves sand on Mars has to be blowing about ten times stronger than one on Earth.

Tiny dust particles may be hazardous to future visitors to Mars. The dust will gum up spacesuits, scratch helmet visors, cause electrical shorts, sandblast instruments and clog motors. On the Moon, which is similarly dusty, spacesuits lasted only two days before they began to leak, and unlike the Moon, intense dust storms can arise on Mars, creating blizzards that will limit vision and drive dust into astronauts' clothing and equipment.

The icy Martian winds have swept up vast dunes of sand and fine-grained dust (Fig. 8.6). Rippled dunes have piled up in basins, craters and chasms (Fig. 8.7), but the dunes cover only a small percentage of the land on Mars, probably less than one percent. Both dark and light sand dunes are found on Mars (Figs. 8.8, 8.9). Starkly beautiful patterns are created when the polar caps warm up in local spring and summer, exposing dark sand dunes (Figs. 8.10, 8.11). Extensive dunes form a dark collar around the north polar cap in the local summer, while global winds tend to blow fine dust from the south to the north.

Dust devils and global dust storms

As the powerful winds disrupt an otherwise silent world, they occasionally stir up small, local dust storms, in much

Fig. 8.6 Rippled dunes and eroding hills The surface of Mars is dominated by features created and shaped by the wind, such as these sand dunes in Hebes Chasma. They seem to resemble sand dunes in the Sahara desert on Earth, but probably include finer dust. This image was taken from the *Mars Global Surveyor* on 13 March 1998. It is 2.3 kilometers wide, and centered near 0.8 degrees south and 76.3 degrees west. (Courtesy of NASA, JPL, and Malin Space Science Systems.)

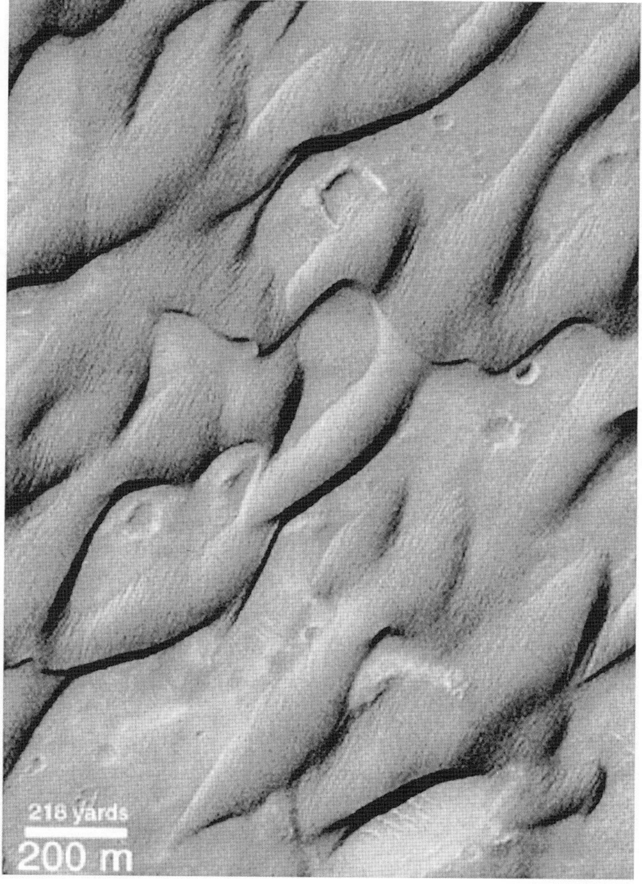

Fig. 8.7 Frozen desert The rough, grooved surfaces of these dunes, found in the Herschel basin, indicates that the sand is cemented into place. Winds had to scour the surface to remove material. The Herschel basin is named after the British astronomers William H. Herschel (1738–1822) and his son John F. Herschel (1792–1871). This image was taken from the *Mars Global Surveyor* on 5 May 1999. It is about 1.2 kilometers wide, and centered near 15 degrees south and 228 degrees west. (Courtesy of NASA, JPL, and Malin Space Science Systems.)

the same way that winds sometimes whip the terrestrial soils into towering columns called dust devils. The local dust storms form when the ground heats up during the day, warming the atmosphere immediately above the surface. The warm gas rises in a spinning column that moves across the landscape like a miniature tornado, sweeping up dust that makes the vortex visible and leaving a dark streak behind. They have scratched tangled paths across some parts of Mars, often crossing hills and running across large sand dunes and through fields of house-sized boulders (Fig. 8.12).

Large dark areas, such as the elevated plateau Syrtis Major, apparently develop when the surface rocks are scoured by powerful, seasonal winds. On close inspection, these dusky areas dissolve into swarms of elongated light and dark streaks, often tens of kilometers long, pointing in the direction of strong prevailing winds (Fig. 8.13). The light streaks consist of fine dust deposited on a darker terrain by the prevailing winds, on the downwind, or leeward, side of craters and hills. The dark streaks result from the removal of dust by strong winds to expose the underlying rock. When all the steaks in a given area are integrated and superimposed by the human eye or at the detector of a telescope, they form the larger, global features visible from Earth, in much the same way as dots in newsprint combine to make a picture.

Strong winds carry dust from the surface high into the atmosphere, forming yellow dust clouds that have been reported by telescopic observers for centuries. Numerous

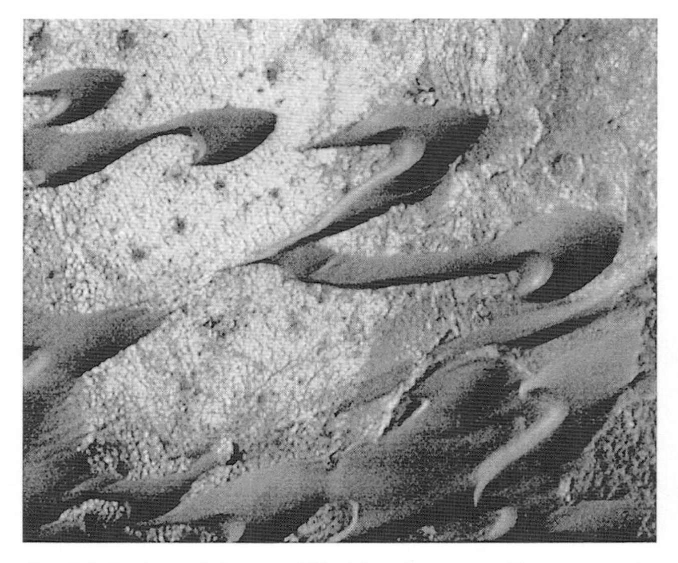

Fig. 8.8 Dark sand dunes Wind has been steadily transporting dark sand across the Nili Patera region of Syrtis Major. This image, 2.1 kilometers wide, was taken from the *Mars Global Surveyor* in March 1999. (Courtesy of NASA, JPL, and Malin Space Science Systems.)

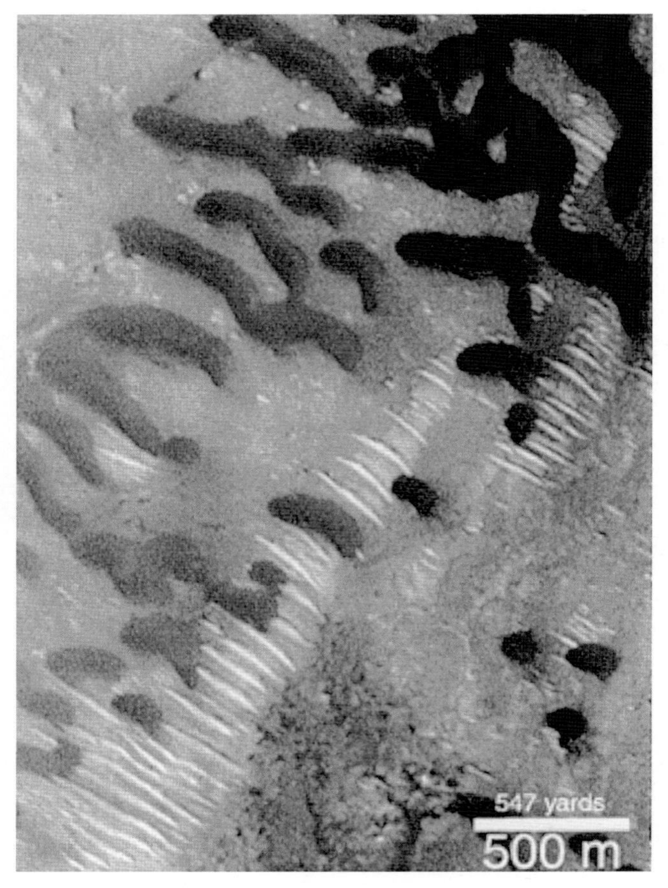

Fig. 8.9 Dark dunes overriding bright dunes In this picture, wind has caused the dark and somewhat crescent-shaped dunes to move toward the lower left, across sets of smaller, bright ridges that also formed by wind action. All of these dunes are located on the floor of an impact crater in western Arabia Terra. This image was taken from the *Mars Global Surveyor* in February 2000. It is about 2.1 kilometers wide, and centered near 10.7 degrees north and 351.0 degrees west. (Courtesy of NASA, JPL, and Malin Space Science Systems.)

fleeting and localized dust storms occur each Martian year (Fig. 8.14). They can occur at any season, but are more frequent in southern spring and summer. Small dust storms can form simultaneously at several points on the planet, and then coalesce with each other, producing dust storms larger by far than any seen on Earth. They sometimes grow and spread across the planet, engulfing the entire globe and shrouding it in an opaque yellow cloud.

Only a handful of these global dust storms have been observed, beginning with telescopic observations during the oppositions of 1922 and 1956. *Mariner 9* arrived at Mars during the slow decay of one of them; until the dust settled, only the summits of volcanoes were visible to the spacecraft cameras. Two planet-encircling dust storms were observed during the *Viking* missions, when the sky above the landers turned dark red and the Sun was greatly dimmed. A global dust storm next hid the surface of Mars from view during the summer of 2001 (Fig. 8.15).

These awesome, globe-covering storms occur during the hot summers in the southern hemisphere when Mars comes closest to the Sun, and also moves the fastest along its orbit. The rapid temperature increase generates hurricane-speed winds that can sweep fine dust particles high into the atmosphere. As more dust is carried aloft, it absorbs sunlight, further heating the atmosphere and strengthening the winds. They eventually cover the planet with a deep cloud of dust that you would not be able to see through. But as the darkening cloud blots out the Sun, the lower layers of the atmosphere cool, the winds diminish, and the dust settles back down to the surface.

Such global dust storms are not generated during cool, long summers in the northern hemisphere, when Mars is 20 percent further away from the Sun than in southern summer, receiving less heat while also moving at a slower speed. The extra sunlight provided during the southern summer, when Mars is close to the Sun, provides the heat and winds that energize the most powerful storms.

8.5 The polar regions

Seasonal polar caps

As on Earth, there are large white caps at both poles of Mars, first observed with telescopes centuries ago. These

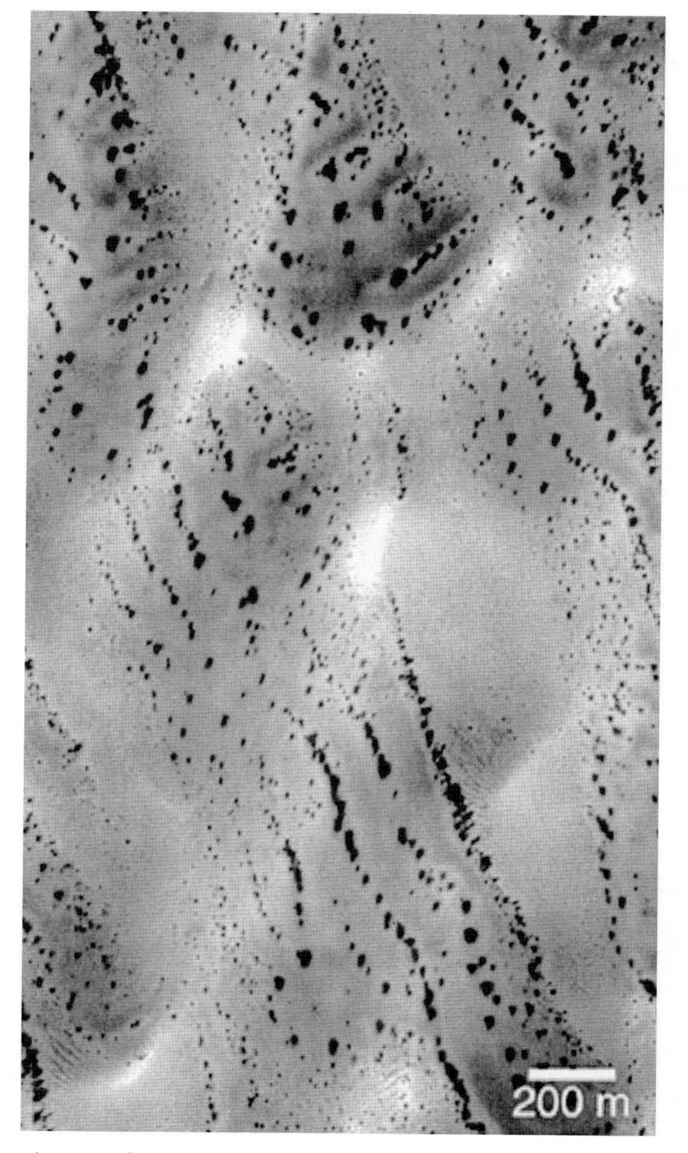

Fig. 8.10 The snow leopard These strange, beautiful dark spots were created when frost was evaporating from the south polar dunes on Mars. The spots are areas where dark sand has been exposed from beneath bright frost as the south polar cap begins to sublimate and retreat. This image was taken from the *Mars Global Surveyor* on 1 July 1999. It is about 1.2 kilometers wide, and located at 61.5 degrees south and 18.9 degrees west. (Courtesy of NASA, JPL, and Malin Space Science Systems.)

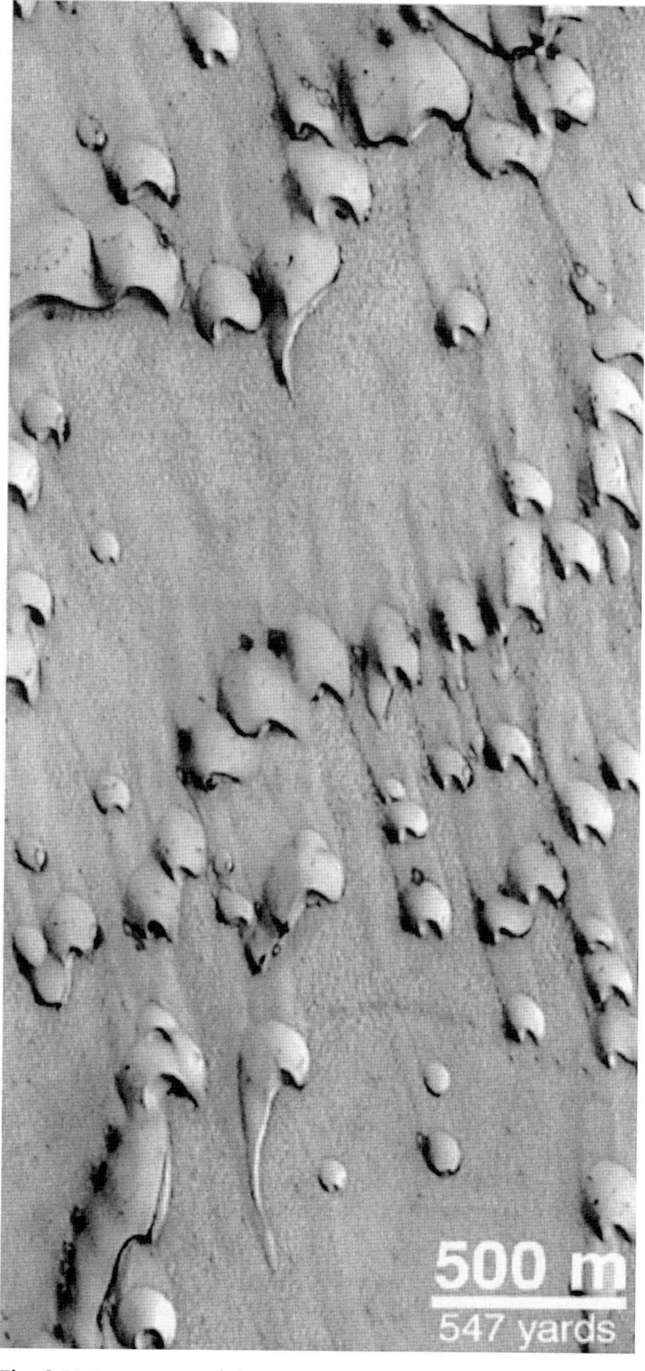

Fig. 8.11 Frost-covered dunes A giant trough in the north polar cap shows dark sand emerging from beneath a veneer of bright frost left over from the northern winter. This image of dunes in the Chasma Boreale was taken from the *Mars Global Surveyor* in September 1998. It is about 1.6 kilometers wide. (Courtesy of NASA, JPL, and Malin Space Science Systems.)

seasonal deposits change in size during the Martian year, growing in the cold of local fall and winter and shrinking in the summer heat. But these varying caps are poles apart in size and seasonal change. The seasonal polar cap in the southern hemisphere is larger in the winter (extending halfway to the equator) than the seasonal northern cap ever gets. The southern cap also disappears more rapidly in the spring, and is smaller in the summer than is the northern cap.

This asymmetry is a direct consequence of the planet's eccentric orbit, which carries Mars furthest away during the southern winter, which is longer and colder than the northern one. The north polar cap grows to a lesser extent in the

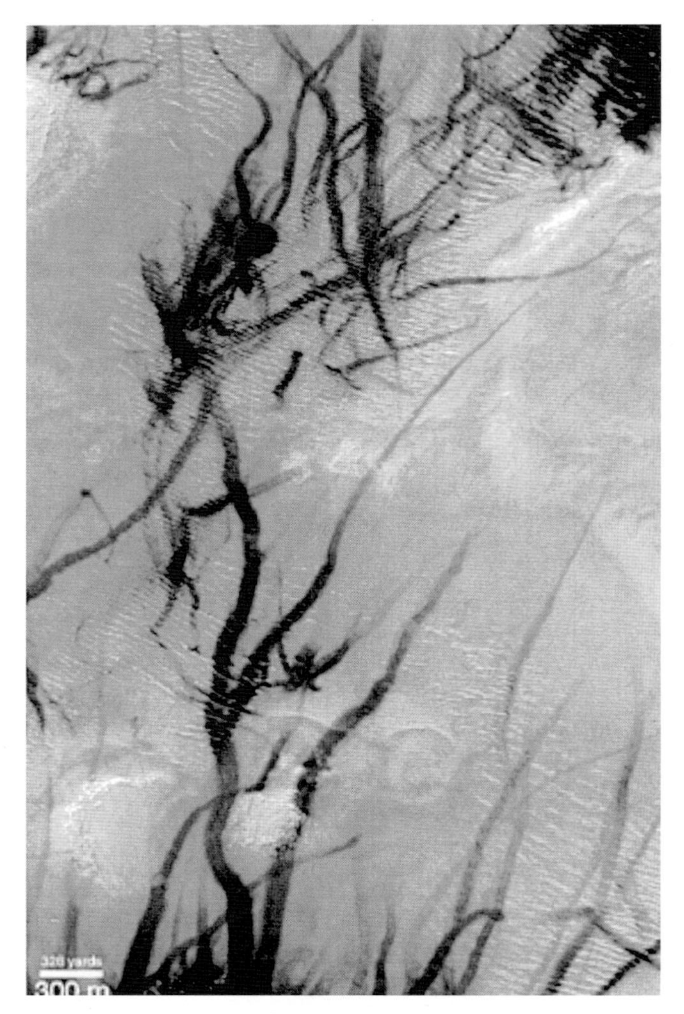

Fig. 8.12 Twisted paths of dust devils Spinning columns of warm air, called dust devils, rise above the Sun-heated surface of Mars. Each tornado-like vortex picks up light-colored dust, exposing the darker surface underneath. Dust devils have created this wild pattern of criss-crossing dark streaks in the rippled flats of Argyre Planitia, covering an area 3 by 5 kilometers at a latitude of 51 degrees south. This image was taken from the *Mars Global Surveyor* in March 2000. (Courtesy of NASA, JPL, and Malin Space Science Systems.)

shorter, colder northern winter, when the red planet is closest to the Sun, near perihelion, and moving at its fastest orbital speed. The relative warmth of southern summer, also near perihelion, causes the south polar cap to almost disappear from sight.

The seasonal caps were long thought to be composed of water ice, by analogy with the Earth's polar caps. But the seasonally varying Martian caps are composed of frozen carbon dioxide, or dry ice. This is the same dry ice that is used on Earth to keep ice cream, lobsters and other things cold for days at a time. The winter on Mars is so cold that the carbon-dioxide gas above each pole freezes and falls down to the ground.

The atmospheric carbon dioxide condenses in the winter when the temperature at the pole drops below 150 kelvin, forming a large seasonal polar cap. It sublimates, or evaporates from solid ice to gas, when the polar temperature rises above 150 kelvin in the spring and summer, returning to the atmosphere. This condensation of carbon dioxide during winter and its subsequent sublimation in the spring is what gives rise to the familiar waxing and waning of the Martian polar caps. The process is entirely analogous to the snowfall that blankets the Earth's polar regions in the winter, and evaporates in the summer, except the "snowfall" on Mars consists of dry ice. It also accounts for the enormous seasonal change in the surface pressure on Mars; about 30 percent of the atmospheric carbon dioxide cycles into and out of the polar regions each year (see Fig. 3.4, Section 3.1).

The residual, remnant, perennial or permanent caps

At both poles, the caps never completely disappear in the heat of the summer, when the temporary, seasonal deposits of dry ice sublimate back into the atmosphere. Residual, or remnant, polar caps are left behind. Since they remain throughout the Martian year, these residual caps have also been called perennial or permanent caps.

The residual caps at the two poles have a split personality, with different sizes and compositions. At the south pole of Mars, the frozen carbon dioxide never entirely disappears, and a residual deposit of dry ice persists throughout the summer's warmth (Fig. 8.16). All the seasonal dry ice disappears during the northern summer, and the part that survives is about three times larger than the southern residual cap. Instead of frozen carbon dioxide, the residual cap at the north pole is composed of water ice (Fig. 8.17).

The discovery of water ice at the top of Mars was one of the great surprises of the *Viking* missions. Dry ice could not survive the relatively high temperature measured by the *Viking* orbiters at the north pole in the summer. The orbiting spacecraft also showed that the Martian atmosphere above the summer north polar cap is somewhat more humid than the atmosphere over the rest of the planet, as would be expected if it was sublimating water ice. The evaporating water molecules would rise above the cap, increasing the humidity of the atmosphere in this region.

The *2001 Mars Odyssey* spacecraft has detected water ice buried in both the Martian northern and southern hemispheres in areas that are far larger than Mars' permanent polar caps, with a total volume that probably exceeds ten thousand billion (10^{13}) cubic meters. Because

Fig. 8.13 Martian wind streaks The wind is always blowing the lighter particles around, leaving light and dark streaks across the Martian surface. With changing seasons, the winds alter direction, blowing dust away from some regions to reveal darker rocks beneath and covering other dark rocky regions with the brighter dust. Wind streaks have been recorded at mid-latitudes of Mars using cameras aboard the *Mariner 9*, *Viking 1* and *2*; and *Mars Global Surveyor* spacecraft. As seen from Earth, these streaks can combine to form large bright or dark patches on Mars. (Courtesy of NASA, JPL, and Malin Space Science Systems.)

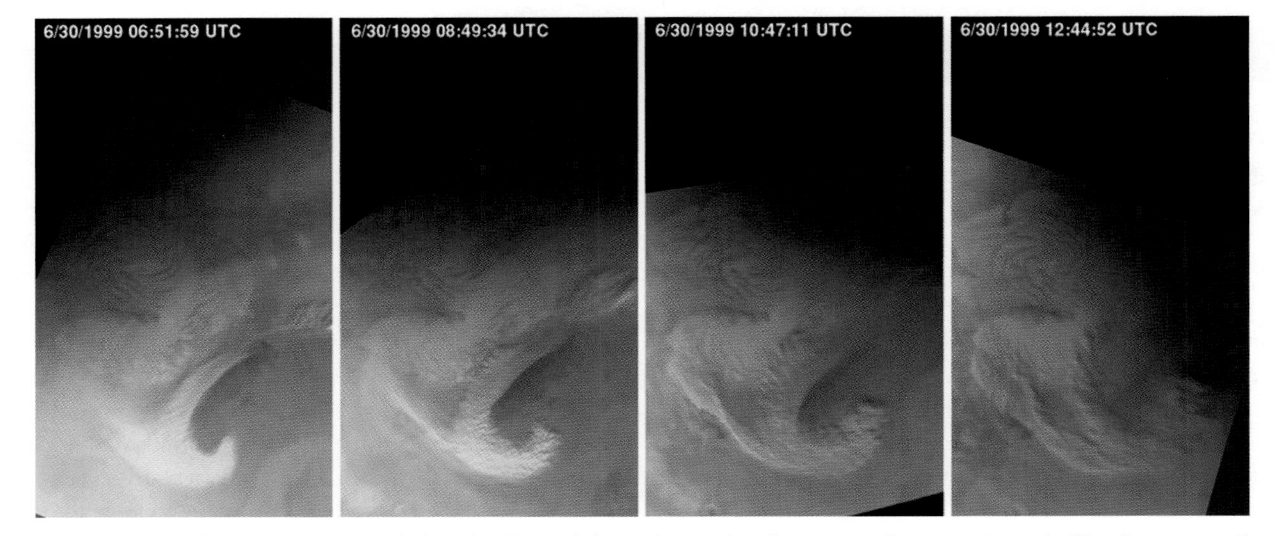

Fig. 8.14 Stormy weather on Mars Dust-laden clouds swirl above the north polar region of Mars at the end of local summer. Clouds that appear white consist mainly of water ice, while the orange-brown clouds contain dust. These images were taken at two-hour intervals from the *Mars Global Surveyor* on 30 June 1999. Storms similar to those shown here continued throughout the month of July and into August. (Courtesy of NASA, JPL, and Malin Space Science Systems.)

Fig. 8.15 Dust storm clouds out Mars Nothing on our world matches the global dust storms on Mars, dramatically displayed in this pair of natural-color *Hubble Space Telescope* images. Surface features that were crisp and clear when the first picture was taken (*left*) were rapidly covered with blinding dust by the time of the second picture (*right*). (Courtesy of James Bell, Michael Wolf, the Hubble Heritage Team, NASA, JPL, and STSI.)

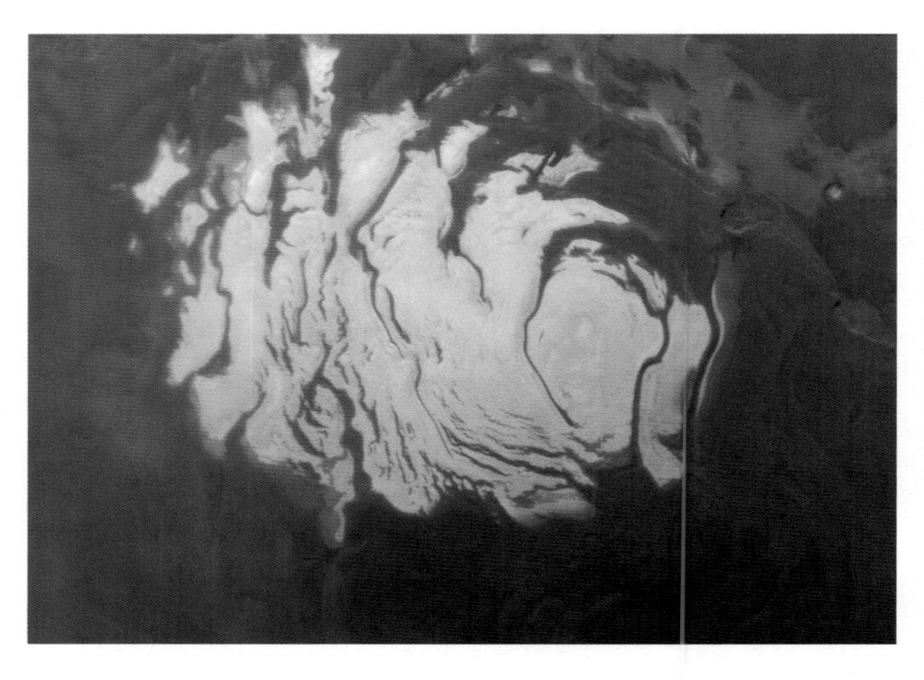

Fig. 8.16 Residual south polar cap In winter the southern polar cap is covered with extensive deposits of solid carbon-dioxide frost. In the southern summer, the cap shrinks to its minimum size shown here, about 400 kilometers across. Even though it is summer, the south polar cap remains cold enough that the residual polar frost consists of frozen carbon dioxide, or dry ice. The polar deposits lie on top of sediments that have been carved by wind into a spiral shape. This image was taken from the *Mars Global Surveyor* on 17 April 2000. (Courtesy of NASA, JPL, and Malin Space Science Systems.)

the spacecraft can detect water ice to no more than a meter under the surface, it could extend much further down.

The observed water-ice cap at the north pole could mark the tip of a vast iceberg. Its volume is estimated at about 1.2 million billion (1.2×10^{15}) cubic meters, which is less than half that of the Greenland ice cap and about four percent of the Antarctica ice sheet. If the surface pressure and temperature increased, permitting liquid water to exist on Mars, the melted cap could potentially cover the Martian surface to a depth of 10 to 40 meters. Still, the estimated volume of the water-ice cap is about ten times less than the minimum volume of an ancient ocean that some scientists believe once existed on Mars (Section 2.4). If a large body of water once existed on the red planet, the remainder of the water was either lost to space or is now stored below the surface of the freeze-dried planet.

Each residual polar cap also has a unique surface texture, suggesting that these regions have had differing climates and histories for thousands and perhaps even millions of years. Closely spaced pits, cracks and small bumps and knobs cover the flat water ice in the north residual cap, something like cottage cheese. The cold, dry carbon-dioxide ice in the south residual cap has larger pits, troughs and flat-topped mesas that make it look something like Swiss cheese. Long, narrow depressions and a waffle-like pattern of intersecting ridges are found in other parts of the south polar regions.

Evaporation of carbon-dioxide ice in the summer reveals laminated terrain that extends horizontally for several

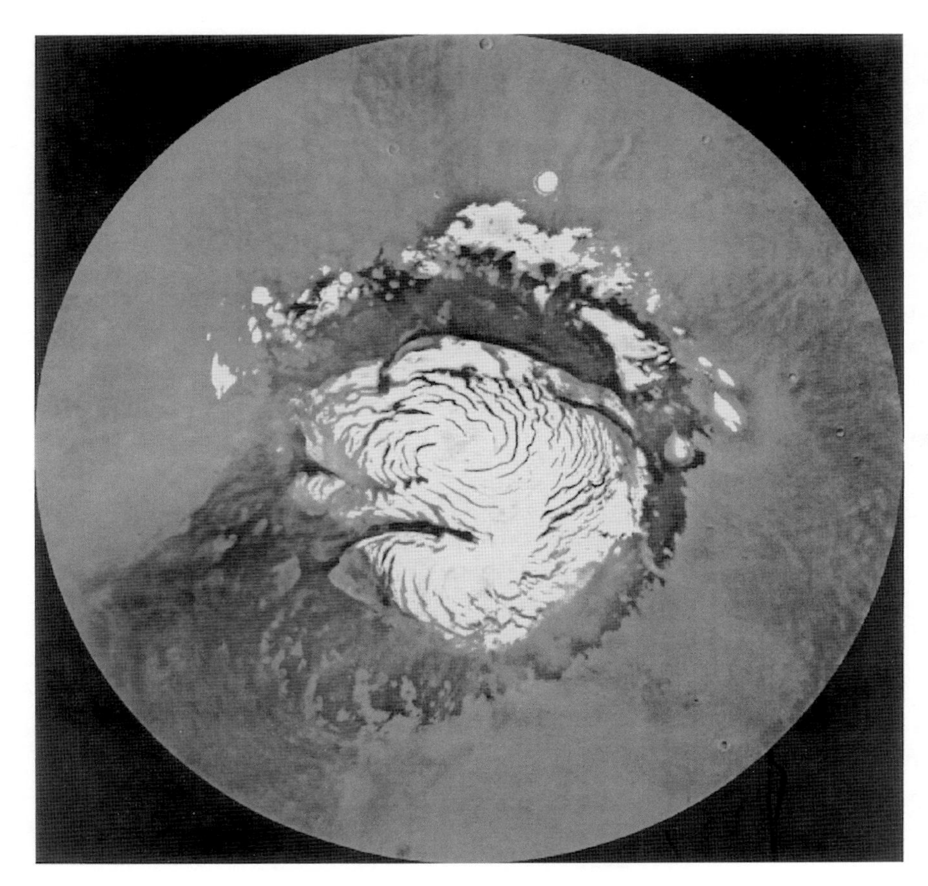

Fig. 8.17 Residual north polar cap The portion of the north polar cap that remains in summer is a towering mountain of water ice, about 1.2 thousand kilometers across. The summit, which nearly corresponds with the planet's spin axis, stands about 3 kilometers above the flat surrounding plains. The north residual cap is surrounded by a nearly circular band of dark sand dunes formed and shaped by wind. This image, which was acquired by the *Viking* orbiters during the northern summer of 1994, strongly resembles that taken by the *Mars Global Surveyor* in the northern summer of 1999. Both the north and south residual caps contain deep valleys that curl outward in a swirled pattern that has been cut and eroded into the icy deposits, like a giant pinwheel. But the residual ice cover in the south (Fig. 8.16) is made of frozen carbon dioxide and is about one-third the size of that in the north. (Courtesy of NASA, JPL, and the U. S. Geological Survey.)

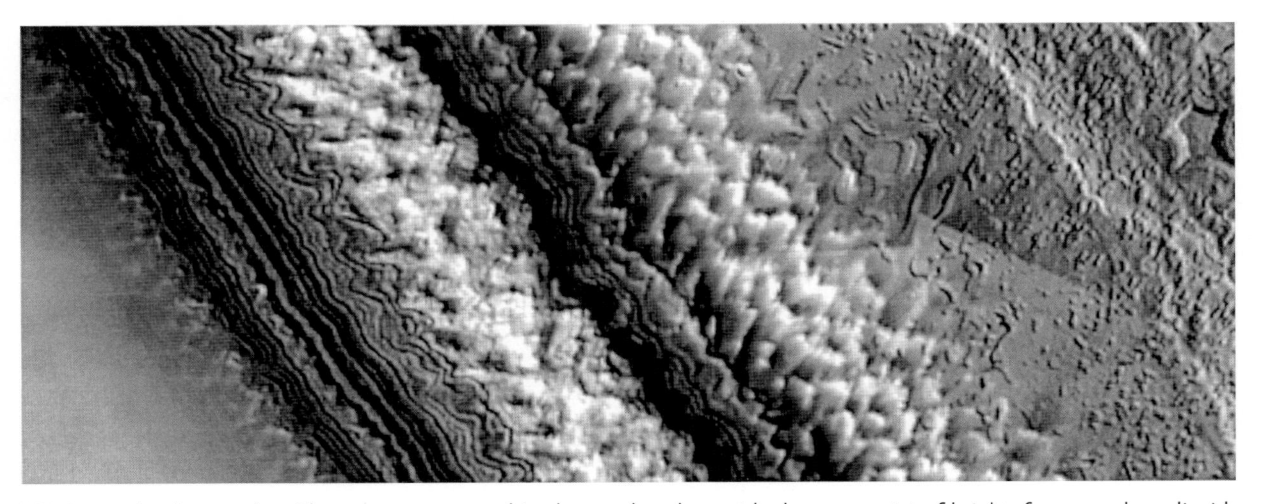

Fig. 8.18 Layered polar terrain These layers, exposed in the south polar residual cap, consist of bright, frozen carbon dioxide and dark, fine dust deposited over millions of years. Water ice could also be mixed in, but no one knows for sure. The layered terrain in both the north and south residual caps is thought to contain detailed records of the climate history of Mars. This image covers an area of 10 by 4 kilometers. It was taken from the *Mars Global Surveyor* in October 1999 at 87 degrees south and 10 degrees west, near the central region of the residual south polar cap. (Courtesy of NASA, JPL, and Malin Space Science Systems.)

kilometers along the edges of both residual caps (Fig. 8.18). Up to 20 layers have been exposed, each a few tens of meters thick, alternating between dark dust and bright ice. The extensive, regular polar layers were probably deposited during periodic climate changes. It is estimated that the layered material was laid down at the rate of about 0.001 meters (1 millimeter) per year. So a layer that is several tens of meters thick took ten thousand to one hundred thousand years to accumulate, which is roughly comparable to the periodicity of the great ice ages on

Earth (Section 4.4). The laminated terrain on Mars might be attributed to astronomical rhythms that have similarly created long-term, periodic changes in the climate of Mars, at least for the past few million years and perhaps longer.

The global climate of Mars may be changing even now, from year to year and decade to decade. A comparison of *Mars Global Surveyor* images, taken one Martian year apart, indicates that the residual southern cap is wasting away. The ice cap's circular pits are growing while its high ridges and mounds are shrinking. The changes are attributed to the evaporation of carbon-dioxide ice, with a loss that could be increasing the mass of the Martian atmosphere at the rate of 1 percent every decade. So the Martian climate seems to be in flux, changing dramatically on yearly time-scales.

The difference between the north and south extends from the polar regions to the equator, resulting in two strikingly different hemispheres on Mars.

Table 8.5 Impact basins and large craters on Mars[a]

Basin	Latitude (degrees)	West Longitude (degrees)	Diameter (km)
Hellas	−43.0	291.0	2300
Isidis	16.0	272.0	1900
Argyre	−49.5	42.0	1200
Polar	−82.5	267.0	850
Chryse	24.0	45.0	800
Renaudot	38.0	297.0	600
Ladon	−18.0	29.0	550
Sirenum	−43.5	166.5	500
Hephaestus Fossae	21.1	237.5	500
Schiaparelli	−2.7	343.3	471
Huygens	−14.0	304.4	456
Leverrier	−37.0	356.0	430
Cassini	23.8	328.2	412
Tikhonravov	13.5	324.2	386
Antoniadi	21.5	299.2	394
Nilosyrtis Mensae	33.0	282.5	380
Newcomb	−22.5	3.0	380
Herschel	−14.9	230.3	304
Al Qahira	−20.0	190.0	300
South	−73.0	344.0	300

[a] Large impact craters and basins are mainly located at southern latitudes. Adapted from P. Cattermole, *Mars, The Story of the Red Planet*, Chapman and Hall, New York, 1992.

8.6 Highland craters, lowland plains, and remnant magnetism

A world divided

The two hemispheres of Mars have distinctly differing terrains. The planet's southern half is extremely ancient, generally elevated, and highly cratered. Like the lunar highlands, most of the craters in the southern landscape on Mars probably date back to an intense bombardment by meteorites early in its history, estimated to be about 3.9 billion years ago. The northern hemisphere, by contrast, consists mainly of younger, lower-lying, flat plains that are predominantly of volcanic origin and have been greatly transformed over the eons. It is as if the two hemispheres somehow fused together to form a divided world.

Most of the north is depressed by a few kilometers below the mean level on Mars, while the majority of the south is elevated by a few kilometers. But there are exceptions, the north includes towering volcanoes that rise as much as 25 kilometers above the surrounding terrain and the south includes giant impact basins, such as the Hellas basin whose floor marks the lowest point on Mars. The northern half is also distinguished by the presence of volcanoes, canyons and flood channels.

The difference between the northern and southern hemispheres has been referred to as the great crustal dichotomy; it most likely originated during the planet's formation processes.

Heavily cratered highlands

The cratered scars of impacting meteorites can be found all over Mars, but the craters are more densely concentrated in the southern hemisphere where the terrain resembles the lunar highlands (Fig. 8.19). The largest meteorites have gouged huge impact basins out of the Martian surface, throwing up mountains along their rims (Table 8.5). The smooth floors of the biggest impact basins are light-colored, circular features that have been observed with ground-based telescopes. They retain their classical designations made more than a century ago – such as Argyre for the "silver" island at the mouth of the Ganges river and Hellas, the Greek word for Greece. The giant Hellas basin, some 2300 kilometers across, is covered with white frost in southern winter, forming a brilliant white disk seen from Earth.

The inquisitive, close-up eyes of spacecraft detect numerous craters that are smaller than the impact basins. Some of these craters become frosted in the southern

Fig. 8.19 Heavily cratered highlands This mosaic of *Viking* images displays the southern hemisphere of Mars. This half of the Martian surface retains the cratered scars of an ancient bombardment dating back to the first 500 million years of the solar system, as well as the marks of a continued bombardment since then. The conspicuous, light-colored, circular depression (*lower right*) marks the Hellas impact basin. Several large craters are located directly northeast (*upper left*) of Hellas, including those named after Giovanni Cassini (1625–1712), Christiaan Huygens (1629–1695), and Giovanni Schiaparelli (1835–1910). The residual south polar cap is located near the bottom. An enlarged version of the central part of this image is given in Fig. 8.21. (Courtesy of NASA, JPL, and the U.S. Geological Survey.)

autumn and winter (Fig. 8.20). Large craters on Mars are named after astronomers and scientists who have studied Mars, including the Schiaparelli crater (Fig. 8.21), named after Giovanni Virginio Schiaparelli (1835–1910); smaller craters are named after villages on Earth.

Although the cratered highlands of the Moon and Mars resemble each other, the Martian craters are somewhat different from their lunar counterparts. The craters on Mars are flatter and more subdued, with greater degradation and a less pristine appearance than craters on the Moon. When compared with the lunar highlands, the Martian highlands have fewer small craters, less than 30 kilometers across. The worn-down appearance of large Martian craters and the missing small ones are attributed to early, intense erosion by wind, water and perhaps even glaciers on Mars.

The ejected material surrounding some craters on Mars flows out in a splashed pattern, somewhat like that formed when you drop a pebble in the mud (Fig. 8.22). It can be explained by supposing that the Martian ground contained liquid water or water ice when the craters were formed. On Mars, the heat of impact may have melted or vaporized water ice frozen in the Martian ground just below the surface, like the layers of permafrost underlying the Arctic landscapes of Earth. Or the impact might have released liquid water from the ground beneath the permafrost. The steam and liquid water then acted as a lubricant for the flowing debris and the muddy material sloshed outward like a wave until it dried and stiffened, or became cool and refroze.

Lowland volcanic plains

The extensive lowland plains that dominate the northern hemisphere of Mars are relatively flat and relatively sparsely cratered. Most of the plains appear to be covered with lava flows, and are thus most likely of volcanic origin.

The plains of Mars are designated by the names of lands, followed by the Latin *planitia*, meaning "a level surface or plain". But they are not completely smooth. Volcanoes rise up in some of them, mesas and buttes in others; boulders or dunes give them a small-scale texture.

The Latin term *planum*, meaning "plateau or high plain" follows the name of flat elevated regions, in contrast to a low-lying planitia. Most of the planitiae are located in the northern hemisphere, while the plana are found just south of the equator (Fig. 8.23). Another Latin name, *terra*, is used to designate an extensive landmass in the older, heavily cratered highlands.

The plains of Mars are sparsely cratered, and therefore post-date the period of heavy bombardment that gave rise to the profusely cratered Martian highlands. Different plains

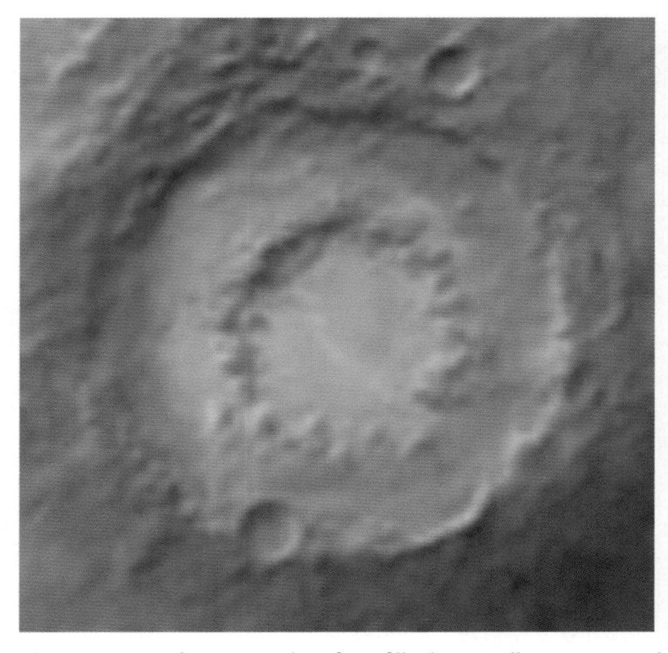

Fig. 8.20 Frosted crater White frost fills the Lowell crater, named after the American Percival Lowell (1855–1916). This image was taken from the *Mars Global Surveyor* during autumn in the southern hemisphere of Mars. The crater is 20 kilometers in diameter, and is located at a latitude of 52.3 degrees south and at 81.3 degrees west longitude. (Courtesy of NASA, JPL, and Malin Space Science Systems.)

Fig. 8.22 Yuty The lobate, layered material surrounding this crater may have been ejected when an impacting object melted the permafrost, or frozen ground, on Mars. Multiple layers of successive flows resemble the overlapping petals of a flower. The thin flow partly buries one crater, and is halted and deflected by the rim of another. The muddy sludge seems to have sloshed across the surface, and was then refrozen. Such ejected features have not been found around craters on the Moon or the other planets. This crater, named Yuty after a town in Paraguay, is 19.9 kilometers in diameter, and located in Chryse Planitia at 22.4 degrees north and 34.2 degrees west. (A *Viking* image courtesy of NASA.)

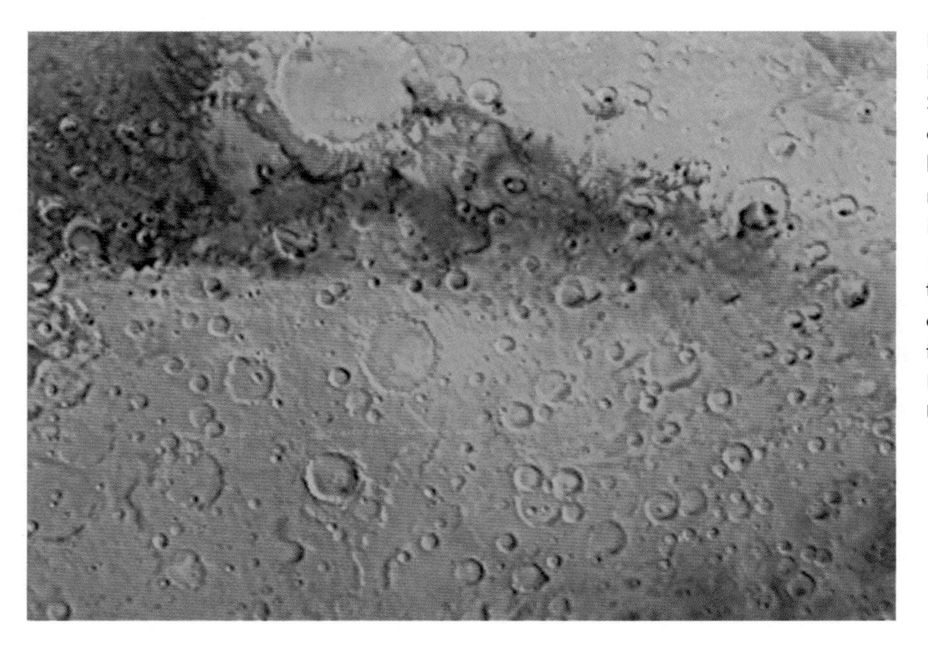

Fig. 8.21 Sinus Sabeus quadrangle Heavily cratered highlands dominate the Sinus Sabeus region of Mars, located just south of the equator between 0 and −30 degrees latitude. A large impact crater, 471 kilometers in diameter, and named after the Italian astronomer Giovanni Schiaparelli (1835–1910), marks the northern part of this mosaic image, taken from the *Viking* orbiters. A full disk image that shows this feature in lower resolution is shown in Fig. 8.19. (Courtesy of NASA, JPL, and the U.S. Geological Survey.)

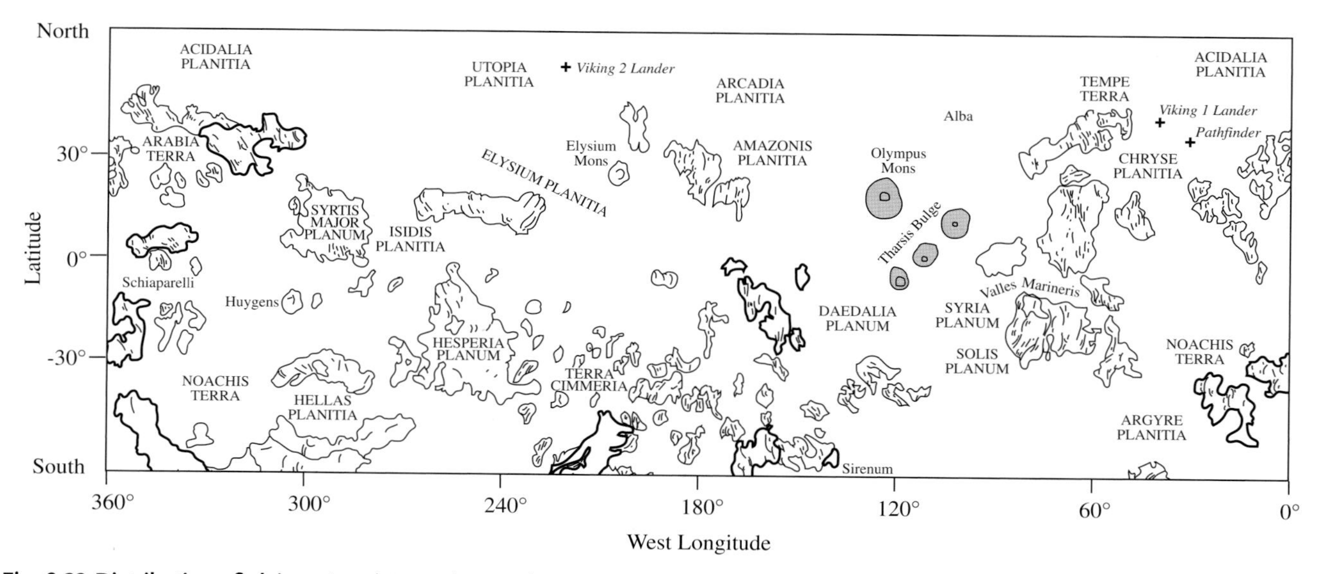

Fig. 8.23 Distribution of plains Low-lying volcanic plains, each designated as a planitia, are located throughout the northern hemisphere of Mars. Other relatively smooth regions are found at the top of elevated plateaus, each called a planum; they are located in the southern side of the equator. Shaded, circular regions denote volcanoes, including Olympus Mons that rises out of the Tharsis bulge, or uplift. The small crosses designate the landing sites of *Viking 1* and *Pathfinder* in Chryse Planitia (*upper right*) and *Viking 2* in Utopia Planitia (*top center*).

nevertheless exhibit varying crater densities, and this provides a method for tracing the planet's development.

The more craters there are in a given area, the older that terrain is, and the fewer the craters the younger the terrain. Volcanic flows of lava probably helped fill in and erase older craters in some regions (Fig. 8.24), while other processes such as wind-blown deposits also contributed. The cratered highlands on Mars probably were formed about 3.9 billion years ago, as the Moon's highlands were, and younger surfaces can be roughly dated by comparisons of the number of impact craters with the lunar record.

Such comparisons show that different regions on Mars span a large range in ages, from the ancient, heavily cratered highlands to very lightly cratered volcanoes that may be younger than a million years. The residual polar caps are practically crater free, suggesting that they are younger still and may now be undergoing modification.

Scientists have used this method to map out the broad periods in the history of Mars, defining three periods known as the Noachian, Hesperian and Amazonian eras, spanning the ages of 4.00 to 3.80, 3.80 to 3.55, and less than 3.55 billion years, respectively (Table 8.6). Each era has been further divided into its early, middle and late parts, but this is not needed to follow the main steps in the planet's geological history.

The Noachian era was a time of intense bombardment that created the large impact basins and the heavily cratered highlands. Early volcanic resurfacing also occurred during this period. The formation of craters had declined by

Table 8.6 The geological development of Mars

Era	Approximate Age (10^9 yr = 1 byr)	Major Geological Events and Representative Terrain
Noachian	4.00 to 3.80	Formation of large impact basins such as Isidis, Hellas and Argyre. Oldest landmasses, like Noachis Terra.
Hesperian	3.80 to 3.55	Catastrophic floods from outflow channels, faulting of Noctis Labyrinthus and Valles Marineris.
Amazonian	3.55 to 1.80	Vast lava flows form lowland plains such as Amazonis and Elsium Planitia, shield volcanoes produced, formation of polar ice caps and their layered deposits.

Fig. 8.24 Lava-flow Extensive volcanic plains and towering volcanoes are found on Mars. This image, taken from the *Mars Global Surveyor*, captures lava frozen into the surface of Daedalia Planum, southwest of the Arsia Mons volcano, probably between 1 and 2 billion years ago. An area of 1.5 kilometers by 2.0 kilometers is covered in this image. (Courtesy of NASA, JPL, and Malin Space Science Systems.)

the Hesperian era, when vast canyons were formed and catastrophic floods carved out huge channels. The extensive, lowland volcanic plains were emplaced across the Martian surface during the Amazonian era, leading to the formation of the Lunae Planum, Chryse Planitia, Syrtis Major Planum, Amazonis Planitia, and Utopia Planitia, with estimated ages of 3.5, 3.0, 2.9, 2.8 and 1.8 billion years, respectively. Towering volcanoes were also formed throughout this era, from 3.5 billion years ago to relatively recently, perhaps even now.

Leftover magnetic fields

Unlike Earth's global, dipolar magnetic field, the magnetism on Mars is now detected only in local regions on the surface. Most likely, the local fields are fossil remnants of an early time when the Martian dynamo was sufficiently vigorous to magnetize the planet's crust and maintain a dipolar field that is now imprinted in the ancient rocks.

The dichotomy between the northern and southern hemispheres of Mars is retained in its leftover magnetism (Fig. 8.25). The magnetic fields that have survived from the planet's active youth are located within the heavily cratered southern hemisphere, and the northern lowlands are now largely free of magnetism. Magnetic fields were therefore present when the highlands were formed. But when the impacts stopped and the lowland plains originated there was no magnetic field left.

The remnant magnetic fields preserved in the ancient highland crust are also missing from the very large, southern impact basins, such as Hellas and Argyre (Fig. 8.25). When these impacts occurred, the internal dynamo must have been shut down, and the global fields were no longer present. Otherwise, the impacted material in the planet's crust would have been magnetized or else something related to the impact process would have removed magnetized crustal material.

So the global magnetic fields, and the dynamo that generated them, had a relatively short life on Mars. They were most likely gone after the first few hundred million years of the planet's life. One explanation is that Mars cooled down enough to form a thick, solid, unbroken outer shell, or lithosphere, that inhibited the internal churning of its dynamo.

The magnetic fields that have survived on Mars are banded into a striped, bar-code pattern of alternating magnetic polarity, each stripe extending up to 2000 kilometers from east to west (Fig. 8.26). The magnetic field in one band points out of the planet, and the adjacent one points in; next to that it points out again and so forth.

Fortunately for life on Earth, the terrestrial magnetic field shields our planet from fast-moving particles in the Sun's winds, as well as energetic, cosmic-ray particles. In contrast, the current lack of a global magnetic field on Mars, combined with its very thin atmosphere, leaves the surface virtually defenseless against the onslaught. Life could not survive when exposed to the relentless bombardment of the Martian surface by energetic solar and cosmic particles. Any living things, from ancient microbes to future astronauts, would have to be protected from the lethal particles, as well as from the dangerous ultraviolet rays that pass right through the tenuous air. Perhaps primitive Martian organisms once took refuge deep within the planet's hot interior or under lakes or seas that are no longer there.

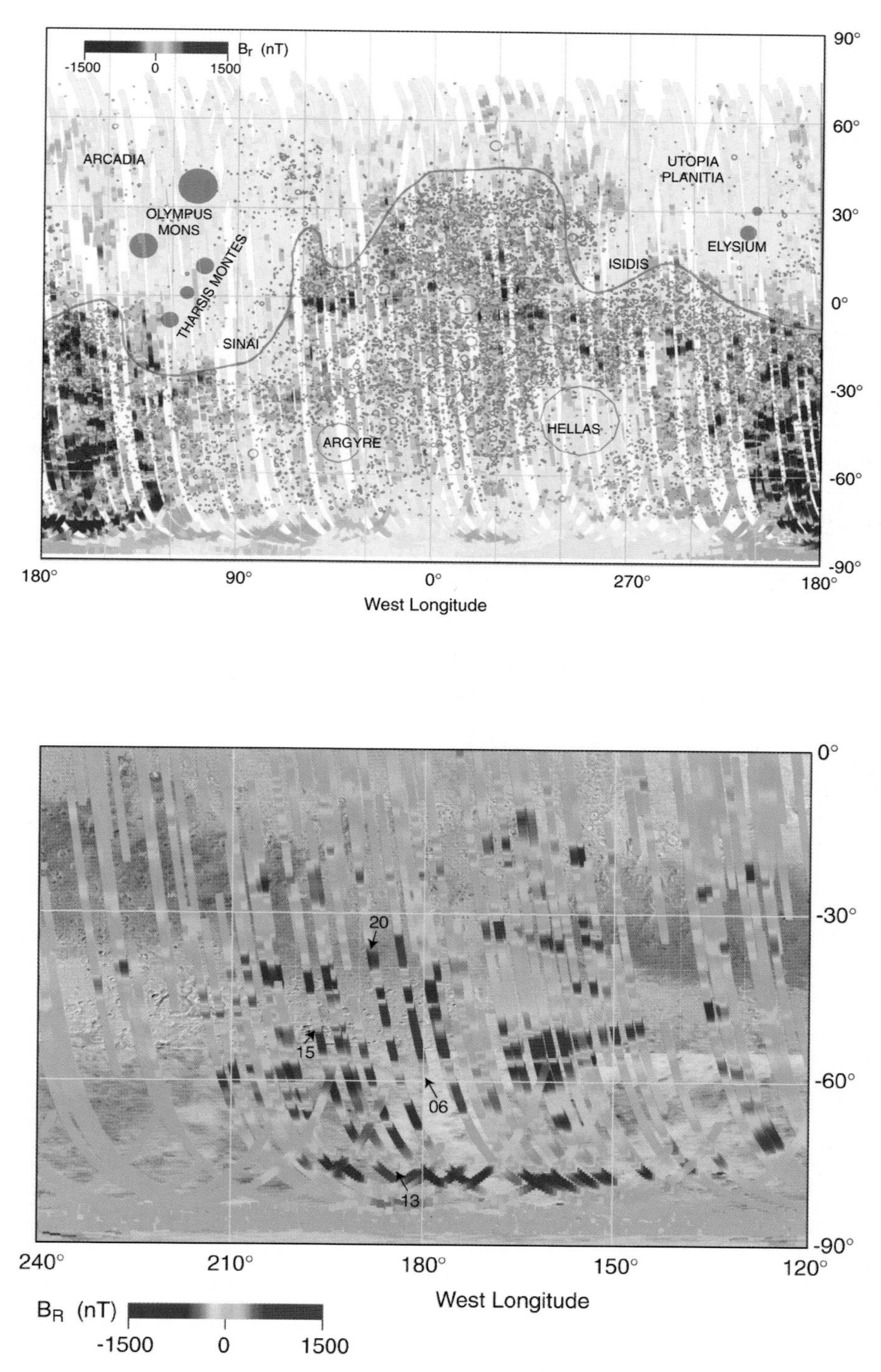

Fig. 8.25 Crustal dichotomy and remnant magnetism The magnetic fields that are currently on Mars are located mainly in the older, heavily cratered southern hemisphere, below the dichotomy boundary (*solid line*) that separates the highlands (*bottom*) from the younger, smoother lowland plains in the north (*top*). A global magnetic field, which no longer exists, was probably imprinted in the highland rocks when they formed more than 4.0 billion years ago, before the lowland plains existed. Hellas and Argyre are two deep impact craters blasted out of the Martian surface after the global magnetism was no longer present. The blue and red colors represent magnetic fields pointing in opposite directions, in and out of the planet, as measured by an instrument aboard the *Mars Global Surveyor*. They form a banded pattern shown in greater detail in Fig. 8.26. (Courtesy of NASA, JPL, and GSFC.)

Fig. 8.26 Magnetic stripes Alternating bands of magnetic polarity are most prominent in this part of the southern highlands, near Terra Cimmeria and Terra Sirenium. The magnetic data were obtained from an instrument aboard the *Mars Global Surveyor*. This map is color-coded red for a positive magnetic field pointing out of the planet and blue for a negative one pointing in, with a strength up to 1500 nanotesla, or 1.5×10^{-6} tesla. Stripes of alternating polarity, or direction, extend up to 2000 kilometers across the planet in the east–west direction. They are similar to the magnetic patterns seen in the Earth's crust at both sides of the mid-ocean ridge, where the spreading crust has recorded flip-flop reversals in the Earth's dipolar magnetic field. (Courtesy of NASA, JPL, and GSFC.)

8.7 Towering volcanoes, immense canyons

Eons of volcanic activity

When *Mariner 9* and *Viking 1* and *2* surveyed Mars in the 1970s, they revealed a mind-boggling world with kaleidoscopic beauty and variety. They replaced the bleak, drab, Moon-like view of a frozen, lifeless Mars, obtained during the partial, fleeting glimpses of previous spacecraft, with a new picture of a dynamic, living planet. Powerful forces have molded the face of Mars at an unsuspected scale, including giant volcanoes and immense canyonlands that dwarf their terrestrial counterparts (Fig. 8.27).

Fig. 8.27 Mars revealed Thousands of *Viking* images of Mars have been pieced together to provide the first detailed map of the entire globe of another planet. This mosaic, centered on 20 degrees latitude and 60 degrees west longitude, is most like the view seen by a distant observer looking through a telescope. The projection includes the great equatorial canyon system, Valles Marineris (*center, below middle*), the four huge Tharsis volcanoes (*left*), and the north polar cap (*top*). Also note the heavy impact cratering of the highlands (*bottom* and *right*), and the younger, less heavily cratered terrains elsewhere. (Courtesy of NASA, JPL, and the U.S. Geological Survey.)

Though about half the Earth's size, Mars is a big world. Rising above the landscape is a towering volcano, Olympus Mons, twice as tall as Mount Everest. A gigantic canyon dubbed Valles Marineris after *Mariner 9*, could stretch from New York City to San Francisco, putting our Grand Canyon to shame.

As *Mariner 9* settled into orbit around Mars, in November 1971, the planet was buried beneath a global dust storm. After circling the planet for a couple of months, the winds abated, the dust settled, and the spacecraft watched four high mountains emerge from the pall, each with craters at their summit (Fig. 8.28).

The volcanoes are perched on top of a vast uplift, known as the Tharsis bulge, which straddles the ancient uplands and lowland plains near the equator and overlies them both. The Tharsis bulge extends more than 2500 kilometers across, and it was formed roughly 2 billion years ago, after the division between the highlands and lowlands. Olympus Mons, the tallest mountain in the solar system, lies on the western edge of the Tharsis bulge (Fig. 8.29). Three other tall volcanoes, named Arsia Mons, Pavonis Mons and Ascraeus Mons, crown Tharsis. They run diagonally across the equator along a ridge known as Tharsis Montes (Fig. 8.28).

These other three volcanoes are smaller than Olympus, but still much larger than any terrestrial volcano. With their gently sloping flanks and roughly circular summit calderas (Fig. 8.30), the Martian volcanoes resemble the shield volcanoes of Hawaii, such as Mauna Loa, which has a similar slope but one-third the height and one-twentieth the volume of Olympus Mons. Such volcanoes are formed by the repeated eruption of lava that cascades down the flanks in thousands of individual flows.

In addition to the four large Tharsis shield volcanoes, Mars has hundreds of smaller shields, concentrated primarily in the general area of Tharsis and in the older volcanic region of Elysium. Volcanoes of the Tholus type have steeper slopes than the shields, perhaps because of a more viscous lava or lower eruption rate. Ancient volcanoes called patera resemble gigantic shield volcanoes that have collapsed and eroded.

Why are some Martian volcanoes so much higher than their terrestrial counterparts? Perhaps it has to do with the thickness of the outer shell, or lithosphere, of Mars. Because Mars is smaller than the Earth, it probably cooled faster, and its lithosphere became relatively stronger and thicker, and not prone to breaking up. This gives Martian volcanoes a longer chance to grow in one spot.

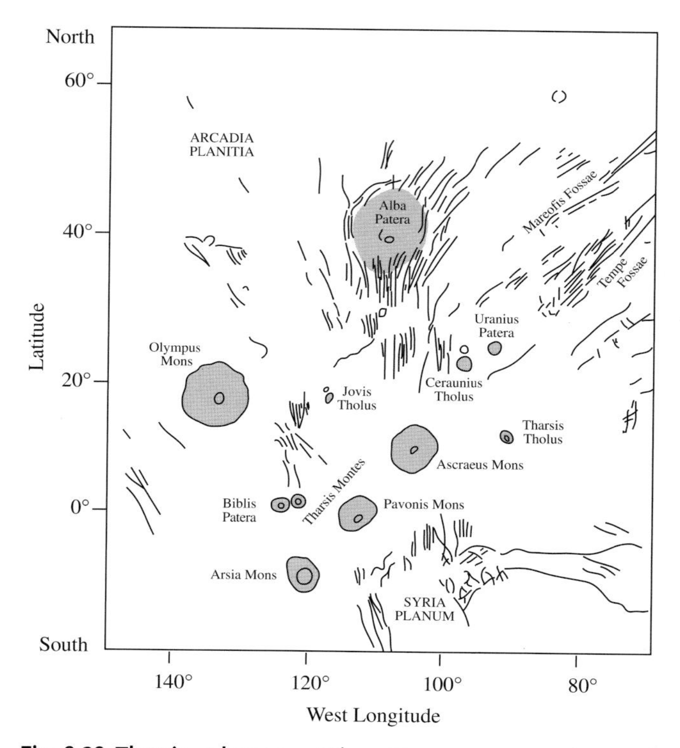

Fig. 8.28 Tharsis volcanoes When Mars was young, it experienced internal adjustments that resulted in a swollen and cracked surface. The most prominent bulge, known as the Tharsis uplift, has four gigantic shield volcanoes on top of it. They are the Olympus, Arsia, Pavonis and Ascraeus Montes. Lava has flowed down the flanks of these volcanoes for perhaps a billion years, but the most recent eruptions could be 300 million years old, or less. Very old, presumably extinct volcanoes are also found on the Tharsis uplift, such as Ceraunius Tholus (Section 2.3, Fig. 2.31) with an estimated age of 2.4 billion years.

Fig. 8.29 Oblique view of Olympus Mons Clouds rest against Olympus Mons, which is about 25 kilometers high. The volcano's flanks have a gentle slope, with a diameter at their base of about 600 kilometers. Thus, Olympus Mons is about 24 times broader than it is high. (Courtesy of NASA, JPL, and Malin Space Science Systems.)

The lithosphere on Mars is one thick, solid plate, so crustal movements are mainly vertical rather than horizontal. In contrast, the Earth has a relatively thin outer shell, and internal currents of hot rock have broken this solid skin into plates. They move with respect to one another and slide horizontally over the deep-seated, hot-spot sources of magma. This motion limits the growth of individual volcanoes on Earth, and produces chains of smaller volcanoes, such as the Hawaiian chain in the Pacific Ocean. On Mars, the crust does not move across the internal hot spots, so the lava can erupt from them for billions of years, building up volcanoes far larger than any volcano on the Earth.

The difference in gravity between Mars and the Earth also contributes to the size of Martian volcanoes. As a volcano grows in height, it eventually becomes too heavy for the underlying rock to support, and the added weight causes the entire mountain to spread outwards. Because the force of gravity on Mars is only about one-third as great as that on Earth, the Martian volcanoes can grow more than twice as tall as their terrestrial counterparts before reaching the limiting height caused by too much weight.

Since the Martian lithosphere did not break apart and move sideways, its volcanoes continued to erupt for a very long time, thereby growing to their immense size. While the latest lava-flows on their flanks may be relatively recent,

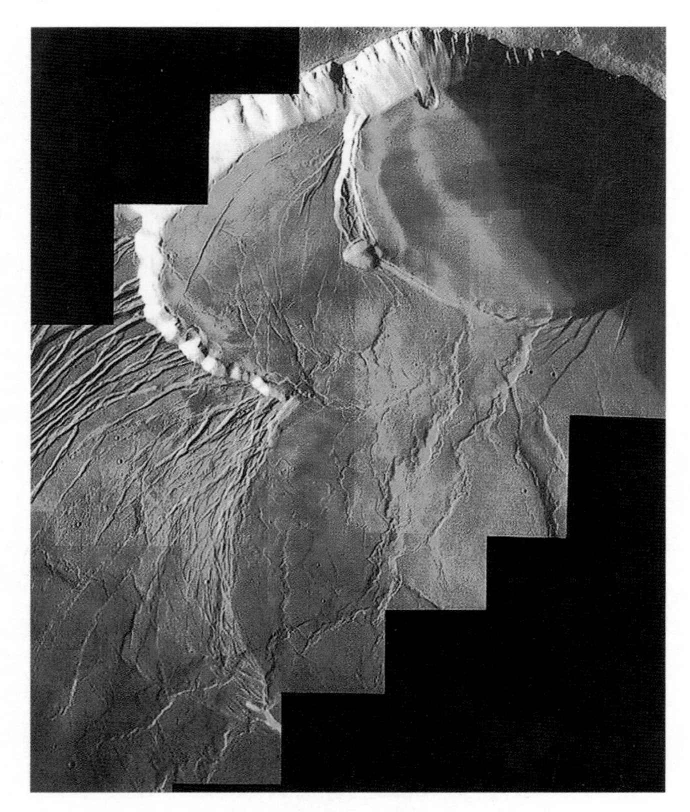

Fig. 8.30 Summit caldera of Olympus Mons The summit region of Olympus Mons contains nested calderas and radiating lava-flows. The crater-like pits have been formed by repeated collapses after eruption, each one marking the point where lava has withdrawn from a chamber within the volcano. The largest collapse feature in this *Viking* image is about 25 kilometers across and nearly 3 kilometers deep. (Courtesy of NASA and JPL.)

lava may have been flowing down some of them for at least two billion years before that. By way of comparison, the oldest volcanoes on Earth have ages of just a few million years.

The lack of craters on some surfaces indicates volcanic activity in the recent geological history of Mars. Close-up images of several volcanic landforms exhibit very few impact craters, suggesting that recent lava-flows have smoothed them over. As an example, the summit of Arsia Mons has impact crater densities only about 2 to 6 percent of those on the lunar maria, over a range of sizes. The latest lava-flows in the volcano's caldera floor were probably emplaced during the past 100 million years or so. The lowland plains located near Elysium Mons, a large and relatively youthful shield volcano, have less than 1 percent of the craters found on the lunar maria. Thin volcanic flows may have spread across this part of the Elysium Planitia within the last 10 million years. And some northern conical volcanoes are so low and fresh looking that their lava was probably deposited within the past 1 to 20 million years.

Yet, Martian volcanism dates back billions of years. Some volcanoes erupted 2 or 3 billion years ago; Ceraunius Tholus is an example. The sparsely cratered, young flows in Elysium Planitia lie atop stratified deposits whose bottom layers are most likely a few billion years old. And the high crater density on the outer flanks and outer edges of the Martian colossus, Olympus Mons, indicate a similar old age for the edifice, even though the paucity of impact craters at its summit imply that lava-flows may have occurred less than 300 million years ago.

Since it is bigger than Mars, the Earth has remained much more internally active than Mars, continually renewing the terrestrial surface and destroying most of the Earth's older terrain. And the internal heat of the Moon and Mercury, which are smaller than Mars, cooled off long ago; they have not been active for billions of years. Mars is in between these extremes. It is just large enough to have remained active for most of the solar system's history, but not so active that all record of its early history has been erased.

Mars has therefore been alive with internal heat for most of the planet's history, and surface volcanism has persisted for much of it. The planet's internal furnace is probably still turned on, and Mars has not yet become a dead world. Although no spacecraft has yet detected a volcanic eruption on the red planet, there may still be internal stirrings just beneath the planet's surface.

Vast canyons

The turmoil associated with the formation of the Tharsis uplift and associated volcanoes opened up a network of vast canyons that are collectively known as Valles Marineris (Valleys of the Mariner – named for the *Mariner 9* spacecraft). The colossal system of interconnected canyons, or chasmata, extends down the eastern flanks of the Tharsis bulge and along the Martian equator for 4000 kilometers, one-quarter the way around the planet (Fig. 8.31). In places the chasms are as wide and deep as Mount Everest is high. Their formation may be similar in origin to the rift valley in Africa (Section 4.3), but on a much vaster scale. For Mars, it was as if a cosmic sculptor was trying to split the planet asunder.

The deep rifts opened up early in the planet's history, a few billion years ago. Erosive forces then took over. Landslides fell into the newly created voids, powerful winds blasted through the canyons, and water apparently flowed through parts of the abyss. You can see the effects in the canyon walls, which have been widened by gigantic landslides and are dissected by gullies.

Valles Marineris originates close to the summit of the Tharsis uplift, at Syria Planum, where the surface

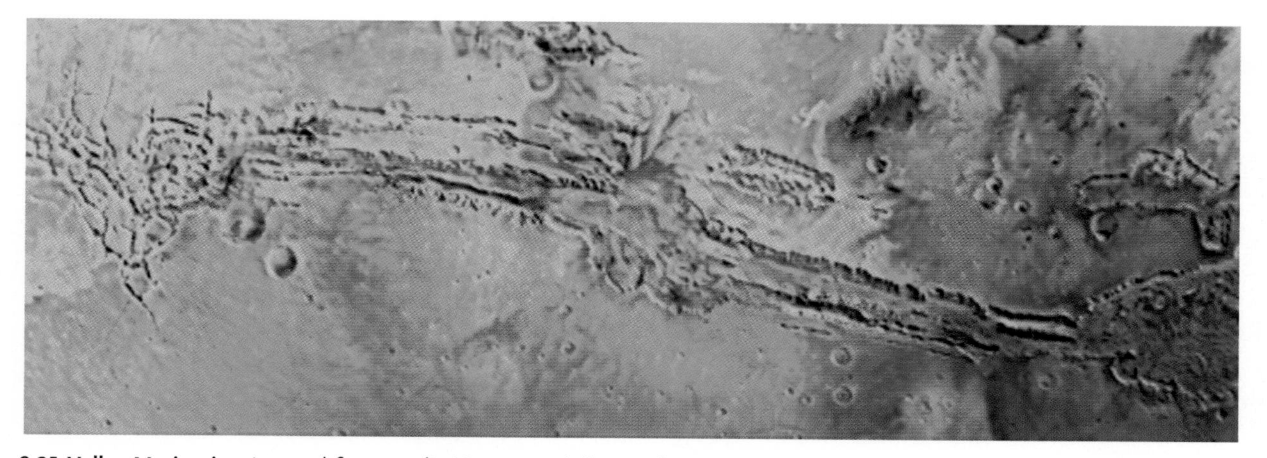

Fig. 8.31 Valles Marineris Internal forces split Mars open billions of years ago, creating a vast system of connected canyons. Landslides, winds and water have subsequently modified the terrain. Collectively known as Valles Marineris, the canyons extend from the Noctis Labyrinthus in the west (*left*) to the Chaotic Terrain in the east (*right*), spanning more than 4 thousand kilometers and averaging 8 kilometers in depth. This panorama is a mosaic of images taken from *Viking 1*. (Courtesy of NASA, JPL, and the U.S. Geological Survey.)

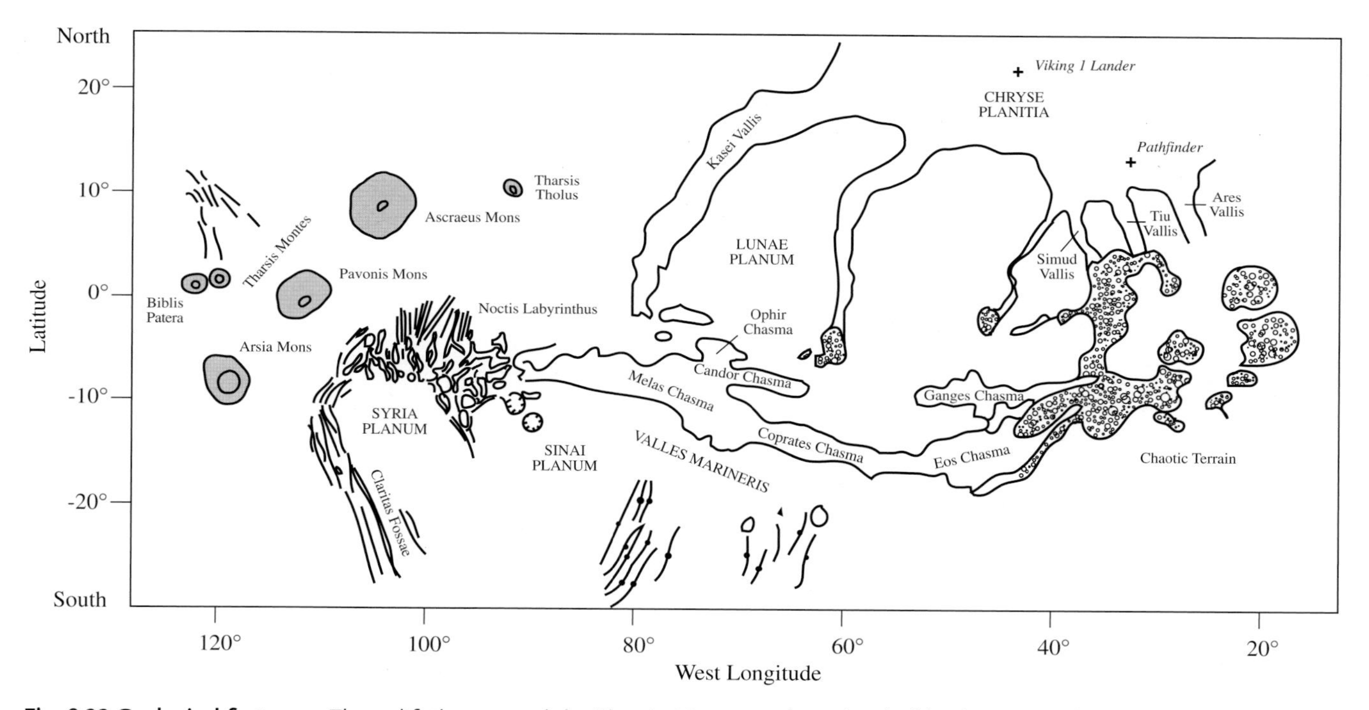

Fig. 8.32 Geological features The uplift that created the Tharsis Montes and nearby shield volcanoes (*left*) seems to have fractured the terrain and opened up an enormous network of chasmata, or canyons, knows as Valles Marineris. Catastrophic floods originating in the vicinity of these canyons flowed north (*top*) into Chryse Planitia, which contains the site of the *Viking 1* lander and the *Mars Pathfinder* lander.

expansion and consequent stretching has produced the intricately fractured Noctis Labyrinthus (Labyrinth of the Night). As the name suggests, it is a maze of short, deep gashes intersecting at all angles (Fig. 8.31).

Further east the depressions become deeper, wider and more continuous. In the middle section of Valles Marineris they branch into three parallel canyons, the Ophir Chasma, Candor Chasma and Melas Chasma, which are separated by intervening ridges (Fig. 8.32). These canyons connect with the single long Coprates Chasma, which runs eastward and joins Eos Chasma. Even further east the canyons become shallower, with evidence of past water flow, and finally the canyons terminate in the jumbled, blocky region called the Chaotic Terrain. The system for naming all of these newly discovered surface features is described in Focus 8.2 and Table 8.7.

Focus 8.2 Naming features on Mars

There are two names that describe every feature on Mars. They are a particular name that identifies it, plus a descriptive name that says what it looks like. Some of the descriptor terms, or feature types, commonly used on Mars are given in Table 8.7.

The particular names for large craters on Mars are those of deceased astronomers, scientists and writers who have contributed to the study or lore of Mars. Small craters receive the names of small villages of the world with a population less than 100 000. Large valleys, described by the Latin term *Vallis*, are given the name of Mars in various languages. Examples include the Ares, Kasei, Nirgal, Simud and Tiu Valles, named respectively for the word "Mars" in Greek, Japanese, Babylonian, Sumerian and old English. Small valleys receive the classical or modern names of terrestrial rivers.

Other features on Mars are designated for the nearest named albedo (light and dark) feature on the maps of Giovanni Virginio Schiaparelli (1835–1910) or Eugène Marie Antoniadi (1870–1944), whose appellations were drawn from classical literature and the Bible. The bright areas were named for continents or islands, such as Argyre, Arabia, Chryse, Elysium, Hellas and Tharsis. The main dark areas were given the names of bodies of water, such as Sinus Sabaeus, Solis Lacus, and Syrtis Major. The complete list is available in the International Astronomical Union's *Gazeteer of Planetary Nomenclature*, web site: http://planetarynames.wr.usgs.gov/

The origins of the particular names for some of the surface features shown in other parts of this chapter are:

Amazonis	Land of the Amazons, on the island Hesperia
Arabia	Country bordering on Aeria (Egypt)
Arcadia	Mountainous region in southern Greece, imagined rural paradise
Argyre	"Silver" island at mouth of Ganges river
Candor	Latin for "blaze" or "white"
Chryse	Island rich in gold
Ceraunius	"Thunderclap", Ceraunii Mountains on coast of Epirus, Greece
Coprates	Old name for Persian river Ab-I-Diz
Elysium	Home of the blessed on the western edge of the world
Hellas	Greek name for Greece
Hesperia	The land where the Sun sets
Nix Olympica	Snows of Olympus, the mountain home of gods in Greece
Noachis	Biblical region of Noah
Ophir	Biblical land to which King Solomon sent naval expedition, probably India
Sinai	Biblical, named for area next to Mare Erythraeum (Indian Ocean)
Sinus Sabaeus	Today's Red Sea
Solis Lacus	Lake of the Sun, the so-called "Eye of Mars"
Syria	Province in Near East including Phoenicia; or one of the Cyclades
Syrtis Major	Libyan Gulf, now Gulf of Sirte
Tharsis	Connecting link between East and West
Utopia	An ideal and perfect place or state, Greek for "no place" and modern Latin for "imaginary island"

8.8 Liquid water on Mars

Cold, parched and wrapped in a thin, carbon-dioxide atmosphere, Mars today is a frozen, desiccated and inhospitable world. It cannot now rain on Mars, and liquid water cannot now remain on its surface. Yet dry riverbeds, deep winding channels, and streamlined, washed-out landforms all provide compelling evidence for abundant liquid water on Mars in the distant past. In the planet's early history, rivers ran across its surface and powerful floods coursed down its valleys, emptying into the russet plains and perhaps forming ancient lakes or seas.

The ancient, water-cut features take two main forms, dubbed the valley networks and outflow channels. The valley networks seem to have been derived from the gradual flow of liquid water. The outflow channels were gouged out of the surface by the powerful rush of short-duration floods.

Table 8.7 Common features on the surface of Mars

Descriptor Term (singular, plural)	Feature Type	Example
Chasma, Chasmata	Deep, elongated steep-sided depression	Candor Chasma (6°S, 71°W),
Labyrinthus, Labyrinthi	Complex of intersecting valleys or canyons	Noctis Labyrinthus (7°S, 101°W)
Mons, Montes	Mountain, Volcano	Olympus Mons (18°N, 133°W)
Planitia, Planitiae	Low plain	Elysium Planitia (20°N, 230°W)
Planum, Plana	Plateau or high plain	Sinai Planum (15°S, 87°W)
Terra, Terrae	Extensive landmass	Noachis Terra (35°S, 335°W)
Vallis, Valles	Valley	Ares Vallis (10.4°N, 25°W)

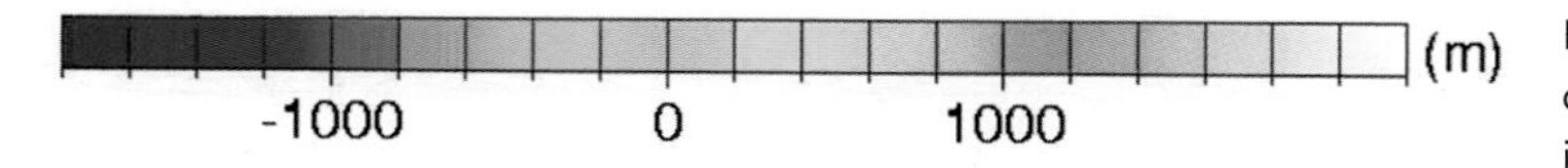

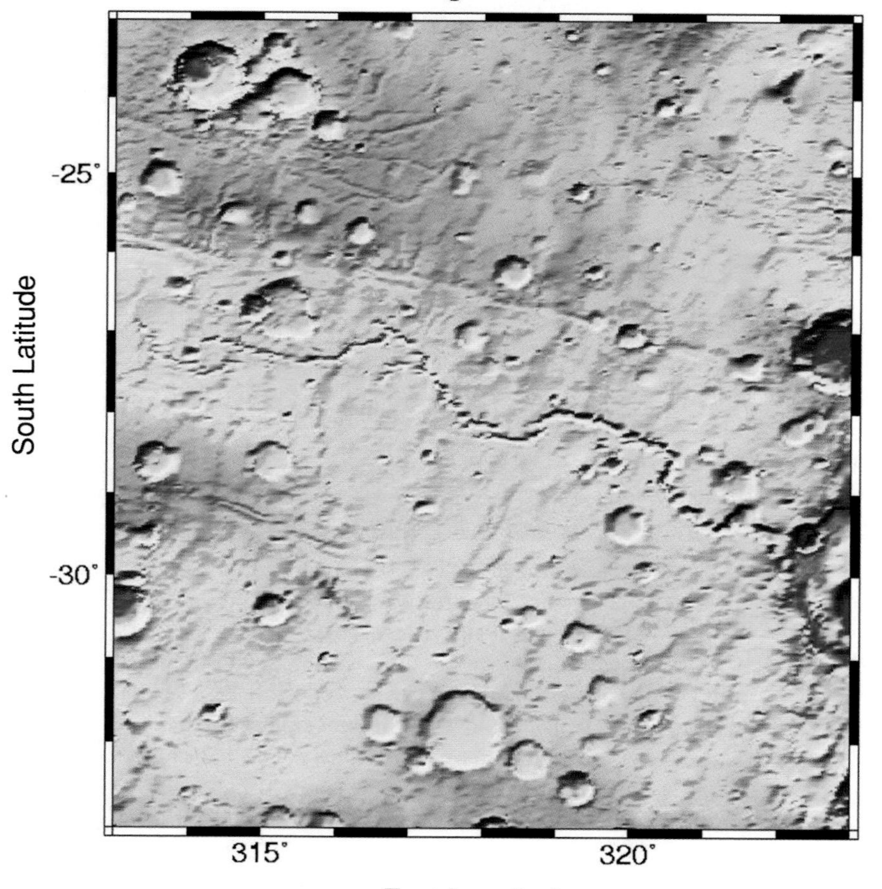

Fig. 8.33 Nirgal Vallis This valley meanders over 500 kilometers across the heavily cratered terrain on Mars. It is shown in topographic relief with a vertical accuracy of approximately 1 meter provided by the Mars Orbiter Laser Altimeter on the *Mars Global Surveyor*. Nirgal is the word for "Mars" in Babylonian. (Courtesy of NASA, JPL, and GSFC.)

Dry riverbeds

The valley networks have branching tributaries that connect into larger flows, so they look like dry riverbeds. The meandering flows increase in size downstream, and follow the local topography (Fig. 8.33). They have lengths of up to hundreds of kilometers and widths of one to a few kilometers, and they resemble river valleys on Earth. Of course, there is now no liquid water in sight on Mars and these dry riverbeds were created long ago. They lie almost entirely

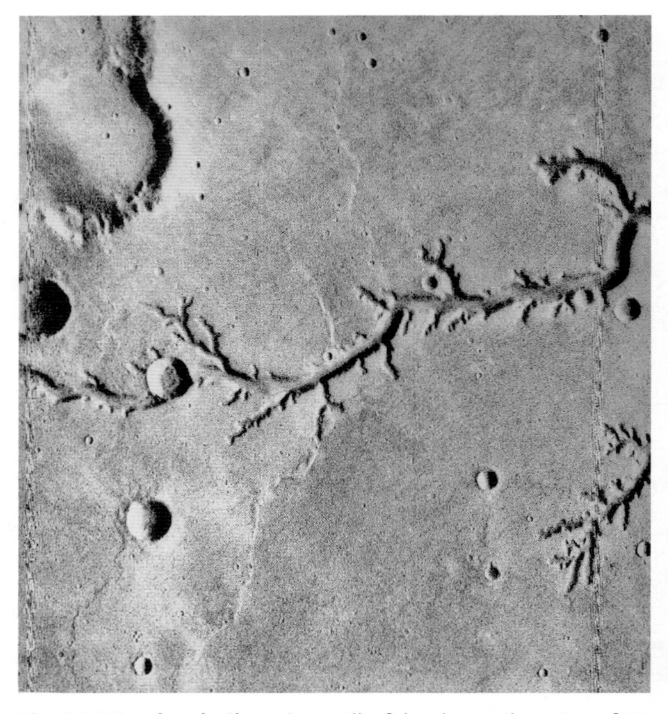

Fig. 8.34 Dead end tributaries All of the deep tributaries of Nirgal Vallis have blunt ends and steep walls, with no indication of erosion between them. These characteristics suggest an origin by collapse into cavities that were eroded by underground streams rather than surface rivers. This *Viking* image is 60 kilometers across. (Courtesy of Michael Carr and NASA.)

in the ancient, heavily cratered highlands, and are rarely found in the younger, lowland plains. So, the valley networks are about as old as the highlands, dating back 4.0 to 3.8 billion years ago.

As the valley networks wind and meander their way downhill, they coalesce with several well-developed tributaries. This suggests that they, like most terrestrial river valleys, formed by the slow and prolonged erosion of running water rather than by rapid, surging floods.

Scientists have debated whether the running water fell as rain, during a sustained period of warm, wet climate, or flowed just below the Martian surface, warmed by internal heat. Recent evidence supports the view that the valley networks formed by collapse into cavities formed by water running under the frozen, ice-rich surface. The branching tributaries of some valley networks end abruptly in box canyons, which indicate underground cavities rather than surface rain (Fig. 8.34). Valleys that seem to have been deeply cut by the continual flow of water also have features that suggest formation by collapse rather than rainfall (Fig. 8.35). Yet, other evidence points to a time when Mars was wrapped in a thicker and warmer atmosphere than now, with possible rainfall, rivers, lakes and maybe even an ocean (Focus 8.3; Fig. 8.36).

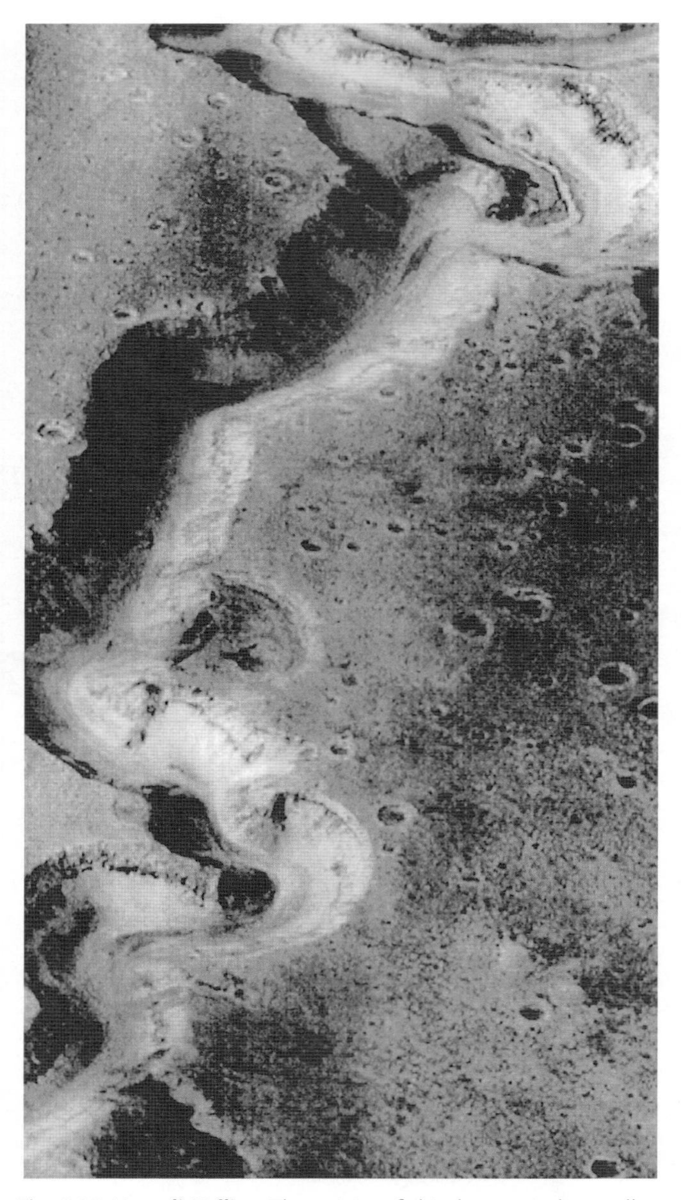

Fig. 8.35 Nanedi Vallis The origin of this long, winding valley is enigmatic. Some features, such as the terraces and the small "stream" in its floor (*top*), suggest that water flowed continuously across the surface for an extended period of time, deeply eroding the rock layers like rivers do on Earth. Such conditions would suggest that liquid water was once stable on Mars, which would have required a denser atmosphere and a higher ground temperature than exists now. Other features, including the dearth of tributaries, suggest that ground collapse could have also contributed to the valley's formation. Both continual flow and collapse probably played a role in Nanedi's formation. This image, taken from the *Mars Global Surveyor*, covers an area of 9.8 by 18.5 kilometers; the valley is about 2.5 kilometers wide, while the small "stream" near the top is just 200 meters wide. Nanedi Vallis is located at 4.9 degrees north and 49.0 degrees west. Nanedi is the word for "planet" in Sesotho, national language of Lesotho, Africa. (Courtesy of NASA, JPL, and Malin Space Science Systems.)

Focus 8.3 The ancient, warm, wet climate on Mars

Today Mars is cold, dry and inhospitable. The desolate, frozen and arid surface is seared by lethal ultraviolet sunlight and bombarded by killer energetic particles from space. Yet the geological record suggests that the planet was not always so uninviting. It might have been warmer, wetter and more Earth-like a few billion years ago, with a thicker atmosphere that would warm the surface by the greenhouse effect and shield it from the ultraviolet rays and cosmic particles. The geological evidence for a more clement environment early in Martian history is given in Table 8.8.

Several lines of evidence point to running water on Mars 3 or 4 billion years ago, when huge amounts of water swept through outflow channels into flood plains, and stately rivers slowly carved the valley networks. There may even have been lakes or shallow seas on early Mars. The floods, rivers and possible lakes or seas all suggest that the Martian atmosphere was once warm and dense enough to allow water to remain liquid on or near the surface of Mars. But then something changed. The atmosphere nearly disappeared and most of the water turned into ice.

Early volcanic eruptions may have released large amounts of carbon dioxide, creating a dense, thick atmosphere during the planet's youth. If the pressure of that atmosphere was just a few times that of Earth's air today, the heat trapped by its greenhouse effect could have warmed the surface enough to sustain water flow for long distances. That water might have seeped out of the frozen ground or it could have resulted from volcanic activity. It is estimated, for example, that volcanoes once released enough water to cover Mars with a global ocean 46 meters deep, most of it during the first 2 billion years of Martian history.

But if Mars once had a warm dense atmosphere, where did its gases go? By a few billion years ago, the originally dense atmosphere must have evolved into a colder, thinner one, similar to the Martian atmosphere we see now. Perhaps much of the carbon dioxide dissolved in the primeval water and became fixed in carbonate rocks. Due to the low gravity on Mars, some of the atmosphere could have also drifted off into space.

Thus, even if Mars did start out warm, it ended up in a deep freeze. As the hypothetical early dense atmosphere became thinner and colder, Mars entered an ice age. The rivers dried up and the liquid water froze into the ground and the remnant polar caps.

Yet, it is also possible that Mars never had a wet and warm atmosphere, and that something else accounts for the features that suggest liquid water in the past. The dry riverbeds, or valley networks, are best explained by collapse into features formed by underground rivers, so they do not require rain or running surface water. Perhaps, the critics argue, raging winds in a thick, dry atmosphere formed the scoured outflow channels, smooth surfaces and surface deposits without the need for liquid water.

Table 8.8 Summary of evidence for a warmer, wetter Mars[a]

Geological Feature	Probable Origin	Implication
River-like valley networks	Water flow out of ground or from rain	Either geothermal heating or thicker atmosphere
Central channel in broader valleys	Fluid flow down valley center	Valleys formed by water flow
Lake-like depressions with layered deposits	Flow through channels into lake	Water existed on surface
Extended low-altitude topography	Possible ocean basin	Northern ocean
Rimless craters	Highly eroded ancient terrain	Water eroded surface
Rounded pebbles and possible conglomerate rock	Rock formation in flowing water	Water was stable; thicker, warmer atmosphere
Abundant sand	Action of water on rocks	Water was widespread
Highly magnetic dust	Magnetic minerals embedded in dust by water	Iron leached from crust

[a] Adapted from an article entitled "The Mars Pathfinder Mission" in the July 1998 issue of *Scientific American*.

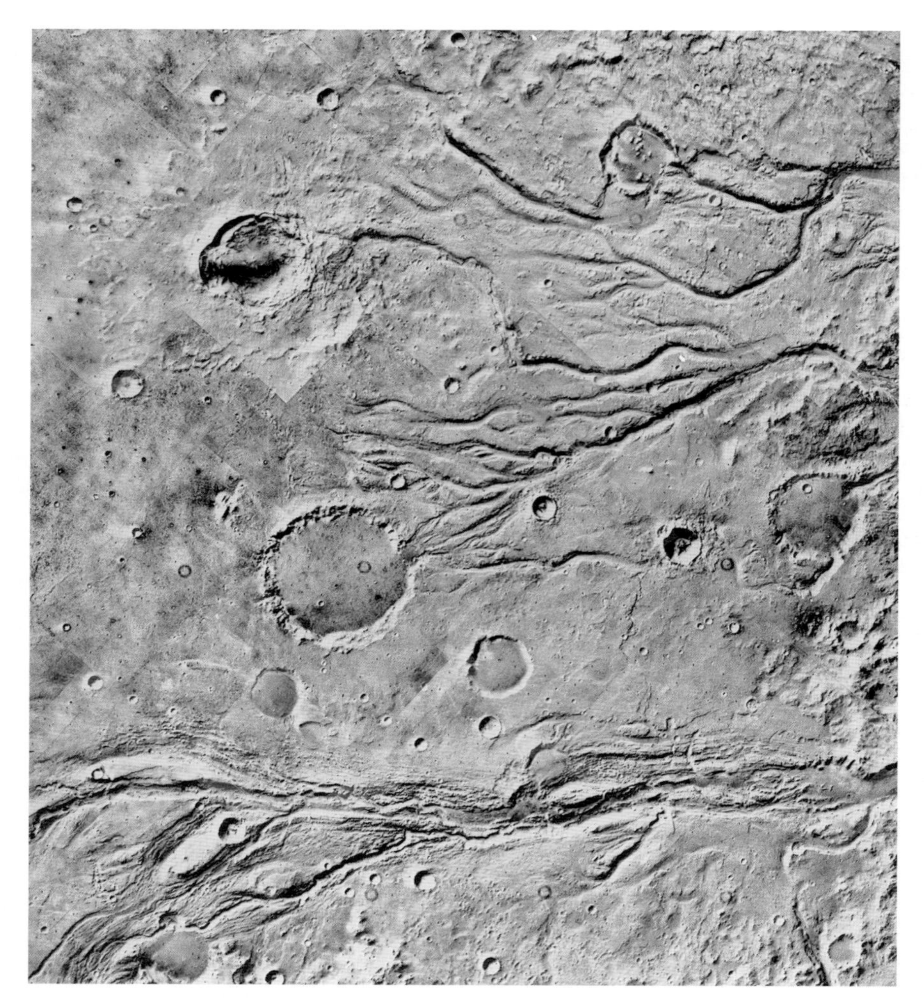

Fig. 8.36 Channels with tributaries Massive floods of water from the highlands into the Chryse basin in the lowlands may have carved these channels, located in the region of Mangala Vallis. The tributaries are rather shallow features, and join their main channels at quite acute angles – evidence of their rapid formation. This image, taken from the *Viking 1* orbiter, has a width of 400 kilometers. (Courtesy of NASA.)

Ancient, water-charged torrents

Long, wide grooves have been gouged out of the equatorial regions of Mars, running downhill from the equatorial highlands to the lowland plains, measuring up to a thousand kilometers long. They tend to be narrow and deeply incised near their origins in the highlands, and broad and shallow in the volcanic plains. Unlike valley networks, these enormous channels lack tributaries and are characterized by sculptured landforms such as scoured surface features, streamlined hills, and teardrop-shaped islands where the flowing water encountered an obstacle (Fig. 8.37).

In at least one instance, the channels are braided together, resembling silt-laden rivers on Earth. The silt drops to the river bottom and builds up until it blocks the flow, forcing the river to redirect its course into a new channel nearby.

The gigantic furrows bear all the marks of catastrophic outpourings of water, and are hence known as outflow channels. They were formed by impulsive, short-lived and catastrophic floods of liquid water, somewhat like flash floods on Earth but on an even more monumental scale.

It is estimated that water flowed through the outflow channels as rapidly as 75 meters per second, about a thousand times faster than the Mississippi River on Earth. Once released, the surging Martian torrents could not be stopped. Such high discharges would not freeze even under present conditions on Mars.

A vast quantity of water was required to create the enormous outflow channels. It would have been enough to fill a global Martian ocean that was 500 meters deep, although it didn't all flow at once.

The outflow channels exhibit a wide spread in ages. Although some of them date back 3.8 billion years ago, almost to the end of the heavy bombardment, others were formed 2 or 3 billion years ago. That is relatively young on a cosmic scale, but still early in the planet's history and old in terms of geological time-scales.

Where did all this water come from? All of the outflow channels emerge from discrete sources in areas that

Fig. 8.37 Streamlined island in outflow channel A raised crater rim, about 4 kilometers in diameter acts as a barrier to the catastrophic floods that discharged from the outflow channel Ares Vallis. The water flowed from the southwest (*bottom left*) with a peak discharge more than 2000 times that of the Mississippi River. (Courtesy of Michael Carr and NASA.)

have undergone collapse, suggesting that the rapid melting of subsurface ice filled the outflow channels with raging floods. Three of the largest outflow channels, the Ares, Simud and Tiu Valles, originate in the Chaotic Terrain, regions of fractured, jumbled rocks that apparently collapsed when groundwater suddenly poured out.

And what triggered the sudden release of such huge volumes of water? Liquid water might have been trapped beneath a thick frozen expanse of permafrost. When volcanic eruptions or the formation of an impact crater breached the overlying frozen seal, the underground water would be suddenly released under great pressure. The rapid surge of water would create dramatic and sudden floods, each lasting only a few days, weeks or month. The overlying surface layer would then collapse, creating Chaotic Terrain.

Possible ancient lakes and seas

As we all know, water collects within holes in the ground, ranging in size from potholes in winter roads to stream-fed lakes and ocean basins. And if water once flowed across the surface of Mars, it would similarly pool in low-lying depressions, such as impact craters and basins, deep canyons and the northern lowland plains (Fig. 8.38). At one time, they could all have been filled with water, forming ancient lakes and seas with perhaps a thin layer of ice on top, but all that now remains is their dried-out floors and sediment.

Widespread, stratified rock structures, found in topographical lows on Mars, could have been deposited by standing bodies of water. The layered material is located in impact craters, on parts of the Hellas impact basin, and on the floors of canyons in Valles Marineris, such as Candor Chasma (see Fig. 2.40 in Section 2.4). Hundreds of individual, horizontal, regularly layered deposits have been laid down, apparently compressed and cemented into rock. Scientists estimate that most of the layered sediments were deposited 3 to 4 billion years ago, but more recent ones are possible, perhaps within the past few hundred million years. The layered sediments seem to have been laid down in large bodies of water, such as ice-covered lakes or shallow seas, during a warmer, wetter, early era on Mars (Focus 8.3).

The low, flat northern regions of Mars could mark the dried-out bottom of a former ocean that once occupied up to one-third of the surface area of Mars. Massive rivers that flowed through the outflow channels and into the northern plains might have fed the ocean. The smooth northern terrain could be due to ancient sediments, and there are some hints of a possible shoreline and parallel terraces carved out by lapping waves.

However, many do not accept that there ever was a true ocean on Mars. Intense searches have, for example, failed to identify confirming details of the supposed shoreline. And if Mars was once bathed in an ocean, where has all the water gone? The north polar cap now contains less than ten times the amount of water required to fill the putative ocean. Perhaps the lowlands were just covered by a shallow, muddy deposit that seeped into the ground and froze there.

Where did all the water go?

Everyone agrees that water once flowed on the surface of Mars in large quantities, but the exact fate of all that water remains unknown. If spread uniformly over the surface, the amount of water involved in the flooding of the outflow channels would, by itself, cover the entire planet to a depth of 500 meters. Yet, there is no liquid water residing on Mars today, and the amount of water vapor in its atmosphere is negligible.

Water ice is locked into the residual north polar cap (Section 8.5), but it cannot explain the missing water. It now contains no more than 1.2 million cubic kilometers of ice. If liquid water could exist on Mars, and the entire polar cap melted, it would create a global ocean less than 20 meters deep. Even if both polar caps have water

Fig. 8.38 Highs and lows In these images, red denotes elevated regions roughly 8 kilometers above the mean level, represented by green. Blue corresponds to 10 kilometers below the mean elevation. Flowing water would run downhill and collect in such low-lying regions. They include the Hellas basin (*upper image*), about 2.8 thousand kilometers across and the largest impact basin in the solar system, the Valles Marineris (*lower image*), shown as a horizontal gash beside the Tharsis volcanoes (*pink*), and the extensive lowland plains in the north (*top of both images*). These maps of the global topography of Mars were obtained with the laser altimeter on *Mars Global Surveyor* – also see Fig. 2.39 of Section 2.4. (Courtesy of NASA, JPL, and GSFC.)

ice buried within them, they represent a small fraction, less than 10 percent, of the water thought to have been present on the surface of Mars.

Instruments on the *2001 Mars Odyssey* spacecraft found evidence in early 2002 for large amounts of subsurface water ice in the upper meter of broad regions surrounding the planet's north and south poles. The water's presence was inferred from measurements of hydrogen-enriched material that produces distinctive spectral signatures, called gamma rays, when bombarded by cosmic rays. The instrument is similar to the one on the *Lunar Prospector* spacecraft, which detected signs of water ice near the poles of the Moon (Sections 2.4, 5.6).

The amount of hydrogen detected in the upper meter of the polar regions of Mars indicates more than 50 percent water ice by volume. Altogether the observed amount is enough to fill Lake Michigan twice over, but not enough to fill even a small ocean. Still, there could be much more liquid water or water ice at greater depths.

Some of the water might have evaporated in more clement times, to be later lost to space, but most of the water that once flowed on Mars probably persists today, frozen just beneath the surface or remaining liquid at greater, warmer depths. Some of the water is most likely buried as ice, frozen into the soil as permafrost. That would account for the muddy debris ejected from some impact craters (Section 8.6). Vast stretches of ice could be hidden under dust and sand far outside the polar caps, perhaps frozen into the places where the outflow channels emptied their flows. The rest might be liquid water buried beneath the ice. Spring-like seeps and small gullies suggest that liquid water has been released from under the frozen surface of Mars in recent times, perhaps even today.

Running water in modern times

Until the early-21st century, most scientists believed that water has not flowed on Mars for billions of years, but features identified in close-up images suggest that liquid water may still flow on Mars. The small, unassuming gullies have

Fig. 8.39 Running water in recent times The inside cliff faces of this crater are etched with gullies that may have been cut by water. Thick sequences of layered rock are also present, attesting to a Martian past of substantial geological activity. Water has apparently seeped from between layers of rock high on the wall, and flowed downhill in deep, channels that have merged together. The lack of small craters superimposed on the gullies and their deposits indicates a geologically young age. They could have formed hundreds, thousands or millions of years ago, and they might even be contemporary. The crater wall shown in this image, taken from the *Mars Global Surveyor*, is located in central Noachis Terra at 47 degrees south and 355 degrees west. (Courtesy of NASA, JPL, and Malin Space Science Systems.)

been carved into the steep, inside walls of some craters or valleys (Fig. 8.39), with shapes that resemble gully washes on Earth. The Martian flow features emerge high up on the wall, run downhill in deep, winding channels, and fan out with an abrupt ending in an apron of dirt and rock (Fig. 8.40).

The amount of water involved in each event can be estimated by measuring the volume of deposits with a conservative assumption of a 2-meter thickness, and by assuming that water accounted for just 10 percent of that volume. The rest of the debris flow is attributed to the dirt and rocks detected at the end of the gullies. It is thereby estimated that about 2500 cubic meters of water (Fig. 8.41) have shaped each small gully, about the volume of an Olympic-sized swimming pool. That is equivalent to 2.5 million liters, or 660 thousand gallons, of water for each gully, enough to sustain future visitors to the planet.

Geologically speaking, the spring-like seeps and flows are as fresh as newly fallen snow. They cut through a terrain that is relatively young, including sand dunes and crater-free landscapes. The lack of small craters superimposed on the gullies and their debris indicates that they are no more than a few million years old, and possibly a lot younger; older regions are pock-marked with craters. Some of the gullies appear to be dust free, which may imply that they happened within the last year or two. Otherwise, the planet's perennial dust storms would have partly covered or even buried them in dust. The gullies might even be forming today, from liquid water that might be now present under the surface of Mars.

A peculiar thing about these gullies is their location. More than one hundred of them have been identified, and nearly all of them occur in places that are well below freezing all year round. Like moss on trees, the gullies form on the coldest slopes facing away from the Sun, where they are usually in shadow and rarely warmed by sunlight. Practically all of the seeps and flows occur at high latitudes, more than 30 degrees from the equator where the surface temperatures range from 173 to 203 kelvin, or −100 to −70 degrees Celsius.

A possible mechanism for recent gully formation involves repeated discharge from behind a dam of ice. Underground water emerging from the initial seepage site would freeze, plugging up the escape route and trapping liquid water behind the ice plug. Pressure would build up until there was enough force to beak the barrier open, releasing the water trapped behind it. The cascade of water would flow hard and fast enough to form the muddy rivulets before freezing, while also carrying rocks and other debris down the steep slopes.

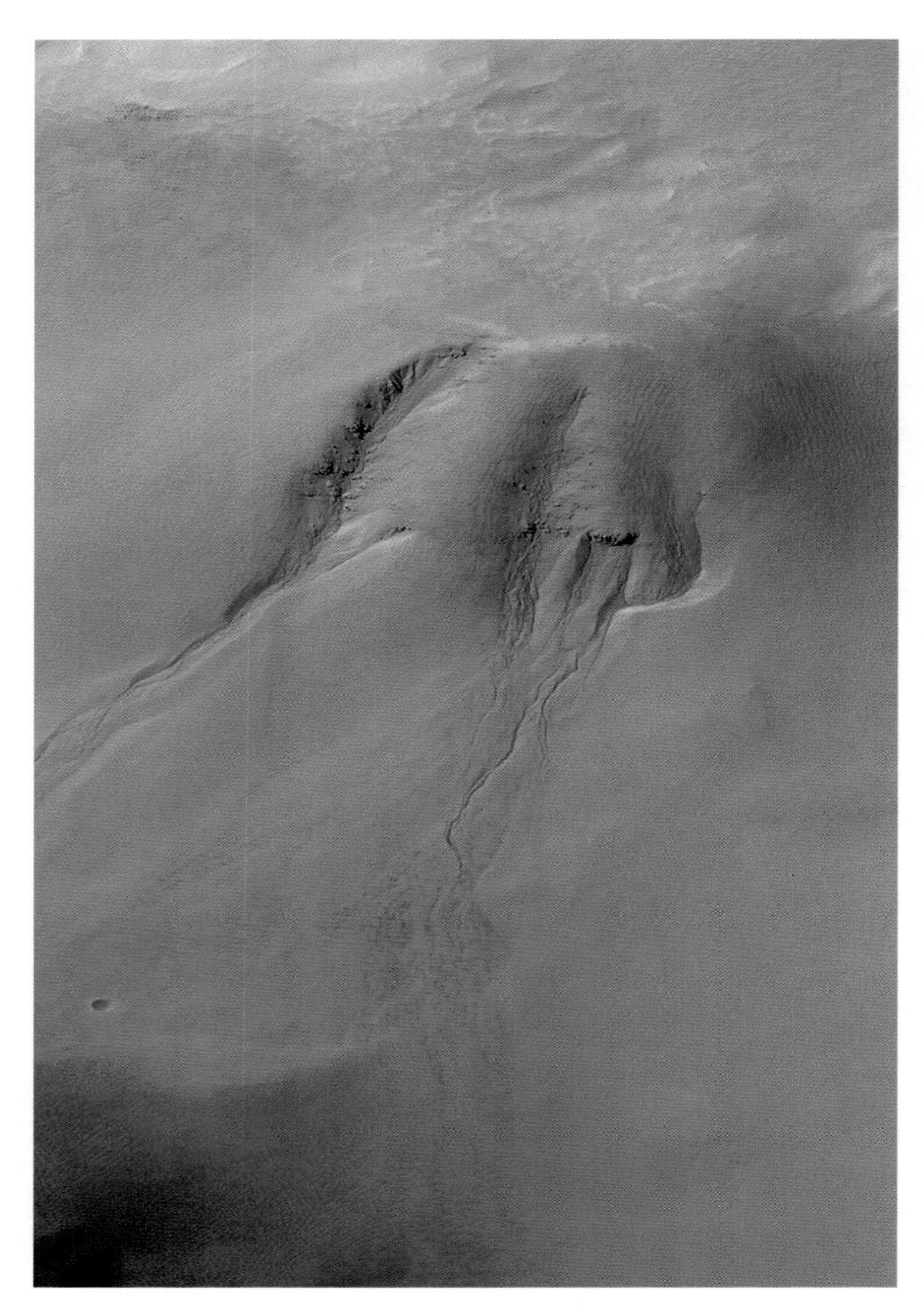

Fig. 8.40 Gully features The channels and associated aprons of debris that are shown here are interpreted to have formed by groundwater seepage, surface runoff, and debris flow. The lack of small craters superimposed on the channels and apron deposits indicates that these features are geologically young. It is possible that these gullies indicate that liquid water is present within the Martian subsurface today. This image, taken from the *Mars Global Surveyor*, covers an area of 1.3 kilometers wide by 2.0 kilometers long. It is located on the south-facing wall of an impact crater in Noachis Terra near 54.8 degrees south and 342.5 degrees west. (Courtesy of NASA, JPL, and Malin Space Science Systems.)

This model might explain why the outbursts only occur in the coldest places on Mars. Any water emerging from the warmer places would be more likely to evaporate, so no ice plug would be formed and there would be no water running downhill.

If the ice-plug model is correct, reservoirs of liquid water lie hidden just a few hundred meters below the icy surface of Mars. Yet, contemporary Mars ought to be deeply frozen, as far down as a few kilometers below ground level. Mars is just too incredibly cold to have liquid water on or anywhere close to its surface at the present time, unless some unexpected heat source is warming the planet up.

In a different explanation, the gullies were created when the polar regions tilted toward the Sun, obtaining enough sunlight to warm the topmost ground layers above freezing. The known swaying tilt of Mars' rotation axis moves the polar regions toward the Sun and away from it, oscillating back and forth between 15 and 35 degrees every 100 thousand years, with wider excursions occurring every 1 to 10 million years. The planet's axis is currently tilted at 25.2 degrees to its orbit, and when it becomes more than about 30 degrees the top layer of the Marian permafrost will thaw. Gullies might then be created by the melt of water mixed with ice. This would also explain why the Martian

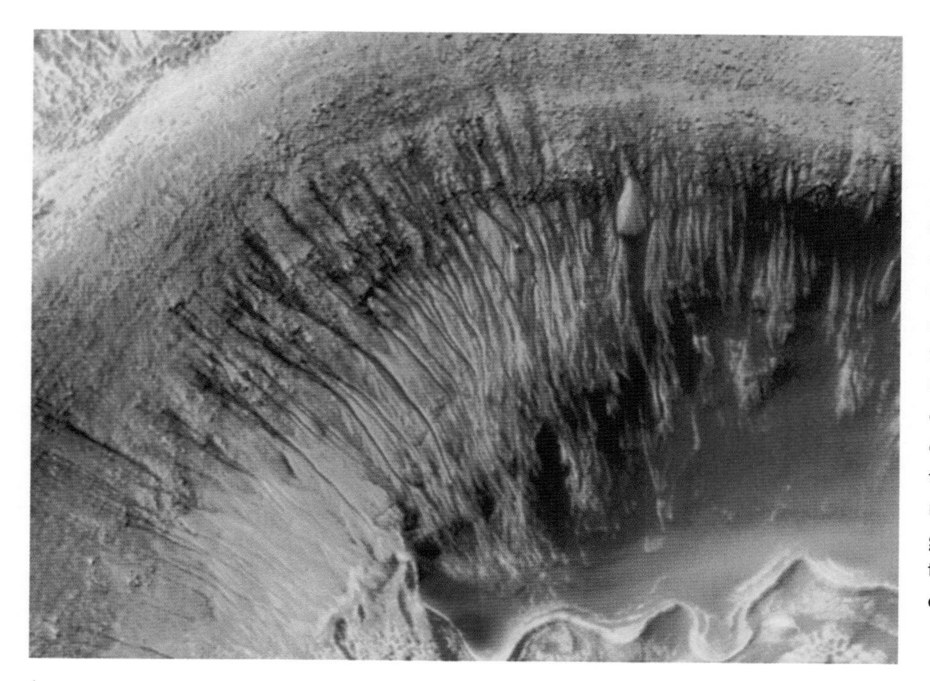

Fig. 8.41 Martian gullies Narrow gullies are eroded into the north wall of a small crater, about 7 kilometers across, that is itself located on the floor of the larger Newton Crater, named after the British scientist Isaac Newton (1643–1727). Flowing water may have caused these gullies and transported debris downhill, creating the lobed and finger-like deposits on the floor and at the base of the crater wall (*bottom right*). The individual deposits were used to estimate that 2.5 million liters, or 660 thousand gallons, of water flowed down each gully. This image, taken from the *Mars Global Surveyor*, is about 4 kilometers across, and centered at 41.1 degrees south and 159.8 degrees west. (Courtesy of NASA, JPL, and Malin Space Science Systems.)

gullies are found at high latitudes and on poleward-facing slopes.

According to another theory, the Martian gullies were carved by water from melting snow deposits. The water flowed beneath the snow packs, where it was sheltered from rapid evaporation in the planet's thin atmosphere. The snow is most likely to accumulate on the coldest slopes where the gullies are now found.

Everyone agrees that something is coming down the walls of the craters to create the gullies, and many suspect it is liquid water.

Liquid water and life

Liquid water is a key ingredient of life as we know it. Without water, there was no such life on Mars in the past, and there would be no life surviving now. Life might have thrived on Mars billions of years ago, when abundant water flowed down its valleys and across its plains. And the possibility of recent liquid water on Mars enhances the prospects for finding microscopic life there now.

No one is suggesting that the Martians are out there relaxing in warm springs or hot tubs, but microscopic life could be thriving in wet, warm areas on Mars. They could have thrived in the muddy gullies, as recently as a million years ago or just the other day. Some of them might be hanging on at the bottoms of former lakes, or maybe they have moved deep underground, where there is enough warmth and water to survive.

Or the microbes might have existed billions of years ago, when water flooded the terrain, and only fossils remain. And maybe there has never been a living thing on Mars. Of course, it is all speculation, until we land on the planet and find out. That is exactly what NASA is planning, sending robotic spacecraft to the sites of past water flow in the search for past or present life.

8.9 Search for life on Mars

Could life originate on Mars?

The question of whether or not life exists on Mars is intimately related to hypotheses for the origin of life on Earth. It could have arisen more than 3.5 billion years ago as the result of chemical reactions in tidal pools, on mineral surfaces, or in the primeval oceans. Ultimately molecules capable of reproducing themselves formed within tiny cells bounded by membranes, and life began.

But could life originate on Mars? The Earth and Mars were formed out of similar material at about the same time, and at a similar distance from the Sun. There is evidence for an ancient period of flowing water on Mars and suggestions of recent subsurface water and heat. Martian life may have breathed carbon dioxide, and the Martian air may even have contained substantial amounts of oxygen that is now locked into its soil. In short, all the basic ingredients for life *may* have once been present on Mars, and life *could* have arisen there.

If life did arise on Mars, how could we find evidence for it? Martian life ought to be based on the chemistry of the cosmically abundant atoms, including carbon, which is a key substance in building complex molecules. Carbon atoms can form large molecules by combining with other atoms, including other carbon atoms. Complex molecules based upon carbon are called organic molecules. Every living thing on Earth is composed of organic molecules, providing them with the capacity to evolve, adapt, and replicate. The discovery of organic molecules on Mars could therefore provide evidence that life might exist, or perhaps once existed, on the red planet.

Real live organisms, or fossils of former ones, might even be detected on Mars, but if anything now lives there it would have to be very strong, tough and exceptionally small. After all, an ice age now prevails on Mars.

Table 8.9 Adapting to hostile conditions now on Mars

Obstacle to Life	Adaptation Required
No liquid water now on surface.	Use subsurface water. Extract water from ice or rock by heat or chemical processes.
Surface temperature usually below freezing point of water.	Live underground where it is warm, develop internal anti-freeze.
Very little oxygen in atmosphere.	Breathe carbon dioxide.
Lethal ultraviolet sunlight and solar and cosmic particles.	Live inside rocks, or underground, hide under rocks and sand, develop protective shell.

Can living creatures survive on Mars?

Chemical evidence and ancient fossils indicate that primitive life existed on Earth 3.8 billion years ago, when our planet was still a fairly inhospitable place for current terrestrial life. The Earth was then cooling off from the intense bombardment that marked the last stages of its formation, and our planet lacked oxygen in its atmosphere. Since there was no oxygen, the Earth had little or no ozone to protect life from harmful ultraviolet sunlight. Yet, life developed in this apparently harsh environment.

Similar conditions existed on Mars at about the same time, up until about 3.5 billion years ago. Both the Earth and Mars probably had protective magnetic fields and thick atmospheres back then, and water seems to have flowed across the Martian surface 3 or 4 billion years ago. Thus, life could have developed on early Mars, at about the same time that life was establishing a foothold on Earth, and fossils of ancient life might be found on Mars now.

Subsequently, however, Mars lost its magnetic field and most of its atmosphere, making the Martian surface an extremely hostile place by Earthly standards. Today Mars does not have enough oxygen, liquid water or heat for most, or possibly any, forms of terrestrial life. Its atmosphere is nearly all carbon dioxide, with very little oxygen or water vapor, and it is a hundred times thinner than our air.

If any hypothetical Martian creatures avoided being asphyxiated by carbon dioxide, or frozen to death at night, they would be faced with damaging ultraviolet sunlight during the daytime. Most organisms that are found on Earth today would be quickly killed if they were exposed to the Martian surface levels of ultraviolet rays. On Earth the lethal ultraviolet is absorbed in the ozone layer and never reaches the ground. On top of that, Mars has no magnetic field or thick atmosphere to keep potentially destructive solar particles or cosmic rays from reaching the ground. Many scientists therefore remain skeptical about the chances of locating living organisms in the dry, cold Martian world.

Optimists, on the other hand, have thought of plausible ways by which primitive Martian life might survive when things turned hard on Mars, taking refuge from increasingly threatening conditions (Table 8.9) and perhaps surviving until the present day. After all, they argue, single-celled life dominated the surface of the Earth for nearly 3 billion years, well before the rise of oxygen in its air, and microbial organisms now thrive on Earth under extreme conditions that were once thought to be lethal. The most notorious of these so-called extremophiles exist at temperatures near and above the boiling point of water; they would freeze to death at temperatures that would result in severe burns to humans. Other microbes now exist at freezing temperatures underneath and within Antarctica sea ice. They even proliferate under dark, pressure-cooker conditions at the bottom of the ocean, feeding on materials emerging from volcano vents, and some of them dwell deep within the Earth's rocky interior, thousand of meters down.

All of these extreme life-forms require water to survive, so the search for life on Mars ought to begin at places there might have been water. Even so, to send a spacecraft to Mars in search of life was an exciting long-shot gamble.

Viking 1 and *2* search for life

One of humanity's most daring and imaginative experiments involved landing spacecraft on the surface of Mars (Table 8.10), and searching for evidence of life there. The

Table 8.10 Locations of Mars landing sites

	Viking 1	*Viking 2*	*Pathfinder*
Location Name	Chryse Planitia	Utopia Planitia	Chryse Planitia
Latitude (degrees)	22.5	48.0	19.3
Longitude (degrees)	48.0	225.7	33.6

one-billion-dollar gamble, in 1976 dollars, began on 20 July 1976, when the *Viking 1* lander came to rest on the western slopes of Chryse Planitia, the Plain of Gold, region of Mars. It appeared to have once been inundated by a great flood and was thus a promising place for life to have arisen. Six weeks later, the *Viking 2* lander settled down in the Utopia Planitia region on the opposite side of the planet, near the maximum extent of the north polar cap, again a favorable site for water and possible life.

How did the *Viking* landers test for life on Mars? The first, most obvious, moving creature test consisted of looking to see if any creatures were frolicking on the Martian surface. The cameras could detect anything down to a few millimeters in size if it came within 1.5 meters of the landers. Pictures were taken of all the visible landscape, from the stubby lander-legs to the horizon, for two complete Martian years, but the view was always one of a desolate, rock-strewn, wind-swept terrain (Fig. 8.42). A careful inspection of all these pictures failed to reveal any motions or shapes that could suggest life, not a single wiggle or a twitch, or an insect or worm. Unless they look like rocks, there are probably no forms of life on Mars larger than a few millimeters in size.

Of course, no one really expected that the *Viking* eyes would see living things, and each of the landers carried a $50 million biology laboratory designed to detect tiny, invisible microbes. Computerized devices inside the landers measured samples of the Martian soil for organic molecules and for signs of growth that might signal the presence of living micro-organisms.

The presence of microbes could be inferred if the Martian soil contained organic molecules with carbon in them. Such a test for organic molecules might be called a dead-body test, for soil would be expected to contain a higher proportion of organic molecules derived from dead bodies than from living ones. But not a single carbon compound was detected, even though the instruments could have spotted organic molecules at a concentration of one in a billion.

The *Viking* experiments were so sensitive that they would have easily detected organic molecules from the most barren and desolate environments on Earth, and even carbon compounds deposited by meteorites as observed in the soils returned from the Moon. Thus, something on Mars had to be destroying carbon compounds. The carbon, denoted by C, was probably oxidized, or combined with oxygen, O, to make carbon-dioxide gas, CO_2, that has escaped into the atmosphere.

The other experiments on board the *Viking* landers searched for the vital signs of living microbes. They did this by exposing the soil to various nutrients, and sniffing the atmosphere to see if microbes ate the food and released gas. Something did emit carbon-dioxide and oxygen gas, but it wasn't alive.

When samples of the Martian soil were exposed to liquid food laced with radioactive carbon, large amounts of radioactive carbon dioxide poured out from the soil. This certainly suggested that animal-like microbes were

Fig. 8.42 Rock-strewn surface This image, taken from the *Viking 2* lander, shows angular rocks that have been tossed across the Martian surface, perhaps as the debris of a nearby crater-forming impact. They cast razor-sharp shadows in the cold, thin atmosphere. Most of the rocks have numerous small holes due to the bursting of bubbles of volcanic gas and pitting by small meteorites. Fierce winds have also eroded the rock surfaces, and piled up fine-grained soil in their lee. The rock in the right foreground is about 0.25 meters (25 centimeters) across. (Courtesy of NASA and JPL.)

Table 8.11 Composition of the Martian surface[a]

Oxide	Mars (percent)	Earth (percent)	Oxide	Mars (percent)	Earth (percent)
SiO_2	51.6	50.7	SO_3	5.3	0.2
FeO	13.4	8.4	Na_2O	2.0	0.2
Al_2O_3	9.1	15.0	TiO_2	1.1	0.8
CaO	7.3	12.0	Cl	0.7	0.2
MgO	7.1	11.0	K_2O	0.5	0.1

[a] The composition of the average Martian "soil" at the *Mars Pathfinder* landing site is somewhat different from that of typical basalt from the Earth's sea floor. Although the Martian material has about the same percentage of silicon dioxide, SiO_2, it appears enriched in iron, Fe, as well as volatile elements such as sodium, Na, potassium, K, and sulfur, S. Adapted from M. H. Carr's article in J. K. Beatty, C. C. Petersen and A. Chaikin (eds.), *The New Solar System*, Cambridge University Press, New York 1999.

digesting the food and exhaling carbon-dioxide gas. But when additional nutrients were added to the soil, there was no additional increase in radioactive gas. Living creatures would have continued to ingest the food.

When the Martian soil was exposed to water, a burst of oxygen flowed from it. At first, the surprised scientists thought that plant-like microbes were emitting the oxygen, but they soon realized that the release was too fast and brief. Microbes would grow and produce more oxygen as time went on, thereby releasing oxygen at a steady rate. Moreover, the oxygen was released when the experiment was performed in the dark, and this behavior would not be expected from plant-like photosynthesis that depends on sunlight.

After further experiments, scientists concluded that the biological tests failed to detect any unambiguous evidence for life on Mars. Instead of being produced by organisms of any kind, all of the results were attributed to non-biological, chemical interactions. Highly oxidized minerals in the Martian soil were reacting with the nutrients, breaking them up and liberating some oxygen gas and even more carbon dioxide.

Since Mars has no ozone layer, the planet's surface is exposed to the full intensity of the Sun's ultraviolet radiation, and these rays have turned the surface into an antiseptic form. The lethal soil has apparently destroyed any cells, living, dormant or dead, wiping them out with chemical reactions, somewhat like pouring hydrogen peroxide into a cut. The only safe haven for any living organism would thus be under the chemically reactive ground.

The highly oxidizing material has even turned Mars red. Its distinctive hue looks like rust, and for a good reason. Analysis with *Viking* lander instruments has shown that the reddish soil has a high content of iron, denoted by Fe, which has been oxidized to produce a form of rust known as ferric oxide, or FeO (Table 8.11). It accounts for the planet's pronounced red color.

The suspension of fine reddish dust can color the Martian sky pink. Because the atmosphere is so thin, its light-scattering properties, which determine color, are dominated by dust particles lofted into the atmosphere by periodic dust storms. In contrast, abundant molecules in the Earth's thick air preferentially scatter blue sunlight down to the surface, making our sky blue.

As it turned out, the pioneering *Vikings* paved the way for the next spacecraft to land on the reddish-brown surface under a cold pink sky.

Mars Pathfinder and *Sojourner Rover*

The primary objective of the next spacecraft to land on Mars, the *Mars Pathfinder* lander and its diminutive roving vehicle, *Sojourner*, was to demonstrate a low-cost means of landing a small payload and mobile vehicle on Mars. Unlike previous spacecraft, that went into stately orbits around Mars and descended gently to the surface, *Mars Pathfinder* was shot directly at the planet, using a small parachute on arrival and cushioning the impact with air bags.

Landing on 4 July 1997, in celebration of Independence Day and America's 221st birthday, the mission became a celebrated example of NASA's new mantra of "faster, better, cheaper". *Mars Pathfinder* and the *Sojourner Rover* were developed in four years, at a total cost of $250 million for launch, lander and rover. That is comparable to the budget of a major motion picture, and a mere fraction of the $1 billion cost of the twin *Viking* missions in the 1970s – which amounts to more than $3 billion in 1997 dollars. Still, *Mars Pathfinder* was not designed to do anywhere near the amount of science the *Vikings* were capable of; science

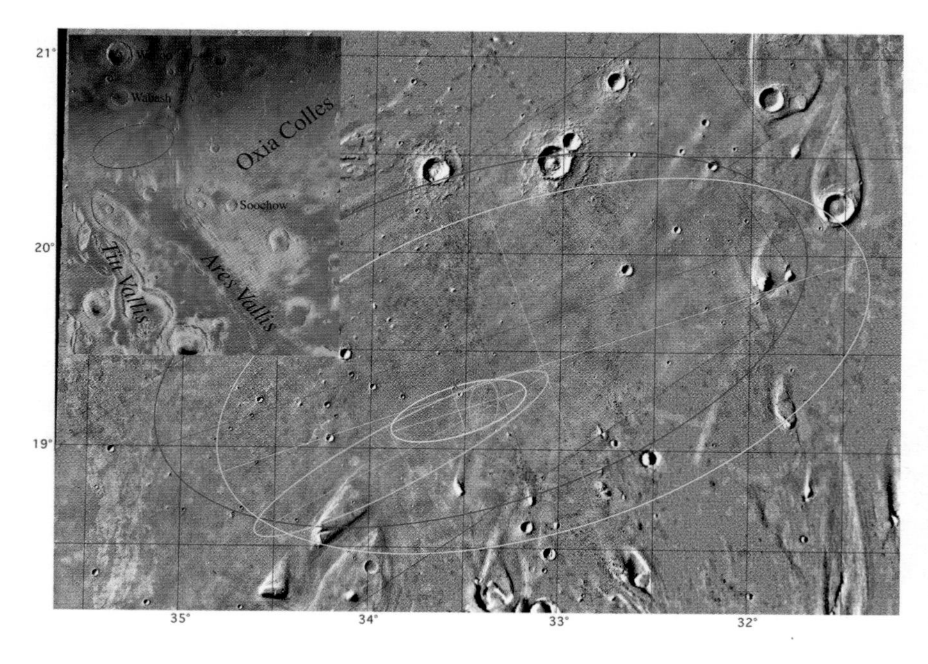

Fig. 8.43 *Mars Pathfinder* lands on an ancient flood plain Billions of years ago, when water flowed on Mars, great floods rushed out of the outflow channel, Ares Vallis, and emptied into the Chryse Planitia, or Plain of Gold, region of Mars (*color inset*). The flowing water carved out streamlined islands around craters (*top right*). This area was chosen as the *Mars Pathfinder* landing site for three reasons: it seemed safe, with no steep slopes or rough surfaces; it had a low elevation, which provided enough air density above the surface for a parachute to work; and it appeared to offer a variety of rock types deposited by the floods. The ellipses mark the area targeted for landing of *Mars Pathfinder*, as refined several times during the final approach to Mars. The site is about 850 kilometers southeast of the location of the *Viking 1* lander. (Courtesy of NASA and JPL.)

was not even a factor allowed to drive the *Mars Pathfinder* design.

Although the *Pathfinder* mission was not designed to look for direct signs of life, it did search for very indirect evidence of a formerly warm, wet Mars that might have supported life. The landing site in Chryse Planitia, the Plain of Gold, was chosen because it lies at the mouth of a large outflow channel, called Ares Vallis, apparently carved by catastrophic floods in the distant past (Fig. 8.43). It is thought that running water flowed down the Ares Vallis and flooded the plain at the landing site between 4.5 and 3.6 billion years ago.

One of the major unexpected results of *Mars Pathfinder* was the inability to detect chemical diversity in the rocks. The data are consistent with all measured rocks being chemically the same and covered by different amounts of the same dust. This was a surprise, because the landing site was expected to contain a wide variety of ancient rocks washed down by the flooding waters (Fig. 8.44).

The distant, streamlined hills, known as Twin Peaks, appear to have been smoothed by water (Fig. 8.44), and evidence of layered sedimentation shows up in both the nearby rocks and the Twin Peaks. Even the presence of sand, as opposed to smaller dust particles, suggests the widespread action of flowing water. The sand detected from *Pathfinder* was light in color, just like beach sand on Earth; and sand on Earth is formed by running water. In addition, magnets on *Pathfinder* found that the airborne dust is very magnetic, which can be explained if liquid water helped embed magnetic minerals in the dust.

Thus, there is abundant evidence that liquid water once flowed across Mars, and that the climate must have once

Fig. 8.44 View from the surface On 4 July 1997 *Mars Pathfinder* landed safely on a wind-swept plain littered with the debris of catastrophic floods from early in the planet's history, including a diversity of rock sizes and types. Between and partially covering some of the rocks is reddish iron-oxide dust, the result of chemical weathering of exposed rock surfaces here and elsewhere. There are also bare rocks left uncontaminated by the seemingly ubiquitous dust. In the foreground the dust or sand is partly cemented. The two modest-sized hills in the distance have been dubbed the Twin Peaks. They are roughly 30 meters tall and about 1 kilometer from the lander. (Courtesy of NASA and JPL.)

Fig. 8.45 *Sojourner* Rover The tiny rover *Sojourner* hit the ground with all six wheels running; energized by a solar panel on its top which delivered up to 16 watts of power. The diminutive rover is just over half a meter long, but equipped with three cameras and an instrument that determined the chemical composition of rocks and soil. *Sojourner* has demonstrated that a small, unmanned rover is an effective vehicle to explore another planet, and that such exploration can be done quickly and relatively cheaply. (Courtesy of NASA and JPL.)

been warmer and wetter than at present. Perhaps the planet had a thicker atmosphere in its early history, and conditions were then conducive to the survival of life.

The immediate vicinity of the *Pathfinder* landing site, however, appears to have been dry and unchanged for eons. The region seems to have remained almost unaltered since catastrophic floods sent rocks tumbling across the plain more than three billion years ago. It has apparently been untouched by water ever since the ancient deluge. Only the winds remained to erode and shape the surface.

The small rover, called *Sojourner*, added an important scientific component to the *Pathfinder* mission, along with elements of drama and excitement that captivated the public. The tiny vehicle weighed just 10.6 kilograms, about the same as a house pet, and had the overall size and rectangular shape of a small microwave oven (Fig. 8.45). Equipped with six-wheel drive, *Sojourner* explored about 250 square meters of the Martian surface, measuring the chemical makeup of the rocks and soil. In contrast, each of the two *Viking* landers were shackled to one location, unable to roam across the surrounding terrain, somewhat like getting sick in bed on vacation.

Sojourner moved slowly across the terrain, at less than one-hundredth, or 0.01, of a meter per second, like a baby learning to walk. It could "think" with behavior programmed into its computerized "brain", albeit with an intelligence quotient less than that of an insect. Five laser beams, sent out in different directions, were used to sense obstacles and determine how far away they were, enabling the rover to find and analyze rocks, and to either steer around or climb over them. A gyroscope, spinning like a top inside *Sojourner*, kept the course steady.

Sojourner's behavior captivated the imagination of the public, which followed the mission with great interest via the World Wide Web; at its peak the web site recorded 47 million hits in just one day. In addition, the diminutive rover operated 12 times its design lifetime of seven days in the unforgiving cold, earning the title "the little engine that could". The tried-and-true technology embodied in the pioneering *Sojourner* will be incorporated in larger roving vehicles that NASA plans to send to Mars, probably to scout the planet's surface for signs of water.

The rover examined an array of Martian rocks of different shapes, sizes and texture (Fig. 8.46), provided with colorful names like Barnacle Bill, Yogi, Scooby Doo, Casper, Wedge, Shark and Half Dome. It also analyzed the Martian soil in the vicinity, showing that it is very similar to the soil at the *Viking 1* and *2* landing sites.

Fig. 8.46 Rock garden This image, taken from *Mars Pathfinder*, shows the rocks named "Shark" and "Half Dome" (*top, left* and *center, respectively*). Between these two large rocks is a smaller one, about 0.2 meters wide and 0.1 meters high, that was observed close up with the *Sojourner Rover*. The rocks in this area are inclined and stacked, as if deposited by rapidly flowing water. The smooth, angular rocks are relatively dust free and apparently uncontaminated, allowing unambiguous measurements of their chemical constituents by the *Sojourner Rover*. (Courtesy of NASA and JPL.)

Scientists expected to find a type of volcanic rock known as basalt, which forms by partial melting of the mantle. This kind of igneous rock is typical of lava found on the Earth's ocean floor, as well as on the maria of the Moon and on the surface of Venus. Many of the rocks analyzed by *Sojourner*, however, contained much more silicon, or quartz, than pure basalt. On Earth, such volcanic rocks are called andesites – named after the Andes Mountains that are made out of a mixture of quartz and basalt.

The Martian rocks tell of a heated, tumultuous internal history. To form quartz, the crustal material must have been heated, cooled and reheated many times by volcanic reprocessing, much like quartz rocks on Earth but unlike rocks returned from the Moon. The implication is that Mars has been convulsed by internal heat through much of its 4.6-billion-year history, marked by repeated internal melting, cooling and remelting that produced an abundance of tell-tale quartz.

Moreover, Mars, like the Earth is an internally layered globe, with a crust, a mantle and an iron core. The evidence that the red planet is not merely a solid ball of rock comes from changes in radio communication with *Pathfinder* as Mars rotated about its axis. By combining these signals with similar measurements from the *Viking* landers, scientists have determined the density at various depths in Mars, inferring the presence of a dense, metallic core. Early in the planet's history, the molten rock on Mars became differentiated, with heavy elements like iron sinking to the bottom and light ones rising to the top. Analysis of radio tracking data from *Mars Global Surveyor* indicates that the iron core is now at least partly liquid.

So *Mars Pathfinder* and *Sojourner Rover* have shown that Mars was Earth-like in its infancy. Both planets sustained enough internal heat to produce surface quartz and distinct internal layers. This, in turn, suggests that ongoing volcanic activity could have produced a thick atmosphere during the planet's youth, becoming warm and wet enough to sustain life.

Possible life in a rock from Mars

Rocks that arrive from space and survive their fiery descent to the ground are given the name meteorites. Thousands of them have been recovered from the ice sheets of Antarctica, and most are chipped fragments of asteroids, a ring of rubble located between the orbits of Mars and Jupiter (Section 13.7).

Only about a dozen meteorites have been identified as coming from Mars. They have been collectively named the SNC meteorites after the initials of the locations where they were first observed to fall from the sky – near Shergotty, India in 1865, Nakhla, Egypt in 1911, and Chassigny, France in 1815.

How do we know that the SNC meteorites came from Mars? These igneous rocks have characteristics similar to terrestrial lava, but the ratios of certain elements, the oxygen isotopes, are distinct from all rocks on Earth. This means that they came from a different planet, a moon, or the asteroid belt. Most of the SNC meteorites solidified from molten material less than 1.5 billion years ago, so they had to come from a place that was volcanically active long after the origin of the solar system 4.6 billion years ago. This rules out Mercury and the Moon, as well as the asteroids, which are not large enough to be volcanically active anyway. The minerals in some of the SNC meteorites have been chemically altered by water, so they originated in a place with water, ruling out the Moon and Venus. That leaves Mars as the only place with the required attributes that is close enough for a rock to be propelled to Earth. The clinching argument for an origin on Mars comes from analysis of pockets of gas trapped in the SNC meteorites. This gas has a unique composition that exactly matches that of the same gases in the Martian atmosphere, as measured by the *Viking* landers.

The origin of each meteorite from Mars can be traced back to the jolt of a much larger impacting object. Most of the debris of this violent collision would fall back to the Martian surface to form the rim of a crater, but some of it would be blasted off at a high enough speed to escape the weak tug of Martian gravity. That material would move in its own orbits around the Sun. These orbits would be gradually skewed by the gravitational pull of the distant planets, and very occasionally redirected on a collision course with Earth. Over an interval of 10 to 100 million years, a small fraction of the ejected debris would eventually strike the Earth.

Interest in the possibility of Martian life was heightened when scientists found possible signs of ancient, primitive bacteria-like structures inside just one of these rocks from Mars. This meteorite crash landed on the blue ice near the South Geographic Pole toward the end of the last ice age, resting there in frigid isolation for millennia, most likely compressed in the snow and later exposed. Then in 1984 a geologist spotted it in the Allan Hills region of Antarctica, bagged it, and sent it to the United States for analysis. Since it was the first meteorite to be processed from the 1984 expedition, the unusual rock has been designated ALH (for Allan Hills) 84001.

The resumé of ALH 84001 has been established from laboratory analysis of different sets of radioactive chemical elements or isotopes. This Martian meteorite differs from the other ones in being very ancient, having solidified from molten material about 4.5 billion years ago. The rock then stayed on the surface of Mars more or less undisturbed for eons. Then 16 million years ago, the blast of

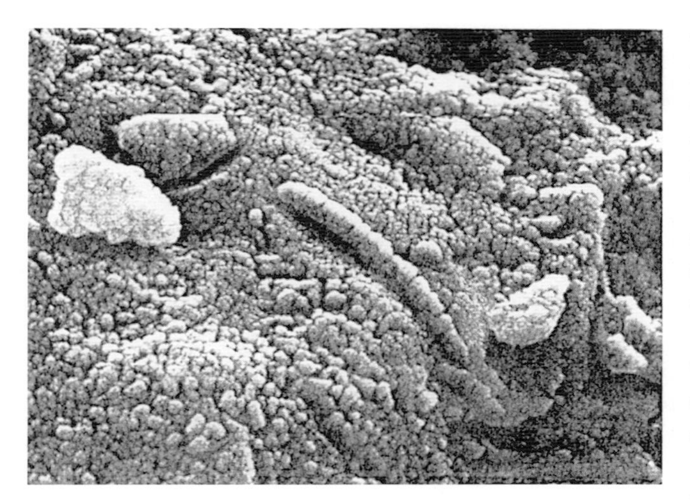

Fig. 8.47 Possible Martian microfossil This image, taken with a scanning electron microscope, shows an unusual structure located in a carbonate globule within meteorite ALH 84001 from Mars. The tube-like feature in the center of the image is only about 200 nanometers, or 0.000 000 2 meters, long, which is about one-hundredth the width of a human air. It appears segmented, as if it were a filament composed of separate elements. The minute structure looks like some very small, fossilized bacteria found on Earth. Some scientists interpret it as a possible Martian microfossil of exceedingly small bacteria that may have lived on Mars about 3.6 billion years ago. Other scientists dispute this interpretation. (Courtesy of NASA and JPL.)

an impacting object excavated the rock from the surface and flung it into space. There it wandered in an increasingly eccentric orbit, bombarded with cosmic rays whose byproducts tell how long the rock was out there. It eventually moved close enough to be captured in Earth's gravity and fell on the Antarctica ice sheet 13 thousand years ago. That much everyone agrees on. ALH 84001 originated on Mars soon after the planet formed, and was found in Antarctica 16 million years after being blasted away from the red planet and 13 thousand years after arriving at Earth.

The controversy and excitement centers on suggestions that microbial life took refuge in cracks within ALH 84001 a very long time ago. The microscopic and chemical evidence includes: globules of calcium carbonate; the first organic molecules thought to be of Martian origin; several mineral features characteristic of biological activity; and what looks like fossils of extremely small bacteria-like organisms that lived on Mars billions of years ago (Figs. 8.47, 8.48). Some scientists argue that these observations collectively provide strong circumstantial evidence for past, primitive life on Mars; others reason that the evidence is not conclusive and that there are non-biological explanations for all of it.

Regardless of the outcome of the ongoing controversy, the evidence only concerns very ancient structures. The carbonates that cover the walls of cracks in the meteorite are probably about 3.6 billion years old. Thus, even if there was something once living in ALH 84001, we are now looking at fossils of long-dead corpses and have no reason to think life might still be around.

The life-like structures in ALH 84001 are also very, very small. The globules are smaller than the period at the end of this sentence, and the putative microfossils within them are less than one-hundredth the width of a human hair. So, we are talking about structures that can only be seen with the most powerful electron microscopes, and there certainly isn't any evidence that any higher life-form ever existed on Mars.

Still, even the possibility that rudimentary life once existed on another planet is so important that we should carefully examine the pros and cons of the controversy, and we first discuss the supporting evidence for alien life on Mars. All of that evidence is found within the meteorite's round carbonate globules. They were apparently formed billions of years ago when water-bearing fluid percolated through the crustal rocks of Mars. The carbonate globules that are found in ALH 84001 are similar in size and texture to carbonate features within fossils of ancient bacteria on Earth. The organic molecules that are located in and on the meteorite's carbonate globules provide the first evidence that such molecules ever existed on Mars. Certain magnetic minerals are also found within the globules, and they resemble those created by terrestrial bacteria. Highly magnified images reveal structures resembling bacterial fossils found on Earth. These tube-like features seem to have formed in colonies within the confines of the globules.

But there are alternative, non-biological explanations for all of the supporting evidence. Geological and geochemical processes can explain the carbonate globules and the complex molecules, magnetic minerals and apparent fossils within them. Carbonates can be deposited in changing chemical environments in the absence of life. Organic molecules are formed by non-biological processes as well as by biological ones. Hundreds of organic molecules are distributed throughout interstellar space and in meteorites known to have come from the asteroid belt, created by lifeless processes in these sterile locations.

Detailed study of the structure of the magnetic crystals found in ALH 84001 indicates that they were probably produced by non-living, inorganic processes rather than by living bacteria. The fossils may also be too small to contain all the genetic essentials of a living cell, for a good-sized protein cannot fit inside them. And the so-called fossils do not seem to contain cells, or the cell walls and membranes that could protect a cell from its environment while also extracting energy from it.

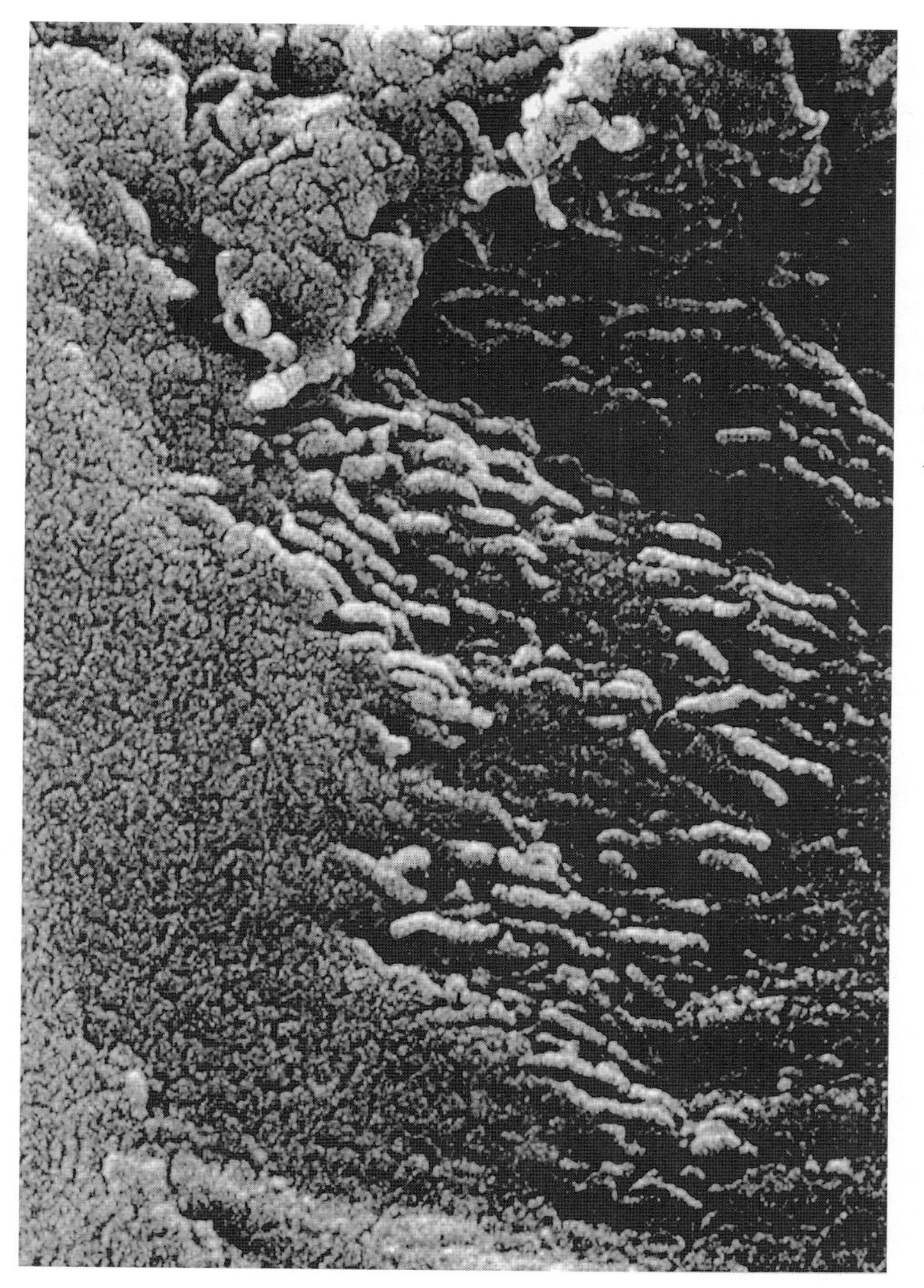

Fig. 8.48 Primitive, ancient Martian life? This electron microscope image shows many rod-shaped objects in a carbonate globule within meteorite ALH 84001 from Mars. The minute structures could be microscopic fossils of a colony of primitive, bacteria-like organisms that lived on Mars more than 3.6 billion years ago. Even the largest structures are very small, about 100 nanometers, or 0.000 000 1 meters, in length; nanobacteria on Earth are the same size. Nevertheless, the tubular structures could also have a non-biological origin that has nothing to do with life on Mars. (Courtesy of NASA and JPL.)

An informed, objective jury would probably examine the different interpretations of the evidence, and conclude that no one has proven, beyond a shadow of doubt, that ancient microscopic Martian life existed in ALH 84001. Indeed, the majority of scientists now think there is nothing in the meteorite that conclusively indicates whether life once existed on Mars or exists there now. They aren't saying that there couldn't be life on Mars, or doubting the arguments for it, but just that there is enough uncertainty to cast doubt on the claim that microscopic life once lived in ALH 84001. This does not refute the general arguments for possible life on Mars, either past or present, and many scientists subscribe to the belief that life will be discovered in some location other than Earth.

The continuing hunt for extraterrestrial life

It seems that modern civilization has always anticipated finding life on Mars, perhaps because of its Earth-like seasons, clouds, past flowing water, ice caps and similar daily rhythm. And Mars remains the most likely, nearby place in the solar system to find extraterrestrial life. We have visited the Moon and concluded that there is no life there. The intense heat on the surface of Venus boiled away any water long ago; it would fry and vaporize all living things that we know of. Mercury has essentially no protective atmosphere, and its temperature extremes from day to night would alternately boil and freeze anything on its surface, except perhaps at the polar regions.

Thus, as dry and cold as it might be, Mars remains the most plausible *nearby* home for life in the solar system outside Earth itself, and future voyages to search for life on Mars are inevitable (Focus 8.4). Robotic spacecraft, and eventually humans, will visit the most likely places to contain life, returning rocks and soil to be scrutinized in the terrestrial laboratory.

Terrestrial microbes will surely accompany astronauts to Mars, perhaps contaminating the planet and complicating the search for life. Conversely, rocks and soils brought back to Earth from Mars by a future space mission could be full of deadly microbes that might cause a global catastrophe on Earth. Visiting astronauts could conceivably get infected with an alien Martian plague, and be forced into quarantine if they make it back home. Precautions should also be taken against an unexpected crash landing of the returning spacecraft, which could release dangerous alien organisms. Since the surface of Mars is now sterilized, these are unlikely possibilities, but precautions are being taken just in case the improbable occurs.

As a matter of fact, it is possible that Earth and Mars have regularly exchanged microbes over the years, without any modern spacecraft or astronauts being involved. Life-forms may have arisen on Mars first and then migrated to Earth, hitching a ride on a meteorite; or it might have been the other way around, with life originating on Earth and traveling to Mars. Cosmic impacts with Mars have sent hundreds of tons, or hundreds of thousands of kilograms, of nomadic Martian rocks to Earth over the centuries, and much more during the past eons. The rain of impacting debris was hardest in the early days of both planets, increasing the likelihood of biological exchange between them. And even now, two tons of Martian rocks are thought to rain down on Earth each year, and about the same amount of terrestrial rocks annually smashes into Mars.

So, life might not have emerged spontaneously on Earth. It could have come from Mars, and in that case we might all be Martians. Or life might have originated on Earth and was then delivered to Mars. Maybe life arose on the two planets independently and spontaneously or perhaps impacting comets or asteroids pollinated both planets. And, just maybe, Earth is the only place to harbor life in the entire solar system. It's all a lot of fun to think about, but every one of these possibilities is mainly speculation with little hard scientific evidence.

To get back to everyday reality, how can we best widen the search for life, especially for extant and dormant life on Mars? To begin, we should understand what characterizes living things, and determine how to find these characteristics on an alien planet. Then we should decide what locations on Mars are likely to have sustained life, and go there and look for it.

What is life anyway? Perhaps the most important key to life is energy, which flows through life and powers it. All living things have to extract energy from their environment to fuel themselves. The most powerful source of this energy is sunlight striking the surface of a planet, but heat energy from the planet's hot interior will also suffice. Tides provide an additional, but usually less powerful, source for energizing life.

Focus 8.4 Future missions to Mars

NASA's Mars Exploration Program intends to ultimately address the question of whether life is or ever was present on Mars. The strategy begins with a global inventory of the Martian surface, with the high-resolution images already taken from the *Mars Global Surveyor*, and then with instruments on the *2001 Mars Odyssey* spacecraft, safely inserted into Martian orbit in October 2001. It is mapping the chemical and mineralogical makeup of Mars, looking for signs of liquid water near the surface and shallow buried ice, while also evaluating potential radiation risks to future human explorers.

Then, in 2003, NASA expects to land two *Mars Exploration Rovers* at "hot spot" locations specified from the orbital reconnaissance missions as locations where liquid water was probably once present on Mars. Each rover will behave as a robot geologist looking for water-affected minerals on the surface. This will be followed in 2005 with the *Mars Reconnaissance Orbiter* whose instruments will narrow the focus, using high-resolution images and spectroscopy to search for the most likely places suitable for past or present life, including signs of liquid water. One of these locations will be chosen for NASA's first sample return mission, hopefully launched during the second decade of the century.

In the meantime, the European Space Agency (ESA) is expected to use the 2003 launch opportunity to send its *Mars Express* orbiter carrying the British *Beagle 2* lander. A Japanese orbiter called *Nozomi* is scheduled to arrive at Mars in 2003, studying the interaction of the solar wind with the Martian upper atmosphere.

On Earth, the major source of available energy is sunlight. The geothermal energy provided by the planet's hot inside accounts for less than one-thousandth of the amount of available power as sunlight. Tidal energy accounts for one-tenth of the available power provided by geothermal energy.

Chemical processes convert the available energy into fuel that is consumed by biological organisms, which return waste products to the environment. Photosynthesis by plants, for example, uses the energy of sunlight to extract carbon, C, from carbon dioxide, CO_2, in the atmosphere, releasing oxygen, O_2, into the air. With the help of water, the carbon is incorporated into organic molecules within the plants. Chemists say that the carbon dioxide, CO_2, has been reduced, or broken apart into its simpler constituents, C and O_2.

Despite its overwhelming energy input to Earth, the Sun is not the only source of life-giving energy on our planet. Geothermal energy powers chemical processes that produce inorganic compounds without carbon, such as hydrogen, hydrogen sulfide, ferrous iron, and iron sulfide. Organisms also use these compounds for fuel, effectively eating rocks, and sometimes living in complete darkness.

We have found giant clams and fields of tubeworms living in the blackness of the deep sea, near underwater volcanoes on the ocean floor. The Earth's inner heat releases chemicals in the volcanic vents, providing food for tiny microbes, which are eaten by larger organisms. But they are ultimately dependent on photosynthetic oxygen, which enters the deep ocean from the surface water. Other microbes survive at least a kilometer inside the Earth's crust, sometimes living in solid formations of granite. Like the deep underwater microbes, these subterranean organisms have substituted volcanic fires for solar ones as an energy source, living off the Earth's inner heat and inorganic chemicals.

Living things also have to be protected from their own waste products, generated while consuming either organic or inorganic substances. Cells, whose membranes wall off the cell interior from its own chemical byproducts, can do this. And cells involve structure, shape and form. In order to survive as time goes on, life must also replicate itself and eventually adapt to a changing environment.

So, what should we look for when returning to Mars, seeking the signs of life? Either extant or fossil life will reside in a structure, and it will have to reproduce itself to survive. Martian life will convert either the energy of sunlight or geothermal energy into a biologically useful form, involving chemical processes that can include either organic or inorganic molecules. The surviving structures are likely to be very small, similar to the tough and tenacious, single-celled creatures that dominated the surface of the Earth for three billion years and are now resistant to, and even thrive on, extreme conditions.

It's likely that hypothetical Martian microbes could similarly thrive on geothermal or geochemical energy, and tests should be designed to detect the chemistry involved in sustaining organisms with this type of energy. Since they require sunlight, other plant-like microbes would have to be at or near the sterile surface of Mars, where they cannot now live; and the *Viking* experiments seem to have ruled out this type of mechanism for obtaining energy.

And where should we look for the putative Martian organisms? The *Viking* tests literally just scratched the surface, the top layer of soil. Since the surface has been exposed for eons to freezing cold, lethal solar ultraviolet, killing cosmic particles and a lengthy parched drought, you might have expected that no life would be found on it. If simple life-forms once existed on the surface of Mars, they might have survived by burrowing inside rocks or deeper underground for warmth and sustaining moisture.

The first place to look for signs of Martian life will be places where liquid water once existed, and where subsurface liquid water might still reside. The layered beds of ancient lakes, the sites of past catastrophic floods, and the possible locations of recent water flow or volcanic activity are all logical places to land a spacecraft. Robotic vehicles or astronauts could retrieve rocks from these locations, and drill core samples from deeper down. Living microbes might still be hiding inside the rocks or cores, or we might find fossilized remnants there.

The discovery of life on Mars, even primitive life in the very distant past, would have profound implications. It would give us companionship in a vast and lonely Universe, and it would also be a little humbling. The discovery would raise the likelihood that life might be found elsewhere in the Universe as well, perhaps on one of the countless planets that surely exist in our Galaxy (Focus 8.5). Of course, the enduring idea of life on Mars could prove as illusory as the Martian canals, but even in that event many humans will still retain a passionate conviction that there must be life somewhere else in the Universe. So the quest will continue, and whatever happens the human spirit will remain as beautiful and glorious as ever.

8.10 The mysterious moons of Mars

Discovery and prediction

Mars has two little moons that are so dark and small that they were undetected for centuries, even after the invention of the telescope. They were discovered in 1877 by Asaph Hall (1829–1907) using a 0.66-meter (26-inch) telescope at

Focus 8.5 Widening the search for life

Looking beyond Mars, there are other locations that contain two of the ingredients essential for terrestrial life – water and energy. Europa, a large moon of Jupiter, seems to have a global sea of liquid water just beneath its thin, cracked icy crust (Sections 2.4, 9.4). Though far from our home star, and therefore bathed in diminished sunlight, alien life might thrive in Europa's dimly illuminated seas, warmed by inner tidal heat and powered by chemicals. Underwater volcanoes, analogous to those discovered in the deep-ocean floors of Earth, could exist at the bottom of Europa's oceans, similarly teeming with microbial life. Ultimately, we would like to obtain samples from the fractured ice, but we don't know how to accomplish a round-trip to Europa, including a stopover, with any current technology.

Saturn's giant moon Titan, larger than the planet Mercury, possesses a thick, Earth-like nitrogen atmosphere with abundant methane and other organic molecules (Sections 3.3, 10.5). A rich chemistry powered by sunlight in the upper atmosphere produces hazes that hide the surface below. The *Cassini–Huygens* spacecraft, scheduled to arrive at Saturn in July 2004, will send a probe down to examine Titan's surface, to see if it offers a sanctuary to life-forms coming in out of the cold. We may then gain new perspectives on what happened before or during the origin of life on Earth. Unfortunately, it is even harder to get to Titan than Europa, and a round-trip with a landing is not feasible using today's technology.

And even if life within our solar system is only found on Earth, the discovery of planets around far-away stars suggests that life might exist in the cosmos at large. These stars and their planets were formed in interstellar clouds that contain vast amounts of water, one of the key molecules of life, as well as all kinds of organic, carbon-bearing molecules, from formaldehyde to benzene.

As yet, not one of these planets has been seen through a telescope. Most have been detected indirectly by the gravitational pull they exert on their star, causing it to apparently "wobble" as the planet orbits the star. Moreover, the orbiting planet has to be at least as large and massive as Jupiter to produce a detectable effect. Smaller planets, comparable in mass and size to the Earth, cannot now be detected by this method.

Telescopes must be lofted into space, outside the Earth's obscuring atmosphere, and linked together to discover Earth-sized planets around stars other than the Sun. NASA and ESA are actively, and independently, planning such a mission, that may eventually discover planets more akin to Earth and Mars, scrutinizing them for the chemical signatures of life. Planetary atmospheres that contain mixtures of gases that are out of equilibrium, and should not be present together, would be one example of a clear signature of life. Large amounts of molecular oxygen in a planet's atmosphere could, for example, indicate the presence of plant life.

Many scientists think that somewhere out there, among the 100 billion (10^{11}) stars in our Galaxy, there ought to be at least one habitable planet similar to Earth, swarming with organisms and perhaps with something more advanced than us. Others discount the possibility; but now we are back in the realm of speculation that may only be clarified by future scientific investigations.

Table 8.12 The satellites of Mars[a]

Property	Phobos	Deimos
Mass (kg)	1.08×10^{16}	1.8×10^{15}
Radii of Triaxial Ellipsoid (km)	$13.3 \times 11.1 \times 9.3$	$7.6 \times 6.2 \times 5.4$
Mean Mass Density (kg m^{-3})	1905 ± 53	1700 ± 500
Mean Distance from Mars (10^3 km)[b]	$9.378 = 2.766\,R_M$	$23.479 = 6.926\,R_M$
Sidereal Period (Mars days)[c]	0.3189	1.262 44

[a] Adapted from H. H. Kieffer, B. M. Jakosky, C. W. Snyder and M. S. Matthews (eds): *Mars*, University of Arizona Press, Tucson Arizona 1992.

[b] The mean radius of Mars is $R_M = 3389.9$ kilometers.

[c] One Mars day = 24 hours 37 minutes 22.663 seconds, so the orbital periods of Phobos and Deimos are 7 hours 39 minutes 13.84 seconds and 30 hours 17 minutes 54.87 seconds, respectively. Both satellites are locked into synchronous rotation with a rotation period equal to their orbital periods.

Fig. 8.49 Phobos A close-up image of the Martian moon Phobos, taken from the *Mars Global Surveyor*. It shows the moon's highly irregular shape and battered, cratered surface. The largest crater, Stickney (*top left*), is 10 kilometers in diameter, and about half the size of Phobos. This crater is named after Angeline Stickney (1830–1892), wife of Asaph Hall (1829–1907), who discovered the two Martian moons. Individual boulders can be seen near the rim of the crater, presumably ejected by the impact that formed Stickney. (Courtesy of NASA, JPL, and Malin Space Science Systems.)

the United States Naval Observatory. He named the inner moon *Phobos*, the Greek word for "fear" and the outer one *Deimos*, Greek for "flight, panic or terror", after the attendants of the Greek god of war, Ares, in Homer's *Illiad*.

Johannes Kepler (1571–1630) suggested the possible existence of two Martian moons as early as 1610. Since Venus has no moons, the Earth has one, and Galileo had discovered four large moons orbiting Jupiter, it seemed logical to Kepler that Mars, with an intermediate orbit between Earth and Jupiter, would have two moons.

Then Jonathan Swift (1667–1745) endowed Mars with two fictional moons in his *Gulliver's Travels*, published in 1726. He placed them close to Mars, at 6 and 10 Mars radii, close to the 2.7 and 6.9 of Phobos and Deimos. Swift's prediction had to be a lucky guess, for there was no telescope at the time powerful enough to detect the two moons.

These are not the same kind of object as the Earth's large and spherical Moon. Both moons of Mars are very small, with insufficient gravity to mold them into a spherical shape (Table 8.12). They have a battered appearance, with a profusion of craters large and small (Fig. 8.49). In fact, the surface of Phobos has been pounded into a layer of insulating dust at least a meter thick, enough to bury anything that tried to land on the small moon. Eons of meteorite impacts have apparently sandblasted it.

A maverick moon

Both Martian moons move within the planet's equatorial plane, but they orbit so near to the surface of Mars that an observer at the poles of Mars could see neither moon. Phobos is the real maverick. It moves around Mars at 2.766 Mars radii, so close that it rises and sets three times in a single day. That is, the orbital period of Phobos, of 7 hours and 39 minutes, is less than one-third of the planet's rotation period of 24 hours 37 minutes. From the surface of Mars, the small moon Phobos would be seen to move backward across the sky, rising in the west and setting in the east.

Phobos is about as close to Mars as it can get. If it came much nearer to Mars, the planet's gravitational forces would tear Phobos apart. In fact, the orbit of Phobos is steadily shrinking. If it continues to move toward Mars at the present rate, Phobos will either smash into the Martian surface or be torn apart by the planet's gravity to make a ring around Mars in 50 million years. Because Phobos is about 4.6 billion years old, we are, astronomically speaking, catching a fleeting glimpse of the last few moments of its life.

On the other hand, Deimos is near the outer limit for an object to be orbiting Mars. If it moved much further away, the Martian gravity would be too weak to hold on to the moon.

Phobos' suicidal motion has resulted in some fascinating, but untrue, speculations about the maverick moon (Focus 8.6). We now know that tidal forces are pulling the small moon toward unavoidable destruction. Phobos produces two tidal bulges in the solid body of Mars, in much the same way that the Moon produces ocean tides on the Earth (Section 5.8). As Phobos moves ahead of Mars, the closest tidal bulge pulls gravitationally on the moon, causing it to lose energy and move inexorably toward self-destruction. Because Phobos orbits so swiftly, this tidal action pulls the moon inward, instead of pushing it outward as tidal interaction does for the Earth's Moon (Section 5.8).

Fig. 8.50 Comparisons of an asteroid and the Martian moons Three irregularly shaped, cratered objects, each about the same size, are shown at the same scale and nearly identical lighting conditions. They are Phobos (*lower right*), the largest moon of Mars, Deimos (*lower left*), the other Martian moon, and the asteroid 951 Gaspra (*top*) which is about 17 kilometers across. The Phobos and Deimos images were obtained from one of the *Viking* orbiter spacecraft in 1977; the Gaspra image was taken from the *Galileo* spacecraft in 1991 on its way to Jupiter. (Courtesy of NASA and JPL.)

Origin of the Martian moons

The irregular shapes, small sizes, and low mass densities of Phobos and Deimos (Table 8.12) closely resemble those

Focus 8.6 Speculations about Phobos

Phobos is speeding up in its orbit as it gradually falls toward Mars. Some astronomers thought that atmospheric friction, or air drag, was responsible, because air drag causes the orbits of all artificial, terrestrial satellites to move slowly toward the Earth and inevitable destruction. The tenuous Martian air can, however, only produce the required drag if Phobos also has a low density of about one-thousandth the density of water. Since no known solid material is that light, the Soviet astrophysicist Iosif Samuilovich Shklovskii (1916–1985) concluded that Phobos is not solid. He claimed that it might be a hollow artificial satellite launched by a past Martian civilization. This remarkable conjecture was given the stamp of approval when the American astronomer, Carl Sagan (1934–1996), included it in his book with Shklovskii entitled *Intelligent Life in the Universe* (Holden-Day, San Francisco 1968). They described the two moons of Mars as artificial satellites that were sent into orbit by an ancient, dying civilization whose other edifices are today covered by the sands and dust of Mars. Spacecraft observations have instead shown that Phobos is actually a battered rock with a mass density nearly twice that of water, and that tidal interaction with Mars is responsible for the moon's peculiar motion.

of the numerous asteroids that orbit the Sun between the orbits of Mars and Jupiter. The two Martian moons and the asteroids have a battered appearance, with a profusion of craters large and small (Fig. 8.50). Moreover, the surfaces of the Martian moons are as dark as some asteroids, known as the carbonaceous C-type, and nowhere near as lightly colored as the surface of Mars. Scientists therefore speculate that Phobos and Deimos were adopted from the asteroid belt.

One idea is that two asteroids wandered close by Mars and were pulled in by its gravity, perhaps at different times. In another variant on this theme, one larger asteroid was captured by Mars, and subsequently broke apart during a collision with another larger body, becoming Phobos and Deimos. In both explanations, the captured asteroids would have to lose energy during their encouter with Mars. Otherwise they would hurtle back into space rather than going into orbit around Mars. Perhaps it all happened early in the planet's history, when frictional forces in a thick, extended atmosphere slowed them down.

Alternatively, Phobos and Deimos could have been created during the formation of Mars. They might, for example, be the last surviving remnants of a ring of debris blasted into orbit by a huge meteorite that collided with Mars during its formative years, about 4 billion years ago. Such a giant impact is now the favorite explanation for the origin of the Earth's Moon (Section 5.9). It would provide a good explanation for the nearly circular orbits of Phobos and Deimos, which lie in the equatorial plane of Mars; such orbits are difficult to explain by the captured-asteroid hypothesis.

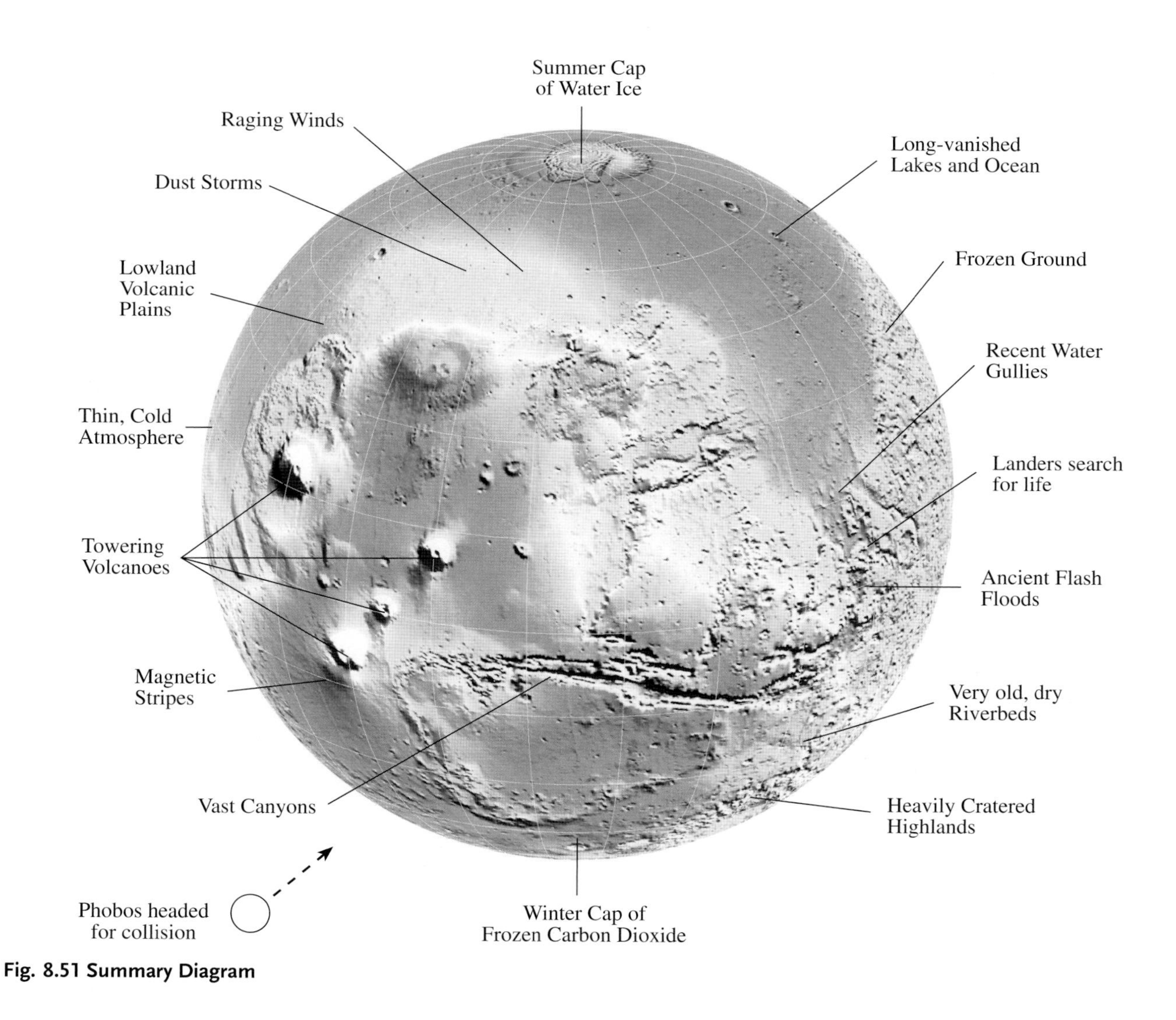

Fig. 8.51 Summary Diagram

Part 3 The giant planets, their satellites and their rings – worlds of liquid, ice, and gas

9 Jupiter: a giant primitive planet

- All we can see on Jupiter are clouds, swept into parallel bands of bright zones and dark belts by the planet's rapid rotation and counter-flowing, east–west winds.
- Jupiter turns to liquid under high pressures within its interior, so the cloudy atmosphere has no distinct bottom and Jupiter's weather pattern is free to flow in response to the giant planet's rapid spin.
- Jupiter's Great Red Spot and white ovals are huge shallow anti-cyclonic storms, which can have diameters larger than the Earth's and last for centuries.
- Large whirling storms on Jupiter gain energy by merging with, and engulfing, smaller eddies. The little storms pull their energy from hotter, lower depths.
- White clouds of ammonia ice form in the coldest, outermost layers of Jupiter's atmosphere. Water clouds are expected to form at greater depths, and ammonium-hydrosulfide clouds should condense between the water and ammonia clouds.
- All of the clouds on Jupiter ought to be white; their colors are attributed to an active chemistry that produces complex compounds in small amounts.
- Bolts of lightning illuminate deep, wet storm clouds on Jupiter.
- When the *Galileo* spacecraft parachuted a probe into Jupiter, the entry site, a region of downdraft, was missing the expected three layers of clouds and it was far drier and windier than anticipated.
- The fierce winds that give rise to Jupiter's banded appearance run deep, indicating that Jupiter's ever-changing weather patterns are driven mainly from within, by internal energy rather than by external sunlight.
- When compared to the outer layers of the Sun, the outermost atmosphere of Jupiter is slightly depleted in helium, and enriched in carbon, nitrogen and sulfur by a factor of about three.
- Jupiter is a primitive incandescent globe that radiates 1.67 times as much energy as it receives from the Sun, probably as heat left over from when the giant planet formed.
- Jupiter originated together with the Sun, and both the giant planet and the star are mainly composed of the lightest element, hydrogen.

- If Jupiter was about 80 times more massive, it could have become a star.
- Jupiter has a non-spherical shape with a perceptible bulge around its equatorial middle.
- The visible cloud tops and outer atmosphere of Jupiter form a very thin veneer that covers a vast global sea of liquid hydrogen.
- Most of Jupiter's interior consists of fluid metallic hydrogen formed under the extreme pressures that exist inside the planet.
- Jupiter probably has a dense core with a mass that is less than or equal to twelve times the mass of the Earth.
- By re-creating extreme conditions like those inside Jupiter, modern laboratory experiments have compressed liquid hydrogen so that it becomes highly conductive like a metal.
- Jupiter's powerful magnetic field is generated by rotationally-driven electrical currents inside its vast internal shell of liquid metallic hydrogen.
- The volcanoes on Jupiter's innermost large moon Io have turned the satellite inside out. It is the most volcanically active body in the solar system.
- The pyrotechnic red and yellow colors on Io are due to sulfur at various temperatures. Its volcanoes emit plumes of sulfur-dioxide gas that freeze onto the surface as a white frost.
- Volcanic vents on Io are filled with melted silicate rocks that are hotter than any place on any planet's surface, even Venus.
- Changing tidal forces squeeze Io's rocky interior in and out, making it molten inside and producing volcanoes.
- A vast current of 5 million amperes flows between the satellite Io and the poles of Jupiter, generating 2.5 trillion watts of power and producing auroral lights on both the satellite and the giant planet.
- Jupiter's magnetic field sweeps past Io, picking up a ton, 1000 kilograms, of sulfur and oxygen ions every second and directing them into a doughnut-shaped ring known as the plasma torus.
- There are no mountains or valleys on the bright, smooth, ice-covered surface of Europa; it has few impact craters, indicating a relatively young age.
- Long, deep fractures run like veins through Europa's icy covering, apparently filled by the upwelling of dirty liquid water or soft water ice. Warmer, slushy material just beneath the crust also lubricates large blocks of water ice that float like rafts across Europa's surface.
- An electrically conducting, subsurface sea within Europa may be responding to Jupiter's magnetic field, generating a time-varying magnetism in the satellite.
- Scientists speculate that subsurface liquid water in Europa may harbor alien life that thrives in the dark.
- Jupiter's moon Ganymede is bigger than the planet Mercury. The satellite's icy surface has been fractured and pulled apart, producing a grooved terrain, and surface depressions that have been filled by eruptions from volcanoes of water ice.
- Ganymede has an intrinsic magnetic field, and it is the only satellite that now generates its own magnetism.

- Jupiter's moon Callisto has one of the oldest, most heavily cratered surfaces in the solar system. Yet, the satellite is covered by fine dark, mobile material and it has a lack of small craters when compared to the surfaces of the Moon and Mercury.
- Like Europa, the outermost large moon Callisto has a borrowed magnetic field, apparently generated by electrical currents in a subsurface ocean as Jupiter's powerful field sweeps by. But Callisto has a largely homogeneous interior without any apparent dense iron core, and the buried sea has to lie deep enough to not affect its unaltered, cratered surface.
- Jupiter's faint, insubstantial ring system is made of dust. The ring particles can last for only a few thousand years, and they must be replenished if the ring system is a permanent feature.
- When interplanetary meteoroids, attracted by Jupiter's powerful gravity, pound into the small inner moons of Jupiter, they chip off dust fragments that go into orbit around the planet, forming its ring system.

9.1 Fundamentals

Jupiter's exceptional brightness and stately motion among the stars earned it the reputation as king of the planets to ancient astronomers. The giant planet outshines everything in the night sky except the Moon and Venus, and it revolves around the Sun at a leisurely pace with an orbital period of 11.86 Earth years.

Jupiter's complete orbital journey across the background stars is close enough to 12 years that the Chinese adopted it for their 12-year astrological cycle, using the giant planet's motion to mark out the years. Named *Sui Xing*, for the "Year Star", Jupiter passes through patterns of stars representing an ordered sequence of a dozen animals. The arrival of the Chinese New Year at the end of January 2000 marked the beginning of the Year of the Dragon, and the succeeding years are designated as Snake, Horse, Sheep, Monkey, Cock, Dog, Pig, Rat, Ox, Tiger and Rabbit. This system dates at least as far back as Marco Polo's (1254–1324) visit to the Mongol rulers of China. Jupiter also passes eastward through one of the twelve constellations of the Greek zodiac each year, but these stellar configurations are not related to the Chinese menagerie.

Jupiter's orbital radius is 5.2 times the radius of the Earth's orbit, so the planet's distance from Earth changes relatively little in the course of a year. As a consequence, its apparent size and brightness are fairly constant, unlike the behavior of Mars and Venus. When these nearby planets are on the same side of the Sun as the Earth, they appear much bigger and brighter than when they move to the opposite side of the Sun.

Jupiter is a true monarch of the planets, the largest planet in the solar system with a radius of about eleven times that of the Earth. The giant is so large that it could contain more than 1300 Earth-sized planets inside its volume. Yet, Jupiter is only 318 times as massive as our planet. So Jupiter must be composed of something lighter than the rock and iron that constitute the Earth.

If we divide the mass by the volume, we find a mean density of 1326 kilograms per cubic meter, only about one-quarter the mean mass density of the Earth. In fact, the mass density of Jupiter is only slightly greater than that of water, at 1000 kilograms per cubic meter, and this implies that Jupiter, like the Sun, is composed primarily of hydrogen. No other element is light enough to account for the low density of the planet.

Despite its great size, Jupiter rotates so fast that daylight and night-time each last about 5 hours and its full day is less than one-half of an Earth day. The precise rotation rate is found by tracking radio bursts that are linked to the planet's spinning magnetic field, which emerges from deep within the planet. A rotation period of 9 hours 55 minutes 29.7 seconds = 9.9249 hours is obtained from the repeated passage of the radio storm centers (Table 9.1). This rapid rotation can easily be detected in an hour or so with a small telescope if the planet's cloud markings are carefully watched, but not measured with precision.

9.2 Stormy weather

Wind-blown zones and belts

The only features we can see on Jupiter are multi-colored clouds drawn into adjacent bands or stripes by the planet's rapid rotation. There are alternate dark bands, called belts, and light ones known as zones. The zones and belts

Table 9.1 Physical properties of Jupiter[a]

Mass	$18\,986 \times 10^{23}$ kilograms = $317.70 M_E$
Equatorial Radius at 1 bar	71 492 kilometers = $11.209 R_E$
Polar Radius at 1 bar	66 854 kilometers
Mean Mass Density	1326 kilograms per cubic meter
Rotation Period	9.9249 hours = 9 hours 55 minutes 29.7 seconds
Orbital Period	11.86 Earth years
Mean Distance from Sun	7.7833×10^{11} meters = 5.203 AU
Age	4.6×10^9 years
Atmosphere	86.4 percent molecular hydrogen, 13.6 percent helium
Energy Balance	1.67 ± 0.08
Effective Temperature	124.4 kelvin
Temperature at 1-bar level	165 kelvin
Central Temperature	17 000 kelvin
Magnetic Dipole Moment	$20\,000 D_E$
Equatorial Magnetic Field Strength	4.28×10^{-4} tesla or $14.03 B_E$

[a] The symbols M_E, R_E, D_E and B_E denote respectively the mass, radius, magnetic dipole moment, and magnetic field strength of the Earth. One bar is equal to the atmospheric pressure at sea level on Earth. The energy balance is the ratio of total radiated energy to the total energy absorbed from sunlight, and the effective temperature is the temperature of a black body that would radiate the same amount of energy per unit area.

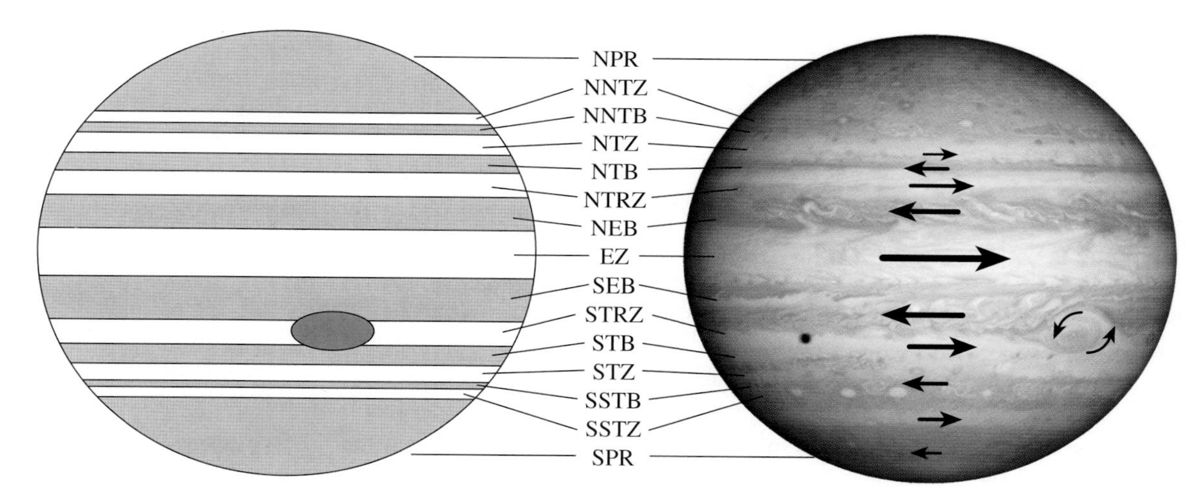

Fig. 9.1 Banded wind-blown clouds on Jupiter The traditional nomenclature of Jupiter's light and dark bands of clouds (*left*) is given in abbreviated form (*center*). The dark bands are called belts, denoted by "B", the light bands are known as zones, or "Z", and the rest of each name is based on climatic regions at the corresponding latitudes on Earth. North, letter "N" is at the top, and south, denoted by "S", is at the bottom. The equatorial, or "E", bands are in the middle, the tropical, "TR", bands on each side of the equator, and the temperate, "T", ones at mid-latitudes. Far northern latitudes are denoted by "NN", far southern latitudes by "SS", and the polar regions by "P". The image of Jupiter (*right*) was taken from the *Cassini* spacecraft on 7 December 2000, when Jupiter's moon Europa cast a shadow on the planet. The arrows point in the direction of wind flow, and their length corresponds to the wind velocity, which can reach 180 meters per second near the equator. The Great Red Spot, abbreviated GRS, is seen just below the equator at the right side of the *Cassini* image (also see Fig. 9.3). (Image courtesy of NASA and JPL.)

surround the planet, running parallel to the equator at different speeds (Fig. 9.1).

Small eddies and larger vortices interrupt the smooth profile of the belts and zones on Jupiter. The biggest whirlpools are visible from Earth using small telescopes of just 0.08-meter (3-inch) aperture, so they have been observed for centuries. The earliest sightings of a large spot in the atmosphere of Jupiter have been credited to Robert

Hooke (1635–1703) in 1664 and to Giovanni Domenico (Jean Dominique) Cassini (1625–1712) the following year. The most prominent one – the Great Red Spot – appears in records and drawings dating back to 1831, and might have coincided with the earlier sightings.

The giant planet's belts and zones are the sites of counter-flowing winds that are sometimes called zonal, or east–west, jets (Fig. 9.1). Where the Earth has just one westward air current at low latitudes – a trade wind, and one nearly eastward current at mid-latitudes – a jet stream, at the cloud-top level Jupiter has five or six of these alternating jet streams in each hemisphere.

The cloud-top winds on Jupiter have higher speeds than the winds on Earth, and storms last longer on the giant planet. Its raging winds are powerful jets, moving at speeds of up to 180 meters per second, more than four times faster than any jet stream on Earth. And unlike most winds on our planet, the east–west winds on Jupiter are remarkably steady. They vary in strength and direction as a function of latitude with a regular pattern that apparently has remained unchanged for at least a century.

On Earth, heating by sunlight results in a large temperature difference between the poles and equator, which drives our winds and circulates the air. But Jupiter's pole and equator share about the same temperature, at least near the cloud tops, so its winds are not just due to solar heating. An internal heat source most likely drives Jupiter's turbulent weather system from below.

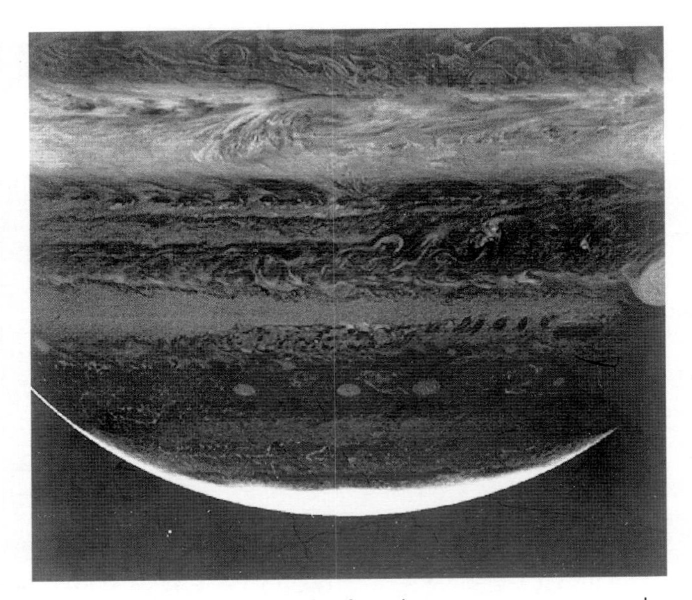

Fig. 9.2 Jupiter's turbulent clouds This image accentuates the detailed cloud structures and movements at the top of Jupiter's atmosphere. It was taken from the *Cassini* spacecraft at a red wavelength where methane gas absorbs light at high altitudes within the ammonia and ammonium-hydrosulfide cloud layers. A polar stratospheric haze (*bottom*) makes Jupiter bright near the south pole. (Courtesy of NASA, JPL, and the University of Arizona.)

Storm clouds

The colorful spots and stripes that dominate Jupiter's face mark patterns of stormy weather, as clouds billow, churn and seethe. Red, white and brown spots stare out of Jupiter's atmosphere like gigantic eyes (Fig. 9.2). Huge storms larger in size than the Earth swirl across the planet, while smaller eddies chase each other, whirling and rolling about.

Jupiter's famous Great Red Spot is essentially a huge weather system, with an east–west dimension greater than two Earth diameters (Fig. 9.3). Because of rapidly increasing pressure with depth it cannot extend deeply into the planet. It is simply an enormous, shallow eddy trapped between counter-flowing jets, so large that the strong prevailing winds are forced to flow around it. The winds, in turn, funnel smaller eddies toward the Red Spot, helping to roll it around.

Winds are swirling inside the awesome vortex in the counter-clockwise direction, at speeds up to 110 meters per second. Since it is in the southern hemisphere of Jupiter, this rotational direction indicates that the Red Spot is a swirling high-pressure vortex, known as an anti-cyclone. A low-pressure cyclone would spin in the opposite direction. Jupiter's Great Red Spot is similar to the high-pressure systems that drift across the terrestrial surface, but unlike much of the Earth's stormy weather that is often associated with low-pressure cyclones, including hurricanes. As air moves freely above the ground, the low-pressure systems on Earth draw surrounding air in and push it down, while high-pressure ones have an upward flow of air in the center and push the air up.

The Great Red Spot is not unique, but just the largest of hundreds of different storms on Jupiter, including three egg-shaped white ovals, located at about 30 degrees south latitude, that recently merged. These weather systems formed in 1939. At least at cloud-top levels, ninety percent of these long-lived vortices are high-pressure, anti-cyclones that rotate counter-clockwise in the planet's southern hemisphere and clockwise in the northern hemisphere, with counter-flowing winds on their sides. Instead of wandering unpredictably like terrestrial hurricanes, the titanic whirlpools on Jupiter drift at a steady rate in either the eastward or westward direction, apparently rolling between and with the winds, like a giant ball bearing. In contrast, storms that whirl in the other cyclonic direction on Jupiter are short-lived, lasting several days or less before being torn apart by the action of shearing winds. Lightning

Fig. 9.3 Great Red Spot The largest and oldest known weather system in the solar system is Jupiter's Great Red Spot. The spot's east–west diameter is more than twice that of the Earth, and one-sixth the diameter of Jupiter itself. The red vortex swirls in the counter-clockwise direction in Jupiter's southern hemisphere, showing that it is a high-pressure anti-cyclone. It has been observed for centuries, probably ever since astronomers turned the first telescopes toward the giant planet. Some small eddies are sucked into the Great Red Spot, helping to sustain it, while other eddies roll around the perimeter, probably reinforcing its circulation. A long-lived white oval is seen just below the Red Spot, also rotating in an anti-cyclonic manner. This oval is one of three systems that formed in 1939. They had slowly contracted in length and drifted about the planet approaching and moving apart. Recently the three storms merged. (Courtesy of NASA and JPL.)

observed by the *Galileo* spacecraft was associated with these cloud systems.

One reason that the storms can last so long on Jupiter is that there is no solid surface directly below the clouds to interfere with the flow. The atmosphere has no distinct bottom, just shearing due to greater pressure. Thus, the weather pattern is free to flow in response to the giant planet's spin. A nearby solid surface would dissipate the energy of the storm clouds, as happens to hurricanes that make landfall on Earth.

But how can the biggest storms on Jupiter last for centuries within a constantly changing atmosphere? The large ovals and spots can survive by sucking in and engulfing smaller eddies that pass in their vicinity, like leaves in a whirlpool of water, consuming them and extracting their energy. The little, short-lived storms feed their energy into the larger storms, just as smaller fish nourish larger ones. And the food chain continues to the very top, with giant white ovals, each half the size of the Earth, occasionally merging together to become one.

The big storms engulf small ones, and the small storms probably pull up their energy from hotter, lower depths. As the moist, internal heat rises in the stormy updrafts, it can sustain the whirling clouds, supplying the energy that drives much of Jovian weather. Terrestrial weather systems can similarly include hot rising air as well as cool downdrafts, but the heat on Jupiter is generated from a completely different source – from deep in the planet and not from sunlight.

Cloud layers and colors

Radio signals can penetrate Jupiter's clouds and probe the giant planet's outer atmosphere, just as radio waves travel from a distant transmitter to your car radio on a cloudy day. Since the weight of overlying layers compresses the gas to greater density at lower depths, the radio signals experience more pronounced refractive alterations when passing through deeper regions of the Jovian atmosphere. These changes have been observed by monitoring homeward-bound radio transmissions from *Voyager 1* and *2* when the spacecraft passed behind the planet, and they have been used to deduce the density, pressure and temperature as a function of altitude in and below the clouds.

If we could descend through Jupiter's thin cloud layer, we would find that the temperature and pressure increase with depth (Fig. 9.4). As in any planetary atmosphere, the atoms and molecules collide more frequently in the increasingly compressed, denser regions of Jupiter's atmosphere so the pressure and temperature increase there. At the cloud tops the temperature is a freezing

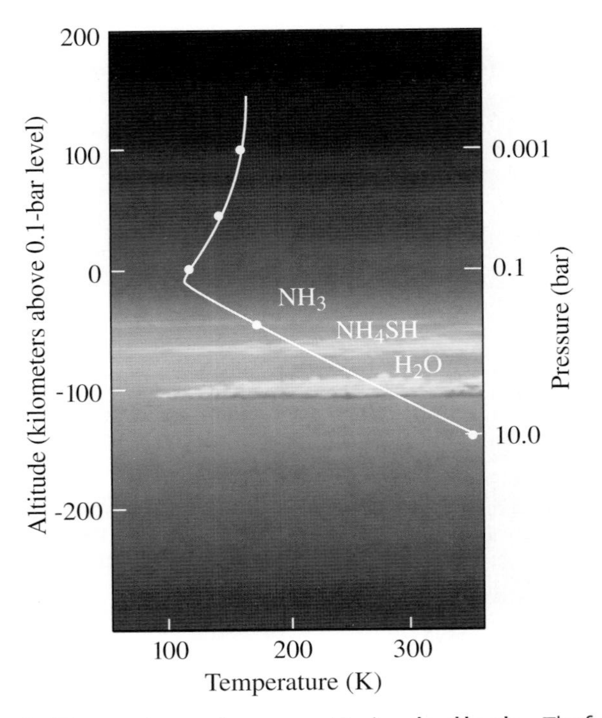

Fig. 9.4 Temperature and pressure at Jovian cloud levels The fading radio signals when the *Voyager 1* and *2* spacecraft passed behind Jupiter in 1979 have revealed the temperatures (*bottom axis*) and pressures (*right axis*) in its upper atmosphere. The temperature reaches a minimum of about 114 kelvin at a level called the tropopause where the atmospheric pressure is 0.1 bar, or 100 millibar. By way of comparison, the pressure of the Earth's atmosphere at sea level is 1.0 bar. The altitudes (*left axis*) are relative to the 0.1-bar level, and the dots are spaced to indicate tenfold changes in pressure. Solar radiation causes the temperature to increase with height just above the tropopause. At lower levels, the temperature and pressure increase systematically with depth. Three expected clouds layers of ammonia, NH_3, ammonium hydrosulfide, NH_4SH, and water ice, H_2O, are shown. The altitudes of the predicted cloud layers are based on a gaseous mixture that is of solar composition. An increase of the abundance of a condensable gas by a factor of three would lower the altitude of the cloud base by about 10 kilometers.

Focus 9.1 Speculations about life in Jupiter's atmosphere

Once they established the temperatures and pressures in the outer atmosphere of Jupiter, scientists could speculate about the possibility of primitive life existing there. The outer atmosphere may be too cold for life to exist, for it would freeze to death. Deeper down inside the planet it is too hot to even allow organic molecules to exist; they would break apart into their constituent atoms. The molecular constituents of life might nevertheless survive in the region between, where Earth-like temperatures and pressures exist, perhaps being synthesized from simpler molecules by the action of Jovian lightning.

But the warm part of Jupiter's extended atmosphere contains no solid surface on which primitive creatures could creep or crawl, and strong atmospheric currents would most likely either cycle them up into the frigid heights or drag them down to scalding depths. Heavy organisms would just sink down into the lethal heat. Some imaginative astronomers have therefore argued that buoyant inflated organisms could be floating in Jupiter's outer atmosphere, bobbing up and down like terrestrial jellyfish to seek more clement conditions without sinking too deeply into the planet.

Other researchers have argued that life on Jupiter is very implausible. They reason that biological compounds could not survive the harsh environment. This conclusion was reinforced when the *Galileo* spacecraft dropped a probe into the giant planet, showing that sophisticated organic molecules were not present at the entry site.

Nowadays, many scientists have forgotten about these early speculations about possible life inside Jupiter, and have turned their attention to Jupiter's moon Europa, which might have a life-sustaining ocean beneath its icy surface (Section 9.4).

114 kelvin and the atmospheric pressure is about 0.1 bar, or one-tenth that of the Earth's air at sea level. In slightly deeper layers, about 130 kilometers down, the temperature rises to a balmy 300 kelvin, well above the freezing point of water, at 273 kelvin. In these warmer regions, the pressure is comparable to the air pressure at the surface of the Earth, leading to speculations that, if vertical mixing were small, living things might reside there (Focus 9.1). And above them it is cold enough to freeze various gases into ice to form the clouds.

Given the profile of temperature and pressure with altitude, it is possible to infer the altitude at which clouds of various types should form. The early calculations, initiated by John S. Lewis (1941–) in 1969, assumed that the gas mixture is in chemical equilibrium and has a uniform composition like that of the Sun. And although there have been many refinements since then, three distinct cloud layers are always predicted. As one proceeds upward from the interior of Jupiter, the temperature and pressure fall to the point where the gases of water, ammonium hydrosulfide and ammonia are expected to condense to form clouds. Water clouds similarly form in the colder, higher parts of the Earth's atmosphere, which has only one layer of cloudy, stormy weather.

The visible cloud tops of Jupiter consist of ammonia-ice crystals, which condense out of the atmosphere at the very

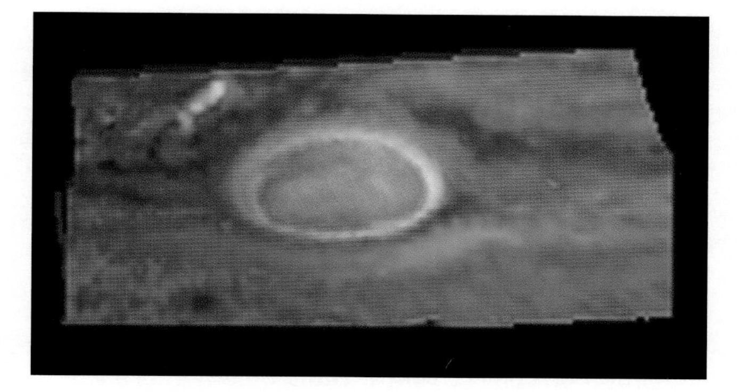

Fig. 9.5 Jupiter's ammonia-ice clouds A cloud of ammonia ice (*light blue*) is shown at the northwest (*upper left*) of the Great Red Spot (*middle*), inside its turbulent wake. The cloud was most likely produced by powerful updrafts of ammonia-laden air from deep within Jupiter's atmosphere. Reddish-orange areas show high-level clouds, yellow regions depict mid-level clouds, and green areas correspond to lower-level clouds. Darker areas are cloud-free regions. Light blue depicts regions of middle-to-high altitude ammonia-ice clouds. This near-infrared image was taken on 26 June 1996 from the *Galileo* spacecraft. (Courtesy of NASA and JPL.)

low temperatures and pressures there. They create graceful white clouds that probably make up the cold, light-colored zones observed from Earth (Fig. 9.5). This is consistent with spectroscopic measurements of abundant ammonia at the cloud tops of Jupiter.

Below the ammonia layer, the models predict two other cloud layers. At a depth where the pressure is slightly higher, at about 2 bar, the ammonia combines with hydrogen sulfide to form clouds of ammonium hydrosulfide, that would smell something like rotten eggs if you could get near them. The lowest clouds are predicted to contain water ice, formed at a pressure of about 5 or 6 bar. The water clouds are obscured by the higher cloud cover of ammonia, and are almost never seen from outside the giant planet.

The layered cloud model does not explain Jupiter's highly colorful appearance. Ammonia, ammonium hydrosulfide and water form white ices, so all the expected clouds should be white. Scientists speculate that complex molecules formed by interaction of gases with solar ultraviolet radiation in the outer atmosphere might coalesce and grow to form brown and yellow smog particles. They could account for some of the colors found in the belts – and remind us of a bad smoggy day in Beijing, Los Angeles or Mexico City.

The origin of the red color in Jupiter's clouds is something of a mystery. It might arise when the chemical equilibrium is disturbed by something that energizes an active chemistry. Lightning bolts, energetic ultraviolet sunlight, high-speed particles, or extreme temperature variations might be responsible. One or more of these sources of energy probably breaks down molecules to produce coloring compounds of sulfur or phosphorus that are present in the atmosphere in only minute quantities. Alternatively, the red colors might be dredged up from greater depths by whirling storms, or perhaps attributed to rare organic, carbon-bearing compounds.

Lightning bolts in wet spots

Ancient mythology was close to the mark when it designated Jupiter as master of the rains, hurling thunderbolts at those who displeased him. Lightning flashes were discovered in *Voyager 1* and *2* images of the dark night side of Jupiter, apparently illuminating the clouds in massive thunderstorms, and the lightning was confirmed by instruments aboard the *Galileo* spacecraft. Both missions showed that the lightning is concentrated near oppositely directed winds where storm clouds are found.

How deep the lightning occurs can be estimated from its diameter. The larger the flash, the deeper the lightning discharge. The observed sizes of Jupiter's lightning flashes suggest that they originate from layers in the atmosphere where water clouds are expected to form, at about 100 kilometers down. Only water could condense at these depths. When the *Galileo* cameras followed the night-side lightning sources into the day side, they confirmed that the lightning originates in deep moist clouds (Fig. 9.6).

If the lightning bolts on Jupiter are similar to those on Earth, then they probably occur in water clouds where partially frozen water particles become electrically charged. The electrified rain or ice particles rise and fall in the turbulence, causing positive and negative charges to separate. A high current discharge can then flow through the atmosphere, producing the lightning flashes.

So the lightning bolts on Jupiter most likely point to places where there are rapidly falling raindrops and quickly rising gas currents. Such a moist convection might transport heat upward, carrying energy into the outer Jovian atmosphere. The circulating atmosphere seems to have been detected when *Galileo* sent a probe into the planet's clouds.

Plunging into a dry and windy spot

A pioneering descent into Jupiter's atmosphere took place on 7 December 1995 when a 339-kilogram probe was

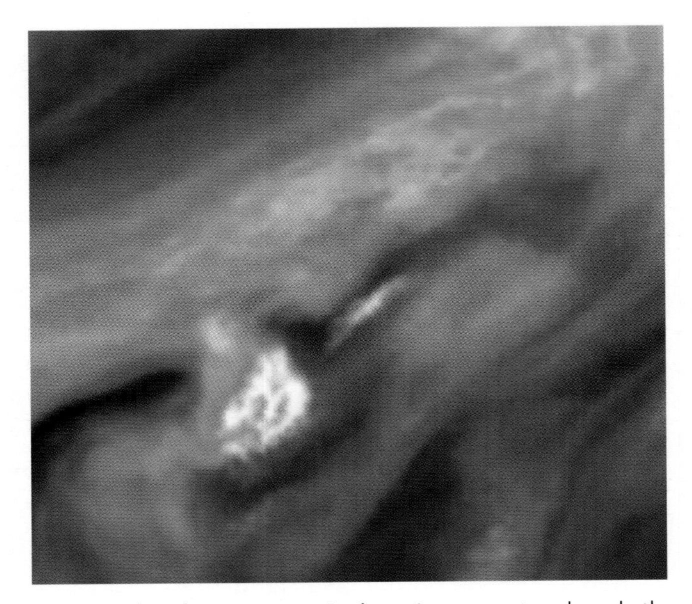

Fig. 9.6 Thunderstorm on Jupiter Instruments aboard the *Galileo* spacecraft determined the altitudes of this tall, thick thunderstorm (*mottled white region*), which towers 25 kilometers above the surrounding clouds and extends 50 kilometers below them (*red base*). On Jupiter, water is the only substance that can form a cloud at this depth, where the pressure is four or five times the sea-level pressure on Earth. Other towering thunderstorms on Jupiter have been shown to produce powerful lightning bolts, suggesting that these clouds contain falling raindrops and rising gas columns. Massive storm cells like these probably transport heat from Jupiter's interior into its long-lived cloudy weather patterns of bands and ovals. Similar water-rich thunderstorms with lightning exist on Earth, but their Jovian counterparts are about five times broader and taller. (Courtesy of NASA, JPL, Cornell University, and the California Institute of Technology.)

dropped from the *Galileo* spacecraft on a suicide plunge into the planet. The instrument-laden *Galileo Probe* was parachuted through the thickening gas, taking the measure of composition and winds to well below the visible clouds. The intrepid capsule returned data for just over an hour, down to the 20-bar pressure level, until the rising temperatures and crushing pressures wiped the probe out and it disappeared without a trace.

Scientists had expected that the *Galileo Probe* would pass through three cloud layers, composed of different chemicals that condense from tenuous gases at successively higher and colder levels. Below the bottom, water-cloud layer, which for Jupiter is formed at the 5-bar level, the atmosphere was expected to be well stirred, and therefore more representative of the planet's uniform, global composition. But contrary to expectations, the clouds were not where everyone thought they would be.

It was apparently a clear day at the probe's entry point. When the capsule plummeted into the maelstrom, its instruments saw almost no evidence for clouds. All of the expected cloud constituents were still in the gaseous state, and were found in increasing amounts through and well below the condensation levels where the cloud's ice particles should have been found.

Moreover, the planet was a lot drier than anticipated, at least in the vicinity of the probe entry site. Extrapolating from the Sun's makeup, researchers had expected at least five or ten times as much water, under the assumption that Jupiter coalesced out of similar material with the same proportion of oxygen, which in the outer atmosphere of Jupiter should all be in the form of water molecules. The amount of lightning detected during the probe's hour-long descent was also far less than expected, supporting the conclusion that this part of the upper atmosphere contains little water.

The missing clouds and water might be explained if the probe descended into an unusually clear spot of dry, downwelling gas and reduced cloud cover. In fact, observations from Earth-based telescopes indicated that the entry site was a region where emerging infrared, generated at deeper, warmer levels, shines through a gap in the clouds, and dry, wrung-out gas from high altitudes might have been forced downward into this region. Jupiter has several of these clear hot spots, that alternate with cloudy places in a band extending around the planet at eight degrees north latitude (Fig. 9.7).

The previous models of Jupiter's clouds were probably too simplified, for they assumed a uniform layering with depth and ignored deep, vertical, up-and-down weather-related activity. Both wet and dry regions are found in the outer atmosphere of Jupiter, just as Earth has tropics and deserts, and they may be related to the circulation of rising and falling gas. Winds that rise from the deep atmosphere could dredge up material that lacks water and other cloud-making ingredients. When these winds converge and drop back down, nothing is left to condense back into clouds and a dry clearing is created. Similar downdrafts occur over subtropical deserts on Earth, but, unlike our planet, Jupiter has no firm surface to quickly stop the falling gas.

One part of the weather forecast that proved correct below Jupiter's cloud tops was "windy". Instead of decreasing to a dead calm as the probe descended, the zonal winds stayed strong and even increased with depth. The zonal winds that create the planet's banded appearance continued to whip around the planet at speeds of up to 200 meters per second, until the capsule fell silent and stopped sending readings at about 600 kilometers down. Because little sunlight can penetrate to such depths, the winds must be driven mainly from below. Internal heat, probably left over from

Table 9.2 Element abundance in the outer layers of Jupiter and the Sun[a]

Element	Symbol	Chemical Form	Jupiter	Sun	Jupiter/ Sun
Helium	He	Helium	0.078	0.097	0.804
Carbon	C	Methane, CH_4	1.0×10^{-3}	3.6×10^{-4}	2.78
Nitrogen	N	Ammonia, NH_3	4.0×10^{-4}	1.1×10^{-4}	3.64
Oxygen	O	Water, H_2O	3.0×10^{-4}	8.5×10^{-5}	3.53
Sulfur	S	Hydrogen sulfide, H_2S	4.0×10^{-5}	1.6×10^{-5}	2.50
Deuterium	D	Deuterium	3.0×10^{-5}	3.0×10^{-5}	1.0
Neon	Ne	Neon	1.1×10^{-5}	1.1×10^{-4}	0.10
Argon	Ar	Argon	7.5×10^{-6}	3.0×10^{-6}	2.50
Krypton	Kr	Krypton	2.5×10^{-9}	9.2×10^{-10}	2.72
Xenon	Xe	Xenon	1.1×10^{-10}	4.4×10^{-11}	2.50

[a] Number of atoms per atom of hydrogen, designated by the symbol H.

Fig. 9.7 Hot spot on Jupiter The dark region near the center of this image is an equatorial "hot spot", similar to the site where the *Galileo* spacecraft parachuted a probe into Jupiter's atmosphere in December 1995. Jupiter has many such regions, and they continually change, so the probe could not be targeted to either hit or avoid them. The dark hot spot is a clear gap in the clouds where infrared radiant energy from the planet's deep atmosphere shines through. Although hotter than the surrounding clouds, these so-called "hot spots" are still colder than the freezing temperature of water. Dry atmospheric gas may be converging and sinking in these regions, maintaining their cloud-free appearance. The bright ovals, shown in other parts of this image, may be examples of upwelling moist air. The images combined in this mosaic were taken on 17 December 1996 from the *Galileo* spacecraft. (Courtesy of NASA and JPL.)

planetary formation or contraction of the planet as it slowly cools, is therefore the most likely driving force for Jupiter's powerful winds and ever-changing weather patterns.

Composition of Jupiter's upper atmosphere

When the *Galileo* spacecraft sent a probe into Jupiter, the relative amounts of several elements were measured in the planet's outermost atmosphere. The abundance of these ingredients has been compared to that of hydrogen, by far the most abundant element in Jupiter, and to the relative amounts found in the outer layers of the Sun (Table 9.2).

Instruments aboard *Galileo* found that helium, the second most abundant element in both Jupiter and the Sun, is just a bit depleted from solar amounts, as had been suggested by previous measurements from the *Voyager 1* and *2* spacecraft. The *Galileo* results were more accurate, indicating somewhat more helium than the previous missions had, but still less than the outer layers of the Sun. Since some original helium in the outer solar atmosphere has been lost to the Sun's internal fires, helium must also be removed from Jupiter's outermost atmosphere, by helium raining into the interior of the planet. This slow helium removal process operates on an awesome scale in neighboring Saturn, whose outer atmosphere is severely depleted in helium (Section 10.3).

Heavier elements, such as carbon, nitrogen and sulfur, were enriched in the Jovian atmosphere by a factor of about three when compared to a mixture of solar composition. This result had also been anticipated, for the carbon, C, and nitrogen, N, combine chemically with hydrogen, H, to form methane, CH_4, and ammonia, NH_3, and scientists have long known that Jupiter has about three times as much carbon and nitrogen in the form of these gases as the

Sun. Still, even with this enrichment, hydrogen and helium comprise 99 percent of Jupiter's substance by volume.

The *Galileo* probe's instruments detected surprisingly high concentrations of argon, krypton and xenon. These three chemical elements are called noble gases because they are very independent and do not combine with other chemical elements. They are enriched compared with solar composition by about the same factor as carbon, nitrogen and sulfur. In contrast, another noble element, neon, was starkly depleted with about ten times less than the solar amount.

These composition results are related to the formation and subsequent evolution of Jupiter. The apparent enrichment of elements like carbon, relative to light hydrogen, probably occurred when some of the light gas was blown out of the solar system by the active young Sun. In order to have caught and retained the noble gases, Jupiter would have had to freeze them – which is not possible at the giant planet's current distance from the Sun. Nobles gases do not condense until they encounter temperatures much lower than the freezing point of water and the current temperatures of the Jovian satellites. Perhaps the noble gases were brought in from colder, more distant regions by cometary bodies that helped build up young Jupiter, or the entire planet might have originated further away from the Sun and gradually migrated inward to where it is now.

The anomalous depletion of neon was even explained before its discovery. Scientists predicted that the neon would dissolve in helium, which rains down inside the planet and takes the neon with it.

9.3 Beneath Jupiter's clouds

Educated guesses about Jupiter's internal constitution

We cannot see beneath the clouds of Jupiter, but we can use external measurements to place constraints upon its internal properties. As an example, we now know that the giant planet emits its own heat radiation, which means that it is hot inside. Since Jupiter and the Sun originated from similar material at the same time, a good initial assumption is that they have the same ingredients with similar proportions. The planet's low average mass density indicates that it is in fact composed largely of hydrogen, just as the Sun is. The planet's oblate shape and rapid rotation also tell us something about the way it is constructed inside. Due to the enormous pressures inside Jupiter, most of the planet's hydrogen is compressed into a liquid metallic form, which helps account for the giant's strong magnetic field. All of these constraints have been pieced together to make a picture of Jupiter's invisible interior.

An incandescent globe

With the advent of ground-based infrared measurements of the planets, pioneered by Frank J. Low (1933–) and his colleagues in the 1960s, astronomers were surprised to discover in 1969 that the giant planet is an incandescent globe with its own internal source of heat. This result was confirmed in greater detail with instruments aboard the *Voyager 1* and *2* spacecraft, that determined precisely how much infrared heat radiation was emerging from inside the planet. They showed that Jupiter is radiating 1.67 times as much energy as the atmosphere absorbs from incoming sunlight. In other words, the giant planet radiates nearly twice as much energy as it receives from the Sun, and almost half of the total energy that Jupiter loses must come from its interior. This essentially means that the planet has to be unexpectedly hot inside.

Jupiter must have been still hotter when it formed, thanks to the energy released during the gravitational collapse that accompanied its growth. When the new born planet coalesced from a larger primordial cloud, gravitational energy was converted into heat as particles and small bodies fell inward and collided with each other. Such a process ignited the internal fires of the Sun and other stars, but Jupiter was not quite massive enough to become a star (Focus 9.2).

The compression inside Jupiter also excites the gas and leads to radiation that can carry off some of the energy. But so much heat is still left inside Jupiter since its time of formation that it is still cooling off, pumping out about twice as much energy as it receives from the Sun. In contrast, a small planet like Earth would have radiated away the heat of its formation long ago. The giant planet's enormous size makes it a much better heat-trap than the terrestrial planets.

The Earth is now heated internally by the decay of radioactive elements contained in its rocks, but this has nothing to do with Jupiter's excess heat radiation. The giant planet contains relatively little rocky material, and the observed heat flow is one hundred times too large to be explainable by radioactive heat production, even if Jupiter were composed almost entirely of terrestrial rocks rather than mostly hydrogen.

Jupiter could also be shrinking slightly today, converting gravitational energy into heat and supplementing the heat left over from its initial formation. But we cannot determine if that is happening. The giant planet needs to contract by only one meter per century to supply the

Focus 9.2 Stars that do not quite make it

Planets are supposed to shine only by reflected sunlight, while most stars generate their own radiation by thermonuclear reactions in their exceptionally hot cores. Jupiter has its own internal energy source, so it resembles a star in this respect, but it shines by heat left over from the planet's formation rather than by nuclear fusion.

The process of gravitational contraction that warmed the inside of young Jupiter to its present central temperature of about 17 thousand kelvin also heated the center of the Sun to about 15 million kelvin. That was hot enough to ignite the thermonuclear reactions that make our star shine. The Sun became much hotter inside because it is more massive, weighing in at one thousand times the mass of Jupiter. So Jupiter could have collapsed to form a star if it had more mass, and some astronomers therefore like to call Jupiter a star that didn't quite make it. In fact, calculations indicate that Jupiter might have become a star if it had been only eighty times more massive than it is now.

There are certain stellar objects, known as brown dwarfs, which originate from the gravitational collapse of a cloud of gas in space, but do not have enough mass to trigger nuclear fusion reactions in their core. They can shine faintly for about 100 million years as gravitational energy is converted into heat. Some of these objects have been detected in binary stellar systems by the means of the gravitational pull they have on their primary stars. One of them, cataloged as Gliese 229B, is roughly forty times as massive as Jupiter and is very similar to it in radius. Objects with masses between thirteen and eighty Jupiter masses are thought to be brown dwarf stars, while those with masses between about one-fiftieth and thirteen Jupiter masses are classified as giant planets.

observed amount of internal energy radiated by the planet, and astronomers could never measure its radius with that kind of precision.

Ingredients at formation

According to the widely accepted nebular hypothesis, the Sun and planets formed together during the collapse of a rotating interstellar cloud called the solar nebula. Most of it fell into the center, until it became hot enough to ignite the Sun's nuclear fires. Further out, the planets formed from a whirling disk of the same material.

If the nebular hypothesis is correct, and the whole solar system originated at the same time, then you might expect Jupiter to have a similar chemical composition to the Sun. To a first approximation, the abundance of the elements in the giant planet does indeed mimic that of the Sun, with a predominance of the lightest element hydrogen. It is the most abundant element in most stars, in interstellar space, and in the entire Universe. The second most abundant element in both Jupiter and the Sun is helium, and hydrogen and helium together account for the low mean mass density of both objects, at 1326 and 1409 kilograms per cubic meter respectively.

Spectroscopic observations indicate that the outer atmosphere of both Jupiter and the Sun also contain heavier elements, like carbon, oxygen and nitrogen. The giant planet has a bit more than the star, but even with this enrichment these elements together comprise less than 1 percent of the planet's composition by volume – all the rest is hydrogen and helium.

At the frigid temperatures where Jupiter and the other giant planets originated, the carbon, C, oxygen, O, and nitrogen, N, atoms would have bonded with the abundant hydrogen, H, to form methane, CH_4, water, H_2O, and ammonia, NH_3, respectively. These compounds are known as "ices" because they would have condensed into solid ice at the freezing temperatures far from the Sun. Rocky material, containing atoms of silicon and iron, would also have been present in lesser amounts. The ice and rock are now located at the center of Jupiter, probably because they coalesced to form a massive nucleus that pulled in the surrounding hydrogen and helium, or perhaps because they settled gravitationally into the planet's core after it formed.

At the higher temperatures closer to the Sun, where the Earth and other terrestrial planets formed, the icy material would be vaporized and could not condense. That left only rocky substances to coalesce and merge together to form the terrestrial planets. Their modest mass and proximity to the Sun would not allow them to capture and retain the abundant lighter gases, hydrogen and helium, directly from the solar nebula.

An equatorial bulge

Observations with even a small telescope show that Jupiter is not a sphere. It has a perceptible bulge around its equatorial middle and is flattened at the poles. This elongated oblate shape is caused by Jupiter's rapid spin. The outward force of rotation opposes the inward gravitational force, and this reduces the pull of gravity in the direction of spin. Since this effect is most pronounced at the equator, and least at the poles, the planet expands into an oblate shape that is elongated along the equator. The same thing happens to all the giant planets, and even to the solid Earth (Table 9.3).

Table 9.3 Oblateness of the Earth and giant planets[a]

Planet	Equatorial Radius, R_e (km)	Polar Radius, R_p (km)	Oblateness $(R_e - R_p)/R_e$
Earth	6 378.140	6 356.755	0.003353
Jupiter	71 492	66 854	0.0649
Saturn	60 268	54 364	0.0980
Uranus	25 559	24 973	0.0229
Neptune	24 766	24 342	0.0171

[a] The radii of the giant planets are those at the level where the atmospheric pressure is equal to 1 bar, the pressure of air at sea level on Earth.

Table 9.4 Range of pressures

Location	Relative Pressure
Beneath the foot of a water strider[a]	0.000 01
Inside a light bulb	0.01
Earth's atmosphere at sea level	1.0
Inside a fully charged scuba tank	100.0
Deepest ocean trench	1 000.0
Pressure at which hydrogen becomes metallic	1 000 000.0
Center of Jupiter	70 000 000.0

[a] A water strider is an insect with long legs that walks on water and feeds on dead insects.

The amount of Jupiter's non-spherical extension depends on both its rate of rotation and the internal distribution of its material. The faster the spin, the more the outward push and the greater the elongation. And given its rotation, with the rapid period of 9.9249 hours, the size of the equatorial bulge depends on how Jupiter's mass is distributed inside. The more massive the planet's dense core, the smaller the equatorial bulge.

Scientists have measured the properties of Jupiter's equatorial bulge by accurately determining the motion of Jupiter's natural satellites, and closely tracking the *Pioneer 11* and *Voyager 1* and *2* spacecraft as they flew close to the giant. If Jupiter were a perfectly spherical planet, it would act as if all its mass was concentrated in a single central point and the motions of natural satellites or spacecraft would not depend on their orientation with respect to the planet's equator. In contrast, an oblate planet produces an extra force that tugs the moving object toward its equatorial bulge, and also toward any internal core. When combined with the known mass, volume and rotation rate of Jupiter, observations of these effects indicate that Jupiter has a dense core containing up to 12 Earth masses. Such a central object, presumably composed of non-gaseous rocky and icy material, was apparently required to initiate the accumulation of the giant planet's extensive hydrogen shell, which now compresses the core to high temperatures and pressures, melting the ice and the rock.

Enormous pressures and strange matter

The pressure inside Jupiter grows with depth, just as the pressure increases when you dive into the depths of an ocean. The particles at greater depths are compressed into smaller volumes by the weight of overlying material, so they collide with each other more frequently. This causes an increase in pressure at deeper levels inside the giant planet (Table 9.4).

Near Jupiter's cloud tops, the pressure is about the same as the air pressure at sea level on Earth, designated as 1 bar or 10^5 pascal, and the pressure increases to an astonishing 70 million times that amount at the center of Jupiter. Higher pressures are associated with hotter temperatures, so the temperature also increases with depth inside Jupiter. Out at the one-bar level, the temperature is a freezing 165 kelvin. Deep down at the center, the temperature has risen to 17 thousand kelvin, just over three times as hot as the visible disk of the Sun – at 5280 kelvin.

To understand the internal constitution of Jupiter, we need to know what happens to its most abundant ingredient, hydrogen, as the pressure increases. At the low pressure in the outer, visible parts of Jupiter, hydrogen forms a molecular gas, but this atmosphere is just a thin veneer. In proportion, the outer layer of gaseous hydrogen molecules resembles the skin of an apple. Deeper down, where the pressures and temperatures are higher, the hydrogen is liquefied. Indeed the planet is almost entirely liquid (Fig. 9.8). It is mostly just a vast, global sea of liquid hydrogen.

As suggested by Rupert Wildt (1905–1976) in 1938, the intense pressures and temperatures deep inside Jupiter will cause hydrogen molecules to break down forming something he called "metallic hydrogen". Jupiter is indeed so massive that most of the fluid hydrogen inside is believed to be squeezed into metallic form. Below about one-seventh of Jupiter's radius, from the cloud tops, the internal pressure exceeds one million bars (1 Mbar) and the liquid molecular hydrogen is transformed into liquid

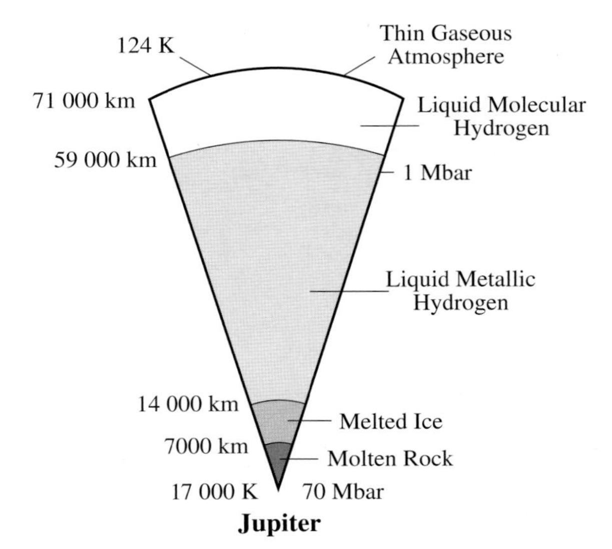

Fig. 9.8 Inside Jupiter Giant Jupiter has a thin gaseous atmosphere covering a vast global ocean of liquid hydrogen. At the enormous pressures within Jupiter's interior, the abundant hydrogen is compressed into an outer shell of liquid molecular hydrogen and an inner shell of fluid metallic hydrogen. The giant planet probably has a relatively small core of melted ice and motlen rock.

metallic hydrogen. It is said to be in a metallic state because, like a metal, it is an excellent conductor of electricity.

The hydrogen molecules are pushed so closely together that the electrons are squeezed free of any single atom or molecule. These mobile electrons can travel freely from one place to another, moving about and conducting electricity like the electrons in normal metals such as copper or iron. Note, however, that, the metallic hydrogen inside Jupiter has been melted at the high temperatures down there, and therefore behaves like a molten metal rather than a solid.

Most of Jupiter is in the form of liquid metallic hydrogen. And underneath it all are relatively small amounts of molten rock and ice, up to twelve times heavier than the Earth (Fig. 9.8). Theoretical considerations suggest that Jupiter's hydrogen probably accumulated around such a massive, pre-existing central object.

The vast liquid shell of metallic hydrogen is no longer just a theoretical conjecture. By re-creating extreme conditions like those inside Jupiter, modern laboratory experiments have turned liquid hydrogen into a liquid metal. A two-stage gas gun has succeeded in pressurizing liquid hydrogen to 1.4 million bars and 3 thousand kelvin, squeezing the hydrogen into a liquid metallic state that conducts electricity just like any solid metal.

Electrical currents, driven by Jupiter's fast rotation within its liquid metallic shell, apparently generate the planet's strong magnetic field, in much the same way that electricity in the Earth's molten metallic core produces our planet's magnetism. Jupiter's magnetic field is much more powerful than Earth's magnetism, with a magnetic moment that is 20 000 times as large and a cloud-top strength that is about 14 times Earth's surface magnetic field strength (Section 3.6). The greater strength of Jupiter's magnetism could be attributed to the planet's faster rotation, more extensive metallic region, and the relative proximity of the internal electrical currents to the cloud tops. By way of comparison, Earth's magnetic field is produced within a much smaller metallic core, which extends only halfway to the surface.

9.4 The Galilean satellites

Introduction to the larger moons

Galileo Galilei (1564–1643) discovered Jupiter's four largest moons in January 1610, using the newly invented telescope. They are bright enough to be seen with a pair of binoculars or a small telescope, and if it were not for the glare of Jupiter, these moons would be visible to the unaided eye.

These objects are now collectively called the Galilean satellites, even though Galileo wanted to name them the Medician planets after the Medici family of Firenze. They retain the individual names given to them by Simon Marius (1573–1624), also in 1610. In order of increasing distance from the giant planet, they are Io, Europa, Ganymede, and Callisto, all the names of mythological consorts of Zeus, the Greek equivalent of the Roman god Jupiter.

Zeus changed the mortal Io, a beautiful river nymph, into a cow to hide her from his jealous wife. The Ionian Sea is named after the sea that Io the cow swam during her wanderings. Europa, a Phoenician princess, bore Jupiter three sons, including Minos legendary ancestor of the Minoan civilization. Charlemagne subsequently named the continent which he had conquered Europe, after the young lady. Ganymede was a beautiful Trojan boy, carried off by an eagle to be cupbearer to the gods. The nymph Callisto also conceived one of Jupiter's sons, but Jupiter's enraged wife turned her into a bear. Callisto and her son were placed in the heavens as the constellations Ursa Major and Ursa Minor, the big and little bears. Parts of these constellations are also known as the Big Dipper and the Little Dipper.

Many of the surface features on the Galilean satellites are named after persons or places in world-wide mythology, including those in the myths of Io, Europa, Ganymede and Callisto. Gods of fire and volcanoes from many cultures were also used for Io – this was introduced when scientists realized the extent of the volcanism on Io.

Table 9.5 Properties of the Galilean satellites[a]

Satellite	Distance from Jupiter Center (Jovian radii)	Orbital Period[b] (days)	Mean Radius (km)	Mass (10^{22} kg)	Mean Mass Density (kg m^{-3})
Io	5.90 R_J	1.769	1821.6	8.932	3528
Europa	9.48 R_J	3.551	1560.8	4.80	3014
Ganymede	15.0 R_J	7.155	2631.2	14.82	1942
Callisto	26.3 R_J	16.69	2410.3	10.76	1834

[a] The mean distances from the center of Jupiter are in units of Jupiter's equatorial radius, $R_J = 71\,492$ kilometers. The radii are given in units of kilometers, abbreviated km, the mass is given in kilograms, abbreviated kg, and the mass density is in units of kilograms per cubic meter, denoted kg m^{-3}. By way of comparison, the radius of our Moon is 1738 kilometers and the Moon's mean mass density is 3344 kg m^{-3}. The planet Mercury has a radius of 2440 kilometers and a mean mass density of 5427 kg m^{-3}.
[b] The orbital period of Europa is about twice that of Io, and the orbital period of Ganymede is nearly twice that of Europa.

The Galilean satellites provided the first clear example of objects moving about a center other than the Earth, and for this reason they played an important role in the eventual acceptance of Copernicus' model of the solar system (Section 1.2). They move in nearly circular orbits near Jupiter's equatorial plane with periods of days, moving around the planet so quickly that their positions can be seen to change from hour to hour.

The first quantitative physical studies of these worlds became possible during the 19th century when Pierre Simon Marquis de Laplace (1749–1827) derived the satellite masses from their mutual gravitational perturbations. When large ground-based telescopes were constructed, astronomers could measure the sizes of these moons, and make approximate estimates of their mean mass densities, accurate to roughly ten percent. Precise determinations of these physical parameters became possible as the result of the *Voyager 1* and *2* flybys of Jupiter in 1979 (Table 9.5). The radius of each satellite was measured from the spacecraft images, and the mass of each was derived from its perturbation of the spacecraft trajectories as they passed close by.

The smaller, inner Galilean satellites, Io and Europa, have the higher mass densities, which are comparable to that of the rocks found on Earth. These satellites are about the same size as our Moon, and have about the same mean mass density (Table 9.5). In comparison, the larger, outer satellites Ganymede and Callisto are about the size of Mercury, but much less dense. Their low mean mass densities indicate that they are composed in part of water ice. A mean mass density of 2000 kilograms per cubic meter would, for example, be explained if an object consists of half silicate rock and half water ice, with respective mass densities of 3000 and 1000 kilograms per cubic meter.

The compositions of the Galilean satellites were most likely affected by their relative proximity to Jupiter, in much the same way that the ingredients of the planets are related to their distances from the Sun. In both instances, the relatively small, dense objects are close to the center and larger, less dense objects are found further out. It was probably too warm near the new born Jupiter for water to condense, explaining why Io and Europa are largely composed of rock. Europa is now encased in ice, which may cover an ocean of liquid water, but this blanket is a relatively thin veneer.

In contrast, the relative cold of regions further from Jupiter permitted Ganymede and Callisto to retain their ice and become mixtures of ice and rock. This would explain their low mass densities, as well as the fact that they are more massive than Io and Europa.

The high reflectivities of the Galilean satellites, combined with the very cold temperatures at their remote distances from the Sun, has long suggested that ice might be present on their surfaces. In fact, Europa is so bright, and reflects so much incident sunlight, that it ought to be covered by pure water ice. Spectroscopic observations in the 1970s, using Earth-based telescopes at infrared wavelengths, indeed identified the expected water ice on the surfaces of Europa, Ganymede and Callisto. Although Io has the high reflectivity one might expect from an ice-covered sphere, the infrared observations failed to detect any signs of water ice. Instead, sulfur dioxide frost was identified on Io's surface by its spectroscopic signature. The *Voyager* reconnaissance wasn't very helpful in this regard; although the infrared spectrograph was capable of detecting water ice, it was not directly detected on the Galilean satellites, due to the rough surfaces and strong absorption of infrared light. But the cameras on *Voyager 1* and *2* accomplished much more – they turned the four moons into real places with an astonishing range of geological landforms and surface activity.

When the two *Voyager* spacecraft flew past Jupiter in 1979, they got only a brief look at the Galilean satellites. However, it was time enough for their cameras to discover active volcanoes on Io, smooth ice plains on Europa, grooved terrain on Ganymede, and the crater-pocked surface of Callisto (Fig. 9.9). The incredible complexity and rich diversity of their surfaces, which rival those of the terrestrial planets, are only visible by close-up scrutiny from nearby spacecraft. Ground-based telescopes provide only a blurred view of the tiny, distant moons.

The volcano-ravaged surface of Io is being transformed before our very eyes, as spacecraft catch volcanoes in the act of erupting and watch lava flowing across its surface. Io must be at least partly molten inside to account for this rampant volcanism. The remarkably smooth surface of Europa, which is nearly devoid of large craters, also suggests a warm, active interior for this satellite. It has emitted material onto the surface that has erased crater-forming impacts in the recent past. An ocean of liquid water, or at least slushy ice, apparently exists at shallow depths beneath Europa's frozen surface, lubricating overlying ice rafts and oozing out into cracks in its icy covering.

Fig. 9.9 Giant Red Spot and Galilean satellites This "family portrait" includes the edge of Jupiter with its Great Red Spot and, on the same scale, the planet's four larger moons, known as the Galilean satellites. From top to bottom the moons are Io, Europa, Ganymede and Callisto. The Great Red Spot is larger than two Earth diameters, while Europa is about the size of Earth's Moon. Ganymede is the largest moon in the solar system. (Courtesy of NASA and JPL.)

Between 1997 and 2002, the orbiting *Galileo* spacecraft sharpened our view of the Galilean satellites, providing high-resolution images of their surfaces that are presented in subsequent parts of this section. Both *Galileo* and the *Hubble Space Telescope* also looked above the satellite surfaces, making the first observations of the very tenuous atmospheres that envelop them (Section 3.4). Their wispy atmospheres include hydrogen and oxygen knocked off surface water ice by energetic particles and radiation.

The *Galileo* spacecraft effectively looked beneath the surfaces of the four largest moons as well, inferring their interior structure from their gravitational forces and magnetic fields. It flew close enough to each satellite to measure small changes in the spacecraft's trajectory produced by each moon's gravitational forces, and this provided information about how the satellite's mass is distributed inside. There is less gravitational distortion of the spacecraft's motion if the heavy material is concentrated in a core.

The gravitation measurements indicate that Io, Europa and Ganymede are all denser toward the center than the surface. They all have a metallic heart, and only the outermost, Callisto, remains internally uniform. Additional evidence for a metallic core inside Ganymede was provided by the *Galileo* spacecraft's discovery of its intrinsic magnetic field, which ought to be generated within a massive molten core. The metallic cores of Io, Europa and Ganymede probably originated when the satellites were molten inside, the lighter material rising toward the surface and the heavier material sinking toward the center.

Scientist's have created three-layer models for the interiors of Io, Europa and Ganymede, based on the *Galileo* spacecraft's gravitation and magnetic data and constrained by the satellites' surface properties and overall mass density (Fig. 9.10). They all have a large metallic core, a rocky silicate mantle, and an outer layer of either water ice, for Europa and Ganymede, or rock, for Io. In contrast, Callisto is a relatively uniform mixture of ice and rock.

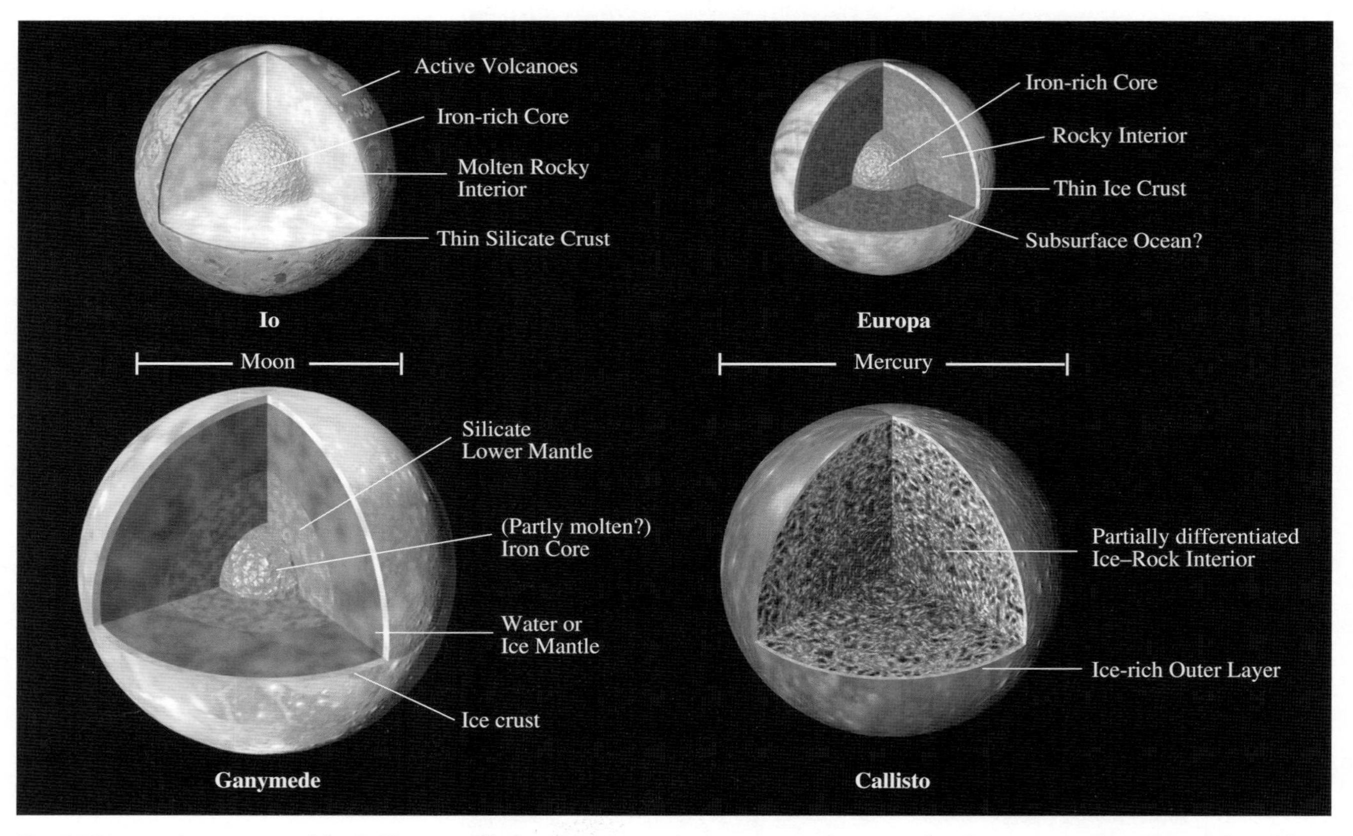

Fig. 9.10 Internal structures of the Galilean satellites The interior characteristics of Jupiter's four larger moons have been inferred from gravitational and magnetic field measurements from the *Galileo* spacecraft. The satellites are shown according to their actual relative size, and compared with those of the Moon and Mercury (*horizontal bars*). With the exception of Callisto, the metal, rock and ice have separated, or differentiated, into distinct layers. Io, Europa and Ganymede all have metallic, or iron and nickel, cores surrounded by rock, or silicate, shells. Io's rocky shell extends to the surface, while outer layers of water, in ice or liquid form, surround Europa and Ganymede. The surface of Io contains numerous active volcanoes, and Europa's thin, frozen crust of water ice probably covers a liquid ocean. Callisto seems to have a relatively uniform internal mixture of ice and rock. (Courtesy of NASA and JPL.)

Fig. 9.11 Volcanic deposits on Io Massive eruptions continually disfigure the surface of Jupiter's volcanically active moon Io. Its entire surface is peppered with volcanoes, some of which are erupting as you read this. A prominent, bright-red ring, 1400 kilometers across, surrounds the volcano Pele, marking the site of sulfur compounds deposited by its volcanic plumes. A dark circular area intersects the upper-right part of the red ring, 400 kilometers in diameter, that surrounds another volcanic center named Pillan Patera. The dark deposits suggest a high silicate content. Deposits of sulfur-dioxide frost appear white and gray in this view, while other sulfurous materials probably cause the yellow and brown shades. Pele is the Hawaiian goddess of the volcano, and Pillan Patera is named for the Araucanian thunder, fire and volcano god. This image was taken on 19 December 1997 from the *Galileo* spacecraft. (Courtesy of NASA and JPL.)

Io: a world turned inside out

The innermost Galilean satellite, Io, has a radius and mass density that are nearly identical to those of our Moon, but contrary to expectation, there are no impact craters on Io. The dramatic landscape is instead richly colored by hot flowing lava and littered with the deposits of volcanic eruptions (Fig. 9.11). The active volcanoes emit a steady flow of lava that fills in and erases impact craters so fast that not a single one is left.

Io's volcanoes are literally turning the satellite inside out. Each volcano can churn out 100 cubic meters of lava every second, fast enough to fill an Olympic-sized swimming pool every minute, and the active volcanoes collectively provide 45 000 tons, or 45 million kilograms, of lava every second. They eject enough material to cover the satellite's surface to a depth of 100 meters in a short span of a million years or less. So all of the material that we now see on Io was probably deposited there less than a million years ago. Evidently Io's mantle and crust have been recycled many times over the span of Io's history.

Sulfur and sulfur dioxide give rise to Io's colorful appearance. Its red and yellow hues are attributed to different forms of sulfur, probably formed at different temperatures. Volcanic plumes of sulfur-dioxide gas fall and freeze onto the surface, forming white deposits that were first detected by ground-based infrared spectroscopy in the 1970s.

Whereas our Moon has been geologically inactive for eons, Io is the most volcanically active body in the solar system (Fig. 9.12). It exhibits gigantic lava-flows, fuming lava lakes, and high-temperature eruptions that make Dante Alighieri's (1265–1321) *Inferno* seem like another day in paradise. Scientists estimate that Io has about 300 active volcanoes, and the intense heat from at least 100 of them has been observed.

The cameras aboard the *Voyager 1* spacecraft discovered nine active volcanoes during its flyby in 1979, and the most active volcanoes, such as Prometheus, Loki and Pele, were observed from the *Galileo* spacecraft two decades later. Prometheus is the "Old Faithful" of Io's many volcanoes, remaining active every time it has been observed. Loki is the most powerful volcano in the solar system, consistently putting out more heat than all of Earth's active volcanoes combined. And Pele, the first volcano to be seen in eruption on Io (Fig. 9.13), has repeated the performance for *Galileo* and the *Hubble Space Telescope*. Like other volcanic centers on Io, these active volcanoes have been named for gods of fire, the Sun, thunder and lightning (Table 9.6).

Fig. 9.12 Volcanoes erupting on Io Several volcanic plumes are shown in this diagram, derived from *Galileo* observations of Io when Jupiter eclipsed the satellite. Bright-blue glows were observed emanating from the volcanic plumes, including the volcano Prometheus whose diffuse glow extended 800 kilometers from the edge of Io. The glow is probably molecular emission from sulfur dioxide, denoted SO_2, known to be abundant in volcanic plumes. The eruptive centers are named after mythological figures such as Amirani, the Georgian god of fire, Culann, the Celtic smith god, Prometheus, the Greek god of fire, and Zamana, the Babylonian Sun, corn and war god. (Courtesy of PIRL, University of Arizona,)

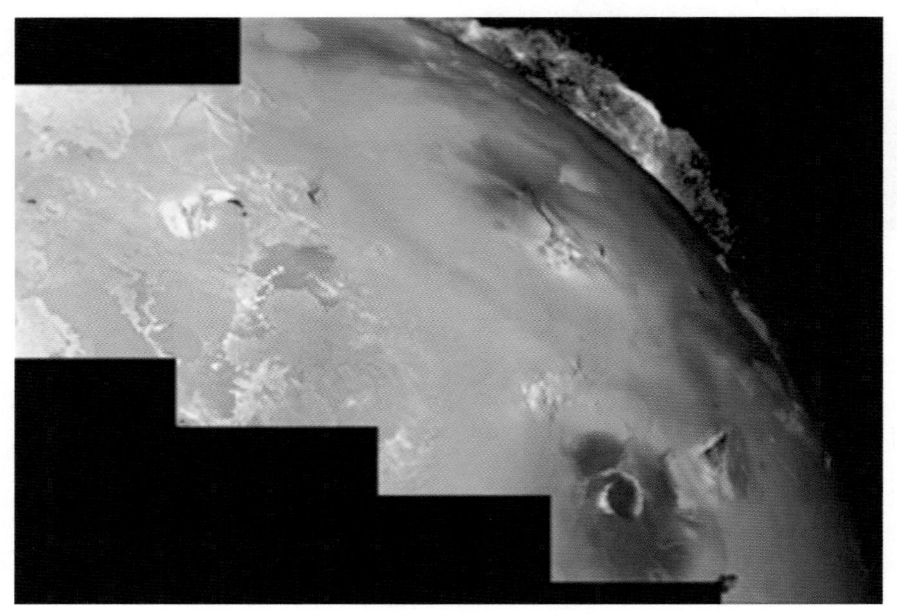

Fig. 9.13 Eruption of Pele During its flyby on 4–5 March 1979, the *Voyager 1* spacecraft captured this image of an active volcano on Jupiter's moon Io. The volcano has been named Pele, after the Hawaiian goddess of the volcano. Its erupting plume is visible at the upper right, rising to a height of about 300 kilometers above the surface in an umbrella-like shape. In this enhanced color image, we see the plume fallout as concentric brown and yellow rings, the largest stretching across 1400 kilometers and covering an area the size of Alaska. Pele remained active for at least two decades, and new deposits from its plumes have been imaged by the *Galileo* spacecraft in the late-1990s (Fig. 9.11). (Courtesy of NASA, JPL, and the U.S. Geological Survey.)

Plumes of volcanic gas erupt from Io's active volcanic vents, rising up to 500 kilometers above the surface. They spread out in graceful, fountain-like trajectories (Fig. 9.14), depositing circular rings of material about a thousand kilometers in diameter. Instruments aboard *Galileo* have practically smelled the hot, sulfurous breath of the eruptions, monitoring the sulfur-dioxide gas as it rises, cools and falls. Diatomic sulfur, S_2, gushing out of the active volcanoes, has also been detected by instruments on the *Hubble Space Telescope*.

Although the kaleidoscopic colors on Io's surface are attributed to sulfur, the bulk of the lava is melted silicate rock. Instruments on the ground and in space have taken the temperature of the searing lava, showing that it sizzles at temperatures of 1700 to 2000 kelvin. That is hotter than any surface temperature of any planetary body in the solar system, even Venus at 735 kelvin. These temperatures rule out substances that melt at lower temperatures, such as liquid sulfur.

So the volcanic vents on Io must be spewing out melted rock, somewhat like terrestrial volcanoes whose lava is rich in iron, magnesium and calcium silicates. But lava this hot has not been common on Earth for more than three billion years, so Io may be giving us an unexpected glimpse into Earth's geological youth.

The ubiquitous high-temperature volcanism on Io has nothing to do with water, a common propulsive agent for some terrestrial volcanoes. There is no evidence that any

Table 9.6 Major volcanic centers on Io

Name	Latitude (degrees)	Longitude (degrees west)	Origin of Name
Amirani	25.9N	114.5	Georgian god of fire
Aten Patera	47.9S	310.1	Egyptian Sun god
Chaac Patera	11.0N	158.0	Mayan thunder and rain god
Culann Patera	19.9S	158.7	Celtic smith god
Kanehekile	18.0S	40.0	Hawiian thunder god
Loki	17.9N	302.6	Norse blacksmith, trickster god
Malik Patera	34.3S	128.5	Babylonian, Caananite Sun god
Marduk	27.1S	207.5	Sumero–Akkadian fire god
Maui	20.0N	122.0	Hawaiian demigod who sought fire
Monan Patera	19.0N	106.0	Brazilian god who destroyed the world with fire and flood
Pele	18.6S	257.8	Hawaiian goddess of the volcano
Pillan Patera	12.0S	244.0	Araucanian thunder, fire and volcano God
Prometheus	1.6S	153.0	Greek fire god
Ra Patera	8.6S	325.3	Egyptian Sun god
Tupan Patera	18.0S	141.0	Thunder god of the Tupi-Giaramo Indians of Brazil
Tvashtar Catena	62.8N	123.0	Indian Sun god and smith who forged the thunderbolt of the thunder god Indra
Surt	45.5N	337.9	Icelandic volcano god
Volund	25.0N	184.2	Germanic supreme smith of the gods
Zamana	18.0N	174.0	Babylonian Sun, corn, and war god

water now exists on Io, and there may never have been any there. Io is close enough to Jupiter that heat received from the young planet during the satellite's formation may have kept water from condensing. And if any water managed to collect in Io, the interior heat would have probably boiled it away long ago.

Io's tides of rock

What is keeping Io hot inside, warming up its interior, melting its rocks, and energizing its volcanoes? The heat released during the moon's formation and subsequent radioactive heating of its interior should have been lost to space long ago, just as our Moon has lost the internal heat of its youth and become an inert ball of rock. Unlike the Earth, whose volcanoes are energized by heat from radioactivity and friction due to diving plates, it is tidal distortions, created by massive Jupiter and its other moons, which sustain Io's molten state.

Just as the gravitational force of the Moon pulls on the Earth's oceans, raising tides of water, the gravitational force of massive Jupiter creates tides in the rocks of Io. Since the pull of gravity is greatest on the closest side to Jupiter, and least on the farthest side, Io's solid rocks are drawn into an elongated shape. But this tidal distortion does not melt the rocks by itself. If Io remained in a circular orbit, one side of the moon would always face Jupiter, its tidal bulges would not change in height, and no heat would be generated.

Shortly before the *Voyager* spacecraft encountered Io in 1979, Stanton Peale (1937–) Patrick Cassen (1940–) and Raymond Reynolds noticed that Io's orbit is slightly out of round, and predicted that the resultant tidal flexing of Io would cause "widespread and recurrent volcanism". The three Galilean satellites Io, Europa and Ganymede resonate with each other in a unique orbital dance, known as the Laplace resonance, in which Io moves four times around Jupiter for each time Europa completes two circuits and Ganymede one. This congruence allows small forces to accumulate into larger ones. The resultant gravitational tug-of-war between Jupiter and the satellites distorts the circular orbits of all three moons into more oblong elliptical ones. The effect is greatest for Io, which revolves nearest to Jupiter, but there is a noticeable consequence for Europa and perhaps even Ganymede.

During each lap around its slightly eccentric orbit, Io moves closer to Jupiter and then further away, wobbling back and forth slightly as seen from Jupiter (Fig. 9.15). The strong gravitational forces of the planet squeeze and stretch Io rhythmically, as the solid body tides rise and fall. Friction caused by this flexing action heats the material in much the same way that a paper clip heats up when rapidly bent back and forth. This tidal heating melts Io's interior rocks and produces volcanoes at its surface.

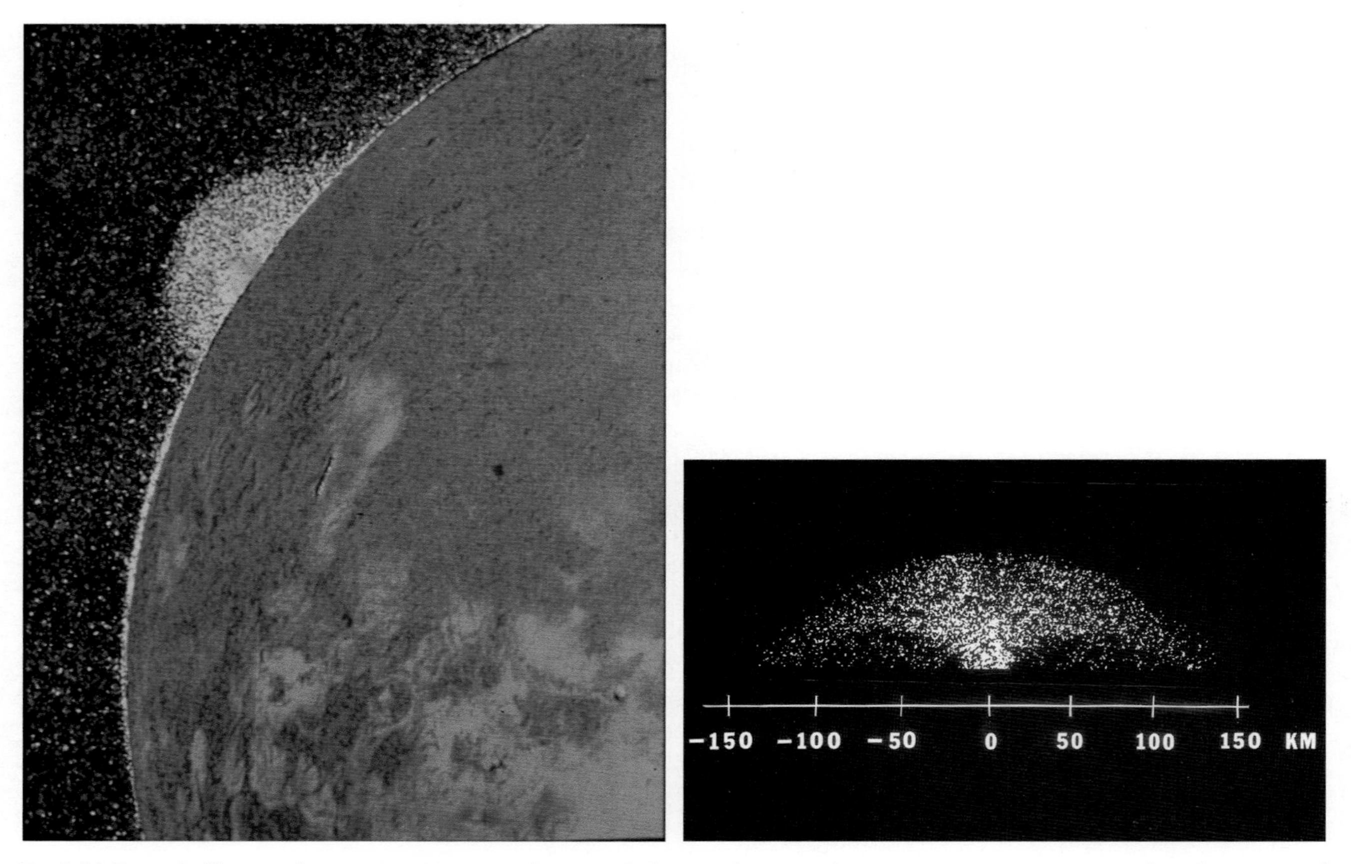

Fig. 9.14 Fountain-like eruptions on Io *Voyager 1* discovered plumes of active volcanoes on Jupiter's moon Io when the spacecraft flew by the planet on 4–5 March 1979 (*left*). A computer simulation of a volcanic eruption on Io shows the fountain-like trajectories of the volcanic plume (*right*). The symmetry and umbrella shape of the volcanic eruption is due to Io's low gravity and the lack of winds or substantial atmosphere on the satellite. Courtesy of NASA and JPL (*left*), and Nicholas M. Schneider, Lunar and Planetary Laboratory, Tucson (*right*).

Magnetic connections with Io

Earth-based observations in the 1970s revealed a vast cloud of sodium atoms that envelops Io, forming an extended atmosphere that is nearly as big as Jupiter (Fig. 9.16). The sodium cloud stretches backward and forward along Io's orbit, until the sodium atoms become ionized and no longer emit the light that makes them visible. The neutral, or un-ionized, sodium atoms have probably been chipped off the surface of Io by the persistent hail of high-energy particles found near the giant planet.

The volcanoes on Io provide the raw material for the satellite's tenuous atmosphere of sulfur dioxide, designated SO_2, that gathers above the erupting vents like localized umbrellas. The volcanic plumes are like fountains, with eruptions that arch gracefully back to Io's surface, and the gas is not propelled with sufficient velocity to escape the satellite's gravitational pull. Nevertheless, atoms of sulfur, S, and oxygen, O, can escape from Io once they are ionized by exposure to radiation from the Sun or from the hail of energetic particles in Io's vicinity. These ions have been detected from the *Voyager* and *Galileo* spacecraft by their ultraviolet glow.

Since charged particles cannot cross magnetic field lines, Jupiter's spinning magnetic field confines and directs the sulfur and oxygen ions into a doughnut-shaped ring known as the plasma torus (Fig. 9.17). As the giant planet rotates, it sweeps its magnetic field past Io, stripping off about a ton, or 1000 kilograms, of sulfur and oxygen ions every second. This material is lost from Io forever, and is continuously replenished by its volcanic activity, albeit indirectly through subsequent ionization.

Once coupled to the Jovian magnetic field, the sulfur and oxygen ions are accelerated to high velocity. Carried by the field, which is anchored inside Jupiter, the ions revolve around the planet once every 9.9249 hours, while Io orbits Jupiter in a leisurely period of 42.46 hours. So, the ions are always catching up with the satellite, and some of them slam into its surface, dislodging and energizing material and lifting it into the thin atmosphere.

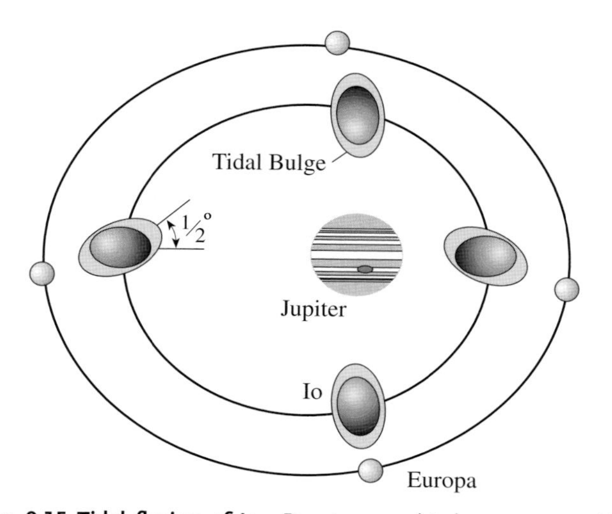

Fig. 9.15 Tidal flexing of Io Due to an orbital resonance with nearby Europa, Jupiter's satellite Io has a non-circular orbit. The forced eccentricity makes Io travel at different velocities along its orbit and the side facing Jupiter nods back and forth slightly, as seen from the planet. Although only half a degree in extent, this movement causes varying tidal forces inside the satellite, flexing it in and out like squeezing an exercise ball with your hand. This, in turn, generates internal friction and heat, leading to the active volcanoes seen on Io with instruments aboard the *Voyager 1* and *2* and *Galileo* spacecraft. In this drawing, Io's size and the eccentricity of its orbit are exaggerated when compared with Jupiter.

As Jupiter's magnetic field sweeps past Io, it generates an enormous electrical potential of 400 000 volts, allowing a powerful electric current of 5 million amperes to flow from Io to the poles of Jupiter and back again. The electrons move along Jupiter's magnetic field lines, within a magnetic flux tube that attaches the moon to its planet like a giant electromagnetic umbilical cord. Instruments aboard the *Galileo* spacecraft have detected beams of electrons flowing along the flux tube. They generate an awesome natural power of 2.5 trillion (2.5×10^{12}) watts, vastly exceeding that of any terrestrial energy-generation plant.

When the electrons in this huge electrical circuit collide with the atoms in Io's tenuous atmosphere, they generate a dazzling light show of red, green and blue emissions. And when the electrons are directed into the atmosphere of Jupiter, at the opposite end of the circuit, they trigger its bright auroral emissions (Section 3.7), marking the glowing foot of the flux tube. Currents in this cosmic power station also generate powerful bursts of radio noise, noticed since 1964, which are strongly controlled by Io's orbital position.

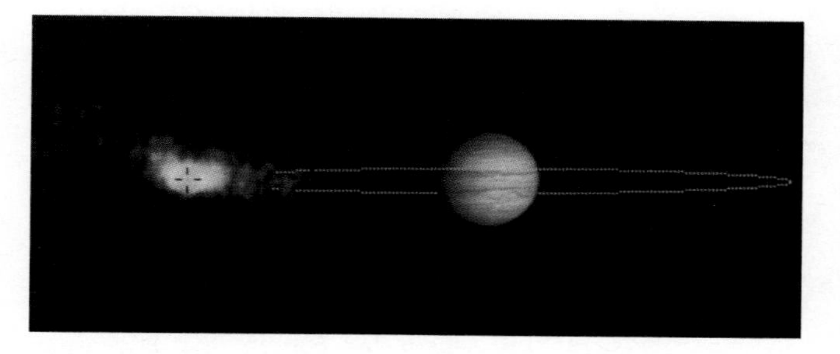

Fig. 9.16 The sodium cloud Io's neutral sodium cloud (*left*) as seen from the Earth, shown together with Jupiter (*center*) and a schematic drawing of Io's orbit to scale. Excited sodium atoms that come from Io (*location denoted by a cross*) emit radiation at spectral lines that can be detected from ground-based optical telescopes. These observations of the sodium cloud were made with the 0.61-meter (24-inch) telescope at Table Mountain Observatory in California; the image of Jupiter was taken at a different time and place. (Courtesy of Bruce Goldberg and Glenn Garneau, JPL.)

Europa's bright, smooth icy complexion and young face

The smallest, and yet brightest, of the Galilean satellites, Europa, has a density comparable to that of rock, but its surface is as bright and white as ice. In fact, it is water ice! With surface temperatures of 110 kelvin or less, the water ice on Europa is frozen as hard and solid as granite on Earth.

Sunlight and charged particles cause some of the water ice to vaporize, and reactions caused by ultraviolet sunlight split the molecules of water vapor into hydrogen and oxygen atoms. The hydrogen escapes into space, leaving behind a very tenuous atmosphere of oxygen (Section 3.4).

Europa's surface is nearly devoid of impact craters, and there are no mountains or valleys on its bright smooth surface. No features extend as high as 100 meters, making Europa the smoothest body in the solar system.

The paucity of cratered impact scars indicates that Europa has a comparatively young surface, showing few signs of age. Since its surface has been accumulating impact craters as time goes on, Europa must have been resurfaced in recent times, geologically speaking, probably in the past few hundred million years. Whatever is keeping Europa smooth is doing it from beneath the frozen crust, as eruptions of liquid water to the surface or flows of soft water ice. But it is unknown if the eruptions or flows are still occurring.

Long cracks, ice rafts and dark places on Europa

A veined, spidery network of long dark streaks marks Europa's young face, suggesting great inner turmoil. The

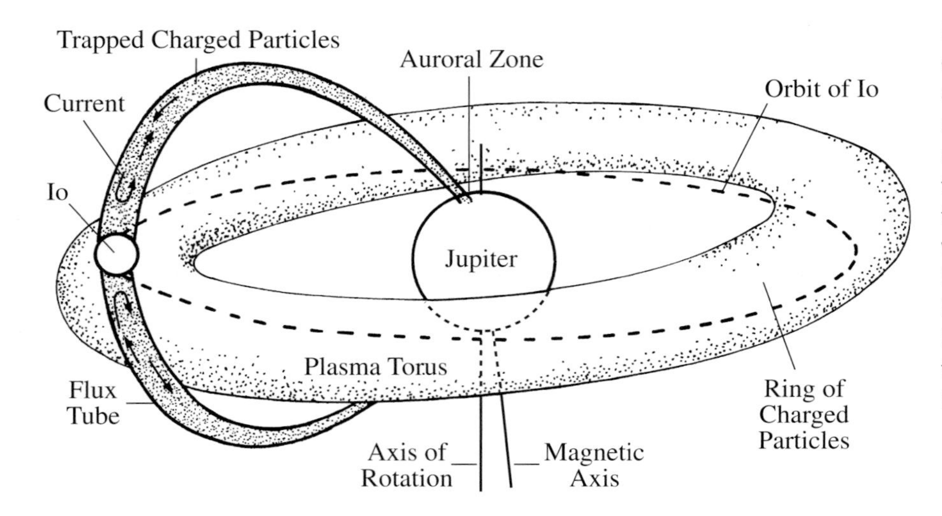

Fig. 9.17 Flux tube and plasma torus An electric current of 5 million amperes flows along Io's flux tube. It connects Io to the upper atmosphere of Jupiter, like a giant umbilical cord. The plasma torus is centered near Io's orbit, and it is about as thick as Jupiter is wide. The torus is filled with energetic sulfur and oxygen ions that have a temperature of about 100 thousand kelvin. Because the planet's rotational axis is tilted with respect to the magnetic axis, the orbit of the satellite Io (*dashed line*) is inclined to the plasma torus.

fine lines run for thousands of kilometers, intersecting in spider-web patterns (Fig. 9.18). They give Europa a broken appearance that resembles a cracked mirror or an automobile window that has been shattered in some colossal accident. The dark lines are most likely deep fractures formed when that part of the ice cracked open, separated, and filled with darker, warm material seeping and oozing up from below. Dirty liquid water or warm dark ice has apparently welled up and frozen in the long cracks, producing the lacework of dark streaks.

As two adjacent pieces of ice pull apart slightly, warm soft ice might push up and freeze to form long ridges that parallel the cracks (Fig. 9.19). Other ridges may have originated when the sides were pushed together, closing the crack and crumpling its edges to form a ridge.

The surface of Europa is fragmented everywhere, as if pieces of ice have broken apart, drifted away and then

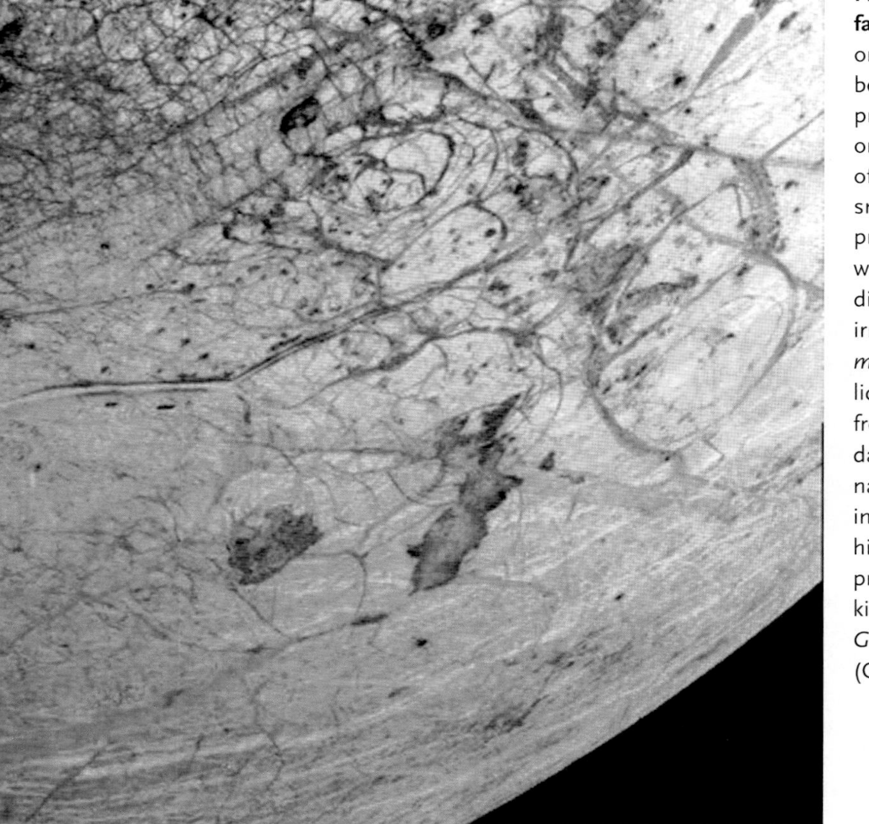

Fig. 9.18 Europa's frozen, disrupted surface Old impact craters are not visible on Jupiter's moon Europa. They must have been erased, perhaps by fresh water ice produced along cracks in the thin crust or by cold glacier-like flows. The number of impact craters found on the bright, smooth surface indicates an age of approximately 100 million years. The thin, water-ice crust has undergone extensive disruption from below (*upper left*). Two irregular, chaotic dark features (*just below middle*) were most likely formed when liquid water or warm ice welled up from underneath Europa's icy shell. These dark spots, technically called macula, are named Thera and Thrace after two places in Greece at which Cadmus stopped in his search for Europa, the Phoenician princess. This image, approximately 675 kilometers across, was taken from the *Galileo* spacecraft on 20 February 1997. (Courtesy of NASA and JPL.)

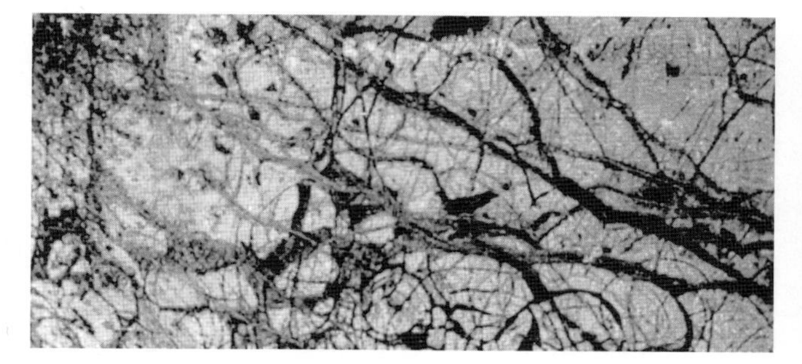

Fig. 9.19 Broken ice on Europa Dark, linear crack-like features extend for thousands of kilometers across Jupiter's moon Europa. They are believed to have formed when the satellite's thin, icy crust fractured, separated and was filled by a dirty slush from a possible ocean below. The long cracks were most likely caused by tides raised on Europa by the gravitational pull of Jupiter. This image, about 770 kilometers wide, was taken from the *Galileo* spacecraft on 27 June 1996. (Courtesy of NASA and JPL.)

frozen again in slightly different places (Fig. 9.19). Large blocks of ice have floated like rafts across the moon's surface, shifting away from one another like moving pieces of a jigsaw puzzle. Some of them are tilted; others rotated out of place, like plastic toys bobbing in a bathtub. This shows that the ice-rich crust has been or still is lubricated from below by either slushy ice or liquid water.

The size and geometry of the ice floes on Europa suggest that internal heat has melted the ice just a few kilometers below the surface. The warmth and currents have broken the thin crustal ice into pieces that slide over the underlying watery slush. They resemble disrupted pack ice seen on Earth's polar seas during springtime thaws. But the thaw on Europa is coming from heat below, not from sunlight above.

Explosive ice-spewing volcanoes and geysers may erupt from the buried seas, reshaping the chaotic surface of the frozen moon and leaving dark scars behind. Extended dark regions may, for example, have formed when the underground ocean melted through Europa's icy shell, exposing darker material underneath, or when upwelling blobs of dark, warm ice broke through the colder near-surface ice (Fig. 9.20).

Europa's underground sea of melted ice

Tidal distortions could explain how water ice has melted in the frigid environment surrounding Europa. The satellite has a slightly eccentric orbit due to gravitational interactions with Io and Ganymede, which revolve closer and further away, respectively, from Jupiter than Europa. Over the course of one trip around Europa's elongated path,

Fig. 9.20 Europa under stress When viewed at high resolution, many sets of parallel and crosscutting ridges and fractures are detected on Jupiter's moon Europa. These features are the frozen remnants of surface tension and compression, probably produced by heating and upwelling from below. The icy crust has also been broken into plates or "rafts", ranging up to 13 kilometers across, which have separated and moved into new positions, somewhat like pack ice in the Earth's polar seas. Soft ice or liquid water below the surface most likely lubricated the moving ice rafts on Europa at the time of disruption. This image, approximately 42 kilometers across, was taken from the *Galileo* spacecraft on 20 February 1999. (Courtesy of NASA and JPL.)

Jupiter's strong gravity stretches and compresses the satellite, in a process called tidal flexing. Frictional heat associated with similar tidal flexing melted the rocks inside Io, and it operates on Europa as well – to a smaller extent since Europa is further from Jupiter. But the warmth generated by tidal heating may have been or may still be enough to soften or liquefy some portion of Europa's internal ice, perhaps sustaining a subsurface ocean of liquid water.

The tidal flexing that warms Europa's interior may also crack the blanket of ice that traps the hypothetical liquid water below. The varying distance of Europa from Jupiter causes the tides in the underground sea to rise and fall as much as 30 meters. The pressure of this continual, rhythmic in-and-out motion probably cracks the brittle crust apart.

Magnetic measurements from the *Galileo* spacecraft provide more evidence for an otherworldly ocean inside Europa. The satellite's magnetism changes direction as Jupiter's magnetic field sweeps by in different orientations to the satellite, owing to the tilt between the planet's rotation axis and magnetic axis. This means that the magnetic field at Europa is not generated in a core, but is instead induced by the passage of Jupiter's field in

Focus 9.3 Life in Europa's ice-covered ocean

The possibility of liquid water just below Europa's surface has led to speculation that life could have gained a foothold there. Tidal flexing might make the internal seas warm enough, and they would be wet enough; there might even be organic molecules down there. A global sea of liquid water could seethe with alien microbes hidden beneath Europa's gleaming ice-covered surface. After all, we know that the heat, minerals and chemical energy of underwater volcanoes on Earth's sea floors sustain life in the dark without sunlight. Tubeworms one foot (0.3 meters) long and giant white clams thrive near the hot underwater vents on Earth, and micro-organisms even live inside them.

But this does not mean that there is life inside Europa, and there is no direct evidence for it. We will not know until robotic spacecraft go there to see if complex organic molecules can be found on the ice. Future spacecraft might even melt a path through an unknown thickness of ice to the possible ocean below, investigating if alien life is swimming through its dark seas.

an electrically conducting liquid, such as salt water, beneath the ice. Although this evidence for a subsurface liquid ocean is indirect, it is the only indication that buried water is there now, rather than in the geological past.

So, it is highly likely that Europa had liquid water near its surface at one time, and it might still be there. Gravitation data tells us that the water moved to the top long ago, within an outer layer about 100 kilometers thick, and the cracked surface, floating icebergs, and changing magnetic field provide strong circumstantial evidence for internal seas just below the crust of ice. If the liquid water is still there, we can stretch our imagination and speculate that life might reside within its lightless depths (Focus 9.3). Still, we cannot unequivocally prove that Europa has a subsurface ocean of liquid water, and most of its water could be frozen solid. This brings us to Ganymede, which is also covered by a cracked, mobile crust of ice.

Cratered, wrinkled Ganymede

Ganymede, the largest moon in the solar system, has a radius that exceeds that of the planet Mercury, but the satellite's density is so low that it must contain substantial quantities of liquid water or water ice. Its icy surface has experienced a violent history involving crustal fractures, mountain building and volcanoes of ice.

Bright regions on Ganymede's surface contain sets of parallel ridges and valleys, termed grooved terrain, which look like the swath of a giant's rake (Fig. 9.21). The grooved terrain was most likely formed when the moon's water-ice crust expanded and stretched, cracking and rifting open as it was pulled apart. The crustal expansion might have happened when the satellite's rocks melted and moved into its interior while its water migrated to the top where it froze.

Sets of intersecting mountain ridges overlap and twist into each other. Some of the ridges cut across craters, while craters appear on other ridges (Fig. 9.22). Ganymede evidently experienced several epochs of mountain building. These crustal deformations may have continued for a billion years.

Water-ice volcanism played a role in creating the bright terrain on Ganymede. Prominent depressions were apparently flooded with liquid water or icy slush, and then froze into bright smooth bands that now cover much of the moon (Fig. 9.23). Craters found in these areas indicate that this also happened early in the satellite's history, at least a billion years ago.

Darker regions on Ganymede are older and more heavily cratered. Some of these large polygonal blocks rise about a kilometer above the bright, grooved terrain, and look as if they have moved sideways for tens of kilometers along the moon's surface (Fig. 9.23).

Ganymede, a moon with its own magnetic field

One of the major surprises of the *Galileo* mission was the discovery that Ganymede has its own intrinsic magnetic field. The moon is generating a magnetic dipole similar to those of most planets, and roughly one-thousandth of the strength of Earth's. No other satellite now has such a magnetic field, but our Moon might have had one in the distant past.

Currents stirred inside Ganymede's large, dense molten iron core may have produced its self-generated magnetic field, but a molten core would cool down and the inner flows might last for only a million years or so. Scientists have therefore speculated that the satellite's orbit has shifted over time, and that strong tidal flexing once heated its interior more than it does now. If the moon moved in a closer or more eccentric orbit many millions of years ago, Jupiter's gravity would have squeezed it in and out by greater amounts, heating it up inside and generating a strong magnetic field that lingers today. The wrinkled, grooved terrain on Ganymede's icy surface might record this earlier period of intense heating. The hot core might

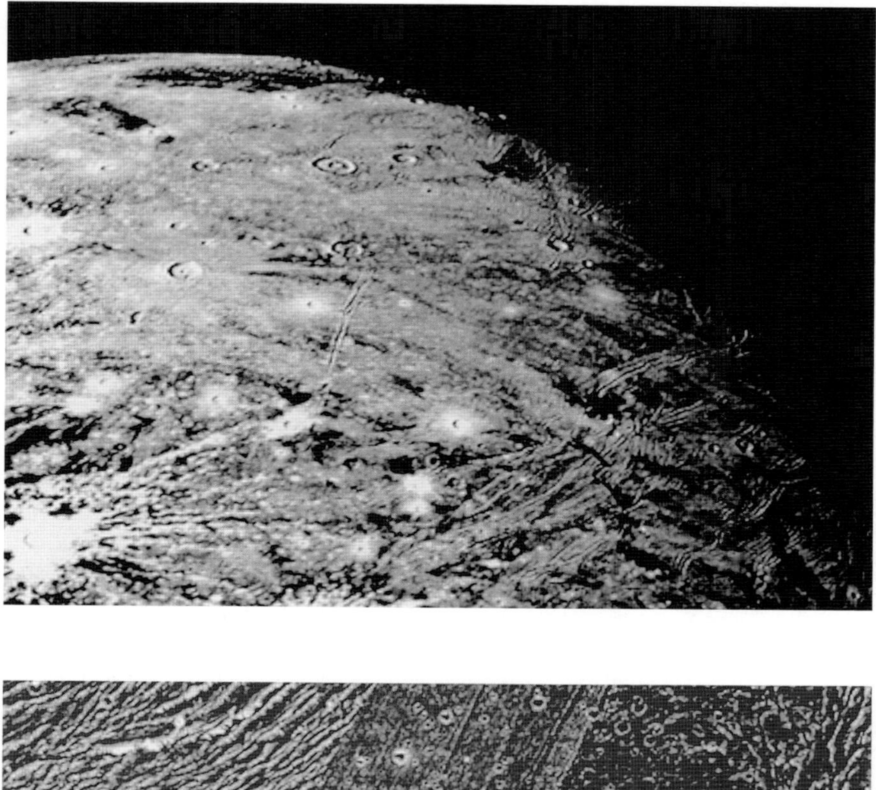

Fig. 9.21 Ganymede The surface of Ganymede, Jupiter's largest satellite, includes impact craters that suggest an age of a few billion years. The bright rays that surround many craters (*lower left*) probably consist of icy material thrown out by the impacts. Sinuous ridges and grooves traverse the surface (*lower right*) most likely caused by deformation of the thick ice crust from below. The *Voyager 1* spacecraft took this image, about one thousand kilometers wide, on 5 March 1979. (Courtesy of NASA and JPL.)

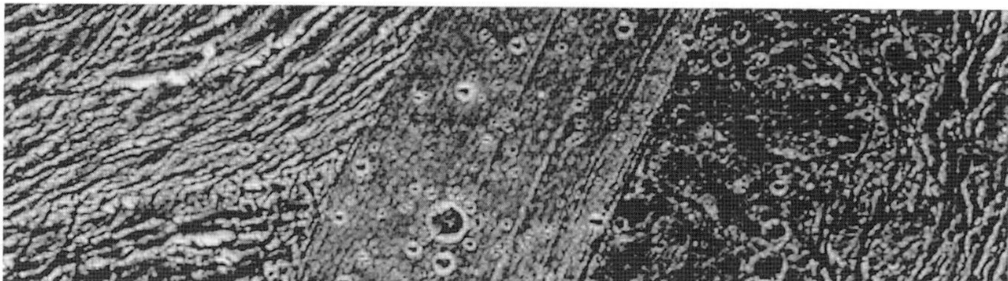

Fig. 9.22 Ganymede close up The bright icy crust on Jupiter's moon Ganymede contains both young and old terrain with bright grooves, caused by internal stress, and craters due to external impact. The youngest terrain (*center*) is finely striated and relatively lightly cratered. The oldest terrain (*right*) is rolling and relatively heavily cratered. The highly deformed grooved terrain (*left*) is of intermediate age. This image, approximately 89 kilometers across, was taken from the *Galileo* spacecraft on 20 May 2000. (Courtesy of NASA and JPL.)

still be cooling off, with internal currents that generate the magnetism seen today.

Since Ganymede's intrinsic magnetic field is nestled within Jupiter's stronger and more extensive one, a changing magnetism is also induced within Ganymede when the giant planet's magnetic field sweeps past the moon. Although this induction magnetic signature is much weaker than the satellite's intrinsic dipolar field, it has the important implication that a thick layer of salty water lies somewhere deep beneath the satellite's frozen crust. Electric currents coursing through Ganymede's internal shell of salt water can also contribute to its intrinsic magnetic field.

Callisto, an ancient, battered world

Remotest of the Galilean moons, Callisto has had a much more sedate and peaceful history than the other large satellites of Jupiter, with little sign of internal activity. It is a primitive world whose surface of ice and rock is the most heavily cratered in the solar system (Fig. 9.24). Unlike nearby Ganymede, the moon Callisto has no grooved terrain or lanes of bright material, and it exhibits no signs of icy volcanism. So, Callisto is a long-dead world unaltered by resurfacing since it formed and ancient impacts molded its face, a fossil remnant of the origin of planets and their moons.

Yet, when seen close up by the *Galileo* spacecraft there are indications of subdued, youthful activity on Callisto's surface. It is blanketed nearly everywhere by fine, mobile dark material, interrupted only where bright crater rims poke up through it (Fig. 9.25). Small impact craters are mostly absent, and those that are found sometimes appear worn down and eroded. Thus, the smaller craters seem to have been filled in and degraded over time, perhaps by the dark blanket of debris that might have been thrown out by the larger impacts. Ice-flows may have alternatively

Fig. 9.23 Varied terrain on Ganymede The surface of Jupiter's largest moon, Ganymede, contains old, dark polygonal blocks frozen within its icy surface. They resemble brown, frozen-over continents floating on a background of translucent ice. The ancient blocks have apparently separated like the moving pieces of a huge mosaic or giant jigsaw puzzle, perhaps because the satellite's crust has expanded. The brilliant, relatively young white material that surrounds some craters is probably fresh, clean water ice that splashed out from inside the satellite. The rays extending from the bright crater in the northern (*top*) part of this picture are up to 500 kilometers long. This image was acquired from the *Voyager 1* spacecraft on 5 March 1979. (Courtesy of NASA and JPL.)

deformed and leveled many craters, because ice, which is rigid to sharp impact, can flow gradually over long periods of time, as glaciers do on Earth. The lack of small craters on Callisto might also be explained if the ancient population of impacting objects near the remote satellite had relatively few small objects when compared to the population near the Moon and Mercury.

Perhaps because Callisto is further from Jupiter than the other Galilean moons, its ingredients are somewhat separated but still largely mixed, like a half-baked potato that is hard on the inside and soft on the outside. Unlike the other three Galilean satellites, Callisto has a homogeneous interior, without a dense metallic core, and it has no magnetic field of its own. But Callisto does have a crust of ice that may cover a subsurface ocean of liquid water.

Like Europa and Ganymede, the battered Callisto has a variable magnetic field, apparently generated by electrical currents as Jupiter's powerful field sweeps by. A shell of liquid water can explain the internal conductivity if it has the salinity of terrestrial seawater, but it would have to be deep enough inside the moon that the water could not rise to the surface, keeping it unaltered.

Since it does not participate in the orbital push and pull of Io, Europa and Ganymede, tidal flexing by Jupiter has not kneaded or heated Callisto inside. So Callisto's internal ocean can only be heated by radioactive elements. The lack of tidal flexing may also help explain the unwrinkled nature of Callisto's pockmarked face.

9.5 Jupiter's mere wisp of a ring

The rings of Saturn were discovered in the 17th century (Section 1.5). About three centuries later, in 1977, several faint and unsuspected narrow rings were discovered about the planet Uranus (Section 11.4). Jupiter was next to join the group of ringed planets, but this time the discovery was not a complete surprise. In 1974, the *Pioneer 11* spacecraft had encountered an unexpected reduction in the amount of high-energy charged particles when the spacecraft passed near Jupiter. Not much was made of this anomaly, although some scientists thought that the falloff could be due to a previously unknown satellite or ring that blocked the energetic particles.

Finally, in 1979 – after much debate about the likelihood of finding a ring – a search was carried out with a camera on *Voyager 1*, and a narrow faint belt of material was found encircling the planet in its equatorial plane at a distance close to where the energetic particles had disappeared. The ring was not previously observed from Earth because it was too faint and close to the bright planet. Since its discovery, Jupiter's main ring has been detected by Earth-based telescopes sensing infrared radiation, and more fully explored by the inquisitive eyes of the *Galileo* spacecraft (Fig. 9.26).

When the retreating *Voyager 1* camera looked back at the shadowed side of Jupiter, the main ring became brighter. The Sun was backlighting the tiny particles that make up the main ring, making it shine brightly. This behavior is typical of very small particles that scatter light in the forward direction, like tiny salt grains on the windshield of an automobile, the smoky haze in some movie theaters, or the condensation trails of airplanes.

The size of the ring particles can be inferred from the way they scatter light, and the conclusion is that they are a few millionths of a meter across or about the same size as flour dust or grains of pollen. The particles that make up cigarette smoke and the hazes in the Earth's atmosphere have similar sizes. Numerous, larger ring particles would

Fig. 9.24 Callisto Jupiter's outermost large moon Callisto exhibits more craters and older terrain than seen on any of the Galilean satellites. It is a battered world, pock-marked with impact craters dating back to the final stages of planetary formation over 4 billion years ago. Because Callisto's icy surface is as rigid as steel, it retains the scars of an ancient bombardment similar to the one that created the heavily cratered terrain on the Moon and Mercury. The bright regions probably contain fresh crustal ice thrown out from relatively young impact craters, and splashed upon the older, dirtier surface ice. This image was acquired in May 2001 from the *Galileo* spacecraft. (Courtesy of NASA, JPL, and DLR – the German Aerospace Center.)

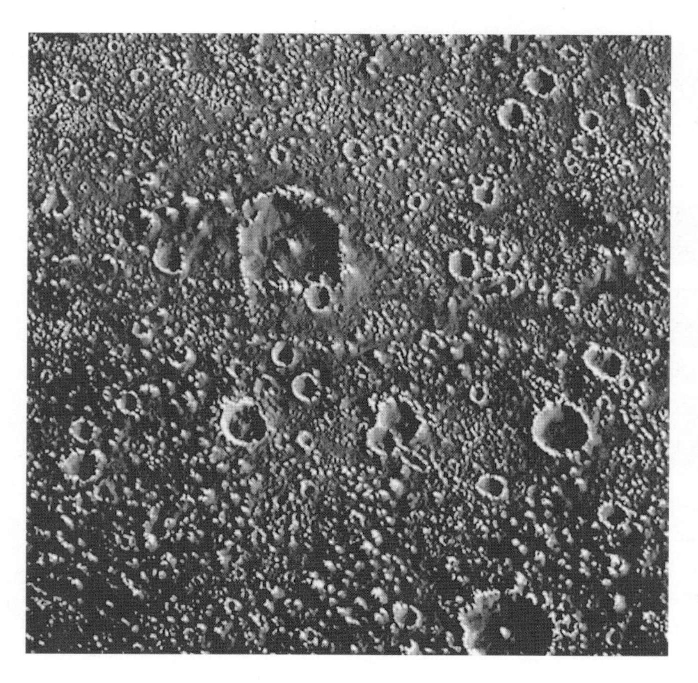

Fig. 9.25 Dark material and few small craters on Callisto A dark, mobile blanket of fine material covers Callisto's surface, sometimes collecting within crater walls. While Jupiter's moon Callisto is saturated with large impact craters, it has fewer very small craters when compared with the Moon and Mercury. One explanation is that the smaller craters have been filled by dark material that has moved down surface slopes. An alternative explanation for the paucity of little craters on Callisto is that there were fewer small impacting objects in its vicinity when compared with the amount within the inner solar system. This image, about 74 kilometers across, was taken from the *Galileo* spacecraft on 17 September 1997. (Courtesy of NASA and JPL.)

have reflected sunlight, making the rings appear brightest when approaching them, which did not happen.

Voyager 1 and *2* viewed a three-ring system around Jupiter, consisting of a flattened main ring, an inner, extended cloud-like ring, called the halo, and a third outer ring, known as the gossamer ring because of its transparency. Observations from the *Galileo* spacecraft in 1996 and 1997 showed that the gossamer ring is, in fact, two rings, one embedded in the other.

The most plentiful particles of all four parts of Jupiter's insubstantial ring system are the size of dust. It is practically made of nothing at all, no dustier than a typical living room. And the individual dust particles only reside temporarily in Jupiter's rings, just as the dust in the air of your room settles onto the room's furniture and bookshelves.

The dust particles can last no more than a few thousand years before being tossed out of the ring plane or spiraling down into Jupiter's upper atmosphere. Given this short lifetime, the fine particles must be continuously replenished if the Jovian rings are permanent features.

They are replenished by dust that is blasted off Jupiter's four small, innermost moons by interplanetary meteoroids, the fragments of comets and asteroids. The meteoroids are drawn in by Jupiter's very strong gravity, which also greatly increases their speed. When the high-velocity cosmic debris slams into one of the small inner moons, it creates a dust cloud, resembling the puff of chalk dust that arises when an eraser is banged against a chalkboard. The dust is thrown off at such high velocity that it escapes the moon's relatively small gravitational pull, orbiting Jupiter and contributing to one of its rings.

The outer edge of the main ring lies just inside the orbit of the tiny moon Adrastea, just 15 kilometers in size and too small to be seen from Earth. It was discovered by the *Voyager* spacecraft, as was another tiny moon, named Metis, which is embedded near the bright mid-point of the main ring. The dust generated by meteoritic impact on Adrastea and Metis can easily escape the small gravity of these moons, accounting for the dense accumulation of particles in the main ring. Some of the microscopic particles are small enough that, if they are slightly electrically charged, electromagnetic forces can overpower the effects of Jupiter's gravity, pumping them into the inner halo that is seen above and below the main ring (Fig. 9.26).

The two much fainter gossamer rings, which are more distant from Jupiter than the main ring, lie just inside the orbits of the small moons Amalthea and Thebe. Detailed observations from the *Galileo* spacecraft indicate that dust

Fig. 9.26 Jupiter's main ring and halo Jupiter's bright, flat main ring (*bottom*) is a thin strand of material encircling the planet with an outer radius of 128.94 thousand kilometers, or about 1.8 Jovian radii, located very close to the orbit of the giant planet's small moon Adrastea, at 128.98 thousand kilometers. The brightness of the main ring drops markedly very near the orbit of another moon, Metis. A faint mist of particles, known as the ring halo, surrounds the main ring and lies above and below it (*top*). The vertically extended halo is unusual for planetary rings, which are normally flattened into a thin plane by gravity and motion. The halo probably results from the "levitation" of small charged particles that are pushed out of the main-ring plane by electromagnetic forces. These images were obtained from the *Galileo* spacecraft on 9 November 1996 when it was in Jupiter's shadow, looking back toward the Sun. The rings of Jupiter proved to be unexpectedly bright when seen with the Sun behind them, just as motes of dust or cigarette smoke brighten when they float in front of a light. A third gossamer ring, which consists of two components, is not shown here; it lies beyond the main ring, at greater distances from Jupiter. (Courtesy of NASA and JPL.)

particles knocked off Amalthea and Thebe feed the two gossamer rings, with a thickness that corresponds to each satellite's elevation above the planet's equatorial plane. Amalthea was discovered using a powerful terrestrial telescope long ago, in 1892 by the sharp-eyed astronomer Edward Emerson Barnard (1857–1923), while Thebe was found with cameras aboard *Voyager 1* and *2*. Because these two satellites orbit in planes tilted slightly from that of Jupiter's equator, the gossamer rings are not precisely flat, but have a vertical extent that matches the satellite orbital inclinations.

This ends our space-age visit to Jupiter. Following the *Voyager* spacecraft, we now step out to Saturn, lord of the rings – and beyond.

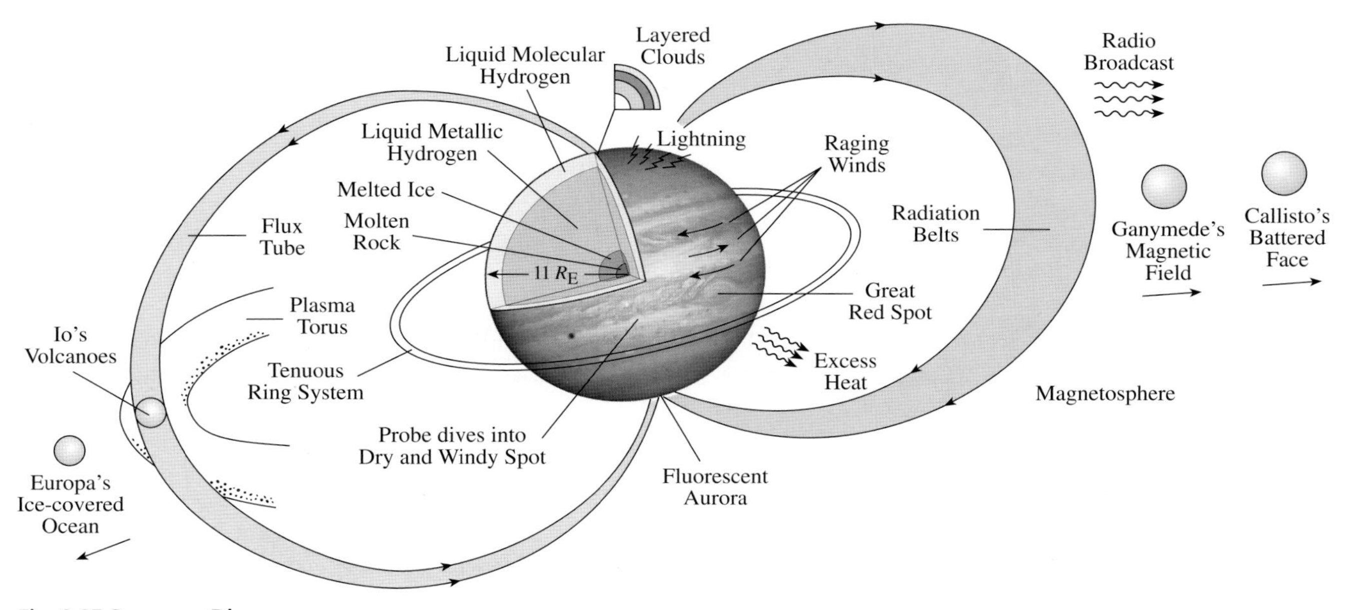

Fig. 9.27 Summary Diagram

10 Saturn: lord of the rings

- Saturn has the lowest mass density of any planet in the solar system, low enough for the planet to float on water, and this means that Saturn is primarily composed of the lightest element hydrogen.
- The lightweight material inside Saturn has been flung out by the planet's rotation, creating the most pronounced equatorial bulge of any planet.
- Saturn is just a great big liquid drop, covered by a thin atmosphere of gas, slightly smaller than Jupiter, and less than a third the mass of Jupiter.
- Liquid hydrogen is compressed inside Saturn's depths to form an electrically conducting, liquefied metal.
- There is no solid surface anywhere inside Saturn, though it might have a core, composed of melted ice and molten rock, that is about ten times as massive as the Earth.
- Saturn radiates almost twice as much energy as it receives from the Sun, and most of the planet's excess heat is generated by helium raining down into its inner metallic hydrogen shell.
- Saturn's rings are completely detached from the planet and separated from each other.
- The rings of Saturn are not solid, but instead are composed of innumerable small particles of water ice.
- The particles in Saturn's main A, B and C rings are as big as hailstones, snowballs and even icebergs; there are more smaller particles than bigger ones, but the big ones supply most of the ring mass.
- The total mass of the rings of Saturn is comparable to that of its satellite Mimas.
- Saturn has a retinue of diffuse, tenuous, and nearly transparent rings, designated the D, E, F and G rings, that are most likely composed of microscopic water-ice crystals, smaller than snowflakes and about the size of the dust particles in your room.
- Saturn's narrow F ring is kept together by two small moons that skirt its edges and confine and shepherd the ring particles between them.

- The icy material in the prominent rings of Saturn has been marshaled into thousands of individual ringlets, resembling ripples on a pond, but with circular, oval and even spiral shapes.
- Gravitational interaction with nearby, external satellites can sculpt the ring material into numerous ringlets and produce waves in it; the small moons can also sharpen ring edges and clear gaps in the rings.
- The gaps within Saturn's rings are not completely empty; one of them, the Cassini Division, contains about 100 ringlets.
- Enigmatic dark spokes stretch radially across the rings, moving at constant speed and keeping their shape in apparent violation of the laws of gravitation.
- Saturn's dark ring spokes consist of microscopic dust-sized particles that may become electrically charged and levitated above the larger ring particles. They might then be swept around Saturn by its rotating magnetic fields.
- Planetary rings lie closer to a planet than its large satellites, within the Roche limit where the planet's tidal forces will rip a large satellite to pieces and prevent small bodies from coalescing to form a larger moon.
- Saturn's rings are thought to be less than 100 million years old, or less than three percent of Saturn's age, so they cannot be left over from the formation of the planet.
- The rings of Saturn could have formed when a moon was pulled toward the planet by tidal forces and eventually ripped apart.
- Small moons embedded in the rings of Saturn might sustain them.
- Saturn's largest moon, Titan, is a planet-sized world with a substantial atmosphere whose surface pressure is about 1.5 times the air pressure at sea level on Earth.
- Nitrogen molecules are the main constituents of Titan's atmosphere, as they are in the Earth's air.
- Clouds of methane and raining ethane might exist beneath the high-altitude smog on Titan.
- Shallow seas of methane and ethane could lap the shores of Titan; they can exist as liquids at the cold surface temperature of only 94 kelvin.
- A veil of orange smog hides the surface of Titan from view at visible wavelengths, but radar signals and infrared images have revealed some surface detail.
- The *Cassini* spacecraft is expected to arrive at Saturn in July 2004. Its *Huygens Probe* will be parachuted into Titan's murky atmosphere, measuring the clouds, gases, smog particles and winds around the probe and taking pictures of the surface. The main spacecraft will continue to orbit Saturn for the next four years, scrutinizing the planet's atmosphere, magnetic environment, rings, Titan and icy satellites.
- Saturn has six mid-sized icy moons that retain impact craters dating back to the early history of the satellites; some of them exhibit signs of internal activity and ice volcanism. Impacting objects almost broke the moons Mimas and Tethys apart. Enceladus has a bright icy coating with volcanoes that belch ice and water from its warm interior. Two-faced Iapetus is bright and icy on one side and dark and dusty on the other.
- A number of unique small, irregularly shaped moons revolve around Saturn with remarkable orbits. The co-orbital moons have almost identical orbits, the Lagrangian moons share their orbit with a larger satellite, and the shepherd moons confine the edges of rings.

10.1 Fundamentals

Majestic Saturn, the sixth planet from the Sun, was the most distant world known to the ancients, and it moved least rapidly around the zodiac. The Greeks identified the planet with Kronus, the father of Zeus, while the Romans named the planet Saturn after their god of sowing. Both the Greeks and the Romans associated Saturn with the ancient god of time, who later became Father Time.

You can see Saturn's oblong, golden disc with a small telescope, girdled by its beautiful rings, unattached to the globe (Fig. 10.1). They set Saturn apart from all the other planets. Even though we now know that all four of the giant planets possess ring systems of some kind, Saturn's rings easily outclass the others.

Saturn's orbital radius is 9.5 times the radius of the Earth's orbit, and it takes 29.458 Earth years for Saturn to complete one revolution around the Sun. Perhaps because of its remote orbit and slow motion, the planet's name has been adopted for the word "saturnine", to describe a cool and distant temperament.

At its large distance from the Sun, the ringed planet and its satellites receive only about one percent as much sunlight and solar heat as the Earth does. The surfaces of many of Saturn's satellites are therefore covered with water ice. And even though Saturn generates some of its own heat, its cloud tops have a temperature of only 95.0 kelvin.

Saturn is the second largest planet in the solar system, overshadowed only by Jupiter. The radius of Saturn, without the rings, is about four-fifths the radius of Jupiter and slightly more than nine times the radius of the Earth.

The volume of Saturn is great enough to encompass 764 Earth-sized planets. But Saturn's mass is only 95 times greater than the Earth's mass, so the giant planet must be composed of material that is much lighter than rock and iron, the primary ingredients of the Earth.

From Saturn's mass and volume, we calculate its average mass density to be only 687 kilograms per cubic meter, the lowest of any planet and less than that of liquid water. If Saturn were placed in a large enough ocean of water, it could float. It has a low average mass density because it is mainly composed of the lightest element, hydrogen, in the gaseous and liquid states.

Saturn rotates with a day of only 10.6562 hours – only 44 minutes longer than the Jovian day of 9.9249 hours (Table 10.1). This is the rotation period of Saturn's magnetic field that is anchored inside the planet, and it is inferred from the observed periodic modulation in Saturn's radio emission, generated in the spinning magnetic fields. The visible clouds spin at different speeds, faster at the equator and slower at the poles.

Fig. 10.1 Ringed planet with icy moons Saturn's yellow-brown clouds are swept into bands by the planet's rapid rotation. Two of its white moons (*left*), Tethys (*above*) and Dione, are covered with water ice. The shadows of Saturn's main rings and Tethys are cast onto the cloud tops. The outer A ring is separated from the central B ring by the Cassini Division. This gap is so tenuous that the edge of Saturn can be seen through it. The faintest of Saturn's main rings, the C ring or crepe ring, is barely visible against the planet. This image was obtained from the *Voyager 1* spacecraft on 3 November 1980. (Courtesy on NASA and JPL.)

Table 10.1 Physical properties of Saturn[a]

Mass	5684.6×10^{23} kilograms = $95.162\,M_E$
Equatorial Radius at 1 bar	60 268 kilometers = $9.449\,R_E$
Polar Radius at 1 bar	54 364 kilometers
Mean Mass Density	687.3 kilograms per cubic meter
Rotation Period	10 hours 39 minutes 22.3 seconds = 10.6562 hours
Orbital Period	29.458 Earth years
Mean Distance from Sun	1.4294×10^{12} meters = 9.539 AU
Age	4.6×10^{9} years
Atmosphere	97 percent molecular hydrogen, 3 percent helium
Energy Balance	1.79 ± 0.10
Effective Temperature	95.0 kelvin
Temperature at 1-bar level	134 kelvin
Central Temperature	13 000 kelvin
Magnetic Dipole Moment	$600\,D_E$
Equatorial Magnetic Field Strength	0.22×10^{-4} tesla or $0.72\,B_E$

[a] The symbols M_E, R_E, D_E, B_E denote respectively the mass, radius, magnetic dipole moment, and magnetic field strength of the Earth. One bar is equivalent to the atmospheric pressure at sea level on Earth. The energy balance is the ratio of total radiated energy to the total energy absorbed from sunlight, and the effective temperature is the temperature of a black body that would radiate the same amount of energy per unit area.

10.2 Saturn's winds and clouds

The wind speeds of Saturn's equatorial jet streams reach 500 meters per second, almost four times the speed of Jupiter's fastest winds and ten times hurricane force on the Earth. The dominant winds on Saturn blow eastward, in the same direction as the planetary rotation, at almost all latitudes, with the most powerful nearest to the equator. Reversals in wind direction are only found near Saturn's poles, where the clouds counter flow in the eastward and westward direction (Fig. 10.2). They form banded belts and zones similar to those observed almost everywhere on Jupiter.

Despite its raging winds, Saturn lacks the dynamic and colorful storm clouds of Jupiter. Stormy weather on Saturn is apparently masked by an upper deck of dirty, smog-coated particles that give the planet a pastel, butterscotch hue. Jupiter, being warmer than Saturn, has less of this smoggy haze, and its cloud features are more distinct.

On rare occasions, a gigantic storm cloud of fresh, clean white ammonia ice warms up, rises and punches through the opaque upper cloud deck, somewhat like the upwelling of warmer air in terrestrial thunderheads. The swirling, white equatorial ovals have been recorded by ground-based telescopic observers, but only three times in two centuries. The oval spotted in 1990 was at least three times the size of Earth. The previous one appeared in 1933 and survived only a few years, while the last one before that was in 1870. No other major storms were seen in intervening years. High-resolution images from the *Hubble Space Telescope* are clear enough to pick out the details of large storm systems and to record minor storms that look like bright cloud features.

When *Voyager 2* passed behind Saturn, its homeward-bound radio signals penetrated the planet's upper atmosphere, and alterations in these transmissions have been used to deduce the pressure and temperature below the clouds (Fig. 10.3). Because there is no solid surface directly below the clouds, altitudes are referred to the level in the atmosphere where the pressure is equal to 0.1 bar, or one-tenth the sea-level pressure on Earth. This is the approximate level where the temperature bottoms out, at about 82 kelvin, and the obscuring veil of haze may be formed.

Under the assumption that Saturn's gas mixture is in chemical equilibrium, with a uniform composition like that of the Sun, ammonia is expected to condense and form clouds at about 100 kilometers below the reference level, where the pressure has risen to about 1 bar. These clouds of ammonia ice presumably rise to form the bright, white storms that are occasionally seen above the global haze. Water clouds may form much lower in the atmosphere, where the pressure rises to almost 10 bar, but no one has ever seen them.

Fig. 10.2 Clouds and hazes in Saturn's atmosphere Enhanced colors bring out the details of Saturn's banded clouds in this image, taken in infrared light. The blue color indicates a clear atmosphere down to a main cloud layer. Different shadings of blue indicate variations of the cloud particles, in size or chemical composition. The cloud particles are believed to be ammonia-ice crystals. The red and orange colors mark clouds reaching up high into the atmosphere, and the dense parts of the two storms near Saturn's equator appear white. The green and yellow colors indicate a haze above the main cloud layer. Saturn's counter-flowing east–west winds have aligned the clouds and haze within fixed latitude bands that become more pronounced near the planet's polar regions. The dark region around the south pole (*bottom*) marks the location of a large hole in the main cloud layer. The rings are made up of chunks of water ice, with a white color that has been browned in some cases, somewhat like dirty snow on a winter road. Two of Saturn's satellites were also recorded, Dione (*lower left*) and Tethys (*upper right*). They appear in different yellow and green colors, indicating different conditions on their icy surfaces. This false-color image was taken on 4 January 1998 using the *Hubble Space Telescope*. (Courtesy of NASA and STSI.)

10.3 Beneath the clouds of Saturn

The internal constitution of Saturn

Saturn's low mass density indicates that the lightest element, hydrogen, is the main ingredient inside the planet, just as it is for Jupiter and the Sun. The lightweight material, just 68.8 percent as dense as water, is hurled outward in its equatorial regions by the planet's rapid 10.6562-hour rotation, making Saturn the most oblate planet in the solar system. Its equatorial bulge amounts to about ten percent of the radius, and is about as big in extent as the Earth. Or, as some view it, the polar regions of Saturn are squashed and flattened by this amount.

The oblong shape of Saturn can be seen with a small telescope, and measured precisely from its satellite orbits and ring positions as well as from the trajectories of the passing *Voyager* spacecraft. When these measurements are combined with Saturn's known mass, volume and rotation rate, scientists can obtain information about its internal distribution of mass.

The model the experts come up with is just a scaled-down version of Jupiter, with a small, dense core of melted ice and molten rock surrounded by a vast shell of liquid hydrogen and topped by a thin gaseous atmosphere (Fig. 10.4). Like Jupiter, giant Saturn is not a solid world, and is essentially a great big drop of liquid.

Deep down inside, the liquid hydrogen is compressed to such high pressures that it conducts electricity like a metal. But since Saturn is less than three times as massive as Jupiter, and only slightly smaller, the internal pressure at a given depth is less, and the liquid hydrogen turns into a metal further down in the ringed planet. Saturn therefore has a smaller shell of liquid metallic hydrogen.

Also like Jupiter, the giant planet Saturn has a magnetic field generated by rotationally driven electric currents in the planet's liquid metallic shell. But since Saturn has a thinner shell, the strength of its magnetic field is about one-twentieth of Jupiter's, despite the fact that both planets rotate with about the same period. Auroras are produced in Saturn's polar regions when charged particles spiral down along the magnetic field lines and collide with gases in the atmosphere (Section 3.7).

One unusual characteristic of Saturn's magnetism is that its magnetic axis is almost precisely aligned with its rotation axis. No other planet has such an alignment of the two axes, and it is difficult to explain how the magnetic field can be maintained in such a way.

Interior heat and helium rain

Precise measurements from the *Voyager 1* and *2* spacecraft indicate that Saturn is radiating 1.78 times more energy in visible and infrared light than it absorbs from incoming

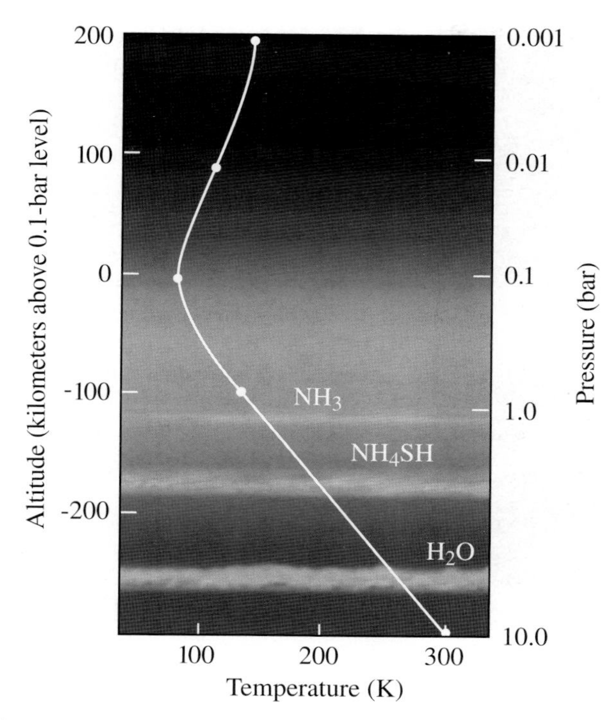

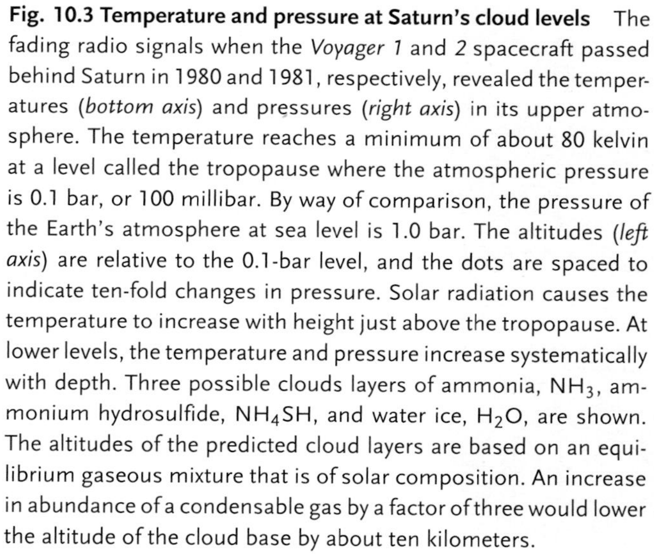
Fig. 10.3 Temperature and pressure at Saturn's cloud levels The fading radio signals when the *Voyager 1* and *2* spacecraft passed behind Saturn in 1980 and 1981, respectively, revealed the temperatures (*bottom axis*) and pressures (*right axis*) in its upper atmosphere. The temperature reaches a minimum of about 80 kelvin at a level called the tropopause where the atmospheric pressure is 0.1 bar, or 100 millibar. By way of comparison, the pressure of the Earth's atmosphere at sea level is 1.0 bar. The altitudes (*left axis*) are relative to the 0.1-bar level, and the dots are spaced to indicate ten-fold changes in pressure. Solar radiation causes the temperature to increase with height just above the tropopause. At lower levels, the temperature and pressure increase systematically with depth. Three possible clouds layers of ammonia, NH_3, ammonium hydrosulfide, NH_4SH, and water ice, H_2O, are shown. The altitudes of the predicted cloud layers are based on an equilibrium gaseous mixture that is of solar composition. An increase in abundance of a condensable gas by a factor of three would lower the altitude of the cloud base by about ten kilometers.

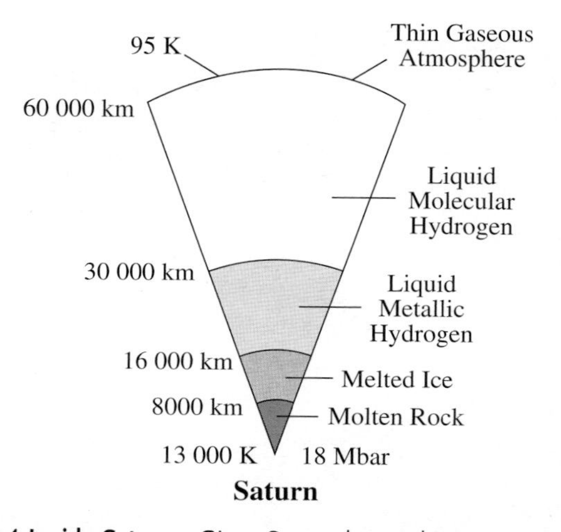

Fig. 10.4 Inside Saturn Giant Saturn has a thin gaseous atmosphere covering a vast global ocean of liquid hydrogen. At the enormous pressures within Saturn's interior, the abundant hydrogen is compressed into an outer shell of liquid molecular hydrogen and an inner shell of fluid metallic hydrogen. The giant planet may have a relatively small core of melted ice and molten rock.

sunlight. This excess energy must be coming from within the planet. It implies that Saturn, like Jupiter, is an incandescent globe with an internal source of heat (Fig. 10.5).

Both Jupiter and Saturn radiate almost twice as much energy as they receive from the Sun, but the dominant source of internal heat is different for the two giant planets. Jupiter's internal heat is primarily primordial heat liberated during gravitational collapse when it was formed, and Saturn must have also started out hot inside as the result of its similar formation. But being somewhat smaller and less massive than Jupiter, the planet Saturn was not as hot in its beginning and has had time to cool. As a result, Saturn lost most of its primordial heat and there must be another source for most of its internal heat.

Saturn's internal heat is generated by the precipitation of helium into its metallic hydrogen shell. The heavier helium separates from the lighter hydrogen and drops toward the center, somewhat like the heavier ingredients in a bottle of salad dressing that hasn't been shaken for awhile. Small helium droplets form where it is cool enough, precipitate or rain down, and then dissolve at hotter deeper levels. As the helium at a higher level drizzles down through the surrounding hydrogen, the helium converts some of its energy to heat. In much the same way, raindrops on Earth become slightly warmer when they fall and strike the ground; their energy of motion – acquired from gravity – is converted to heat.

The helium rain theory has apparently been confirmed by *Voyager* measurements of a lower abundance of helium in the outer atmosphere of Saturn than in Jupiter or the Sun. The number of helium molecules in Saturn is only 3 percent, and hydrogen 97 percent, while the number is 16 percent in the Sun and 13.6 percent for Jupiter. Since Jupiter is just slightly depleted in helium when compared to the Sun, helium rain is also probably operating inside this giant planet, but in more modest amounts because of Jupiter's greater mass and internal temperature.

10.4 The remarkable rings of Saturn

Billions of whirling particles of water ice

The austerely beautiful rings of Saturn are so large and bright that we can see them with a small telescope. And

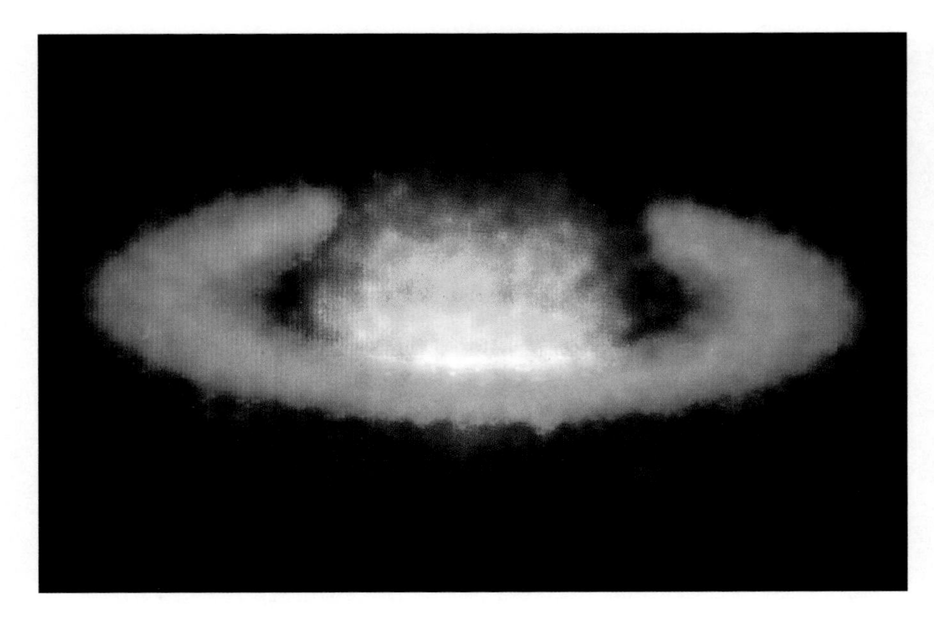

Fig. 10.5 Saturn and its rings at infrared wavelengths Because Saturn's rings are made of water ice, they reflect relatively large amounts of sunlight at an infrared wavelength of 3.8 micrometers (*blue*). Methane in Saturn's outer atmosphere absorbs radiation at this wavelength, but the incandescent globe has its own internal source of heat that makes it shine brightly at the longer infrared wavelength of 4.8 micrometers (*orange*). The heat welling up from within Saturn is partly due to helium rain, that is falling inside the planet. (Courtesy of David Allen, Anglo-Australian Telescope Board © 1983.)

Fig. 10.6 Saturn's rings open up These images were taken from the *Hubble Space Telescope* during a four-year period, from 1996 to 2000 (*left to right*), as Saturn moved along one-seventh of its 29-year journey around the Sun. As viewed from near the Earth, Saturn's rings open up from just past edge-on to nearly fully open as it moves through its seasons, from autumn towards winter in its northern hemisphere. (Courtesy of NASA and STSI.)

because the glittering rings are tipped with respect to the ecliptic, the plane of the Earth's orbit about the Sun, they change their shape when viewed from on or near the Earth. The rings are successively seen: edge-on, when they can briefly vanish from sight in a small telescope; from below, when they are wide open; edge-on again; and then from above. The complete cycle requires 29.458 Earth years, the orbital period of Saturn, so the rings nearly vanish from sight every 15 years or so. The last disappearance took place in 1995. One-seventh of the cyclic change in appearance, from 1996 to 2000, is illustrated in Fig. 10.6.

The three main rings of Saturn have been observed for centuries (Section 1.5). There are the outer A ring and the central B ring, separated by the dark Cassini Division, and an inner C, or crepe, ring that is more transparent than the other two. They all remain suspended in space, unattached to Saturn, because they move around the planet at speeds that depend on their distance, opposing the pull of gravity.

The motions of Saturn's rings can be measured using spectral features in their reflected sunlight. When part of a ring moves toward or away from an observer, the spectral

Table 10.2 Saturn's rings

Ring	Width (km)	Closest Distance (km)	Distance Range[a] (R_S)	Particle Size (m)	Optical Depth	Mass (kg)
D	7540	66 970	1.11–1.235	$< 10^{-6}$	0.0001	
C	17 490	74 510	1.235–1.525	0.01–3.0	0.05–0.35	1×10^{17}
B	25 580	92 000	1.525–1.949	0.01–5.0	0.4–2.5	2×10^{19}
A	14 610	122 170	1.949–2.025	0.01–7.5	0.05–0.15	4×10^{17}
F	50	140 180	2.324	10^{-7}–10^{-5}	0.1	
G	500–3000	170 180	2.82	3×10^{-8}	2×10^{-6}	
E	302 000	181 000	3 to 8	1×10^{-6}	1.5×10^{-5}	7×10^{8}

[a] The distance range is given in units of Saturn's apparent equatorial radius, $R_S = 60\,330$ kilometers. At the 1-bar pressure level, the radius is 60 268 kilometers.

features are displaced in wavelength by an amount that depends on the velocity of motion. There is a shift toward shorter wavelengths when the motion is toward the observer, while motion away produces a shift toward longer wavelengths. Observations of this Doppler effect, by the American astronomer James Edward Keeler (1857–1900) in 1895, showed that the inner parts of the rings move around Saturn faster than the outer parts, all in accordance with Kepler's third law for small objects revolving about a massive, larger one. They orbit the planet with periods ranging from 5.8 hours for the inner edge of the C ring, to 14.3 hours for the outer edge of the more distant A ring. Since Saturn spins about its axis with a period of 10.6562 hours, the inner parts of the main rings orbit at a faster speed than the planet rotates, and the outer parts at a slower speed.

The difference in orbital motion between the inner and outer parts of the rings means that they are not a solid sheet of matter, for they would be torn apart by the differential motion. As demonstrated by James Clerk Maxwell (1831–1879) in 1867, the rings are instead made up of vast numbers of particles, each one like a tiny moon in its own orbit around Saturn. Billions of ring particles revolve about the planet. They have been flattened and spread out to a thin, wide disk as the result of collisions between particles.

The rings of Saturn are flat, wide and incredibly thin (Table 10.2). Measured from edge to edge, the three main rings span a total width of 62 200 kilometers, so they are a little wider than the planet's radius, at 60 300 kilometers. When observed edge-on, from on or near the Earth, the rings practically disappear from view (Fig. 10.7). They look about a kilometer thick, but this is an illusion attributed to warping, ripples, embedded satellites and a thin, inclined outer ring. When instruments on *Voyager 2* monitored starlight passing through the rings, they found that the ring edges extend only about 10 meters from top to bottom. If a sheet of paper represents the thickness of Saturn's rings, then a scale model would be two kilometers across.

What are the ring particles made out of? At visible wavelengths, the rings are bright and reflective, but at infrared wavelengths they are dark and less reflective. This suggests that the particles are cold and made of ice. In fact, they are composed of water ice.

Detailed Earth-based infrared spectroscopy of the main rings in the 1970s showed that incident sunlight is absorbed by water ice at the surfaces of the particles. Subsequent spectral investigations indicated that the frozen water is exceptionally clean and pure, with few impurities of dust or rock. They are also poor absorbers and emitters

Fig. 10.7 Edge-on view of Saturn's rings When the Earth is in the plane of Saturn's rings, an observer on the Earth views the rings edge-on. Because the rings are so thin, they are then barely visible. Saturn's largest satellite, Titan, is seen just above the rings (*left*); it is enveloped in a dark brown haze and casts a dark shadow on Saturn's clouds. Four other moons are clustered near the other edge of Saturn's rings (*right*), appearing bright white because their surfaces are covered with water ice. These icy satellites are named Mimas, Tethys, Janus and Enceladus. This image was taken on 6 August 1995 from the *Hubble Space Telescope*. (Courtesy of NASA and STSI.)

of microwaves, which implies that more than 99 percent of the mass of the rings is water ice, and that less than 1 percent consists of dirty contaminants.

The total mass of the prominent A, B and C rings is about equal to that of Saturn's satellite Mimas, which weighs in at 4.5×10^{19} kilograms, and such a mass is consistent with particles composed of water ice. To check that, just multiply the mass density of water, at 1000 kilograms per cubic meter, by the total volume of the main rings – 10 meters thick, 60 000 kilometers wide, and a circumference of 600 000 kilometers. Since the particles are not jammed tightly together, and probably separated by five to ten times their size, the resulting mass has to be diluted by a corresponding factor.

Typical chunks of ice in the main rings vary in size from hailstones to fist-sized snowballs, some are as small as snowflakes and a few are icebergs as large as houses. In other words, the ring particles range in size from a hundredth of a meter to ten meters across (Table 10.2). There are more and more particles of smaller and smaller size within this range, so the main rings consist primarily of the smaller particles about 0.01 meters in size. Though far less numerous, the larger particles greater than 1 meter across contain most of the ring mass.

The ring particles are too small for spacecraft cameras to see individually, but scientists can infer their size from radio measurements. Since the rings are very reflective to ground-based radar transmissions, we know that their particles are comparable to, or larger than, the radar wavelength of about 0.1 meters. The particle size distribution has been determined from the way the rings blocked the radio signals from *Voyager 1* and *2* when the spacecraft passed behind the rings. This method showed that there are remarkably few particles larger than 5 to 10 meters in size or smaller than 0.01 meters. Within these bounds, the number of particles in the main rings decreases with increasing size, in proportion to the inverse square of their radius.

However, four additional rings, designated the D, E, F and G rings (Fig. 10.8), consist of much smaller, microscopic particles. These rings, discovered using ground-based or spacecraft observations, are all very diffuse, tenuous and nearly transparent. The way that their particles scatter light indicates that they are very small, roughly a micron in size – a micron is one-millionth, or 10^{-6}, of a meter.

Lying between the C ring and the planet is the D ring. Although terrestrial observers had reported a faint ring between the C ring and the globe as early as 1969, these reports remained controversial until the *Voyager* spacecraft definitely verified the existence of the D ring. It is so tenuous and transparent that it is probably impossible to see from the Earth using the best telescopes. According to one hypothesis, splintered chips from colliding ice particles drift down into Saturn's atmosphere and form the D ring.

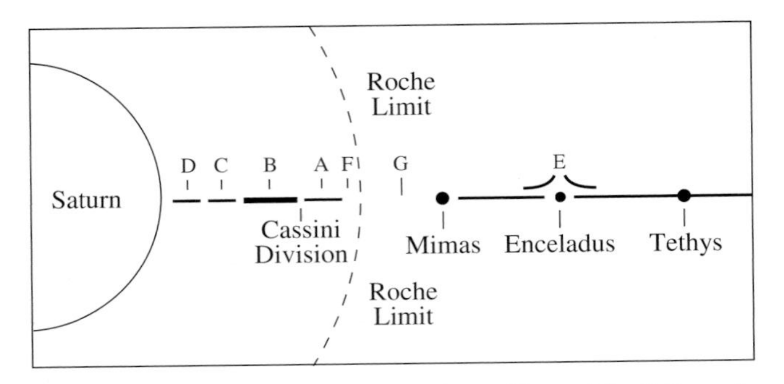

Fig. 10.8 Cross-section of rings and satellites All of Saturn's main rings lie inside the Roche limit (*dashed curve*) within which the planet's gravity will tear a large satellite apart. The A and B rings have been observed for centuries. The more tenuous C ring was discovered in the 19th century, and definite observations of the transparent D ring awaited the arrival of the *Voyager 1* spacecraft on 12 November 1980. The icy satellite Enceladus probably feeds the tenuous E ring, also revealed from *Voyager*. For clarity, the thickness of the rings has been exaggerated.

Outside the traditional system lies the huge, sparse E ring, a broad tenuous band of small particles that is five times the combined breadth of rings A, B and C, and roughly centered on the orbit of Enceladus. Discovered with ground-based telescopes in 1966, the E ring becomes visible when the ring system is viewed approximately edge-on. It is composed almost exclusively of small grains just one micron, or 10^{-6} meters, in size.

Because they have relatively short lifetimes of several thousand years, these tiny particles must be continually replenished if the E ring is a permanent feature. They could be thrown out during meteoritic impacts with Enceladus or other nearby icy satellites. Watery eruptions from ice volcanoes or geysers on Enceladus might also feed small bits of ice into Saturn's E ring. Once lofted into space, the pressure of sunlight, the gravitational tug of Saturn, and possibly electromagnetic effects spread the particles out into space.

Initially discovered by *Pioneer 11* and verified in *Voyager* images, the tenuous G ring lies between the A and E rings. The G ring consists of similar micron-sized particles to the E ring, and may be renewed from small moons embedded within it.

Pioneer 11 discovered the incredibly narrow F ring, that lies just outside the A ring, by its absorption of energetic particles; images from the *Voyager* spacecraft showed the F ring in great detail, demonstrating that its width varies from a few to tens of kilometers. Moreover, it is not just a single ring; *Voyager 1* spotted a contorted tangle of narrow strands that had smoothed out by the time *Voyager 2* arrived about nine months later. Because the F-ring particles are brighter when backlit by the Sun, and fainter in reflected sunlight, we know that the particles are also micron-sized, much smaller than snowflakes and comparable in size to the dust particles in your room.

But how can this ring retain such narrow boundaries? In the absence of other forces, collisions between ring particles should spread them out, causing the particles to fall inward toward Saturn and expand outward from it, thus creating a broader and more diffuse ring. Two tiny moons, named Pandora and Prometheus, flank the F ring and confine it between them, thereby keeping the particles of the F ring from straying beyond the ring's narrow confines (Fig. 10.9).

These shepherd satellites, discovered by the *Voyager* spacecraft, chase each other around the narrow F ring and keep it from spreading, as though they were two gravitational sheepdogs herding sheep into a narrow path. Each shepherd tends one edge of the ring. The moon outside the ring moves more slowly than the ring particles, which in turn are outpaced by the faster inner moon. The faster-moving inside satellite gravitationally pulls the inner F-ring particles forward as it passes, causing them to accelerate and spiral outward. The slower-moving outer shepherd exerts a net backward force on the outer ring particles, causing them to move inward. The result is a very narrow ring. Such confining moons were originally proposed to account for the narrow rings surrounding Uranus, constraining their edges from the otherwise inevitable spreading, and such a pair of satellites was eventually found astride one of the rings of Uranus (Section 11.4).

The shepherd satellites that flank the F ring, and possibly one or more other satellites embedded in the ring system, gravitationally interact with the ring material and distort its normal, circular shape, producing temporary kinks and twists in it. Because their orbits are slightly eccentric, these two satellites produce a varying perturbation of ring particles, perhaps accounting for the changing appearance of the F ring. Through similar gravitational interactions, small moons can produce ripples and waves on the surface of Saturn's main rings, and clear gaps within them.

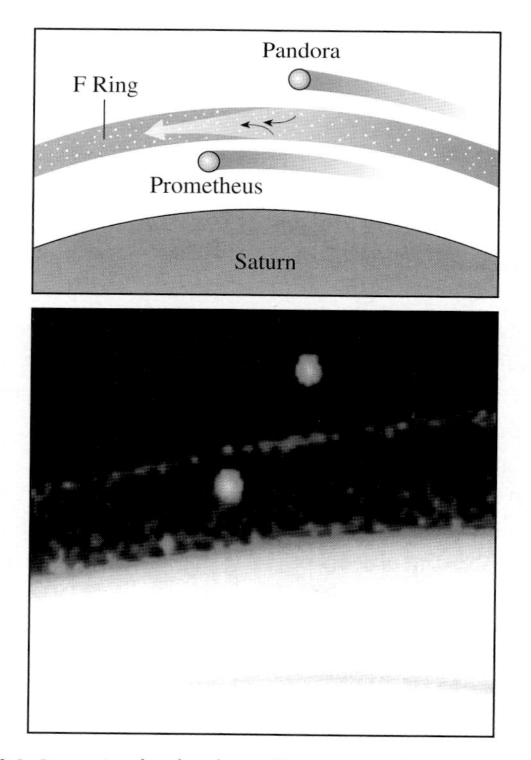

Fig. 10.9 Saturn's shepherd satellites Two shepherd satellites confine Saturn's narrow F ring. The outer shepherd gravitationally deflects ring particles inward, and the inner shepherd deflects ring particles outward. (Courtesy of NASA and JPL.)

Ringlets, waves, gaps and spokes

From a distance, the principal rings of Saturn look like smooth, continuous structures. Up close, however, from the views provided by the *Voyager 1* and *2* spacecraft, the icy material is marshaled into thousands of individual ringlets (Fig. 10.10). Some of the ringlets are perfectly circular, others are oval-shaped and a few seem to spiral in toward the planet like the grooves on an old-fashioned vinyl record. In some places, the flat plane of the rings is slightly corrugated, and ringlets are seen at the crests and dips of the corrugations, like ripples running across the surface of a pond.

An outside hand is at work sculpting at least some of the intricate ring structures through the force of gravity. The combined gravitational pull of Saturn and the accumulated pull of nearby moons can redistribute the ring particles, concentrating them into many of the observed shapes. Although small nearby moons have only a weak gravitational pull on the particles in the rings, the pull is repeated over and over again at certain resonant locations. Just as we can make a child on a swing arc high above the ground with a gentle, repeated push in the same place of the swing, so the repeated gravitational pull of a small external moon during each orbit can give an unexpectedly large perturbation. The interplay of this effect and Saturn's inward gravitational pull can repel and attract the ring particles, pushing and pulling them into localized concentrations such as ringlets. Detailed structures in Saturn's A ring have, for example, been identified with alternating compressed and rarefied ring material in orbit around Saturn, in a rippling driven by interactions with nearby external satellites.

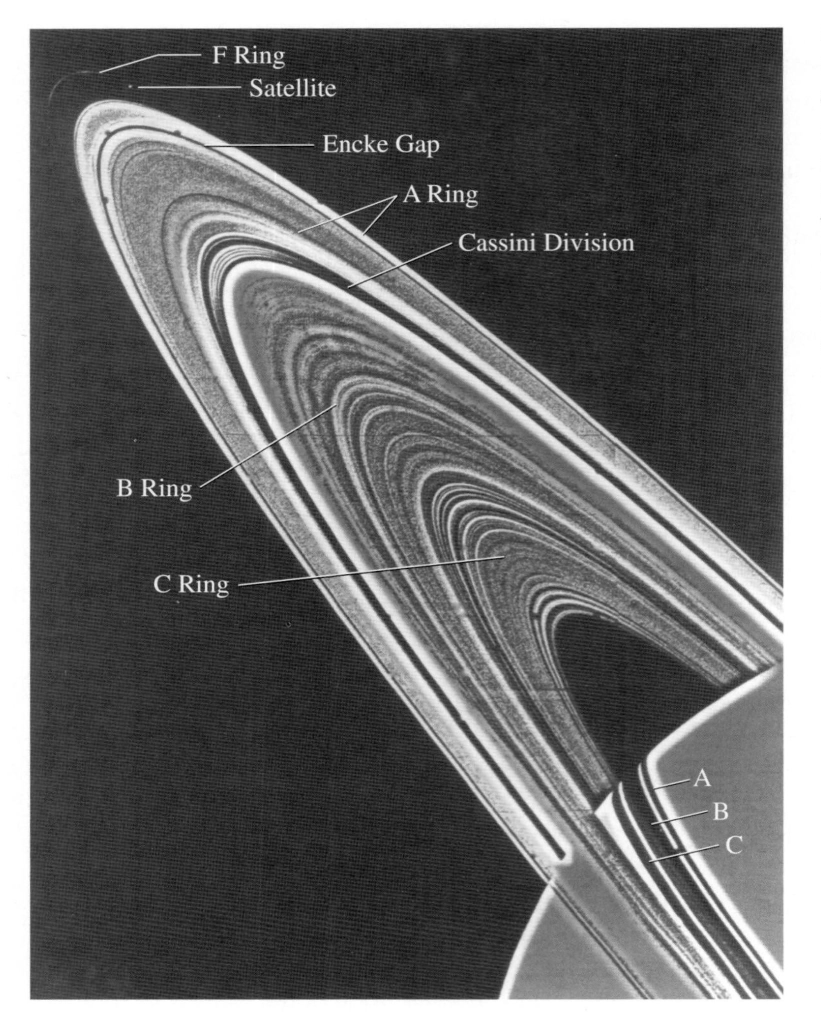

Fig. 10.10 Ringlets When viewed with high resolution, approximately 100 concentric features are seen within Saturn's rings, including some in the Cassini Division. The ring system would probably separate into countless ringlets if we could detect fine enough detail. A small satellite, discovered by *Voyager 1*, is seen (*upper left*) just outside the narrow F ring. The *Voyager 1* spacecraft took this mosaic of Saturn's rings on 6 November 1980. (Image courtesy of NASA and JPL.)

Repeated gravitational resonance interactions occur when the particle orbital period and the satellite orbital period are exact fractions or whole-number ratios, such as 1/2 for the 2:1 resonance. The particle then experiences a repeated additive perturbation during each orbit that can set up gravitational waves that propagate radially through the rings. The particles tend to congregate at the crests and troughs of these density waves, like automobiles in traffic intersections or a crowd starting a "wave" in a stadium. Similar spiral density waves on a vastly greater scale are thought to create the stellar arms of spiral galaxies.

Resonance can also confine ring edges and clear gaps. Particles straying into a gap at a resonant location are removed by repetitive interaction with a particular moon. The ring particles at the outer edge of the B ring and the inner edge of the Cassini Division, for example, are traveling almost twice as fast as Saturn's largest inner satellite, Mimas, with a period one half as large and occupying the 2:1 resonance with this moon. The razor-sharp outer edge and scalloped hem of the A ring similarly result from a 7:6 resonance with the external co-orbital satellites Janus and Epimetheus. In addition, many low-density regions in Saturn's A ring are located at positions resonant with small moons that lie just exterior to the ring.

Tiny satellites embedded within the rings can also sweep out a tidy gap with neat edges. The small moon Pan, embedded within Saturn's A ring, apparently plows its way through Encke's gap, keeping it open and creating wavy radial oscillations around the inner and outer edges of the gap.

But simple interactions with known moons have not been completely successful in accounting for all of the intricate detail found in Saturn's rings. The apparent gaps in the system are not completely empty. The Cassini Division, for example, contains perhaps 100 ringlets (Fig. 10.11), with particles just as large as those in the neighboring ring. Some gaps do not even occur at known resonant positions or contain detected moons embedded within them. Unseen moons might influence the clumping and removal of material in these locations.

Perhaps the most bizarre *Voyager* discovery was the long, dark streaks, dubbed spokes, that stretch radially

Fig. 10.11 Beneath the rings of Saturn When the *Voyager 1* spacecraft moved beneath Saturn's rings, it could view sunlight transmitted through the rings, presenting a reversed image of the sunlit side. Both the C ring and Cassini's Division appear bright because they are sparsely populated with small particles that efficiently scatter light in the forward direction, whereas the A and B rings appear dark because their densely-packed particles absorb all the incident sunlight. This perspective is not available from Earth, where we always see the sunlit side of the rings. (Courtesy of NASA and JPL.)

across the rings, keeping their shape like the spokes of a wheel (Fig. 10.12). These ephemeral features are short-lived, but regenerated frequently. They are found near the densest part of the B ring, that co-rotates with the planet at a period of 10.6562 hours. But the inner and outer parts of Saturn's dark spokes also whirl around the planet with this period, at constant speed, in apparent violation of Kepler's third law and Newton's theory of gravitation. If the spokes consisted of dark particles embedded in the rings, the particles would move with speeds that decrease with increasing distance from Saturn, and the spokes would quickly stretch out and disappear.

Fig. 10.12 Dark spokes Several dark spoke features streak across the central third of Saturn's B ring. They sweep around Saturn with a uniform velocity in apparent defiance of Kepler's laws. Electromagnetic forces probably levitate the dark, electrically charged particles above the main rings, permitting Saturn's magnetic field to carry them around the planet. This image was taken from the *Voyager 2* spacecraft on 22 August 1981. (Courtesy of NASA and JPL.)

From their reflective properties, we know that these enigmatic features consist of microscopic grains, much smaller than the dominant particles in the prominent A, B and C rings. As the *Voyagers* approached Saturn, the spokes appeared dark in reflected sunlight, against a bright ring background. As the spacecraft departed, sunlight passed through the spokes, which appeared brighter than the surrounding ring areas. This means that the spokes are composed of very small dust-like particles that are comparable in size to the wavelength of light, about a millionth of a meter, or one micron, across. And since the spoke particles move with velocities equal to the velocity of Saturn's rotating magnetic field, they are probably carried around by it. But the exact mechanism for generating and sustaining the mysterious spokes remains obscure.

According to one hypothesis, the small dust particles may become charged, perhaps as the result of collisions with energetic electrons. Electromagnetic forces then raise or levitate the tiny, charged particles off the larger ring bodies, and the spokes are swept around Saturn by its rotating magnetic field. It sounds bizarre, but subtle forces are required to overcome gravity.

Why do planets have rings?

One might expect the particles of a ring to have accumulated long ago into larger satellites. But the interesting

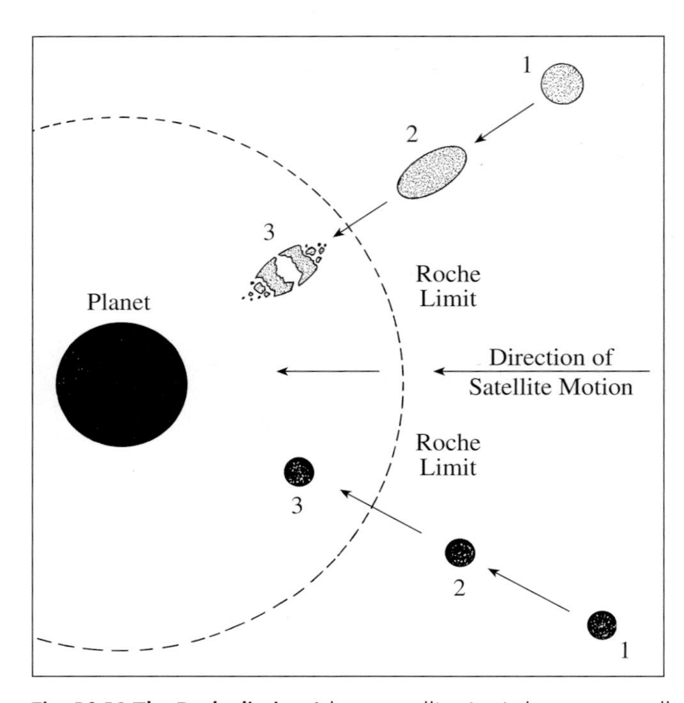

Fig. 10.13 The Roche limit A large satellite (*top*) that moves well within a planet's Roche limit (*dashed curve*) will be torn apart by the tidal force of the planet's gravity. The side of the satellite closer to the planet feels a stronger gravitational pull than the side farther away, and this difference works against the self-gravitation that holds the body together. A small solid satellite (*bottom*) can resist tidal disruption because it has significant internal cohesion in addition to self-gravitation.

feature of rings – and a clue to their origin – is that they do not coexist with large moons. Planetary rings are always closer to the planets than their large satellites.

The rings are confined to an inner zone where the planet's tidal forces would stretch a large satellite until it fractured and split, while also preventing small bodies from coalescing to form a larger moon. The outer radius of this zone in which rings are found is called the Roche limit (Fig. 10.13) after the French mathematician Eduard A. Roche (1820–1883), who described it in 1848. For a satellite with no internal strength and whose density is the same as the planet, the Roche limit is 2.456 times the planetary radius, or about 147 000 kilometers for Saturn.

Jupiter's mere wisp of a ring, the icy snowballs of Saturn's rings, and the dark boulders in the narrow rings encircling Uranus and Neptune all lie within the Roche limit for the relevant planet. Anywhere inside this distance a large satellite can no longer remain intact, but instead gets torn apart by planetary tides. Nevertheless, because of their material strength and great internal cohesion, small moons less than 100 kilometers across can exist inside the Roche limit without being tidally disrupted, just as the ring particles can.

To visualize the physical significance of the Roche limit, try to imagine what happens when two ring particles approach each other slowly while orbiting a planet. As they come closer together, their gravitational attraction for each other increases, and the maximum attraction occurs when the particles are touching each other. Larger, more massive particles will feel greater attraction. At the moment of contact, the planet pulls harder on the particle that is closer to it and less hard on the particle that is further away. The difference between the planet's gravitational pull on the inner and outer particles is the tidal force, and if it exceeds the mutual gravitational attraction of the particles toward each other, the particles will not stay together. The outcome of the tug-of-war between the tidal force and the mutual attraction is primarily decided by the particles' distance from the planet. At distances less than the Roche limit, particles are pulled apart and this prevents the accumulation of larger moons.

And where did Saturn's rings come from? There are two possible explanations for their origin. In the first explanation, the rings consist of material left over from Saturn's birth about 4.6 billion years ago. This hypothesis assumes that the rings and moons originated at the same time in a flattened disk of gas and dust with large, new born Saturn at the center. According to the second explanation, a former moon or some other body moved too close to Saturn and was torn into shreds by the giant planet's tidal forces, making the rings. In this case, the rings could have formed after Saturn, its satellites and much of the rest of the solar system.

It has long been thought that Saturn's rings and satellites are both primordial leftovers of the planet's formation process. Any disk material initially within the Roche limit soon after Saturn formed would have been prevented by the giant planet's tidal stresses from gathering or accreting into a large satellite. But outside the Roche limit, satellites could have coalesced from smaller bodies. Thus, any primordial, circumplanetary disk should develop into rings near the planet and exterior satellites, as we see today around Saturn. This could also help explain why only the giant planets have rings and a retinue of satellites, while the rocky terrestrial planets have no rings and either no satellite or just one or two of them. But there are recent objections to this long-standing explanation, centered on the fact that the planetary rings are relatively young and therefore cannot be as ancient and enduring as the planets themselves.

Astronomers now estimate that Saturn's rings are less than 100 million years old, or less than three percent of Saturn's life-span. The dazzling, sparkling brightness of Saturn's rings provides evidence for this youth. They glisten with clean particles of pure water ice, unsullied by the constant pelting by cosmic dust. The rings would look much darker if they were very old, just as new-fallen snow becomes dirty over time. Calculations indicate that in 100 million years Saturn's bright rings will be darkened by the pervasive cosmic debris to the same extent as the older, coal-black rings of Uranus and Neptune (Section 11.4).

The gravitational tugs of Saturn's moons on the rings will shorten the lives of the rings, providing another indication of their youth. When setting up density waves in the rings, nearby moons extract momentum from the ring particles, causing them to slowly spiral toward Saturn; to conserve momentum in the overall system, the moons gradually move away from the planet. The A ring will eventually be dragged down into the B ring, and all the rings should collapse as the result of this moon–ring interaction in about 100 million years.

So, it is now thought that Saturn's resplendent rings cannot be older than 100 million years or they would not be there now, and it is just a matter of time before the current rings disappear. If we had lived at the time of the dinosaurs, we might not have seen any rings around Saturn, and the rings we see today are just temporary embellishments destined to disappear from sight in 100 million years or so.

Saturn's rings might have formed and dissipated many times since the beginning of the solar system, the byproduct of short-lived processes of creation and destruction. Today's rings could be just the most recent incarnation of a process that keeps going on. The birth process is probably related to the planet's moons, which might also play the main role in their death.

This brings us back to the second explanation for Saturn's rings, in which a pre-existing body strayed too close to Saturn and was torn apart by tidal forces. It might have been one of Saturn's moons, or an interloper from another region of the solar system. A satellite could form outside the Roche limit and move inward due to the pull of tidal forces that would eventually rip the satellite to pieces. As previously mentioned, the total mass of all the ring particles is similar to the mass of Saturn's relatively small satellite, Mimas, so it seems reasonable that the rings could have formed from such a moon, or from a few smaller ones. After all, the Martian moon Phobos is now being drawn inexorably toward the red planet by its tidal forces, and Neptune's largest satellite Triton is also headed on a collision course toward its planet.

A former satellite of Saturn could have been broken up by a collision with a comet, hurling smaller pieces in toward the planet. Or the rings might have been created from small moons that were already inside the Roche limit. Some astronomers think that the remains of the moons that spawned Saturn's rings are buried somewhere inside them, and that the bright rings hide other moons that are destined to create rings in the future.

10.5 The moons of Saturn

Titan – moon of mystery

Titan is the largest of Saturn's satellites, much larger than the planet's other moons. It is the second largest satellite in the solar system, and the only satellite possessing an extensive, dense atmosphere with a surface pressure comparable to that of the Earth's air.

When the *Voyager 1* and *2* spacecraft passed behind Titan, as seen from Earth, the homeward-bound radio signal penetrated Titan's atmosphere, permitting an accurate determination of the surface radius from the time the signal disappeared. Under its thick atmosphere, the solid surface of Titan has a radius of 2575 kilometers, just slightly smaller than Jupiter's satellite Ganymede, at 2631 kilometers. Titan is a little larger than Mercury, whose radius is 2440 kilometers, except Titan is in orbit around Saturn. So Titan is a moon, not a planet.

The trajectories of the *Voyagers* were deflected by a small amount due to Titan's gravitational pull. The size of the deflection permitted an improved determination of the satellite's mass. From the mass and radius we can determine the mean mass density of Titan – 1880 kilograms per cubic meter. That is almost twice the mass density of water ice. If Titan were solid rock, like the Earth's Moon, its average density would be about three times that of water ice. So Titan may be composed of nearly equal amounts of ice and rock.

By way of comparison, Mercury has an average mass density greater than five times that of water. It has a dense iron core in addition to a rocky mantle, and a magnetic field generated by currents in the core. Titan does not have such a core or any detectable magnetic field.

Visible light cannot penetrate the veil of orange smog that covers Titan's surface (Fig. 10.14). In the satellite's dry, cold atmosphere, the smog builds up to an impenetrable haze that extends to altitudes as high as 300 kilometers. On Earth, smog similarly forms by the action of sunlight on hydrocarbon molecules in the air. The urban smog usually forms within a kilometer of the Earth's surface. Titan's atmosphere extends far above its surface because of the high atmospheric pressure and the relatively low mass and gravitational pull of Titan.

Instruments aboard the *Voyager* spacecraft determined the composition of Titan's atmosphere. The dominant gas

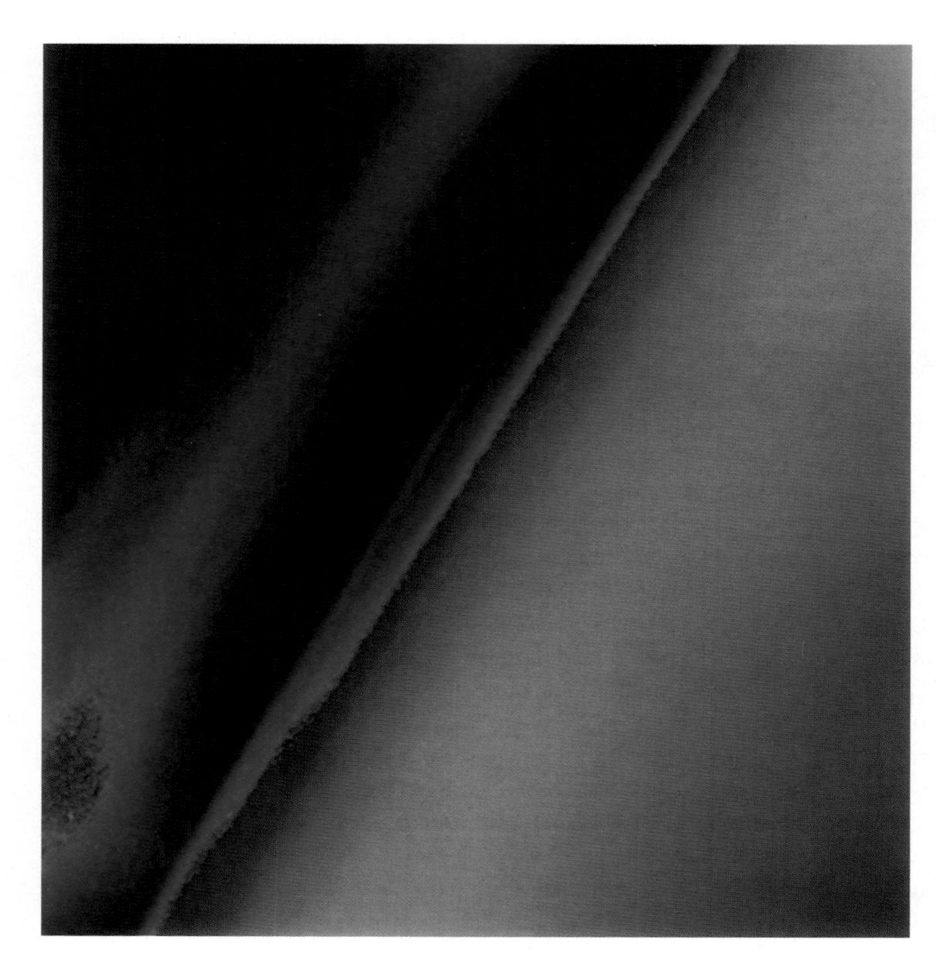

Fig. 10.14 Titan's smoggy haze The surface of Saturn's largest moon, Titan, is hidden beneath a thick haze that completely envelops the satellite. Divisions in the layer of haze (*blue*) occur at 200, 375 and 500 meters above the edge, or limb, of the satellite's atmosphere (*orange*). This false-color image was taken from the *Voyager 1* spacecraft on 12 November 1980. (Courtesy of NASA and JPL).

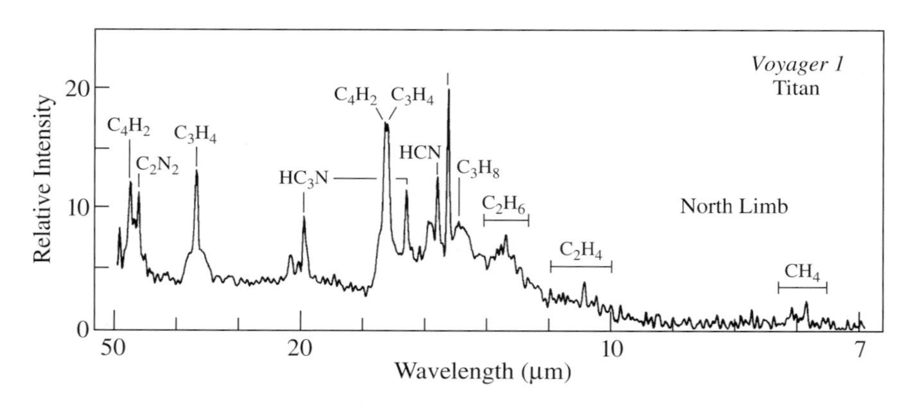

Fig. 10.15 Molecules in Titan's atmosphere Emission features in the infrared spectrum of Titan's reflected sunlight identify the molecular constituents of its atmosphere. Sharp peaks in this spectrum, acquired from the *Voyager 1* spacecraft in 1980, are attributed to methane, CH_4, acetylene, C_2H_2, ethane, C_2H_6, and more complex hydrocarbon molecules, as well as nitrogen compounds. The wavelength is given in units of microns, or 10^{-6} meters, abbreviated μm.

surrounding the satellite is molecular nitrogen, between 82 and 99 percent. Methane, the one gas identified with certainty before the *Voyagers* arrived, turned out to be a minor constituent, with an abundance of 1 to 6 percent. So, nitrogen molecules account for the bulk of Titan's atmosphere as they do on Earth – 77 percent of our air is molecular nitrogen.

But unlike Earth, the atmosphere of Titan contains no molecular oxygen, which accounts for 21 percent of our air. The freezing temperature on Titan is way too low for any living things, such as plants, to supply oxygen.

High above Titan's surface, abundant nitrogen and methane molecules are being broken apart continuously by the Sun's energetic ultraviolet light and by the bombardment of electrons trapped in Saturn's magnetic environment. Some of the fragments then recombine to form more complex molecules that have been detected in small amounts by *Voyager*'s infrared spectrometers (Fig. 10.15, Table 10.3). In addition to methane, CH_4, the list includes hydrocarbons like ethane, C_2H_6, acetylene, C_2H_2, and propane, C_3H_8, and nitrogen compounds such as hydrogen cyanide, HCN. Many of these molecules can join

Table 10.3 Composition of Titan's atmosphere

Molecule	Symbol	Amount
Major constituents		*Percent*
Nitrogen	N_2	82–99
Methane	CH_4	1–6
Minor constituents		*Parts per million*
Hydrogen	H_2	2000
Hydrocarbons		
Ethane	C_2H_6	20
Acetylene	C_2H_2	4
Ethylene	C_2H_4	1
Propane	C_3H_8	1
Methylacetylene	C_3H_4	0.03
Diacetylene	C_4H_2	0.02
Nitrogen Compounds		
Hydrogen Cyanide	HCN	1
Cyanogen	C_2N_2	0.02
Cyanoacetylene	HC_3N	0.03
Acetonitrile	CH_3CN	0.003
Oxygen Compounds		
Carbon Monoxide	CO	50
Carbon Dioxide	CO_2	0.01

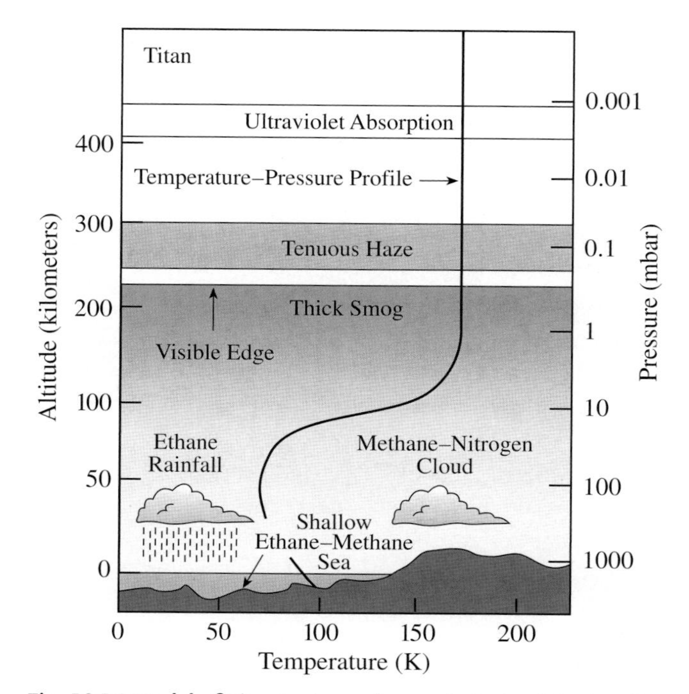

Fig. 10.16 Model of Titan's atmosphere A study of the bending and fading of homeward-bound radio signals when the *Voyager 1* spacecraft passed behind Titan led to this plot of the temperature and pressure in the satellite's atmosphere. The temperature (*bottom axis*) decreases with height until about 40 kilometers altitude, and then increases rapidly at higher altitudes (*left axis*). The entire atmosphere is well below the freezing temperature of water, 273 kelvin, but the lower atmosphere is just warm enough to allow the condensation of liquid nitrogen. The high-altitude smog might cover clouds of methane, and liquid ethane and methane could rain down to the surface. The pressure (*right axis*) is given in units of millibars or 0.001 bar, and it reaches 1500 millibar or 1.5 bar near the surface of Titan. The air pressure at sea level on Earth is 1 bar.

together in chain-like, polymer structures that contribute to Titan's dark, smoggy haze.

Although you cannot see beneath the smog-covered globe, *Voyager*'s radio signals have been used to infer the pressure and temperature down to the surface (Fig. 10.16). The surface pressure is an ear-popping 1.5 bar. That is one-and-a-half times the 1-bar air pressure at sea level on Earth, and equivalent to the pressure experienced by a deep-sea diver at about six meters under the ocean's surface.

Titan's temperature profile is very similar to that of the Earth's atmosphere, where the temperature initially drops with increasing altitude above the surface and then rises again due to heating by ultraviolet sunlight. But unlike Earth, the temperatures in Titan's atmosphere are everywhere below the freezing point of water, at 273 kelvin. Besides being about a billion kilometers from the Sun's heat, the surface of Titan lies below a haze that blocks out about 90 percent of the incident light. As a result, the surface temperature is 94 kelvin, and peaks at 175 kelvin at about 40 kilometers above the surface. The surface temperatures are far too cold to permit life on Titan, but its atmosphere may nurture chemical reactions similar to those at work on Earth before life began there (Focus 10.1).

Although liquid water cannot now lap the shores of Titan, it might contain shallow hydrocarbon seas. In fact, ethane or methane could play the role of water on Earth. The methane can condense in Titan's cold atmosphere to produce thick clouds that lie beneath the haze. Infrared observations that penetrate the smog suggest the presence of short-lived methane clouds in Titan's lower atmosphere, which form briefly and irregularly. Since the atmosphere is not fully saturated with methane, there cannot be extensive oceans of pure methane on the surface, but both ethane and methane can rain out of the atmosphere. They can exist as a liquid rather than a solid at the surface temperature of 94 kelvin. Evaporation of the liquid seas can resupply the hydrocarbons to the atmosphere, completing the cycle.

Thus, we expect seas, lakes and ponds of liquid hydrocarbons on Titan, consisting of ethane, methane and even propane. They are all highly flammable. The entire satellite could go up in flames, but it won't ignite because of the absence of molecular oxygen in the atmosphere.

We now know that Titan is not completely covered by a global hydrocarbon sea. Radar signals that penetrate the

Focus 10.1 Titan could be an early Earth in a deep freeze

The chemistry in Titan's atmosphere may be similar to that in Earth's atmosphere several billion years ago, before living things released molecular oxygen into the air and modified it. So Titan could serve as a time machine, taking us back to a simpler era on Earth before life began to "contaminate" the planet. Titan could even provide clues to how life got started when the Earth was young. Nevertheless, current life on Titan is ruled out by the exceedingly low surface temperature, which must slow chemical reactions to unproductive rates.

All the life that we know about depends on molecules that contain carbon and hydrogen atoms, and such hydrocarbon molecules have been found in Titan's atmosphere. The chemical study of these compounds is known as organic chemistry – but it has nothing to do with organic foods, which are those grown without artificial fertilizers. And on Titan the organic chemistry is going on without concurrent life.

So Titan is a frozen moon that resembles the early Earth in a deep freeze. About seven billion years from now, the Sun will near the end of its life and swell up to become a bright giant star. The intense heat from the aged and swollen Sun will warm Titan's surface and may bring it to life. The moon could become an oasis of liquid water and organic chemicals, ready to initiate life or to serve as a haven for interplanetary immigrants.

haze indicate that different regions of the surface reflect radar by varying amounts, so any liquid would have to be pooled in lakes or small seas rather than in a homogeneous covering. Titan's thick orange smog is also transparent enough at infrared wavelengths to allow mapping of the surface. Images obtained with the *Hubble Space Telescope* indicate an inhomogeneous landscape with bright and dark features that reflect infrared radiation by different amounts (Fig. 10.17). They could be continents and oceans, but no one knows for sure. We might find out when the *Cassini* spacecraft arrives at Saturn in July 2004, and parachutes the *Huygens Probe* down to Titan's surface (Focus 10.2).

Saturn's medium-sized icy moons

Saturn has only one large satellite, Titan, comparable in size to Jupiter's four Galilean satellites, but it has an extensive family of smaller moons, including six mid-sized

Focus 10.2 The *Cassini* mission to Saturn

The *Cassini* spacecraft was launched on its seven-year journey to Saturn on 15 October 1997, with arrival expected in July 2004. It consists of an orbiter, that will spend the next four years taking data while in orbit around Saturn, and the *Huygens Probe* that will be parachuted into the hazy, dense atmosphere of Saturn's intriguing moon Titan, determining the properties of its Earth-like atmosphere and its mysterious surface below.

The *Cassini Orbiter* carries a sophisticated set of imaging, spectroscopic, photometric and field and particle instruments that will study the planet's atmosphere, magnetic environment, rings, Titan, and icy satellites.

The orbiting spacecraft contains radar equipment that will look right through Titan's smog and map its surface. Spectrometers will determine the detailed composition of Titan's atmosphere. In addition to about 40 planned flybys of Titan, the *Cassini Orbiter* will have approximately six close flybys of Saturn's medium-sized icy satellites, obtaining detailed images of their surfaces. They could determine if ice volcanoes now exist on Enceladus, and help us understand the mysterious dark face of Iapetus.

Perhaps the most anticipated results will come from the *Huygens Probe.* If all goes well, the probe will be parachuted into Titan's murky atmosphere in late-November or December 2004. Descending slowly through the atmosphere for almost three hours, the probe will measure the clouds, gaseous molecules, smog particles, and winds around it. A sensitive camera will record the scene below at different wavelengths, returning images of the surface on the way down. The probe could survive landing, and make additional measurements at the surface. It is even designed to float on the liquid ethane and methane that are almost certainly present somewhere on the surface.

Imaging instruments on the orbiter will study Saturn's energetic winds and high-flying clouds. Radio signals will be sent through the atmosphere to examine its vertical structure, and gravitational and magnetic measurements will reveal more about Saturn's interior.

The rings will also be examined by sending radio signals through them to Earth, or by looking at stars that pass behind them. Orbiter instruments will also scrutinize the dark stuff in the enigmatic spokes, determining what holds them together. They will additionally determine the positions of the small inner moons, discovered at the time of *Voyager* encounters in 1980 and 1981, to see how their orbits have been altered by generating ring waves. The high-resolution images will almost certainly reveal new, previously unknown moons that help shape and sustain the rings.

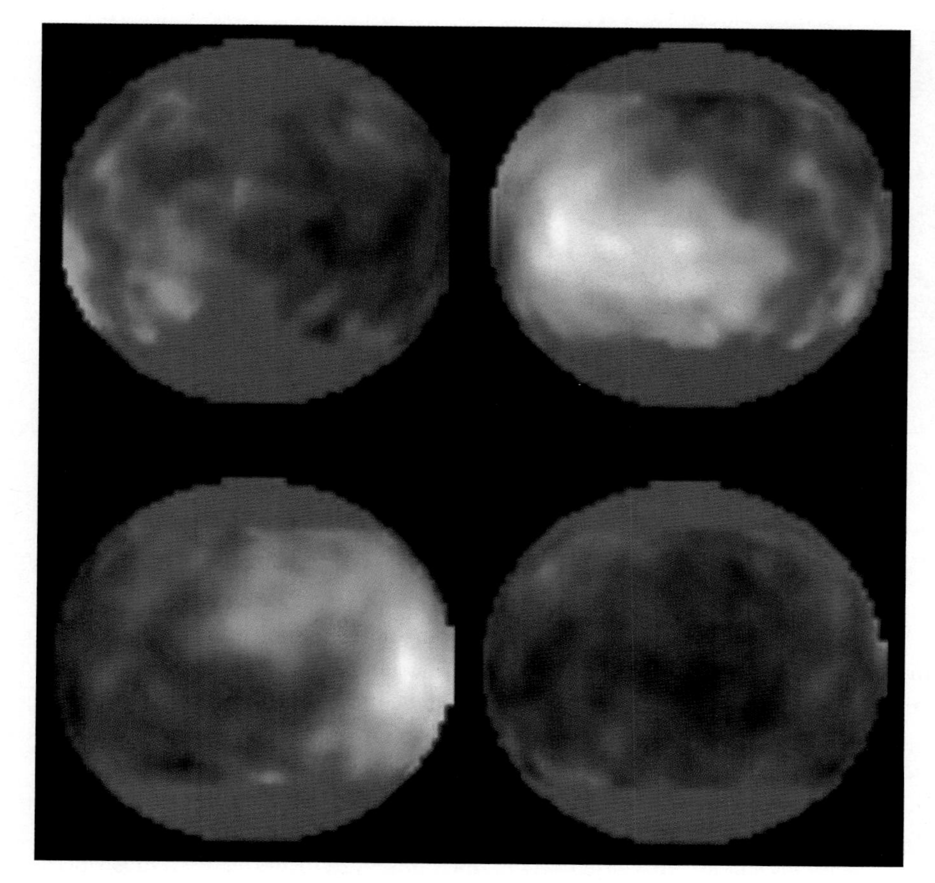

Fig. 10.17 Surface of Titan The obscuring haze in Titan's atmosphere is transparent enough at some infrared wavelengths to allow rough mapping of its surface. Dark and bright surface features are seen in these infrared images, taken from the *Hubble Space Telescope* at wavelengths near a millionth, or 10^{-6}, of a meter as Titan circled Saturn in its 16-day orbit. The bright area is about 4 thousand kilometers across, roughly the size of Australia, suggesting that Titan has regions that are elevated above its possible oceans, but it is not certain what the landforms represent. (Courtesy of NASA and STSI.)

Table 10.4 Properties of Saturn's largest moons[a]

Name	Mean Distance from Saturn (Saturn radii)	Orbital Period (days)	Mean Radius (km)	Mass (10^{21} kg)	Mean Mass Density (kg m^{-3})
Mimas	3.08	0.942	199	0.04	1142 ± 21
Enceladus	3.95	1.370	249	0.08	1000 ± 30
Tethys	4.88	1.888	530	0.76	1006 ± 11
Dione	6.26	2.737	559	1.05	1498 ± 40
Rhea	8.73	4.518	764	2.5	1236 ± 38
Titan	20.22	15.945	2575	134.57	1881 ± 4
Hyperion	24.53	21.277	$205 \times 130 \times 100$	–	1250
Iapetus	59.01	79.331	718	1.9	1025 ± 103

[a] The orbital distances are given in units of Saturn's radius, which is 60 268 kilometers, nearly ten Earth radii. The satellite radii are given in units of kilometers. By way of comparison, the mean radius of the Earth's Moon is 1738 kilometers. The mass is given in units of 10^{21} kilograms – our Moon's mass is 73.5 in these units, and the mass density is given in units of kilograms per cubic meter, abbreviated kg m^{-3}.

icy bodies that range from 196 to 765 kilometers in radius. In order of increasing orbital distance from Saturn, and also of increasing size, they are Mimas, Enceladus, Tethys, Dione, Rhea and Iapetus (Table 10.4). They were all discovered long ago (Section 1.5, Table 1.7) – four by Christiaan Huygens (1629–1695) and two by William Herschel (1738–1822).

The mean mass densities of these moons are low, between 1100 and 1400 kilograms per cubic meter, which suggests that they are mainly composed of pure water

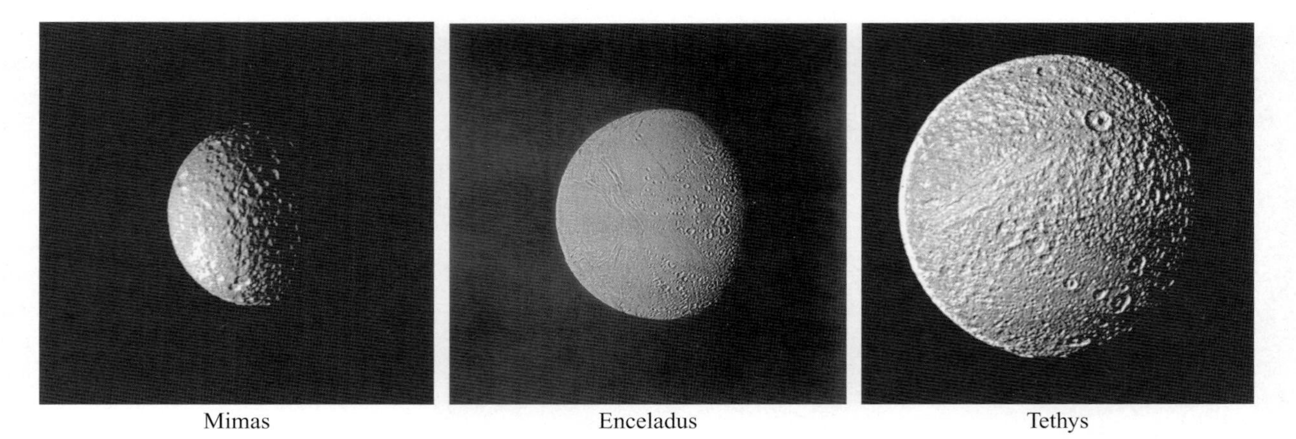

Fig. 10.18 Three icy moons of Saturn Cameras aboard the *Voyager 2* spacecraft obtained these pictures of three ice-covered moons of Saturn in 1980. They are Mimas (*left*), Enceladus (*center*) and Tethys (*right*) with respective radii of 199, 249 and 530 kilometers. The surfaces of all three satellites are covered with water ice. They contain the cratered scars of past impacts as well as relatively young, smooth regions. (Courtesy of NASA and JPL.)

ice. This is consistent with their highly reflective surfaces. With the exception of Iapetus, they all reflect more than 50 percent of the incident sunlight, and one of them, Enceladus, reflects almost 100 percent of the sunlight that strikes it. Water ice was also identified by infrared spectroscopy in the years prior to the *Voyager* missions.

The cold surfaces of all of Saturn's mid-sized icy satellites preserve ancient impact craters dating back to their formative years. Several of them also exhibit evidence for melting ice and active geology after their formation.

There are several ways to group these objects. Mimas, Rhea, and Iapetus, for example, have surfaces dominated by craters with no signs of internal activity. Or the moons can be grouped into pairs of approximately the same size – Mimas and Encleadus, Tethys and Dione, and Rhea and Iapetus. Here we will discuss each of them in the order of increasing distance from Saturn.

Saturn's innermost, medium-sized moon, Mimas, has a surface that is saturated with overlapping impact craters (Fig. 10.18 – *left*), including one crater that is about one-third the diameter of the moon itself. The impact that made this crater was nearly powerful enough to completely shatter Mimas.

Though of comparable size, Enceladus is a very different world from Mimas. Parts of the smooth, nearly crater-free surface of Enceladus (Fig. 10.18 – *center*) have been coated with fresh icy material that rose from the warm interior of the satellite. Other parts of the surface contain cracks and grooves, suggesting that internal stresses may have discharged liquid water that froze into smooth ice. As the satellite moves around its eccentric orbit, produced by Dione's gravitational tugs, tidal flexing by Saturn probably heats the interior of Enceladus, melting the water ice and permitting its eruption. Active ice volcanoes may be even now be erupting on Enceladus.

Tethys is about twice the size of Enceladus with nearly the same mean mass density, but Tethys is more akin to Mimas, with a large number of impact craters and one enormous impact that nearly broke the moon apart (Fig. 10.18 – *right*). A gigantic fracture covers three-quarters of the moon's circumference, suggesting internal activity early in its history. But the satellite shows no evidence for current activity.

Dione is nearly the same size as Tethys but denser, and shows a wide variety of surface features (Fig. 10.19). Next to Enceladus among Saturn's moons, it has the most extensive evidence for an active interior. It has enough

Fig. 10.19 The Saturnian moon Dione Many large impact craters (*left*) and bright regions (*right*) are found on Saturn's satellite Dione. The bright areas could be attributed to impact debris, ridges and valleys, or even surface frost deposits. This image was taken from the *Voyager 1* spacecraft on 12 November 1980.

Fig. 10.20 Saturn's moon Rhea The icy, cratered surface of Saturn's satellite Rhea is shown in this *Voyager 1* image, taken on 12 November 1980. The craters and landscape resemble those on Mercury and the Earth's Moon. Rhea probably froze and became rigid, behaving like a rocky surface, very early in its history. (Courtesy of NASA and JPL.)

Fig. 10.21 Two-faced Iapetus Saturn's outermost large moon, Iapetus, has a bright, heavily cratered terrain and a dark landscape, as shown in this *Voyager 2* image taken on 22 August 1981. The bright regions are probably made of water ice, and the dark substance may be composed of organic substances. Iapetus apparently plows through the dark material, covering the leading, frontal hemisphere of the satellite, as it orbits Saturn. (Courtesy of NASA and JPL.)

rocky material in its makeup to produce internal heat from natural radioactivity. Most of the surface is heavily cratered, but differences in the number of craters within various regions indicate that several periods of resurfacing occurred during the first billion years of Dione's existence. Bright, wispy streaks, which stand out against an already-bright surface, are believed to be the result of internal heat and subsequent flows of erupting material.

The surface of Rhea is completely saturated with impact craters (Fig. 10.20). It appears to be a dead world, geologically inert and without signs of internal heat. Yet, it is the largest of Saturn's icy moons. Compression resulting from its greater size and mass may have closed any volcanic vents, shutting off the outward flow of warm material.

Curiously, early astronomers could only observe the icy moon, Iapetus, on one side of Saturn. The satellite seemed to disappear when its orbit carried it to the other side of the planet. The reason for this strange behavior is that Iapetus is a divided world; half its surface is as bright as ice, and the other is as dark as asphalt or coal and is thought to contain complex organic compounds (Fig. 10.21). Like the Earth's Moon, the satellite Iapetus keeps one side toward its planet, and as it revolves around Saturn the bright and dark parts are successively turned toward the Earth. When the dark half is pointed at the Earth, the moon becomes very difficult to observe.

The face that Iapetus puts forth as it moves forward in its orbit, called its leading side, is coated with dark material, while the trailing hemisphere is brighter. So one explanation for the moon's divided appearance is that Iapetus is sweeping up the dark stuff as it revolves around Saturn. The satellite Phoebe, in a more distant and retrograde orbit, has been suggested as a possible source of infalling material, though the dark part of Iapetus is redder than Phoebe is. An alternative explanation is that the dark material originates in the satellite's interior. This would account for craters with dark floors, found on the trailing side of Iapetus. Perhaps they contain methane ice that has become darkened by solar ultraviolet radiation.

Small, irregularly shaped satellites of Saturn

Instruments on the *Voyager 1* and *2* spacecraft have discovered a host of small, irregularly shaped satellites that reside within the inner parts of Saturn's satellite system.

They are all bright objects, probably composed of ice, and many of them have orbits that are remarkable in one way or another.

Six of these tiny moons are associated with the rings: Pan, Atlas, Pandora, Prometheus, Janus and Epimetheus. Pan disturbs particles in the A ring to form the Encke division. Pandora and Prometheus shepherd the F ring; Atlas shepherds the outer margin of the A ring.

Saturn's two co-orbital satellites, Janus and Epimetheus, are even more bizarre. Janus and Epimetheus move in almost identical orbits. The satellite on the inner orbit that is closest to Saturn moves slightly faster, overtaking the outer satellite every four years. But the bodies' diameters are greater than the distance between their orbital paths, so they cannot pass without some fancy pirouetting. They avoid a collision at the last moment by gravitationally exchanging energy and switching orbits. The inner one is pulled by the outer one and raised into the outer orbit, and *vice versa*. They then move apart, only to repeat this *pas-de-deux* four years later, and exchange again.

Three so-called Lagrangian satellites move along the orbits of Saturn's larger satellites Tethys and Dione. The satellite Tethys shares its orbit with two small companions, Telesto and Calypso, one about 60 degrees ahead and the other about 60 degrees behind. These two positions are locations within an object's orbit in which a less massive body can move in an identical stable orbit, first specified by the Italian-born French mathematician Joseph Louis (Giuseppe Lodovico)Lagrange (1736–1813) in the 19th century. One additional miniature moon, named Helene, shares Dione's orbit, leading it by 60 degrees.

Fig. 10.22 Leaving Saturn When the *Voyager 1* spacecraft sped past Saturn on 16 November 1980, it looked back to take this picture of the ringed planet from a perspective that cannot be enjoyed from Earth. Saturn's shadow falls upon the rings, and the planet's bright crescent can be seen through the dark Cassini's Division and other parts of the rings. (Courtesy of NASA and JPL.)

The curious small moon Hyperion, discovered in 1848, has an irregular, flattened shape, and it moves in an eccentric orbit just outside that of Titan. This combination produces a chaotic tumbling motion – subject to fits and starts. This random, rotational speeding up and slowing down is explained by a chaotic interaction resulting from tidal forces during close passages to nearby Titan.

This concludes our survey of Saturn, the most distant planet known to the ancients (Fig. 10.22). We will now travel out beyond this enchanting world to the next wanderer, Uranus.

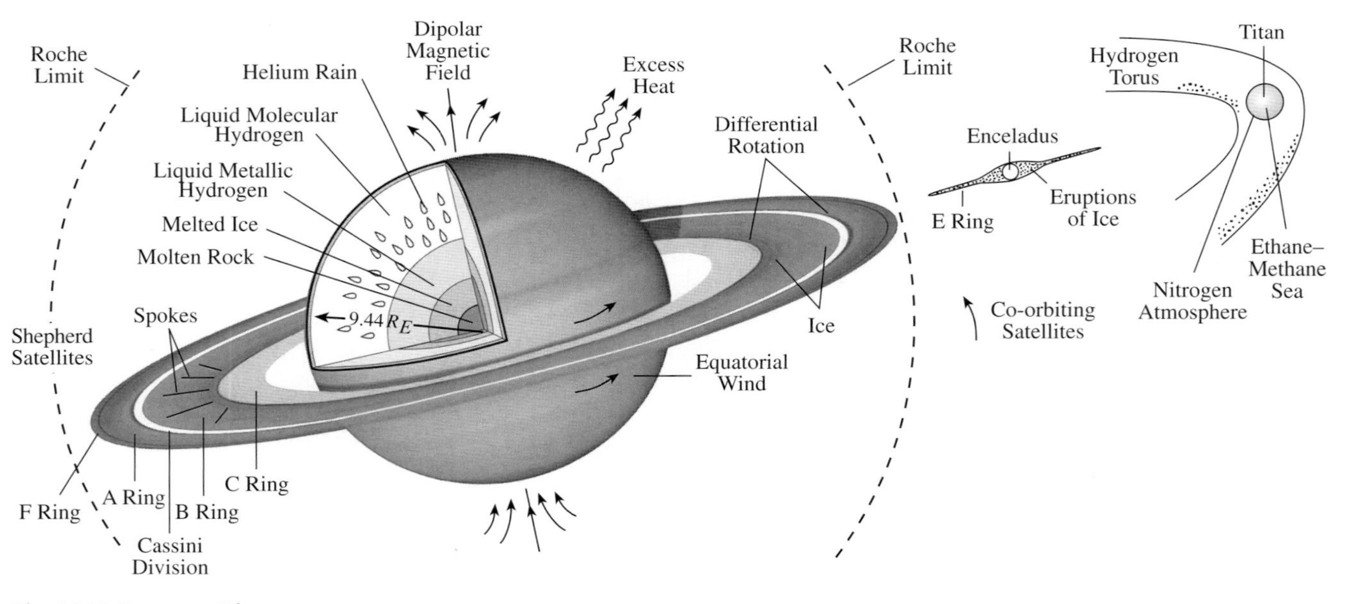

Fig. 10.23 Summary Diagram.

11 Uranus and Neptune

- Uranus and Neptune have a similar size, mass, composition and rotation, and these physical properties lie between those of the Earth and the giants Jupiter and Saturn.
- In contrast to all the other planets in the solar system, Uranus is tipped sideways so its spin axis lies nearly within the planet's orbital plane; Uranus also rotates in the opposite direction to that of most of the other planets.
- Although Uranus apparently has no strong internal source of heat, Neptune is so hot inside that it radiates 2.7 times as much energy as it receives from the Sun.
- During the *Voyager 2* encounter, Uranus had one pole pointing toward the Sun, but the cloud-top temperatures of its dark, winter pole were no colder than its sunlit, summer one.
- The cloud bands and winds on Uranus blow parallel to the planet's equator, apparently controlled by the planet's rapid spin rather than by direct heating from the Sun.
- Neptune's atmosphere is surprisingly active and dynamic with a large storm system and high-speed winds that may be driven by internal heat.
- The raging winds on Neptune are second only to Saturn's in speed, but blow westward at the equator and eastward at both poles.
- Uranus and Neptune consist mainly of vast internal oceans of water, methane and ammonia "ices", melted at the high temperatures inside.
- Unlike Jupiter and Saturn, there is no liquid metallic hydrogen inside Uranus and Neptune.
- The magnetic fields of Uranus and Neptune are askew, tilted by large amounts from their rotation axes; they could be generated by rotation-driven currents in ionized water.
- The austere, skeletal rings of Uranus are very narrow and widely spaced from each other, and made of very dark material.
- The rings around Uranus are not quite circular, do not lie exactly in Uranus's equatorial plane, and vary in width; these irregularities are attributed to the gravitational interaction of ring particles with small nearby moons.
- The material in one narrow ring around Neptune has been concentrated into three clumps, probably by the gravity of a nearby moon.

- The sparse rings around Neptune contain no more material than that found in a single small moon only a kilometer across.
- It is now thought that all the planetary rings are younger than the age of the solar system, that they are not permanent features dating back to its origin, and that they may be only a fleeting stage in a continuing saga of creation and loss.
- Most of the planetary rings that we see now will be ground into dust by collisions and meteoritic bombardment in a few hundred million years, eventually being consumed by their central planet and vanishing from sight. But the rings can easily be replaced by debris blasted off small moons already embedded in them.
- The five major moons of Uranus are darker, denser, and rockier than Saturn's mid-sized satellites, with icy surfaces that contain impact craters, huge fractures and volcanic flows of water ice.
- Miranda, the innermost mid-sized satellite of Uranus, exhibits a bizarre variety of surface features. It may have been shattered by a catastrophic collision and reassembled, or perhaps it became frozen in an embryonic stage of growth.
- Neptune's largest satellite, Triton, revolves about the planet in a direction opposite to that in which Neptune spins.
- Triton has a very tenuous, nitrogen-rich atmosphere, bright polar caps of nitrogen and methane ice, frozen lakes flooded by past volcanoes of ice, and towering geysers that may now be erupting on its surface.
- Triton may have formed in orbit around the Sun and was subsequently captured by Neptune, whose tidal forces kept Triton molten for much of its early history and are now pulling the satellite toward a future collision with the planet.

11.1 Fundamentals

Planetary twins

Saturn was the most distant planet known to the ancients. Uranus and Neptune are both so far away, and so faintly illuminated by the Sun, that telescopes were required to discover them. Uranus was discovered in 1781 by William Herschel (1738–1822) during his telescopic survey of the heavens, and Neptune was found in 1846, as the result of a mathematical prediction based on its gravitational effect on the motion of Uranus (Section 1.2).

Uranus is about 19 times as far away from the Sun as the Earth is, and Neptune is about 30 times as distant. As a result, it takes 84.0 Earth years for Uranus to complete one revolution about the Sun and nearly twice that for Neptune. With an orbital period of 164.8 Earth years, Neptune has not yet completed its first full orbit since discovery.

These two distant planets remain little more than dim, fuzzy spots of light in even the most powerful telescope. From the Earth, the planet Uranus subtends an angle of just 3.5 seconds of arc, and Neptune 2.0. Since the Earth's atmosphere blurs features smaller than about 1.0 second of arc, ground-based observers can distinguish nothing in the outer atmospheres of Uranus and Neptune.

One can still infer enough about Uranus and Neptune from telescopic observations to know that they have similar physical properties. The size, mass, composition and rotation of Uranus and Neptune are in fact so similar that they are often called planetary twins (Table 11.1). These parameters lie between those of the Earth and the giants Jupiter and Saturn.

Both Uranus and Neptune are intermediate in size as far as planets go, four times bigger than the Earth but less than half the size of Jupiter or Saturn. Their mass has been inferred from the orbital periods and distances of their largest satellites. They both weigh about the same as the ice–rock cores of Jupiter or Saturn, just 14.53 and 17.14 times the Earth's mass for Uranus and Neptune, respectively.

Table 11.1 Some comparisons of Uranus and Neptune[a]

	Uranus	Neptune
Mass (Earth masses)	14.535	17.141
Equatorial Radius at 1 bar (Earth radii)	4.007	3.883
Mean Mass Density	1318	1638
Rotation Period (hours)	17.24	16.11
Orbital Period (Earth years)	84.01	165.8
Mean Distance from Sun (AU)	19.19	30.06
Atmosphere	83 percent hydrogen	79 percent hydrogen
	15 percent helium	18 percent helium
Energy Balance	less than 1.4	2.7 ± 0.3
Effective Temperature	59.3 kelvin	59.3 kelvin
Temperature at 1-bar level	76 kelvin	73 kelvin
Central Temperature	5000 kelvin	5000 kelvin
Magnetic Dipole Moment.	$50 D_E$	$25 D_E$
Equatorial Magnetic Field Strength	0.23×10^{-4} tesla	0.14×10^{-4} tesla

[a] The Earth's mass is 59.736×10^{23} kilograms, the Earth's equatorial radius is 6378 kilometers, the astronomical unit, denoted AU, is the mean distance between the Earth and the Sun with a value of 1.496×10^{11} meters. The energy balance is the ratio of total radiated energy to the total energy absorbed from sunlight, the effective temperature is the temperature of a black body that would radiate the same amount of energy per unit area, a pressure of one bar is equal to the atmospheric pressure at sea level on Earth, and D_E is the magnetic dipole moment of the Earth.

From each planet's mass and size, we calculate its mean mass density, which is intermediate between the low-density giants, Jupiter and Saturn, and the dense rocky Earth or Mars. The bulk of Uranus and Neptune must therefore be composed of something less dense than rock, but more substantial than the hydrogen and helium that dominate the composition of Jupiter and Saturn. The main ingredient of Uranus and Neptune is probably liquid water and other melted ice, with no solid surface.

The blue-green color of Uranus and the deeper blue of Neptune are attributed to methane. At the low temperatures prevailing at their cloud tops, the methane condenses to form a top layer of clouds made of methane-ice crystals. And since methane absorbs red light quite strongly, the sunlight reflected off the clouds of Uranus and Neptune has a blue color.

The blue cloud deck of methane ice forms where the atmospheric pressure is about one bar, the same as the air pressure at sea level on Earth. At this level, the equatorial radius of Uranus is 25 559 kilometers while that of Neptune is 24 766 kilometers. Rapid internal rotation produces an equatorial bulge on both planets; the polar radius is 586 kilometers shorter on Uranus and 424 kilometers on Neptune.

The rotation periods of Uranus and Neptune lie between the roughly 10-hour day of Jupiter and Saturn and the 24-hour rotation period of Earth and Mars. Periodic variations in the radio emission of Uranus and Neptune indicate that their magnetic fields, which are anchored deep inside the planets, rotate once every 17.24 hours and 16.11 hours, respectively.

Thus, Uranus and Neptune bear an uncanny resemblance to each other, but there are two remarkable differences. Uranus is tipped sideways and has no significant heat inside, while Neptune is more upright and has a lot of internal energy.

Uranus is tipped on its side and has no strong source of internal heat

Blue-green Uranus lies sideways, with its poles where its equator should be. This knowledge comes not from watching the small, featureless ball rotate, but instead from observing the orbits of its major moons. The orbits are all circular, and they lie in one plane, which is turned at right angles to the plane of Uranus' orbital motion about the Sun. As a result, the moons form a bull's-eye pattern,

Fig. 11.1 Uranus and its rings Thin, spidery rings encircle the methane-rich atmosphere of Uranus in this artist's rendition. The planet is tipped on its side, so its equator, rings and direction of rotation are almost perpendicular the plane of its orbital motion around the Sun. The north pole of Uranus points toward the east (*right*), and toward the Sun during one-quarter of its 84-year orbit around the star. In contrast, the equatorial planes of all the other planets lie near their orbital planes. (Courtesy of NASA.)

revolving around Uranus like a Ferris wheel. Since these satellites should be orbiting within the plane of Uranus' equator, the entire planet has to be tipped on its side (Fig. 11.1). One speculation is that Uranus was knocked sideways during a massive collision, perhaps when the planet was still forming.

The equatorial plane of Uranus is inclined 97.9 degrees from its orbital plane, with a tilt that is just a bit more than a right angle. So Uranus spins in the opposite, retrograde direction from most of the other planets.

Because the rotational axis of Uranus lies near its orbital plane, first one pole and then the other points toward the Sun as the planet slowly progresses around its orbit. Each pole faces the Sun for 21 Earth years, with a corresponding period of darkness at the opposite pole. When either pole points at the Sun, we see the moons moving in circles around Uranus. Between the long summer and winter at each pole, the equator turns toward the Sun, and we observe the moons traveling vertically up and down as they move around the equator.

When *Voyager 2* arrived at Uranus, on 24 January 1986, the spacecraft's infrared detectors found that the planet is radiating about as much energy as it receives from the Sun. This means that Uranus lacks a strong internal heat source, in contrast to Jupiter and Saturn that produce heat in their centers. These two giants each radiate away about twice as much energy as they receive from the Sun. But like Jupiter and Saturn, and unlike Uranus, the planet Neptune glows in the infrared with its own internal heat, discovered when the hardy spacecraft encountered Neptune on 24 August 1989, twelve years after launch.

The temperature at Neptune's cloud tops is 59.3 kelvin, about the same as that of Uranus. But since Neptune is 50 percent further from the Sun, it should have been a lot colder than Uranus; the temperature that would result from sunlight alone is 46 kelvin at the cloud tops of Neptune. The hotter measured temperature implies that Neptune radiates 2.7 times as much energy as it absorbs from the Sun, and its outer atmosphere must therefore receive energy from the interior.

This makes Uranus unique among the giant planets in having no strong internal heat source that warms its outer atmosphere today. Perhaps Uranus lost its internal heat during the collision that knocked the planet on its side. Or maybe it is still hot inside, with an interior that is insulated from the outside in some, as yet unknown, manner.

11.2 Storm clouds on the outer giants

Mild weather on Uranus

At the low temperatures prevailing near the top of Uranus' atmosphere, methane gas freezes and forms a methane cloud deck; haze particles are also formed there due to the action of ultraviolet sunlight on methane. The haze and methane clouds hide the lower atmosphere from view. Ammonia and water clouds probably form deeper in the atmosphere and are difficult to see. On warmer Jupiter and Saturn the topmost light-colored clouds that we see are composed of ammonia-ice crystals rather than frozen methane.

With no appreciable heat rising from the interior to drive the weather system, Uranus presents a dull and placid face to the world. In addition, a cold and hazy atmosphere obscures our view. Even at close range, looking at Uranus is something like gazing down into the depths of a vast sea.

Yet, some zonal banding was extracted from the *Voyager 2* images (Fig. 11.2). The clouds are arranged in bands that circle the planet's rotation axis, running at constant latitudes parallel to the equator like the more vivid bands seen at Jupiter and Saturn. The features at different latitudes on Uranus move in the same east–west direction as the planet rotates, but at faster speeds. The difference is greatest at high latitudes, where the clouds circle the poles in 14 hours, and it gets progressively smaller toward the equator, closer to the internal rotation period of 17.24 hours.

Since the high-latitude clouds are rotating faster than the interior of Uranus, the clouds cannot be simply carried by the planet's rotation. They are being blown by winds in the same direction as the planet rotates, just as clouds on Earth, Jupiter and Saturn are. But unlike these planets, the winds on Uranus are primarily determined by the planet's rotation instead of either solar or internal heating.

The long, alternating periods of sunlight and darkness have little effect on the winds of Uranus. During the *Voyager 2* encounter, the south pole of Uranus was facing the Sun almost directly. The equator was in constant twilight, and the north pole had been in darkness for 20 years. So you might expect the south pole to be the warmest place on the planet and the north pole the coldest, with a temperature difference that would drive winds from pole to pole. But the thin clouds on Uranus move parallel to the equator, orthogonal to the expected poleward direction.

The *Voyager* thermometer showed that the atmospheric temperatures are much the same everywhere on Uranus. The temperature at the pole facing the Sun and the equator were about the same, and the dark winter side may have even been a few degrees hotter. So something has to be redistributing the solar heat, exchanging it between warm and cold places.

Stormy weather on Neptune

The *Voyager 2* flyby in 1989 forever changed our view of Neptune's weather. Despite its great distance from the Sun, the dimly lit atmosphere of Neptune is one of the most turbulent in the solar system, with violent winds, large

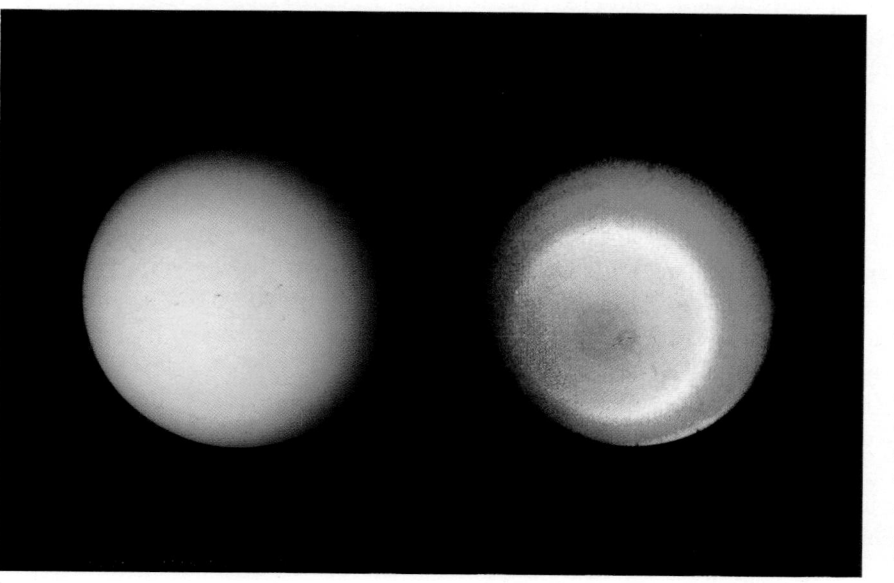

Fig. 11.2 Uranus' clear atmosphere When viewed through a telescope, Uranus is a clear, featureless ball (*left*). Its blue-green color results from the absorption of red light by methane gas in the deep, cold atmosphere. An exaggerated false-color image (*right*) brings out subtle details in the polar regions of Uranus. A thin reddish-brown haze obscures the south pole, perhaps due to chemical reactions of ultraviolet sunlight with methane. Progressively lighter concentric bands may be attributed to zonal motions in the upper atmosphere, which circulates in the same direction as the planet rotates. These *Voyager 2* images were taken on 17 January 1986. (Courtesy of NASA and JPL.)

Fig. 11.3 Neptune's dynamic atmosphere White clouds are seen overlying the swirling Great Dark Spot (*right*), and raging winds reach speeds of 300 meters per second, creating a global banding in the atmosphere of Neptune. The faint sunlight at Neptune's great distance cannot provide the energy of such winds; they are probably energized by heat from the interior of the planet. This image was sent from the *Voyager 2* spacecraft on 14 August 1989, after a 12-year journey to the planet, using a 20-watt transmitter with less power than an ordinary light bulb. Traveling at the speed of light, the signals took more than 4 hours to reach Earth. (Courtesy of NASA and JPL.)

dark storms and high-altitude white clouds that come and go at different places and times (Fig. 11.3).

Neptune has strong zonal winds driven and defined by the planet's rotation. The clouds near the equator circulate slower than Neptune's interior rotates so the prevailing winds blow in the westward direction, opposite to the rotation of the planet. The equatorial wind speed on Neptune is about 325 meters per second relative to the core, almost as fast as Saturn's equatorial wind and faster than those on Jupiter or Uranus.

The wind pattern on Neptune lacks Jupiter's multiple zonal winds that flow in opposite directions. Neptune has just one westward air current at low latitudes, like the Earth's trade winds, and one meandering eastward current at mid-latitudes in each hemisphere, resembling the Earth's jet streams. And like Uranus, the polar and equatorial temperatures on Neptune are nearly equal.

You wouldn't want to forecast the lively, variable and unpredictable weather on Neptune. When the *Hubble Space Telescope* took another look at the planet, in 1994 to 1996, the violent storms seen by the *Voyager 2* cameras had vanished without a trace, and other storms had appeared (Fig. 11.4).

The largest dark storms on Neptune are probably high-pressure systems that come and go with atmospheric circulation. The most prominent one was the Great Dark Spot, a vast, circulating storm almost as large as Earth (Fig. 11.5). It is called the Great Dark Spot because it resembles the Great Red Spot of Jupiter. Both storms are found in the planetary tropics at about one-quarter of the way from the equator to the south pole, both rotate counter-clockwise, in the direction of high-pressure anticyclones, and both are about the same size relative to their planet. As on Jupiter, some of the small dark spots on Neptune may be whirling in the opposite direction to the bigger one, perhaps indicating that they are little cyclones with descending material at their centers (Fig. 11.6).

There are some important differences between the two Great Spots. The Jovian Great Red Spot has survived for centuries, while Neptune's dark one disappeared from view within a few years of its sighting from *Voyager 2*. And Jupiter's whirling storm lies above the clouds while Neptune's seems to form a deep well in the atmosphere, providing a window-like opening to the deeper, darker clouds below.

White, fleecy cirrus-like clouds cast shadows on the blue cloud deck below, indicating that they are high-altitude condensation clouds that rise about 100 kilometers above the surrounding ones. They form as atmospheric gas flows up, over and around the storm center, without being consumed by it. When the rising methane gas cools, it forms white clouds, fashioned from crystals of frozen methane. Water in the Earth's atmosphere freezes in a similar way into ice crystals that form cirrus clouds. When strong upwelling carries the wispy white clouds to great heights in Neptune's atmosphere, they are sheared out, producing anvil-like shapes similar to those observed in terrestrial thunderstorms.

Although the global wind pattern on Neptune resembles the Earth's trade winds and jet streams, they cannot be energized in the same way. Solar heating of the atmosphere and oceans drives the terrestrial winds. At Neptune's distance, the Sun is 900 times dimmer than at the Earth's distance and the winds should be correspondingly weaker if they are driven by the feeble sunlight. The fast winds on Neptune and the planet's complex stormy weather must instead be energized by heat generated in the planet's core. The internal heat warms Neptune from the inside out, producing convecting currents of rising and falling material, somewhat like a pot of boiling water on a stove. Uranus, on the other hand, shows no signs of

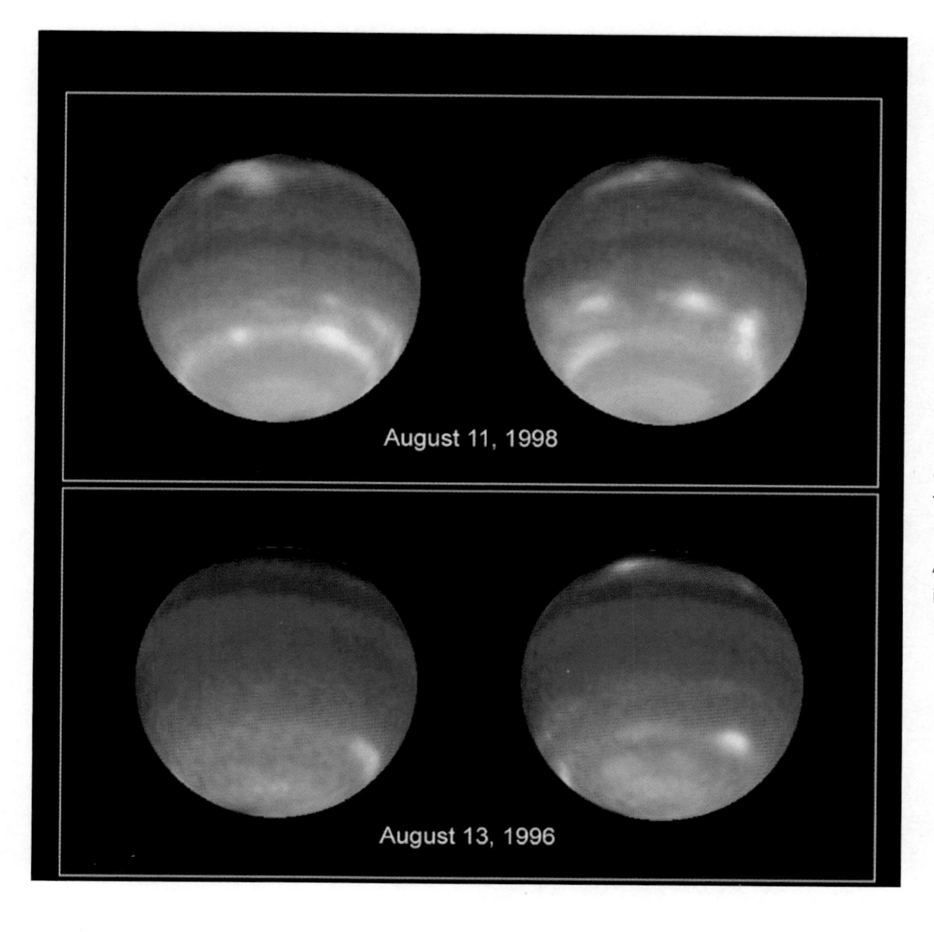

Fig. 11.4 Neptune's stormy disposition The weather on opposite hemispheres of Neptune is recorded in these images that combine simultaneous observations made with the *Hubble Space Telescope* and NASA's Infrared Telescope Facility on Mauna Kea, Hawaii. The predominant blue color of the planet is a result of absorption of red and infrared light by Neptune's methane atmosphere. Clouds elevated above most of the methane absorption appear white. Neptune's powerful jet stream, where winds blow at over 300 meters per second, is centered at the dark-blue belt near Neptune's equator. The Great Dark Spot detected when the *Voyager 2* spacecraft visited Neptune in August 1989 has completely disappeared in these images, taken seven year later. (Courtesy of NASA and STSI.)

substantial internal heating that rises to cloud level, and this may explain why its atmosphere is relatively benign and inactive.

11.3 Interiors and magnetic fields

Water worlds

The atmosphere above the cloud tops of Uranus and Neptune consists mainly of molecular and atomic hydrogen, warmed by the Sun's ultraviolet rays. The tenuous gas forms an extensive hydrogen corona around Uranus, but is held closer to the cloud tops above Neptune. The overwhelming abundance of hydrogen in the outer atmospheres of Uranus and Neptune resembles that in Jupiter, Saturn and the Sun.

Unlike Jupiter and Saturn, however, Uranus and Neptune cannot consist mostly of the lightest element hydrogen, or they would have lower mean mass densities than are inferred from observations. For their size, Uranus and Neptune are too massive for hydrogen to be their main ingredient, and their bulk must instead be composed of heavier abundant elements. To put it another way, both planets are too small for their mass to be mainly composed of hydrogen and helium, and must consist mainly of heavier material (Fig. 11.7).

Since Uranus and Neptune have similar masses, sizes, composition and rotation, their interiors are also expected to be alike. But they must be quite different from Jupiter and Saturn inside. The hydrogen in Uranus and Neptune is confined within a thin atmosphere and liquid molecular shell that do not extend to great depths and contribute only about 15 percent of the planetary mass (Fig. 11.8). These two planets do not have enough hydrogen, or sufficient mass and internal pressure, to squeeze the hydrogen into a metallic state. So there is no internal shell of liquid metallic hydrogen inside Uranus and Neptune.

Most of their interior probably consists of a vast internal ocean of water, H_2O, methane, CH_4, and ammonia, NH_3. Although customarily denoted as ices, since they would be frozen at the cloud tops of these planets, these substances are kept liquid by the high temperatures, up to 8000 kelvin, deep in the planetary interiors. These molecules will form from atoms of hydrogen, H, oxygen, O, carbon, C, and nitrogen, N, the most abundant heavy elements in the material from which the Sun and giant planets originated.

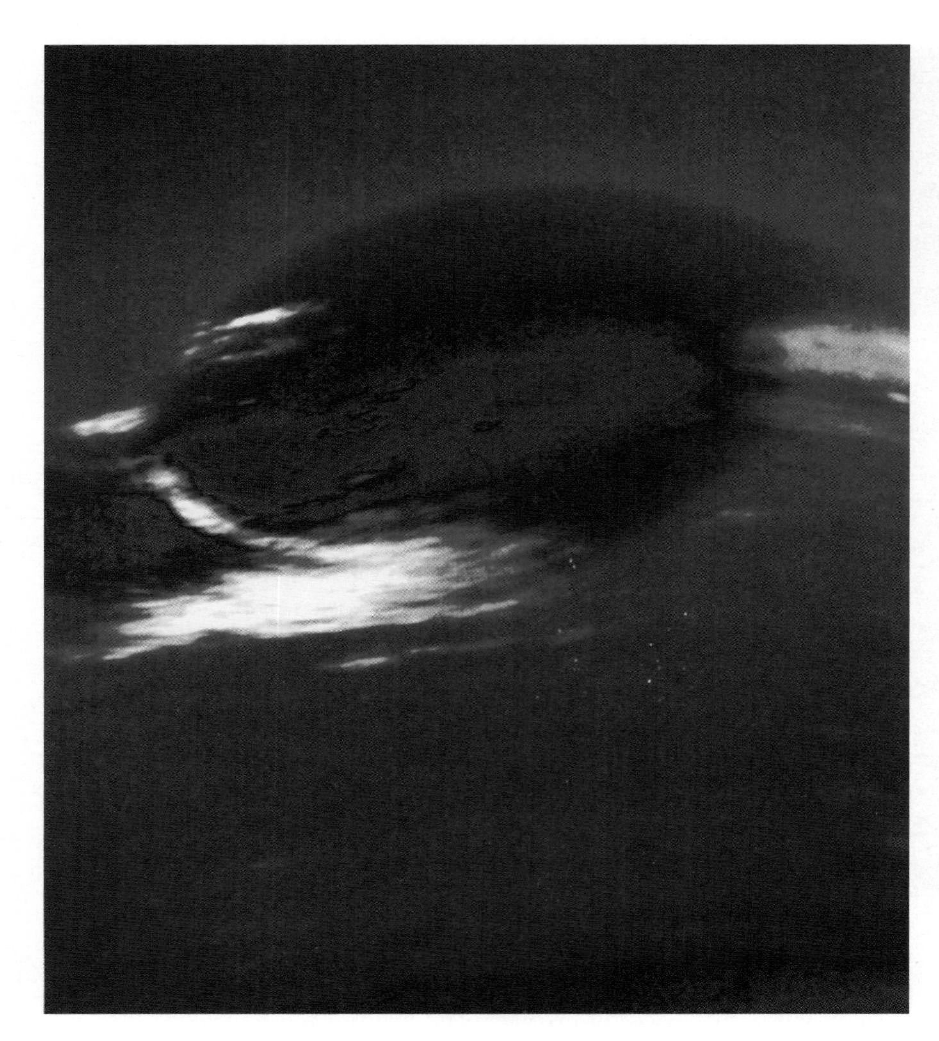

Fig. 11.5 Neptune's Great Dark Spot The Great Dark Spot of Neptune, rotating in an anti-cyclonic, counter-clockwise direction, is as large as the Earth, and about half as large as the Great Red Spot of Jupiter. Unlike the Red Spot, which has lasted for centuries, Neptune's Dark Spot vanished within a few years after the *Voyager 2* spacecraft took this image, in August 1989. (Courtesy of NASA and JPL.)

Uranus and Neptune are not unlike the cores of Jupiter and Saturn, which similarly contain 10 to 20 Earth masses of melted ice and molten rock. But Uranus and Neptune are almost all core, without the deep envelope of hydrogen and helium that make up most of the mass of Jupiter and Saturn. The differences between these four planets apparently derive primarily from the amounts of hydrogen and helium that they were able to attract and hold as they formed.

Why don't Uranus and Neptune contain extensive internal shells of hydrogen when Jupiter and Saturn do? All four of these planets must have formed in the colder outer reaches of the primeval solar nebula that enveloped the young Sun, where there was plenty of hydrogen around. According to one explanation, the cores of Jupiter and Saturn accreted, or gravitationally gathered in, the surrounding gas before it dissipated. The solar nebula was dispersed into a larger volume and lower density at the more distant orbits of Uranus and Neptune. So their cores of ice and rock accumulated slowly and took a longer time to grow. Little hydrogen was left to capture by the time they had grown large enough to start gravitationally collecting the surrounding gas. In another scenario, a blast of radiation from a luminous nearby star, other than the Sun, boiled away any hydrogen or helium that may have collected around Uranus and Neptune, while Jupiter and Saturn were protected by the greater density of the nearby material.

Tilted magnetic fields

Like the Earth, Jupiter and Saturn, both Uranus and Neptune have strong magnetic fields. But the resemblance ends there. Here on Earth our magnetic pole is very near our North Geographic Pole, which is very useful for navigation with a compass. The magnetic and rotational axes of Jupiter and Saturn are also closely aligned. But they are way off kilter on both Uranus and Neptune.

The magnetic axis is tipped by 58.6 degrees to the rotation axis of Uranus, and the two are tilted by 46.8 degrees for Neptune. By way of comparison, the displacement is

Fig. 11.6 Neptune's Small Dark Spot Winds swirl around this dark spot in Neptune's atmosphere, viewed from the *Voyager 2* spacecraft in August 1989. Bright, white methane-ice clouds stare out of its center like the pupil of an eye. The spot may be rotating in the cyclonic clockwise direction, opposite to that of the Great Dark Spot on Neptune and the Great Red Spot on Jupiter. If so the material in the dark oval was descending. (Courtesy of NASA and JPL.)

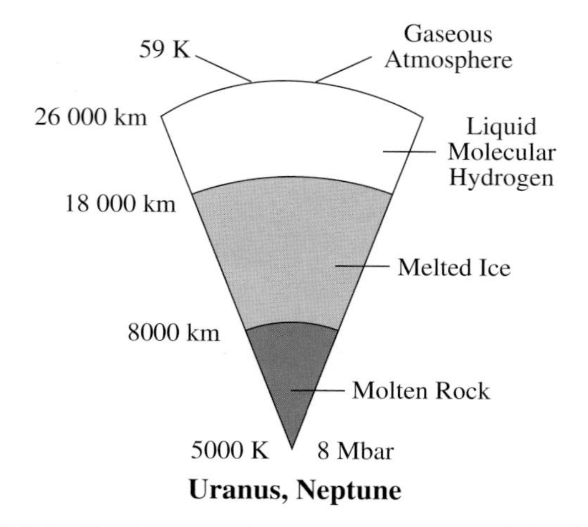

Fig. 11.8 Inside Uranus and Neptune An outer shell of liquid molecular hydrogen covers a thick inner shell of melted ice within the interior of both Uranus and Neptune. Because of their relatively low mass and hydrogen abundance, neither planet contains an inner shell of liquid metallic hydrogen, as Jupiter and Saturn do.

11.7 degrees on Earth, 9.6 degrees on Jupiter, and less than 1 degree on Saturn. If the disparity on Uranus and Neptune existed on Earth, a compass needle would point somewhere near Cairo, Egypt instead of the North Geographic Pole. Theoreticians expected a closer alignment between rotation and magnetism on Uranus and Neptune.

It is almost certain that the same dynamo process as that responsible for Earth's magnetic field generates the magnetic fields of Uranus and Neptune (Section 3.6). In this mechanism, swirling currents in a fluid conduct electricity, generating and sustaining a planet's magnetism. This happens in Earth's molten metallic core, and it occurs within the liquid metallic hydrogen shell inside Jupiter and Saturn. Unlike these two giants, there is no shell of liquid metallic hydrogen inside Uranus and Neptune, but electrical currents within their vast internal oceans might generate the magnetic fields. It is probable that the electrical conductivity within Uranus and Neptune is provided by water-rich material that has a conductivity that is about two orders of magnitude less than that of metals. It is also likely that this conductivity comes from protons, not electrons, within the ionized waters.

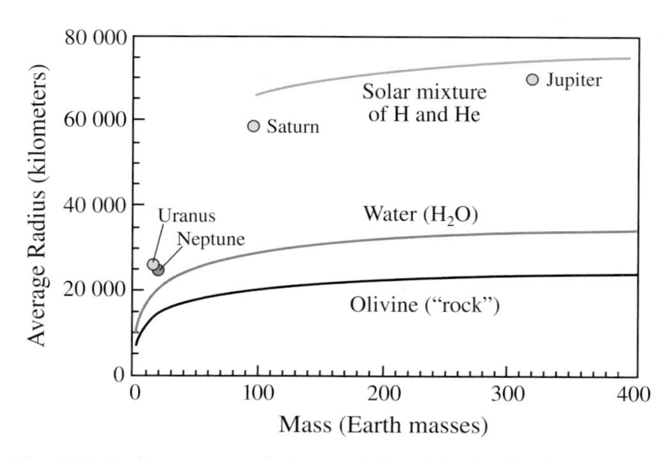

Fig. 11.7 Radius–mass relations A liquid body of solar composition describes a radius–mass relation (*top curve*) that approximates the mass and radius of Jupiter and Saturn. They consist mainly of the lightest element hydrogen, denoted by H, and next lightest abundant element helium, abbreviated He. Uranus and Neptune contain little hydrogen or helium, for their radii are much too small to be consistent with a solar composition. Instead, they lie only slightly above the radius–mass relation for liquid water, so they probably contain large quantities of water. For comparison purposes, the bottom curve is one calculated for a planet composed entirely of rock, as the Earth is.

11.4 Rings of Uranus and Neptune

Narrow, widely spaced rings around Uranus

Astronomers have had a history of happy accidents concerning Uranus, starting with William Herschel's (1738–1822) serendipitous discovery of the planet in 1781 (Section 1.2). Another lucky incident occurred on 10 March 1977, when the planet was scheduled to pass in front of a faint star. By observing such a stellar occultation, astronomers hoped to determine properties of the planet's atmosphere, and to accurately establish its size from the duration of the star's disappearance behind it.

Because of uncertainties in the predicted time of the star's disappearance, one telescope was set into action about 45 minutes early. Soon after the recording began, the starlight abruptly dimmed but then it almost immediately returned to normal, producing a brief dip in the recorded signal. At first, the dip was attributed to a wisp of cloud on Earth or to an unexpected change in the telescope's orientation. But the star blinked on and off several times before and after the planet covered it. Moreover, each dip on one side of Uranus was matched by another dip on the other side, at the same distance from the planet (Fig. 11.9). The symmetrical, brief dips indicated that Uranus is surrounded by a family of narrow rings that blocked out the star's light but could not be seen directly from the Earth.

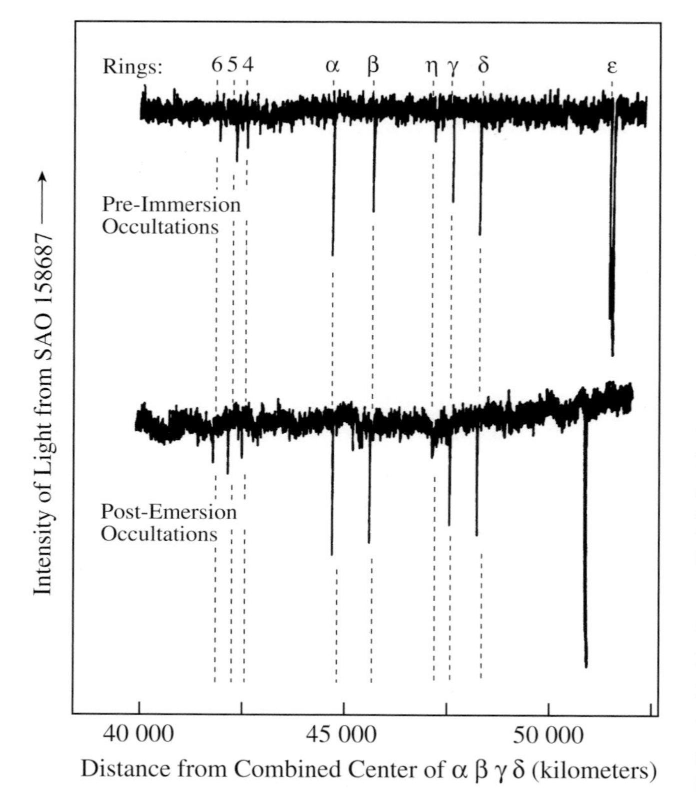

Fig. 11.9 Discovery of the thin rings of Uranus Astronomers recording the light of a star that was expected to disappear behind Uranus, on 10 March 1977, unexpectedly recorded short dips in the starlight before the star passed behind the planet (*top*). The same pattern was repeated when the star reappeared (*bottom*), indicating that Uranus is surrounded by narrow rings that briefly block out the light at the same distance on opposite sides of the planet. The strong and abrupt absorption of starlight indicates that the narrow rings are quite opaque and have well-defined edges. These observations were taken from high above the Indian Ocean aboard the Kuiper Airborne Observatory. (Courtesy of James L. Elliot.)

During the next few years, observations of more than 200 stellar occultations by Uranus revealed the details of nine narrow rings. In order of increasing distance from Uranus, the rings are named 6, 5, 4, α, β, η, γ, δ, and ε, following the differing notation of the discoverers. From the brief duration of the dips of blocked starlight, astronomers concluded that all but one of the individual rings could be no wider than 10 kilometers. The relatively long time between the dips indicated that the thread-like rings are separated by hundreds of kilometers of nearly empty space. These skeletal, web-like rings are unlike any seen before, all very narrow and widely spaced from each other.

The size and orbits of the rings could be established with extraordinary precision by the occultation technique. Features as small as one kilometer could be distinguished from the Earth by rapid recording of the changing starlight. In contrast, the best ground-based telescopic views of the rings might have resolutions of thousands of kilometers. Since the rings are much smaller than this, and separated by wide spaces of almost nothing, they are extremely difficult to see with even a large telescope on Earth.

When *Voyager 2* arrived at Uranus in 1986, nearly a decade after the discovery of its narrow rings, instruments on the spacecraft confirmed all the known rings, and added at least two more (Fig. 11.10). They found the λ ring, a narrow strand between the δ and ε rings, and another one interior to ring 6. The spacecraft also discovered at least ten small moons that are located just outside the ring system.

Both the ring particles and the nearby moons are very dark, quite unlike the bright particles and tiny moons found in Saturn's wide rings. In fact, the particles of Uranus' rings reflect only about two percent of the sunlight falling on them, making them as dark as charcoal. Most investigators agree that the material is dark because it is rich in carbon, like soot or carbon-enriched meteorites and asteroids.

Why are the rings of Uranus so dark? According to one hypothesis, their ancient surfaces contain methane ice that has been darkened by prolonged exposure to energetic electrons trapped within the planet's magnetic environment. In this explanation, Saturn's bright rings are either pure water ice without any carbon-bearing methane, or they are relatively young, exposed to electron bombardment for a comparatively short time. Alternately, the dark material may be of primordial origin, consisting of unaltered material that is already rich in carbon and quite different from the water-ice particles in Saturn's rings.

Fig. 11.10 Uranus' rings This *Voyager 2* image, taken on 23 January 1986, shows five of the nine rings of Uranus that had been previously inferred from Earth-based observations of their brief occultation of a star's light. In this view, sunlight striking the rings' particles was reflected back toward the camera, showing that the denser parts of the ring system consist of narrow rings with wide gaps. In contrast, Saturn's main rings are broad with narrow gaps. From bottom to top, the rings are designated by the Greek letters alpha, beta, eta, gamma and delta. The eta ring orbits Uranus at a distance of 47 000 kilometers from the planet centre and has a width of about 1.5 kilometers. (Courtesy of NASA and JPL.)

Fig. 11.11 Between the rings of Uranus *Voyager 2* took this image while in the shadow of Uranus and looking back at only a small angle away from the Sun. This made the fine dust particles between the rings appear very bright. About 100 very diffuse, nearly transparent bands of microscopic dust particles are seen surrounding the known narrow rings.The bright wide ring at the bottom lies at the distance of the lambda ring that is difficult to discern from the sunlit side of the rings. The lambda ring has a width of about 2 kilometers, lies just inside the epsilon ring, and is located at a distance of about 50 000 kilometers from the planet's center. The long, 96-second exposure produced streaks due to trailed stars. (Courtesy of NASA and JPL.)

The particles in the main narrow rings of Uranus are both dark and large. They range between a softball and an automobile in size, or between 0.1 and a few meters across. And they contain very few smaller particles in the millimeter to centimeter, or 0.001 to 0.01 meters, range, with surprisingly small amounts of micron-sized dust about 10^{-6} meters across.

Broad sheets of dust were nevertheless detected in the wide gaps between the rings. When *Voyager 2* entered Uranus' shadow and looked back at the rings, about 100 very diffuse, nearly transparent bands of microscopic dust were seen with sunlight streaming past them (Fig. 11.11). The dust is lit up when the Sun shines through the rings, in the same way that grime on a car's windshield becomes visible when struck by the lights of an oncoming car.

The slender rings are not perfectly circular and of uniform width. They are slightly out of round and thickened in places. The outermost epsilon, or ε, ring varies in width from 20 to 96 kilometers, increasing in thickness in proportion to its distance from Uranus. At least six of the rings around Uranus are also slightly inclined with respect to the planet's equatorial plane.

The irregular orientation and shapes of the Uranian rings are attributed to small moons that lie just outside them (Fig. 11.12). The repeated gravitational tugs of two of them, Cordelia and Ophelia, pull the epsilon ring into its oval shape and restrain its edges. These tiny moons flank the ring, controlling its shape in much the same way that the shepherd satellites, Pandora and Prometheus, constrain Saturn's F ring (Section 10.4). Nearby moons probably sharpen the edges of the other rings, keeping them from spreading out as the result of particle collisions, but many of the expected moons have not been found. They may have been too dark or too tiny for *Voyager*'s cameras to record.

Neptune's sparse thin rings and arcs

After the discovery of the rings of Uranus by watching a distant star pass behind the planet, astronomers hoped to

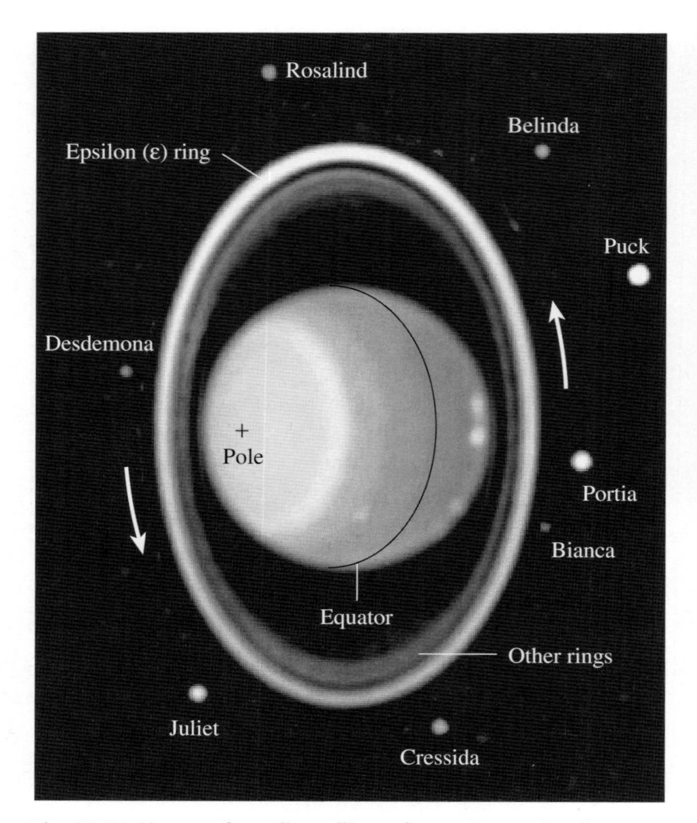

Fig. 11.12 Rings and small satellites of Uranus Eight of Saturn's small satellites circle the planet just outside its bright epsilon ring. This image, taken with the *Hubble Space Telescope* on 28 July 1997, is a false-color composite of three images taken at different infrared wavelengths in which Uranus appears relatively dim but the rings and moons do not. The satellites range in size from 40 kilometers across, for Bianca, to 150 kilometers for Puck. The arrows denote their direction of revolution about Uranus. White clouds are seen just above the planet's blue-green methane atmosphere. (Courtesy of NASA and STSI.)

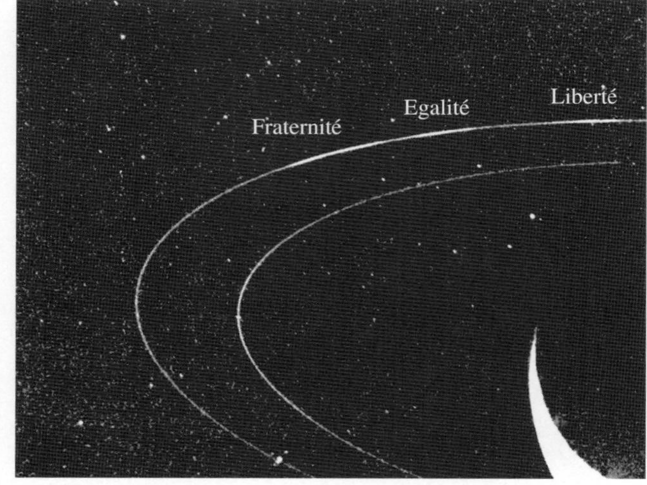

Fig. 11.13 Neptune's rings As *Voyager 2* left Neptune in August 1989, the planet's narrow rings were backlit by the Sun, enhancing the visibility of the rings' dust particles. The outer ring consists of at least three dense clumps of orbiting debris, named Liberté, Egalité and Fraternité, that stand out from the thinner remainder of the ring. Astronomers on the ground had only detected the clumps during some stellar occultations, and assumed that the ring was incomplete. (Courtesy of NASA and JPL.)

repeat the achievement by observing stellar occultations by Neptune, but the results were inconclusive. Sometimes the starlight would remain unchanged before and after the planet directly occulted the star. At other times the star would blink on and off, but always on just one side of the planet. Because the brief dimming of starlight was not symmetrical about the planet, and not all stellar occultations produced a blinking signal, the hypothetical rings became shortened, in the minds of the astronomers, to ring-arcs that only reached part way around the planet. Chance might then dictate which astronomers would detect the obscuration.

Voyager 2 clarified the problem. Neptune's ring-arcs turned out not to be isolated segments, but rather three thicker portions of one very thin ring (Fig. 11.13). The ring is narrow and continuous, stretching all the way round the planet just like any well-behaved ring. Its material is generally spread so thinly that it does not noticeably dim a star's light. The ring is only dense enough to hide a star in three arc-like concentrations, subsequently named Liberté, Egalité and Fraternité after the French revolutionary slogan. It was these high-density clumps that had been detected from Earth, blocking starlight and giving the impression of disconnected arcs. The rest of the ring couldn't be seen from Earth because it is so transparent, and hence below the threshold of detectability.

The existence of such clumps, or concentrations, in the rings was an enigma. Every time that the ring particles collide and bump together, they must change speed, and gradually spread around the ring away from the clumps. Unless something is confining the material in the arc-like concentrations, it should spread uniformly around the entire ring in just a few years. Some external force must therefore be holding the material in place, and keeping it within the three arcs. A small moon might hold the clumps together by its gravity, and astronomers think they have found it – the satellite Galatea orbits Neptune just inside the arc-containing ring, producing a wavy distortion in it.

Altogether *Voyager 2*'s cameras found six rings around Neptune – two relatively bright narrow rings, two faint narrow ones, and two broad, dim rings. Only the outermost narrow, clumpy ring had been detected during stellar occultations from Earth. The rest of the rings are so diffuse and the material in them so fine that starlight passed right through them without a detectable change in intensity. In fact, these tenuous rings could not even be seen with the unaided eye of an imaginary astronaut who might fly out there to get a closer look.

The most conspicuous outer and inner rings are respectively named Adams and Leverrier, after John Couch Adams (1819–1892) and Urbain Jean Joseph Leverrier (1811–1877) who independently predicted the existence of the then unknown planet Neptune (Section 1.2). The faint, innermost ring is broad, roughly 1700 kilometers wide; it has been named Galle after Johann Gottfried Galle (1812–1910) who found Neptune close to both of Adams' and Leverrier's predicted positions using a 0.23-meter (9-inch) refractor. Some astronomers think that the Galle ring could extend all the way down to the top of Neptune's atmosphere.

Another broad band of material, called a plateau, extends about 4000 kilometers out from the inner narrow ring, Leverrier, ending in a narrow, sharp-edged ring. The satellite Galatea moves in another faint, narrow ring, just inside the Adams ring.

Unlike the main narrow rings of Uranus, the Adams and Leverrier rings of Neptune contain a vast amount of microscopic dust particles, apparently produced by grinding down or eroding the larger particles. In fact, the sparse, dusty rings of Neptune amount to almost nothing at all. They contain only about a thousandth as much matter as the rings around Uranus, and a million times less matter than Saturn's rings. Put together, all the particles in Neptune's rings would make a body only a few kilometers across. So, small nearby satellites could have been the source of the Neptune rings. Meteoritic bombardment over the eons could, for example, have pulverized such a moon into rubble, producing all that dust. Even now, three small moons are located deep within Neptune's ring system, between the Galle and Leverrier rings.

Formation and evolution of the rings of Uranus and Neptune

It is now thought that all the planetary rings are younger than the age of the solar system, so they cannot be permanent features dating back to its origin. And the present rings are now viewed as a passing stage in an ongoing process of creation and loss.

The austere rings that now circle Uranus and Neptune may have had a violent and chaotic past, arising from catastrophic collisions of moons or when one larger satellite moved inward by tidal interaction with the planet until it was close enough to be ripped into pieces. The inner small moons and larger particles in the rings were then probably gradually broken up into smaller ones by collisions. And all the ring particles we see today will eventually be eroded away by meteoritic bombardment, ground into fine dust by particle collisions, or displaced by gravitational interaction with neighboring satellites.

Once the rings have been turned into dust, as they have in parts of Uranus' ring system and most of Neptune's, they must eventually spiral down into the planet's atmosphere where they will burn up. Neptune's rings are, in fact, already old and decrepit, containing almost no material at all. The rarefied atmosphere of Uranus extends into the ring region, and its gaseous molecules will collide with the ring dust and drag it out of its orbits in just 100 million years or so. Even without this atmospheric drag, the tiny dust grains recoil from the action of the planet's reflected sunlight, moving slowly and inexorably toward a planet in less than 500 million years.

Thus an entire ring system will eventually be turned into dust. And because all the dust is dragged into the planet's atmosphere or ejected from the system, the rings will inevitably decay and disappear over astronomical times.

This need not imply that rings will vanish from the solar system, for satellites can provide the seeds of new rings for a long time to come. The rings of Uranus and Neptune now contain so little material that their transient particles could easily be replaced by pieces of debris blasted off small satellites by meteoritic bombardment. More substantial rings could be replenished as time goes on by the collisional breakup or meteoritic erosion of tiny moons already embedded in the rings. If you broke up all the satellites now within Neptune's rings, you might produce rings as magnificent as Saturn's present ones. This now brings us to the larger satellites of these two planets.

11.5 The large moons of Uranus and Neptune

Five major moons of Uranus

Uranus possesses five major satellites discovered telescopically from Earth before the space age, and named Miranda, Ariel, Umbriel, Titania and Oberon (Section 1.5). As a group, they are similar in size to the mid-sized satellites of Saturn. The two larger and outer Uranian moons, Titania and Oberon, are roughly half the size of the Earth's Moon; the smallest and innermost, Miranda, is about one-seventh the lunar size (Table 11.2). Infrared spectroscopy

Table 11.2 The five large moons of Uranus[a]

Name	Mean Distance from Uranus (Uranus radii)	Orbital Period (days)	Mean Radius (km)	Mass (10^{21} kg)	Mean Mass Density (kg m^{-3})
Miranda	5.08	1.41	236	0.07	1201
Ariel	7.47	2.52	579	1.44	1665
Umbriel	10.41	4.15	585	1.18	1400
Titania	17.07	8.70	789	3.43	1715
Oberon	22.82	13.46	761	2.87	1630

[a] The orbital distances are given in units of the radius of Uranus, which is 25 559 kilometers, or about four Earth radii. The satellite radii are given in units of kilometers. By way of comparison, the mean radius of the Earth's Moon is 1738 kilometers. The mass is given in units of 10^{21} kilograms – our Moon's mass is 73.5 in these units, and the mass density is given in units of kilograms per cubic meter, abbreviated kg m^{-3}.

from Earth indicated that they all have water ice on their surfaces, but their icy surfaces are darker and less reflective than Saturn's moons.

Accurate masses for the larger moons of Uranus were obtained by observing their gravitational effects on the trajectory of the *Voyager 2* spacecraft during its flyby on 24 January 1986. Combined with the size of the moons, these masses yield mean mass densities for the four larger ones of between 1400 and 1700 kilograms per cubic meter, higher than their Saturnian counterparts, between about 1100 to 1400 in the same units.

The high mass densities imply that the larger moons of Uranus are about half rock and half water ice. Thus, with the exception of Miranda, the major moons of Uranus are rocky on the inside, as well as dirty on the outside. Their dark surfaces and rocky interiors may be related to an ancient collision that might have knocked Uranus on its side before the satellites were fully formed.

Because they are relatively small and very cold outside, these moons were expected to be heavily cratered ice-balls, devoid of any signs of internal activity. When *Voyager 2* photographed their icy surfaces at close range, the expected craters were seen, but the surfaces also displayed a surprising measure of geological activity and diversity. They have been fractured and cracked open, probably as the result of surface expansion, and covered by flowing ice or perhaps even liquid water that subsequently froze. Internal heat supplied by radioactive elements may have produced the surface expansion, melted the ice, and generated icy volcanic flows. Because radioactive elements are embedded in rock, the relatively large amount of rock in the large moons of Uranus, as compared to Saturn's satellites of about the same size, produces greater internal heat.

It was Prospero's daughter Miranda who, in Shakespeare's *The Tempest*, proclaimed:

> O, Wonder!
> How many goodly creatures are there here!
> How beautious mankind is!
> O, brave new world!
> That has such people in't
>
> *The Tempest*, V, i, line 183

and her Uranian counterpart certainly has all the earmarks of a "brave new world".

The landscape on Miranda is one of the most amazing yet observed in the solar system (Figs. 11.14, 11.15). It includes old cratered plains, bright younger terrain, and an eclectic mixture of ridges, grooves, mountains, valleys, fractures, and faults.

There are two possible explanations for Miranda's jumbled surface. One explanation supposes that the satellite was once blasted apart by a catastrophic collision and then pulled itself together again, reforming into a born-again moon. It may even have been broken up and reassembled several times. The alternative explanation holds that Miranda is a half-grown world whose evolution was stopped in mid-course. It supposes that the moon heated up inside, soon after formation, and that the dense rocky material began to sink toward the center while the lighter icy substances started rising to the surface. But the internal heat dwindled before this differentiation process could be finished, and the satellite has remained frozen in an embryonic stage of development ever since.

In either explanation, Miranda is just the right size to lie in a twilight zone between two possible worlds. Any smaller, and the satellite could not pull itself together after collisional disruption or generate sufficient internal heat to

Fig. 11.14 Uranus from its moon Miranda This montage of *Voyager 2* images, obtained in January 1986, shows the blue-green cloud tops of Uranus as they would be viewed from the icy surface of its satellite Miranda. An artist has added the planet's dark rings as they might appear from this vantage point. (Courtesy of NASA and JPL.)

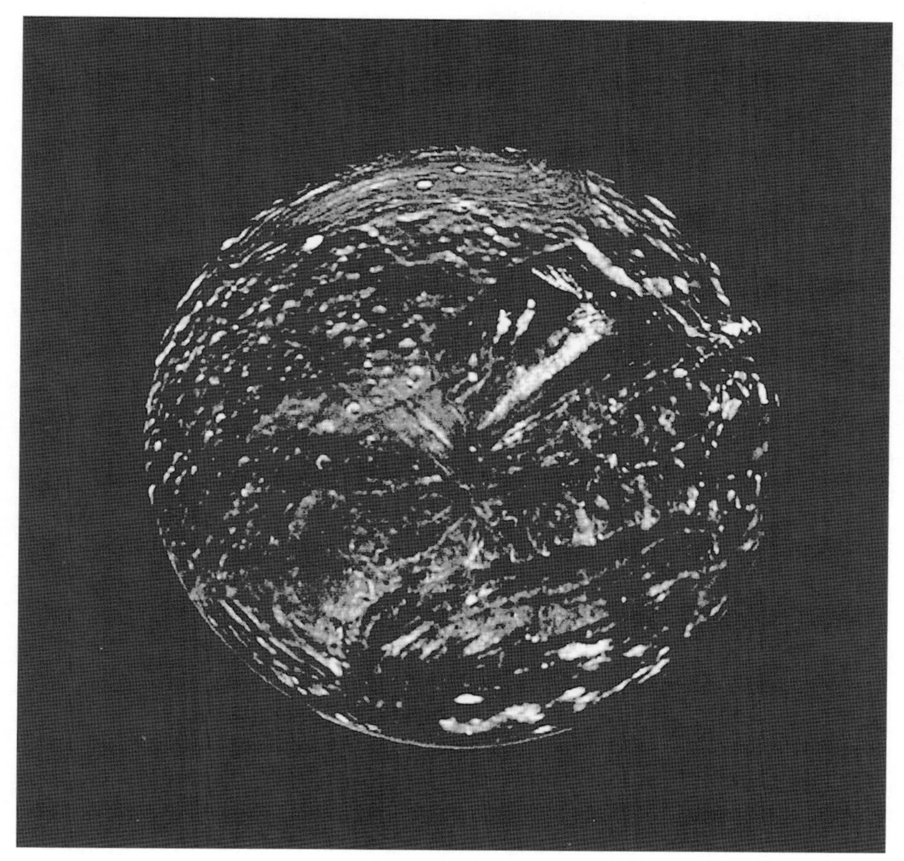

Fig. 11.15 Miranda Two strikingly different types of terrain are found on Miranda, the innermost and smallest of the five major Uranian satellites. It is 472 kilometers in diameter. There is an old, heavily cratered, rolling terrain with relatively uniform reflectivity (*left*), and a young, complex terrain (*right*) characterized by sets of bright and dark bands, jagged cliffs, ridges and grooves, as well as the distinctive chevron feature (*above and right of center*). This image was taken from the *Voyager 2* spacecraft on 24 January 1986. (Courtesy of NASA, JPL, and the U.S. Geological Survey.)

begin differentiation. Any bigger and gravitational forces would have pulled it into a perfect sphere and it might have remained hot enough inside for all its rock to sink to the core.

The four larger moons of Uranus can be divided into two pairs, Ariel and Umbriel, and Oberon and Titania. The members of each pair have similar mass and size, but very different surfaces (Fig. 11.16). The reason that they have

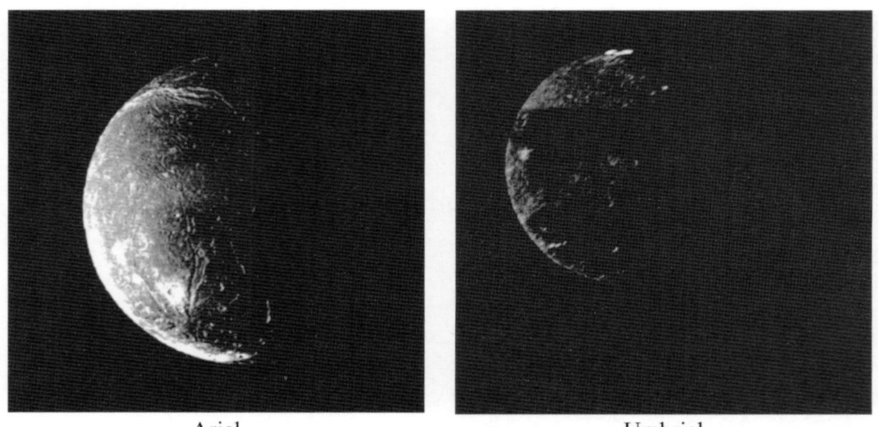
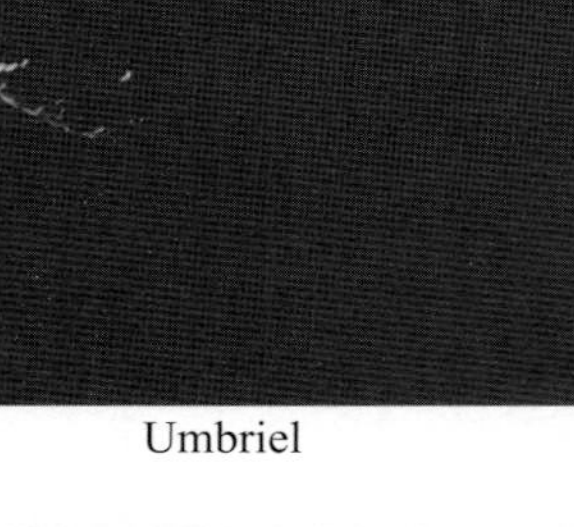

Fig. 11.16 Four icy satellites of Uranus Uranus has four large ice-covered moons that display impact craters, bright rays of ejected material, smooth regions and large rifts or grooves. They range in size from 579 kilometers in radius, for Ariel, to a radius of 789 kilometers for Titania. Numerous valleys and faults cross the terrain of Ariel (*top left*). The darkest large moon is Umbriel (*top right*), which reflects only 16 percent of the sunlight striking its surface. Its heavily cratered surface resembles the lunar highlands, but lacks the numerous bright-ray craters and evidence for geological activity seen on the other large satellites. Titania (*lower left*) retains the numerous scars of impacts, with bright icy ejected material, as well as a trench-like feature that suggests tectonic activity (*right edge*). The icy surface of Oberon (*lower right*) exhibits several large impact craters surrounded by bright rays. One of them has a bright central peak and a floor that is partially covered with very dark, possibly carbon-rich, material that may have erupted onto the crater floor sometime after the crater formed. These images were taken from the *Voyager 2* spacecraft on 24 January 1986. (Courtesy of NASA and JPL.)

similar bulk properties, but look so different, may be related to the differences in the way that the moons produced internal heat and eventually froze inside. Nevertheless, the details remain a perplexing mystery.

Ariel has the brightest, smoothest, and apparently youngest surface of the major moons of Uranus. The bottoms of its canyons and valleys have been resurfaced by icy material that welled up through fractures and spread across the floors. They resemble terrestrial volcanic eruptions of lava at the rifting center of the Earth's spreading ocean floor (Section 4.3), except the rift volcanoes on Ariel produce glacial flows of ice and rock.

Umbriel, with about the same mass and size as Ariel, has an ancient and dark surface. It is among the oldest and most cratered in the Uranian system, with no signs of internal activity.

Oberon and Titantia, the outer and larger of the moons, are both heavily cratered with gray, icy surfaces that reflect about 30 percent of the incident sunlight. Titania resembles a bigger version of Ariel, with large cracks running across its surface and smooth areas that may have been formed by the volcanic extrusion of ice and rock along the faults. Like Umbriel, the older surface of Oberon shows little evidence of internal activity, and is distinguished only by an exceptionally high mountain that was probably created by an enormous impact.

Neptune's Triton, a large moon with a retrograde orbit

Jupiter, Saturn and Uranus have a flock of satellites whose orbits mimic those of the planets around the Sun. Their larger moons revolve in regularly spaced, circular orbits in the same direction as the rotation of the planet and close to the planet's equatorial plane, presumably because they share the rotation of the nebular disk from which the planet and its satellites formed. The radii, orbital distances and other characteristics of these regular satellites also tend to differ in smooth progression.

In sharp contrast to the other major outer planets, Neptune lacks a system of regular larger satellites. Its largest moon, Triton, is the only large satellite in the solar system to circle a planet in the retrograde direction,

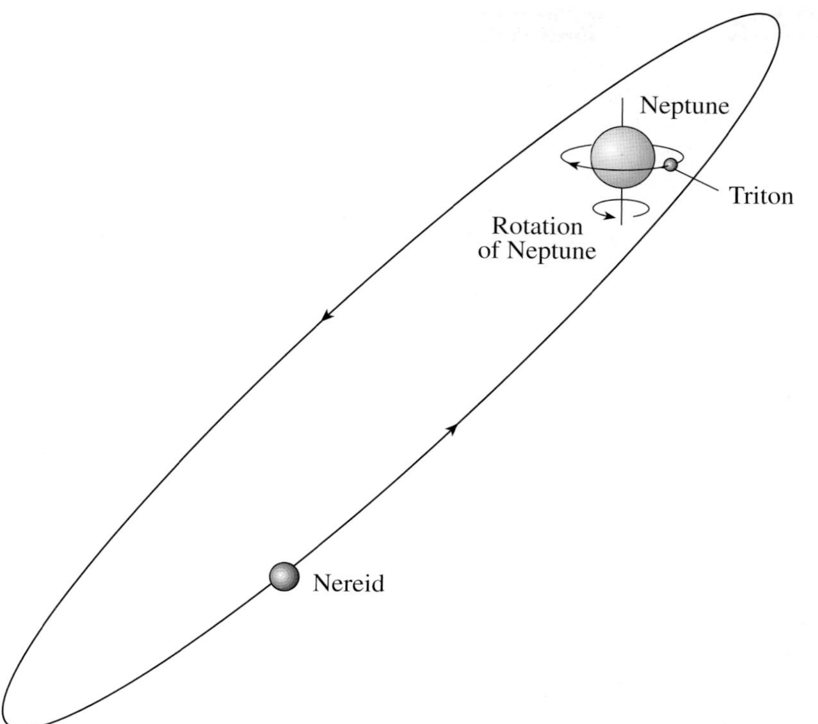

Fig. 11.17 Neptune's odd satellites Nereid, the outermost of Neptune's satellites, travels in a highly inclined, eccentric orbit, in the same direction as that of the planet's rotation. Triton, the largest satellite of Neptune, travels around Neptune in a circular orbit, but, unlike any other large satellite of the giant planets, it travels in the opposite retrograde direction to the rotation of Neptune. In addition, careful analysis of Triton's motions shows that the satellite is in a decaying orbit and is slowly being pulled toward Neptune.

opposite to the planet's direction of rotation (Fig. 11.17). This oddity is compounded by the high orbital inclination. The satellite's orbital plane is tilted at an enormous 157 degrees from the planet's equator. The tilted orbit gives rise to dramatic seasonal variations, for each pole of Triton in turn faces the Sun for nearly half of Neptune's 165-year orbit about the Sun, and the planet appears to move along the satellite's horizon (Fig. 11.18).

Nereid, the outermost moon of Neptune, adds to the mayhem, with the most elongated orbit of any planetary satellite, seven times as distant from the planet at its furthest compared with its closest approach.

Triton's unusual retrograde orbit was quickly established, soon after its discovery – by William Lassell (1799–1880) in 1846, just three weeks after the discovery of Neptune. But due to its great distance from Earth, the radius, mass and reflectivity of Triton remained uncertain until 25 August 1989 when instruments on the *Voyager 2* spacecraft measured them. With a radius of about 1352 kilometers, Triton is about three-quarters of the radius of the Earth's Moon, at 1738 kilometers. The slight gravitational tug exerted on *Voyager* by Triton yielded a mass of 2.141×10^{22} kilograms for the satellite. The size and mass are combined to give a mean mass density of about 2060 kilograms per cubic meter. This is significantly more dense and rock-rich than the medium-sized icy moons of Saturn, between 1100 and 1400 kilograms per cubic meter, or Uranus, between 1400 and 1700 kilograms per cubic meter. But Triton does have nearly the same size and mass density as Pluto (Section 1.5).

Triton's frozen surface, thin atmosphere and geyser-like eruptions

Measurements from *Voyager 2* indicated that Triton is the ultimate icebox, with a daytime surface temperature of 38 kelvin at the time of encounter. That is only thirty-eight degrees above absolute zero, when nothing can move, not even an atom. In fact, Triton has the coldest measured surface of any natural body in the solar system! It is so cold because it is so far away from the Sun, therefore receiving little sunlight, and also because Triton reflects more of the incident sunlight than most satellites – only Enceladus and Europa are comparable. As a result, the total amount of sunlight absorbed by Triton's surface is less than that of any other planet or satellite.

A frosty coating of nitrogen and methane ices overlies all of the surface features, reflecting the incident sunlight (Fig. 11.19). The brilliant ice has a salmon-pink tint with peach hues, possibly due to organic compounds derived from methane by the bombardment of energetic particles from the solar wind and Neptune's radiation belts.

Although nitrogen and methane frosts are apparently the dominant constituents of Triton's visible disk, water ice is needed to support and preserve the observed topography,

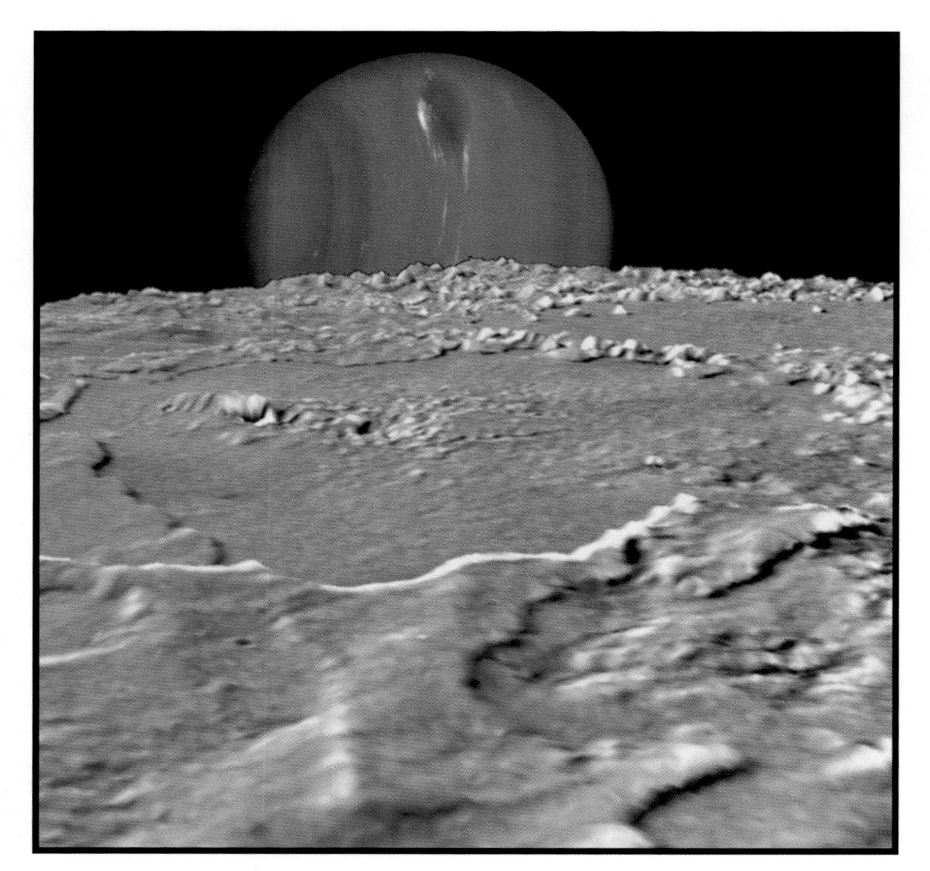

Fig. 11.18 Neptune on Triton's horizon From its satellite Triton (*foreground*), the planet Neptune would appear to move laterally along the horizon. Terraces on the moon's surface indicate multiple exposures of icy volcanic flooding, while a huge anti-cyclonic storm system, the Great Dark Spot, is visible in Neptune's atmosphere. The foreground is a computer-generated view of Triton's maria as they would appear from a point approximately 45 kilometers above the surface. (Courtesy of NASA and JPL.)

including cliffs and ridges that exceed one kilometer in height. At the frigid temperature of Triton, water ice is as strong as steel, and behaves like hard rock on Earth; the methane and nitrogen ice do not have sufficient strength to support the elevated features, which would deform and collapse under their own weight. Thus, thin, brilliant veneers of nitrogen and methane ice apparently overlie a rigid crust of water ice.

An exceedingly tenuous atmosphere envelops the satellite. It consists mainly of nitrogen molecules, the same gas that makes up most of the atmospheres of Earth and Saturn's moon Titan. But the cold, thin atmosphere on Triton has a surface pressure of only 15 microbar, or 15×10^{-6} bar, 15-millionths of the air pressure at sea level on Earth. Triton's surface remains fully visible under its very thin atmosphere. In contrast, Titan has a denser nitrogen-rich atmosphere with a surface pressure 1.5 times that at Earth, and its surface cannot be seen at visible wavelengths.

Voyager 2's cameras showed that Triton's southern polar cap was in the process of dissipating. Nitrogen ice was sublimating, or changing directly from the solid to vapor form, and supplying the atmosphere with gas. The vaporized nitrogen gas is probably carried by atmospheric winds to the dark hemisphere. Thus, the mass of the atmosphere is expected to vary as the polar caps wax and wane with the four long seasons, each of 41 years duration. On Mars, carbon dioxide is similarly recycled between the polar caps and atmosphere during its seasons (Sections 3.1 and 8.3).

Measurements from the *Hubble Space Telescope* indicate that the frozen world may be heating up, with a surface temperature that increased by about five percent between 1989 and 1998. That is a temperature rise of only two kelvin, but it might be enough to melt parts of Triton's surface. The thin shroud of nitrogen molecules could be thickened slightly when the surface ice thaws, but it is still much too tenuous to hide the surface.

Even in the middle of southern summer, a bright ice cap extends from the south pole three-quarters of the way to Triton's equator (Fig. 11.19). It is so cold that some of Triton's atmosphere freezes out at its poles, coating them with a huge ice cap of frozen nitrogen. By way of comparison, the Earth's polar caps contain frozen water ice, for it is too warm for nitrogen to freeze at our planet's poles, and it is too cold for water ice to vaporize from Triton's surface and enter its atmosphere.

Numerous dark streaks run parallel in the midst of the bright southern cap, apparently blown by the prevailing winds in Triton's tenuous atmosphere and strewn over the ice (also see Section 2.3, Fig. 2.36). They emanate abruptly

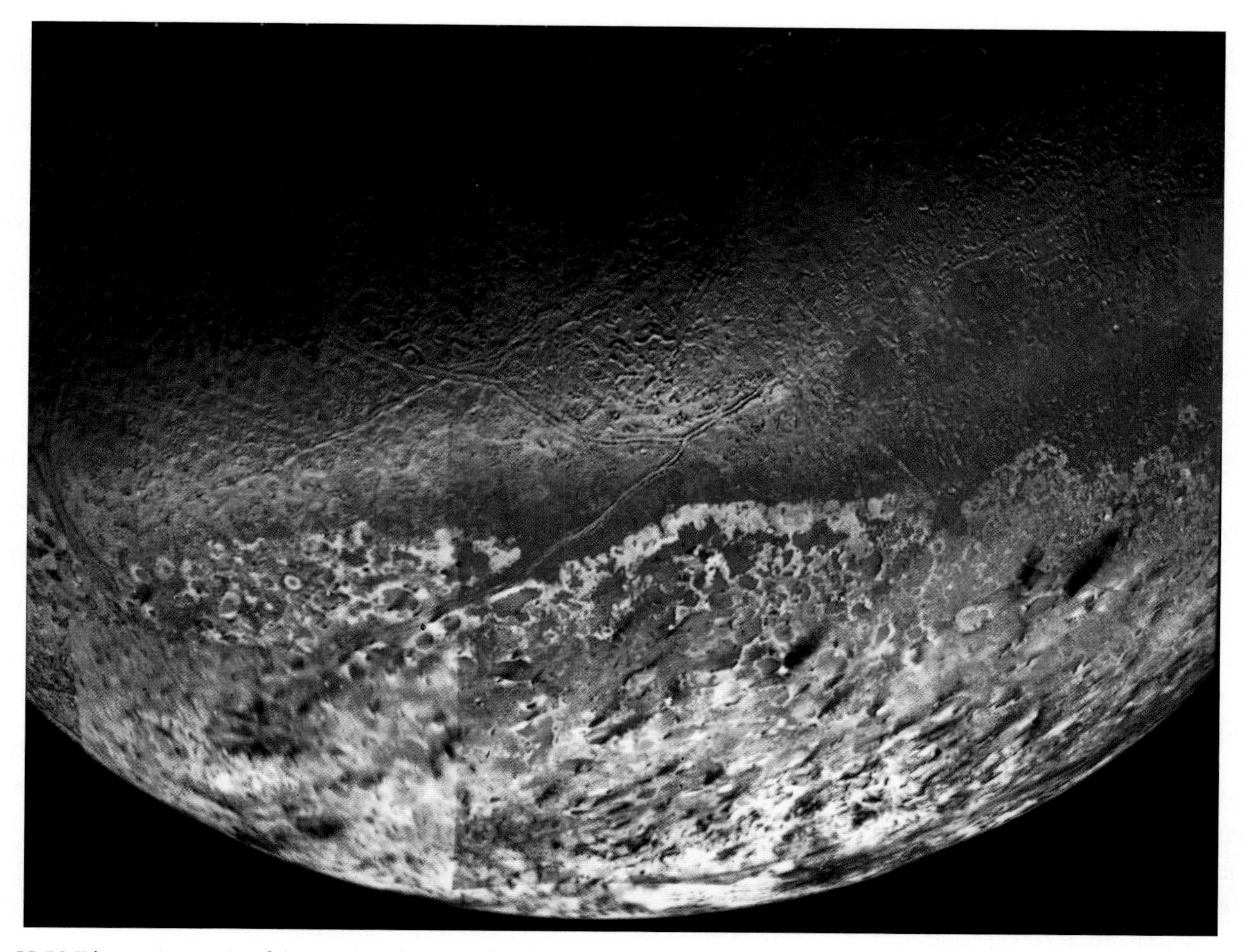

Fig. 11.19 Triton A mosaic of the south polar cap of Triton, the largest satellite of Neptune, taken in August 1989 from the *Voyager 2* spacecraft. At the time of this flyby, Triton was the coldest measured object in the solar system with a surface temperature of 38 kelvin. It is so cold that most of Triton's nitrogen atmosphere is condensed as frost, making it the only satellite in the solar system known to have a surface made mainly of nitrogen ice. Highly reflective methane ice has been colored pink by the action of energetic radiation. (Courtesy of NASA and JPL.)

from specific points, and thin away toward their ends. Some are only a few kilometers in length, while others extend more than 100 kilometers. They must have formed recently, for they seem to overlie deeper ice deposits, and it is unlikely that they could survive sublimation of the polar ice. In fact, they are probably related to erupting plumes that were seen during the *Voyager 2* encounter with Triton.

On a world that is literally frozen solid, astronomers were amazed to find at least four erupting plumes near the center of Triton's sunlit polar cap. These plumes rise in straight columns to an altitude of eight kilometers, where dark clouds of material are left suspended and carried downwind for over 100 kilometers, like smoke wafted away from the top of a chimney. Since the active plumes occur where the Sun is directly overhead, the solar heat might energize them.

Scientists have not reached a consensus about what produces the plumes, but one likely explanation is that geysers are sending up plumes of nitrogen gas laced with extremely fine dark particles. Triton is far too cold for geysers to spout steam and water, like geysers on Earth, but Sun-powered geysers might expel dark material when pent-up nitrogen gas becomes warm and breaks through an overlying seal of ice.

On Triton, the subterranean heat might be accumulated from sunlight, which passes through the translucent ice and is absorbed by darker methane or other carbon-rich material encased beneath. The overlying nitrogen ice would trap the solar heat, for it is opaque to thermal infrared radiation, producing a solid-state greenhouse effect. Nitrogen gas, pressurized by the subsurface heat, then explosively blasts off the iced-over vents or lids, launching volcanic plumes of gaseous nitrogen and ice-entrained darker material into the atmosphere, just as the water in an overheated car radiator is explosively released when the radiator cap is removed.

Astronomers expected to find Triton's surface covered by craters, but there are almost no craters in sight on the satellite. Much of the visible surface outside the polar cap resembles the skin of a cantaloupe, containing numerous pits or dimples with low raised rims and ridges that snake their way through them. But the dimpled pits are too similar in size and too regularly spaced to be impact craters. They may have formed when the satellite heated up inside and volcanic extrusions of slushy ice wiped out the preexisting craters, or even after Triton melted completely and solidified again.

Long cracks or faults on Triton seem to have been partially filled with oozing ice, as they are on Ariel, one of the moons of Uranus. Vast frozen basins found within Triton's equatorial regions have apparently been filled by icy extrusions flowing out from the warm interior, like a squeezed slush cone. These frozen lakes of ice look like inactive volcanic calderas, complete with smooth filled centers, successive terraced flows and vents (Fig. 11.20).

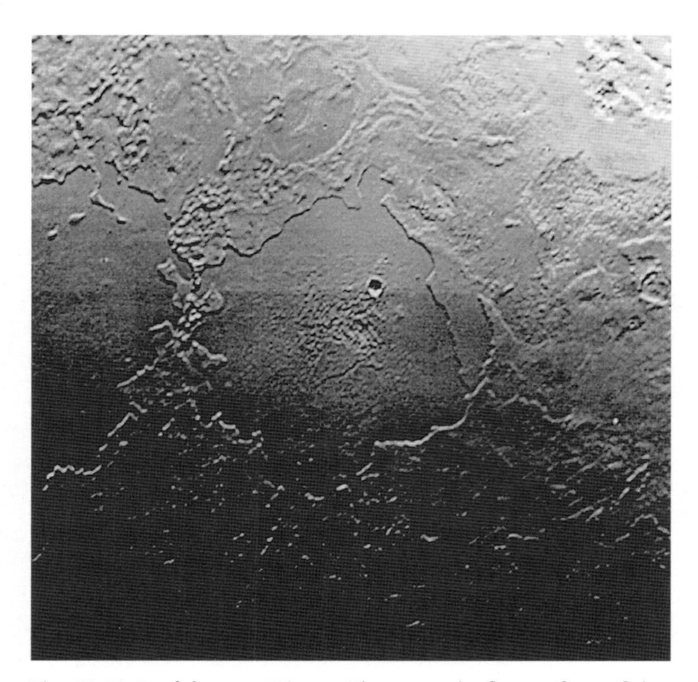

Fig. 11.20 Ice lakes on Triton The smooth, flat surface of this feature on the surface of Triton may have been filled with water ice. It is edged with overlapped terraces, as though the level of the lakes changed as the result of repeated flooding, partial removal and freezing. In Triton's frigid world volcanoes erupt ice, rather than molten rock. The small pits and finely textured areas near the center of the lake-like features are probably associated with the most recent eruption of ice. This image, taken from the *Voyager 2* spacecraft in August 1989, is about 500 kilometers across. (Courtesy of NASA and JPL.)

The warm liquid or slushy ice on Triton apparently acted like lava on Venus, resurfacing the globe and filling low flat areas with smooth deposits that subsequently froze.

It is likely that this volcanism was driven by tidal interaction with Neptune, in somewhat the same manner that Jupiter produces Io's volcanoes (Sections 2.3 and 9.4). This interaction is related to Triton's origin and fate.

Origin and evolution of Triton

Triton's retrograde and inclined orbit suggests that it is not a true satellite of Neptune, dating back to the planet's formation, but that Triton was once a separate world, which was captured into an eccentric, backward and tipped orbit in the remote past. Neptune's lack of an ordered family of large satellites might also be explained if Triton originated in its own independent orbit around the Sun, and was subsequently captured by Neptune, destroying any regular satellite system the planet may have had in the process.

Neptune could only pull Triton into its gravitational sphere of influence if the passing body lost some energy as it went by. Otherwise, Neptune and Triton would just continue to go on their separate ways. One possibility is that Triton smashed into one of the planet's more substantial satellites, losing enough energy that it could not escape Neptune's gravitational pull and went into orbit around it. Or, Triton might have grazed the fringes of Neptune's atmosphere where friction slowed it down enough to be captured.

When Triton began revolving around Neptune, its orbit would have most likely been very elliptical. But tides raised on Triton by Neptune could have circularized the satellite's orbit. While its orbit was evolving, Triton could have cannibalized or ejected satellites it collided with, thereby removing any large regular satellites Neptune may have once had, while also knocking Nereid into its unusual, highly eccentric orbit.

Even today, tidal interaction between Neptune and Triton is gradually drawing the satellite inward. Some time in the very distant future, Triton will be pulled close enough to Neptune for the planet's tidal forces to rip the moon apart, forming rings as bright and magnificent as Saturn's are today.

As tidal forces from Neptune circularized Triton's orbit, they would have squeezed and stretched the satellite, keeping it molten for about a billion years. In the process, Triton's denser rocky material sank to form a core while lighter substances like water rose to form an icy mantle. Volcanoes would have spewed water out of the

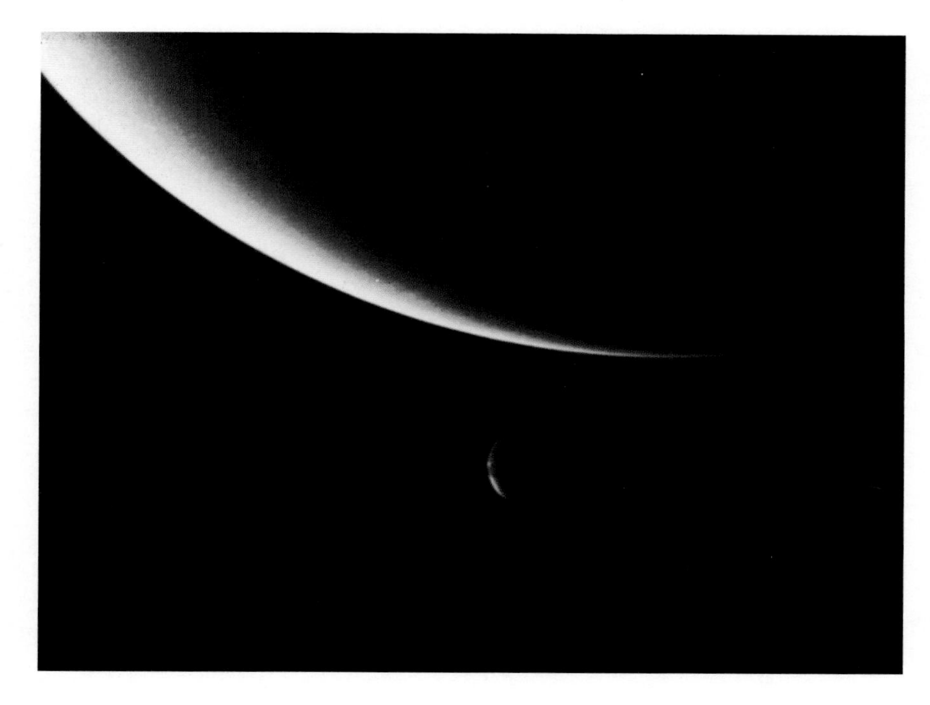

Fig. 11.21 Farewell to the planetary system In parting, *Voyager 2* provided this last picture show, with Neptune's rim in the foreground and Triton appearing as a thin crescent in the distance. The hardy spacecraft, launched in 1977, recorded the swirling face of Jupiter in July 1979, the myriad rings of Saturn in August 1981, pale Uranus in January 1986 and stormy Neptune in August 1989 before heading out of the solar system. (Courtesy of NASA and JPL.)

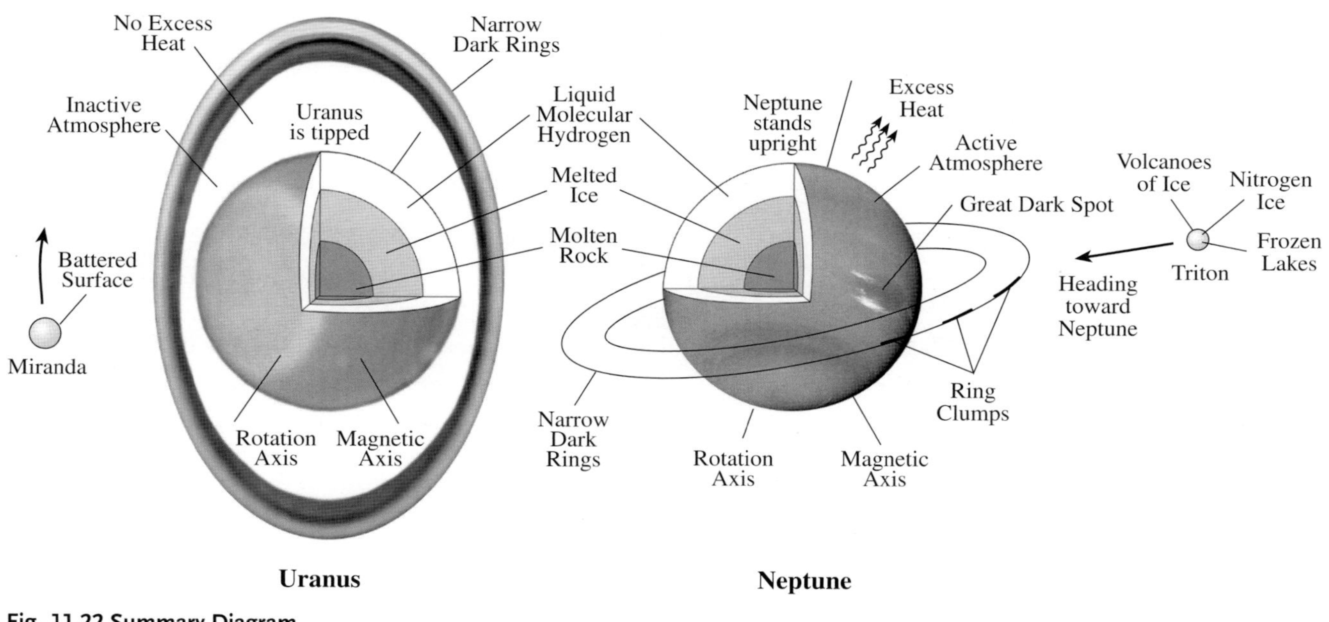

Fig. 11.22 Summary Diagram

satellite's interior, eradicating its craters and other preexisting features, and forming a smoothed icy surface. But this type of water-ice volcanism most likely ended a long time ago when Triton's orbit became circular and the interior tidal-melting stopped.

All in all, *Voyager 2*'s journey past Neptune was a brilliant success. As it drew away, the intrepid explorer took one last picture of Neptune and Triton as neighboring crescents (Fig. 11.21), continuing out to where Triton might have originated and billions of comets now hibernate.

Part 4 Remnants of creation – small worlds in the solar system

12 Comets

- The sudden apparition, changing shapes, and unpredictable movements of comets have puzzled humanity for centuries.
- Comet Halley has returned to fascinate and frighten the world for more than 2000 years.
- Long-period comets, with orbital periods greater than 30 years, have been tossed into the planetary realm from a remote, spherical shell, named the Oort cloud, located about a quarter of the way to the nearest star other than the Sun.
- A million million, or 10^{12}, invisible comets have been hibernating in the deep freeze of the Oort comet cloud since the formation of the solar system 4.6 billion years ago.
- Many short-period comets, with orbital periods less than 30 years, probably came from the Kuiper belt, which lies in the outer disk of the planetary system beyond the orbit of Neptune.
- The Kuiper belt may contain more than a billion comets; the larger ones can be detected with powerful telescopes and electronic detectors.
- Jupiter's immense gravity can propel a comet into a planet-like orbit, or hurl it into the interstellar void.
- Comets light up and become visible for just a few weeks or months, when their orbits bring them near the Sun. The solar heat then vaporizes some of the comet ices, permitting the comets to grow large enough to be seen.
- The solid comet nucleus is just a gigantic ball of water ice, other ices, dust and rock. Some of the comet nuclei are about the size of Paris or Manhattan and roughly one-billionth the mass of the Earth, others are much smaller.
- No two comets ever look identical, and every comet changes shape and form as it whips around the Sun, but they all develop a glowing spherical cloud of gas and dust, known as the coma, when moving close enough to the Sun.
- The comet coma can be larger than the Earth and as big as the Sun, and around the coma there is an even larger envelope of atomic hydrogen, known as the hydrogen cloud, that shines in ultraviolet light.

- Some comets develop tails that flow away from the Sun, briefly attaining lengths as large as the distance between the Earth and the Sun, but other comets have no tail at all.
- Comets can have two kinds of tails, the long, straight ion tails, that re-emit sunlight with a faint blue fluorescence, and a shorter, curved dust tail that shines by reflecting yellow sunlight.
- Spacecraft have peered into the icy heart of two comets, 1P/Halley and 19P/Borrelly, showing that their nuclei are blacker than coal and reflect just 4 percent of the incident sunlight.
- Instruments aboard the *Giotto* spacecraft detected gas and dust jetting out from the sunlit side of the nucleus of Comet Halley, from fissures in its dark crust, but nearly 90 percent of the surface of its nucleus was inactive at the time of the spacecraft encounter.
- When a bright comet nears the Sun, it turns on its celestial fountain, spurting out about a billion kilograms, or a million tons, of water each day.
- The recoil effect of jets of matter ejected from a comet's spinning, icy nucleus can push a comet along in its orbit or oppose its motion, causing the comet to arrive closest to the Sun earlier or later than expected.
- Most of the comets seen during recorded history will vanish from sight in less than a million years, either vaporizing into nothing or leaving a black, invisible rock behind.
- About 40 million kilograms, or 40 thousand tons, of small, cosmic dust particles fall onto the Earth in a typical year, wafting gently through the atmosphere to the ground.
- Visible comets are in their death throes.
- Meteor showers, commonly known as shooting stars, are produced when sand-sized or pebble-sized pieces of a comet burn up in the atmosphere, never reaching the ground.
- Comets strew particles along their orbital path as they loop around the Sun, and when the Earth passes through one of these meteoric streams a meteor shower occurs, recurring at the same time every year.
- Some burned-out comets look like asteroids, with no trace of a coma or tail, and a few asteroids behave like comets, blurring the distinction between these two types of small solar-system bodies.

12.1 Unexpected appearance

Every few years, on average, an unusually bright comet will blaze forth in the night sky, becoming visible to the unaided eye and sporting a graceful tail resembling long hair blowing in the wind (Figs. 12.1, 12.2). In fact, the word comet is derived from the Greek name *aster kometes*, meaning "long-haired star". But comets are nothing like stars; their dramatic display emanates from a relatively small, blackened chunk of ice and dust, comparable to a large city in size.

Unlike the planets, the comets can appear almost anywhere in the sky, remain visible for a few weeks or months, and then vanish into the darkness. Astronomers call this period of visibility an "apparition". During its apparition, a comet changes its shape, often from night to night.

Many of the enigmatic comets travel far outside the paths of the planets and move in every possible direction around the Sun. Their orbits are inclined at all possible angles to the ecliptic, the plane of Earth's orbit, and different comets move in either the same direction around the Sun as the planets or in the opposite retrograde direction. Some comets move in tight orbits, with periods comparable to those of the planets and with low inclinations to the ecliptic.

Fig. 12.1 The Great Comet of 1577 This drawing by a Turkish astronomer appeared in the book *Tarcuma-I Cifr al-Cami* by Mohammed b. Kamaladdin written in the 16th century. The yellow Moon, stars and comet are shown against a light-blue sky. (Courtesy of Erol Pakin, Director, Istanbul Universitesi Rektorlugu.)

Fig. 12.2 Comet Kohoutek A modern photograph of a comet's flowing tail. It was taken on 12 January 1974 with the 1.2-meter (48-inch) Schmidt telescope of the Hale Observatories using a 3-minute exposure in blue light. (Courtesy of the Hale Observatories.)

Comets used to frighten people, filling ancient minds with awe and terror. By their unexpected arrivals, these celestial intruders seemed to upset the natural order of the otherwise placid firmament, and to presage changes in the order of things on Earth, such as the death of rulers, wars and other disasters (Fig. 12.3). One showed up in 44 BC, the year that Julius Caesar (100–44 BC) was assassinated. The fallen emperor's adopted son declared the comet to be Caesar's soul rising to heaven, and used the apparition to gain control over the entire Roman Empire as Augustus (Octavius) Caesar (63 BC–14 AD). William Shakespeare (1564–1616) wrote about the comet's link to Julius Caesar's death 15 centuries later, with:

Fig. 12.3 The eve of the deluge People believed for centuries that unexpected appearance of comets was a premonition of war, death and other disasters. Here the arrival of a comet foretells the great flood at the time of Noah. The 1835–6 apparition of Comet Halley may have influenced the artist, John Martin (1789–1854), for he finished this painting a few years later in 1840. (Collection of Her Majesty, Queen Elizabeth II of the United Kingdom.)

> When beggars die, there are no comets seen;
> The heavens themselves blaze forth the death of princes.
> Cowards die many times before their deaths;
> The valiant never taste death but once.
>
> *Julius Caesar II, ii, 30*

Another famous example was the Norman conquest of England in 1066, which was coincident with the appearance of what is now called Comet Halley. This comet was also seen in 1456 when the Turks conquered Constantinople, and some in Europe prayed for protection from "the Devil, the Turk and the Comet". More recently, the Great Comet of 1811 was supposed "to portend all kinds of woes and the end of the world" in Leo Tolstoy's (1828–1910) *War and Peace.*

The unexpected appearance of a bright comet was also considered an omen of doom and the harbinger of epidemics and disasters. John Milton (1608–1674) imagined pestilence and war raining from a comet's tail. Even in 1910 there were speculations that Comet Halley would impregnate the air with poisonous vapors and wipe out life on Earth, but there were no noticeable effects on humans or other living things when the Earth passed near the comet's tail.

Awe-inspiring Great Comets still arrive without warning today, becoming brighter than the most brilliant stars. At night, the Great Comets can remain visible to the unaided, or naked, eye for months, and they sometimes become visible in daylight. Some of them have tails that stretch up to halfway across the sky. As noted in Table 12.1, they all travel closer to the Sun than the Earth, and sometimes much closer, but a decade or more can pass between the unanticipated discoveries of truly Great Comets.

Amateur astronomers often discover the brightest comets, diligently searching for them with small telescopes or even large binoculars. Professional astronomers sometimes accidentally come across one while using a large telescope for another purpose. In accordance with a tradition that has gone on since the time of the French comet hunter Charles Messier (1730–1817) more than two centuries ago, new comets are given the name of the discoverer, or independent discoverers. Throughout most of the 20th century, the comet name was followed by the year of discovery and a Roman numeral indicating the order

Table 12.1 Some Great Comets of the 19th and 20th centuries[a]

Name	Perihelion Date	Days Visible[b]	Perhelion Distance[c] (AU)	Brightest Apparent Magnitude[d]
Great Comet (1807 R1)	19 Sept. 1807	90	0.65	1 to 2
Great Comet (1811 F1)	12 Sept. 1811	260	1.04	0
Great March Comet (1843 D1)	27 Feb. 1843	48	0.006	1
Comet Donati (1858 L1)	30 Sept. 1858	80	0.58	0 to 1
Great Comet (1861 J1)	12 June 1861	90	0.82	0 (or −2?)
Great Southern Comet (1865 B1)	14 Jan. 1865	36	0.03	1
Comet Coggia (1874 H1)	09 July 1874	70	0.68	0 to 1
Great September Comet (1882 R1)	17 Sept. 1882	135	0.008	−2
Great Comet (1901 G1)	24 Apr. 1901	38	0.24	1
Great January Comet (1910 A1)	17 Jan. 1910	17	0.13	1 to 2
1/P Halley (1910 reappearance)	20 Apr. 1910	80	0.59	0 to 1
Comet Skjellerup–Maristany (1927 X1)	18 Dec. 1927	32	0.18	1
Comet Ikeya–Seki (1965 S1)	21 Oct. 1965	30	0.008	2
Comet Bennett (1969 Y1)	20 Mar. 1970	80	0.54	0 to 1
Comet West (1975 V1)	25 Feb. 1976	55	0.20	0
Comet Hyakutake (1996 B2)	01 May 1996	30	0.23	1 to 2
Comet Hale–Bopp (1995 O1)	01 Apr. 1997	215	0.91	−0.7

[a] Adapted from Donald K. Yeomans', *Great Comets in History*, at the web site http://ssd.jpl.nasa.gov/great_comets.html

[b] Days visible to the naked eye unaided by binoculars or a telescope.

[c] The perihelion distance is the distance from the Sun at the closest approach to the star, given in astronomical units, or AU, roughly the mean distance between the Earth and the Sun.

[d] The apparent magnitude is a measure of the apparent brightness of a celestial object, in which brighter objects have smaller magnitudes. Sirius A, the brightest star other than the Sun, has an apparent visual magnitude of −1.5. Proxima Centauri, the nearest star other than the Sun is about 0 on the magnitude scale, while Venus has an apparent magnitude of −4 when brightest and at its brightest Jupiter appears at magnitude −2.7.

of discovery that year, both placed in parentheses. For example, Comet Ikeya–Seki (1965 VIII) was the eighth comet discovered in 1965, and it was discovered by two Japanese comet hunters, Kaoru Ikeya (1943–) and Tsutomu Seki (1930–).

A prefix "P/" is now used for a periodic comet, defined to have a revolution period of less than 200 years or confirmed observations at more than one perihelion passage, and "C/" for a comet that is not periodic in this sense. A number is also added before the prefix P to designate the order of discovery. For instance, 1P/Halley was the first periodic comet known and 2P/Encke the second. Newly discovered comets are also now designated by the year of observation, followed by an upper-case letter identifying the half-month of the observation during that year, and a consecutive numeral to indicate the order of discovery announcement during that half-month. The letters "I" and "Z" are not used, to make a total of 24 half-months. For example, the third comet reported as discovered during the second half of February 1995 would be designated 1995 D3.

12.2 The return of Comet Halley

Many of the brightest comets seem to come out of nowhere, suddenly moving past the Sun, and are never seen again. Yet, one famous comet has come back for repeat performances, fascinating the world for more than 2000 years. It is now known as Comet Halley, named for the British astronomer Edmond Halley (1656–1742).

Halley demystified comets by showing that at least one of them travels in an elongated orbit around the Sun. He found that the orbit of the comet of 1682 was similar to those of comets observed in 1607 (by Johannes Kepler, 1571–1630) and in 1531 (by Petrus Apianus, 1495–1552). All three comets moved around the Sun in retrograde orbits with a similar orientation. Halley also knew that the Great Comet of 1456 had traveled in the retrograde direction, and he concluded that all four comets were returns of the same comet in a closed elliptical orbit around the Sun with a period of about 76 Earth years. Halley confidently predicted its return in 1758, noting that he would not live to see it. After the comet was rediscovered, on Christmas night of the predicted year, Halley's achievement was acknowledged, albeit posthumously, by naming it Comet Halley (Fig. 12.4).

Halley's is the most famous of the comets because it was the first to arrive on schedule. Its fame is deserved on other counts as well. It displays a complete range of comet fireworks including an exceptionally long tail, a bright head, and jets, rays, streamers and haloes. Moreover, it has been observed for a longer period of time than any other comet. The earliest apparition established with confidence from Chinese chronicles dates back to 240 BC; since then, all its perihelion passages have been retraced in the ancient or modern records of astronomers (Table 12.2).

After its 1910 apparition (Fig. 12.5), Comet Halley moved away from the Sun into the outer darkness, arriving

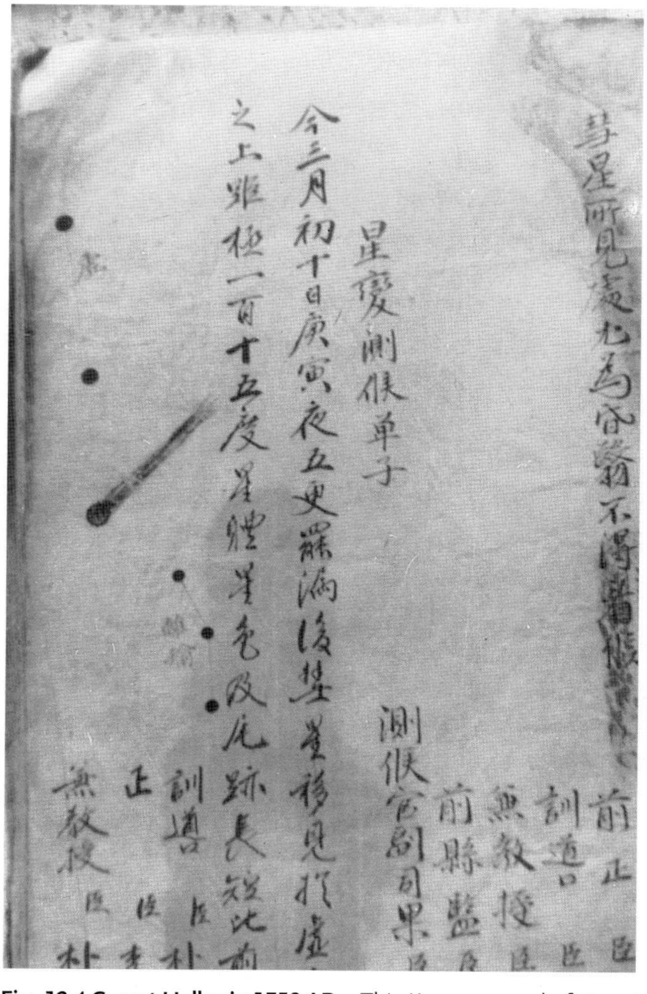

Fig. 12.4 Comet Halley in 1759 AD This Korean record of Comet Halley was made during the comet's first predicted return in 1759 AD. The Korean astronomers have been recording the appearance of comets and other unusual celestial objects for more than 3000 years. (Courtesy of Il-Seong Na, Yonsei University, Seoul.)

Table 12.2 Thirty-two perihelion[a] passages of Comet Halley

Year	Date
240 BC	25 May
164	13 November
87	6 August
12 BC	11 October
66 AD	26 January
141	22 March
218	18 May
295	20 April
374	16 February
451	28 June
530	27 September
607	15 March
684	3 October
760	21 May
837	28 February
912	19 July
989	6 September
1066	21 March
1145	19 April
1222	29 September
1301	26 October
1378	11 November
1456	10 June
1531	26 August
1607	28 October
1682	15 September
1759	13 March
1835	16 November
1910	20 April
1986	9 February
2061[a]	28 July
2134[a]	27 March

[a] The perihelion of a comet is the point in its orbit that is closest to the Sun. The future two perihelion passages of Halley's comet are predicted dates; all of the others have been recorded.

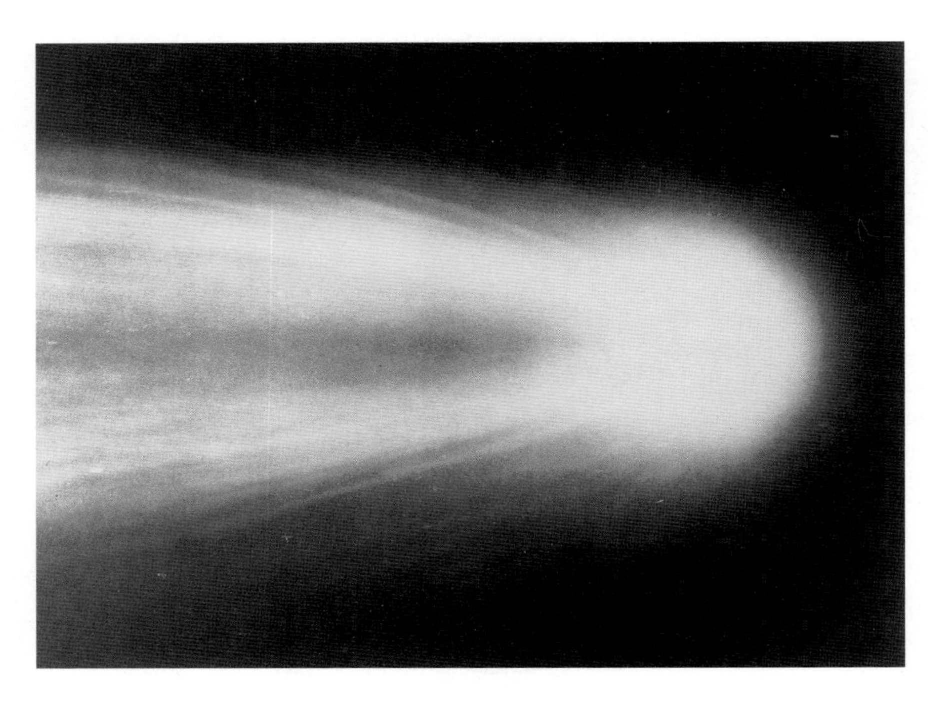

Fig. 12.5 Apparition of Comet Halley in 1910 The head region or coma of Comet Halley observed on 8 May 1910 with the 1.52 meter (60-inch) telescope on Mount Wilson. The comet's tail flows to the left, away from the Sun. (Courtesy of the Hale Observatories.)

in 1948 at the remotest part of its orbit at 35 AU, or at 35 times the Earth's distance from the Sun. The comet then turned the direction of its course, and began falling back toward the heart of the solar system with ever-increasing speed. It reached perihelion, or its closest distance from the Sun, on 9 February 1986. Comet Halley and the Earth were then on opposite sides of the Sun, so this was among the least favorable apparitions for observing the comet with the unaided eye. Nevertheless, it still became one of the most thoroughly studied apparitions in the history of comet research (Fig. 12.6), including visits by six spacecraft (Section 12.6).

After its 1986 visit, Halley's comet headed for the cold reaches of space beyond Saturn, to return in the Sun's neighborhood in 2061. But the comet will not keep on coming back forever. A body such as Comet Halley is caught in a life of continual decay, evaporating and blowing part of itself away each time it comes near the Sun. Scientists estimate that the comet loses about one-thousandth of its mass on each visit to the inner solar system, so Comet Halley will survive for only about another 1000 orbits, lasting roughly a further 76 thousand years.

12.3 Where do comets come from?

Comets are primitive bodies that formed at the same time as the Sun and planets about 4.6 billion years ago. But once they come close enough to be seen, comets begin to fall apart and they must eventually vanish from sight, often in less than a million years after first sighting. So comets are very old, but once they swing near the Sun they do not last very long. This means that ancient reservoirs must be furnishing the inner solar system with new comets if they appeared in the remote past and are going to continue to be seen in the distant future. They come from two reservoirs, one that is very far away, at the fringe of the outer solar system, and a nearer one at the edge of the planetary realm. These small icy worlds have been hibernating in the cold outer reaches of space ever since the formation of the solar system.

Long-period and short-period comets

Most discovered comets have arrived near the Sun from distant regions far beyond the major planets. They have very elongated trajectories that take them back to the distant regions they came from. These comets are known as the long-period comets, with orbital periods larger than 30 Earth years. The long-period comets are observed in the inner part of the solar system just once, arriving unannounced and unpredicted. As you might expect, they come from very far away, at the outer fringes of the solar system.

Some known comets, about 150 in all, have appeared more than once during the past two centuries. These are the short-period comets, which revolve around the Sun with orbital periods of less than 30 Earth years. They are sometimes distinguished by putting a number and the letter "P" before their name, with the short-period comet number corresponding to the order of recognition.

Fig. 12.6 The return of Comet Halley in 1986 Rays, streamers and kinks can be seen in the ion tail of Comet Halley during its 1986 return to the inner solar system. The broad, fan-shaped dust tail can also be seen. The radio galaxy known as Centaurus A, or NGC 5128, can be seen in the bottom left corner. It is about ten trillion, or 10^{13}, times further away from the Earth than the comet is. Photograph taken by Arturo Gomez on 15 April 1986 with the Curtis Schmidt telescope of Cerro Tololo. (Courtesy of the National Optical Astronomy Observatories.)

The short-period comets are seen time and again, trapped in tight orbits within the planetary realm. Most of them have low orbital inclinations near the plane of the Earth's orbit, with mean distances from the Sun of just a few times that of the Earth, or a few AU. It is these short-period comets whose regular returns we are able to predict, and which we can examine in detail with spacecraft (Table 12.3).

The comets have been further divided into those with intermediate orbital periods, between 30 and 200 Earth years, and those that circle the Sun in less than 20 years. Members of the former class are called Halley-type comets, after their prototype Comet Halley with its 76-year period. The type with the shorter orbital periods, close to that of Jupiter at 11.86 years, are known as the Jupiter-family comets. The Halley-type comets are probably long-period comets whose orbits have been modified by the giant planets after they entered the planetary region. Some of the Jupiter-family comets originate from a much closer reservoir, located just beyond the orbit of Neptune.

The Oort cloud

The size and orientation of the trajectories of long-period comets can be explained if they come from a remote, spherical shell belonging to the outer parts of the solar system (Fig. 12.7). This vast comet repository is known as the Oort cloud, named after the Dutch astronomer Jan Hendrik Oort (1900–1992) who first postulated its existence. Since the long-period comets approach the Sun from enormous distances, of 50 000 AU or more, the Oort cloud has a diameter of up to twice this size. By way of comparison, the average distance between the Earth and the Sun is just 1 AU, about 150 billion meters. And because long-period comets enter the planetary realm at all possible angles,

Table 12.3 Selected short-period comets[a]

Name	Orbit Period (years)	Perihelion Date[b] (year)	Perihelion Distance (AU)	Orbital Inclination (degrees)	Absolute Magnitude[c]
1P Halley	76.01	2061	0.59	162.2	5.5
2P Encke	3.30	2003	0.34	11.8	9.8
6P d'Arrest	6.51	2002	1.35	19.5	8.5
9P Tempel 1	5.51	20th century	1.50	10.5	12.0
19P Borrelly	6.88	2001	1.36	30.3	11.9
21P Giacobini–Zinner	6.61	2005	1.00	31.9	9.0
26P Grigg–Sjkellerup	5.11	2002	0.99	21.1	12.5
27P Crommelin	27.41	2011	0.74	29.1	12.0
46P Wirtanen	5.46	2002	1.06	11.7	9.0
55P Tempel–Tuttle	33.22	2031	0.98	162.5	9.0
73P Schwassmann–Wachmann 3	5.34	2001	0.94	11.4	11.7
81P Wild 2	6.39	2003	1.58	3.2	6.5
107P Wilson–Harrington[d]	4.30	2001	1.00	2.8	9.0

[a] Adapted from Lang, Kenneth R., *Astrophysical Data: Planets and Stars*. New York, Springer-Verlag 1992, and Gary M. Kronk's comet web site http://cometography.com. Short-period comets that have either been visited by spacecraft in the past or have been considered for encounters in the future, listed in the order of their recognition, or periodic comet number.

[b] The given perihelion date is the first to occur in the 21st century, and the perihelion distance is in AU, the mean distance between the Earth and the Sun.

[c] The absolute magnitude is a measure of the intrinsic brightness of a comet, and a smaller magnitude indicates a brighter comet.

[d] Also asteroid (4015).

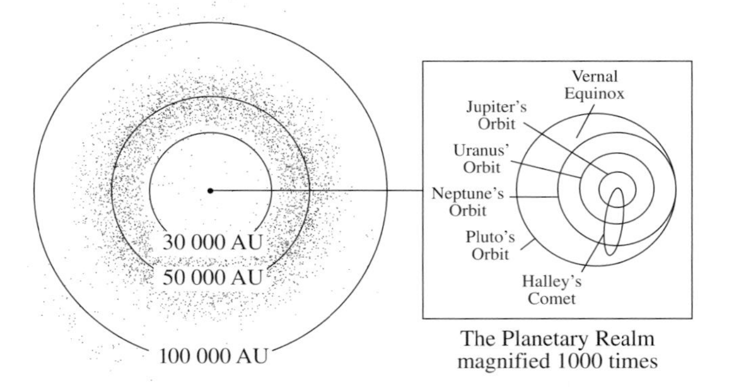

Fig. 12.7 The Oort comet cloud More than 200 billion comets hibernate in the remote Oort comet cloud, shown here in cross-section. It is located in the outer fringes of the solar system, at distances of up to 100 000 AU from the Sun. By comparison, the distance from the Sun to the nearest star, Proxima Centauri, is 271 000 AU, while Neptune orbits the Sun at a mere 30 AU. The planetary realm therefore appears as an insignificant dot when compared to the comet cloud, and has to be magnified by a factor of 1000 in order to be seen. This comet reservoir is named after the Dutch astronomer Jan H. Oort (1900–1992) who, in 1950, first postulated its existence.

with every inclination to the Earth's orbital plane, they must come from a spherical shell. This would also explain the fact that long-period comets move in all directions. Roughly half of them move along their trajectories in the retrograde direction, opposite to the orbital motion of the planets.

The comet cloud extends about one-quarter of the way to the nearest star other than the Sun – Proxima Centauri at 271 000 AU. Even at these enormous distances, the Sun's gravity is powerful enough to hold the unseen comets to gigantic elliptical orbits with the Sun at one focus. Out there, it takes up to 10 million Earth years for a comet to complete one circuit around the Sun. At greater distances the stars in the neighborhood of our solar system compete for gravitational control, and each is imagined to have its own retinue of comets.

Where did the Oort-cloud comets originally come from? They could not have formed in their current position, because the material at such large distances from the young Sun would have been too sparse to coalesce. They instead originally formed in large numbers in the region of the outer planets, between the orbits of Jupiter and Neptune, as the leftover bits and pieces from the formation of the solar system. Once formed, the kilometer-sized cometary bodies were swept out of the region by the nascent giant planets. The nascent giants acted like cosmic street cleaners, either hurling the primeval comets into distant regions or consuming them.

Jupiter and Saturn, the two most massive planets, might have ejected some of them into interstellar space, but they also placed a significant fraction of comets into the Oort cloud. Jupiter may have additionally tossed nearby comets into a collision course with our planet, perhaps supplying some of early Earth's water and organic compounds. Uranus and Neptune, with lower masses, could not easily throw the primitive comets into the space between the stars, but they should have tossed about the same number of comets into the Oort cloud as Jupiter and Saturn did. The strong gravitation of the outer giants would have pulled in and consumed nearby comets, helping the young giants to grow.

Thus, the comets that belong to the Oort cloud are mementos of creation, frozen into the deep freeze of outer space, tumbling unseen in the remote blackness for billions of years. They are much too small and too far away to be seen. So they will remain forever invisible, and will never be directly detected. As the Stoic philosopher Seneca (4 BC–65 AD) put it:

> How many bodies besides these comets move in secret, never rising before the eyes of men? For God has not made all things for man.
>
> *Natural Questions, Book 7, Comets*

Just few of them occasionally return as comets that we see.

But how do comets fall from the Oort comet cloud to the heart of the solar system? The distant comets are only weakly bound to the solar system, and are easily perturbed by the gravitation of nearby, moving objects, which throw some of the comets back into the planetary system. The random gravitational jostling of individual stars passing nearby, for example, knocks some of the comets in the Oort cloud from their stable orbits, either ejecting them into interstellar space or gradually deflecting their paths toward the Sun. Every one million years, about a dozen stars pass close enough to stir up the cometary objects, sending a steady trickle of comets into the inner solar system on very long elliptical orbits. A giant interstellar molecular cloud can also impart a gravitational tug when it moves past the comet cloud, helping to jostle some of them out of their remote resting-place. Tidal forces generated in the cloud by the disk of our Galaxy, the Milky Way, also help to feed new long-period comets into the planetary region.

As time goes on, the accumulated effects of these tugs will send a few comets in toward the Sun – or outward to interstellar space. If the several hundred new comets observed during recorded history have been shuffled into view by the perturbing action of nearby stars or molecular clouds, then there are at least one hundred billion, or 10^{11}, comets in the Oort cloud. There may be a trillion, 10^{12}, or even ten trillion, 10^{13}, of them. This large population of unseen comets can sustain the visible long-period comets and persist without serious depletion for many billions of years, until long after the Sun brightens enough to boil the Earth's oceans away.

The Kuiper belt

The Oort cloud cannot easily explain the comets with the shortest periods, the so-called Jupiter-family comets with periods less than 20 Earth years. These comets have relatively small orbits tilted only slightly from the orbital plane of the Earth, and they usually move in the same prograde direction as the planets. Unlike their longer-period cousins, the motions of the Jupiter-family comets resemble those of the planets.

The main source of these comets is thought to be a ring of small icy objects at the outer edge of the planetary realm, just beyond the orbit of Neptune and a thousand times closer than the Oort cloud. It is known as the Kuiper belt, named after the Dutch–American astronomer Gerard Peter Kuiper (1905–1973) who predicted its existence in 1951. The name Edgeworth–Kuiper belt is used in the United Kingdom, acknowledging Kenneth E. Edgeworth's (1880–1972) proposal of the belt's existence in 1943.

Kuiper believed that comets must have formed throughout the early solar system. Although the giant planets cleared out any comets that were in their vicinity, the comets that formed beyond these planets should still be there.

The material density in this outer region of the primeval planetary disk was so low that the small objects did not coalesce into a single larger planet. They instead formed the flattened Kuiper belt of 100 million to 10 billion, or 10^8 to 10^{10}, small frozen worlds that have remained there for billions of years (Fig. 12.8).

Armed with sensitive electronic detectors and powerful telescopes, astronomers have shown that the planetary system does not end abruptly at Neptune (Section 1.5). They have discovered a substantial population of small, previously unseen bodies in the Kuiper belt, each millions of times fainter than can be seen with the unaided eye. All of these newly found celestial objects travel in trans-Neptunian orbits that are only slightly tilted from the ecliptic, encircling the planetary system somewhat like the ring that wraps around Saturn. In fact, Pluto is probably one of

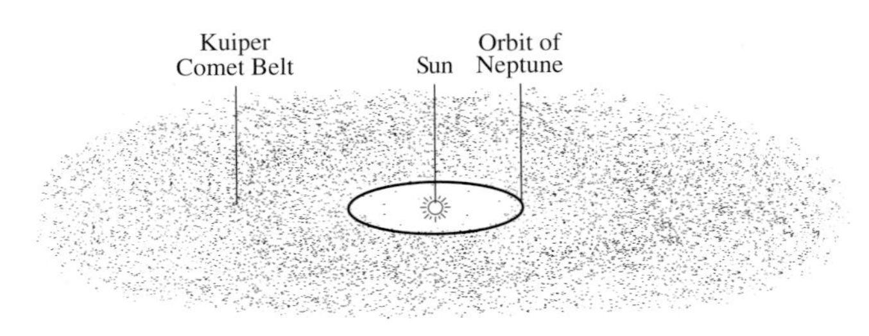

Fig. 12.8 The Kuiper belt A repository of frozen, comet-sized worlds resides in the outer precincts of the planetary system, just beyond the orbit of Neptune and near the orbital plane of the planets. Known as the Kuiper belt, it is thought to contain 100 million to 10 billion, or 10^8 to 10^{10}, comets. Many short-period comets are tossed into the inner solar system from the Kuiper belt.

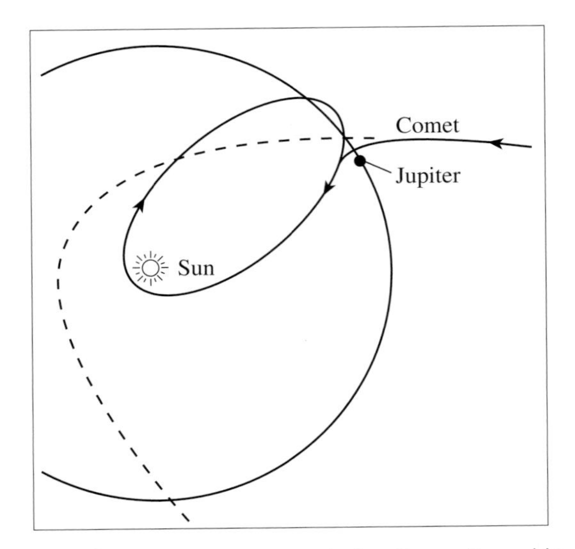

Fig. 12.9 Jupiter captures a comet A planet's gravity could transfer a comet from an extremely elongated orbit (*dashed line*) into a shortened elliptical orbit (*solid line with arrows*). The most massive planet, Jupiter, might have captured some short-period comets in this manner. Some of them have orbital periods that are less than that of Jupiter, or less than 12 Earth years, and their orbits reach out to Jupiter's orbit, 5.2 AU from the Sun, shown as a partial circle.

them. Based on these detections, scientists estimate that the Kuiper belt contains tens of thousands of objects larger than 100 kilometers in size.

The belt ought to contain a much larger population of smaller comet-sized bodies, just a few kilometers across, just as there are more pebbles on a rocky beach than boulders and there are more smaller objects in the solar system than larger ones. There are probably at least a billion pristine comets located just beyond the orbit of Neptune, each so faint and distant that they cannot be seen. Unlike the comets in the Oort cloud, those in the Kuiper reservoir formed at their current locations at the dim horizon of the planetary realm, and have not been significantly perturbed since the origin of the solar system about 4.6 billion years ago. But Neptune's gravity slowly erodes the inner edge of the Kuiper belt, within about 45 AU from the Sun, launching short-period comets from that zone into the inner solar system.

Once a comet is launched into the planetary realm, from either the Kuiper belt or the Oort cloud, it may not stay on the same trajectory. Its orbit can be transformed if it passes near Jupiter, the most massive of planets. The giant planet's gravity can perturb the comet into a new elliptical orbit around the Sun (Fig. 12.9).

12.4 What turns a comet on?

All of the comets in the Oort cloud, and most of those in the Kuiper belt, are invisible. They are the nuclei of comets that can be seen only if they come near the Sun. Each nucleus is the solid, enduring part of a distant comet. It is just a gigantic ball of frozen water ice and other ices laced with darker dust and pieces of rock. Light from the distant Sun is much too feeble to warm the comet ices, which remain frozen solid at the low temperatures in the remote comet reservoirs.

When a comet nucleus emerges from the deep freeze of outer space and moves toward the Sun, the increased solar heat causes the comet's surface material to sublimate, with gases escaping through fissures in the crust of the nucleus. At the low-pressure conditions of space, the solid ice goes directly into gas without passing through a liquid state, in a process called sublimation, just as dry ice does on Earth and water ice does on Mars. The escaping gases also carry along dust particles. The gas and dust make the comet grow in size, enabling it to be seen.

When astronomers mention dust, they mean solid particles between a millionth, or 10^{-6}, and a ten-thousandth, or 10^{-4}, of a meter in size. Originally embedded in the ice of the comet nucleus, the dust particles are driven away from the nucleus as the ices sublime and the gases expand. Larger solid particles, with sizes comparable to sand or even pebbles are also carried away from the comet nucleus into space.

Unlike the planets, a comet lights up and becomes visible for just a brief, fleeting interval during its long journey through space. It can often be detected only when it moves into the inner solar system, within the orbits of the terrestrial planets. Then the comet whips around the Sun and heads outward in more or less the same direction that it came from. As it moves away from the Sun, a comet receives less solar heat, becoming cold and inert and fading into darkness.

The distance at which an invisible comet nucleus turns on, and grows large enough to be seen, varies from comet to comet. Some short-period comets turn on at several AU from the Sun, but new comets that are traversing the planetary system for the first time can turn on at greater distances. For instance, many first-time visitors to the solar neighborhood become unusually bright and extensive at distances of 5 AU or more.

New comets might turn on sooner than repeat visitors because their outer layers more closely reflect primordial conditions in the solar system. Laboratory studies show that ices formed at the extremely low temperatures, of about 15 kelvin, in the outer precincts of the solar system will not be of the usual crystalline form. Instead, they are *amorphous*, from the Greek "without form". When heated by passage near the Sun, this amorphous ice is especially volatile and produces active jets.

In contrast, old comets that are not on their first passage through the inner solar system, including most short-period comets, have made many passages close to the Sun. Their outer layers have been "cooked" and partially stripped off, leaving behind a dark insulating crust composed largely of dust. Solar radiation has a more difficult time penetrating this material than the fresh, icy surface of a new comet. This explains the limited loss of material from periodic comets that have been repeatedly exposed to the Sun. They also have to travel nearer the Sun than a new comet does in order to grow large enough to become visible.

Of course, a warm comet emits all sorts of noxious substances. There is cyanogen, a poisonous, flammable and colorless gas, and hydrogen cyanide, otherwise known as prussic acid. Methane, ethane and acetylene have been identified in the infrared spectra of some comets, together with a host of other organic molecules detected at radio wavelengths. Nevertheless, these complex molecules are almost certainly only minor ingredients of comets.

If a comet nucleus provides such huge quantities of gas and dust, and still survives for hundreds or thousands of trips near the Sun, it ought then to be mainly composed of water ice. The fact that the outer layers of periodic comets start releasing material near 3 AU from the Sun suggests that water ice dominates their nucleus, since the temperature of the Sun's radiation at 3 AU is approximately that required to vaporize water ice. Ices of other possible molecules begin to sublimate off the nucleus at much lower temperatures and greater distances from the Sun. The vaporization of these more volatile substances, such as carbon dioxide, methane, and ammonia, may initiate the production of gas and dust in the newer (fewer solar trips) comets.

12.5 Anatomy of a comet

No two comets ever look identical (Fig. 12.10), just as no two snowflakes are alike, but most comets have basic features in common. When they emerge from the deep freeze of outer space and move toward the Sun, the increased solar heat eventually causes their ices to sublimate and blow dust away with the escaping gas. The comet then becomes visible as an enormous moving patch of light. This glowing,

Fig. 12.10 Comet forms Three kinds of comet shapes. Comet Perrine (1902 III) shows a transparent coma and tail (*left*), Comet Finsler (1937 V) exhibits a coma and tail that are unsymmetrical (*center*), and Comet Morehouse (1908 III) is remarkable for the rapid variations in the structure of its tail (*right*). When a telescope follows a comet, stars move across the field of view, producing numerous short star trails. [Courtesy of the Royal Observatory, Greenwich (*left* and *right*) and the Norman Lockyer Observatory (*center*).]

misty ball of light is called the *coma*, the Latin word for "hair". One or more tails can eventually stream from the coma, in a direction away from the Sun.

Comet gas and dust are initially ejected primarily in the general direction of the Sun; solar forces push them into tails that flow away from the Sun. As a result, a comet travels head first when approaching the Sun and tail first when moving away from it (Fig. 12.11).

A comet has very weak gravity, so the gas and dust that are blown off the surface of the central nucleus easily escape to interplanetary space, forming the coma and comet tails. They are continuously vanishing into space on the breath of the Sun's radiation and winds. All this material is lost to the comet forever, and must be continuously replenished from the solid comet nucleus.

At the heart of a comet's coma lies a nucleus of solid material, no more than 10 kilometers across. The nucleus has been directly imaged from spacecraft that pass near it, measuring oblong shapes of about this size. If a comet moves close enough to the Earth it can be detected by transmitting radio waves to it and detecting the echo. Although this radar technique has not yet been used to directly image a comet nucleus, the echoes detected from several comets provide cross-sections that are consistent with nucleus diameters of a few kilometers.

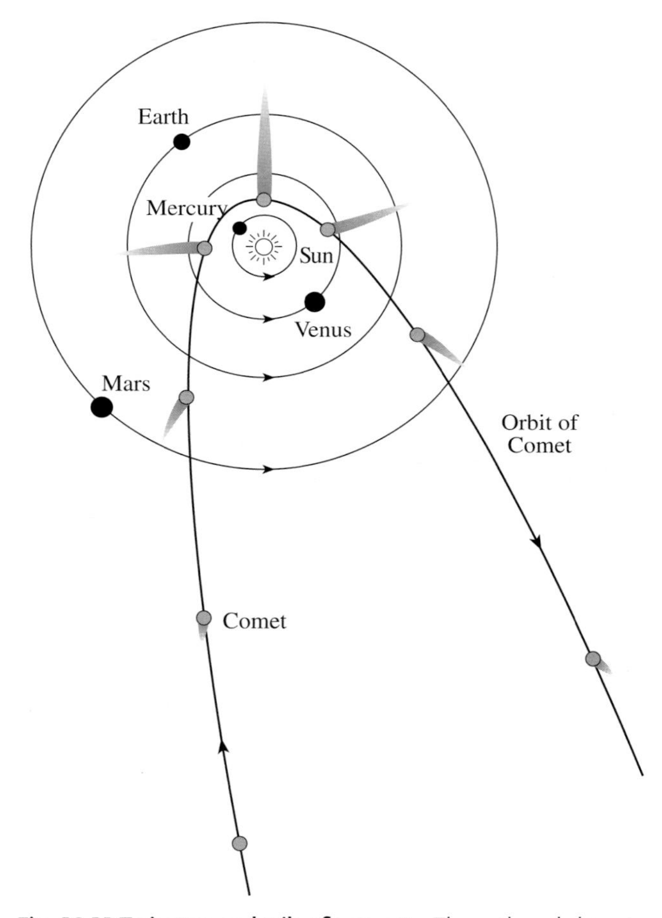

Fig. 12.11 Trajectory and tails of a comet The path and changing shape of a typical comet as it enters the inner solar system. Note that the tail of the comet is oriented away from the Sun, independent of the direction of travel of the comet.

With a mass density somewhere between that of water ice and rock, or between 1000 and 3000 kilograms per cubic meter, the comet nucleus would have a mass of just 10^{15} kilograms or less. This is less than one-billionth the mass of the Earth, which weighs in at 5.97×10^{24} kilograms.

The visible coma, or head, is a spherical cloud of gas and dust that has emerged from the nucleus, which it surrounds like an extended atmosphere. Large comas sometimes reach a billion meters in size, which is about as large as the Sun, and most comas usually become bigger than the Earth (Table 12.4).

A vast hydrogen cloud, containing hydrogen atoms that emit ultraviolet radiation, envelops the coma and nucleus. Observations of this glow – invisible to the eye – indicate that the hydrogen halo can be ten billion, or 10^{10}, meters across (Fig. 12.12), or about ten times bigger than the Sun. The atomic hydrogen is produced when water molecules, released from the comet nucleus, are torn apart by energetic sunlight. The lightweight hydrogen atoms travel at high speed to great distances before they are also ionized by the Sun's energetic light and swept away by its winds.

If the comet travels closer to the Sun than about 1.5 AU, it usually develops a gossamer tail; but many comets that stay outside this distance have no tail. The wispy, ghost-like tails are paler and more tenuous than the coma. The long, flowing tails sweep across the sky in regal splendor, attaining lengths of ten billion, or 10^{10}, meters, and even one hundred billion, or 10^{11}, meters, about 1 AU. Thus, the tails of comets can briefly become the largest objects in the solar system.

Yet the comet tails appear much more substantial than they really are. You can sometimes see stars shining through the tails as if they were not there at all. It is therefore no wonder that the Earth has passed through many comet tails unscathed. So, comet tails look awesome, but they contain so little matter that they are very close to being nothing at all.

Some comets show two types of tail at the same time. There are the long, straight blue ion tails and the shorter, curved yellow dust tails. The gases liberated by a comet nucleus become ionized by the action of solar ultraviolet radiation and emit a faint blue light by fluorescence. The dust tail shines only by reflecting yellow sunlight. Since the individual dust particles enter slightly different orbits of their own, the dust tail often spreads out into a fan shape

Table 12.4 Structural features of a comet

Feature	Size	Composition	Appearance
Nucleus	100–100 000 meters	Dust, ice and rock	Very dark
Coma	Up to 0.01 AU[a]	Neutral (un-ionized) molecules and dust	Slightly yellow
Hydrogen Cloud	Up to 0.1 AU	Hydrogen atoms	Ultraviolet radiation
Dust Tail	Up to 0.1 AU	Dust particles	Yellow, curved
Ion Tail	Up to 1 AU	Ionized molecules	Blue, straight

[a] One AU, roughly the average distance between the Earth and the Sun, is about 150 billion, or 1.5×10^{11}, meters.

(Fig. 12.13). An individual comet may have a dust tail, an ion tail, both types of tail, and no tail at all.

But what are the solar forces that blow the gas and dust into comet tails? The gentle pressure of sunlight pushes the tiny, solid dust grains along curved paths as the comet moves through space. When the Sun's light bounces off the dust particles, it gives them a little outward push, called radiation pressure, and this forces them into a dust tail. For larger solid particles, comparable in size to sand or pebbles, the Sun's gravitational pull overcomes the radiation pressure, and so these particles stay near the orbital path of the comet and they do not enter the dust tail.

A solar wind of electrically charged particles and magnetic fields propels and constrains the ions on straight paths away from the Sun. The solar wind, which continuously flows away from the Sun's surface, also accelerates the ions to high velocities.

Thus, the ion tail acts like a windsock and, in fact, the existence of the solar wind was hypothesized from observations of comets before the age of space exploration. Spacecraft have now confirmed these predictions, and have permitted measurements of the electrons and protons that are blown away from the Sun, carrying the solar magnetic fields with them (Section 3.5).

The gas lost from a comet is ionized by ultraviolet sunlight, producing an ionosphere that envelops the comet nucleus. Magnetic fields carried by the solar wind are unable to penetrate the ionosphere, so they pile up in front of it and drape around it to form nearly parallel, adjacent magnetic field lines that point toward and away from the Sun (Fig. 12.14). Guided and constrained by these folded magnetic field lines, the comet ions are pushed away from the Sun by the much faster solar-wind particles, forming a straight, blue ion tail. These features of the interaction

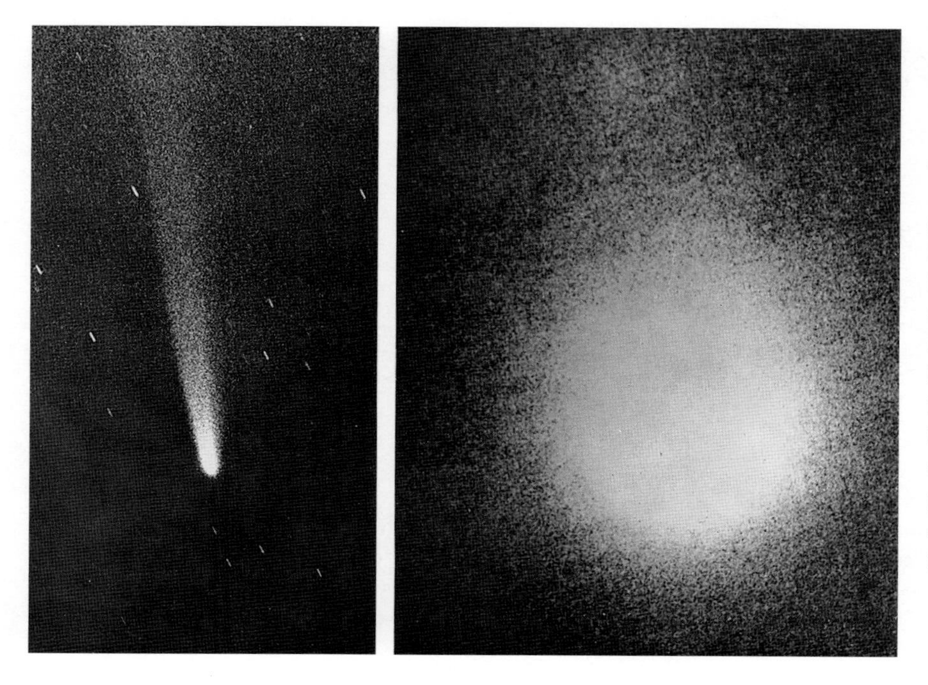

Fig. 12.12 Hydrogen cloud A comparison of the visible image (*left*) of Comet Kohoutek with a far-ultraviolet image (*right*) on the same scale, taken from Aerobee rocket flights on 4 and 7 January 1974. The far-ultraviolet image shows a gigantic cloud of hydrogen nearly 10 billion, or 10^{10}, meters in size, or eight times bigger than the Sun. It is being fed by the comet nucleus at the rate of 500 billion billion billion, or 5×10^{29}, atoms of hydrogen every second. The large size of the hydrogen cloud is due to the fact that hydrogen atoms are much lighter than the other atoms, ions, molecules and dust particles which produce the visible light of the coma. (Courtesy of Chet B. Opal, Naval Research Laboratory.)

Fig. 12.13 Dust tail This photograph of Comet West (1976 VI) shows a broad, curved, pearly-hued dust tail. Because dust particles scatter sunlight, the dust tail has a slightly yellow color. It has a delicate lacy structure, created by countless dust particles shed from the comet nucleus over many days. (Courtesy of Stephen Larson, LPL, University of Arizona.)

between the comet and the solar wind have been confirmed by instruments aboard the *Giotto* spacecraft for Comet Halley, and the *Interplanetary Cometary Explorer*, or *ICE*, spacecraft for Comet Giacobini–Zinner.

But the interplanetary magnetism extending from the Sun is divided into sectors that point in opposite directions, toward and away from the star. When a comet crosses from one sector to another, the magnetism that envelops its ion tail becomes pinched and the comet loses the tail, somewhat like a tadpole (Fig. 12.15). But unlike a tadpole, the comet soon grows another ion tail.

To sum up, a comet's anatomy consists of a concealed nucleus, an Earth-sized or Sun-sized coma, a vast hydrogen cloud, and two types of tail, the dust and ion tails (Fig. 12.16). But a comet's anatomy is not a static thing, for comets are always changing shape. All comet tails grow when the comet approaches the Sun, and shrink when the comet moves away from the Sun. There is no such thing as a typical comet tail. They differ in shape, size and structure.

12.6 Spacecraft glimpse the comet nucleus

An international flotilla of six spacecraft, belonging to four space agencies, flew by Comet Halley in March 1986, to examine the gas and dust in the vicinity of the comet and to photograph its nucleus. Japan launched the *Sakigake* and *Suisei* spacecraft, meaning "pioneer" and "comet",

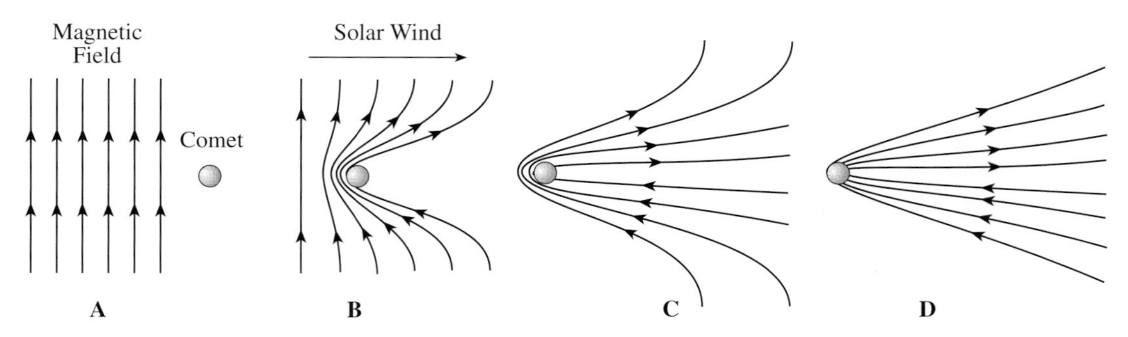

Fig. 12.14 Interaction between a comet and the solar wind Magnetic field lines entrained within the solar wind (A) are unable to penetrate the sphere of ions that envelop a comet nucleus, and so they pile up in front of it and drape around it (B). An ion tail forms on the side of the comet facing away from the Sun (C and D). The ions flow away from the Sun between the oppositely directed magnetic field lines in the tail. When a comet enters a region where the original magnetic field lines in the solar wind (A) change direction, the comet loses its ion tail and soon grows another.

Fig. 12.15 Losing a comet tail Photograph of Comet Halley taken on 6 June 1910, showing part of its ion tail that has become disconnected and does not run into the coma. (Courtesy of the Yerkes Observatory.)

respectively, which observed the comet from a safe distance and measured the interaction of the solar wind with the comet's atmosphere. The American probe *International Cometary Explorer*, or *ICE* for short, also examined the solar wind upstream from the comet. *ICE* had already flown through the tail of the short-period comet 22P/Giacobini–Zinner on 11 September 1985. The two Soviet probes, named *Vega 1* and *Vega 2*, penetrated to 8890 and 8030 kilometers of the sunlit side of the nucleus, and the European *Giotto* approached to within 596 kilometers. These spacecraft were taking a trip back through time to the epoch when comets were born, determining conditions prevailing when the solar system was formed.

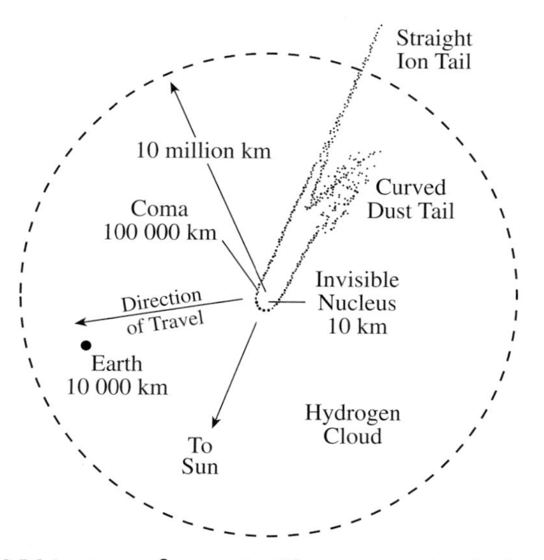

Fig. 12.16 Anatomy of a comet What you see when looking at a comet depends on how you look at it. The nucleus of a comet is usually invisible, unless a spacecraft is sent in to take a glimpse. A comet first becomes visible when it develops a coma of gas and dust. When the comet passes closer to the Sun, long ion and dust tails become visible, streaming out of the coma in the direction opposite to the Sun. When looking at a comet in ultraviolet light, the hydrogen atoms in its huge hydrogen cloud are detected.

The major objective of the cameras on board both the two *Vegas* and *Giotto* was to image the bare surface of a comet nucleus, which no one had ever seen. When viewed from a distance, the nucleus is hidden within the brilliant glare of the coma's fluorescing gases and reflected sunlight. All three spacecraft penetrated the coma of Comet Halley and detected its nucleus, despite damage by the hail of comet particles.

Giotto obtained the best images, with the highest resolution (Fig. 12.17). The surface of the nucleus was charcoal-black, reflecting only about 4 percent of the sunlight that falls on it. Bright jets were spewing gas and dust from the comet's sunlit side, but there wasn't much white ice in sight.

The ice has evaporated away from the outer layers of the nucleus, to leave a dark, tar-like crust all over the surface. When Halley passes close to the Sun, and the solar heat activates the comet, gas jetting out from cracks and holes in its surface carries dust with it. But when the comet is far away from the Sun at the other end of its orbit, some of the black dust settles back down over the surface to form a dark crust, rich in carbon compounds.

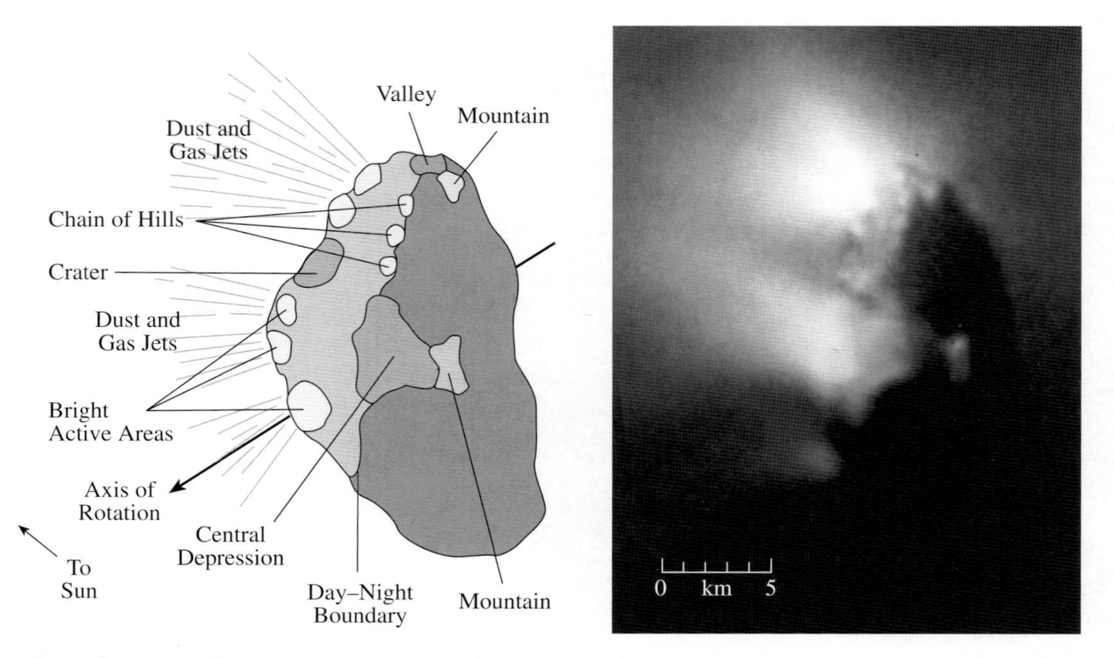

Fig. 12.17 Nucleus of Comet Halley A composite image of the nucleus of Comet Halley (*right*) obtained using images taken in March 1986 with the camera on board the *Giotto* spacecraft, from a distance of 6.5 thousand kilometers before comet dust destroyed the camera. It is compared with a schematic drawing (*left*) that highlights the major features recognizable in the photograph. The nucleus is about 16 kilometers long and 8 kilometers wide. Dust and gas geyser out of narrow jets from the sunlit side of the nucleus, but about 90 percent of the surface is inactive. The gas is mainly water vapor sublimed from ice in the nucleus, while a significant fraction of the dust may be dark carbon-rich matter. A dark surface crust, which insulates most of the underlying ice, is blacker than coal, reflecting about 4 percent of the incident sunlight. "Mountains" rise about 500 meters above the surrounding terrain, while a broad "crater" is depressed about 100 meters. (Image courtesy of Harold Reitsema of the Ball Aerospace Corporation, and Horst Uwe Keller.)

Gas and dust can now only get out in vents where the crust has broken to expose the underlying ice, rather than from the whole sunlit hemisphere. The sunward jets are emitted from roughly ten percent of the total surface area. Nearly 90 percent of the surface was inactive at the time of observation.

The images obtained with the camera on *Giotto* showed that the nucleus of Comet Halley has an elongated, irregular shape with dimensions of $16 \times 8 \times 7$ kilometers, about the size of Paris or Manhattan. For a mass density about that of water, this volume corresponds to a mass of about 10^{15} kilograms or 1000 billion tons. A varied, lumpy topography was seen, with craters, valleys, hills, and mountains (Fig. 12.17).

When the nucleus of Comet Halley was at its maximum rate of gas emission, near its closest approach to the Sun, it released up to 1.7×10^{30} water molecules per second, or up to 56 tons (56 thousand kilograms) of water every second. On average, the comet nucleus was emitting 20 tons of gas every second during the *Giotto* encounter, and about 80 percent of that was water. This is comparable to a loss of about 1.4 million tons of water every day, the approximate amount of water need to supply the hydrogen haloes observed around other comets. The nucleus of Comet Halley was also ejecting about 10 tons of dust every second. This fully confirmed the icy conglomerate model in which a comet nucleus consists mainly of water ice and dust particles.

If the comet suffers such a high rate of loss of material during the few months when it is close to the Sun, then it must lose on the order of 100 million tons, 100 billion kilograms, of material during the course of each orbit. With a total mass of 100 billion tons, Comet Halley can survive a further one thousand orbits, or 76 thousand years, before it is wasted away.

We now know that Comet 19P/ Borrelly also has a black heart, with a nucleus that is just as dark and unreflective as that of Comet Halley. The *Deep Space 1* spacecraft was directed toward this comet after completing its primary mission of flight-testing an ion engine and eleven other advanced technologies. On 22 September 2001, the spacecraft whizzed by Comet Borrelly at a distance of just 2.2 thousand kilometers, revealing an irregular chunk of rock and ice, about 8 kilometers long and perhaps 4 kilometers wide (Fig. 12.18). It is covered with a dark, carbon-rich slag that reflects only about four percent of the incident sunlight, on average, comparable to the reflectivity of the powdered toner used in laser printers. The surface of this nucleus

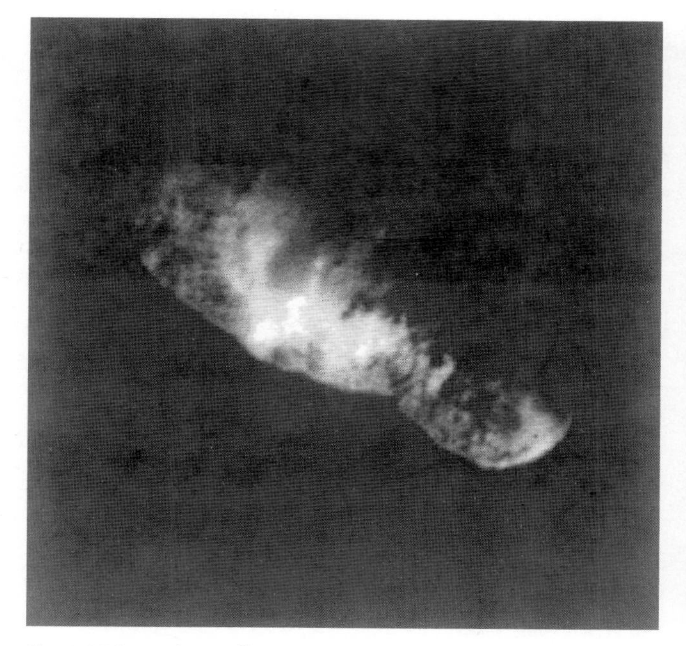

Fig. 12.18 Nucleus of Comet Borrelly A camera on board the *Deep Space 1* spacecraft peered into the icy heart of Comet Borrelly on 22 September 2001, taking this image from a distance of 3.4 thousand kilometers. The nucleus is shaped like a gigantic bowling pin, with a length of about 8 kilometers and a width of roughly half that size. A dark veneer of material covers most of the nucleus, reflecting only 4 percent of the incident sunlight, on average. Rugged terrain is found on both ends of the nucleus, while bright smooth plains are present in the middle. Jets of gas and dust shot out from all sides of the comet's nucleus as it rotated, producing a flow of ions that was not centered on the nucleus. (Courtesy of NASA and JPL.)

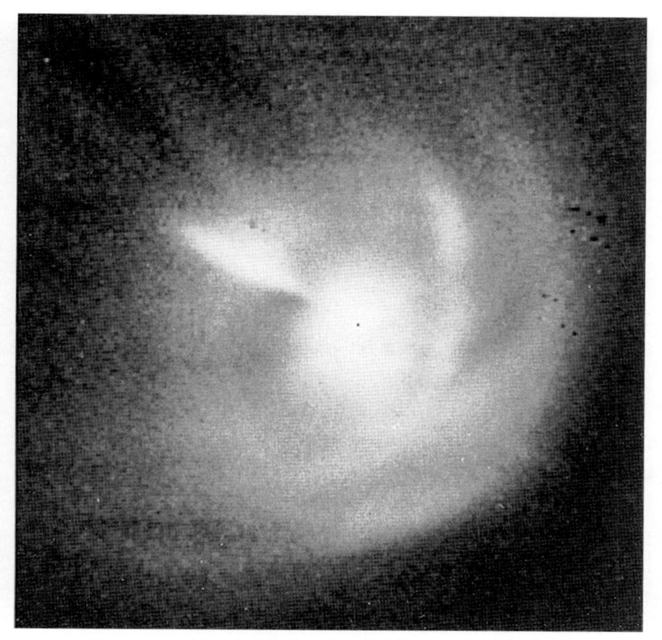

Fig. 12.19 Rotating comet This photograph of Comet Halley shows jets of dust ejected from a rotating nucleus. Stephen Larson and David Levy took the image on 6 January 1986 with the 1.55-meter (61-inch) Catalina reflector on Mount Lemmon. They used a CCD camera and a red filter to enhance the dust component of Halley's light. (Courtesy of Stephen Larson, LPL, University of Arizona.)

also has a rugged terrain, with mountains, valleys, deep fractures, and smooth rolling terrain where jets of gas and dust have apparently polished the surface. Some of the jets vastly exceed the nucleus in length, resulting in an asymmetric coma that was offset from the center of the nucleus by up to two thousand kilometers.

Because Borrelly is a short-period comet, circling the Sun in just 6.9 years, it has built up a thick surface layer, left behind each time the comet's ices sublimate and then cool. The terrain viewed by *Deep Space 1* exhibited different shades of black, with no traces of water ice in sight. The comet was near enough to the Sun that any surface ice would change quickly to a gas, leaving the dark crust behind, like a dirty street a few days after a snowstorm. The ice is there, but hidden beneath a dark, dry surface.

A comet's nucleus also rotates (Fig. 12.19). Typical rotation periods are a few hours to a few days. Observations of Comet Halley, for example, indicate that it rotates around its longest axis once every 7.4 days, and that it wobbles about its shortest axis once every 2.2 days or 53 hours. As the nucleus rotates, new regions turn to face the Sun, heat up and become active, while others face away from the Sun and momentarily turn off their activity.

Jets from a rotating comet nucleus can help explain comets that seem to defy gravity by arriving at perihelion before or after the time calculated using Newton's theory of gravitation. Some comets are speeding up while others are slowing down. Halley's comet, for example, last returned closest to the Sun four days later than it would under the influence of gravity alone, so its latest orbit lasted four days longer than the previous one. Sunward jets of matter ejected from a spinning icy nucleus can explain these non-gravitational motions.

The gas and dust streaming off the sunlit side of a comet nucleus initially head toward the Sun, before being swept back into the comet tails. The expelled material pushes the comet in the opposite direction, making it arrive sooner or later than expected (Fig. 12.20). A similar recoil effect explains the darting action of a small balloon when it is released, as well as the forward thrust of a rocket engine. The thrust of the rocket-like, comet jets either pushes the comet along in its orbit or slows it down.

The extraordinary success of the *Giotto* mission to the dark core of Comet Halley and the *Deep Space 1* encounter

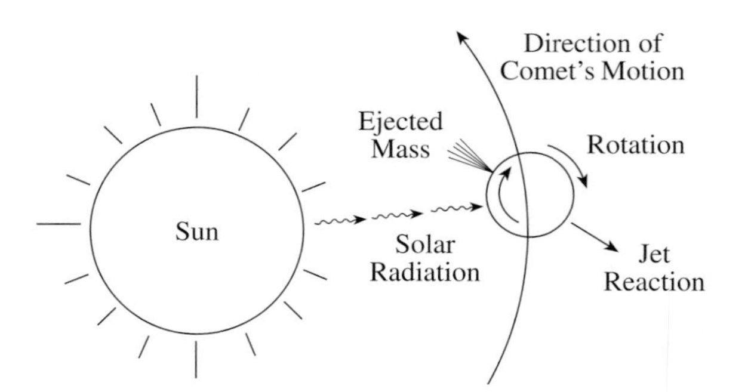

Fig. 12.20 How to make a jet engine out of a dirty ball of ice Unexpected cometary motions are attributed to non-gravitational forces caused by jets of matter ejected from a comet's spinning, icy nucleus. In this illustration, the ejected material pushes the comet in the opposite direction to its motion, causing the comet to arrive closest to the Sun at a time later than expected. If the comet had been rotating in the opposite direction, the jets would have pushed the comet along in its original direction, resulting in an early arrival time.

with the icy heart of Comet Borrelly have stimulated future missions to the nucleus of other comets (Focus 12.1).

12.7 Comet decay and meteor showers

Once a comet enters the inner solar system, it returns again and again on a relentless voyage of continual decay and disintegration. Most comets are consumed by their own emissions, blowing themselves away as they sublimate the ice that holds them together. Sooner or later most short-period comets will either fall apart or turn into a dark rocky corpse that looks like an asteroid. It's just part of the aging process. The only way to avoid this fate is to be thrown out of the solar system through close passage to Jupiter, or to collide with one of the planets or the Sun.

Comets are very fragile, with little internal strength and a very low mass density. Some comet nuclei are crumbly, fluffy structures with mass densities less than that of solid ice and far less than solid rock. The central pressure of their nucleus is probably comparable to that under a thick layer of blankets. So it is little wonder that some comets have been observed to break up as the result of tidal forces from either the Sun or Jupiter. These massive objects pull on the near side of the comet a little more than the far side, tearing the comet apart.

As an example, shortly after its closest approach to the Sun, the Great September Comet (1882 II) divided into four or more pieces stretched along nearly the same orbit like a string of pearls. The nucleus of Comet West (1976 VI) also split into pieces when it passed near the Sun in 1976

Focus 12.1 Future missions to the comets

NASA's *Stardust* spacecraft, launched on 7 February 1999, will be the first mission to collect samples from a comet and return them to Earth. The pioneering spacecraft will encounter Comet Wild 2 in January 2004, and swing by the Earth two years later, parachuting the comet samples to a desert in Utah. This comet has made only three passes around the Sun since Jupiter's gravity pulled it into its present orbit, leaving much of its core material intact. *Stardust* will catch the pristine comet dust on a paddle covered with an exotic substance called aerogel, and commonly referred to as frozen smoke for its hazy appearance. One version of aerogel has such a low mass density, of just 3 kilograms per cubic meter, that the *Guinness World Records* has recognized it as the least-dense solid known.

The dust particles would evaporate when striking a conventional sheet of metal, but the tenuous aerogel will capture them intact, somewhat like a small pellet shot into a Styrofoam container. When the comet dust is examined in the terrestrial laboratories, scientists expect to learn more about the origin of the solar system and possibly to identify the carbon-rich molecules that might have been a necessary precursor to life.

ESA's *Rosetta* spacecraft will rendezvous with Comet Wirtanen in 2011, eight years after its launch, expected in 2003. *Rosetta* will be the first spacecraft to orbit a comet, obtaining a detailed look at its nucleus. Its instruments will determine the detailed shape and composition of the nucleus, and examine organic molecules and other substances released by it. After encircling and studying the comet remotely for about 20 months, a landing vehicle will swoop down to inspect the surface of the nucleus from close up. Prior to arrival at Comet Wirtanen, the *Rosetta* spacecraft will fly by asteroid 4979 Otawara in July 2006 and asteroid 140 Siwa in July 2008, so it will also obtain information about the surface and composition of these asteroids. The mission is named after the Rosetta stone, an ancient tablet that was discovered in 1799 near the small Egyptian town of Rosetta (Raschid), providing a major key to the translation of Egyptian hieroglyphic writing.

NASA's *Deep Impact* spacecraft is scheduled for launch in January 2004, encountering Comet Tempel 1 in July 2005. *Deep Impact* will be the first mission to look beneath the surface of a comet by sending a 370-kilogram projectile into it, creating a fresh crater that is expected to be larger than a football field and deeper than a seven-story building. This will determine if large amounts of ice remain sealed inside this comet, or if the subsurface ice has been largely exhausted by past sublimation from the surface.

(Fig. 12.21). The pieces of the split nucleus have too little mass to pull themselves together gravitationally. So once a nucleus splits, its pieces remain forever separated. The jets of escaping gas kick them away from each other, and they continue to drift further apart.

Another dramatic example of comet disintegration was Comet Shoemaker–Levy 9, which broke into more than a dozen fragments when the comet passed near Jupiter in July 1992, and collided with the giant planet in 1994 (Section 14.1). Other comets are headed on a kamikaze collision course with the Sun. They plunge directly into the star, never to return (Section 14.2).

Any comet that we can see is slowly wasting away from the outside in. Each time it approaches the Sun, the solar heat causes the outer ice layers of the comet to sublimate and escape into space, dragging along grains of dust. An example is Comet Encke that revolves about the Sun with a 3.3 Earth year orbital period; it loses about a meter of ice and dust each time it passes near the Sun. If its nucleus were 10 kilometers across, then this comet would disappear in just 33 thousand years. Most of the comets seen during recorded history will probably vanish from sight in just a million years. They will either vaporize into nothing or their gas and dust will be preferentially removed, leaving behind a black, rocky, burned-out corpse that is indistinguishable from an asteroid – unless they are thrown out of the solar system or collide with one of the planets or the Sun.

Earth's cosmic dusting

Studies of tiny dents made in an Earth-orbiting satellite indicate that about 4×10^{7} kilograms, or 40 thousand tons, of cosmic dust falls from space onto Earth in a typical year. On average, each of the impacting particles has an insignificant mass, just 1.5×10^{-8} kilograms, and a miniscule diameter of 200-millionths, or 0.0002, of a meter. Some of these dust-like particles originate from comets, and some come from the asteroid belt between Mars and Jupiter, where billions of small rocks are constantly bumping into and pulverizing each other (Section 13.1).

Computer simulations and satellite observations indicate that the Earth's gravitational pull traps some of the dust particles into a ring that revolves around the Sun in

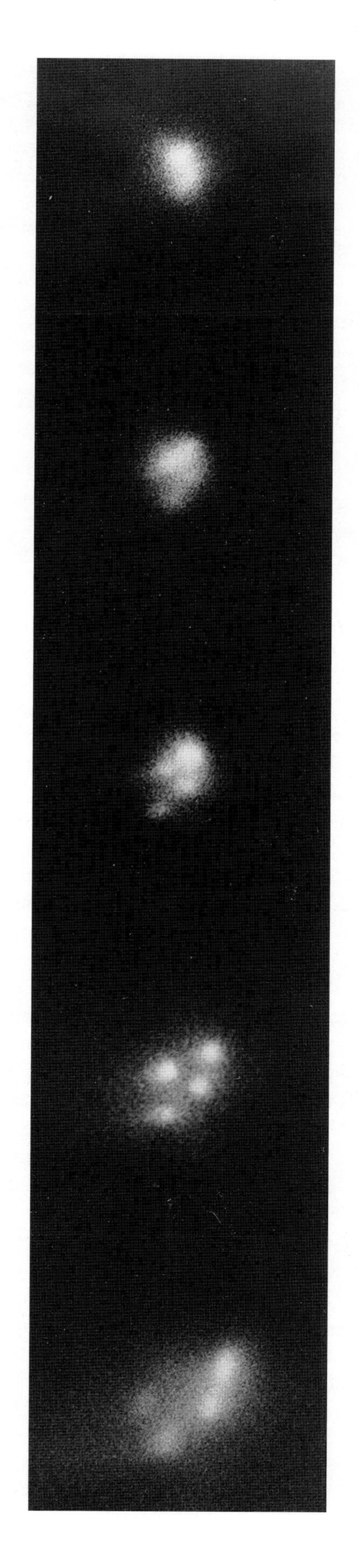

Fig. 12.21 Comet clones The splitting of the nucleus of Comet West (1976 VI) photographed (*top to bottom*) on 8, 12, 14, 18 and 24 March 1976, in yellow-green light using a 0.60-meter (23.6-inch) Cassegrain reflector. On 18 March, the diameter of the four features was about 10 thousand kilometers. (Courtesy of C. Knuckles and S. Murrell, New Mexico State University Observatory.)

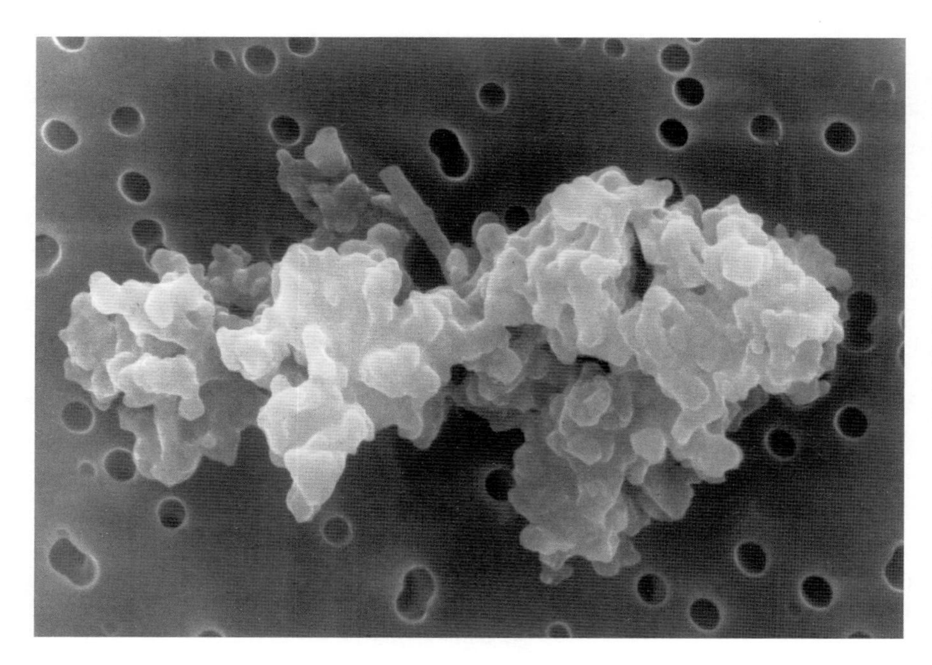

Fig. 12.22 Cosmic dust This interplanetary dust particle is probably of cometary origin. It is a mere one-ten-thousandth, or 0.0001, of a meter long. The particle was collected at 20 kilometers altitude by a U-2 aircraft, and then photographed at a magnification of 14 000 using a scanning electron microscope. The embedded crystals were probably formed in the primeval solar nebula from which the planets formed, or perhaps in the presolar interstellar environment. (Courtesy of Donald E. Brownlee.)

the Earth's orbit. Sunlight scattered off the interplanetary dust creates the zodiacal light, a luminous pyramid of light visible in the tropics. And much of the dust falls to the ground, accounting for most of the material raining down from the sky.

Cosmic dust is everywhere. It is in the air we breathe, the food that we eat, and the water we drink. Some of the dust that has been spawned by comets and asteroids even enters our own hair. The smallest dust particles are so tiny that the air slows them down rather than burning them up. Hundreds of them have been collected from the stratosphere and examined in the terrestrial laboratory, often with the fragile, porous structure expected of comet debris (Fig. 12.22).

The cosmic dust particles that we examine today are time capsules that may date back to the origin of the solar system. They have probably not been significantly altered from the moment of their creation. The delicate primordial comet dust is therefore thought to preserve a record of chemical conditions at the time of planet formation, and it may even contain the ashes of stars that existed before the Sun was born.

A few astronomers have endowed comets with life-giving and death-giving properties, arguing that they might breathe life into sterile planets or spread disease on life-bearing ones. A rain of comet dust may have brought early Earth not only water but the self-replicating, carbon-rich molecules necessary for life. The tiniest dust particles would be small enough to fall gently through the atmosphere, reaching the ground intact. Bigger pieces of comets may have wafted down through a thicker, denser atmosphere 3 or 4 billion years ago, permitting fertile comets to impregnate the ancient oceans with life. Other comets might have injected deep-frozen viruses into the Earth's air, bringing the sudden, widespread appearance of ancient plagues and modern flu. Nevertheless, most of these ideas are discredited speculations ungrounded in hard scientific evidence, viewed as wild conjectures of the fringe minority.

Nights of the shooting stars

Spectacular meteor showers, spawned by passing comets, have periodically returned for at least a thousand years, inspiring fear, wonder and admiration. William Blake (1757–1827) caught some of the excitement of the August shower, the Perseids, in his poem "The Tyger":

> When the stars threw down their spears
> And water'd heaven with their tears . . .
>
> *Songs of Experience (1794)*

Robert Frost (1874–1963) provided a different appraisal of the November shower, the Leonids, in "A Loose Mountain":

> Did you stay up last night (the Magi did)
> To see the star shower known as Leonid
> That once a year by hand or apparatus
> Is so mysteriously pelted at us?
> It is but fiery puffs of dust and pebbles,
> No doubt directed at our heads as rebels
> In having taken artificial light
> Against the sovereignty of night.

Fig. 12.23 The shooting stars Two couples portray meteor showers or falling stars in this picture painted by Jean-Francois Millet (1814–1875) in 1847. They soar through the skies, perhaps illustrating the transcendental nature of erotic love. (Courtesy of the National Museum of Wales, Cardiff.)

A fusillade of blanks and empty flashes,
It never reaches Earth except as ashes
Of which you feel no least touch on your face
Nor find in dew the slightest cloudy trace . . . .

A Witness Tree. New York, Henry Holt, 1942

Although meteor showers are commonly called shooting stars (Fig. 12.23), they are not stars, but fragile material from comets. In addition to spewing off small dust particles, which can drift down to the ground, comets also expel larger particles ranging in size from sand grains to pebbles. This debris burns up when it enters our atmosphere, producing visible meteors.

The meteors associated with comets are fragile and they vaporize completely in flight. In fact, no meteor associated with a comet has ever reached the ground. There is a great difference between the fragile meteoric material associated with comets and the tough meteorites that survive the atmospheric flight to the ground (Section 13.7).

When just one of the comet particles rubs against the air, it vaporizes in a streak of light, producing the luminous trail of a meteor (Fig. 12.24). And when many fall into the dark night sky, they produce meteor showers.

From the luminous path of a meteor, it is possible to determine the incoming particle's orbital path around the Sun, and in most cases the orbits are similar to those of comets (Table 12.5). A comet ejects the particles along its orbital path as it loops around the Sun, and this material continues to revolve around our star, something like the ice particles that circle Saturn in its ring. The swarm of comet material is called a meteoroid stream. And when the Earth passes though one of these streams, it intercepts some of the orbiting particles that enter our atmosphere and create a meteor shower (Fig. 12.25).

When a meteor shower includes large numbers of "shooting stars", the trails appear to intersect and emanate

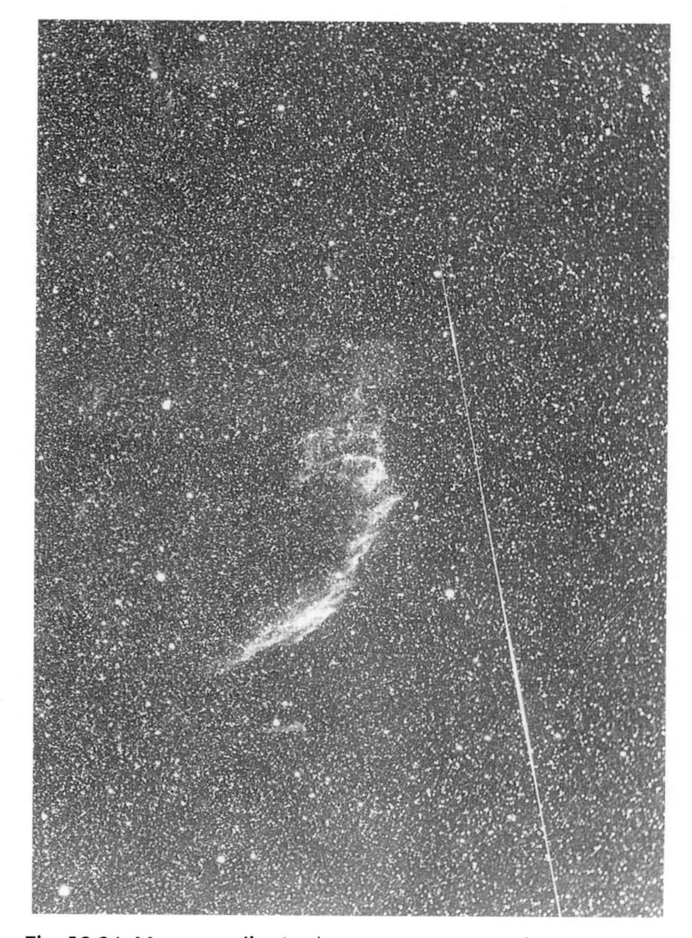

Fig. 12.24 Meteor trail A glowing meteor streaks across the backdrop of the stars near the constellation Cygnus. The straight trail was produced by a sand- or pebble-sized piece of a comet, burned up by friction as it entered the Earth's atmosphere. The curved structure (*left middle*), known as the Cygnus Loop, is an expanding shell of material thrown off during the supernova explosion of a massive, dying star. (Courtesy of the Yerkes Observatory.)

Table 12.5 Comets associated with meteor showers[a]

Meteor Shower[b]	Comet
Lyrids	Thatcher C/1861 G1
Eta Aquarids	1P/Halley
Scorpiids–Sagittariids	2P/Encke
Bootids	7P/Pons–Winnecke
Perseids	109P/Swift–Tuttle
Aurigids	Kiess C/1911 N1
Draconids	22P Giacobini–Zinner
Orionids	1P/Halley
Epsilon Geminids	Ikeya C/1964 N1
Taurids	2P/Encke
Adromedids	3P/Biela
Leonids	55P/Temple–Tuttle
Geminids	3200 Phaethon[c]
Monocerotids	Mellish C/1917 F1
Ursids	8P/Tuttle

[a] Adapted from Lang, Kenneth R., *Astrophysical Data: Planets and Stars*. New York, Springer-Verlag, 1992.
[b] For the visibility dates of these showers, see Table 12.6.
[c] 3200 Phaethon is cataloged as an asteroid, but it may be an inactive comet nucleus.

from a distant point called the radiant (Fig. 12.26). But meteors that appear to diverge from a point are actually moving on parallel paths, just as parallel railroad tracks seem to come from a point on the distant horizon. Meteor showers are named after the constellation in which their radiant appears (Table 12.6).

Because the Earth may pass through a comet's orbit just once a year, a meteor shower often appears at yearly intervals. The Lyrids, for instance, appear in April, the Perseids are seen every August, and the Leonids light up the night sky in November. Since the distribution of material along a comet orbit is generally non-uniform, the numbers of meteors in a particular shower can vary from year to year.

The Earth has been passing through some comet orbits for hundreds of years. The Lyrids were known in 687 BC, the Perseid stream was first recorded in 36 AD and the Leonids were first recorded in 902 AD. Nevertheless, the comet that produces a meteor shower will eventually disintegrate completely, and that shower will eventually vanish from sight. But this may not happen for a million years from now.

12.8 Crossover objects

Small bodies of the solar system

Astronomers have separated the small objects in the solar system into two main categories based on their telescopic appearance. Known as the comets and asteroids, they differ in composition, orbits, and beginnings. Most comets are dirty balls of ice with very elongated orbits that take them out to the distant reaches of the solar system, moving at every possible angle to the Earth's orbital plane. The asteroids are lumps of rock confined mostly to nearly circular orbits relatively close to the Sun, in the main belt between Mars and Jupiter, moving in the same plane and direction as the planets. Comets sublimate surface ice when they travel

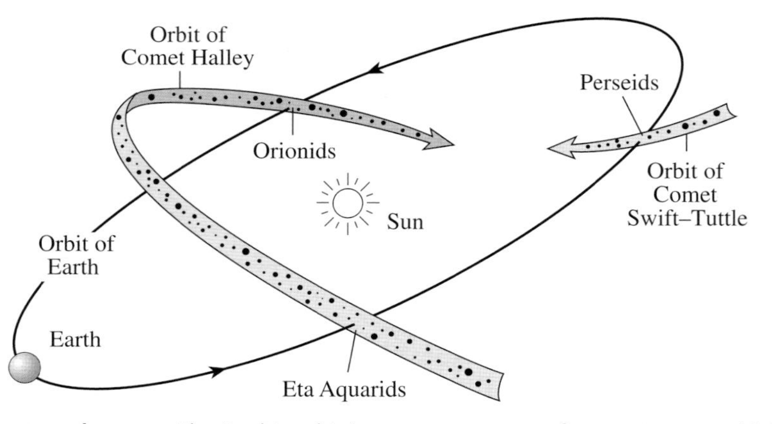

Fig. 12.25 Comets produce meteor showers The Earth's orbit intersects a stream of meteoric material left along the orbit of Comet Halley, producing two meteor showers, the Eta Aquarids in May and the Orionids in October. Other comets intersect the Earth's orbit just once during their trip around the Sun. Annual meteor showers are created when the Earth enters the intersection point, such as the August Perseids produced by debris from Comet 109P/ Swift–Tuttle. The orbit of Comet Halley is inclined by 162 degrees with respect to the ecliptic, the plane of the Earth's orbit, while the orbit of Comet 109P/ Swift–Tuttle has an inclination of 114 degrees.

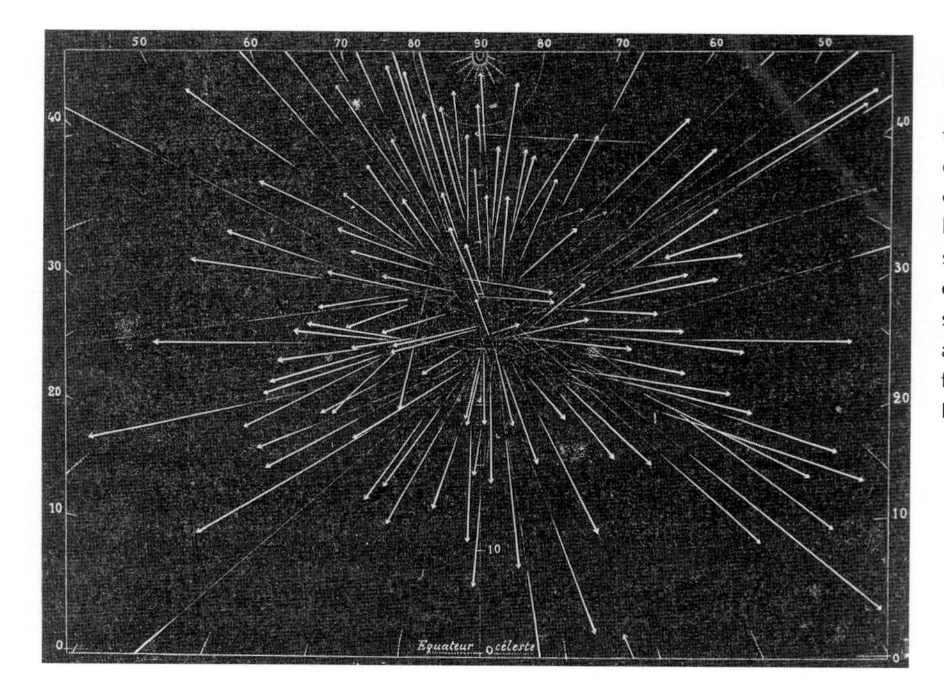

Fig. 12.26 Radiant meteors The apparent paths of "shooting stars" on 27 November 1872. Meteor showers are named after the constellation in which their radiant appears. This meteor shower is called the Andromedids meteor shower because its radiant appears in the constellation Andromeda. The shower occurs every November when the Earth intersects the debris that has been scattered along the orbit of Biela's comet. [Adapted from Amedee Guillemin's *Le Ciel*, Librairie Hachette, Paris (1877).]

close to the Sun, emitting gas and dust, but asteroids do not emit anything, wherever they are.

Both the comets and asteroids are thought to be leftover bits and pieces from the formation of the solar system, but differing in their place of origin. The long-period comets formed in the frozen region between the orbits of Jupiter and Neptune, and were subsequently thrown out to the distant Oort cloud, while many of the short-period comets formed just beyond the orbit of Neptune where they now reside, in the Kuiper belt. In contrast, the asteroids formed much closer to the Sun, within the main belt where the solar heat should have vaporized the ices away from most of them.

Recent discoveries have blurred the boundaries between these two classes of small objects. There are comets that behave like asteroids, and asteroids that act like comets. A handful of bodies have a dual personality, behaving like both a comet and an asteroid. So there is some overlap between the two categories, and our strict definitions have to be relaxed to take the anomalies into account.

Burned-out comets that look like asteroids

Although comets are distinguished by their ability to emit gas and dust when near the Sun, some comets may be either dormant or extinct. In fact, there are a few small objects that move like comets, but emit no gas and dust and display neither a coma nor a tail. Some of them have eccentric, elongated comet-like orbits that stretch into the vast outer reaches of the solar system, beyond the most-distant planets. Yet, they show no trace of comet activity when approaching the warmth of the Sun, sometimes nearer to it than the Earth. So these objects look like asteroids and are usually designated that way. Other apparent asteroids move into the sunward side of the main belt, approaching the Earth, but with orbits that resemble short-period comets.

All of these small bodies may be inert comets, which have turned into inactive objects that are indistinguishable from asteroids. These comets either exhausted all the volatile ices they once had to feed a coma and tail, or their ice might be completely shrouded in a thick, insulating cover of dust and dirt, preventing the ice from sublimating into luminous material.

The near-Earth asteroid 3200 Phaethon is a well-known example. It moves in a highly eccentric orbit – once thought to be the hallmark of comets. In fact, it follows the orbit of the meteoroid stream that produces the Geminids meteor shower. Since the large majority of meteor showers are caused by debris scattered along a comet's orbit, it is likely that this "asteroid" is a defunct comet that has now lost its ability to emit gas and dust.

In other words, some comets can turn into asteroids. The short-period Comet Encke could be a future example. It circles the Sun once every 3.3 Earth years, moving between 4.1 and 0.34 AU from the Sun. Once this comet runs out of gas, in about 33 thousand years, it will be indistinguishable from an asteroid.

Table 12.6 The principal annual night-time meteor showers[a]

Shower	Maximum Date	Radiant RA	Position[b] Dec.	Visibility Dates	Meteors Per Hour[c]
Quadrantids	3–4 January	15h 28m	+50°	1–6 January	110
Alpha Aurigids	6-9 February	04h 56m	+43°	Jan.–Feb.	10
Virginids	12 April	14h 04m	−09°	March–April	5
		13h 36m	−11°		5
Lyrids	21–22 April	18h 08m	+32°	19–25 April	15
Eta Aquarids	5 May	22h 20m	−01°	24 April–20 May	35
Alpha Scorpiids	28 April	16h 32m	−24°	20 April–19 May	5
	13 May	16h 04m	−24°		
Ophiuchids	9 June	17h 56m	−23°	May–June	5
	19 June	17h 20m	−20°		
Alpha Cygnids	21 July	21h 00m	+48°	June–August	5
Capricornids	8–15 July	20h 44m	−15°	5 July–20 Aug.	5
Alpha Capricornids	2 August	20h 36m	−10°	15 July–20 Aug.	5
Delta Aquarids	29 July	22h 36m	−17°	15 July–20 Aug.	25
	6 August	23h 04m	+02°		10
Iota Aquarids	6 August	22h 10m	−15°	July–August	10
Piscis Australids	31 July	22h 40m	−30°	July–August	5
Perseids	12 August	03h 04m	+58°	25 July–20 Aug.	80
Alpha Aurigids	28 August	04h 56m	+43°	August–October	10
Piscids	9 September	00h 36m	+07°	Sept.–Oct.	10
Orionids	21 October	06h 24m	+15°	15 Oct.–2 Nov.	30
Taurids	3 November	03h 44m	+14°	15 Oct.–25 Nov.	10
Leonids	17 November	10h 08m	+22°	15–20 November	45
Geminids	13–14 December	07h 28m	+32°	7–15 December	70
Ursids	22–23 December	14h 28m	+78°	19–24 December	10

[a] Adapted from Bone, N.: *Meteors*, Cambridge, Mass., Sky Publishing Co. 1993 and Lang, Kenneth R., *Astrophysical Data: Planets and Stars*. New York, Springer-Verlag 1992.

[b] The celestial coordinates of the radiant are right ascension, abbreviated as RA, in hours, h, and minutes, m; and declination, or dec. for short, in degrees,°.

[c] The maximum hourly rate of meteors is under the assumption that the radiant is at the zenith, but this rate can vary from year to year because of the non-uniform distribution of the relevant meteoroid stream.

Small objects that move like asteroids and look like comets

Some small bodies travel in low-inclination, low-eccentricity orbits that resemble those of asteroids, but develop comet-like activity. The main-belt asteroid 7968 Elst–Pizarro, for example, displayed a temporary dust tail.

There are other asteroids whose orbits take them outside the main asteroid belt, further away from the Sun and into the realm of the outer planets. They can move near the Kuiper belt of comets, and may have originated there. A few dozen of these so-called Centaurs circle the Sun within the orbits of the giant planets. They may be transition objects from the Kuiper belt of short-period comets, but with asteroid-like orbits.

The first known Centaur was designated 2060 Chiron, the two-thousand-and-sixtieth numbered asteroid. It travels between the orbits of Saturn and Uranus. But then it was discovered that Chiron has a dual personality, developing a luminous shroud of gas and dust just like a comet. So it is now also known as the ninety-fifth periodic comet, designated 95P/Chiron. A few of these objects have been discovered, receiving a dual designation as both asteroids and comets.

So, we now have comets in asteroid-like orbits, asteroids in comet-like orbits, and small objects that can behave like both comets and asteroids. The distinction between the two classes is therefore not as sharp as once believed. Yet, these anomalous bodies are in a definite minority, like the small numbers of deviant adults in the overall human population. They account for a very modest fraction of the thousands of known comets and asteroids. We therefore now turn our full attention to the ubiquitous asteroids.

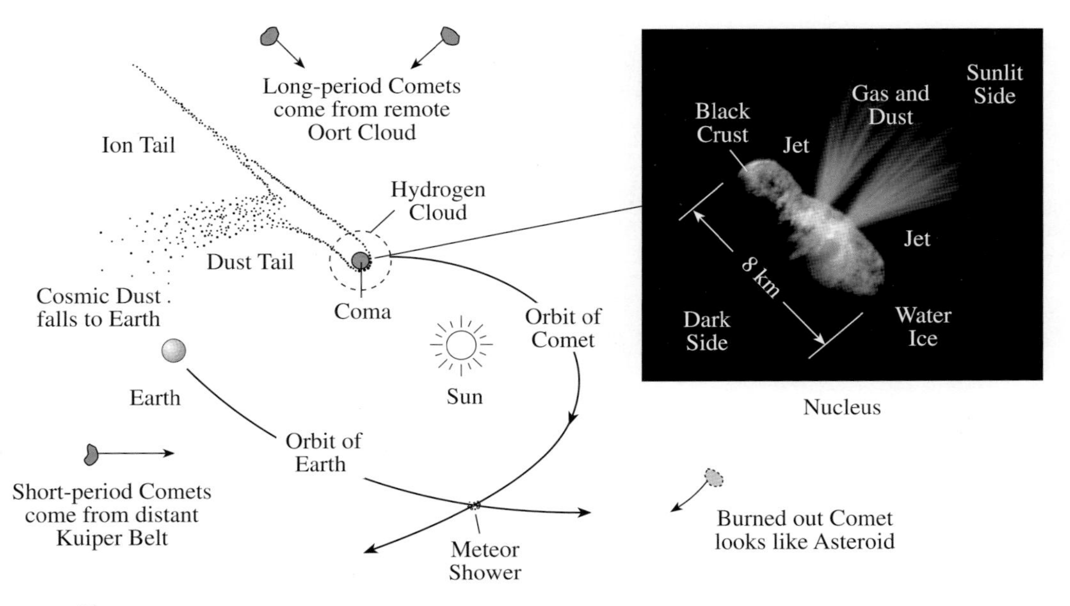

Fig. 12.27 Summary Diagram

13 Asteroids and meteorites

- There are billions of asteroids in the main asteroid belt, located between the orbits of Mars and Jupiter.
- The asteroid belt is largely empty space, and a spacecraft may safely travel through it.
- Hundreds of Trojan asteroids circle the Sun in the same orbit as Jupiter. These asteroids are located near the two Lagrangian points where the gravity of the Sun balances that of Jupiter.
- The Earth resides in a swarm of asteroids. Many of these near-Earth asteroids travel on orbits that intersect the Earth's orbit, with the possibility of an eventual devastating collision with our planet.
- Asteroids can be chaotically shuffled out of certain orbits in the main belt, and redirected into the inner solar system.
- The asteroids are the pulverized remnants of former worlds that failed to coalesce into a single planet.
- Groups of asteroids, known as families, have very similar orbits. The members of each family are the collisional fragments of a larger object, which was itself much smaller than a major planet.
- The biggest asteroid, 1 Ceres, is only about a third the size of the Moon and has less than two percent of the Moon's mass.
- The combined mass of billions of asteroids is less than five percent of the Moon's mass.
- An asteroid's gravity is too weak to hold onto an atmosphere or to pull asteroids into a spherical shape.
- The colors of sunlight reflected from asteroids indicate that they formed under differing conditions that prevailed at varying distances from the Sun.
- Roughly 75 percent of the main-belt asteroids are the dark, black carbonaceous C-type orbiting the Sun in the outer half of the belt; about 15 percent of asteroids are bright, red, silicate S-type, residing on the sunward side of the main belt.
- In years to come, asteroids may be mined for minerals or water.

- Periodic brightness variations tell us that most asteroids are non-spherical objects spinning with periods of a few hours.
- The close-up view, obtained by passing spacecraft, indicates that asteroids have been battered and broken apart during catastrophic collisions in years gone by.
- The *Near Earth Asteroid Rendezvous*, abbreviated *NEAR* and subsequently named *NEAR Shoemaker* was the first spacecraft to orbit an asteroid and the first to land on one. *NEAR Shoemaker* circled the near-Earth asteroid 433 Eros for a year, examining its dusty, boulder-strewn landscape in great detail, obtaining an accurate mass for the asteroid and showing that much of it is solid throughout.
- Radar images of asteroids reveal diverse shapes and surfaces ranging from solid metal to loose meteoritic rock, as well as binary asteroids.
- Asteroid rotation periods range from a couple of minutes to a few months, and a few asteroids wobble instead of rotating uniformly.
- Some asteroids are thought to be rubble piles, the collected fragments of past collisions held together by gravity; other asteroids are solid rocks of uniform composition.
- Meteorites are rocks from interplanetary space that survive their descent to the surface of the Earth or of any other object in the solar system.
- The number of meteorites recovered on Earth more than doubled when scientists discovered a large number of them on the blue ice of Antarctica.
- A few meteorites found on Earth may have been blasted off the Moon or Mars, but most of them are chips off asteroids.
- The organic matter found in meteorites recovered on Earth predates the origin of life on Earth by a billion years; but the meteoritic hydrocarbons are not of biological origin.

13.1 The orbits of asteroids

The main belt

Millions of asteroids are confined within a wasteland between the orbits of Mars and Jupiter, like so much rubble left over from the creation of the solar system. Most of them occupy a great ring, known as the asteroid belt, at mean distances of 2.2 to 3.3 AU from the Sun and with orbital periods of 3 to 6 Earth years. For comparison, the mean distance between the Earth and the Sun is roughly 1 AU, or one astronomical unit, about 150 billion, or 1.5×10^{11}, meters. Not all asteroids lie in this belt, but those that do are said to belong to the main belt.

The Sicilian astronomer Giuseppe Piazzi (1749–1826) discovered the first asteroid accidentally, on 1 January 1801, the first day of the 19th century. He named the tiny object Ceres, in honor of the patron goddess of Sicily. Another asteroid, named Pallas, was located the following year, and the third and fourth, designated Juno and Vesta, were found in 1804 and 1807, respectively. Ceres is the largest asteroid, with a radius of 475 kilometers, about twice as large as Pallas and Vesta.

We now know that there are very many smaller asteroids. After a gap of nearly a century, astronomical photography enabled them to be found by the hundreds. During a long-exposure photograph made through a telescope tracking the stars, the faster-moving asteroids make short trails while the stars look like dots.

When an asteroid is discovered, it receives a temporary designation, consisting of the year of discovery followed by two letters. The first letter indicates the half-month of the asteroid discovery – the letters "I" and "Z" are not used, to make a total of 24 half-months. The second letter shows the order of discovery within that half-month, but the letter "I" is not used for this second letter. Thus, the asteroid 1998 KY was the 24th (letter Y) asteroid found during the second half of May (letter K).

After an asteroid has been observed often enough for an accurate orbit to be established, it receives a number corresponding to the chronological order of reliable orbit determination – but not the order of discovery. The number is often followed by a name provided by the discoverer. There are about 100 000 asteroids with reliably known orbits, and the list keeps growing.

With a million asteroids whirling around the Sun, the main belt was once thought to be a hazard to space travel. In reality, however, the volume of space they occupy is so large that any one asteroid is typically several million kilometers away from its nearest neighbor, and the asteroids are so small that it would be hard to hit one even if you tried to do so. Thus, despite their vast numbers, the asteroids leave plenty of room for space flight. The *Pioneer 10* and *11*, *Voyager 1* and *2*, *Ulysses*, *Galileo*, and *Cassini* spacecraft have all passed through the asteroid belt unharmed.

And since the asteroid belt is largely empty space, collisions between asteroids are relatively rare over short timescales. It takes tens of millions of years for shattering collisions to occur, so they happen, but it takes a long time.

Asteroids are also known as minor planets, but most of them differ noticeably from the familiar major planets. Because of their small size, asteroids have very little mass or gravity. Unlike the larger planets, the asteroids are not big enough to hold onto an atmosphere. Most asteroids are also too small to retain a spherical shape.

The majority of asteroids are not large enough to have ever had a molten core, and magma has never flowed out from the inside of most of them. So, except for ongoing collisions, most asteroids have been unaltered since the formation of the solar system about 4.6 billion years ago. In contrast, the surfaces of large planetary bodies have been destroyed by eons of planetary evolution, including towering volcanoes and extensive lava-flows across their surfaces.

Trojan asteroids

Not all asteroids are found in the main belt. An especially interesting type is further away from the Sun than the asteroid belt. These asteroids move along Jupiter's orbit, keeping pace with the giant planet. The first known one, 588 Achilles, was discovered photographically by the Heidelberg astronomer Maximilian Franz Joseph Cornelius Wolf (1863–1932) in 1906. Hundreds of them are now known, traveling on both sides of Jupiter in two clouds, one preceding the giant planet and one following it (Fig. 13.1). As with Achilles, they are all named after heroes of the Trojan War and they are therefore collectively known as the Trojan asteroids.

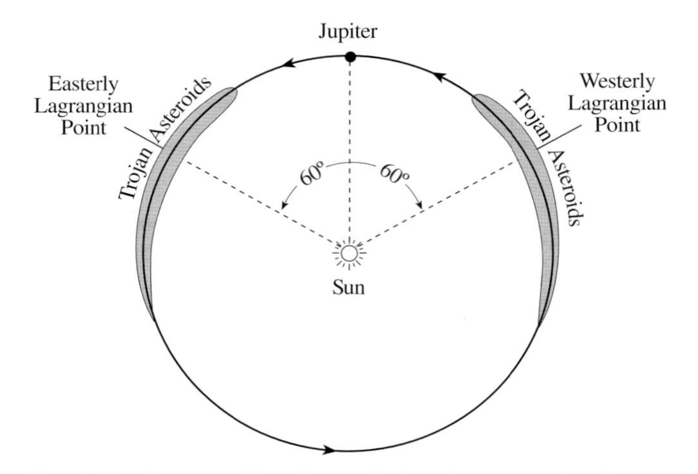

Fig. 13.1 Trojan asteroids Asteroids that share Jupiter's orbit are known as the Trojan asteroids. They are located near the point where the gravitational force of Jupiter and the Sun are equal. The gravitational perturbations of the inner planets produce slight swinging motions, so the Trojan asteroids oscillate within the two shaded regions. Some of the Trojan asteroids may occasionally move close enough to be captured by Jupiter's gravity, thereby accounting for the planet's unusual outer satellites.

The Trojan asteroids are held captive by the gravity of both Jupiter and the Sun. They are found near two of the five Lagrangian points, named after the Italian-born French mathematician Joseph Louis (Guiseppe Lodovico) Lagrange (1736–1813) who predicted their existence 134 years before the discovery of 588 Achilles. At these points, the gravitational force of Jupiter is equal to that of the Sun, which is much more massive than Jupiter but also a lot further away from the asteroids. These Lagrangian points lie in the corners of equilateral triangles that have Jupiter and the Sun at the other corners (Fig. 13.1).

The Trojan asteroids do not stay precisely at a Lagrangian point, but instead oscillate around it. They pace back and forth along Jupiter's orbit in paths that take them toward and away from the planet over a cycle lasting hundreds of years.

Once locked into their haven near the Lagrangian points, the Trojan asteroids move at slow speeds along well-defined paths. The Trojans therefore suffer fewer collisions than their counterparts in the main asteroid belt, and their surfaces have probably gone unchanged for billions of years. Thus, the Trojan asteroids may be pristine remnants of the early solar system.

Near-Earth asteroids

Although the vast majority of asteroids travel in the main belt lying between the orbits of Mars and Jupiter, there are some notable exceptions that reside within the inner solar system. Known as the near-Earth asteroids, they move inward toward our planet as they travel around the Sun.

One of them could hit our planet someday, with devastating consequences. So astronomers are still actively trying to find all of these interlopers and monitor their motions, with the hope of avoiding the collision (Section 14.6). Since these asteroids move closer to the Sun than Mars, they travel at faster speeds than asteroids in the main belt, and thus make longer trails on images taken to discover asteroids.

There are three populations of near-Earth asteroids, called the Atens, Apollos and Amors (Fig. 13.2). Both the Aten and Apollo asteroids move on eccentric orbits that can cross the Earth's path in space. The Atens are always close to the Sun, never moving out as far as the orbit of Mars. The elongated orbits of the Apollo objects loop in from the main belt to within the Earth's orbit. The Amors travel around the Sun between the orbits of Mars and the Earth, and often cross the orbit of Mars.

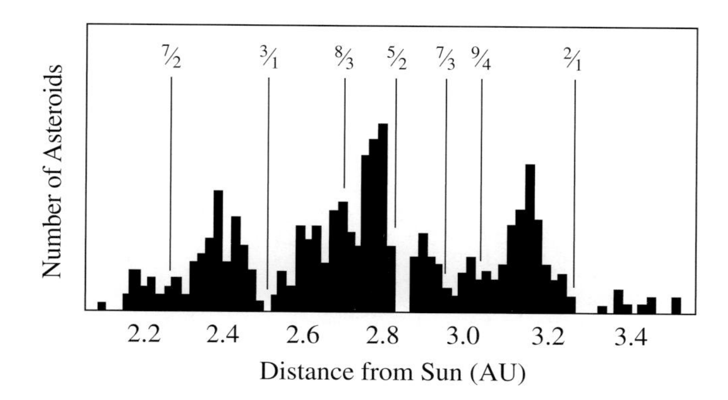

Fig. 13.3 Orbital resonance with Jupiter There are very few asteroids at certain distances from the Sun, due to resonance with the orbit of Jupiter. Vertical lines in this diagram of the first 400 numbered asteroids denote these vacancies, called Kirkwood gaps. The ratio 3/1 means that an asteroid at that distance makes 3 revolutions around the Sun for each 1 revolution completed by Jupiter. Asteroids are tossed out of such resonant orbits by Jupiter's repeated gravitational perturbations. (Adapted from Charles T. Kowal's *Asteroids – Their Nature and Utilization*, Second Edition. John Wiley, New York 1996.)

Chaotic orbits

Why do some asteroids move near the Earth, while most of them stay in the asteroid belt? Gaps of missing asteroids in the main belt provide some clues to these wandering interlopers. These are the Kirkwood gaps (Fig. 13.3), discovered long ago by the American astronomer Daniel Kirkwood (1814–1895) – also see Section 1.2.

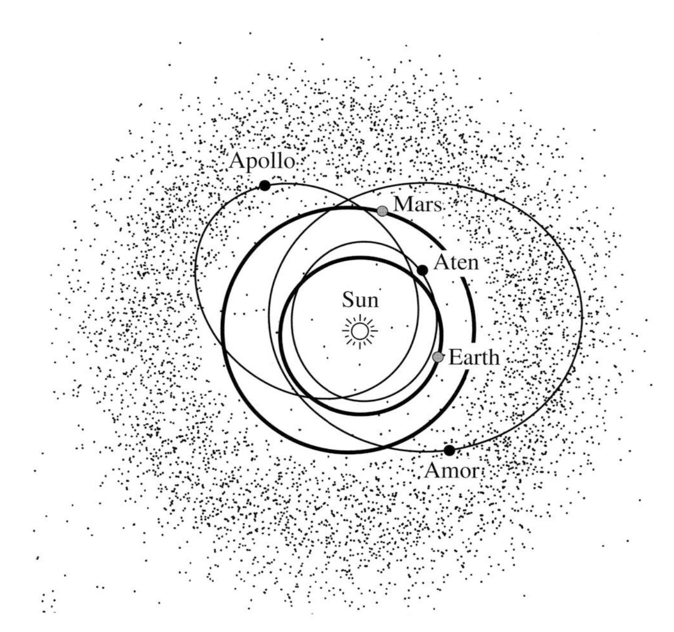

Fig. 13.2 Near-Earth asteroids The paths of three representative near-Earth asteroids, 1221 Amor, 1862 Apollo and 2062 Aten, all come closer to the Sun than most asteroids, located in the main belt beyond the orbit of Mars. Amor crosses the orbit of Mars, and almost reaches the Earth's orbit. Apollo crosses the orbits of Mars, Earth and Venus (not shown). Aten is always fairly close to the Earth's orbit.

The locations of these clearings correspond to orbital resonance with Jupiter, in which the orbital periods are exact fractions of the giant planet's period. Any object that orbits the Sun at the 3:1 resonance, for example, would have exactly one-third, or 1/3, the orbital period of Jupiter, and it would complete three circuits around the Sun for every one that Jupiter completes. Such an asteroid would revolve around our star at a distance of 2.5 AU with a period of 3.95 years, compared with Jupiter's orbital distance of 5.2 AU and orbital period of 11.86 years. An asteroid that happened to stray into this resonance would come close to Jupiter at almost the same part of the asteroid's orbit at regular 11.86-year intervals and the accumulated gravitational interaction with Jupiter could dislodge the asteroid from its orbit.

So the net effect of the resonance is to clear asteroids out of the resonant locations, and some of them might eventually be brought into Earth-crossing orbits. Asteroids that are not in a resonance are affected by Jupiter at completely random time intervals and places along their orbit, so the giant planet's gravitational perturbations tend to cancel each other over long times. It is somewhat analogous to repeated pushes on a swing. If the pushes occur at the same point in each swing, they can amplify the change, but haphazard pushes would produce little net effect.

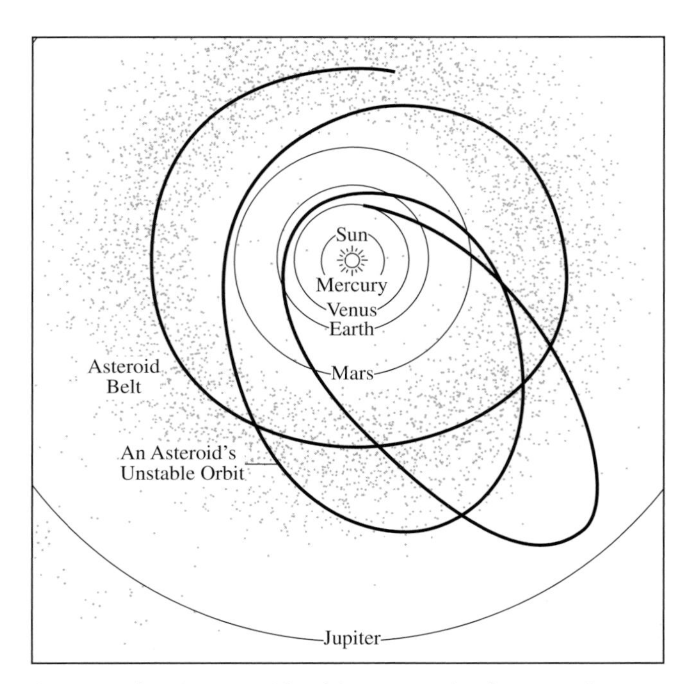

Fig. 13.4 Chaotic asteroid orbit Asteroid orbits can become chaotic under the gravitational influence of nearby massive Jupiter. Asteroids at certain locations in the main belt follow a trajectory that becomes increasingly off-center over thousands of orbits. They may eventually become interlopers with orbits that cross the Earth's orbit. Some of these near-Earth asteroids may eventually collide with our planet.

A satisfactory explanation of the Kirkwood gaps was not achieved until the 1980s when Jack Wisdom (1953–) used powerful computers to study Jupiter's influence on the motion of asteroids. The computer simulations showed that Jupiter induces a chaotic zone in the vicinity of an orbital resonance, and that an asteroid that moves into the resonant orbit will eventually be tossed out of it. An asteroid in the chaotic zone can spend tens of thousands of years in a well-behaved, near-circular orbit. But that ordered, placid behavior can be unexpectedly interrupted after 100 thousand years or so, when the orbit is suddenly stretched and elongated in a chaotic way (Fig. 13.4).

The increasingly elongated orbit can become so elliptical that it crosses the orbit of Mars or the Earth, and gravitational interaction with these planets can fling the asteroid into a totally different trajectory that removes it from the asteroid belt. Thus, orbits that initially fall within the chaotic zone around a resonance with Jupiter would be gradually cleaned out, creating a gap in the asteroid belt. Asteroids that have been thus redirected could abruptly end their voyage through space in a collision with Earth.

The entire solar system is an odd mix of order and chaos. You might call it a constrained chaos. Computer simulations for a time-span of nearly 100 million years show that planetary orbits cannot be accurately predicted in the very distant future. Nevertheless, all of the major planets have been around for 4.6 billion years without flying out of their orbits. So, it is perhaps most likely that the planets will wander and stray just a little over the coming eons.

13.2 Origin of the asteroids

Former worlds

In the past, there have been two extreme theories for the origin of asteroids. According to the first, the asteroids represent the fragments of a former planet that has been torn apart. The second theory proposes that the asteroids are the pieces of a planet that never formed. Today, astronomers favor a theory that lies between the extremes.

It is now known that the combined mass of all asteroids is far too small to make up a major planet. If all the known asteroids were brought together, they would create a body less than five percent the mass of the Moon. So the first extreme must be discarded. On the other hand, the second extreme can also be excluded because there is strong evidence that many – and probably most – asteroids were once collected into a relatively small number of slightly larger parent bodies.

Remnants of a planet that never formed

Why did the asteroids fail to coalesce into a single planet? It is likely that gravitational forces from the rapidly forming and massive Jupiter took charge of its neighborhood, stirring it up and keeping the original asteroids from growing too large. Numerous asteroids in resonant orbits with youthful Jupiter probably permeated the region of the asteroid belt, between 2 and 4 AU. Chaotic zones in the vicinity of these resonances would have pumped up the eccentricities of initially circular orbits, flinging the resident bodies into elongated and inclined orbits, accelerating them to high velocities, and causing them to crash into each other. The colliding objects would be moving too fast to stick to each other. Instead, they would break apart into fragments.

Collisions therefore pulverized the early asteroids, grinding them down to the numerous smaller asteroids we see today. Almost every asteroid we see today must be a fragment of a larger original body.

In contrast, a swarm of small, solid bodies in the inner solar system, with orbits closer to the Sun than the asteroid belt, were located far from Jupiter's gravitational influence.

These so-called planetesimals therefore remained in nearly circular orbits, moving at slow velocity around the Sun. This permitted the neighboring planetesimals to merge gently together and form larger ones. They eventually coalesced into the four terrestrial planets – Mercury, Venus, Earth and Mars.

If it wasn't for Jupiter's chaotic interference, a similar terrestrial planet might have formed in the asteroid belt. Gravitational interaction with the giant planet removed objects in this region, throwing them into collisions with the other planets or out of the solar system. These events cleared the asteroid zone of 99.9 percent of its original mass. The missing material could have coalesced to form an Earth-sized planet. What remain today are the relic building blocks of a planet that failed to form.

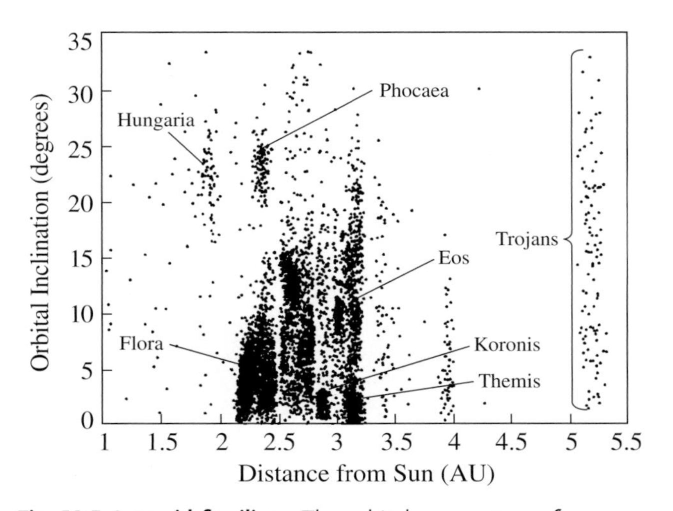

Fig. 13.5 Asteroid families The orbital parameters of many asteroids are very similar, as shown in this diagram of orbital inclination (*vertical axis*) plotted as a function of orbital distance from the Sun (*horizontal axis*). Three families with common orbital elements are the Koronis, Eos and Themis families. The Flora family is sometimes subdivided into several separate ones. The Hungaria and Phocaea groups of asteroids at high inclinations are separated from the main belt by secular resonance with Jupiter, that clears asteroids out of certain locations, and they are not true families. The Kirkwood gaps, also cleared by resonance, are noticeable by vertical white spaces; the one at 2.5 AU corresponds to an orbital period of 4 years and the 3:1 resonance. Another sharp break is present at 3.3 AU, corresponding to a period of 6 years and the 2:1 resonance. (Adapted from Charles T. Kowal's *Asteroids – Their Nature and Utilization*, Second Edition. John Wiley, New York 1996.)

A lifetime of catastrophic collisions

Billions of years ago, before Jupiter began to disrupt nearby objects, a few large bodies probably inhabited the asteroid belt. The largest of these would-be planets accumulated enough internal heat to differentiate, their dense material sinking to form iron cores and leaving rocky residues in their outer layers. Volcanoes may have erupted from a few of them.

Most of the initial asteroids never grew large enough to form cores, so they preserved matter typical of the region in which they formed. The majority of asteroids, both large and small, are, in fact, undifferentiated asteroids like Ceres.

Encounters among the earliest asteroids became increasingly violent as Jupiter stretched and twisted their orbits into eccentric and inclined orientations. These orbits criss-crossed each other, resulting in violent collisions at the place where they meet. Instead of continuing to grow, the largest asteroids were chiseled and blasted apart by mutual collisions.

So the original asteroids never grew larger than about one thousand kilometers across, and they never accumulated into a major planet. The pulverized remnants of these former worlds became the present asteroids, often orbiting the Sun in families with common orbital characteristics and spectral properties.

The Japanese astronomer Kiyotsugu Hirayama (1874–1943) discovered asteroid families. He noticed that groups of main-belt asteroids share very similar orbits, with common orbital inclinations and distances from the Sun (Fig. 13.5), suggesting that they are the broken fragments of larger objects. Hirayama called each group a family because he believed the members shared a common origin as the children of a bigger parent body. He also named a number of families after their largest member asteroids, such as the Eos, Koronis and Themis families. Many, and probably most, asteroids belong to families of collisional fragments of a larger parent object.

Within each of these families, the orbits are so similar that the members must have originated from a single object. Hundreds, and perhaps thousands, of small asteroids making up each family are probably the debris of a collision that disrupted the once-larger parent body. These parents may have been several hundred kilometers in diameter.

And not only are the orbits similar within each family, the colors and surface compositions are also often alike. These similarities imply that the families are real physical groupings. The Koronis family is, for example, composed of bright, silicate asteroids and the Themis family of dark, carbonaceous ones. The asteroid families can be easily recognized by color, providing independent confirmation of their reality.

The asteroid belt has been dominated by collisions for billions of years. Major impacts between asteroids larger than one kilometer across occur every 10 to 100 million years. So each one of them must have suffered roughly

Table 13.1 Principal characteristics of three larger asteroids[a]

Asteroid	Type	Radius (km)	Mass[b] (10^{21} kg)	Mass Density (kg m^{-3})	Rotation Period (hours)	Semi-major Axis (AU)
1 Ceres	C	475	0.95	2120 ± 40	9.1	2.77
2 Pallas	B	269[c]	0.214	2710 ± 110	7.8	2.77
4 Vesta	V	265	0.267	3440 ± 120	5.3	2.36

[a] The radius is in units of kilometers, abbreviated km, the mass is in units of kilograms, abbreviated kg, and the mass density is in units of kilograms per cubic meter, abbreviated kg m^{-3}.
[b] The mass determinations are from the work of E. M. Standish at the Jet Propulsion Laboratory where he uses observations of Mars to solve for the masses of the larger asteroids.
[c] Pallas is not quite spherical. A triaxial ellipsoid fit to occultation observations has diameters of 559 km, 525 km, and 532 km with a mean diameter of 538 ± 12 km.

a hundred devastating collisions over the past 4 billion years.

Over time, the asteroids crashed into one another and modified their shapes. When a small asteroid hit a larger one, it gouged a crater out of the surface of the larger one. If the two colliding objects were of comparable size, they could have broken each other apart into smaller fragments. As a result, some present-day asteroids are the metal-rich cores of larger, former parents, stripped of their rocky mantles by the ongoing collisions. Others are the shattered remains of bodies that have remained homogeneous. Such destructive encounters occur even now, dominating the evolution, shapes and sizes of the asteroids.

13.3 Size, color and spin

The size of asteroids

Due to its small size, an asteroid remains an unresolved point of light in even the best telescopes on Earth, just like a faint star. This explains the name *asteroid*, which comes from a Greek word that means "starlike". Although the name describes the visual appearance of these objects in a telescope, it is totally inappropriate to their physical nature. Using our instruments on Earth, we can determine the sizes of these objects, and they are much smaller than either a star or a major planet.

One method of measuring the size of an asteroid is to watch it pass in front of a star. During such a stellar occultation, the asteroid casts a shadow on the Earth, and the width of this shadow is equal to the asteroid's diameter. As the Earth rotates and the asteroid moves across the sky, the asteroid's shadow is swept by an observer on Earth, and its width can be determined by measuring the length of time the star is invisible. Observations of stellar occultations have established accurate values for the sizes of the largest asteroids, such as 1 Ceres and 2 Pallas (Table 13.1).

Even the bigger asteroids are smaller and less massive than the Moon. Ceres is by far the largest asteroid, having a radius of about 475 kilometers and a mass of 0.95×10^{21} kilograms. That is about a third the radius of the Moon and only 0.013 the mass of the Moon. The brightest asteroid, Vesta, is even smaller, a little less than half the size of Ceres.

There are many more small asteroids than big ones. About 1000 asteroids are larger than 15 kilometers in radius. Surveys of the faintest asteroids suggest that there are about half a million asteroids in the main belt larger than 1.6 kilometers. Yet, despite their vast numbers, the total mass obtained by adding up the contributions of all asteroids, of all sizes, is far less than the mass of any major planet. The entire asteroid belt is less than five percent the mass of the Moon and less than 0.0006 the mass of the Earth.

Another way of estimating an asteroid's size is to measure its apparent brightness. Bigger asteroid's reflect more sunlight and are therefore brighter. The amount of sunlight reflected from an asteroid nevertheless depends on both its size and reflectivity, or albedo – the ratio of light reflected by a surface to the incident light, a measure of the efficiency of the reflection process. As an example, 1 Ceres is the biggest asteroid, but 4 Vesta can appear brighter because it has a higher albedo. Vesta reflects 42 percent of the sunlight that strikes it, compared with Ceres' reflectivity of only 11 percent.

Both the size and the albedo of an asteroid can be determined by observing it in visible and infrared light.

When sunlight strikes an asteroid, some of the light is reflected, but most of it is absorbed. The absorbed radiation makes the surface heat up and emit infrared light. By measuring the infrared radiation from the asteroid and comparing it to the amount of reflected visible light, we can determine its albedo. And by knowing the asteroid's apparent brightness and distance as well, we can infer its size.

An added complication is caused by the rotation of a non-spherical asteroid, which can bring various surfaces into view, reflecting different amounts of sunlight. This produces a brightness variation, with periods of hours, that can be used to infer an asteroid's irregular shape. These rotational modulations of the light reflected from the largest asteroids, such as 1 Ceres and 2 Pallas, are very small, indicating that these asteroids are practically spherical in shape. Most asteroids, however, have very irregular shapes with pronounced rotational brightness variations. They are too small to retain a spherical shape during their lifetime of disruptive collisions.

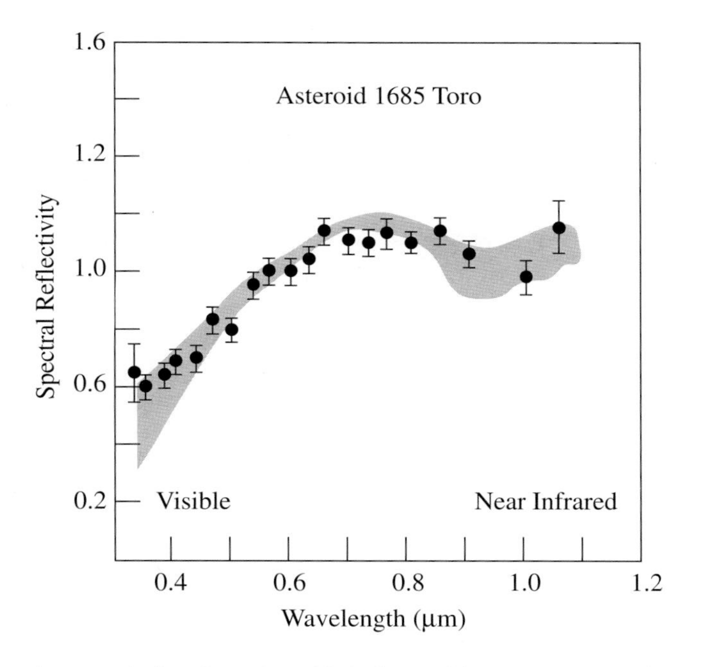

Fig. 13.6 Finding the composition of asteroids A prominent silicate absorption feature is present in the spectrum of sunlight reflected from the S-type asteroid 1685 Toro (*filled circles with error bars*). Asteroids like Toro may be the source of the stony meteorites recovered on Earth. The shaded spectrum is the reflection spectrum of a stony-chondrite meteorite. The wavelength is in units of micrometers, or 10^{-6} meters, designated as μm for short.

An asteroid's color

Because asteroids display no visible disk, Earth-based observers must infer their physical characteristics from the intensities and spectral properties of their reflected sunlight. By comparing an asteroid's reflected light, wavelength by wavelength, with that of the incident sunlight, it is possible to deduce its surface composition. Astronomers divide the amount of incident sunlight at each wavelength by the amount of reflected sunlight at that wavelength, and the ratio tells them how much light of each color is reflected compared to any other color. Such spectral measurements have revealed the physical diversity of the asteroids, and shown that their compositional differences tend to depend on distance from the Sun.

The bulk of main-belt asteroids can be divided into two broad spectral categories, known as the S, for silicate, and C, for carbonaceous, types. The bright S-types have a reddish color and exhibit spectral dips identified with absorption by silicate minerals (Fig. 13.6). They prevail in the inner part of the asteroid belt, orbiting closer to the Sun than the belt mid-point (Fig. 13.7).

In contrast, the C-type asteroids are darker, bluer and richer in carbon, with relatively flat and featureless spectra at visual wavelengths. The C-type asteroids far outnumber all types, possibly composing three-quarters of the main belt. The C-type asteroid 1 Ceres is a representative example; it has a smooth visible spectrum with an infrared absorption feature attributed to water embedded in its mineral structure.

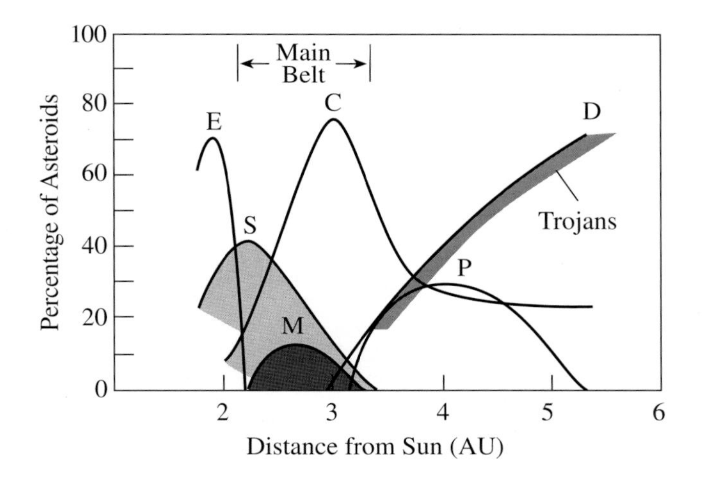

Fig. 13.7 Asteroid distribution with distance The color, or surface composition, of the asteroids is correlated with distance from the Sun. In order of increasing distance, there are the white E asteroids, the reddish S or silicate ones, the black C or carbonaceous ones, and the unusually red D asteroids. This systematic change has been attributed to a progressive decrease in temperature with distance from the Sun at the time the asteroids formed. Simple temperature differences within the primeval solar nebula cannot, however, explain the rare metallic M asteroids found in the middle of the asteroid belt; they are probably the cores of former, larger parent bodies.

Focus 13.1 Mining asteroids

The asteroids are rich storehouses of valuable materials such as iron, nickel and water. The utilization of their minerals could overcome growth limits imposed by dwindling natural resources on Earth. The water bound up in the carbonaceous asteroids could supply future space travelers with this vital commodity.

Prospecting spacecraft might be sent to an asteroid in search of a cosmic El Dorado. Because of its low gravity, valuable metals could be easily removed, and the extracted material might be shipped back by *Space Shuttle* from an asteroid that traveled near to the Earth. Imaginative engineers speculate that a more distant asteroid might be brought closer to Earth using a "mass driver", a device that would chew off pieces and fling them into space, propelling the asteroid like a rocket.

A metallic asteroid that is one kilometer in diameter contains about 8 billion tons, or 8 trillion kilograms, of metal, with an estimated market value of 10 thousand trillion (10^{16}) dollars, so mining asteroids could be a lucrative business. The cost analysis must nevertheless include the expense of bringing the metals back to Earth, which is not insignificant. Mining asteroids for interplanetary structures and habitats would not be so expensive in this regard.

Metals are relatively scarce in the Earth's outer layers because most of them sank inside to the central core where they will remain forever sequestered, buried under our planet's crust and mantle. In contrast, some metallic asteroids may be pure core. The outer mantle and crust of their former parents were probably chipped away by eons of cosmic collisions. So the metals on an asteroid could be relatively abundant and easy to extract. Moreover, the residual gas and dust would be swept from the solar system by the solar wind, thereby avoiding the problems of industrial pollution when recovering valuable metals on Earth.

You might not even have to mine an asteroid to retrieve its valuable material. Close-up pictures of the near-Earth asteroid 433 Eros show that it is covered with thousands of boulders that have already been mined and thrown out by impact. Robotic vehicles could just go there and pick up what's lying about.

Water may be the more valuable commodity, especially in space, making it more practical to mine than metal. But water in asteroids is chemically bound and must be extracted by some process, and only the more primitive carbonaceous asteroids have water. Heating their material would release water and leave a residual.

The S-type asteroids probably account for up to 15 percent of all asteroids. Some of the less common M-type asteroids reflect sunlight in a way that suggests that their surfaces are composed of nickel and iron, hence the designation M for metallic for at least some of them. These objects could be the metal cores of larger parent bodies, stripped bare by collisions. The M-types are most common in the middle of the asteroid belt. Future space entrepreneurs may want to mine them for valuable metals (Focus 13.1).

There is an intriguing connection between the composition of asteroids and their distance from the Sun. The innermost asteroids, with orbits closest to the Sun, are rocky, siliceous and dry, while the outer ones are carbonaceous with water-rich, clay-like minerals. The silicate asteroids, found closer to the Sun, have fewer volatile compounds and less water, and they have been subject to greater heating. The primitive asteroids, which are located furthest from the Sun, are primarily rich in carbon and water.

There is a related, progressive decrease in an asteroid's reflecting power with increasing distance from the Sun. The brighter asteroids that reflect the most sunlight tend to lie near the inner edge of the main belt, closest to the Sun, while the most distant asteroids are, on the average, the darker ones with the lower reflecting power. The very darkest are found in the remote regions near Jupiter's orbit.

These differences in the composition and reflecting power of asteroids are probably related to conditions in the primeval solar nebula – the interstellar cloud of gas and dust from which the solar system originated. They may be a consequence of a decrease in temperature with increasing distance from the young Sun when the asteroids were formed.

The middle of the main asteroid belt marks the boundary dividing the cold, outer, water-condensing regions and the hot, inner, water vapor parts of the primeval nebula. Dark material, rich in carbon and water, could condense in the colder regions further from the Sun, but not in the hot regions near to our star where the early water was probably vaporized. But the bright rocky material was less volatile and could remain within the hotter regions closer to the Sun. The igneous asteroids contain material that was formerly molten, perhaps in the intense heat within a large parent body. In this way, the temperature of the solar nebula may have led to the pattern of materials and colors now seen in the asteroids. This separation of asteroid type with distance from the Sun is also an argument against the theory that the asteroids are the remnant of a single disintegrated planet.

Astronomers have identified a plethora of other, less common classes of asteroids, based upon the shape and slope of their reflectance spectra. In addition to the most common S-, C- and M-types, there are at least eleven other classes denoted by different letters, such as the red, possibly organic D-types found in the outer belt, and the white, igneous E-types that are closer to the Sun. In addition, rare, individual asteroids exhibit unique, well-identified absorption features in their spectra. As an example, 4 Vesta, brightest and third largest of the asteroids, shows the distinct absorption signature of volcanic basalt, indicating that its surface is covered with lava-flows. Its parent body spawned a family of smaller V-type asteroids with similar spectra and basaltic composition.

The spin of an asteroid

Asteroids do not shine like a steady beacon with constant brightness. They instead reflect a varying amount of sunlight toward the Earth. The observed brightness variation, also known as a light curve, is periodic, often with two maxima and two minima. The overall repetition is due to rotation, while the double pattern of variability results from alternating side views of an asteroid's elongated shape. When we see the biggest side of an asteroid, with the greatest area, the asteroid is brightest, while the smaller area reflects less sunlight and the asteroid is dimmer.

Almost all asteroids spin about a single axis. The period of rotation is inferred from the amount of time that it takes for the complete pattern of brightness variation to repeat itself. The rotation periods are usually between 2.4 and 24 hours (Fig. 13.8), although a few of them rotate with longer periods, such as 253 Mathilde with a rotation period of 17.4 days and some have periods of only a few minutes. Frequent oblique collisions can increase the rate of rotation or decrease it, depending on whether the collision is in the direction of rotation or opposite to it.

Some asteroids are probably rotating as fast as they can. If an asteroid is not solid, and is thus bound only by its own gravity, it can only spin at a certain maximum rate before material is whirled off it. Asteroids larger than 200 meters seem to have reached this limit, for most of them do not rotate faster than once every 2.2 hours (Fig. 13.8), suggesting that there is nothing stronger than gravity holding them together. If it lacks the material strength of a solid, an asteroid with a faster rotation rate, and a shorter period, will throw material off its surface and fly apart. Such an asteroid might resemble a rubble pile or gravel heap, formed after collisions have blasted a larger asteroid to bits, with

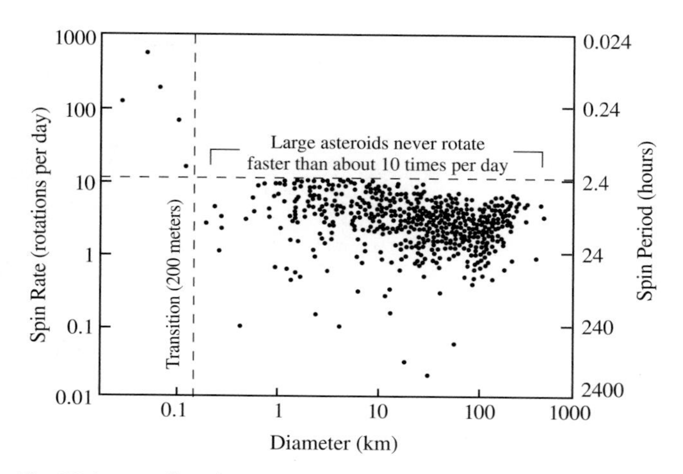

Fig. 13.8 How fast do asteroids spin? Two groups of asteroids are indicated on this plot of their rotation rate (*vertical axis*) versus size (*horizontal axis*). There are the large asteroids that rotate at a wide variety of speeds, except at the faster rates, and the small asteroids that spin most rapidly. The larger asteroids usually do not rotate faster than once every 2.2 hours, perhaps because these asteroids are piles of rubble that fly apart if spun too fast. Smaller asteroids, which can turn once every few minutes, must be solid rocks.

the fragments reassembling into a loosely bound object with little internal cohesion. Close-up scrutiny by spacecraft has shown that large asteroids can be both rubble piles and solid rocks (Section 13.6).

Asteroids smaller than 200 meters in diameter can rotate at faster rates, some turning once every few minutes. Their rotation is too rapid for these asteroids to consist of multiple components bound together by mutual gravitation. They must instead be rock solid. Some small asteroids rotate so swiftly that their day ends almost as soon as it begins. An example is the 30-meter asteroid 1998 KY26 (Fig. 13.9). Its day and night are only 5 minutes long; sunrises and sunsets on this asteroid take less than one second. By way of comparison, daylight at some places on Earth can last 12 hours or longer, and terrestrial sunrise or sunset usually takes about two minutes.

The direction of an asteroid's rotation axis can be determined by noting the way the brightness variation changes as the asteroid moves about the Sun. A highly elongated asteroid will reveal little or no light variation if its spin axis points directly toward us. As the asteroid's orbital motion carries it away from this alignment, the amount of light reflected toward Earth will increase, reaching a maximum when its broadest, equatorial surface is turned toward Earth. Studies of this effect suggest that the rotation axes of the asteroids are haphazardly oriented, pointing in all directions in space, contrary to the orderly rotations of the major planets except Uranus, whose rotation axis points more or less north and south. Frequent collisions

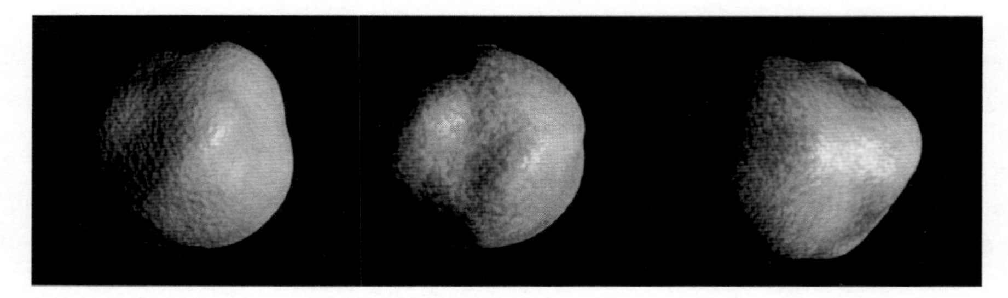

Fig. 13.9 Asteroid 1998 KY26 This lumpy, fast-spinning, near-Earth asteroid is just 30 meters across and it rotates with a period of just 10.7 minutes. These pictures of a computer-generated model were obtained by bouncing radar signals off the object when it passed within 800 thousand kilometers of Earth, or at about twice the distance between the Earth and the Moon. The radar reflectivity showed similarities to carbonaceous chondrites, primordial meteorites that contain 10 to 20 percent water, suggesting that this asteroid might contain about a million gallons, or 3.8 million liters, of water. The asteroid is too small for gravity to play a role in its shape, so its lumpy, round, nearly spherical shape is the result of collisions with other asteroids. (Courtesy of Steven J. Ostro, JPL.)

may have been responsible for the diverse orientations of asteroid rotation axes.

13.4 Shape and form

For more than two centuries, no one knew what an asteroid looks like. Since they are so small and far away, the surfaces of asteroids cannot be distinguished with telescopes on Earth, although modern technology has been used to image one of the larger ones, Vesta, with some success. The shape of an asteroid can nevertheless be inferred from the form and amplitude of its periodic light variations. They have told us that most asteroids are at least slightly elongated, chipped and pummeled into irregular shapes by eons of collisions.

The stretched out, irregular shapes of some asteroids have also been determined from radar observations of near-Earth asteroids that travel close enough for scientists to detect the echoes of radio waves bounced off them. During their close approach, these asteroids speed by the Earth at distances of several hundred thousand kilometers, permitting brief, high-resolution radar images before they move on and fade from view.

RADAR is an acronym for RAdio Detection And Ranging. In this technique, a coded radio signal is sent to the asteroid, and the properties of the return "echo" are compared to those of the transmitted signal. This yields information about the composition, surface texture, shape and rotation of the asteroid. Smooth surfaces reflect more radio energy than rough ones, and the faster an asteroid rotates the broader the range of returned wavelengths, which have been Doppler-shifted by the rotation to longer or shorter wavelengths.

The radar data indicate that the overall shape of some asteroids (an estimated 16 percent) is dominated by two irregular, lumpy components that touch each other, something like a dumbbell. Each of these asteroids is a double object, that is, two bodies in contact. Examples are 4179 Toutatis, pronounced too-TAT-is (Fig. 13.10) and 4769 Castalia. The two pieces probably merged after a past catastrophic collision of a larger body; they may have been thrown apart and subsequently came together under their mutual gravity. Or they might be two former asteroids that joined in a gentle encounter.

The inquisitive eyes of spacecraft were nevertheless required for the full resolution of the surface details of asteroids, and to turn these moving points of light into real places. The first glimpse was provided when the *Galileo* spacecraft flew close by two asteroids, 951 Gaspra and 243 Ida on its way to Jupiter, revealing details of these ravaged, misshapen worlds (Fig. 13.11, also see Section 2.1, Fig. 2.14).

Another sideways glance was obtained when the *NEAR Shoemaker* spacecraft flew past 253 Mathilde on its way to a rendezvous with the asteroid, 433 Eros. Like Gaspra and Ida, asteroid Mathilde has survived blow after blow of cosmic impacts (Fig. 13.12). Its surface is covered with the crater scars of past collisions that have disfigured the asteroid's shape, like the battered and scarred face of a professional boxer who has just lost a fight. Huge pieces have been removed from Mathilde, leaving four enormous craters tens of kilometers across.

13.5 *NEAR Shoemaker* embraces Eros

On Valentine's day 14 February 2000, the *Near Earth Asteroid Rendezvous,* abbreviated *NEAR,* spacecraft became the first to orbit an asteroid, arriving at 433 Eros after a four-year journey from Earth. The *NEAR* craft was the first in NASA's Discovery Program of no-frills, scientifically focused, low-cost missions, designed to

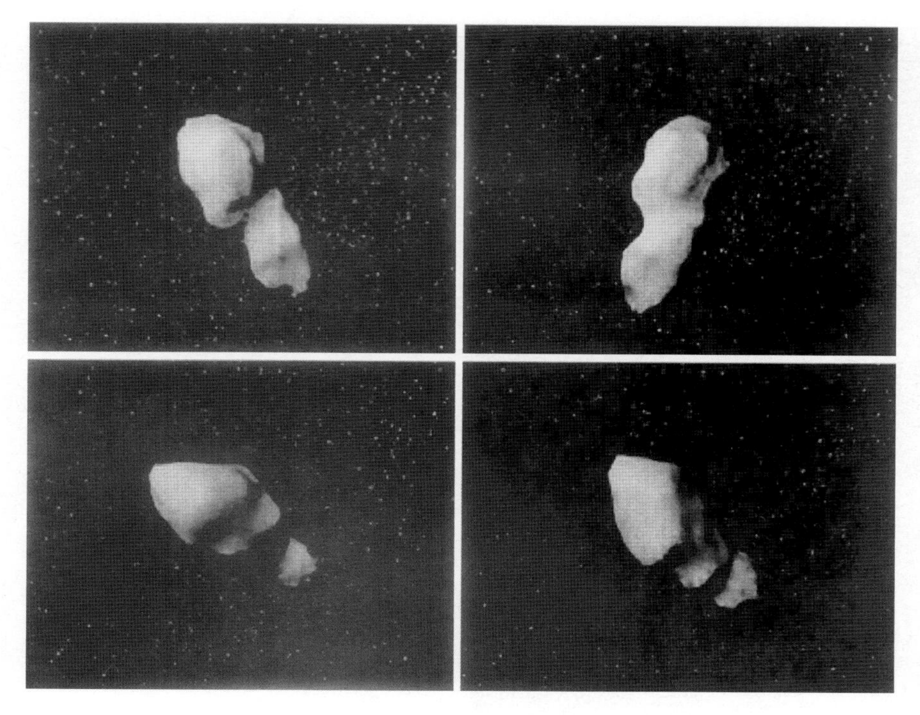

Fig. 13.10 Asteroid 4179 Toutatis These four views of a radar-derived model of 4179 Toutatis show shallow craters, linear ridges and a deep topographic "neck". It may have been sculpted by impacts into a single, coherent body, or this asteroid might consist of two separate objects that came together in a gentle collision. Toutatis is about 4.6 kilometers long. (Courtesy of Steven J. Ostro, JPL.)

do quality science in a "faster, better, cheaper" mode. The mission took just 26 months from start to launch, at a bargain total cost of $224 million. The car-sized vehicle circled Eros for a year, landing on the asteroid on 12 February 2001, another historic first. In the meantime, NASA renamed the spacecraft *NEAR Shoemaker* in honor of the astronomer–geologist Eugene M. Shoemaker (1928–1997), a pioneering expert on asteroid and comet impacts. Other successful missions in the Discovery Program were the *Lunar Prospector* (Section 5.4) and the *Mars Pathfinder* (Section 8.9) missions.

Discovered in 1898, asteroid 433 Eros was one of the first asteroids found to cross the orbit of Mars and come near the Earth's path in space – 132 Aethra also comes nearer the Sun than Mars. When *NEAR Shoemaker* arrived at the asteroid, it was just 1.7 AU from Earth, or less than twice the average distance between Earth and the Sun. Eros is one of the larger near-Earth asteroids, about twice the size of Paris and with a mass thousands of times greater than other asteroids that approach the Earth. So Eros is big

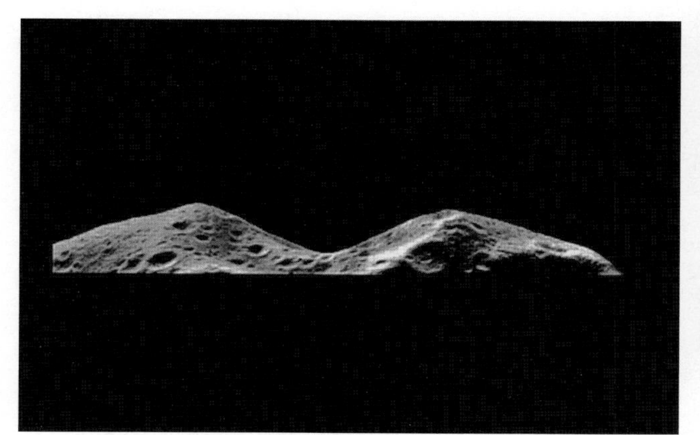

Fig. 13.11 Asteroid 243 Ida A camera aboard the *Galileo* spacecraft captured this picture of the limb, or visible edge, of the S-type asteroid 243 Ida on 28 August 1993. Prominent in this view is a 2-kilometer deep valley seen in profile on the limb. Many small craters and some grooves are also seen on the surface, the scars of past collisions. (Courtesy of NASA and JPL.)

Fig. 13.12 Asteroid 253 Mathilde An image mosaic of the C-type asteroid 253 Mathilde acquired by the *NEAR Shoemaker* spacecraft on 27 June 1997, when the spacecraft flew by the asteroid at a distance of 1.8 thousand kilometers. The part of the asteroid shown is about 59 kilometers by 47 kilometers across. The angular shape of the upper left edge is a large crater viewed edge-on. (Courtesy of NASA and the Johns Hopkins University Applied Physics Laboratory.)

Fig. 13.13 Asteroid 433 Eros This global view of the S-type asteroid 433 Eros was obtained by the *NEAR Shoemaker* spacecraft on 29 February 2000 from a distance of 200 kilometers. This perspective highlights the major features of the asteroid's northern hemisphere. The asteroid's largest crater (*top*) measures 5.5 kilometers wide and sits opposite from an even larger 10-kilometer, saddle-shaped depression (*bottom*). (Courtesy of NASA and the Johns Hopkins University Applied Physics Laboratory.)

and relatively nearby, two important reasons for a visit. It is also named after the Greek god of love – most appropriate for an encounter on Valentine's day.

Radio tracking of the orbiting spacecraft was used to determine the mass of Eros, which weighed in at 6.687 million billion, or 6.687×10^{15}, kilograms, about one-billionth the mass of Earth. That means that most adults would weigh less than a few ounces if standing on Eros, about as much as a bag of airline peanuts. And on Eros you could jump thousands of meters high, never to return. The gravity is so slight that *NEAR Shoemaker* had to keep its speed down to about 5 kilometers per hour to stay in orbit, moving about as fast as a casual bicyclist. If it moved at a faster speed, the spacecraft would escape the asteroid's feeble gravity and move into interplanetary space.

Although previous spacecraft passed close to a few asteroids, none had orbited one. In contrast to these previous, brief flybys, *NEAR Shoemaker*'s cameras scrutinized 433 Eros for a solid year, sending back 160 thousand images of the asteroid and recording its diverse surface from all angles and distances.

Eros is a warped and misshapen world, with heavily cratered expanses abutting relatively smooth areas (Fig. 13.13). The asteroid's biggest crater measures 5.5 kilometers across, and most of the surface is peppered with smaller craters (Fig. 13.14). The *NEAR-Shoemaker* scientists spotted 100 thousand of them.

The craters on Eros have been given names of famous lovers from history and fiction, taken from different cultures. They include Bovary from Gustav Flaubert's (1821–1880) novel *Madame Bovary*, Don Quixote and his Dulcinea from Miguel de Cervantes' (1547–1616) novel *Don Quijote de La Mancha*, Lolita from Vladimir Nabokov's (1899–1977) novel of that name, Don Juan, the legendary Spanish nobleman known for his seduction of women, and both Eurydice and her lover Orpheus, from Greek mythology.

The surface of Eros is saturated with craters, as are the surfaces of Mathilde and Ida. There are as many craters on their surfaces as there can be, so continued cratering would not change their appearance. This suggests that their present surfaces are at least two billion years old, and that they have retained their irregular shapes in their present form for a least that long. For Eros there is the additional

Fig. 13.14 Craters on asteroid 433 Eros The many craters on the surface of the asteroid 433 Eros are attributed to eons of collisions with other asteroids. Large boulders, perhaps broken off Eros during these impacts, are perched on the crater's edge. The largest boulder, on the horizon in the center of the picture, is about 40 meters long. The two overlapping craters shown here were probably formed many millions of years apart. This picture, taken on 7 July 2000 from the *NEAR Shoemaker* spacecraft, is 1.8 kilometers wide. (Courtesy of NASA and the Johns Hopkins University Applied Physics Laboratory.)

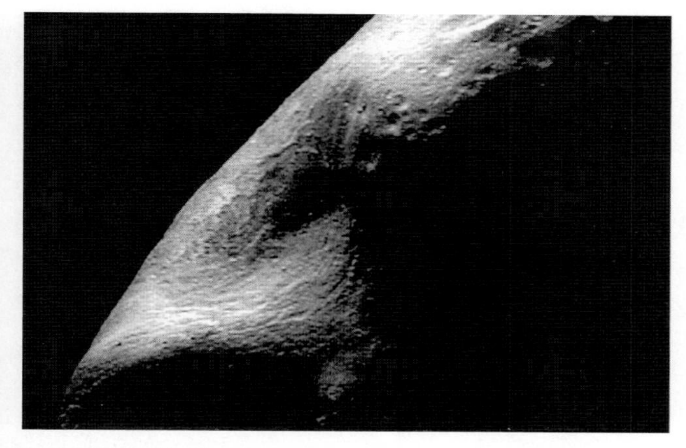

Fig. 13.15 Inside the giant gouge on Eros This mosaic is composed of images taken from the *NEAR Shoemaker* spacecraft on 15 February 2000, while the spacecraft was passing directly over the large gouge in the surface of the asteroid 433 Eros. Many narrow parallel troughs closely follow the shape of the gouge, which is about 7 kilometers across. Most of the asteroid's surface is saturated with impact craters. Inside the gouge, however, only smaller craters are present, indicating that the area within the gouge is younger than the nearby surface. So the event that caused the gouge must have happened more recently than the formation of the rest of the surface of Eros. (Courtesy of NASA and the Johns Hopkins University Applied Physics Laboratory.)

implication that it was not ejected toward Earth's general vicinity by a relatively recent collision within the asteroid belt, because such an impact would likely have erased most surface detail.

Eros has also been smoothed and rounded by glancing blows during its catastrophic past (Figs. 13.15, 13.16). This cosmic sculpture rivals the smaller bronze and marble forms of Constantin Brancusi (1876–1957) and Henry Moore (1898–1986). Equally beautiful is the broad, curved, saddle-like depression that connects two mountains on Eros, each several kilometers high (Figs. 13.17, 13.18).

Far from being a barren lump of rock, Eros has a dusty, boulder-strewn landscape. Despite its weak gravity, the diminutive asteroid has managed to hold on to about seven thousand boulders larger than 15 meters across, forced out of craters and pulled back to the surface during the relentless bombardment of its past. Some of the isolated stones are as large as a house, and up to 100 meters across.

The positions of the boulders on Eros indicate that at least three thousand of them were scoured out of a single crater by a colliding projectile a billion years ago. Some of these boulders went straight up and straight down. Most of the remainder traveled as far as two-thirds of the way around the asteroid, in all directions, before finally coming to rest on the surface.

Fig. 13.16 Over the horizon of Eros This incredible picture of the asteroid 433 Eros, taken from the *NEAR Shoemaker* spacecraft on Valentine's day 14 February 2000, shows the view looking from one end of the asteroid across the gouge on its underside and toward the opposite end. The two ends, which cap the long dimension of the asteroid, are separated by a distance of 31 kilometers. House-sized boulders are present in several places; one lies on the edge of the giant crater separating the two ends of the asteroid. A bright patch is visible on the asteroid in the top left-hand part of this image, and shallow troughs can be seen just below this patch, running parallel to the asteroid's long dimension. (Courtesy of NASA and the Johns Hopkins University Applied Physics Laboratory.)

Fig. 13.17 Back in the saddle again This image of the saddle region on 433 Eros was taken on 22 March 2000 by the *NEAR Shoemaker* spacecraft. The saddle is about 10 kilometers across. It may be the scar of an ancient crater, or somehow related to a different large crater on the opposite side of the asteroid. (Courtesy of NASA and the Johns Hopkins University Applied Physics Laboratory.)

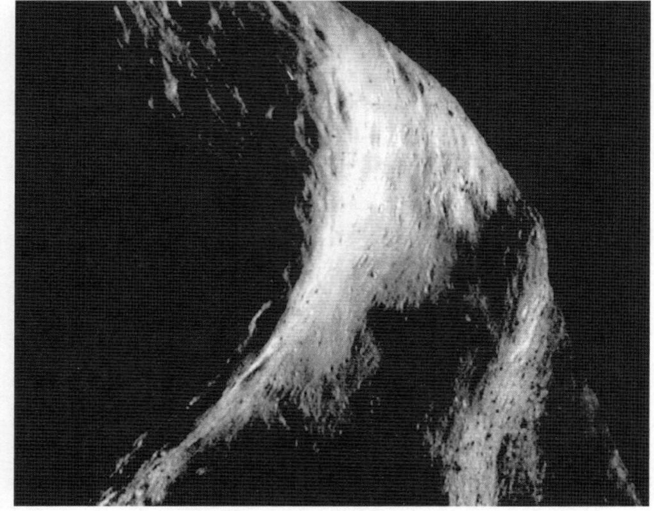

Fig. 13.18 The south saddle of Eros This mosaic of four images, photographed on 26 September 2000, was taken as the *NEAR Shoemaker* spacecraft looked down on the saddle region of the asteroid 433 Eros. The saddle is about 10 kilometers across. A broad, curved depression stretches vertically across the image, as if scooped out by a cosmic sculptor. A boulder-rich area is seen in the lower right. (Courtesy of NASA and the Johns Hopkins University Applied Physics Laboratory.)

Fig. 13.19 Touchdown The tip of the white arrow marks the location of the *NEAR Shoemaker* historic touchdown on the asteroid 433 Eros on 12 February 2001, when it became the first craft to land on an asteroid. The asteroid's overall dimensions are described by a triaxial ellipsoid measuring 31 × 13 × 13 kilometers.This image mosaic was taken from the spacecraft on 3 December 2000 from an altitude of 200 kilometers. (Courtesy of NASA and the Johns Hopkins Applied Physics Laboratory.)

Smaller rocks and a loose layer of dirty debris came into view when the *NEAR Shoemaker* spacecraft moved in to land on the boulder-strewn surface of 433 Eros (Fig. 13.19). It took pictures as close as 125 meters above the surface, showing features as small as a golf ball (Fig. 13.20). They indicate that Eros is something between a very big rock and a planet, large enough to hold onto its pieces yet too small to lose its odd, distorted shape.

Most asteroids are probably covered with a blanket of dust, pebbles and rocks that rests on solid bedrock. This layer of loose rock particles is known as the *regolith*, from the Greek word for "rock layer". The regolith formed during repeated bombardment by small meteorites that broke apart the surface rock. The Moon and Mars contain such dusty surface layers. Small asteroids may have a thin coating of dust, while the larger asteroids could have a thicker, powdery veneer, even thousands of meters deep.

The fine, dusty material on Eros has settled downhill, collecting in ponds that are tens of meters wide and a few meters deep. Hundreds of them are found in low-lying hollows, such as the bottoms of craters. The powdered deposits have flat level surfaces, resembling ponds of water

Fig. 13.20 Close-up view of the surface of 433 Eros This *NEAR Shoemaker* picture of the surface of asteroid 433 Eros was taken from a range of 250 meters on 12 February 2001, just before landing on the asteroid. The image is just 12 meters across, and the cluster of rocks at the upper right measures 1.4 meters across. (Courtesy of NASA and the Johns Hopkins Applied Physics Laboratory.)

on Earth, but there is no water on a rocky asteroid like Eros, and there hasn't been any for billions of years. Something else is causing the material to flow down hill. Since the ponds are found in well-lit areas, scientists speculate that the dust moves when it has been in the Sun too long, but the details of the sorting mechanism are not well understood.

The incredible success of the *NEAR Shoemaker* visit to 433 Eros has given us a reference point for guiding future studies of asteroids, and led to at least one future asteroid mission now on the slate of NASA's Discovery Program (Focus 13.2).

13.6 Rubble pile or solid rock

There are two hypotheses for the internal structure of asteroids. According to the rubble pile hypothesis, an asteroid is the reassembled debris of previous impacts, with a porous interior that is literally filled with holes. In this interpretation, an asteroid consists of smaller pieces loosely held together by their mutual gravitational attraction. As mentioned earlier in this chapter (Section 13.3), investigations of asteroid rotation rates suggest that many of them are not solid, and that there is nothing stronger than gravity holding them together. An alternative scenario proposes that asteroids are solid inside, and held together by their own material strength. It turns out that both ideas may be correct, depending on the asteroid.

Over billions of years, asteroids on intersecting orbits have collided with enough force to shatter and break them into pieces. Instead of dispersing, the fragments might have reaccumulated into a rubble pile. Asteroid 253 Mathilde, a dark, carbon-rich, C-type asteroid is an example of such an asteroid. But the collisional energy may have also been enough to throw the fragments far into space, leaving a solid, chipped rock behind. The bright, silicate, S-type asteroid 433 Eros is this kind of object.

Evidence about the internal constitution of asteroids is provided by measurements of their bulk mass density. The mass of 253 Mathilde was determined by radio tracking of small perturbations in the trajectory of *NEAR Shoemaker* during its close flyby. A similar technique established an accurate mass for 433 Eros when the spacecraft circled that asteroid. These mass determinations were combined with measurements of the asteroid volume to provide their bulk mass density (Table 13.2). *Galileo* flew by too fast and too far away to be affected noticeably by the asteroids it encountered, so masses could not be determined in this

Focus 13.2 Future missions to asteroids

The Japanese Institute for Space and Astronautical Sciences, abbreviated ISAS, plans to launch its *MUSES-C* spacecraft in December 2002, arriving at the near-Earth asteroid (25413)1998 SF36 in September 2005. This asteroid has an elliptical orbit that takes it out a little beyond the orbit of Mars and then back within the Earth's orbit. The spacecraft will touch down briefly on the asteroid, collect up to three surface samples, and return them to Earth in June 2007. It will be the first sample return mission to an asteroid.

NASA's *Dawn* spacecraft will orbit two of the largest known asteroids, 1 Ceres and 4 Vesta, for about one year each. This spacecraft's launch is scheduled for May 2006, arriving at Vesta in July 2010 and at Ceres in August 2014. These two asteroids feature striking contrasts in surface composition. According to current theories, their differing properties are the result of their being formed and evolving in different parts of the solar system. Ceres is, for example, still primitive with some water bound up inside it, while Vesta is evolved and dry. By observing these two asteroids with the same set of instruments, *Dawn* will characterize conditions and processes at the earliest moments in the origin of the solar system, establish the nature of the building blocks from which the terrestrial planets formed, and address the role of size and water in the formation and evolution of asteroids.

Table 13.2 Physical properties of asteroids visited by spacecraft[a]

Asteroid	Type	Overall Dimensions (km)	Mass (10^{15} kg)	Mass Density (kg m^{-3})	Rotation Period
243 Ida	S	$60 \times 25 \times 19$	42 ± 6	2600 ± 500	4.63 hours
253 Mathilde	C	$66 \times 48 \times 46$	103.3 ± 4.4	1300 ± 300	17.4 days
433 Eros	S	$31 \times 13 \times 13$	6.687 ± 0.003	2670 ± 30	5.27 hours
951 Gaspra	S	$18 \times 11 \times 9$	–	–	7.04 hours

[a] The overall dimensions are the diameters of a triaxial ellipsoid fit. The mass is in kilograms and the mass density is in units of kilograms per cubic meter, denoted by kg m^{-3}.

way. But a rough estimate for the mass of 243 Ida was inferred from the motion of its tiny moon Dactyl.

Mathilde has an exceptionally low mass density of just 1300 kilograms per cubic meter, only slightly higher than water, at 1000 in the same units, and about half the average density of the Earth's crust. The asteroid must therefore contain as much empty space as rock in its interior. It is indeed a loose pile of rubble, broken apart and stuck back together again, so pervasively fragmented that no solid bedrock is left.

The porous interior of Mathilde might explain the mysterious absence of visible rims or ejected deposits around its enormous craters. The unusually large craters may have been formed by compression of the surface during impact, like the dents in a beanbag, rather than by excavation of the material. Mathilde might even have swallowed up the colliding objects, like a bullet shot into a sandbag.

In contrast to Mathilde, the S-type asteroid 433 Eros has a bulk mass density of 2670 kilograms per cubic meter, comparable to that of terrestrial rocks. This asteroid is a solid, consolidated body. Millions of range, or distance, measurements with a laser aboard *NEAR Shoemaker* have established the asteroid's complex shape and topography. Comparisons of this shape with radio tracking of the gravitational pull on the orbiting spacecraft show that the mass density of Eros' interior must be nearly uniform. Thus, the asteroid 433 Eros is mostly solid throughout, with a uniform composition and homogeneous internal structure, rather than a loosely bound collection of smaller components. It also never separated into a distinct crust, mantle and core. Although Eros has been whittled away by surface-wrenching collisions, they never completely broke the asteroid apart, and it never became a rubble pile.

Scientists estimate that about 16 percent of near-Earth asteroids larger than 200 meters in diameter may be binary systems. Some of them come close enough to be imaged by radar techniques, enabling the total mass of the system to be inferred from the separation and orbital period of the components, using Kepler's third law. The bulk mass density of the component can also be inferred if its size is known. As an example, the near-Earth binary asteroid 2000 DP107 has a total mass of $4.6 \pm 0.5 \times 10^{11}$ kilograms and a primary component that is 800 meters across with a bulk mass density of 1700 ± 1100 kilograms per cubic meter.

13.7 Meteorites

Space rocks

Most meteors, or shooting stars, are produced by tiny fragments of comets, which burn up in the air and never reach the ground (Section 12.7), but occasionally a stone will fall from the sky, producing a brilliant trail of light flashing across the night sky. A rumbling sound and what appears to be a great burst of sparks may accompany it. These are fireballs and they are produced by tougher chunks of matter from space, resembling rocks (Fig. 13.21).

Extraterrestrial chunks of rock and metal that survive the fiery descent through the atmosphere and reach the ground have been given the name meteorites. And strictly speaking a meteoroid is the solid object in interplanetary space that appears as a meteor when it lights up in the atmosphere and becomes a meteorite if it reaches the ground. The term meteorite is also used for an interplanetary projectile, or meteoroid, that impacts the surface of any moon or planet.

Meteorites found on Earth have long been recognized as celestial objects. The *New Testament*, Acts of Apostles (19:35) refers to a temple dedicated to Artemis in which there is a "sacred stone that fell from the sky". These black objects have also been found in the Egyptian pyramids with a hieroglyph meaning "heavenly iron".

Fig. 13.21 Fireball A great flash of light, called a fireball, is produced when a large meteoroid streaks through the atmosphere. It is often accompanied by sonic booms and rumbling noises. A camera-chopping shutter used for timing and velocity determinations produced spaces between the luminous segments of the fireball's trajectory. The faint curved lines in the background are star trails caused by the Earth's rotation during the three-hour exposure. (Courtesy of the Smithsonian Astrophysical Observatory.)

Until the late-20th century, only about ten meteorites were recovered each year. They were ones that happened to fall near populated regions of the Earth, and many more must be at the bottom of the ocean, lost in the jungles, or buried in desert sand.

The Antarctica lode

In 1969, a group of Japanese scientists discovered a bountiful source of meteorites on the blue ice fields at the bottom of the world (Fig. 13.22), leading to a dramatic increase in the number of recovered meteorites. During the next three decades, about 20 thousand cosmic rocks were harvested from the Antarctic ice. The most productive areas were near the Allan Hills in Victoria Land and the Yamato Mountains in Queen Maud Land.

Many of the objects recovered from Antarctica must be fragments of the same meteorite, but they still represent thousands of new meteorites, more than doubling the number found on Earth. Before scientists traveled to Antarctica, the world's meteorite collections only had about 2600 different specimens, most of them collected during the past two centuries.

A meteorite landing on the Antarctic ice becomes buried in compressed snow, and is quickly frozen into the thickening ice. The cosmic rock soon sinks to great depths, where it remains preserved against corrosion. The tremendous mass of the ice, which reaches a thickness of four kilometers, squeezes the ice downward and pushes it outward toward the edges of the continent. The buried meteorite becomes caught in the flowing ice that moves like a river under the surface, creeping along at rates of several meters per year.

Some of the ice and its enclosed meteorites ultimately reaches the sea and breaks off as icebergs. At other locations the flowing ice encounters an obstacle, such as a mountain range. The moving ice then thrusts the meteorite upward and forces the buried rock to the surface. Strong winds corrode and wear away the surface, removing the ice covering and exposing the meteorite. The dark fusion crusts of the meteorites are then easy to spot against the bright icy background.

Meteorites recovered in Antarctica have stayed there for a long time. Radioactive dating indicates that most of them have spent about half a million years entombed in the ice, remaining virtually unchanged since the time they struck the Earth. By way of comparison, most of the meteorites found elsewhere on Earth fell between 1800 and 2000 AD.

Typical meteorites

Meteorites, together with rocks returned by astronauts from the Moon and grains of dust collected from the high levels of the Earth's atmosphere, are the only samples we know of extraterrestrial material. Although a meteorite's surface is usually coated with dark, glassy material, known as a fusion crust, that melted during its descent through

Fig. 13.22 Antarctica The midnight Sun illuminates the wind-swept ice at the bottom of the world. Numerous meteorites have been found embedded in the ice in this region near Allan Hills, Antarctica. These meteorites are probably fragments of asteroids that once had orbits between those of Mars and Jupiter, but a few of them may have come from the Moon or even Mars. (Courtesy of Ursula Marvin, Harvard–Smithsonian Center for Astrophysics.)

the atmosphere, the heat of friction did not have time to penetrate deeply into the falling rock. So the interiors of most recovered meteorites are unaffected by their rapid fall though the atmosphere. Meteorites may therefore be cut open and examined with microscopes and subjected to chemical analysis that reveals their original constitution.

Most meteorites that have been seen to fall and then recovered are stones, rather than chunks of metal (Fig. 13.23). About 94 percent of the fallen meteorites are stones, 5 percent irons and 1 percent stony-irons (Table 13.3). About 90 percent of the stony meteorites are, in turn, classified as chondrites. So most of the meteorites that fall to Earth are chondrites. The name "chondrite" is derived

Table 13.3 Classes of fallen meteorites

Name	Composition	Mass Density[a] ($kg\ m^{-3}$)	Percent[b]
Stones	Silicates (75–100 percent) and metallic nickel–iron (0–25 percent)	3500–3800	94
Irons	Nickel, iron (100 percent)	7600–7900	5
Stony-irons	Silicates (50 percent) and metallic nickel–iron (50 percent)	4700	1

[a] For comparison, typical rocks on Earth are largely silicates with mass densities in the range 3100 to 3300 kilograms per cubic meter, denoted $kg\ m^{-3}$, so meteorites are usually denser than other rocks found on the Earth's surface.

[b] Percent of total meteorites.

Fig. 13.23 Stony meteorite Fragment of the meteorite that fell near Johnstown, Colorado. The specimen is 0.09 meters wide. (Courtesy of the American Museum of Natural History.)

from the ancient Greek word, *chondros*, meaning "grain" or "seed", and the chondrite meteorites contain small spherical masses of material known as chondrules (Fig. 13.24). The other 10 percent of the stony meteorites are achondrites, which show signs of past igneous activity that once melted their material (Fig. 13.25).

The chondrites have been additionally divided into groups with common properties, suggesting that each group formed in the same region of the solar system. They are the ordinary chondrites, which are the most abundant, the carbonaceous chondrites, and the enstatite chondrites, named for their high abundance of the mineral enstatite.

Most meteorites are denser than terrestrial rock. So, if you find a dark, rather smooth rock that you suspect of being a meteorite, it must weigh at least as much as an ordinary Earth-born rock of the same volume if it is to pass muster as a rock from the sky.

But, of course, every rule has an exception – except that rule. There is a rare class of meteorites, with the ponderous name carbonaceous chondrites, that are fragile and have unusually low densities in the range 2200 to 2900 kilograms per cubic meter, making them less dense than an average terrestrial rock. They often contain appreciable amounts of carbon and water and they are considered to be among the most primitive and least altered samples of solids in the solar system.

Chronology of the meteorites

If we ask, "How old is a meteorite?" the question can have several meanings. Each meaning refers to the time since a significant event in the history of meteorite, and we shall describe them as follows.

1 Formation

Meteorites are as old as the solar system, dating back to its earliest days. Most of them were formed at about the same

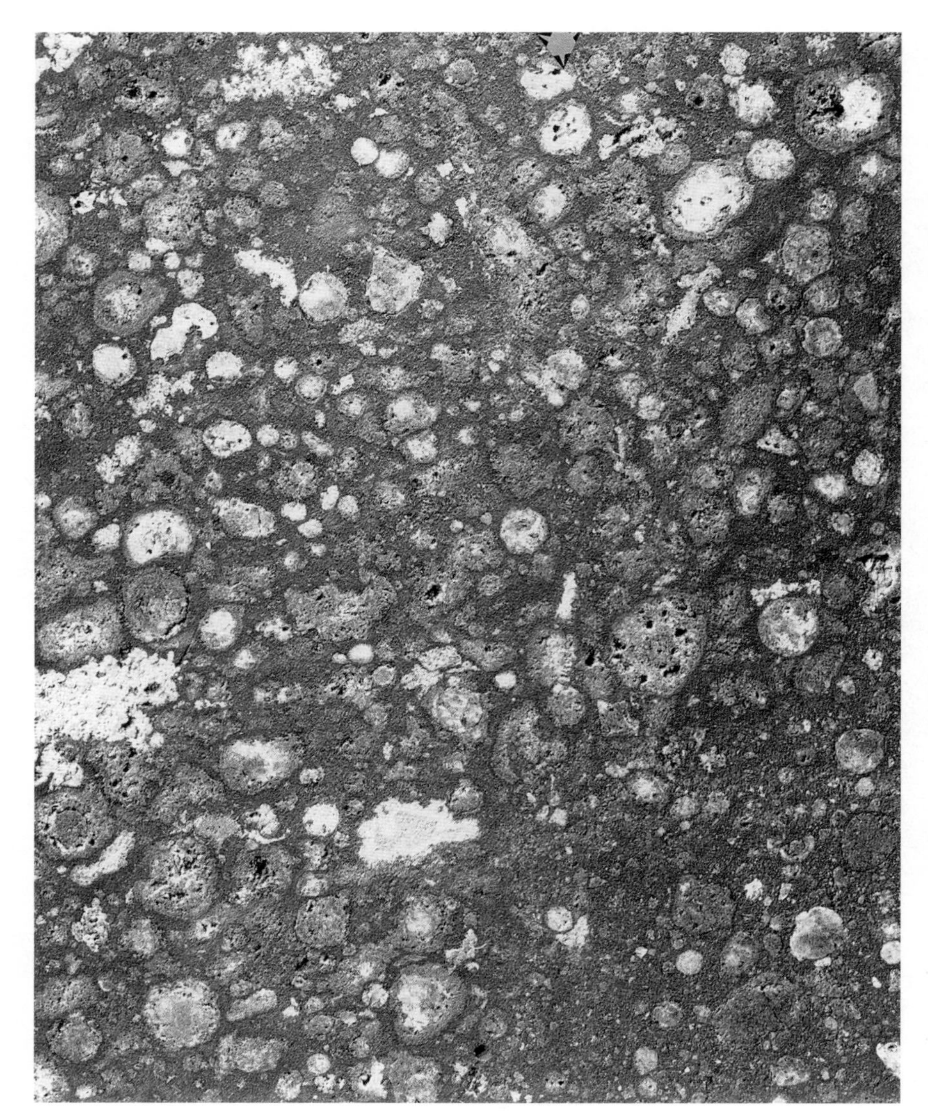

Fig. 13.24 Chondrules in Allende This photomicrograph of a section of the Allende meteorite shows numerous round silicate chondrules together with irregular inclusions. The meteorite section is 0.021 meters across and 0.027 meters high. (Courtesy of the Smithsonian Institution.)

Fig. 13.25 Achondrite meteorite A photomicrograph of a section of the achondrite meteorite that fell near Juvinas, France on 15 June 1821. It contains basalt, resulting from the melting and separation of material inside a relatively large parent body. The section shown here is 0.0032 meters across. (Courtesy of Martin Prinz, American Museum of Natural History.)

time as the planets. They accumulated directly from the primeval solar nebula and they have compositions similar to that of the Sun, except for their lack of hydrogen and helium.

The dates of formation of the meteorites can be determined by radioactive dating, in much the same way that the ages of lunar rocks were determined (Section 5.7). The relative concentrations of the decay products of elements such as rubidium and uranium reveal the time since these rocks were formed. Such measurements indicate that the carbonaceous chondrite meteorites formed 4.566 ± 0.002 billion years ago, and that some stony and iron meteorites are 4.55 ± 0.07 billion years old. At these times, the mineral grains in the meteorites crystallized, and their radioactive clocks started ticking.

Rounding off the numbers, and allowing for possible systematic errors, we obtain an age of about 4.6 billion years for most meteorites. The presence of decay products from short-lived radioactive elements additionally indicates that the meteorites formed in just a few million years back then.

The meteorites are the oldest rocks that we can touch. They are hundreds of millions of years older than the oldest rocks on the surface of the Earth. All of the planets and satellites are thought to have originated together with the meteorites 4.6 billion years ago, but erosion and geological processes have destroyed the original rocks on Earth. As a result, it is meteorites that reveal the age of the solar system and give clues to its origin.

2 Breakup and exposure

Another type of radioactivity also occurs in meteorites – radioactivity that is continually being caused by cosmic rays in the solar system. These "rays" are not rays in the usual sense; they are very energetic particles that bombard the meteorites and penetrate their surface for short distances. This cosmic-ray bombardment performs a bit of alchemy and transforms some of the atoms of the meteorite into radioactive nuclei. These radioactive nuclei slowly disintegrate, creating "daughter" nuclei. As the meteorite continues to be exposed to cosmic rays, the daughter nuclei become more and more abundant, and by a careful measurement of the amount of daughter nuclei in a meteorite it is possible to estimate the duration of this exposure interval. This gives the exposure age of the meteorite.

Now, the exposure ages of the meteorites that have been recovered on Earth are remarkably short in astronomical terms. Typically they are between 5 and 60 million years – just an instant in the life of the solar system and the meteorites. Evidently the meteorites have spent most of their lives shielded from cosmic rays. Astronomers now believe that most meteorites spent a larger portion of their life inside a parent body that was much smaller than the Earth but larger than a typical meteorite. According to this view, an important event in the chronology of most meteorites was the breakup of a parent body, exposing smaller fragments to space and to the bombardment by cosmic rays. The exposure ages measure the time that has elapsed since the breakup took place.

3 Collision with the Earth

Finally, when a meteorite falls through the blanket of the Earth's atmosphere, it becomes protected from cosmic-ray bombardment. No more radioactive atoms are created, and the ones that already exist inside the meteorite begin to decay, like the slow ticking of a clock.

In this way, the atoms of a meteorite carry a record of their chronology that can be unlocked with radiochemistry.

Rare and exotic finds

The frozen cargo of the Antarctica ice includes at least a dozen, greenish-brown meteorites that are strikingly similar to the welded highland rocks from the Moon (Fig. 13.26). The abundance of various elements and gases in these meteorites are virtually identical to those found in lunar rocks; at the same time, they are unlike those found in any other known meteorite or terrestrial rock. These small stones were blasted off the Moon by impacting objects.

Out of the thousands of stony meteorites now found in terrestrial collections, roughly a dozen are believed to be pieces of Mars. They were blasted into space by impacting objects, with such force that they escaped the red planet's gravitational pull and eventually reached Earth. One of them, dubbed ALH (for Allan Hills in Antarctica) 84001 contains controversial evidence for ancient microscopic life on Mars (Section 8.9).

Unlike rocks from the Moon, the meteorites from Mars contain small amounts of water and water-altered minerals. Moreover, gases trapped in bubbles within some of them resemble the gases in the Martian atmosphere. So there can be little doubt that these meteorites came from Mars.

They are often referred to as the SNC, pronounced "snick", meteorites, short for Shergotty (India), Nakhla (Egypt), and Chassigny (France) – three locations where they were observed to fall from the sky. Radioactive dating indicates that many of the SNCs were molten in the relatively recent history of our solar system. Those named for Nakhla and Chassigny hardened into solid rock 1.3 billion years ago, while the ones found near Shergotty solidified from molten lava just 180 million years ago. In

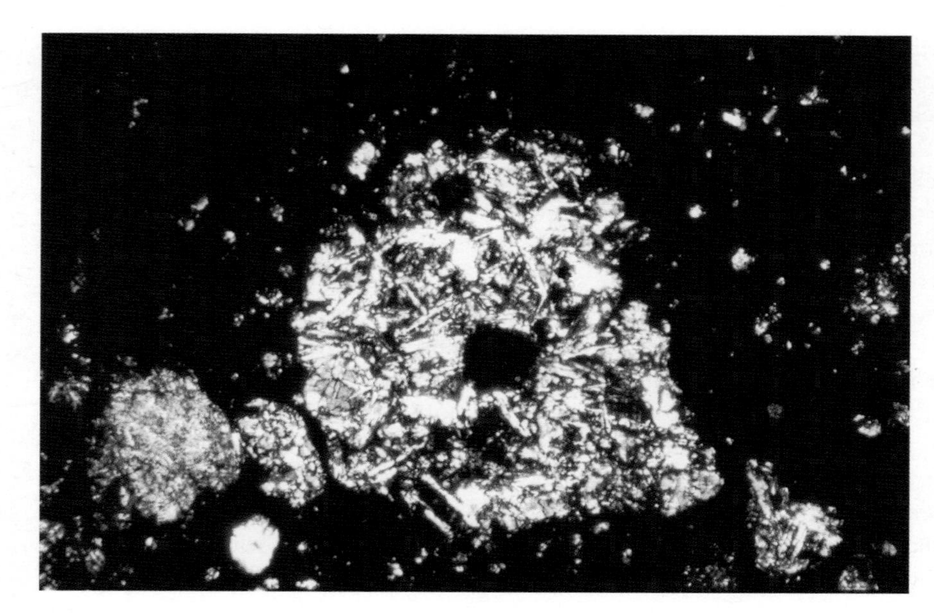

Fig. 13.26 Meteorite from the Moon Polarized light brings out the structure of a thin slice of a meteorite that probably came from the Moon. The abundance of several elements found in this meteorite are virtually identical with those found in rocks returned from the Moon, and unlike those found in any other meteorite or terrestrial rock. (Courtesy of Darrell Henry, NASA.)

comparison, most other meteorites solidified from molten materials between 4.6 and 4.5 billion years ago.

When looking at the structure of the meteorites from Mars, we see bits of glassy substance of the type produced when a rock is struck a hard blow, most likely by the impacting object that tossed them off the planet. Cosmic-ray exposure ages indicate that the meteorites subsequently traveled in space for several million years before their orbits intersected that of the Earth, and they fell to the ground. Some of the meteorites from Mars have been on the Earth for less than 200 years; others have been locked in the Antarctica ice for more than 10 thousand years.

Organic matter in meteorites

For more than a century, organic molecules have been suspected to be inside meteorites, although their presence in newly arrived meteorites led to the suggestion that they were the result of terrestrial contamination. But their existence became a certainty when 20 kinds of amino acids were found in the carbonaceous chondrite meteorites from Antarctica. These meteorites had lain in a sterile environment and were collected using sterile procedures. The organic matter in these carbonaceous chondrites apparently formed with the meteorite parent bodies, probably when water was bound into the structure of its clay minerals.

Many organic compounds, such as the amino acids, can come in two versions that are mirror images of each other. They are identified as left and right handed, based on their ability to rotate light in one direction or another. All living organisms on Earth use only left-handed amino acids. In contrast, the carbonaceous chondrites contain roughly equal amounts of both types, including the right-handed amino acids that are not found in living systems on Earth. This provides convincing evidence that the organic matter in meteorites did not originate on our planet, and that it is not directly related to life as we know it.

The discovery implies that the amino acids and other organic molecules probably existed in the solar system a billion years before the appearance of life on Earth. Does this imply the existence of early life before it came to Earth?

Probably not. The molecules found in carbonaceous chondrites are generally thought to be of non-biological origin. Carbon monoxide and hydrogen can, for example, be converted into organic molecules in the presence of an iron catalyst, and when ammonia is present in a laboratory sample, amino acids can be produced by electrical sparks.

So, the organic molecules found in meteorites are not in themselves vestiges of extraterrestrial life. But they are certainly primitive, and their cousins may have been the precursors to living matter.

The asteroid–meteorite connection

What is the source of meteorites? The primitive nature of most meteorites, called chondrites, indicates that they came from objects that have not experienced geological processes. They are non-igneous, and did not go through a hot, liquid stage. This is most easily understood if their parent bodies were small asteroids. A small fraction of meteorites, known as the achondrites, formed by igneous processes in larger parent bodies, most likely one of the original asteroids.

There is little doubt that most of the meteorites have come from the asteroid belt. They are probably chips off wayward asteroids, and there are two pieces of direct evidence for this conclusion.

1 Orbits

Photography of meteorites as they descend through the Earth's atmosphere can be used to determine their precise speed and direction of motion when they encountered the Earth. From these data, their orbits may be inferred, and many of the objects came from space beyond Mars, in the main belt of asteroids (Fig. 13.27).

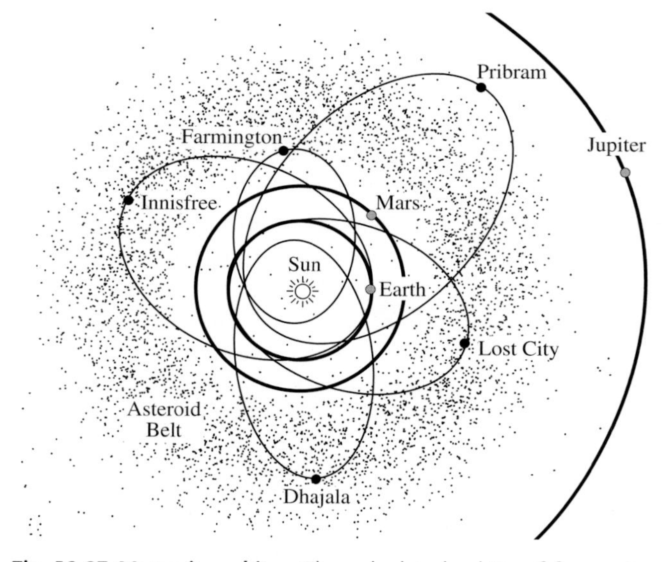

Fig. 13.27 Meteorite orbits The calculated orbits of five meteorites, inferred from their trajectory before hitting the ground. All of them originated in the asteroid belt, indicating that these meteorites are chips off asteroids.

2 Colors

The surface composition of asteroids has been inferred by breaking down their reflected sunlight into its component colors. Such spectral displays are similar to those of meteorites, suggesting that the meteorites are the debris of colliding asteroids. The relative abundance of asteroid types are not like those of the fallen meteorites, but this may simply reflect the ease or difficulty in sending asteroid fragments to Earth.

The sunlight reflected from C-type asteroids, that far outnumber all other asteroid types, closely resembles that of the relatively scarce carbonaceous chondrite meteorites. But the C-type asteroids reside in the remote, outer part of the main belt, furthest away from the Earth. The most common meteorites, the ordinary stony chondrites, display spectral colors that resemble those of the S-type asteroids found in the inner half of the asteroid belt. Due to its proximity, most meteorites are expected to originate from this part of the main belt. The light reflected from rare, bright, M-type asteroids, which reside near the middle of the main belt, matches the spectrum of relatively scarce iron meteorites.

Nevertheless, some of the color coordination is not exact. The S-type asteroids have redder overall colors and subdued light absorption compared with the abundant meteorites. This discrepancy may be explained if solar radiation or small, impacting cosmic particles gradually altered the thin outer layers of asteroids, darkening and reddening

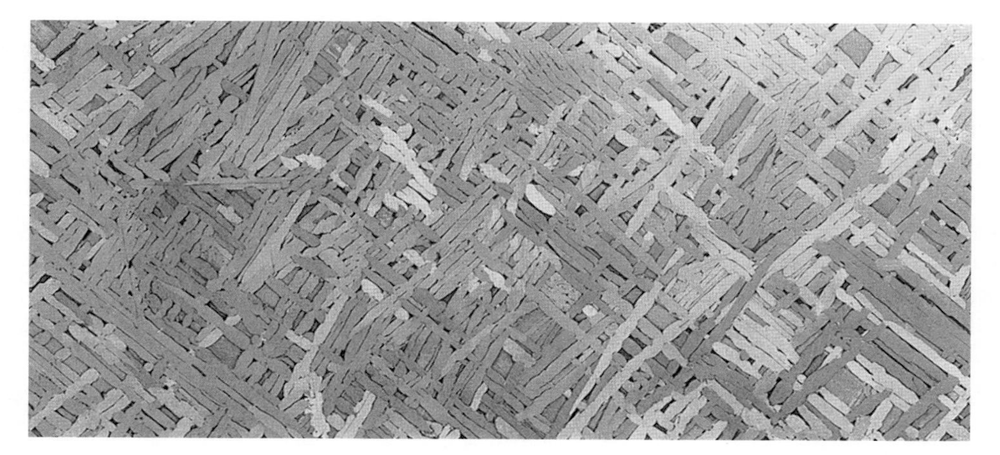

Fig. 13.28 Widmanstätten pattern When polished and etched with acid, an iron meteorite displays this distinctive Widmanstätten pattern produced by crystals of two differing iron-nickel phases. The pattern provides evidence that this meteorite was once buried within a parent body between 50 and 200 kilometers in radius. This sliced specimen is about 0.05 meters across. (Courtesy of the Smithsonian Institution.)

their surfaces. Similar "space weathering" apparently makes the lunar surface much redder than the color of unexposed rocks returned from the Moon. And the *Galileo* spacecraft showed that material near the sharp-edged, relatively young craters on 951 Gaspra and 243 Ida is slightly bluish, while the low-lying, older areas are slightly reddish. So the evidence now suggests that the color of freshly exposed asteroid surfaces gradually reddens with time. The bluish surfaces are recently exposed, while the red areas have been weathered for millions and even billions of years.

Earth-based telescopic observations indicate that many, relatively small S-type asteroids in the inner belt display a range of colors between those of the larger S-type asteroids and the ubiquitous meteorites. And instruments aboard *NEAR Shoemaker* revealed that 433 Eros, also an S-type asteroid, has the same basic composition as some ordinary stony chondrite meteorites.

Powerful collisions most likely excavate meteorites from deeper layers inside asteroids, which retain their pristine, unweathered color. So, the surface appearances of asteroids are slightly misleading.

Deep down inside, the most common, inner main-belt asteroids are probably very similar to the most common meteorites.

In addition, there is some compelling indirect evidence.

3 Crystalline structure

When the majority of iron meteorites are cut and polished and then are etched with acid, a delicate and complex pattern emerges (Fig. 13.28). It is produced by regions of crystalline structure, depending on the local orientation of the crystals in the iron. The sizes and shapes of these crystals indicate that they grew very slowly, and that the meteorite must have been hot, almost to the melting point, for tens of million of years. It probably cooled at the rate of a few degrees in a million years.

Such a slow cooling rate is compelling evidence that the meteorites were once inside a sizeable parent body. If a small meteorite had been exposed to space when it was still hot, it would have cooled in a matter of days. Small meteorites cool rapidly because their material is close to the surface, through which the heat can escape. Large crystal patterns would not have grown in such small bodies.

The meteorites that retained their heat for millions of years must have been buried within parent bodies a few hundred kilometers in radius, and this is just the size of large asteroids.

So, the crystalline patterns of iron meteorites also suggest an asteroid–meteorite connection. Additional suggestive evidence is the relationship among the sizes of asteroids, meteorites, and non-cometary meteoroids (Fig. 13.29). The classes are not mutually exclusive, and there is considerable overlap. The simplest explanation of all the evidence is that meteorites are the debris of collisions among the asteroids.

Whatever their precise history, the asteroids and meteorites are primitive objects that can act as beacons to the past. They represent a tableau of the ancient objects that littered space 4.6 billion years ago, and they carry a code for our understanding of the formation of the solar system.

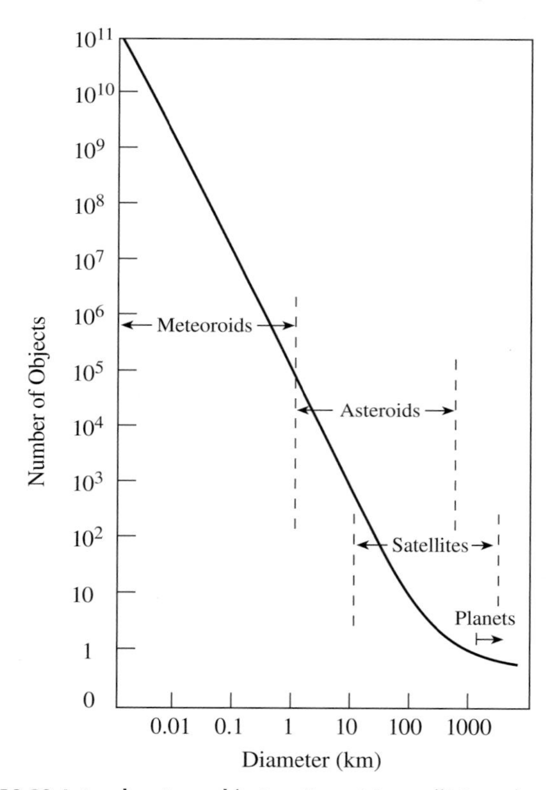

Fig. 13.29 Interplanetary objects Repetitive collisions between interplanetary objects have produced many more smaller meteoroids than larger ones. Some of the larger asteroids are comparable in size to small moons, and ongoing collisions between asteroids have produced numerous smaller meteorites.

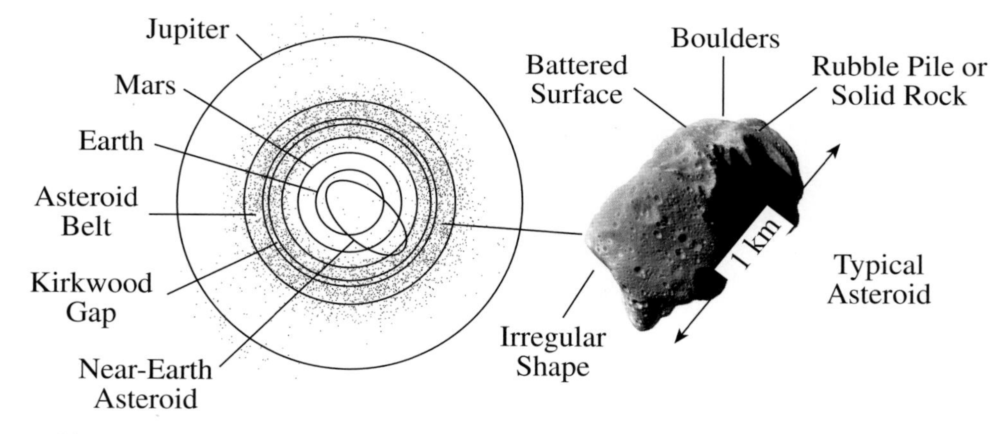

Fig. 13.30 Summary Diagram

14 Colliding worlds

- At least twenty pieces of a comet recently hit Jupiter, producing explosive fireworks and dark scars that fascinated astronomers throughout the world.
- Some comets are on suicide missions to the Sun, diving into our star and being consumed by it.
- Comets or asteroids may have supplied water to the Earth's oceans early in the planet's history when impacts by these objects were more frequent than they are now.
- Most of the impact craters on the Earth were eroded away long ago, but a few of the relatively recent ones have been discovered from airplanes or spacecraft.
- An asteroid wiped out the dinosaurs when it hit the Earth 65 million years ago, and similar cosmic impacts might be responsible for other mass extinctions.
- If an asteroid or comet of about 10 kilometers in size hit the Earth, the horrific blast could generate towering ocean waves, block out the Sun's light and heat, ignite wildfires that might set the world ablaze, drench the land and sea with acid rain, and ignite deadly volcanoes on the side of the Earth opposite from the impact site.
- The Earth is immersed within a cosmic shooting gallery of potentially lethal, Earth-approaching asteroids and comets that could collide with our planet and end civilization as we know it.
- The lifetime risk of your dying as the result of a two-kilometer asteroid or comet striking the Earth is about the same as death from an airplane crash, but a lot more people would die with you during the cosmic impact.
- It is estimated that the Earth receives a direct hit by an asteroid or comet about two kilometers in size every million years or so, resulting in a global catastrophe. It could happen tomorrow or it might not occur for hundreds of thousand of years.
- Astronomers are now taking a census of most of the Near-Earth Objects that are big enough and close enough to threaten us. When the survey is complete, they will have found 90 percent of those larger than one kilometer in size and will be able to determine their future trajectories for at least much of the coming century.
- With enough warning time, we could redirect the course of an asteroid or comet that is headed for collision with the Earth.

14.1 A comet hits Jupiter

Discovery of Comet Shoemaker–Levy 9

Eugene Shoemaker (1928–1997), his wife Carolyn, and the amateur astronomer, David H. Levy (1948–), were involved in routine observations the evening of 23 March 1993, a dark and cloudy night with stormy weather on its way. They were continuing a ten-year search for comets and asteroids that might be headed toward the Earth using the small 0.46-meter (18-inch) wide-field photographic telescope at Palomar Observatory in California. Good and expensive film couldn't be wasted during the poor weather conditions, so some fogged film, which had been partially exposed to light, was used to photograph a clear place in the night sky near Jupiter before the clouds covered it up.

Two days later, when Carolyn Shoemaker (1929–) examined the images taken on the flawed film, she saw an elongated feature that looked to her like a "squashed comet". When the discovery was confirmed with better telescopes, the stretched-out blur of comet light was resolved into several objects aligned along a single straight line projected in the sky, like pearls on a string. In accordance with tradition, it was named Comet Shoemaker–Levy 9, after the discoverers, ninth in a series of objects the trio found traveling around the Sun in short-period orbits.

When a comet nears the Sun, the ices in the comet nucleus turn directly from solid to gas and release dust to form a round, fuzzy coma, sometimes accompanied by a tail that points away from the Sun. But instead of a single coma and tail, powerful telescopes revealed a train of baby comets, each with its own invisible nucleus, nearly spherical coma and elongated dust tail (Fig. 14.1).

Breakup and collision

Comet Shoemaker–Levy 9, abbreviated SL9, consisted of the pieces of a former, single comet that had been trapped in a two-year orbit around Jupiter for decades. But when it traveled too close to Jupiter in 1992, the comet was shredded apart and launched on a trip to oblivion. It passed within about 20 thousand kilometers of Jupiter's cloud tops, and 90 thousand kilometers from the planet's center. So there was a modest difference between the planet's gravitational attraction on the near and far side of the comet, enough to rip the fragile comet into at least twenty observable pieces. Most of these fragments remained visible over the entire subsequent lifetime of the comet.

A few previous comets had been known to orbit Jupiter temporarily, and the disruption wasn't unprecedented. After all, that's how you might make a planetary ring. The differential gravitational forces on the front and back of a satellite that moves too near a planet rip it into pieces to make a ring of smaller objects that revolve around the planet.

What made SL9 unique was that the broken comet was inexorably hurtling along a path to total destruction,

Fig. 14.1 Fragments of Comet Shoemaker–Levy 9 In July 1992, this comet passed so close to Jupiter that the icy material of its nucleus was torn apart by the differential gravitational forces of the giant planet. This panoramic image of the comet fragments was taken from the *Hubble Space Telescope* in January 1994, eight months after they were discovered and six months before they dived into the atmosphere of Jupiter. The length of the string of comet pieces is about 1.1 million kilometers, three times the distance from the Earth to the Moon. The largest of the fragments in the string is about two kilometers across. Each fragment mimics a larger comet with a round coma and a dusty tail. (Courtesy of NASA and STSI.)

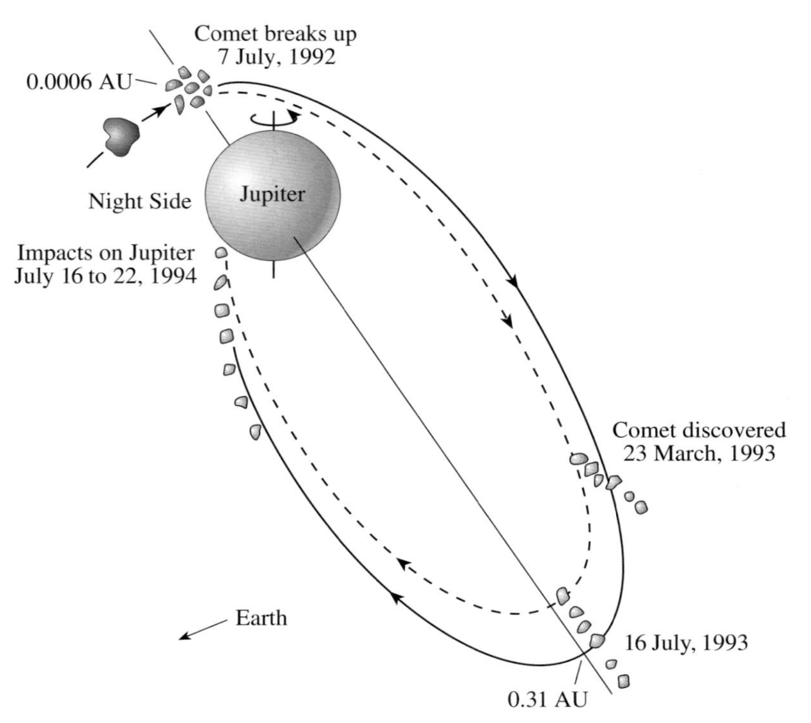

Fig. 14.2 Final orbit of Comet Shoemaker–Levy 9 This comet was orbiting Jupiter for more than half a century, until it was ripped apart during a close encounter with the planet and collided with it two years later. The disruption occurred on 7 July 1992 when the comet passed within 0.0006 AU, or 90 thousand kilometers from the planet's center. Since Jupiter has a radius of just over 70 thousand kilometers, the comet passed within about 20 thousand kilometers of the planet's cloud tops. Jupiter's unequal gravitational pull on the near and far sides of the comet nucleus then tore the object apart. Carolyn and Eugene Shoemaker and David Levy discovered the comet fragments on 23 March 1993, when the broken comet was almost at its furthest distance from Jupiter, at 0.31 AU. One by one the icy fragments exploded in Jupiter's cloud tops during impact week, from 16 to 22 July 1994.

doomed to collide with Jupiter. Orbital calculations indicated that the train of comet fragments would plunge into the giant planet in July 1994, two years after the former single comet's disruption and more than one year after the discovery of its pieces (Fig. 14.2).

Such advance knowledge of a collision with any planet was unprecedented in human history. Numerous craters on the terrestrial planets, as well as the Moon and some of Jupiter's satellites, bear silent witness to cosmic collisions of the past, but now for the first time astronomers could see the collision happening before their very eyes.

The anticipated impact was an incredible opportunity, occurring just once in the lifetime of any astronomer and most likely once in a millennium. Some scientists speculated that Jupiter would swallow up the comet pieces without a trace. Others reasoned that each fragment would dig a "tunnel of fire" into the giant planet's atmosphere, exploding and sending a spectacular fireball back into space. But no one knew for sure what would happen.

Impact

The collision of Comet Shoemaker–Levy 9 with Jupiter in July 1994 was perhaps the most widely witnessed event in astronomical history. Practically every telescope in the world was trained on Jupiter during impact week, between 16 and 22 July 1994. Infrared heat detectors were placed at the focal point of the Keck Observatory's giant 10-meter (394-inch) telescope atop Mauna Kea in Hawaii, and the *Hubble Space Telescope* was poised to record the event at visual wavelengths. Every other major astronomical observatory participated, as did numerous amateur astronomers from their own backyard.

Detailed calculations indicated that the collisions would be hidden on the dark "back" side of Jupiter, where Earth-based astronomers could not see them. So it might be something like watching a World Series ball game from a seat behind a stadium post. The comet fragments would nevertheless strike Jupiter close to the side facing Earth, so astronomers hoped that something would be seen when the planet's rapid rotation, of once every 9 hours 55.5 minutes, brought the impact sites into view. Moreover, the *Galileo* spacecraft, on its way to Jupiter, had a direct view of the actual collisions from its unique position in space.

No one was disappointed! Fantastic fireworks were produced when the comet fragments plowed into Jupiter. Although the chunks of ice and rock were no more than one or two kilometers in size, each fragment exploded in Jupiter's upper atmosphere with an energy equivalent to the simultaneous explosion of hundreds of thousands of nuclear bombs on Earth. As it penetrated the outer atmosphere of Jupiter, each comet fragment heated and compressed the surrounding gas, producing a violent explosion high in the atmosphere. The resultant fireball punched a hole through the overlying material and sent plumes of hot gas rising into space.

Fig. 14.3 Dark impact scars This *Hubble Space Telescope* image shows several dark spots (*right*) on Jupiter, each marking the impact site of a fragment of Comet Shoemaker–Levy 9. The Earth-sized scars remained visible for about five months, until the winds in Jupiter's outer atmosphere pulled them apart. A thin expanding ring of dark material, suggesting waves spreading out from the impact explosion, surrounded some of the dark central spots. Jupiter's Great Red Spot is also prominent, as it has been for centuries. (Courtesy of NASA and STSI.)

Instruments aboard the *Galileo* spacecraft measured temperatures that soared to 10 or 20 thousand kelvin when the fragments plunged into the clouds of Jupiter. That's at least twice as hot as the Sun's visible disk, at 5.28 thousand kelvin. Rising plumes of hot gas were hurled three kilometers above Jupiter's clouds, each generating a bright flash of infrared light.

When an ejected plume fell back into Jupiter, it produced another infrared flash. It took 10 or 20 minutes for each plume to rise and fall again, by which time the impact site had rotated into view from Earth. But the bright fireworks could only be seen by their infrared heat radiation. In visible light there was little to see but the dark ashes.

What incredible luck, to have a comet break into pieces, hit Jupiter, and generate brilliant bursts of infrared light so close to the edge of the planet's back side that they could be seen from on or near our planet. And the good fortune didn't end there, for the arching plumes of hot gas produced great dark scars when they cascaded back down into the giant planet.

Fallout

The comet fragments plunged into Jupiter, one after another, like the cars of a train when its locomotive is derailed. After generating a bright ball of light, each fragment disfigured Jupiter with a black scar that had never been seen before, twice as large as the Earth and spanning tens of thousands of kilometers (Fig. 14.3). Meanwhile waves swept across the impact site and reverberated deep within the planet, which seemed to shudder from the impacts.

The black spots were by far the darkest features on Jupiter, and they were easily visible with small telescopes as the planet's rotation carried them into view. They endured for months, gradually spreading, merging and slowly fading from view as the Jovian winds dispersed the dark material.

The collision of Comet Shoemaker–Levy 9 with Jupiter produced dramatic effects, unleashing incredible violence. It raised awareness of a similar threat to our home planet. Such impacts have happened on Earth in the past and they could happen again, with devastating effects to civilization. And to heighten the concern, astronomers were detecting numerous comets plunging into the Sun, never to return again.

14.2 Consumed by the Sun

Some comets plunge deep into the Sun's thin, million-degree outer atmosphere, or corona. Instruments aboard the *SOlar and Heliospheric Observatory* satellite, abbreviated *SOHO*, have recorded their death-defying trip around the Sun. One of its instruments uses an occulting disk to block out the bright light of the visible solar disk, enabling it to detect the comets as they move through the inner corona (Fig. 14.4). A comet often pays a heavy price for this trip, sometimes breaking apart because of the Sun's forces.

Other comets are hurtling toward complete meltdown, passing so close to the Sun that the encounter is fatal. Though rarely, if ever, hitting the visible solar disk, or photosphere, these comets can come closer than 50 thousand kilometers from it. They are unlikely to survive the Sun's intense heat and gravitational forces at that range. Amateur astronomers from all over the world have examined *SOHO's* real-time images posted on the Internet, discovering hundreds of previously unknown comets on their death-dive into the Sun (Fig. 14.5).

Most of the comets discovered by *SOHO*, about 90 percent of them, are small cometary fragments known as the Kreutz sungrazers, which closely approach the Sun from one direction in space. They are named after the German astronomer Heinrich Kreutz (1854–1907) who found that many of the comets, which had come closest to the Sun in the 19th century, seemed to have a common origin with similar orbits. It turned out that they are all fragments of a single large comet that first broke up when passing

Fig. 14.4 Sunbathing comet The short-period comet 96P/ Machholz passed through the Sun's million-degree atmosphere, or corona, in January 2002, when it approached within about 18 million kilometers of the Sun, or about 0.12 AU where one AU is roughly the average Earth–Sun distance. The comet's coma and foreshortened tail can be seen above the occulting disk of the *SOHO* instrument that recorded this image. In a spectacular coincidence, a giant magnetic bubble, known as a coronal mass ejection, was explosively hurled from the Sun at about the same time (*left*). A CME tosses billions of tons, or trillions of kilograms, of coronal gas into space, expanding to become much larger than the Sun. The white circle denotes the outer edge of the visible solar disk. Venus lies just outside the occulting disk (*below right*). The planet is so bright that its image is marred by a horizontal stripe. (Courtesy of the *SOHO* LASCO consortium. *SOHO* is a project of international collaboration between ESA and NASA.)

Fig. 14.5 Fatal impact This composite image records a comet plunging into the Sun on 23 December 1996. The innermost image (*center*) records the million-degree solar atmosphere, known as the corona. The electrically charged coronal gas is seen blowing away from the Sun just outside the inner dark circle, which marks the edge of one instrumental occulting disk. Another instrument records the comet (*lower left*), as well as the coronal streamers at more distant regions and the stars of the Milky Way. (Courtesy of the *SOHO* EIT, UVCS, and LASCO consortia. *SOHO* is a project of international collaboration between ESA and NASA.)

Fig. 14.6 Death of a sungrazer This *SOHO* image shows a bright sungrazing comet (*left of center*) headed into the inner atmosphere of the Sun on 22 October 2001. The white semi-circle marks part of the outer edge of the visible solar disk, whose intense glare is hidden by the instrument's occulting disk (*opaque semi-circular region at top center*). The million-degree solar atmosphere can also be seen, streaming away from the Sun. (Courtesy of the *SOHO* LASCO consortium. *SOHO* is a project of international collaboration between ESA and NASA.)

very close to the Sun thousands of years ago. The original fragmentation may have been witnessed in 321 BC by the Greek philosopher Aristotle (384–322 BC) and by the Greek historian Ephorus (405–330 BC), but it may have occurred much later. The break up gave rise to two main comets, perhaps with orbital periods of 350 years and about 700 years, and the two parts were split into more pieces during return visits to the Sun.

When a member of the Kreutz sungrazer group moves around its orbit and returns to our vicinity, it can dive into the inner corona and disappear forever (Fig. 14.6). Spectroscopic observations from *SOHO* indicate that each comet fragment can be very small, just 6 to 12 meters across, despite their spectacular display. Such a tiny object, falling so close to the Sun, would vaporize completely away, like the proverbial snowball in hell.

In just six years of service, *SOHO* has spotted more than four hundred comets, making it by far the most prolific comet finder in the history of astronomy. Aside from the numerous Kreutz sungrazers, *SOHO* has found more than forty new comets, which is comparable to the number of comet discoveries during almost any decade throughout the previous two centuries.

The other main contributor to recent comet discoveries has been the LIncoln Near Earth Asteroid Research, abbreviated LINEAR, Program, operated by MIT's Lincoln Laboratory. It consists of a pair of telescopes dedicated to detecting and cataloging Near-Earth Objects that threaten Earth. LINEAR has completed millions of observations, finding more than seven hundred confirmed near-Earth asteroids or comets. One of them, dubbed Comet LINEAR and also known as C/1999 S4, has been caught breaking up on its way into the Sun, vanishing much further out than the Kreutz sungrazers.

When discovered in September 1999, Comet LINEAR was exceptionally bright at a relatively large distance of about 4 AU, which has often happened to other comets during their first trip through the inner solar system. But then something unexpected happened. As it moved closer to the Sun, the comet broke apart into numerous parts (Fig. 14.7). But the comet was as far as 0.8 AU from the Sun when it disintegrated.

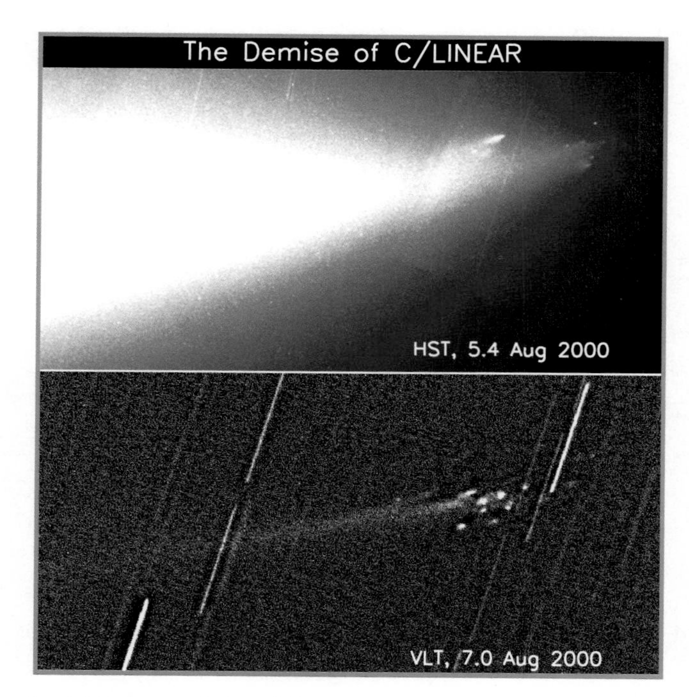

Fig. 14.7 Comet LINEAR falls apart A composite of two *Hubble Space Telescope*, or *HST*, images (*top panel*), taken on 5 August 2000, shows the bright dust tail of Comet C/1999 S4 (LINEAR) that is nearly 58 thousand kilometers across. The tail was created primarily during the complete destruction of part of the comet's nucleus about two weeks earlier. The rest of the nucleus broke up into fragments that are seen clustered near the tip of the tail (*right*). The European Southern Observatory's Very Large Telescope image (*bottom panel*), taken about 35 hours after the *HST* image, was processed using unsharp masking which suppressed light from the tail and enhanced the visibility of the fragments. This image is about 103 thousand kilometers across at the comet, and the nearly vertical streaks are tracks of stars passing through the field near the comet. Since the Earth's orbit will not intersect the comet's orbit, the comet debris will not create future meteor showers on Earth. (Courtesy of Harold Weaver, the *HST*/VLT observing team, NASA/ESA, ESO, and STSI.)

The tidal forces from the Sun should have been negligible at that distance, and previous comets have survived such a relatively distant passage. The explanation for the catastrophic disintegration probably lies in some combination of solar heat, the explosive release of volatile substances, rapid comet rotation and the intrinsically weak physical structure of the comet nucleus. In fact, when the amount of water vapor released during the break up, about 3 million tons, or 3 billion kilograms, is combined with the total volume of the fragments, a remarkably low value for the density was found. It was about 15 kilograms per cubic meter and less than two-hundredths that of water, suggesting the comet was very loosely packed and filled with holes.

This brings us to the main objective of the LINEAR program, a search for cosmic objects that may collide with the Earth in the future. Such a collision wiped out the dinosaurs 65 million years ago.

14.3 Impacts with Earth

The solid surfaces of almost all planets and satellites, from Mercury and the Moon to Jupiter's satellite Callisto, are marked with impact craters; the one exception is Jupiter's moon Io whose volcanic outpourings of lava have erased all the craters from its surface. The impact craters are the scars of past collisions with cosmic objects speeding through space. The terrestrial planets originated by the coalescence of these objects, and one of them tore enough material out of our planet to forge the Moon. Large comets or asteroids might even have brought water to the young Earth. And the cosmic barrage continues today.

A hail of cosmic objects is now pelting the Earth as it sweeps through space. Some of them are tiny, and burn up in the atmosphere. Every month at least one house-sized object is blowing up when it enters our air, producing a blast as forceful as a nuclear bomb. Now and then a bigger one gets through, gouging a crater out of the ground and even threatening the inhabitants of Earth.

Where did the Earth's water come from?

When it formed, the Earth was most likely too hot for water to condense from the primeval solar nebula. Space and planetary scientists therefore argue that much of the Earth's water could have been delivered from outside, by comets or asteroids. Scientists calculate that one big comet impact every millennium during the Earth's first billion years would have provided enough water to fill the oceans and lakes that are now on Earth.

Since the ratio of heavy hydrogen, or deuterium, to ordinary light hydrogen in some comets is about twice the ratio in our seawater, comets might not have brought all the water to the nascent Earth. But colliding asteroids could have supplied it. Many of these flying rocks contain about 10 percent water, in contrast to 50 percent in comets, so the colliding asteroids would have to be bigger than impacting comets, or else collide more frequently with the Earth. But that does not present a problem for this explanation. We know from the Moon's cratering record that the Earth must have been bombarded by large numbers of comets and asteroids soon after it formed.

Many geologists argue that the Earth's water came from inside the planet, during intense bouts of volcanism that filled the oceans. But the space scientists disagree with this explanation. They argue that most, if not all, of the volcanic water is not primordial in origin, and that it has instead been recycled in and out of the terrestrial crust and mantle, over and over again by geological forces.

The controversy over the source of the Earth's water, a prerequisite for life, continues to rage today, and all of the scenarios remain plausible. Some of the water may have been tied up in the rocks that made Earth, and some water had to be supplied during subsequent impacts by comets and asteroids. No one knows for sure exactly how much was provided by these different sources. And in any event, since the Earth was formed by asteroid, and perhaps comet, agglomeration, the Earth's water comes from these objects whether or not the delivery is during the Earth's formation or subsequent to it.

Explosions in the atmosphere

We have dramatic evidence that intruders from space are constantly bombarding the Earth today. Unknown to the public, the U.S. Department of Defense has been detecting them for decades. Military satellites, designed to watch for enemy rocket firings and nuclear explosions, have detected the explosions produced when speeding, house-sized fragments of asteroids or comets enter the air. The incoming projectiles are heated to incandescence and then self-destruct in the upper atmosphere, vanishing without a trace on the ground.

Ground-based defense networks, designed to listen to the sounds generated by man-made nuclear explosions, have confirmed the satellite results. They show that one of these cosmic bombs is bursting overhead every month, each with an energy equivalent to a nuclear bomb. These investigations might be aiding world peace by helping to distinguish between natural explosions and those caused by humans, thereby preventing false warnings of clandestine nuclear tests or of nuclear attack by terrorists.

Somewhat larger pieces of asteroids or comets can survive passage through the Earth's atmosphere more or less intact and strike the terrestrial surface at high velocity. Even a relatively small metallic asteroid, just tens of meters across, is tough enough to penetrate the atmosphere and hit the ground. Stony asteroids have to be more than 50 meters in size to survive passage through the atmosphere, and the fragile comets must be at least ten times as big to make it through.

A classic example is the 1.2-kilometer-wide Meteor Crater in northern Arizona, also known as the Barringer Crater. It was formed about 50 000 years ago by the impact of an iron asteroid just 60 meters in diameter. When it struck the Earth, this relatively small projectile released an estimated 100 million billion, or 10^{17}, joules of kinetic energy, an amount comparable to the energy released by the explosion of 20 million tons, or 20 billion kilograms, of trinitrotoluene, abbreviated TNT.

The largest object to strike the Earth in the 20th century wasn't quite big enough to reach the ground. It disintegrated between 5 and 10 kilometers up, over the Podkamennaya Tunguska River in central Siberia. The shock wave generated by the ensuing explosion leveled trees over 2 trillion (2×10^{12}) square meters of the underlying land, an area larger than New York City and surrounding suburbs (Fig. 14.8). The energy produced was equivalent to the aerial explosion of the nuclear bomb that leveled Hiroshima. So much devastation, yet it failed to produce a crater.

A Tunguska-like atmospheric explosion, relatively near the ground, is the usual fate for a stony asteroid fragment roughly 50 meters across. An asteroid of this size would have had the internal strength to penetrate deeply into the Earth's atmosphere before exploding, but a comet of comparable size would have disrupted too high in the atmosphere to cause much damage to the ground.

Scientists estimate that a Tunguska-like explosion should occur every century or two, on average. If it happens above a large city, the results will be devastating. But then, it is more likely to enter the atmosphere above the oceans or remote land regions, since they cover much more of the planet's surface area. Terrestrial impacts by larger cosmic objects occur less frequently, but they will reach the ground with more powerful consequences.

The terrestrial impact record

Even the largest crater, produced by the biggest comet or asteroid, will gradually disappear from sight with the passage of time. The same forces that erode mountains, deposit sediments, eject lava and shift continents are erasing the craters and removing them from sight. If not for these dynamic forces, the craters accumulated over the ages would be as densely distributed and prominent as the overlapping craters on the Moon.

Only about 160 terrestrial impact craters have managed to survive the ravages of time. They have been identified on images taken from space, using airplanes, the *Space*

Fig. 14.8 The Tunguska explosion On 30 June 1908, a giant blue-white ball of fire streaked across the daytime sky above the Tunguska River in Siberia, apparently becoming brighter than the Sun. The Tunguska fireball then exploded in the atmosphere, felling underlying trees like matchsticks. All of the toppled trees point away from a central location that must have been directly below the point of the explosion. Interpreting the extent and orientation of the tree-fall pattern, scientists concluded that the explosion above Tunguska released an energy equivalent to exploding a nuclear bomb between 5 and 10 kilometers above the Earth's surface.

Fig. 14.9 Wolf Creek impact crater This relatively well-preserved crater near Wolf Creek, Australia, is partly buried under wind-blown sand. Iron meteorites have been found in the vicinity, as well as some impact glass. The rim diameter of this crater is 850 meters and the impact that created it occurred about 300 thousand years ago. (Courtesy of Virgil L. Sharpton and the Lunar and Planetary Institute.)

Fig. 14.10 New Quebec impact crater A colliding asteroid or comet caused this impact crater, known as the New Quebec, or Nouveau Quebec, Crater after the region in Canada where it is located. The crater has a rim diameter of 3.4 kilometers and an estimated age of 1.4 ± 0.3 million years. (Image courtesy of George Burnside, Manotik, Ontario, Canada, and the Lunar and Planetary Institute.)

Shuttle, or satellites such as *Landsat*. These craters can be first identified from aerial photographs, by their circular shapes and uplifted and overturned rims (Figs. 14.9, 14.10, and 14.11). But since other processes, such as volcanism and erosion, can also leave circular holes, confirming evidence of an impact origin must be gathered from rocks in and around the crater.

How do geologists know that some terrestrial craters are due to the explosions of projectiles coming from space? They look for rocks that have been transformed under the conditions of extreme temperature, pressure, and shock associated with a high-velocity, external impact. The most apparent shock effect is the formation of conical structures called shatter cones, which point toward the center of the impact (Fig. 14.12). Other evidence includes glassy, previously molten material formed at high temperature, and minerals with a deformed crystal structure produced by a shattering, high-pressure impact. Roughly ten percent of the craters also contain meteorites that had to come from space.

Geologists have dated some of the craters by radiometric age determinations of previously melted rock, determining the time that has elapsed since the molten rock cooled and solidified. They find ages from a few thousand to two billion years, but most of the observed terrestrial impact craters are younger than 200 million years old. The lunar crater record suggests that the cratering rate on Earth must have been roughly constant during the past three billion

Fig. 14.11 Roter Kamm impact crater Located in the Namib Desert, the raised rim of this impact crater makes a stark contrast with the darker background vegetation. Broad, shifting sand dunes, deposited by the wind, cover the crater floor, but rocks melted by the impact have been found on the crater rim. The rim diameter of this crater is 2.5 kilometers and its age is 3.7 ± 0.3 million years. (Courtesy of Christian Koeberl and the Lunar and Planetary Institute.)

Fig. 14.12 Shatter cones The shatter cones that are found in the vicinity of many terrestrial craters provide evidence for shocks associated with the impact of a cosmic object. They point toward the direction of impact, like the cone-shaped plugs of glass that are often formed when a bullet strikes a window. The shatter cones shown here are about 0.05 meters in height. They are from the Wells Creek Tennessee Basin, a crater that is about 14 kilometers across and roughly 200 million years old.

years, so erosion and other geological processes have worn most of the older ones away.

Only the larger terrestrial craters have been able to survive the wearing effects of time for more than a few million years. One of them, located near Vredefort, South Africa, is three hundred kilometers in diameter and it was formed about two billion years ago (Table 14.1).

The consequences of these bigger impacts are even more sobering than the small ones. If an asteroid of just one kilometer in size hit the Earth, the power of its explosion could not be matched by the world's entire nuclear arsenal. Such an impact is estimated to occur every one million years or so. An asteroid or comet of exceptional size, say 20 kilometers across, might hit the Earth less often, every 100 million years, on average. As we shall next see, such collisions have happened in the past, when they altered the course of biological history.

14.4 Demise of the dinosaurs and other mass extinctions

Catastrophe from the sky

Collisions by objects from outer space have always been a menace to life on Earth. During the planet's first billion years, the barrage was probably so intense that living things could not exist on the Earth's surface. After those early times, the rate of bombardment slowed down, so impacts of exceptionally large cosmic projectiles became less frequent. But these giant impacts continued every once in a while, with devastating consequences. The most recent death rock arrived 65 million years ago, resulting in the wholesale removal of life on Earth. Such an abrupt destruction of an entire species of living things by a force of nature is known as a mass extinction.

Table 14.1 Three of the largest identified terrestrial impact craters[a]

Crater Name	Location	Crater Diameter (km)	Age (million years)
Vredefort	South Africa	300	2023.0 ± 4.0
Chicxulub	Yucatán, Mexico	250–280	64.98 ± 0.05
Manicouagan	Quebec, Canada	100	214.0 ± 1.0

[a] Numerical data and images for these and other terrestrial impact craters can be found on the web at http://cass.jsc.nasa.gov/publications/slidesets/craters.html

A thin layer of clay, just 0.01 meters thick, provided the initial evidence that an asteroid or comet collision wiped out the dinosaurs. The clay layer was deposited across the globe at the right time, and it contained unusual amounts of the rare element iridium that had to come from outer space.

When a team headed by the American geologist Walter Alvarez (1940–) determined the layer's age, from it position among geologically dated strata, they found that it was deposited about 65 million years ago when the dinosaurs and a variety of plants and animals disappeared. The iridium-rich clay layer is sandwiched between strata of rock from two geological time intervals, the Cretaceous period, when the dinosaurs flourished, and the subsequent Tertiary period, which saw the rise of the mammals. It is known as the K–T boundary, after *Krieide* the German word for "Cretaceous" and T for Tertiary, perhaps because C is used for the earlier Cambrian period.

Then Walter's father, Nobel-prize-winning physicist Luis Walter Alvarez (1911–1988), came up with an ingenious way to determine how long it took to form the boundary layer, by measuring the amount of iridium in the clay. Early in our planet's evolution, when the Earth was molten, the iridium clung to the heavy iron and sank down to form our planet's core, leaving the mantle and crust almost devoid of these elements. In contrast, iridium is much more abundant in certain meteorites, those rocks that fall from the sky and survive their fiery passage through the atmosphere.

Since the cosmic iridium rains steadily down through the atmosphere and settles in the soil, the amount of iridium in a layer of sediment can be used as a cosmic clock. In an average century, a certain amount of iridium will mix with the soil and become part of any new layers that are forming. If a layer requires twice as long to form, it will have twice as much iridium.

But the geologists found that the iridium clock had gone wild for a short interval about 65 million years ago. The amount of iridium they found in this layer of clay was far higher than normally found in the Earth's crust and about 30 times higher than that found in the fossilized limestone above and below the clay. Moreover, the same type of iridium enrichment was found in clay at widely scattered points on the Earth. So the entire globe had been drenched with an unusually large amount of iridium for a short time.

Walter, Luis, and their colleagues concluded in 1980 that the iridium deluge came from outside the Earth, delivered by a large asteroid or huge comet that struck the Earth and vaporized about 65 million years ago. According to their hypothesis, the iridium was lofted into the atmosphere along with other debris by the fireball of hot gas created during the collision, and then carried by the winds over much of the globe. The global cloud of iridium-rich dust then slowly filtered back down to the ground where it produced a thin global layer that contained relatively large amounts of iridium. They estimated that a layer 0.01 meters thick covering the entire Earth would be deposited by an asteroid about 10 kilometers in diameter.

Most geologists and biologists must have initially dismissed the idea of a killer asteroid from space. They probably attributed it to spaced-out astronomers or science fiction enthusiasts. Such an abrupt cataclysm conflicted with the prevalent concept of gradual evolutionary change over the eons. Sudden, short-lived events were just not supposed to affect the course of evolution. Yet, the evidence for catastrophic events is found throughout the solar system, including the cratered surfaces of the Moon and Earth.

The idea just would not go away and supporting evidence kept accumulating. Shocked quartz grains were, for example, found in the K–T boundary throughout the world. No known terrestrial process, including volcanic flows or explosions, can generate pressure high enough to alter the grains in the observed way, only the sudden shock of an impact can. Still, the skeptics asked where is the crater of the impact that occurred 65 million years ago and was big enough to obliterate most of the Earth's life-forms.

After years of searching, the tell-tale crater was found straddling the northern coastline of the Yucatán Peninsula (Fig. 14.13). It is located below the Mayan village of Chicxulub (pronounced Cheek-shoe-lube, a Maya phase for "horns of the devil"), and is hence known as the Chicxulub impact basin. The discovery of this crater and the subsequent confirmation of its age at 65 million years led most scientists to accept the impact hypothesis for the demise of the dinosaurs.

At the time of the impact, the Yucatán Peninsula was below sea level, a continental shelf, and the center of the crater now lies buried below one kilometer of limestone laid down in the intervening years. So there is nothing on the surface to betray the crater's existence, and the vast scar cannot be seen directly.

The serendipitous discovery of the submerged crater began with a search for oil in the region. The Mexican national petroleum company, Petróles Mexicano or Pemex for short, commissioned an aerial magnetic survey to asses the thickness of sedimentary – and possible oil-bearing – rocks in the region, which revealed a large, buried

Fig. 14.13 Site of impact that wiped out the dinosaurs The long-sought site of the scar left by a killer asteroid has been found near Chicxulub (*filled circle*), a small village at the tip of the Yucatán Peninsula. Material from the submerged crater has been dated at about 65 million years, coinciding with a blast that triggered the eradication of most life on Earth. Thick sedimentary deposits laid down 65 million years ago in Haiti (*big open circle*) contain exceptionally large amounts of iridium, shocked quartz and glassy debris that are thought to be part of the impact across the Caribbean, almost two thousand kilometers away. The small open circles mark the sites of marine wave deposits associated with the same impact. (Adapted from Alan R. Hildebrand and William V. Boynton, *Natural History* 6, 47–53, American Museum of Natural History 1991.)

Fig. 14.14 Chicxulub impact crater This image shows the variations in the magnitude of the gravity field at sea level in the vicinity of the buried Chicxulub impact basin in the northwestern corner of the Yucatán Peninsula. The impact basin exhibits ring-like variations in the strength of the gravity. One interpretation of the crater's horseshoe shape is that the asteroid struck the land at a low, oblique angle from the southeast (*lower-right*). The pronounced ring that separates the deep annular basin (*magenta and deep-blue regions*) from the basin flanks (*cyan and green*) has a diameter of about 180 kilometers. A discontinuous, subtle outer region with a diameter of 250 to 280 kilometers can be discerned particularly toward the southern (*lower*) part of the basin. The irregular white line marks the shoreline of the Yucatán Peninsula, and the straight white lines mark province boundaries. (Courtesy of Virgil L. Sharpton and the Lunar and Planetary Institute.)

semi-circular structure. Coarse gravity maps showed a similar feature, and exploratory oil drilling in the area revealed an underground layer of broken, melted rock. Subsequent radioactive dating of the drill-core fragments showed that they were exactly contemporaneous with the K–T boundary, with an age of 65 million years.

Moreover, the iridium layer in Haiti, located about two thousand kilometers away from the impact site, contains exceptionally large glassy rocks, called *tektites* from the Greek words "molten" and "to melt", that were probably hurled out during the explosive impact. And the K–T boundary layer in other parts of the Caribbean shows evidence of marine deposits laid down by giant waves created by the impact.

High-resolution gravity maps in the 1990s revealed the size and structure of the buried Chicxulub crater (Fig. 14.14). Regions of high density and greater gravitational pull are distributed in several concentric rings, with an outermost diameter of at least 180 kilometers and perhaps as much as 250–280 kilometers. Its size and bull's eye pattern is similar to the largest impact basins on the Moon and Mercury, created by cosmic collisions during the early days of the solar system about four billion years ago.

Chicxulub is the biggest thing to hit Earth in the past one or two billion years, and of just the right size to have been excavated by an asteroid of 10 kilometers

Focus 14.1 The belt of an asteroid.

The mass of an impacting asteroid or comet can be determined from the mass density and volume. Assuming a mass density, ρ, of about $\rho = 3000$ kilograms per cubic meter, comparable to that of stony asteroids and meteorites, and a radius, R, of $R = 5$ kilometers, a mass of $M = \rho(4/3)\pi R^3 = 1.6 \times 10^{15}$ kilograms is obtained. A comet of the same size would have a mass density and mass of just a few times lower than this amount.

The kinetic energy, abbreviated $K.E.$, of an impacting asteroid or comet is given by the expression $K.E. = (1/2)MV^2$, where M is the mass of the projectile and V is its incoming velocity. Assuming an impacting velocity of 20 kilometers per second, and the mass we have estimated, this expression gives $K.E. = 3.2 \times 10^{23}$ joules.

This amount of energy is equivalent to the explosion of 71 thousand terrestrial bombs, each with a destructive force of 100 megatons, or 100 million tons and 10^{11} kilograms, of trinitrotoluene, abbreviated TNT. The devastation produced by the impact would therefore be many orders of magnitude larger than that caused by the simultaneous explosion of the world's entire arsenal of nuclear weapons. It would be equivalent to the detonation of the blast that destroyed Hiroshima, at just 13 thousand tons, or 13 million kilograms, of TNT, every second for 175 years.

in extent. The energy released during the collision of such an intruder is enough to trigger a mass extinction (Focus 14.1).

The day the dinosaurs died

Most scientists are now convinced that the dinosaurs, which had dominated the Earth for over 160 million years, were destroyed when a marauding asteroid or comet dropped out of the sky and struck the Earth 65 million years ago. The explosive impact generated an enormous ball of fire, which incinerated everything in the immediate area, blasted billions of tons, or trillions of kilograms, of debris into space, and gouged out a crater at least 180 kilometers across. But that was just the beginning.

Deadly, consequences of the collision included:

Enormous sea waves. If a huge impacting projectile plows into an ocean, it will generate a relentless succession of colossal waves that race across the sea at the speed of a jet aircraft and rise several kilometers in height. So the cosmic object that slammed into the water above the current Yucatán Peninsula must have generated great moving walls of water that surged into the East Coast of what is now known as the United States, leveling everything in its path and leaving almost nothing standing. These enormous sea waves are known as *tsunamis*, from the Japanese for "harbor waves", but they have nothing to do with the tides, which is suggested when using the alternative term "tidal waves".

Frightful darkness and freezing cold. The cosmic intruder was completely vaporized during the explosive impact, and a great fireball rose into the stratosphere, carrying with it large amounts of pulverized debris. Vast clouds of dust and ash remained suspended and were circulated by air currents until they encircled the Earth, covering it in total darkness. Since the global shroud blocked the Sun's heat, as well as its light, the surface temperature plummeted and the planet entered a dark chill, lasting for months before the dust eventually settled back to Earth.

The globe-spanning pall was most likely accentuated and prolonged by clouds of sulfuric acid, formed from the debris thrown skyward by the impact. The blast must have vaporized the sulfur-rich rock around Chicxulub, and tossed great quantities of sulfur into the air, where it mixed with moisture to form tiny droplets of sulfuric acid. These drops reflected enough sunlight back into space to drop ground temperatures below freezing, and could have remained airborne for decades.

Plants would have failed to receive enough sunlight to allow photosynthesis to continue, and a prolonged "winter" of unusual cold would have added to the devastation. As plant life withered and froze, plant-eating animals dependent on them would also die, as would meat eating animals once their plant-eating prey were gone. By the time that the dust settled, many land plants and animals were no longer there.

Wildfires. As hot material ejected from the impact rained back down to the ground, it would have ignited wildfires that scorched the Earth in a global inferno. Dead vegetation, killed by a combination of darkness and cold, would have fueled the conflagration. A layer of carbon soot particles, found in the K–T deposits from many parts of the globe, seems to confirm that the world was set ablaze. And since the impacting asteroid or comet rammed into the Earth on an oblique trajectory from the southeast, it would have sent a blowtorch of white-hot debris forward toward North America, along the impactor's path, accounting for the high rate of plant extinction there.

Acid rain. Conditions in the oceans would not have been much better than those on land. Torrents of acid rain would have poisoned the water and destroyed shell-bearing creatures, disrupting the entire food chain. Sulfur ejected into the stratosphere during the collision would have slowly converted to sulfuric acid that rained into the oceans, wreaking havoc on marine animals and plants. And the pulse of heat sent into the atmosphere by the impact would have created lethal nitric acid rain.

Volcanoes. If anything managed to escape the enormous sea waves, extensive wildfires, cold darkness and acid rain, then volcanoes might have got them. According to one hypothesis, the colliding projectile generated shock waves that propagated through the Earth and converged to an area opposite to the impact. The shock energy was focused in this direction by the body of the planet itself, acting like a giant lens. When the concentrated shock waves reached the other side of the planet, they heated the ground and triggered major volcanic eruptions, incidentally accounting for massive outpourings of lava that occurred about 65 million years ago in the region now known as India. These volcanoes might have contributed to the mass extinction by spewing carbon dioxide, deadly ash, sulfur and other substances that disrupted the atmosphere and altered the climate. When combined with the dust, sulfur, and hot debris tossed skyward by the original impact, the volcanic activity could have spelled doom for many living things.

When the cosmic blast and its aftermath were over, the dinosaurs were gone, along with most marine animals and many land plants. Altogether roughly half of all animal and plant species died out. That sounds pretty awful, but from catastrophes there arise opportunities. The biological devastation caused by the impact apparently cleared the way for the rise of the relatively small mammals, so your very distant ancestors may have benefited.

Other mass extinctions

The dinosaur era may have both started and ended with the collisions of asteroids with Earth. The layer of rock laid down 200 million years ago also contains elevated iridium content, suggesting a cataclysmic impact then. It apparently eliminated half the groups of life and opened the evolutionary door for what was then a relatively small group of animals – the dinosaurs.

Another, truly staggering mass extinction occurred about 250 million years ago. About 90 percent of life in the oceans vanished forever, as did 70 percent of the land animals. Scientists have speculated that a killer asteroid or comet also caused this global destruction, known as the Permian–Triassic extinction.

Although it was long believed that the Permian creatures disappeared gradually, over 10 million years, recent investigations suggest that the extinction happened much more quickly. Marine fossils dating back 250 million years apparently indicate that they died off in just 8000 years or less, which is sudden death in terms of geological times. An asteroid or comet impact could be blamed for wiping out life in such a short time.

Moreover, traces of helium and argon found in rocks dating back to the extinction are in proportions that could not have been created on Earth. The isotopic signatures of these gases seem to require a cosmic origin in the extreme conditions within an ancient star billions of years ago. These gases were subsequently blown into interstellar space, and later consolidated into an asteroid or comet that carried them to Earth. The collision site is unknown, so the killer projectile most likely dived into the ocean.

Deadly catastrophes from the sky are now in vogue, perhaps accounting for relatively recent extinctions. Spherules of impact-melted glass have, for example, been found in layered deep-sea sediments dating back to a mass extinction that occurred 35 million years ago. Cosmic impacts have been blamed for four global mass extinctions of the last 100 million years, taking place 91, 65, 35 and 11 million years ago. Some scientists have even argued that killer asteroids or comets may periodically sweep a wave of death across the Earth, thereby ending the rule of the dominant species.

We humans have flourished in the past half million years, developing wonderful civilizations, building great cities, generating profound knowledge, and sending spacecraft throughout the solar system. Yet, it might suddenly come to an end. Such catastrophes have a small probability of occurring during our relatively short lives, but over astronomical times of millions and billions of years, the exceedingly unlikely becomes a virtual certainty. After all, 99 percent of all species that ever lived have gone extinct.

14.5 Probability of death from above

The Earth does not occupy a secure niche in space. Our planet is instead immersed in a cosmic shooting gallery, subject to a steady bombardment by lethal, Earth-approaching asteroids and comets. Somewhere in space, one of them is hurtling toward a future collision with Earth (Fig. 14.15). And if it is large enough, the impact will severely disrupt terrestrial life upon impact. It's only a matter of time.

So when are we going to be hit hard enough to worry about it? Since astronomers have not yet located the doomsday rock that will definitely collide with Earth, we don't

Fig. 14.15 Doomsday rock This computer-generated image depicts a view of the Earth as seen at a distance of 5.2 million kilometers from the surface of an asteroid or comet that might collide with our planet. If the impacting projectile is big enough, then the collision will severely disrupt human life and might even destroy it altogether. (Courtesy of JPL and NASA.)

know exactly when the next impact will take place. But we can calculate the odds, and they depend on the size of the colliding object. Since there are many more small cosmic objects than large ones, the smaller ones hit our planet more frequently. Bigger asteroids or comets strike the Earth less often, but they have greater destructive power.

Thus, to estimate the risk of being hit in a way that matters, the potential impacting projectiles first have to be sorted according to size (Fig. 14.16). Fragments smaller than a few tens of meters across burn up in the atmosphere and rarely reach the ground – except for the exceedingly tiny particles of cosmic dust that drift down into your hair. Asteroids or comets a hundred meters in diameter are expected to strike Earth every thousand years, on average. They could take out a city and cause severe local damage, but pose no threat to the Earth as a whole.

Asteroids or comets about one kilometer in size pose a greater peril. They are large enough to destroy a large country and produce global consequences. Contemporary surveys indicate that there may be about one thousand of these objects now on paths that come near the Earth's orbit. Most of the time, the Earth will be somewhere else if one of them crosses its path, but occasionally they will arrive almost simultaneously at the intersection. The average time between such impacts is about one million years, an interval vastly longer than the history of civilization.

During your lifetime, the chance that you will be wiped out by the impact of a two-kilometer asteroid or comet is 1 in 20 000, the same as death from an airplane crash (Table 14.2). Such a risk is much greater than those of death by snakebite or food poisoning, and just a little less than dying in a car accident. But it is not just one person that dies when a large cosmic projectile hits the Earth. About 100 million people are expected to perish if the object is two kilometers across, and a bigger one could destroy all of us. The other causes of death usually affect just one person at a time, or perhaps a few of them.

Moreover, to declare that cosmic disaster strikes the Earth, on average, just once in a million years or so does not mean that we are guaranteed such a long interval between catastrophic impacts. The very small chance of such a collision is the same today as it will be millions of years from now. It could be tomorrow, or it might be long after you're gone.

In fact, there are all sorts of other things you can worry about if you are in a morbid mood. The chance of dying before age five is 1 in 8 in South Asia, and 1 in 100 in the United States. The risk of dying in childbirth in the United States is about 1 in 12 000, roughly twice the chance of death by the impact of a two-kilometer cosmic projectile, but thousands of times greater than death by dog bite or drinking detergent.

But to get back to the dangers of cosmic impact, it takes an even larger, rarer asteroid, 10 kilometers across or larger, to render the human species extinct. The impacting object would blast out a huge crater and eject billions of tons of pulverized rock and dust into the air. Hot material lofted into the atmosphere could produce widespread wildfires that would scorch the continents. The sulfur tossed into the air would result in acid rain that might lead to an even greater toll in the oceans than on land. A

Table 14.2 The dangers of a lifetime[a]

Cause of Death	Chance of Dying in a 65-year Period	Number of People Killed
Car accident	1 in 100	5
Murder	1 in 300	1
Gun accident	1 in 2500	1
Accidental electrocution	1 in 5000	1
Asteroid or comet impact (Two kilometers in diameter, global catastrophe)	**1 in 20 000**	**100 million**
Airplane crash	1 in 20 000	300
Tornado	1 in 60 000	10
Snake bite or bee sting	1 in 100 000	1
Food poisoning	1 in 3 000 000	1

[a] Adapted from Clark R. Chapman and David Morrison, *Nature* **367**, 33–40 (1994). The risks are for a person living in the United States.

globe-encircling pall of dust and other debris would block out the Sun's light and heat, crippling agriculture, producing widespread starvation, and perhaps leading to a breakdown of our fragile civilization.

Such a mass extinction might occur once in a hundred million years on average, destroying in an instant what it has taken humans millennia to build. The chances of that happening are very low, but they are not zero. The lifetime risk of your being wiped out with the rest of humanity during such a mass extinction is roughly one in a million. So it might happen and a prudent society should prepare for the possibility.

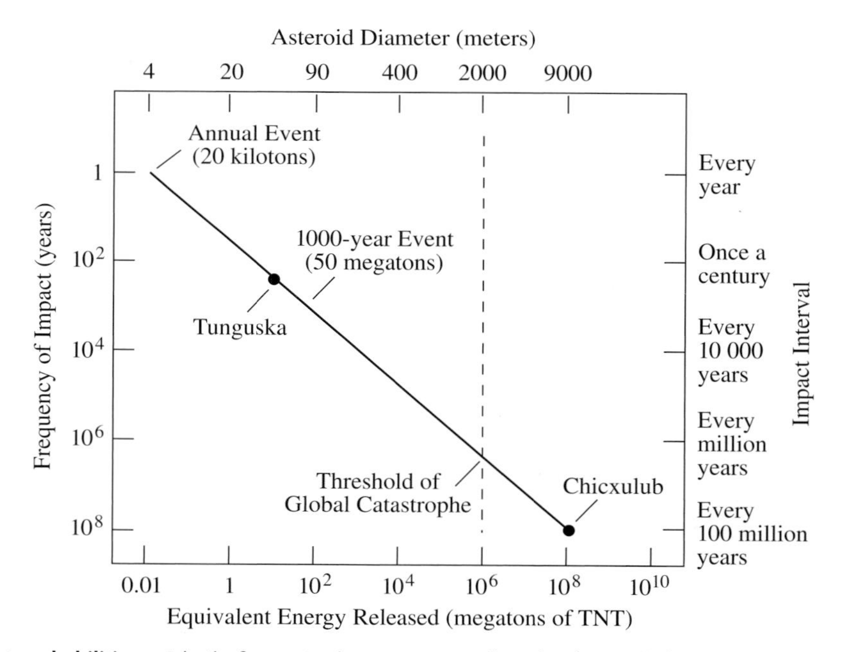

Fig. 14.16 Cosmic impact probabilities A hail of cosmic objects continually pelts the Earth from space. This plot shows the likelihood (*vertical axis*) that a member of the current population of Earth-crossing asteroids and comets will hit our planet. An object 2 kilometers in diameter, capable of producing certain global damage, hits the Earth every million years on average. An impact like the one that wiped out the dinosaurs, giving rise to the Chicxulub crater, is estimated to occur every 100 million years. Since smaller asteroids and comets are much more numerous than larger ones, the smaller objects strike our planet more often. An impact like the Tunguska event, which occurred on 30 June 1908, might occur every 700 centuries or so. The explosive energy of the impact is also given (*horizontal axis*) in units of megatons of exploded trinitrotoluene, or TNT for short. One hundred megatons (10^{11} kilograms) of TNT is equivalent to about 4×10^{17} joules of energy, the amount released by a typical nuclear bomb on Earth.

14.6 Breaking a date with doomsday

Finding the hidden threat

While we know that asteroids and comets have collided with the Earth in the past, and that they will inevitably hit our planet in the future, we do not yet know if any of them are now headed for a deathly collision with our solitary outpost of life. Astronomers are therefore taking a census of everything out there that is big enough and close enough to threaten us (Fig. 14.17). Once all of these Near-Earth Objects are located, and their current trajectories known, astronomers can use computers and refined observations to determine their precise future paths and establish whether and when any of them will strike the Earth.

After decades of scanning the skies, ongoing search programs will find and catalog the most threatening asteroids (Focus 14.2). They will provide the exact positions and orbits of about 90 percent of the Near-Earth asteroids larger than one kilometer in size. That is the minimum diameter of a space rock that could have global consequences if it hit Earth.

Astronomers have found at least half of the menacing space rocks that might wreak global havoc, and none of those discovered so far is on a direct collision course with Earth – at least in the near future. The search is therefore slowly diminishing the chances of our unexpected demise, gradually improving the odds of this cosmic Russian roulette.

We know nothing about the other half of the larger asteroids that are expected to cross the Earth's orbit but are not yet discovered. In addition, there are many more smaller, undiscovered cosmic objects, less than one kilometer across, that are now moving about in the space near our planet. They pose a significant threat, but not a global one. They can hit the Earth with an energy equivalent to a large nuclear bomb. Moreover, large new comets will always come unannounced and unexpected from the remote outer fringes of the solar system, so we can never completely assess the comet hazard.

Today it is beyond our technology to detect and defend against relatively small asteroids or comets, but they can be seen when coming very near the Earth. Some asteroids have come uncomfortably close (Table 14.3), passing the Earth within less than the distance between the Earth and the Moon. By cosmic standards, that is a close call, only a hairbreadth away. These objects are all relatively small, most of them less than 200 meters across, at least so far.

Fig. 14.17 Asteroid streak As the nearby asteroid 2100 Ra-Shalom moved along its orbit around the Sun, it produced an elongated streak or trail on this photograph, while the background stars retained their round shape. This 20-minute exposure was taken with a 0.46-meter (18-inch) telescope on Palomar Mountain in California, during a pioneering search for Near-Earth Objects. (Courtesy of Eleanor F. Helin, JPL, California Institute of Technology.)

Doing something about the threat

Sooner or later we will discover an asteroid or comet headed toward collision with the Earth. But unlike most natural disasters, the impact can be forestalled once the killer object has been fingered and we can see it coming. Evasive action will depend on the lead-time available and the expected physical nature of the object, including whether the object is a binary system.

How much warning will we have? When the inventory of large Near-Earth Objects, or NEOs, is complete, and one is found on its way to strike the Earth, then advanced warning would probably come decades in advance of a collision. The killer rock would most likely swing near the Earth and loop around the Sun several times before hitting our planet. But if we are unlucky, a large comet, entering the solar system for the first time, might be identified just a few months before impact. And since existing surveys do not regularly detect small NEOs, less than 200 meters in size, as efficiently as they discover larger ones, we may not know about them until they are about to collide with us. Thus, we will either have a long lead-time or none at all.

And what do we do if we find a large asteroid or comet headed our way? The Earth cannot be moved out of the way, but we could launch an intercept mission to redirect the object's course. If the impact is many years away and the threatening object relatively far away from us, all we have to do is give it a little nudge. By the time the asteroid or comet reaches the Earth's vicinity, that small change in trajectory will make a big difference, enabling it to bypass the planet.

An orbiting mirror could be used to focus sunlight and vaporize the object's surface, producing a jet of gas and

Focus 14.2 Searching for cosmic bombs headed our way

The large majority of asteroids are in orbits between those of Mars and Jupiter where they pose no threat to Earth. Some of these rocky objects, however, follow a more eccentric course that takes them closer to Earth, sometimes crossing its path. And although most short-period comets do not come closer to the Earth than the nearest planet, some of these icy intruders can pass perilously near to us. It is these threatening asteroids or comets, collectively known as Near-Earth Objects and abbreviated NEOs, which may be on a collision course with the Earth. If one of these cosmic bombs hits our planet it could explode with a violence that far surpasses the world's entire nuclear arsenal, threatening civilization and possibly making humans extinct.

The search for NEOs was pioneered by a small number of dedicated amateur and professional astronomers in the 1970s, primarily with the 0.46-meter (18-inch) telescope on Palomar Mountain in California. The comet that hit Jupiter (Section 14.1) was discovered in 1993 as a byproduct of the ongoing NEO search with this telescope. This relatively small, photographic telescope has a wide field of view that can be used in a dedicated NEO survey, unlike bigger, more powerful telescopes that are in high demand for other purposes and scrutinize tiny pieces of the sky.

As time passed, technology improved, and it was realized that the rate of finding asteroids and inactive comets could be greatly increased by using electronic detection devices and fast computers. Then the collision of a comet with Jupiter, anticipated in 1993 and watched by millions in 1994 (Section 14.1), raised public consciousness of the impact threat to planet Earth. The U.S. Congress held hearings to study the threat, and asked NASA to formulate plans to deal with the problem.

Public awareness of the cosmic bombs was notched up once more with the release, in 1998, of two blockbuster movies, *Deep Impact* which deals with a killer comet and *Armageddon* in which two astronauts save humanity by diverting a rogue asteroid headed toward collision with Earth. In the same year, the U.S. Congress held more hearings about the threat, and NASA initiated the Spaceguard Survey, intended to find 90 percent of the NEOs larger than one kilometer within the next decade. Although the threat has passed out of the science fiction domain to a serious public issue, it's entertaining to realize that Arthur C. Clarke (1917–) advocated a "Spaceguard Project" for protection from future cosmic impacts in his 1973 novel *Rendezvous with Rama*.

By the end of the 20th century, several teams of astronomers were surveying the sky with electronic cameras to complete the inventory of large NEOs, with primary support from NASA and the U.S. Air Force. The search involves finding every object that moves against the background stars down to a magnitude of 19.5, almost 100 000 times fainter than the detection limit of the human eye. The LIncoln Near Earth Asteroid Research, abbreviated LINEAR, project of MIT's Lincoln Laboratory had been especially productive, detecting about 250 confirmed NEOs. Other survey groups include the Spacewatch program at the University of Arizona, which has done much of the pioneering work in the field of NEO detection, the Near Earth Asteroid Tracking, or NEAT, system operated by the Jet Propulsion Laboratory in California, the Lowell Observatory Near Earth Object Search, or LONEOS for short, in Arizona, and the Catalina Sky Survey, abbreviate CSS, also in Arizona.

Once an NEO is discovered by the ongoing surveys, previous unsuspecting observations made before the NEO discovery and follow-up observations with powerful radar telescopes can be used to refine knowledge of its future trajectory and push predictions far into the future. By bouncing radio waves off the object, and examining the return echo, the radar technique can be used to establish an exact orbit, providing us with centuries of advance notice of close encounters or impacts by NEOs. As an example, radar and optical measurements of the asteroid 1950 DA, spanning 51 years, indicate a 20-minute interval in March 2880 when there could be a non-negligible probability of the 1-kilometer object colliding with Earth.

In the early 21st century, about two-thirds of the estimated one thousand NEOs larger than one kilometer had been discovered and confirmed. They are cataloged at the International Astronomical Union's Minor Planet Center, located on the web at http://cfa-www.harvard.edu/iau/mpc.html

dust that might change its course. Alternately, astronauts could rendezvous with it, attaching small explosives, rocket engines, solar sails or mass drivers that could push it into a harmless trajectory. A large solar sail would use the pressure of sunlight to move the body slowly, while a mass driver would scoop up surface material and hurl it away, creating an appropriate recoil reaction just as the small explosives or rocket engines would.

If the warning time is only a matter of months or less, the sole recourse might be to send a high-powered rocket armed with a bomb powerful enough to redirect the object or blow it up. Such a possibility has sparked the interest

Table 14.3 Close approaches of asteroids to Earth[a]

Closest Distance from Earth (AU[b])	Date	Absolute Magnitude[c]	Estimated Diameter (m)
0.0007	9.80 December 1994	28.0	9
0.0010	20.90 May 1993	29.0	6
0.0011	15.70 March 1994	28.5	7
0.0011	18.70 January 1991	28.5	7
0.00205	15.85 January 2001	25.8	25
0.0029	27.20 March 1995	26.5	18
0.00303	19.69 May 1996	20.5	220
0.00306	5.39 December 1991	28.8	6
0.00311	8.04 March 2002	24.4	50
0.00321	8.80 February 2002	26.9	16
0.00332	2.90 June 2000	29.0	6
0.00380	6.14 May 1999	24.5	50
0.00457	22.90 March 1989	20.5	220
0.00484	29.33 November 2001	24.9	30
0.00486	24.85 November 1994	22.0	140
0.0049	30.70 October 1937	18	900

[a] Adapted from lists of the International Astronomical Union's Minor Planet Center, available on the web at http://cfa-www.harvard.edu/iau/lists A list of the closest comet encounters can also be viewed at this site, but even the nearest one, Comet Lexell at 0.015 AU in 1770, was more than 20 times further away from the Earth at closest approach. The Near Earth Object web site of the Jet Propulsion Laboratory, located at http://neo.jpl.nasa.gov/ carries several tables that can be sorted by close approach, date, object name, past and future approaches, and comets or asteroids.

[b] The average distance between the Earth and the Sun is roughly one astronomical unit, abbreviated AU, with a value of 1.496×10^{11} meters. By way of comparison, the average distance between the Earth and the Moon is 384.4 thousand kilometers and 0.002 56 AU.

[c] The absolute magnitude is a measure of the brightness of an asteroid's reflected sunlight, and a smaller value indicates a brighter asteroid. The asteroid diameters are estimated from the absolute magnitudes under the assumption that the asteroid reflects 14 percent of the incident sunlight.

Focus 14.3 Star Wars in outer space

When it was realized that an asteroid or comet might be hurtling toward collision with the Earth, experts in nuclear weapons and missile defense systems had something other than nuclear war to worry about. They proposed that nuclear-tipped missiles could be used to blow the threatening object up or to deflect it *en masse* from its Earthward trajectory.

Of course, the military would want to test the warheads. With the whole world threatened you wouldn't want to use an untested weapon. In 1996, China even advocated underground tests to prepare nuclear warheads as a possible defense against the cosmic threat.

That brings up the question of who is told about plans to protect civilization from the intruder. A madman, or someone making a stupid error, could use the Cosmic Star Wars weapons to deflect a previously benign asteroid or comet into a collision course with Earth. After all, we now know how terrorists can redirect airplanes to destroy thousands of innocent lives.

of bomb designers and some members of the military (Focus 14.3). A conventional nuclear weapon might be used to deflect or destroy a small, solid, rocky asteroid, but a much larger explosion could be needed to divert or pulverize a loosely bound object, like some large asteroids and most comets.

It might not be a good idea to blow the threatening object up, for the Earth might then be struck by a hail of dangerous projectiles rather than a single blast. Some of the pieces could still head straight at us. If they were not small enough to burn up in the atmosphere, the fragments might cause massive devastation in several places on the Earth.

In the absence of early warning and evasive action, advance knowledge of the time and place of an impact would at least allow people to store food and supplies and to evacuate regions near the impact site where damage would be the greatest. It could happen today, or it might not happen for millennia. Some of us will place our trust in God, and many of us have more immediate concerns, but it will not hurt to subscribe to a cosmic insurance policy and fully identify the swarm of asteroids and comets that might be headed our way. The policy has relatively small premiums and enormous potential benefits.

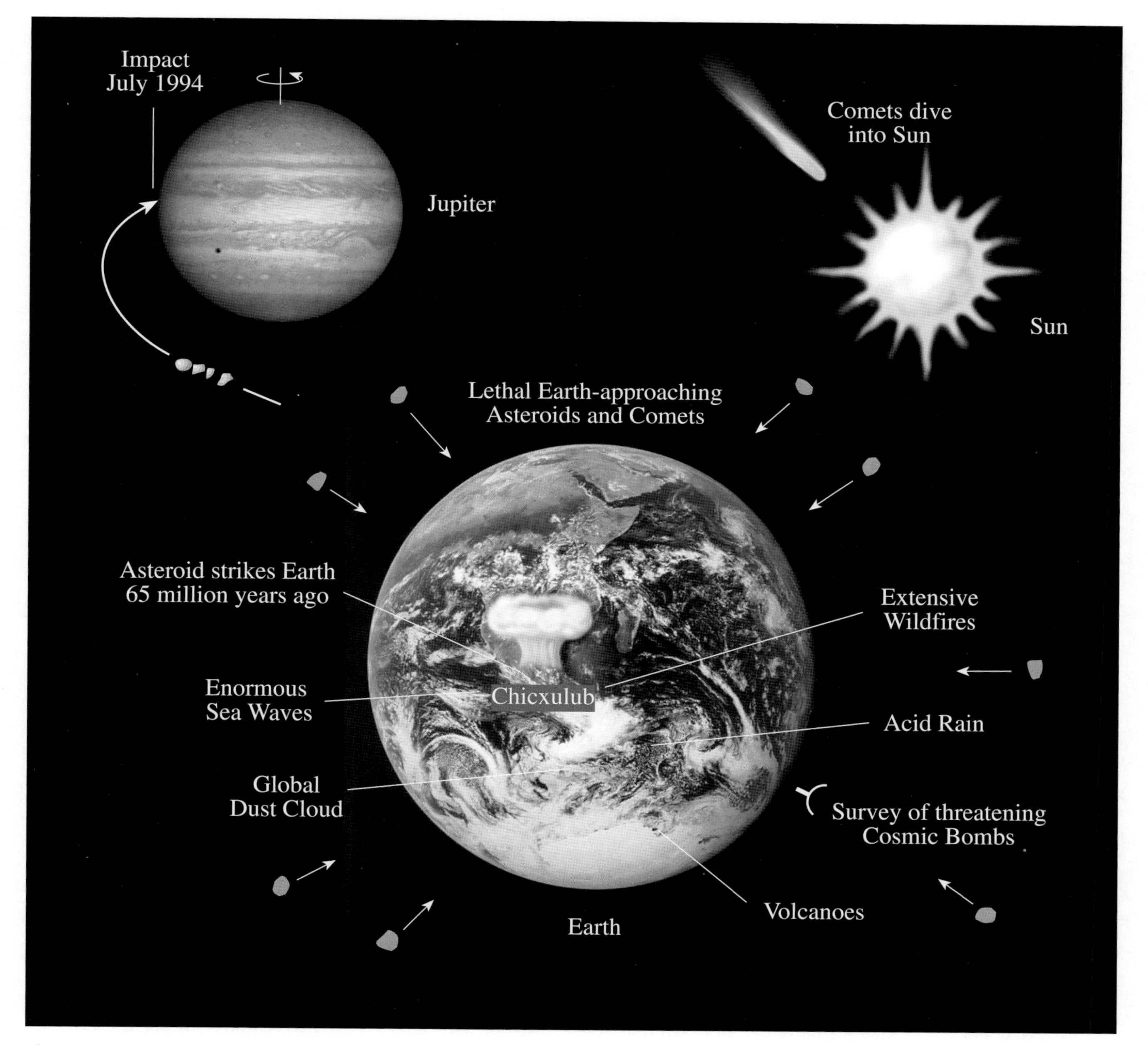

Fig. 14.18 Summary Diagram

Appendix 1

Further reading

An annotated list of books published between 1990 and 2002

Beebe, Reta: *Jupiter: The Giant Planet*, Second Edition. Smithsonian Institution Press, Washington, D.C. 1997.

A good summary of our knowledge of Jupiter, written after the two *Voyager* missions but before the main results of the *Galileo* mission. This book includes discussions of Jupiter's atmosphere, interior, satellites, rings and magnetosphere, as well as the Shoemaker–Levy 9 comet collision with the planet.

Benedick, Richard Elliot: *Ozone Diplomacy: New Directions in Safeguarding the Planet*. Harvard University Press, Cambridge, Massachusetts 1991.

A fascinating discussion of the scientific, political, economical and diplomatic efforts that forged *The Montreal Protocol on Substances that Deplete the Ozone Layer*, adopted in 1987 and revised in 1990. This was the first international treaty to protect the global environment. The participants in the negotiations included scientists, industrialists, and government officials, including the author who was the chief U.S. negotiator of the treaty.

Bone, Neil: *The Aurora: Sun–Earth Interactions*. John Wiley and Sons, New York 1996.

A complete review of all aspects of the aurora, describing the mechanism of auroral displays and their observation, including the history of aurora investigations from the earliest ideas to scientific investigations with satellites. Solar activity, Earth's magnetosphere and solar–terrestrial interactions are also reviewed.

Booth, Nicholas: *Exploring the Solar System*. Cambridge University Press, New York 1995.

More than 300 full-color images are used to highlight the important scientific advances during more than three decades of planetary exploration with spacecraft.

Bougher, S. W., Hunten, D. M., and Phillips, R. J. (editors): *Venus II – Geology, Geophysics, Atmosphere, and Solar Wind Environment*. The University of Arizona Press, Tucson, Arizona 1997.

A collection of 36 technical articles written by scientific experts in the field, including the Venus atmosphere and ionosphere and their interaction with the solar wind, surface processes, geological structures, volcanism, tectonism, impact cratering and geodynamics.

Brandt, John C., and Chapman, Robert D.: *Rendezvous in Space: The Science of Comets*. New York, W. H. Freeman 1992.

An overview of our recent understanding of comets conveyed with a blend of science and history, including the "dirty-snowball" model and the spacecraft encounters with Comet Halley.

Carr, Michael H.: *Water on Mars*. Oxford University Press, New York 1996.

A complete discussion of the past and present water on Mars, written after the *Viking 1* and *2* missions, but before the *Pathfinder, Mars Global Surveyor* and *2001 Mars Odyssey* missions. This book includes discussions of the planet's atmosphere, outflow channels, valley networks, ground ice, climate change, and implications for life.

Cattermole, Peter: *Mars: The Story of the Red Planet*. Chapman and Hall, London 1992.

The scientific results of the *Mariner 9* and *Viking 1* and *2* missions to Mars are described, including the atmosphere, weather, interior, craters, volcanoes, plains, canyons, channels, polar regions and geological history of Mars.

Cattermole, Peter: *Planetary Volcanism: A study of volcanic activity in the solar system*, Second Edition. John Wiley and Sons, New York 1996.

A thorough technical account of volcanism on Earth, Io, Mars, Venus, and our Moon. It also includes discussion of volcanic flows, landforms and plains, as well as the generation, evolution, ascent and eruption of magma.

Cattermole, Peter: *Venus, the Geological Story*. Johns Hopkins University Press, Baltimore, Maryland 1994.

An excellent technical review of the surface of Venus, based mainly upon the latest *Magellan* results. Topics include the planet's erosion, fracture zones, gravity, highlands, hot spots, mountain belts, plains, tectonics, topography, volcanoes and winds.

Cattermole, Peter, and Moore, Patrick: *Atlas of Venus*. Cambridge University Press, New York 1997.

An exceptionally well written account of Venus, ranging from the historical background to our perception of the planet to both the Russian and American Space programs. It includes the latest images from the *Magellan* mission as well as reference maps and the names of Venusian features. Topics include highland massifs, impact craters, the Venusian plains, and the planet's history.

Chaikin, Andrew. *A Man on the Moon*. Viking Press, New York 1994.

An excellent narrative account of the *Apollo* program that sent twenty-four men to the Moon; twelve of them walked on the Moon and the rest flew around it. Based on extensive interviews with the astronauts, this book brings them to life, revealing the bright colors of their personalities, as well as the fears and exhilaration of their remarkable journeys. It also provides a vivid account of the *Apollo 13* mission that suffered a crippling explosion of an oxygen tank while in space.

Christianson, Gale E.: *Greenhouse: The 200-Year Story of Global Warming*. Walker and Company, New York 1999.

A compelling history of our understanding, and growing evidence for, global warming during the past two centuries, including participants in that history such as industrialists, inventors, politicians and scientists.

Cooper, Henry S. F. Jr.: *The Evening Star: Venus Observed*. Farrar Straus and Giroux, New York 1993.

A wonderful popular account of the *Magellan* mission, written by a veteran science and space reporter for *The New Yorker*. This evocative and lucid narrative includes discussions of the people who do science, and of the challenges that confront them, the importance and politics of NASA, and much of our current scientific understanding of Venus.

Gelbspan, Ross: *The Heat is On: The Climate Crisis, The Cover-Up, The Prescription*. Addison-Wesley, New York 1997; Perseus Books, Cambridge, Massachusetts 1998.

A journalist exposes the deliberate campaign by oil and coal interests to confuse the public about global warming and the disruptive weather patterns that mark its initial stages, and the emerging alliance among environmentalists, the insurance agencies, and vulnerable nations to do something about global warming.

Goldsmith, Donald: *The Hunt for Life on Mars*. Penguin Books, New York 1997.

The complete story of a meteorite from Mars that landed on Earth, was recovered from Antarctica, and exhibits controversial evidence for ancient, primitive life on Mars.

Grinspoon, David Henry: *Venus Revealed: A New Look Below the Clouds of Our Mysterious Twin Planet*. Addison-Wesley, New York 1996.

A comprehensive and imaginative discussion of Venus before the telescope, through the telescope and from spacecraft, including comparisons with Earth and even a discussion of life on Venus.

Hanlon, Michael: *The World of* Galileo*: The Inside Story of NASA's Mission to Jupiter*. St. Martin's Press, New York 2001.

A good popular account of some *Galileo* discoveries with lots of color pictures.

Hansson, Anders: *Mars and the Development of Life*, Second Edition. John Wiley and Sons, New York 1997.

An examination of past, present and future life-forms on Mars, written from a biophysical perspective. Areas are identified where space exploration will lead to a better understanding of the conditions for the origin of life, and the possibilities for terraforming Mars are included.

Harland, David M.: *Exploring the Moon: The Apollo Expeditions*. Springer-Verlag and Praxis Publishing, New York 1999.

A lunar travelogue of the *Apollo* expeditions to the lunar surface, concentrating on the last three missions, *Apollo 15, 16* and *17*, that included *Lunar Rover* visits to Hadley–Appenine, Descartes–Cayley and Taurus–Littrow. The *Galileo, Clementine* and *Lunar Prospector* investigations of the Moon are included in the final chapter.

Harland, David M: *Jupiter Odyssey: The Story of NASA's* Galileo *Mission*. Springer-Praxis, New York 2000.

A good scientific account of the discoveries of the *Galileo* mission, concerning the planet Jupiter and its four larger satellites. This book contains a good selection of images, and appendices with nomenclatures for satellite surface features, a glossary, further reading and web sites.

Heiken Grant H., Vaniman, Davit T., and French, Bevan M. (editors): *Lunar Sourcebook: A User's Guide to the Moon*. Cambridge University Press, New York 1991.

A complete reference encyclopedia and technical account of data gathered during the American and Soviet missions to the Moon, written by scientists active in every field of lunar research and veterans of the *Apollo* program. It explores the formation and evolution of the Moon's surface, the chemical and mineralogical nature of lunar rocks and soils, and the current state of scientific knowledge about the nature, origin, and history of the Moon.

Henbest, Nigel: *The Planets*. Viking Penguin Books, New York 1992.

A splendid collection of modern images of the planets from Mercury to Pluto, describing the *Magellan* mission to Venus, the *Viking 1* and *2* missions to Mars and the *Voyager 1* and *2* missions to the giant planets. The non-technical text provides a very readable background to the images. Since the book was written about a decade ago, it is missing images from the *Mars Pathfinder, Mars Global Surveyor* and Jupiter *Galileo* missions.

Hoskin, Michael (editor): *The Cambridge Illustrated History of Astronomy*. Cambridge University Press, New York 1997.

This excellent volume traces the history of astronomy from antiquity through the rise of astrophysics, including the planetary discoveries and theories of the ancient Greeks, Ptolemy, Copernicus, Tycho Brahe, Johannes Kepler, Galileo Galilei, Isaac Newton, William Herschel and Christiaan Huygens. Discoveries concerning stellar evolution, galaxies and the expanding Universe are discussed in the later chapters of this book.

Hufbauer, Karl: *Exploring the Sun: Solar Science Since Galileo*. Johns Hopkins University Press, Baltimore, Maryland 1991.

A historical account that emphasizes the solar wind and the variable Sun. Techniques, theories, scientific communities and patronage are singled out for discussion.

Jakosky, Bruce. *The Search for Life on Other Planets*. Cambridge University Press, New York 1998.

A comprehensive discussion of the search for life in the Universe – including the origin and development of life on Earth; the environmental conditions necessary for life to exist; the possibility that life might exist elsewhere in our solar system, including Mars, Europa and Titan; the discovery of planets around other stars and the habitability of such planets; and the possibility of intelligent extraterrestrial life.

Kieffer, Hugh. H., Jakosky, Bruce M., Snyder, Conway W., and Matthews, Mildred S. (editors): *Mars*. The University of Arizona Press, Tucson, Arizona 1992.

A collection of thirty-eight technical articles including telescopic observations and spacecraft exploration of Mars, and the geophysics, geology, surface, atmosphere, biology and satellites of Mars, based primarily on results from the *Viking 1* and *2* missions.

Kowal, Charles T.: *Asteroids – Their Nature and Utilization*, Second Edition. John Wiley, New York 1996.

A concise description of asteroids, including a brief history, observational techniques, origins, asteroids within the main belt and outside it, asteroids and comets, and mining asteroids. The second edition includes the *Galileo* flybys of 951 Gaspra and 243 Ida, but the *NEAR Shoemaker* spacecraft had not yet arrived at 433 Eros.

Lang, Kenneth R.: *Sun, Earth and Sky*. Springer-Verlag, New York 1995.

A lavishly illustrated book that introduces the Sun, its physics and its impact on life here on Earth, written in a light and friendly style with apt metaphors, similes and analogies, poetry, art, history, and vignettes of scientists at work.

Lang, Kenneth R.: *Astrophysical Formulae. Volume I. Radiation, Gas Processes and High Energy Astrophysics. Volume II. Space, Time, Matter and Cosmology*. Springer-Verlag, New York 1999.

The third, enlarged edition of a comprehensive, widely-used reference to the fundamental formulae employed in astronomy, astrophysics and general physics, including

4000 formulae and 5000 references to the original papers. It includes all aspects of astronomy and astrophysics that are relevant to studies of the Sun.

Lang, Kenneth R.: *The Cambridge Encyclopedia of the Sun*. Cambridge University Press, New York 2001.

A complete modern guide to this fire of life, our nearest star, including the recent discoveries of the *Yohkoh*, *SOHO* and *Ulysses* solar missions. It provides comprehensive accounts of the most recent discoveries, such as the neutrino observations that are revealing a completely new physics. A description of the use of sound waves to peer deep into the Sun's inner regions and measure temperature right down to the central nuclear reactor is included, and the awesome ejections of matter from the Sun's outer regions and their potential threat to Earth are also discussed. A full bibliography, a list of Internet sites, and a glossary so comprehensive as to constitute a dictionary of solar astronomy complete this reference work.

Lang, Kenneth R.: *The Sun From Space*. Springer-Verlag, New York 2000.

A comprehensive account of solar astrophysics and how our perception and knowledge of the Sun have gradually changed. Timelines and hundreds of seminal papers are provided for key discoveries during the past two centuries, but the emphasis is on the last decade which has seen three successful solar spacecraft missions: *SOHO*, *Ulysses* and *Yohkoh*. Together these have confirmed many aspects of the Sun and its output, and provided new clues to the numerous open questions that remain. This generously illustrated book is written in a clear and concise style, appealing to anyone interested in the Sun, from the amateur astronomer to the expert.

Lewis, John S.: *Mining the Sky: Untold Riches from the Asteroids, Comets and Planets*. Addison-Wesley, New York 1996.

If we send spaceships to mine asteroids and/or comets for precious metals and water, we might overcome the problems of depleting Earth's natural resources and unrestrained human population growth.

Lewis, John S.: *Rain of Iron and Ice: The Very Real Threat of Comet and Asteroid Bombardment*. Addison-Wesley, New York 1996.

An excellent review of the scientific evidence for past and possibly future dangers of the impact of Near-Earth Objects with our planet, written for lay readers and scientists alike. It includes the impact history of the solar system, the contemporary impact hazard and possible defense against incoming comets and asteroids.

McNab, David, and Younger, James: *The Planets*. Yale University Press, New Haven, Connecticut 1999.

A well-illustrated book published to accompany the BBC television series on *The Planets*, including the Moon, planetary surfaces, the giant planets, the Sun, planetary atmospheres and life.

McNeill, J. R.: *Something New Under the Sun: An Environmental History of the Twentieth-Century World*. W. W. Norton and Company, New York 2000.

An excellent historical description of environmental change by humans during the 20th century. It includes discussions of the Earth's atmosphere and crust, the hydrosphere, the biosphere, the growth of economies, industries, cities and population, and environmental and international politics.

McPhee, John: *Annals of the Former World*. Farrar, Straus and Giroux, New York 1998.

A collection of essays that provide a wonderful popular survey of geology, and the modern theory of plate tectonics, set within the context of the author's travels across the United States in the company of geologists.

McSween, Harry Y. Jr.: *Meteorites and Their Parent Bodies*, Second Edition. Cambridge University Press, New York 1999.

An overview of meteorites in chapters that pair each main type of meteorite with its asteroid parent body. A final chapter discusses the importance of meteorites in unraveling the origin and evolution of the solar system.

Moore, Patrick: *Eyes on the Universe: The Story of the Telescope*. Springer-Verlag, New York 1997.

A clear and concise history of astronomical telescopes, from the early beginnings by Galileo to the *Hubble Space Telescope*.

Moore, Thomas Gale: *Climate of Fear: Why We Shouldn't Worry about Global Warming*. Cato Institute, Washington, D.C. 1998.

The author argues that increased carbon-dioxide emissions are likely to cause global warming, but most people will benefit from the slightly higher temperatures that it would produce. Agricultural production, heating costs and transportation would all benefit, and the costs of curbing greenhouse gas emissions would far exceed the losses from the expected climate change.

Morrison, David: *Exploring Planetary Worlds*. Scientific American Library, W. H. Freeman, New York 1993.

This well-written account focuses on the processes that shaped each planet's distinctive character, and what each planet tells us about the Earth and the origin and development of our solar system. The planets are grouped by primary traits, comparing and distinguishing their sizes, atmospheres, surfaces, magnetic fields, energy sources, chemical compositions and geological histories. A prologue provides a short history of early planetary astronomy, while an epilogue discusses the role of collisions between comets, asteroids, and the Earth in shaping the history of life on Earth.

National Academy of Sciences: *Policy Implications of Greenhouse Warming.* National Academy Press, Washington, D.C. 1991.

A scientific description of what is needed to make informed decisions about global warming resulting from releases of heat-trapping gases into the atmosphere, providing specific advice for United States policy and addressing the need for an international response to potential greenhouse warming.

North, Gerald: *Observing the Moon: The Modern Astronomer's Guide.* Cambridge University Press, New York 2000.

Comprehensive descriptions of individual lunar craters and seas are provided. By explaining some of the modern scientific understanding of the Moon, including lunar geology and other lunar science from *Apollo* to *Lunar Prospector*, this book makes observing the Moon more interesting.

Norton, O. Richard: *Rocks from Space.* Mountain Press, Missoula, Montana 1994.

An entertaining account of meteorites, including their physical characteristics and origins as well as well-known meteorite hunters and falls.

Norton, O. Richard: *The Cambridge Encyclopedia of Meteorites.* Cambridge University Press, New York 2002.

A beautifully illustrated compendium that includes the classification, internal structure and origin of meteorites, as well as their terrestrial impacts and museum locations. The *NEAR Shoemaker* mission to the asteroid 433 Eros is also included.

Peterson, Ivars: *Newton's Clock: Chaos in the Solar System.* W. H. Freeman, New York 1993.

This science writer traces the growth in our knowledge of dynamics in the solar system, from clockwise precision to chaos and complexity, culminating in recent findings made possible by modern mathematics and computer simulations.

Philander, S. George: *Is the Temperature Rising? The Uncertain Science of Global Warming.* Princeton University Press, Princeton, New Jersey 1998.

An introduction for the non-scientist to the complex and uncertain factors that determine the world's climate and weather, including winds, clouds, sunlight, air, land, sea and human activity that led to the ozone hole and may cause global warming.

Raeburn, Paul: *Uncovering the Secrets of the Red Planet.* National Geographic Society, Washington, D.C. 1998.

A well-illustrated account of the space-age investigation of Mars, from *Mariner 4* and *9* through *Viking 1* and *2*, the *Pathfinder* mission, including 3-D pictures, and on to the earliest days of the *Mars Global Surveyor* mission.

Rogers, John H.: *The Giant Planet Jupiter.* Cambridge University Press, New York 1995.

This book is one of a series of practical astronomy handbooks, especially suitable for amateur and leisure astronomers who want to make productive observations and new discoveries with their telescopes. It reviews more than 100 years of telescopic observations of Jupiter and provides modern observational accounts of the various regions and structures of the planet's atmosphere. Theoretical discussion of the spot dynamics, anti-cyclonic circulation, jet streams and global patterns of activity are provided. Auroras, lightning, the Jovian ring and major satellites are also included.

Schaff, Fred: *Comets of the Century: From Halley to Hale–Bopp.* Springer-Verlag, New York 1997.

A history of all the greatest comets ever recorded, whose publication was timed to coincide with the most recent one, Comet Hale–Bopp.

Sheehan, William: *The Planet Mars: A History of Observation and Discovery.* The University of Arizona Press, Tucson, Arizona 1996.

The history of our scientific understanding of Mars is traced, from the early telescopic observations of the so-called canals on Mars to the close-up results of the *Mariner 9* and *Viking 1* and *2* missions.

Sheehan, William, and O'Meara, Stephen James: *Mars: The Lure of the Red Planet.* Prometheus Books, Amherst, New York 2001.

An engaging history of why Mars has fascinated humankind for 200 years, from early telescopic observations to the *Viking 1* and *2* and *Mars Global Surveyor* missions. The possibilities of water and even life on Mars are emphasized.

Spencer, John R., and Mitton, Jacqueline (editors): *The Great Comet Crash*. New York, Cambridge University Press 1995.

This book contains a fine description of the collision between Comet Shoemaker–Levy 9 and the giant planet Jupiter in July 1994, written by distinguished scientists that participated in observing it.

Spudis, Paul D.: *The Once and Future Moon*. Smithsonian Institution Press, Washington, D.C. 1996.

An expert in the field has written this book, intended for both a general audience and specialists in the planetary sciences. It presents a detailed picture of the Moon's composition and geological history, based on the *Apollo* missions, including the currently accepted giant impact theory of lunar origin. The *Clementine* lunar mission is also discussed, and scientific arguments for returning to the Moon are presented.

Stevens, William K.: *The Change in the Weather: People, Weather and the Science of Climate*. Random House, New York 1999.

An accessible, comprehensive and objective overview of what we know, and do not know, about global warming and climate change. This account, written by a *New York Times* science reporter, describes catastrophic atmospheric changes on Earth over the past four billion years, including the probable acceleration of weather disaster by burning fossil fuels during the last century.

Walter, Malcolm: *The Search for Life on Mars*. Perseus Books, Cambridge, Massachusetts 1999.

This book focuses on the similarities between Mars and the early Earth, and advocates searching for biological organisms under the surface of Mars.

Wilford, John Noble: *Mars Beckons: The Mysteries, The Challenges, The Expectations of Our Next Great Adventure in Space*. Alfred A. Knopf, New York 1990.

A science correspondent for the *New York Times* traces our fascination with Mars from the earliest written Babylonian records of 3000 years ago, through the planetary theories of Johannes Kepler, Galileo Galilei, Giovanni Schiaparelli and Percival Lowell. After a discussion of the *Viking 1* and *2* search for life on Mars, the author focuses on the possibilities of colonization of Mars.

Wilhelms, Donald E.: *To a Rocky Moon: A Geologist's History of Lunar Exploration*. The University of Arizona Press, Tucson, Arizona 1993.

A definitive account of how we came to understand the geological history of the Moon, including the rationale behind the *Apollo* site selections and a detailed discussion of all six landings.

Yeomans, Donald K.: *Comets: A Chronological History of Observation, Science, Myth and Folklore*. New York, John Wiley & Sons 1991.

This book provides a comprehensive and authoritative history of comet observations, written by one of the world's comet experts. It includes biographical details of the astronomers and mathematicians who have observed and studied comets. The first hundred pages review the early history from Aristotle and Seneca to Kepler, Newton and Halley. The physics of comets, the origin of meteor showers, and spacecraft visits to the nucleus of Comet Halley are also included. A lengthy appendix lists every comet visible to the naked eye from the 11th century BC to 1700 AD.

Zubrin, Robert: *The Case for Mars: The Plan to Settle the Red Planet and Why We Must*. The Free Press, New York 1996.

A blueprint for the future human exploration of Mars using present-day technology, including how we can build bases and settlements on Mars and one day alter its atmosphere.

Appendix 2

Directory of web sites

This book is supported by an Internet site located at http://ase.tufts.edu/cosmos/

General information

Images

http://photojournal.jpl.nasa.gov/

Physical Properties

http://ssd.jpl.nasa.gov/

Asteroids

Minor Planet Center

http://cfa-www.harvard.edu/iau/mpc.html

NEAR-Shoemaker Mission to Astroid 433 Eros

http://near.jhuapl.edu/

Comets

Individual Comets

http://cometography.com

Giotto Mission to Comet Halley

http://sci.esa.int/home/giotto/

Deep Space 1 Mission to Comet Borrelly

http://nmp.jpl.nasa.gov/ds1/

Jupiter

Galileo Mission to Jupiter

http://galileo.jpl.nasa.gov/
http://nssdc.gsfc.nasa.gov/planetary/galileo.html

Voyager Missions to Jupiter

http://nssdc.gsfc.nasa.gov/planetary/voyager.html

Mars

Mars Exploration Home Page

http://mars.jpl.nasa.gov

Mars Global Surveyor Mission

http://mars.jpl.nasa.gov/mgs/
http://www.msss.com/

2001 Mars Odyssey Mission

http://mars.jpl.nasa.gov/odyssey/

Mars Pathfinder Mission

http://mars.jpl.nasa.gov/missions/past/pathfinder.html

Viking Missions

http://mars.jpl.nasa.gov/missions/past/viking.html
http://nssdc.gsfc.nasa.gov/planetary/viking.html

Mercury

Mariner 10 Mission to Mercury

http://www.jpl.nasa.gov/missions/past/mariner10.html

Moon

Apollo Missions

http://nssdc.gsfc.nasa.gov/planetary/lunar/apollo.html

Clementine Mission

http://www-phys.llnl.gov/clementine/
http://nssdc.gsfc.nasa.gov/planetary/clementine.html

Lunar Prospector Mission

http://www.lpi.usra.edu/expmoon/prospector/prospector.html
http://nssdc.gsfc.nasa.gov/planetary/lunarprosp.html

Near earth objects

http://neo.jpl.nasa.gov/neo.html
http://cfa-www.harvard.edu/iau/NEO/TheNEOPage.html

Neptune

Voyager Mission to Neptune

http://nssdc.gsfc.nasa.gov/planetary/voyager.html

Saturn

Cassini–Huygens Mission to Saturn and Titan

http://saturn.jpl.nasa.gov/
http://sci.esa.int/home/huygens

Voyager Missions to Saturn

http://nssdc.gsfc.nasa.gov/planetary/voyager.html

Uranus

Voyager Mission to Uranus

http://nssdc.gsfc.nasa.gov/planetary/voyager.html

Venus

Magellan Mission to Venus

http://www.jpl.nasa.gov/magellan/
http://nssdc.gsfc.nasa.gov/planetary/magellan.html

Index

U

V